T0257219

Bernissart Dinosaurs and Early Cretaceous Terrestrial Ecosystems

Life of the Past   JAMES O. FARLOW, EDITOR

Indiana University Press   BLOOMINGTON & INDIANAPOLIS

BERNISSART DINOSAURS

AND EARLY CRETACEOUS TERRESTRIAL ECOSYSTEMS

Edited by PASCAL GODEFROIT

This book is a publication of

Indiana University Press
601 North Morton Street
Bloomington, Indiana 47404-3797 USA

iupress.indiana.edu

Telephone orders   800-842-6796
Fax orders   812-855-7931

♾ The paper used in this publication meets
the minimum requirements of the American
National Standard for Information Sciences—
Permanence of Paper for Printed Library
Materials, ANSI Z39.48-1992.

Manufactured in the
United States of America

Library of Congress
Cataloging-in-Publication Data

Bernissart dinosaurs and early Cretaceous
terrestrial ecosystems / edited by Pascal Godefroit.
    p. cm. — (Life of the past)
    Includes bibliographical references and
index.
    ISBN 978-0-253-35721-2 (cloth :
alk. paper) — ISBN 978-0-253-00570-0
(e-book) 1. Iguanodon—Belgium—Bernissart.
2. Paleontology—Belgium—Bernissart.
3. Paleoecology—Cretaceous. I. Godefroit,
Pascal, [date]
    QE862.O65B485 2012
    567.914—dc23
                            2011049613

1 2 3 4 5   17 16 15 14 13 12

This book is dedicated to

PIERRE BULTYNCK

As the head of the department of paleontology of the Royal Belgian Institute of Natural Sciences between 1991 and 2003, Pierre initiated the renaissance of vertebrate paleontology, including dinosaur research, in Belgium.

Have pity on an Iguanodon
Whose history lacks precision,
On account of which a lot of savants
Have been saying dumb things for thirty years
Ah! If only Barnum had found him!
He would have informed us better;
I'm sure that if he had looked,
He could have shown us a living one.

Marcel Lefevre, "The Bernissart Iguanodon's Complaint" (1912)

Contents

Contributors

Ainara Badiola Grupo Aragosaurus-IUCA (http://www.aragosaurus.com), Paleontología, Facultad de Ciencias, Universidad de Zaragoza, Pedro Cerbuna 12, 50009 Zaragoza, Spain, abadiola@unizar.es.

Jean-Mac Baele University of Mons, Faculté Polytechnique de Mons, rue de Houdain 9, 7000 Mons, Belgium, jean-marc.baele@umons.ac.be.

Yuri L. Bolotsky Palaeontological Museum of the Institute of Geology and Nature Management, Far East Branch, Russian Academy of Sciences, per. Relochny 1, 675000 Blagoveschensk, Russia, dinomus@ascnet.ru.

Niels Bonde Institute of Geography and Geology (Copenhagen University), Øster Voldgade 10, DK-1350 Copenhagen K and Fur Museum, Nederby, DK-7884 Fur, Denmark, nielsb@geo.ku.dk.

Gábor Botfalvai Eötvös University, Department of Paleontology, Pázmány P. s. 1/c, 1117 Budapest, Hungary.

Eric Buffetaut Centre National de la Recherche Scientifique (UMR 8538), Laboratoire de Géologie de l'Ecole Normale Supérieure, 24 rue Lhomond, 75231 Paris Cedex 05, France, eric.buffetaut@sfr.fr.

Evgenia V. Bugdaeva Institute of Biology and Soil Science, Far Eastern Branch of Russian Academy of Sciences, 159 Prosp. 100-letiya, Vladivostok, 690022, Russia, bugdaeva@ibss.dvo.ru.

Pierre Bultynck Royal Belgian Institute of Natural Sciences, Department of Paleontology, rue Vautier 29, 1000 Bruxelles, Belgium, pierre.bultynck@belgacom.net.

José Ignacio Canudo Grupo Aragosaurus-IUCA (http://www.aragosaurus.com), Paleontología, Facultad de Ciencias, Universidad de Zaragoza, Pedro Cerbuna 12, 50009 Zaragoza, Spain, jicanudo@unizar.es.

Patrick Carlier rue Haute 28 5190 Spy, Belgium, carlierpatrick@skynet.be.

Chen Jun Jilin University Geological Museum, Chaoyang Campus, 6 Ximinzhu Street, Changchun, Jilin Province 130062, People's Republic of China, cj@jlu.edu.cn.

Vlad A. Codrea University Babeş-Bolyai Cluj-Napoca, Faculty of Biology and Geology, 1 Kogălniceanu Str., 400084, Cluj-Napoca, Romania, vlad.codrea@ubbcluj.ro; codrea_vlad@yahoo.fr.

Clément Coiffard UMR 7207—Centre de Recherche sur la Paléobiodiversité et les Paléoenvironnements, 43 rue Buffon, CP 48, F-75005 Paris, France, clement.coiffard@ens-lyon.org; Museum für Naturkunde, Invalidenstrasse 43, 10115 Berlin, Germany, clement.coiffard@mfn-berlin.de.

Walter Coudyzer Universitaire Ziekenhuis Gasthuisberg, Radiologie, Herestraat 49, 3000 Leuven, Belgium, cwscan@yahoo.com.

Gloria Cuenca-Bescós Grupo Aragosaurus-IUCA (http://www.aragosaurus.com), Paleontología, Facultad de Ciencias, Universidad de Zaragoza, Pedro Cerbuna 12, 50009 Zaragoza, Spain, cuencag@unizar.es.

Gilles Cuny The Natural History Museum of Denmark, University of Copenhagen, Øster Voldgade 5-7, 1350 Copenhagen K, Denmark, gilles@snm.ku.dk.

Véronique Daviéro-Gomez Université Lyon 1 and CNRS-UMR 5125 Paléoenvironnements et Paléobiosphère, F-69622, Villeurbanne, France, daviero@univ-lyon1.fr.

Jean Dejax Muséum national d'Histoire naturelle, USM 0203, CNRS UMR 5143, Département Histoire de la Terre, case postale 38, 57 rue Cuvier, 75231 Paris Cedex 05, France, dejax@mnhn.fr.

Christian Dupuis University of Mons, Faculté Polytechnique de Mons, rue de Houdain 9, 7000 Mons, Belgium, christian.dupuis@umons.ac.be.

François Escuillié Eldonia, 9 avenue des Portes Occitanes, 3 800 Gannat, France, eldonia@wanadoo.fr.

Peter M. Galton Professor Emeritus, College of Naturopathic Medicine, University of Bridgeport, CT, USA; Curatorial Affiliate, Peabody Museum of Natural History, Yale University, New Haven, CT, USA. Current address: 315 Southern Hills Drive, Rio Vista, CA 94571, USA, pgalton@bridgeport.edu.

Géraldine Garcia IPHEP, UMR CNRS 6046, Faculté des Sciences de Poitiers, 40 avenue du Recteur Pineau, F-86022 Poitiers cedex, France, geraldine.garcia@univ-poitiers.fr.

Emmanuel Gilissen Royal Museum for Central Africa, Department of African Zoology, Leuvensesteenweg 13, B-3080 Tervuren, Belgium; Université Libre de Bruxelles, Laboratory of Histology and Neuropathology CP 620, B-1070 Brussels, Belgium; Department of Anthropology, University of Arkansas, Fayetteville, AR 72701, USA, Emmanuel.Gilissen@africamuseum.be.

Thomas Gillot Université Lyon 1 and CNRS-UMR 5125 Paléoenvironnements et Paléobiosphère, F-69622, Villeurbanne, France, tom6446@hotmail.fr.

Pascal Godefroit Department of Paleontology, Royal Belgian Institute of Natural Sciences, rue Vautier 29, 1000 Bruxelles, Belgium, Pascal.Godefroit@naturalsciences.be.

Bernard Gomez Université Lyon 1 and CNRS-UMR 5125 Paléoenvironnements et Paléobiosphère, F-69622, Villeurbanne, France, bernard.gomez@univ-lyon1.fr.

Péter Gulyás Eötvös University, Department of Paleontology, Pázmány P. s. 1/c, 1117 Budapest, Hungary.

Jin Liyong Jilin University Geological Museum, Chaoyang Campus, 6 Ximinzhu Street, Changchun, Jilin Province 130062, People's Republic of China, cj@jlu.edu.cn.

Olivier Kaufmann University of Mons, Faculté Polytechnique de Mons, rue de Houdain 9, 7000 Mons, Belgium, Olivier.Kaufmann@umons.ac.be.

Pascaline Lauters Department of Palaeontology, Royal Belgian Institute of Natural Sciences, rue Vautier 29, 1000 Bruxelles, Belgium, Pascaline.Lauters@naturalsciences.be, and Département d'Anthropologie et de Génétique humaine, Université Libre de Bruxelles, avenue F.D. Roosevelt 50, 1050 Bruxelles, Belgium, plauters@ulb.ac.be.

Thierry Leduc Department of Geology, Royal Belgian Institute of Natural Sciences rue Vautier 29, 1000 Bruxelles, Belgium, t.leduc@sciencesnaturelles.be; Laboratory of Mineralogy B.18, University of Liège, 4000 Liège, Belgium.

Jean Le Loeuff Musée des Dinosaures, 11260 Espéraza, France, jean.leloeuff@dinosauria.org.

Li Xiao-Bo College of Earth Sciences, Jilin University, 2199 Jianshe Street, Changchun, 130061, People's Republic of China.

Luciane Licour University of Mons, Faculté Polytechnique de Mons, rue de Houdain 9, 7000 Mons, Belgium, Luciane.Licour@umons.ac.be.

László Makádi Eötvös University, Department of Paleontology, Pázmány P. s. 1/c, 1117 Budapest, Hungary.

Valentina S. Markevich Institute of Biology and Soil Science, Far East Branch of Russian Academy of Sciences, 159 Prosp. 100-letiya, Vladivostok, 690022, Russia, markevich@ibss.dvo.ru.

Thierry Martin University of Mons, Faculté Polytechnique de Mons, rue de Houdain 9, 7000 Mons, Belgium, Thierry.Martin@umons.ac.be.

Edwige Masure Université Pierre et Marie Curie, UMR-CNRS 7207, 4 place Jussieu, 75005 Paris, France, edwige.masure@upmc.fr.

Laetitia Nori Association Carrières d'Euville, Villasatel, Hameau des carrières 55200 Euville, France, circuitdelapierre@wanadoo.fr.

David B. Norman Sedgwick Museum and Department of Earth Sciences, University of Cambridge, Downing Street, Cambridge CB2 3EQ, UK, dn102@cam.ac.uk.

Attila Ősi Hungarian Academy of Sciences–Hungarian Natural History Museum, Research Group for Palaeontology, Ludovika tér 2, 1083 Budapest, Hungary, hungaros@freemail.hu.

Xabier Pereda-Suberbiola Universidad del País Vasco/EHU, Facultad de Ciencia y Tecnología, Departamento de Estratigrafía y Paleontología, Apartado 644, 48080 Bilbao, Spain, xabier.pereda@ehu.es.

Stéphane Pirson Direction de l'archéologie DGO4–Département du Patrimoine, Service public de Wallonie, rue des Brigades d'Irlande 1, 5100 Jambes, stephane.pirson@spw.wallonie.be.

Yves Quinif University of Mons, Faculté Polytechnique de Mons, rue de Houdain, 7 000 Mons, Belgium, Yves.Quinif@umons.ac.be.

Márton Rabi Eötvös University, Department of Paleontology, Pázmány P. s. 1/c, 1117 Budapest, Hungary.

Thomas H. Rich Museum Victoria, P.O. Box 666 Melbourne, Victoria, 3001 Australia, trich@museum.vic.gov.au.

Armand de Ricqlès Collège de France, / (UMR 7093 CNRS-UPMC/ ISTEP Biomineralisations-Paleoenvironnements), UPMC Paris Universitas, armand.de_ricqlès@upmc.fr

Francis Robaszynski University of Mons, Faculté Polytechnique de Mons, rue de Houdain 9, 7000 Mons, Belgium, francis.robaszynski@umons.ac.be.

Christoph Roolf Heinrich-Heine-Universität Düsseldorf, Historisches Seminar II (Neuere Geschichte), Universitätsstraße 1, 40225 Düsseldorf, Germany, and Wimpfener Straße 14, 40597 Düsseldorf, Germany, roolf@uni-duesseldorf.de.

José Ignacio Ruiz-Omeñaca Museo del Jurásico de Asturias (MUJA), 33328 Colunga, Spain, and Grupo Aragosaurus-IUCA (http://www.aragosaurus.com), Paleontología, Facultad de Ciencias, Universidad de Zaragoza, Pedro Cerbuna 12, 50009 Zaragoza, Spain, jigruiz@unizar.es.

José Luis Sanz Unidad de Paleontología, Departamento de Biología, Universidad Autónoma de Madrid, C/Darwin 2, 28049 Cantoblanco, Madrid, Spain, dinopepelu@gmail.com.

Johann Schnyder Université Pierre et Marie Curie et C.N.R.S., UMR 7193 ISTeP, case postale 117, 4 place Jussieu, 75252 Paris Cedex 05, France, Johann.Schnyder@upmc.fr.

Thierry Smith Department of Paleontology, Royal Belgian Institute of Natural Sciences, rue Vautier 29, 1000 Bruxelles, Belgium, Thierry.Smith@naturalsciences.be.

Paul Spagna Department of Paleontology, Royal Belgian Institute of Natural Sciences, rue Vautier 29, 1000 Bruxelles, Belgium, Paul.Spagna@naturalsciences.be.

Suravech Suteethorn Paleontological Research and Education Centre, and Department of Biology Mahasarakham University—Faculty of Science, Maha Sarakham 44150, Thailand, suteethorn@yahoo.com.

Zoltán Szentesi Eötvös University, Department of Paleontology, Pázmány P. s. 1/c, 1117 Budapest, Hungary.

Rodolphe Tabuce ISEM, UMR CNRS 5554, Université Montpellier II, c.c. 064, Place Eugène Bataillon, F-34095 Montpellier cedex 5, France, rodolphe.tabuce@univ-montp2.fr.

Philippe Taquet Muséum National d'Histoire Naturelle, 8 rue Buffon, 75005 Paris, France, taquet@mnhn.fr.

Fidel Torcida Colectivo Arqueológico-Paleontológico de Salas (C.A.S.), Museo de Dinosaurios, Plaza Jesús Aparicio 9, 9600 Salas de los Infantes, Burgos, Spain, fideltorcida@hotmail.com.

Xavier Valentin IPHEP, UMR CNRS 6046, Faculté des Sciences de Poitiers, 40 avenue du Recteur Pineau, F-86022 Poitiers cedex, France, xavier.valentin@univ.poitiers.fr.

Sara Vandycke University of Mons, Faculté Polytechnique de Mons, rue de Houdain 9, 7000 Mons, Belgium, Sara.Vandycke@umons.ac.be.

Martine Vercauteren Département d'Anthropologie et de Génétique humaine, Université Libre de Bruxelles, avenue F.D. Roosevelt 50, 1050 Bruxelles, mvercau@ulb.ac.be.

Monique Vianey-Liaud ISEM, UMR CNRS 5554, Université Montpellier II, c.c. 064, Place Eugène Bataillon, F-34095 Montpellier cedex 5, France, Monique.Vianey-Liaud@univ-montp2.fr.

Patricia Vickers-Rich School of Geosciences P.O. Box 28 Monash University, Victoria 3800 Australia, pat.rich@sci.monash.edu.au.

Wu Wenhao Research Center for Paleontology and Stratigraphy, Jilin University, Changchun 130061, P. R. China, wu_wenhao, yahoo.cn.

Johan Yans FUNDP UCL-Namur, Department of Geology, 61 rue de Bruxelles, 5000 Namur, Belgium, johan.yans@fundp.ac.be.

Preface

On May 7, 1878, P.-J. Van Beneden announced to the Belgian Academy of Science that a major new discovery of fossils had been made at a colliery in Bernissart (southwest Belgium). Among the fossils were teeth that could be identified as belonging to the dinosaur named *Iguanodon*. What was to emerge from Bernissart over the next three years of intensive excavation (supervised by the staff of the Royal Museum of Natural History, Brussels) was genuinely spectacular: a large number of virtually complete skeletons of dinosaurs, crocodiles, rare amphibians, and insects, thousands of fossil fish, an abundance of coprolites, and a diverse flora. It seemed that a window had been opened into an Early Cretaceous ecosystem via some laminated clays and silts that had evidently slumped into chasms, or large fissures, that had formed by dissolution of underlying Carboniferous limestone-dominated beds.

These extraordinary discoveries, not surprisingly, attracted the attention of some of the great paleontologists of the time: Albert Seward (Cambridge) described the flora from Bernissart, Charles Eugène Bertrand (Lille) described the coprolites, and Ramsay Traquair (Edinburgh) described the fish. In comparison, Louis Dollo (lately arrived from Lille) was unknown yet; after the departure of George Albert Boulenger in 1881, Dollo was appointed "Aide Naturaliste" at the Royal Belgian Museum of Natural History and was offered the opportunity to describe the dinosaurs collected from Bernissart.

From 1882 until 1923 Dollo produced a stream of insightful and provocative research papers not just on his beloved dinosaurs, but many of the other vertebrates from Bernissart and elsewhere. Dollo's intellectual grasp and energy were impressive, and he almost single-handedly balanced (from a European perspective at least) the prodigious output of Cope and Marsh in the United States during the extraordinary Bone War years. Dollo, however, was not content simply to describe new fossils per se; he was at least equally (if not more) interested in the biology, inferable behavior, and potential ecology of long-extinct creatures. Just as Bernissart provided the tangible record of an ecosystem, so Dollo wished to see how far one could actually reconstruct that ecology using strictly scientific principles. So advanced was Dollo's thinking (we would casually call it "modern") that as a tribute to the pioneering work of Louis Dollo, Othenio Abel coined the new term *palaeobiologie* to signify the integration of the study of fossils with that of our understanding of the living world.

Dollo also contributed to the theory of evolution and is personally remembered through Dollo's Law, which still exerts its subtle influence in the commonly used algorithms used in numerical phylogenetic analysis programs today.

Bernissart and its remarkable geology—Louis Dollo; anatomy; taxonomy; systematics; paleobiology; paleoecology (and local as well as global ecosystem reconstruction)—are topics that are deeply woven into the fabric of this book, as are the intellects of the contributors to this symposium volume. It is truly remarkable to think how great our debt is to that accidental discovery at Bernissart, which was first hinted at in an urgent telegram sent by the colliery's chief mining engineer, Gustave Arnould, to the director of the Royal Museum in Brussels, Edouard Dupont, on April 12, 1878.

David B. Norman (Cambridge, October 27, 2010)

Acknowledgments

This book was conceived during the symposium Tribute to Charles Darwin and Bernissart Iguanodons: New Perspectives on Vertebrate Evolution and Early Cretaceous Ecosystems, which was held February 9–13, 2009, at the Royal Belgian Institute of Natural Sciences (Brussels) and in Bernissart itself. This meeting was partly devoted to the latest results of a multidisciplinary project dedicated to the material collected in the cores drilled in 2002–2003 in and around the Iguanodon Sinkhole at Bernissart. I should like to express my deep gratitude to the coleaders of this project—J.-P. Tshibangu, J. Yans, and C. Dupuis—and to the FRS-FNRS, which provided financial support. I should like also to thank the people who have contributed to the organization of the symposium. O. Lambert was a particularly efficient co-organizer. The general director of the RBINS, Camille Pisani, and the head of the department of paleontology, E. Steurbaut, provided support, advice, and facilities. W. De Vos and the communication staff of the RBINS, as well as all the members of the department of paleontology, tirelessly helped us in various practical aspects of organization. I am greatly indebted to the Belgian Science Policy, the main sponsor of this meeting, and particularly to its president, P. Mettens, who encouraged its organization. The mayor, R. Vanderstraeten, Annette Cornelis, Corinne Detrain, the Tourism Office of Bernissart, the Local Development Agency of Bernissart, and the Geological Circle of Hainaut efficiently provided support and facilities at Bernissart.

Thanks also to R. Sloan (Indiana University Press) and J. O. Farlow for their support of the project from its earliest inception and for their patience. And of course I thank all the authors for their papers, in particular J.-M. Baele, D. B. Norman, Y. Quinif, P. Spagna, and J. Yans, for their important contribution to the realization of this book.

Pascale Golinvaux created the jacket illustration. H. De Potter and E. Dermience helped with the electronic versions of the figures.

New Investigations into the Iguanodon Sinkhole at Bernissart and Other Early Cretaceous Localities in the Mons Basin (Belgium)

1

1.1. The Sainte-Barbe pit and mine buildings in 1878, at the time when the iguanodons were discovered.

2

Bernissart and the Iguanodons: Historical Perspective and New Investigations

Pascal Godefroit*, Johan Yans, and Pierre Bultynck

The discovery of complete and articulated skeletons of *Iguanodon* at Bernissart in 1878 came at a time when the anatomy of dinosaurs was still poorly understood, and thus considerable advances were made possible. Here we briefly describe, mainly from documents in the archives of the Royal Belgian Institute of Natural Sciences, the circumstances of the discovery of the Bernissart iguanodons. We also provide information about their preparation and mounting in laboratories, for exhibitions, and in early studies. We also summarize the latest results of a multidisciplinary project dedicated to the material collected in the cores drilled in 2002–2003 in and around the Iguanodon Sinkhole at Bernissart.

The discovery of the first *Iguanodon* fossils has become a legend in the small world of paleontology. Around 1822, Mary Ann Mantell accompanied her husband, the physician Dr. Gideon Algernon Mantell, on his medical rounds and by chance discovered large fossilized teeth. Her husband found the teeth intriguing. With advice from Georges Cuvier, William Clift, and William Daniel Conybeare, he described them and named them *Iguanodon*, "iguana tooth," because of their superficial resemblance to those of living iguanas (Mantell, 1825). *Iguanodon* was one the three founding members of the Dinosauria—along with *Megalosaurus* and *Hylaeosaurus*—named by Richard Owen in 1842.

For 56 years, little was known about *Iguanodon* and other dinosaurs. Mantell imagined these antediluvian animals to be some kind of giant lizards with elongated bodies and sprawling limbs (Benton, 1989). In 1854, the sculptor Waterhouse Hawkins, following Owen's advice, realized full-size reconstructions of *Iguanodon* and *Megalosaurus* for the Crystal Palace exhibition in London. *Iguanodon* was reconstructed as a rhinoceros-like heavy quadruped with a large spike on its nose. These impressive monsters invoked the first public sensation over dinosaurs (Norman, 1985).

The first partial dinosaur skeleton, named *Hadrosaurus foulkii* Leidy, 1858, was discovered in 1857 in New Jersey. This skeleton was reconstructed in a bipedal gait at the Academy of Natural Sciences of Philadelphia, but many questions were still left unanswered about the general appearance of dinosaurs.

Then, 20 years later, another *Iguanodon* discovery broke the scientific world—and the dinosaur world—wide open (Forster, 1997). The discovery of complete and articulated skeletons of *Iguanodon* at Bernissart in 1878 revealed for the first time the anatomy of dinosaurs, and thus considerable advances were made possible, in combination with the remarkable

The Bernissart Iguanodons: A Cornerstone in the History of Paleontology

discoveries in the American Midwest described by Marsh and Cope (Norman, 1987).

Many manuscripts and plans relating to the original excavations at Bernissart are preserved in the paleontological archives of the RBINS, which allow us to reconstruct the circumstances of the discovery of these fantastic dinosaurs.

Institutional abbreviations. NHMUK, The Natural History Museum, London (formerly the British Museum [Natural History]), U.K.; RBINS, Royal Belgian Institute of Natural Sciences, Brussels (formerly MRHNB, Musée royal d'Histoire naturelle de Belgique), Belgium.

The Discovery and Excavation of the Bernissart Iguanodons

Bernissart is a former coal-mining village in southwestern Belgium, situated 21 km south of Mons and less than 1 km from the Franco-Belgian frontier. Pre-industrial coal extraction began at Bernissart around 1717 (Delguste, 2003). In 1757, Duke Emmanuel de Croÿ grouped together the different coal companies in northern France into the powerful Anzin Company, which started the industrial exploitation of the coal in the Bernissart area during the second half of the eighteenth century (Delguste, 2006). In the nineteenth century, the Bernissart Coal Board Limited Company dug five coal pits on Bernissart territory. The Négresse pit (no. 1, exploited from 1841) and Sainte-Barbe pit (no. 3, exploited from 1849; Fig. 1.1) were used for coal extraction and coupled with the Moulin pit (no. 2, exploited from 1842) for ventilation. The Sainte-Catherine pit (no. 4, exploited from 1864) was the third extraction pit and was coupled with pit no. 5 (exploited from ?1874) for ventilation. The maximum distance between pits 1 and 5 was about 1,600 m. With a depth of 422 m, the Sainte-Barbe pit was the deepest. In spite of a rather archaic technology, the daily production for the three extraction pits was about 800 tons. However, the flood problems were more important than in other coal mines from the Mons area; steam pumps were used to extract the water.

On February 28, 1878, miners digging a horizontal exploration gallery 322 m below ground level suddenly encountered, 35 m to the south of the Luronne seam, disturbed rocks, indicating that they were penetrating inside a vertical cran—a local term meaning a pit formed by natural collapse through the coal seams that was filled especially with clayey deposits normally located above the coal measures.

On March 1, chief overseer Cyprien Ballez, engineer Léon Latinis, and mine director Gustave Fagès went down into the Sainte-Barbe pit to evaluate the situation. It was decided to traverse this cran and to rejoin the coal seam on the other side. Overseer Motuelle and miners Jules Créteur and Alphonse Blanchard were put in charge of continuing the exploration gallery through the perturbed layers of the cran. On March 9, Ballez noticed that the exploration gallery was still in the perturbed zone of the cran.

In March, the miners had already collected dinosaur remains: fragmentary bones and teeth, which are labeled "remains of the first Iguanodon, March 1878" and are housed in the paleontological collections of the RBINS. But they apparently paid little attention to these discoveries, believing that they were just fossil wood.

On April 1, the exploration team again entered nondisturbed but inclined formations. On April 3, Ballez and Latinis went down again together in the exploration gallery. The engineer estimated that they had again reached coal-bearing formations. Latinis's explanations apparently did not satisfy Fagès. Indeed, the mine manager decided to accompany the engineer and the chief overseer in the exploration gallery on April 5. While inspecting the deposits, Fagès found a long object with an oval cross section and a fibrous texture. Latinis believed that it was a fossil oak branch. Conversely, Fagès ironically asserted that it was a rib of Father Adam. Miner Jules Créteur mentioned that he had already found a larger fossil, and the team soon unearthed limb bones in the gallery. In the evening, miners brought several fragments of these fossils to Café Dubruille. There the local doctor, Lhoir, who also worked for the coal mine, burned one of the fragments and confirmed that the fossils collected by the miners were bones, not wood. Many new fossils were discovered by the miners in the night of April 5–6. On April 6, Fagès ordered Ballez to bring all the fragments of bones that the miners had collected to the surface and to lock up the end of the gallery.

On Sunday, April 7, Latinis was commissioned to go to Mons to show the fossils to the well-known geologist François-Léopold Cornet. But Cornet was not home. Latinis thus left the fossils to his young son, Jules (a future renowned geologist), and asked him to tell his father that these bones had been found in the Sainte-Barbe pit at Bernissart.

On April 8, F.-L. Cornet came to Bernissart and briefly discussed the Bernissart discovery with Latinis. He could not meet Fagès, who was with Ballez in the Sainte-Catherine pit. On April 10, Cornet told the zoologist Pierre-Joseph Van Beneden, professor of paleontology at Leuven University, that Latinis, who was a former student of Van Beneden, had discovered fossil bones at Bernissart, and he sent him some of the bones that Latinis had left with his son. Van Beneden quickly identified the teeth as belonging to the dinosaur *Iguanodon*, previously described from Wealden deposits in England.

On April 12, Fagès went to Mons to meet the chief mining engineer, Gustave Arnould, who immediately sent a telegram to Edouard Dupont, director of the Musée royal d'Histoire naturelle de Belgique (MRHNB) at Brussels to inform him of the important discovery at 322m below ground level in the Sainte-Barbe Pit (Fig. 1.2).

On Saturday, April 13, Louis De Pauw, head preparer at the MRHNB and a man who already had extensive experience in the excavation and preparation of fossil vertebrates, met Arnould at Blaton. They went together to Bernissart. Fagès showed them the bones recently found in the gallery; De Pauw recognized two ungual phalanges and one vertebral centrum. It was then decided to go down together into the fossiliferous gallery. De Pauw (1902) reported that the walls of the exploration gallery were completely covered by fossil bones, plants, and fishes. Ballez, Motuelle, Créteur, and Blanchard soon unearthed a complete hind limb that they transported on a plank covered with straw. But after a 300-m walk, the bones began to disintegrate on contact with the fresh air of the mine galleries. De Pauw protected the biggest remaining fragment with his own clothes, and Ballez and Motuelle brought the fossils to the surface. De Pauw realized that the presence of pyrite inside the bones was one of the biggest problems that they had to face if they were to unearth the fossils from the Sainte-Barbe pit.

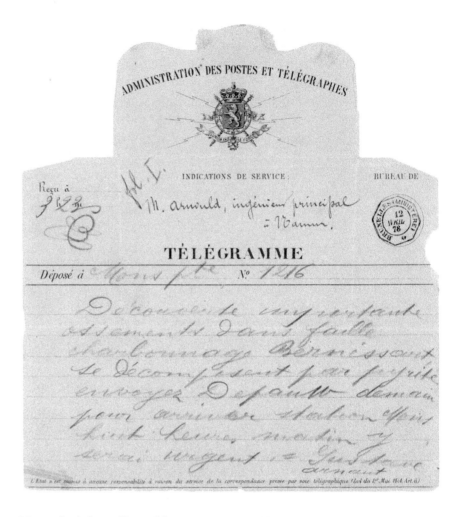

He packed the collected bones in a box full of sawdust and brought them back to Brussels. In MRHNB workshops, he succeeded in solidifying the limb bones from Bernissart with gelatin. In the meantime, Latinis prepared 11 more boxes full of fossils at Bernissart.

Fagès quickly gathered together the board of directors of the Bernissart Coal Board Limited Company. They decided to donate the fossils discovered in the Sainte-Barbe pit to the Belgian state and to notify Charles Delcour, minister of the interior, and Edouard Dupont, director of the MRHNB, about this decision. But the excavations could not immediately begin because the MRHNB team was busy preparing for the Paris World's Fair.

De Pauw settled in Bernissart on May 10, and the excavations began on Wednesday, May 15. The excavation team included one warder (M. Sonnet) and one molder (A. Vandepoel) from the MRHNB, six miners (J. Créteur, A. Blanchard, J. Gérard, E. Saudemont, D. Lesplingart, and Dieudonné), and the overseers Ballez, Mortuelle, and Pierrard. Every day from 5:30 in the morning until 12:30 in the afternoon, the team went down into the Sainte-Barbe pit. The excavation method De Pauw created proved to be efficient and is still used today during paleontological excavations. Each *Iguanodon* skeleton was split into pieces. The exposed bones were first covered by wet paper or liquid clay and coated by a layer of plaster of Paris. The fossils were then undercut in a bed of matrix and the reverse side plastered. The block was then reinforced with either strips of wood or

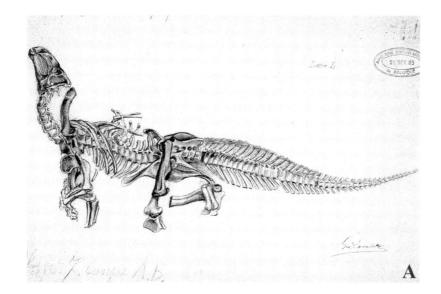

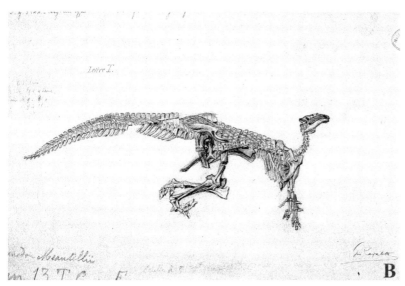

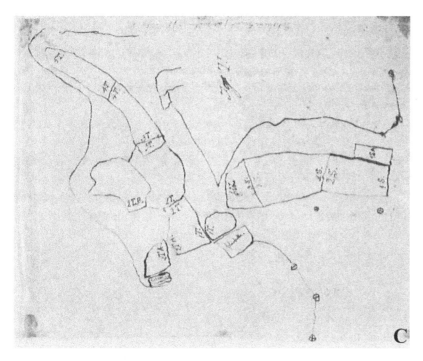

1.3. A, Drawing by G. Lavalette in 1883 of specimen "L" (RBINS R56) of *Iguanodon bernissartensis,* as discovered in the Sainte-Barbe pit. B, Drawing by G. Lavalette in 1882 of specimen "T" (RBINS R57) of *Mantellisaurus atherfieldensis,* as discovered in Sainte-Barbe pit. C, Sketch of the assemblage of plaster blocks containing pieces into which specimen "T" (RBINS R57) was divided for raising to the surface. Block "1T" contains the skull and "5T," the end of the tail of this individual.

steel, then coated with a second layer of plaster. After being sketched and cataloged (Fig. 1.3), the blocks were carried to the surface. Every afternoon, from 3:00, the team collected fossils on the coal tip in sediments previously extracted from the pit (De Pauw, 1902).

In August 1878, a big earthquake blocked the excavation team for 2 hours in the gallery 322 m below ground level. This gallery was subsequently flooded, and on Tuesday, October 22, the team was forced to abandon their work for several months. The tools and the last fossiliferous blocks had to be left behind in the flooded galleries. At that time, five skeletons of *Iguanodon* had already been discovered, although only that of "A" (RBINS VERT-5144-1716) had been excavated completely.

Between October 1878 and April 1879, individual "A" was prepared and mounted in the museum workshop at the St. Georges Chapel of Nassau Palace. The front part of this specimen had been destroyed during the original gallery excavations. This was one of the earliest mounted skeletons of associated dinosaur remains (Norman, 1986; Fig. 1.4).

In the meantime, Antoine Sohier replaced Latinis as engineer in the Bernissart Coal Board Limited Company and received the task of repairing the damaged galleries and replacing the old wooden shaft lining of the Sainte-Barbe pit with a cast iron one (see Sohier, 1880). Since the discovery of the first fossils in the Sainte-Barbe pit, the relations between Fagès and Latinis were characterized by conflict. Latinis was regularly dressed down because he did not regularly inspect the galleries. Latinis was apparently absent without leave when the gallery collapsed after the earthquake, and Fagès held him responsible for the collapse.

The excavations restarted on May 12, 1879. De Pauw was accompanied by four members of the MRHNB team (M. Sonnet, A. Collard, and A. and L. Vandepoel) and by the same miners as in 1878. J. Créteur was the first to find the abandoned tools and blocks in the gallery at −322 m (Fig. 1.5A). The excavations proceeded with great success, resulting in the removal of 14 more or less complete and four partial skeletons of iguanodontids, two *Bernissartia* (a dwarf crocodile) skeletons, one "*Goniopholis*" (larger crocodile) skeleton, two turtles, and innumerable fishes and plant remains. From this first concentration of fossils, the gallery at the 322-m level was extended horizontally for about 50 m in an east–southeast direction across the cran, passing through an area where the stratified sediments were almost horizontal but apparently devoid of large vertebrate remains. On October 22, 1879, another *Goniopholis* specimen was discovered at about 38 m from the entrance of the cran. A further eight well-preserved *Iguanodon* skeletons were discovered between 38 and 60 m from the entrance before reaching its opposite side (Fig. 1.5B).

In 1881, a new horizontal gallery was dug at a depth of −356 m. The miners also encountered fossiliferous clays, but the diameter of the cran was extremely restricted (approximately 8 m) at this level. Three more articulated skeletons were recovered from this third series of excavations (Fig. 1.5C). The clayey layers had completely disappeared 3 m below.

After three years of excavations at Bernissart, about 600 blocks, weighing a total of more than 130 tonnes, were transported to Brussels in furniture removal vans, each of 3 tonnes' capacity.

1.4. Mounting, in 1878, of the first *Iguanodon* specimen (specimen "A," RBINS VERT-5144-1716) in the St. Georges Chapel, or Nassau Chapel, assembly workshop of the MRHNB. This room is now an exhibition hall in the Albert I Royal Library, Brussels. To the left of the iguanodon's hind limb can be seen the skeletons of a kangaroo and a cassowary, used as models in assembling the skeleton.

The excavations at Bernissart were particularly expensive for the Belgian state (about 70,000 francs in the currency of the time), and the government had already allocated two extraordinary grants. In 1881, the expenses involved by this enterprise were considered too high by the Belgian government, and the excavations were stopped. Members of parliament suggested that an *Iguanodon* skeleton should be sold abroad to defray expenses, but public outcry prevented this transaction.

During World War I, the German occupation authorities decided to start new excavations at Bernissart (see Roolf, Chapter 2 in this book). The plans, revealed in documents captured after the liberation of Belgium, indicated that a new gallery was to be excavated at −340 m. The exploration gallery was stopped on October 11, 1918, 30 m in front of the border of

1.5. Plan views of the excavations at Bernissart, with the skeletons restored in their original locations (adapted by Norman 1986 from original archived documents in RBINS). Not all the letter/ number-coded individuals are now identifiable in the collections. A, First series at –322 m; B, second series at –322 m on the east–southeast side of the cran; C, third series at –356 m. *Abbreviations:* Be, the small crocodile *Bernissartia;* Go, the larger crocodile *"Goniopholis"*; ch, turtles.

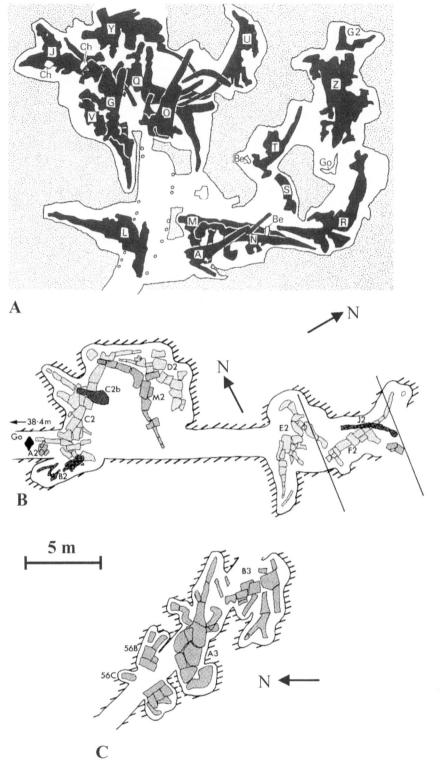

the cran. Unfortunately, the newly excavated tunnel collapsed when the occupying forces withdrew.

On January 20, 1919, Albert Anciaux, then the director general of the colliery at Bernissart, sent a letter to Gustave Gilson (the director of the MRHNB after Edouard Dupont), complaining that the costs of the aborted excavation in Bernissart between 1916 and 1918, which were entirely borne by the colliery owners, amounted to 36,604 francs. The Belgian government reimbursed these expenses in May 1923 (Gosselin, 1998).

After the war, the extraction pits were put again into exploitation at Bernissart. But coal extraction at Bernissart was no longer financially viable, even as the mining activities at the neighboring Harchies colliery became highly profitable.

On September 1921, Anciaux informed Gilson that for reasons of economy, the water pumps and ventilators at the Sainte-Barbe pit would had to be removed and the pit abandoned, unless funds could be found elsewhere. The director of the Bernissart colliery proposed two solutions to maintain the paleontological research activities in Sainte-Barbe pit and estimated the annual maintenance costs: 1,242,700 or 755,310 francs, depending on the solution that was chosen. G. Gilson approached the Belgian government and private sponsors to find financial support. Despite national and international appeals, the Sainte-Barbe pit was definitively closed at the end of October 1921.

At the beginning of the 1930s, Jules Destrée, then minister without portfolio and senator, again requested that the Belgian government release 1 million francs to restart excavations at Bernissart. The moment was badly chosen, thanks to the dramatic consequences of the 1929 stock market crash on the Belgian economy, particularly on the collieries.

According to Gosselin (1998), the German occupation authorities again tried to start excavations at Bernissart between 1940 and 1945. They took maps and documents necessary for a new exploitation of the fossiliferous site away from André Capart (then a section director at the MRHNB) and De Pauw's son. However, we did not find any document in the archives of the RBINS that could corroborate this hypothesis.

From 1882 onward, once the excavations at Bernissart had ceased, museum preparation proceeded rapidly. Once the *Iguanodon* blocks arrived in Brussels, they were stored in the museum workshop, housed in the St. Georges Chapel of Nassau Palace, now preserved as an exhibition hall in the Albert I Royal Library. De Pauw (1902) described in detail the preparation of the *Iguanodon* skeletons. The plastered blocks were exposed on their upper surface (the surface containing the bones exposed in the gallery) by removing the protective casing of plaster of Paris. Then a wall of plaster was constructed around the block and a hot glue mixture, diluted with alcohol and saturated with arsenic, was poured on top. De Pauw believed that the arsenic was able to "kill" the pyrite. Excess glue mixture was cleaned off and the block hardened in a drying room. The reverse side of the block was then prepared with a cold chisel to remove the plaster and the matrix, and the glue mixture was applied on this side. The pyrite was systematically curetted from the bones. Some vertebrae contained more than 1 kg of pyrite. The remaining cavities were filled with *carton-pierre*, a stable mixture of paper, glue, and talc.

It was decided to mount the best preserved *Iguanodon* specimens in a lifelike gait. In 1882, the first complete specimen (individual "Q," RBINS R51, the holotype of *Iguanodon bernissartensis*) was assembled and mounted by L. De Pauw and his team in the St. Georges Chapel. The bones were suspended from scaffolding by ropes that could be adjusted so as to obtain the most lifelike position for the complete skeleton, which

Preparation, Mounting, and Exhibition of the Bernissart Iguanodons

1.6. Mounting in 1882 of the first complete *Iguanodon* specimen (specimen "Q," RBINS R51, the holotype of *I. bernissartensis*) in the St. Georges Chapel. The bearded figure closest to the specimen is L. De Pauw.

was then supported by an iron framework (Fig. 1.6). This first mounted specimen was publicly exhibited in 1883 in a glass cage constructed in the interior court of Nassau Palace. In 1884, the cage was lengthened to accommodate a second specimen (individual "T," RBINS R57, the only complete specimen of *Mantellisaurus atherfieldensis*) and a selection of fossils of the Bernissart flora and fauna (Fig. 1.7).

But the Nassau Palace chapel quickly became too small for the storage, preparation, mounting, and exhibition of these numerous and bulky skeletons. In 1891, the iguanodons were transported to a new location: the Royal Museum of Natural History in Leopold Park. In 1899, five specimens were mounted in a glass cage close to the entrance of the museum. From 1902 onward, the whole Bernissart exhibition was permanently installed in the newly constructed Janlet Wing of the MRHNB. Eleven complete specimens

Godefroit et al.

were exhibited in a lifelike gait, while 12 more or less complete and eight fragmentary individuals were presented as an *en gisement* display (Fig. 1.8).

Between 1933 and 1937, the *Iguanodon* skeletons were dismantled and treated because 30 years of changes in temperature and humidity had damaged them. The bones were soaked in a mixture of alcohol and shellac, a natural lacquer secreted by coccid insects. The specimens were installed into two large glass cages to stabilize the temperature and humidity of their environment (Fig. 1.9).

During World War II, all the specimens were again dismantled and stocked in the cellars of the museum, for fear of aerial bombings. But the humidity was too much for these fragile fossils, which were again mounted in the exhibition hall before the end of the war (Bultynck, 1989).

From 2004 to 2007, the MRHNB's Janlet Wing was renovated. On this occasion, the iguanodon skeletons were again completely restored. All the bones were reinforced by a solution of synthetic polyvinyl acetate in acetone and alcohol (known by the trade name Mowilith). New glass cages were constructed to protect the skeletons (Fig. 1.10).

From the beginning, E. Dupont and G. Fagès had friendly relations. Dupont expressed his gratitude to Fagès, who had accepted the care of the Bernissart fossils for the MRHNB, and Dupont did everything in his power to

1.7. First two complete specimens of Bernissart iguanodons exhibited in the interior court of the Nassau Palace, Brussels, in 1884. Left, holotype specimen of *Iguanodon bernissartensis* (individual "Q," RBINS R51); right, complete specimen of *Mantellisaurus atherfieldensis* (individual "T," RBINS R57).

The Study of the Bernissart Iguanodons

1.8. The Bernissart iguanodons, mounted in life-like gait, in the Janlet Wing of the MRHNB in the early 1930s.

flatter Fagès. In 1883, he suggested to the minister of the interior that Fagès be decorated with the Order of Leopold, the highest distinction in Belgium. Dupont even invited Fagès and his wife to celebrate Christmas 1878 in his home! (They politely refused.) The explanatory label at the feet of the first mounted specimen in the interior court of Nassau Palace mentioned that it was "discovered in 1878 in Bernissart colliery by M. Fagès, director of the society." This label irritated P.-J. Van Beneden, who thought that he ought to be credited with first discovering the iguanodons because it was he who had first identified the fossils as belonging to the genus *Iguanodon* and who had published the first scientific note about these dinosaurs (Van Beneden, 1878). It was the start of an epic, although completely futile, dispute between Van Beneden and Dupont over the authorship of the Bernissart iguanodons during noisy sessions of the Academy of Sciences in 1883. As a consequence of these disputes, Dupont insisted that Van Beneden give the handful of bones that Cornet had sent to him in April 1878 back to the MRHNB. Of course, Van Beneden refused, and the relations between the director of the MRHNB and the professor at Leuven University continued to deteriorate.

A third contentious point, again involving P. J. Van Beneden, concerned the species that had been discovered at Bernissart: did it belong to a new species or to *Iguanodon mantelli*, already described from disarticulated specimens discovered in England? Just after the discovery of the Bernissart iguanodons, Dupont had asked the young naturalist Georges Albert Boulenger to study these specimens. In 1881, Boulenger presented his first

results to the Belgian Academy of Sciences, Letters, and Fine Arts. He described the anatomy of the pelvis of these dinosaurs and proposed that the greater number of sacral vertebrae (six) in the Bernissart form, as opposed to the five sacral vertebrae in *I. mantelli*, merited the establishment of a new species that he named *Iguanodon bernissartensis*. Unfortunately, this paper was refused publication, although a brief, highly critical review of Boulenger's paper was published by Van Beneden (1881), then president of the science section of the academy, who claimed that observed anatomical differences were most probably attributable to sexual dimorphism and that the Bernissart iguanodons belonged to *Iguanodon mantelli*. Shortly afterward, in 1881, Boulenger accepted a post at the British Museum (Natural History), and in 1882, study of the Bernissart iguanodons was entrusted to Louis Dollo, a mining engineer of French origin who eventually became a Belgian citizen and who entirely devoted his career to vertebrate paleontology at the MRHNB.

Between 1882 and 1923, Dollo (1882a, 1882b, 1883a, 1883b, 1883c, 1884, 1885a, 1885b, 1888, 1906, 1923) published many preliminary notes on the Bernissart fauna, especially on *Iguanodon*. While studying in detail several parts of the *Iguanodon* skeleton, Dollo began to adopt a forensic approach to understanding these fossils. He developed a new style of paleontology that became known as paleobiology—paleontology expanded to investigate

1.10. A new cage for the Bernissart iguanodons in 2007.

the biology, and by implication the ecology and behavior, of extinct creatures. The first paper (Dollo, 1882a) examined the basis for the creation of the new species, *I. bernissartensis*, as distinct from *I. mantelli*. Dollo established an overall similarity in anatomy between the smaller and more gracile species from Bernissart (RBINS R57) and the remains of the "Mantelpiece" (NHMUK 3741), and therefore by convention identified RBINS R57 as *Iguanodon mantelli* (Norman, 1993). With respect to the larger species, Dollo (1882a) circumvented the problems of sexual dimorphism in the sacral count by demonstrating a wider range of additional anatomic differences: skull proportions, size of narial openings, shape of the orbit, size and shape of the infratemporal openings, shape of scapular blade, completeness of external coracoid foramen and overall shape of the coracoid, size of the humerus, proportions of the manus and pollex, and shape of the anterior pubic blade. Dollo finally concluded that they merited being considered a separate species.

Dollo's final contribution to the *Iguanodon* story was published in 1923 as a synthetic study to honor the centenary of Mantell's original paper. He identified *Iguanodon* as an ecological equivalent of the giraffe. Its kangaroo-like posture enabled it to reach high into the trees to gather its fodder, which it was able to draw into its mouth with its long, muscular tongue. The sharp beak was used to nip off tough stems, while the teeth served to pulp the food before it was swallowed. This image of *Iguanodon* as a gigantic kangaroo-style creature, as depicted by Dollo, has become iconic during more than 60 years and was reinforced by the distribution of full-size replicas of mounted skeletons of *Iguanodon* from Brussels to many of the great museums around the world (Norman, 2005).

In 1980, British paleontologist D. Norman published a monograph on *Iguanodon bernissartensis*. He described the skeleton with the precision required nowadays. Functional analysis of the skeleton revealed that the vertebral column, stiffened by a network of ossified tendons, was held more or less horizontal while the animal was walking or running. Norman also believed that *I. bernissartensis* was mainly quadrupedal. The structure of the pectoral girdle, the ratios of the forelimb and hind limb lengths, the strongly fused carpal bones, and the presence of hooflike unguals on the middle three digits of the hand suggested that the adult of *I. bernissartensis* spent most of its time in a quadrupedal posture, although juveniles had a predominantly bipedal mode of life.

In 1986, Norman described the small *Iguanodon* species from Bernissart and concluded that it belonged to *Iguanodon atherfieldensis* Hooley, 1925, a species previously described from the Wealden Group of the Isle of Wight. Moreover, he stressed that the former name for it, *Iguanodon mantelli*, is a nomen dubium as a result of the fragmentary preservation of the type material of that species.

On the occasion of the mounting of an *Iguanodon bernissartensis* cast in a quadrupedal position at the RBINS in 1992, Bultynck discussed in a short paper the posture and gait of this species.

It is also worth mentioning that many specialists undertook the study of the other fossils found at Bernissart: C. E. Bertrand (1903) and G. Poinar Jr. and A. J. Boucot (2006) for coprolites, A. Lameere and G. Severin (1897) for insects, R. H. Traquair (1911) and L. Taverne (1981, 1982, 1999) for fishes, Buffetaut (1975) and M. A. Norell and J. M. Clark (1990) for the crocodile *Bernissartia fagesii*, and A. C. Seward (1900), K. L. Alvin (1953, 1957, 1960, 1971), and F. Stockmans (1960) for plants.

New Boreholes within the Iguanodon Sinkhole

In 2002–2003, three new boreholes were drilled within and around the Iguanodon Sinkhole at Bernissart. Initially, the aim of this drilling program was to evaluate the chances of finding more fossils, to understand the genesis of the Iguanodon Sinkhole, and to test a seismic geophysical technique for ground imaging (Tshibangu et al., 2004a, 2004b). In October 2002, the drilling program started with a completely cored well (named BER 3) using the PQ wireline technique. BER 3 reached 349.95 m of Thanetian, Late Cretaceous, Early Cretaceous, and Westphalian sediments (Yans et al., 2005b; Yans, 2007). During these operations, various parameters were recorded: rate of penetration, core recovery, and brief core descriptions (Tshibangu et al., 2004a, 2004b). BER 3 provided exceptional material to improve our knowledge of the iguanodon-bearing Wealden facies, with multidisciplinary research funded by FRS-FNRS (FRFC no. 2.4.568.04.F). Another borehole (BER 2) was also cut into the Wealden facies (Spagna and Van Itterbeeck, 2006).

The formation processes of the Iguanodon Sinkhole were documented by sedimentological studies of the lacustrine Wealden facies (including clay mineralogy, granulometry, and magnetic susceptibility; Spagna et al.,

2008; Spagna et al., Chapter 9 in this book) and by characterization of the organic matter with Rock-eval, palynofacies, soluble alkane content, and carbon isotope and structural analyses (Schnyder et al., 2009). Schnyder et al. (2009) suggested two steps in the life of the lacustrine Wealden facies of Bernissart: a first step with a large supply of plant debris, and a second step with active algal/bacterial activity with amorphous organic matter, which followed the lake's level variations. The paleontological content was studied using paleohistology (de Ricqlès and Yans, 2003; de Ricqlès et al., Chapter 12 in this book) and diagenesis of the bone fragments (Leduc, Chapter 11 in this book), characterization of amber, and preparations for diatom and ostracod analyses, which were unfortunately barren (C. Cornet, pers. comm.; B. Andreu, pers. comm.). A late Late Barremian to earliest Aptian age was estimated for the iguanodon-bearing sediments by both palynology and chemostratigraphy (Yans et al., 2005a, 2006, 2010; Dejax et al., 2007a; Yans et al., Chapter 8 in this book), which permitted a better knowledge of the initial steps of the subsidence in the Mons Basin (Spagna et al., 2007). Moreover, Wealden facies samples from the RBINS collection (historical searches of 1878–1881) and other localities in the Mons Basin (Hautrage, Thieu, Baudour) were also investigated. Rare dinosaur fossils are described from the Baudour Clays Formation (Godefroit et al., Chapter 13 in this book). Palynology and determination of wood and plant mesofossil fragments provide further information about the paleoenvironment of the Mons Basin during the Early Cretaceous (Gerards et al., 2007, 2008; Dejax et al., 2007b, 2008; Gomez et al., 2008; Gomez et al., Chapter 10 in this book). In Thieu, the occurrence of dinoflagellate cysts suggests marine influences in the Wealden facies of the Eastern part of the Mons Basin (Yans et al., 2007). These data were integrated into the Early Cretaceous geological context of Northwest Europe (Thiry et al., 2006; Quinif et al., 2006). Studies are still in progress . . .

References

Alvin, K. L. 1953. Three abietaceous cones from the Wealden of Belgium. Mémoires de l'Institut Royal des Sciences naturelles de Belgique 125: 3–42.

———. 1957. On the two cones *Pseudo-araucaria heeri* (Coemans) n. comb. and *Pityostrobus villerotensis* n. sp. from the Wealden of Belgium. Mémoires de l'Institut Royal des Sciences naturelles de Belgique 135: 3–27.

———. 1960. Further conifers of the Pinaceae from the Wealden formation of Belgium. Mémoires de l'Institut Royal des Sciences naturelles de Belgique 146: 3–39.

———. 1971. *Weichselia reticulata* (Stokes et Webb) Fontaine from the Wealden of Belgium. Mémoires de l'Institut Royal des Sciences naturelles de Belgique 166: 1–33.

Benton, M. J. 1989. On the trail of the dinosaurs. Crescent Books, New York, 143 pp.

Bertrand, C. E. 1903. Les coprolithes de Bernissart. 1. Les coprolithes qui ont été attribués aux iguanodons. Mémoires du Musée royal d'Histoire naturelle de Belgique 4: 1–154.

Buffetaut, E. 1975. Sur l'anatomie et la position systématique de *Bernissartia fagesii* Dollo, L. 1883, crocodilien du Wealdien de Bernissart, Belgique. Bulletin de l'Institut Royal des Sciences Naturelles de Belgique, Sciences de la Terre 51 (2): 1–20.

Bultynck, P. 1989. Bernissart et les Iguanodons. French version by F. Martin and P. Bultynck. Edition Institut royal des Sciences naturelles de Belgique, Brussels, 115 pp.

———. 1992. An assesssment of posture and gait in *Iguanodon bernissartensis* Boulenger, 1881. Bulletin de l'Institut Royal des Sciences Naturelles de Belgique, Sciences de la Terre 63: 5–11.

Dejax, J., D. Pons, and J. Yans. 2007a. Palynology of the dinosaur-bearing Wealden facies sediments in the natural pit of Bernissart (Belgium). Review of Palaeobotany and Palynology 144: 25–38.

Dejax, J., E. Dumax, F. Damblon, and J. Yans. 2007b. Palynology of Baudour Clays Formation (Mons Basin, Belgium): correlation within the "stratotypic" Wealden. Notebooks on Geology e-Journal. http://paleopolis.rediris.es/cg/CG2007_M01/CG2007_M01.pdf, 16–28.

Dejax, J., D. Pons, and J. Yans. 2008. Palynology of the Wealden facies from Hautrage quarry (Mons Basin, Belgium). Memoirs of the Geological Survey 55: 45–52.

Delguste, B. 2003. L'exploitation du charbon dans le Blaton de la première moitié du 17ème siècle. Mercuriale, Mélanges IX, Cercle d'Histoire et d'Archéologie Louis Sarot 11: 29–48.

———. 2006. Les origines de la maison "Canivez." Mercuriale, Mélanges XV, Cercle d'Histoire et d'Archéologie Louis Sarot 17: 7–12.

De Pauw, L. F. 1902. Notes sur les fouilles du charbonnage de Bernissart. Découverte, solidification et montage des Iguanodons. Imprimerie photo-litho, J. H. and P. Jumpertz, Bruxelles, Belgium.

Dollo, L. 1882a. Première note sur les dinosauriens de Bernissart. Bulletin du Musée royal d'Histoire naturelle de Belgique 1: 55–80.

———. 1882b. Deuxième note sur les dinosauriens de Bernissart. Bulletin du Musée royal d'Histoire naturelle de Belgique 1: 205–211.

———. 1883a. Note sur la présence chez les oiseaux du "troisième trochanter" des dinosauriens et sur la fonction de celui-ci. Bulletin du Musée royal d'Histoire naturelle de Belgique 2: 13–20.

———. 1883b. Troisième note sur les dinosauriens de Bernissart. Bulletin du Musée royal d'Histoire naturelle de Belgique 2: 85–126.

———. 1883c. Quatrième note sur les dinosauriens de Bernissart. Bulletin du Musée royal d'Histoire naturelle de Belgique 2: 223–252.

———. 1884. Cinquième note sur les dinosauriens de Bernissart. Bulletin du Musée royal d'Histoire naturelle de Belgique 3: 129–146.

———. 1885a. L'appareil sternal de l'Iguanodon. Revue des Questions Scientifiques 18: 664–673.

———. 1885b. Les Iguanodons de Bernissart. Revue des Questions Scientifiques 18: 5–55.

———. 1888. Iguanodontidae et Camptonotidae. Comptes Rendus hebdomadaires de l'Académie des Sciences, Paris 106: 775–777.

———. 1906. Les allures des Iguanodons, d'après les empreintes des pieds et de la queue. Bulletin Scientifique de la France et de la Belgique 40: 1–12.

———. 1923. Le centenaire des Iguanodons (1822–1922). Philosophical Transactions of the Royal Society of London, Series B 212: 67–78.

Forster, C. A. 1997. Iguanodontidae; pp. 359–361 in P. J. Currie and K. Padian (eds.), Encyclopedia of dinosaurs. Academic Press, San Diego.

Gerards, T., J. Yans, and P. Gerrienne. 2007. Growth rings of Lower Cretaceous Softwoods: some plaeoclimatic implications. Notebooks on Geology e-Journal. http://paleopolis.rediris.es/cg/CG2007_M01/CG2007_M01.pdf, 29–33.

Gerards, T., J. Yans, P. Spagna, and P. Gerrienne. 2008. Wood remains and sporomorphs from the Wealden facies of Hautrage (Mons Basin, Belgium): paleoclimatic and paleoenvironmental implications. Memoirs of the Geological Survey 55: 61–70.

Gomez, B., T. Gillot, V. Daviero-Gomez, P. Spagna, and J. Yans. 2008. Paleoflora from the Wealden facies strata of Belgium: mega- and meso-fossils of Hautrage (Mons Basin). Memoirs of the Geological Survey 55: 53–60.

Gosselin, R. 1998. 1878–1998, 120ème anniversaire. Histoire d'une gigantesque découverte, l'Iguanodon de Bernissart. Administration communale de Bernissart, 118 pp.

Hooley, R. W. 1925. On the skeleton of Iguanodon atherfieldensis sp. nov., from the Wealden shales of Atherfield (Isle of Wight). Quarterly Journal of the Geological Society of London 81: 1–61.

Leidy, J. 1858. Hadrosaurus and its discovery. Proceedings of the Academy of Natural Sciences, Philadelphia 1858: 213–218.

Mantell, G. A. 1825. Notice on the Iguanodon, a newly discovered fossil reptile, from the sandstone of Tilgate forest, in Sussex. Philosophical Transactions of the Royal Society of London 115: 179–186.

Norell, M. A., and J. M. Clark. 1990. A reanalysis of Bernissartia fagesii, with comments on its phylogenetic position and its bearing on the origin and diagnosis of the Eusuchia. Bulletin de l'Institut Royal des Sciences Naturelles de Belgique, Sciences de la Terre 60: 115–128.

Norman, D. B. 1980. On the ornithischian dinosaur Iguanodon bernissartensis from Belgium. Mémoires de l'Institut Royal des Sciences naturelles de Belgique 178: 1–105.

———. 1985. The illustrated encyclopedia of dinosaurs. Salamander Books, London, 208 pp.

———. 1986. On the anatomy of Iguanodon atherfieldensis (Ornithischia: Ornithopoda). Bulletin de l'Institut Royal des Sciences Naturelles de Belgique, Sciences de la Terre 56: 281–372.

———. 1987. On the discovery of fossils at Bernissart (1878–1921) Belgium. Archives of Natural History 13: 131–147.

———. 1993. Gideon Mantell's Mantelpiece: the earliest well-preserved ornithischian dinosaur. Modern Geology 18: 225–245.

———. 2005. Dinosaurs. A very short introduction. Oxford University Press, Oxford, 176 pp.

Owen, R. 1842. Report on British fossil reptiles, part II. Report of the Eleventh Meeting of the British Association for the Advancement of Science, held at Plymouth, July 1841, 66–204. London.

Poinar, G., Jr., and A. J. Boucot. 2006. Evidence of intestinal parasites of dinosaurs. Parasitology 133: 245–249.

Quinif, Y., H. Meon, and J. Yans. 2006. Nature and dating of karstic filling in the Hainaut province (Belgium). Karstic, geodynamic and paleogeographic implications. Geodinamica Acta 19: 73–85.

Ricqlès, A. de, and J. Yans. 2003. Bernissart's iguanodons: the case for "fresh" versus "old" dinosaur bone. Journal of Vertebrate Paleontology 23 (supplement to 3): 45.

Schnyder, J., J. Dejax, E. Keppens, T. Nguyen Tu, P. Spagna, S. Boulila, B. Galbrun, A. Riboulleau, J.-P. Tshibangu, and J. Yans. 2009. An Early Cretaceous lacustrine record: organic matter and organic carbon isotopes at Bernissart (Mons Basin, Belgium). Palaeogeography, Palaeoclimatology, Palaeoecology 281: 79–91.

Seward, A. C. 1900. La flore wealdienne de Bernissart. Mémoires du Musée royal d'Histoire naturelle de Belgique 1: 1–39.

Sohier, A. 1880. Note sur la réparation du cuvelage du puits n° 3, Ste-Barbe, du charbonnage de Bernissart. Revue universelle des Mines (ser. 2) 7: 1–21.

Spagna, P., and J. Van Itterbeeck. 2006. Lithological description and granulometric study of the Wealden facies in two borehole core drilled in the "Cran aux Iguanodons de Bernissart" (N-W of the Mons Basin, Belgium). Geologica Belgica Meeting, Liège, abstract book, 56.

Spagna, P., S. Vandycke, J. Yans, and C. Dupuis. 2007. Hydraulic and brittle extensional faulting in the Wealden facies of Hautrage (Mons Basin, Belgium). Geologica Belgica 10: 158–161.

Spagna, P., C. Dupuis, and J. Yans. 2008. Sedimentology of the Wealden Clays in the Hautrage Quarry (Mons Basin, Belgium). Memoirs of the Geological Survey 55: 35–44.

Stockmans, F. 1960. Guide de la salle des végétaux fossiles. Initiation à la paléobotanique stratigraphique de la Belgique et notions connexes. Les Naturalistes belges et Patrimoine de de l'Institut royal des Sciences naturelles de Belgique, Brussels, 222 pp.

Taverne, L. 1981. Ostéologie et position systématique d'*Aethalionopsis robustus* (Pisces, Teleostei) du Crétacé inférieur de Bernissart (Belgique) et considérations sur les affinités des Gonorhynchiformes. Académie Royale de Belgique, Bulletin de la Classe des Sciences 67: 958–982.

———. 1982. Sur *Pattersonella formosa* (Traquair, 1991) et *Nybelinoides brevis* (Traquair, 1911), deux téléostéens salmoniformes argentoïdes du Wealdien inférieur de Bernissart (Belgique) précédemment décrits dans le genre Leptolepis Agassiz, 1832. Bulletin de l'Institut Royal des Sciences Naturelles de Belgique, Sciences de la Terre 54 (3): 1–27.

———. 1999. Ostéologie et position systématique d'*Arratiaelops vectensis*, gen. nov., Téléostéen élopiforme du Wealdien (Crétacé inférieur) d'Angleterre et de Belgique. Bulletin de l'Institut Royal des Sciences Naturelles de Belgique, Sciences de la Terre 69: 77–96.

Thiry, M., F. Quesnel, J. Yans, R. Wyns, A. Vergari, H. Thévenaut, R. Simon-Coinçon, C. Ricordel, C., M.-G. Moreau, D. Giot, C. Dupuis, L. Bruxelles, J. Barbarand, and J.-M. Baele. 2006. Continental France and Belgium during the Early Cretaceous: paleoweatherings and paleolandforms. Bulletin Société Géologique de France 177: 155–175.

Traquair, R. H. 1911. Les poissons wealdiens de Bernissart. Mémoires du Musée royal d'Histoire naturelle de Belgique 21: 1–65.

Tshibangu, J.-P., F. Dagrain, B. Deschamps, and H. Legrain. 2004a. Nouvelles recherches dans le Cran aux Iguanodons de Bernissart. Bulletin de l'Académie Royale de Belgique, Classe des Sciences 7: 219–236.

Tshibangu, J.-P., F. Dagrain, H. Legrain, and B. Deschamps. 2004b. Coring performance to characterise the geology in the "Cran aux Iguanodons" of Bernissart (Belgium). Lecture Notes in Earth Sciences 104: 359–367.

Van Beneden, P.-J. 1878. Découverte de reptiles gigantesques dans le charbonnage de Bernissart, près de Péruwelz. Bulletin de l'Académie Royale de Belgique (ser. 2) 45: 578.

———. 1881. Sur l'arc pelvien chez les dinosauriens de Bernissart. Bulletin de l'Académie Royale Belge 1: 600–608.

Yans, J. 2007. Lithostratigraphie, minéralogie et diagenèse des sédiments à faciès wealdien du Bassin de Mons (Belgique). Mémoire de la Classe des Sciences, Académie Royale de Belgique (ser. 3) 2046: 1–179.

Yans, J., J. Dejax, D. Pons, C. Dupuis, and P. Taquet. 2005a. Paleontologic and geodynamic implications of the palynological dating of the Bernissart Wealden facies sediments (Mons Basin, Belgium). Comptes Rendus Palevol 4: 135–150.

Yans, J., P. Spagna, C. Vanneste, M. Hennebert, S. Vandycke, J.-M. Baele, J.-P. Tshibangu, P. Bultynck, M. Streel, and C. Dupuis. 2005b. Description et implications géologiques préliminaires d'un forage carotté dans le "Cran aux Iguanodons" de Bernissart. Geologica Belgica 8: 43–49.

Yans, J., J. Dejax, D. Pons, L. Taverne, and P. Bultynck. 2006. The iguanodons of Bernissart are middle Barremian to earliest Aptian in age. Bulletin Institut Sciences Naturelles Belgique 76: 91–95.

Yans, J., E. Masure, J. Dejax, D. Pons, and F. Amedro. 2007. Influences boréales dans le bassin de Mons (Belgique) à l'Albien; in L. G. Bulot, S. Ferry, and D. Grosheny (eds.), Relations entre les marges septentrionale et méridionale de la Téthys au Crétacé. Notebooks on Geology, Mémoire 2007/02, Résumé 06 (CG2007_M02/06).

Yans, J., T. Gerards, P. Gerrienne, P. Spagna, J. Dejax, J. Schnyder, J.-Y. Storme, and E. Keppens. 2010. Carbonisotope of fossil wood and dispersed organic matter from the terrestrial Wealden facies of Hautrage (Mons Basin, Belgium). Palaeogeography, Palaeoclimatology, Palaeoecology 291: 85–105.

The Attempted Theft of Dinosaur Skeletons during the German Occupation of Belgium (1914–1918) and Some Other Cases of Looting Cultural Possessions of Natural History

2

Christoph Roolf

This contribution focuses on the attempted theft of dinosaur skeletons during the German occupation of Belgium in 1914–1918 and addresses some other cases of looting cultural possessions of natural historic value in modern European history. It is not just a question of single incidents that have occurred in the history of science during times of war and occupation. As examples of forced *Kulturtransfer*, they rather turn out to be an integral part of a general history of international science relations, oscillating between cooperation and conflict.

My contribution deals with the hitherto largely unknown attempts of German scientists to seize Belgian cultural possessions during World War I.[1] Probably the most spectacular examples of these are the activities of German paleontologists and German natural history museums at the biggest dinosaur excavation site in Europe, which is located in the Belgian town of Bernissart. These took place during the German occupation of the country between 1914 and 1918. Following a plan of the German paleontologist Otto Jaekel, launched in spring 1915, more *Iguanodon* skeletons were to be excavated and transferred to German natural history museums. The work began in July 1916 and ended, without results, with the retreat of German troops from Belgium in late 1918.[2]

The establishment of German occupation authorities in large parts of Western and Eastern Europe in the late summer of 1914 opened up new and unexpected fields of activity for scientists and scientific institutions. This included direct participation in the governing bodies as well as counselling work, both of which were meant to support German occupation policies and war aims. Most of all, the occupied territories offered ample opportunities to conduct ambitious research projects with the cooperation of the occupation authorities. This would often lead to plans for looting cultural possessions (Roolf, 2009, 141ff., 148, 151). That the German occupation was actually a necessary precondition for some research projects is illustrated by the considerations I discuss below.

To analyze scientific activities during wartime, it is necessary and helpful to combine different approaches. Analytical tools from the history of science can be used in connection with those from the history of particular fields of science and general history, with special emphasis on the analysis of modern societies in times of total war. This is of crucial importance

Introduction

2.1. The German paleontologist Otto Jaekel (1863–1929) in 1890.

Courtesy Humboldt–Universität zu Berlin.

because it is the more or less complete exploitation of the civilian and material resources of occupied countries that is the hallmark of total warfare in the twentieth century.[3]

The case of the excavation site in Bernissart during World War I shows that even a seemingly unpolitical scientific project can become a battleground for different—and often opposing—interest groups within the occupying body. With the *Kulturtransfer* research concept, cases such as that of Bernissart can be understood not only as individual examples of science history in times of war, but can also be integrated into a general concept of history of international science relations. "Perception" (*Wahrnehmung*), "exchange" (*Austausch*), or "transfer" (*Übertragung*), and finally "implementation through productive appropriation, reception and acculturation" (*Implementierung durch produktive Aneignung, Rezeption und Akkulturation*)—the three characteristic stages of investigation of a typical transfer process—allow for an overall perspective of the prevailing history before and after the transfer event.[4] This will be illustrated in the second part of this contribution with some further cases of looting cultural possessions of natural history in modern European history.

Attempted Excavations at Bernissart during the German Occupation of Belgium (1914–1918)

Belgium was occupied by the German Reich during World War I. Without doubt, Otto Jaekel (1863–1929) (Fig. 2.1), a professor of geology and paleontology at Greifswald University who had come to Brügge in Belgium in the spring of 1915 as part of a reserve regiment, was the driving force behind the plans for German-led excavations in Bernissart. The earliest evidence for this is the letter he wrote to Gustav Krupp von Bohlen und Halbach on April 27, 1915. In this letter, he asked the well-known German industrialist for financial support for his project. The idea of showcasing the skeletons in the central natural history museum of the German Reich, the Museum für Naturkunde in Berlin, was already put forth. Together with the *Plateosaurus* skeletons from Halberstadt (excavated by Jaekel himself) and the dinosaur skeletons from the Tendaguru expedition in German East Africa,[5] they could bring the Berlin museum another spectacular increase of prestige among the world's leading museums of natural history. The case of the Bernissart iguanodons highlights the international rivalry of museums, which in turn can be interpreted as a projection of general antagonisms between the time's imperialistic powers into the field of science. Shortly before the war, the Muséum National d'Histoire Naturelle de Paris, an American natural history museum, and probably Jaekel himself[6] had tried in vain to start new excavations in Bernissart. If you bear this in mind, the urge to seize a *günstige Gelegenheit* ("good opportunity") for research activities in the occupied territories after the occupation of Belgium in August 1914 becomes more understandable. Otto Jaekel expressed this in his letter to Krupp, saying that "eine günstige Gelegenheit, diese Funde zu machen, kaum wiederkehren dürfte, auch nicht, wenn wir Belgien dauernd besetzen" ("a better opportunity to find [further dinosaur skeletons] will hardly ever arise again—even if we are to occupy Belgium for good") (Roolf, 2004, 7–9).

In the following months, Jaekel gained the financial support of both the German Zivilverwaltung in Brussels and the responsible ministries in

Exzellenz Dr. von Sandt
Zivilgouverneur in Belgien
mit den Mitgliedern der Zivilverwaltung

Berlin (i.e., the ministry of the interior and the Prussian ministry of culture) for his plans. The German emperor Wilhelm II himself offered to contribute a considerable sum from his private assets. Jaekel had visited the premises of the Société anonyme des charbonnages de Bernissart, where the dinosaur skeletons had first been found in July 1915. He had been accompanied by representatives of the German Zivilverwaltung in Brussels, who were in charge of the mining business within the department of commerce and industry (Fig. 2.2). By September 1915, a report had been filed by Albert Boehm, an employee of the mining section, that recommended digging a tunnel 340 m below the surface, right between the two existing Belgian tunnels (Roolf, 2004, 9–11).

Jaekel succeeded in getting some of the Belgian experts interested in the project of further excavations. Resistance from the mining company and the Musée Royal d'Histoire Naturelle de Belgique in Brussels proved to be strong, however. The museum in particular repeatedly rejected the German plans up until January 1916. The main argument was that there had been a previous agreement with the mining company to transfer any newfound fossils to Brussels. Besides, there had already been plans to continue the excavations at the site once the analysis of the existing fossils from the excavation of 1878–1882 was completed.

The German occupation authorities in Brussels, on the other hand, rejected Jaekel's proposal of breaking the Belgian opposition against the new excavations by force. They were primarily guided by the idea of a so-called tacit collaboration (*lautlose Kollaboration*), which was in line with general policies of German occupation. These emphasized the need of continued cooperation of the local administrations and ministries. The plan was also to gain support of the Flemish movement for postwar plans in Belgium that would result in a German-led Central Europe. In the meantime, it was vital to the war effort to ensure the uninterrupted provision of coal from the mines in the Borinage, Liège, and Verviers. Jaekel's ideas ran contrary to that. He urged for a forced administration to ensure new excavations in Bernissart. This notion was opposed by the German occupation authorities because it would have run counter to the idea of tacit collaboration. There was also some concern that it might cause unrest among the Belgian workforce, and further international protests against it would be perceived as another case of "cultural barbarism" and "Prussian militarism" (Roolf, 2004, 11–13).

It was only the partial consent of the mining company and the intervention of the German Zivilverwaltung that finally lead to an agreement with the Belgian ministry of science, which was in charge of the Musée Royal d'Histoire Naturelle de Belgique in Brussels, on May 10, 1916. The mining company agreed on a preliminary basis to pay for the excavations, and Otto Jaekel was granted unrestricted access to the premises. Any newfound fossils of specimens that were already on exhibit in the Brussels museum were eligible for transfer to German natural history museums, provided they were being paid for. However, both the Belgian and German parties were probably aware that this was a preliminary compromise. In the long run, all questions concerning financial matters and distribution of fossils were likely to depend on the outcome of the war (Roolf, 2004, 13ff.).

The German natural history museums and collections, however, had some ideas of their own. The group of interested museums comprised the Museum für Naturkunde Berlin, the Senckenberg-Museum der Senckenbergischen Naturforschenden Gesellschaft in Frankfurt on Main, the Roemer- und Pelizaeus-Museum in Hildesheim, the Mineralogisch-Geologische Staatsinstitut Hamburg, the Bayerische Geologisch-paläontologische Staatssammlung in Munich, and probably the Institut und Museum für Geologie und Paläontologie der Universität Tübingen. The somewhat pushy intervention of the Bavarian museum made Jaekel and the German mining department especially uncomfortable because it complicated the local situation in 1915–1916 even further. August Rothpletz, head of the Staatssammlung and the Bayerische Akademie der Wissenschaften kept insisting on obtaining a fixed share of the fossils. In the spring of 1916, a high-ranking official of the Bavarian ministry of culture went so far as to demand seizure of the coal mine as a *Faustpfand* ("bargaining counter") for future peace talks and to ensure German excavations in Bernissart. This was emphatically rejected by the German occupation authorities because again, it ran counter to the idea of tacit collaboration (Roolf, 2004, 14–17).

The seizure of cultural possessions also played a crucial role with respect to the *Rückforderungsaktion* (adverse international law), therefore the reclaiming of historical documents and pieces of art that had been stolen from German museums, libraries, and archives during the Napoleonic wars. Since August 1914, 50 German scientific institutions had been working on extensive lists of objects to be reclaimed. These included pieces of art and manuscripts located primarily in France and Russia, but also in Belgium—in this case, mostly in Brussels and Ghent. There was a continual

pressure from the scientists involved to seize exhibits from museums in occupied territories, especially in the north of France. This was supposed to increase the stakes in future peace talks and to ensure the reclamation of German cultural possessions. These plans were regularly thwarted by the German foreign office, which rejected these ideas.[7]

The proceedings and results of the German excavations in Bernissart, beginning in July 1916, can be summed up in a few words. Technical difficulties slowed down the work (Fig. 2.3), and from 1917 on, the massive recruitment of forced laborers in the arrondissement of Mons (which was at the time close to the front line) almost brought the coal mining in Bernissart to a standstill. When the excavations finally promised to unearth new *Iguanodon* skeletons, the German retreat from Belgium began, and the site was abandoned without results. Disappointment and frustration about the sudden end of research activities in the occupied countries was doubtless an important factor that led to nonacceptance of the Versailles peace terms and opposition to the first democracy among the German scientific elite (Roolf, 2004, 18ff.).

2.3. A group of excavation workers of the Bernissart coal mine between 1916 and 1918, who carried out the new German excavation at the dinosaur site, facing the premises of the mine, among them Jules Caulier (1867–1941) and Arthur Bievelez (1880–1941) (after an undated photograph from the family archive of Jean-Paul Caulier).

However, the case of Bernissart during the German occupation of Belgium in World War I is by no means an isolated case in the history of paleontology. In a history book about international scientific communities such as natural history museums, collections, and research institutes—still unwritten—one would probably discover that attempted or realized theft of fossils are often to be found among the behavioral patterns of science institutions. In the continuous anarchic up-and-down of international science

Other Cases of Looting Cultural Possessions of Natural History

relations between institutionalized and personal cooperation, between exchange, competition, alienation, conflict, confrontation, opposition, and rapprochement—all this always set in the context of state, politics, and society as well as war and peace—the looting of cultural possessions of natural history appears to be less an "industrial accident" or a curiosity than an integral part of the history of international science relations.

The campaigns and wars of revolutionary and Napoleonic France between 1794 and 1815—accompanied by systematic looting of art and cultural possessions (Savoy, 2003)—marked the beginning of looting activities of natural history specimens in its modern form. The geologist Faujas and the botanist Thouin, two professors at the Muséum National d'Histoire Naturelle de Paris, were members of the scientific confiscation commissions that followed the French army to the Netherlands, Belgium, and Germany in 1794–1795. In 11 towns and cities, they confiscated several hundred crates with natural history specimens, including plants, petrifications, and fossils—and including items from the castle of Canon Godin in Maastricht, such as the famous fossil skull (with a size of 1.2 m) of *Mosasaurus*, which had been found in the subterranean lime quarries of Maastricht in 1770. However, the subsequent confiscation campaigns in Italy and again in Germany in 1796 and 1797 were less successful. Their aims were, among others, to complete the collection stocks of the natural history museum in Paris, to set up natural history collections in French departments, and to establish a European center of art and natural science in Paris—one with an explicitly republican spirit.[8]

The French occupation of Portugal (approved by Spain) in November 1807 and repeated French attempts at invasion following various revolts—with the aim of enforcing the dissolution of the Portugal–British trade alliance and implementing the continental blockade against England[9]—were accompanied by the confiscation and transportation of fossils to France. A total of 1,959 specimens of mammals, birds, reptiles, and fishes, some originating from Portugal's colony, Brazil, were taken from Lisbon's natural history museum and brought to the Muséum National d'Histoire Naturelle de Paris. Donations of substitute specimen—a collection of birds (313 specimens) and a collection of fossils totaling 1,722 specimens[10]—to the king of Portugal, Peter V, in 1855 (as instructed by Napoleon III) had the function of emphasizing the normalization of relations between Portugal and France and reliving tension from the relations between the natural history museums in Paris and Lisbon (Antunes and Taquet, 2002, 639–647).

During World War I, there were *günstige Gelegenheiten*[11] ("good opportunities") for German paleontologists and natural history museums as well as collections for scientific activity in the extraordinary context of occupation policy by deliberately looting cultural possessions not only in Belgium, but also in other Western and Eastern European territories then occupied by the German Reich. The biography of the paleontologist Friedrich Freiherr von Huene (1875–1969) can serve as a good example of this phenomenon. Since 1902, Freiherr von Huene worked as an assistant, later as a so-called *Hauptkonservator* at the Institut und Museum für Geologie und Paläontologie der Universität Tübingen.[12] He was considered to be an *Altmeister der Dinosaurierforschung* ("pioneer of dinosaur research")

(Probst and Windolf, 1993, 72) in Germany in the first half of the twentieth century. During the prewar years, research stays led him to places in the United States and Canada (among others), and he succeeded in building up a lifelong functioning worldwide network of colleagues (Turner, 2009, 227, 229ff.). During World War I, he was active for two years as a *Kriegsgeologe* ("war geologist"), since late fall 1916 primarily in the Dobrudcha in the occupied territory called the German Militärverwaltung Rumänien, and since March 1918 in occupied northern France (Huene, 1944, 26). Among his official tasks were the professional discussion of and assistance in exploring the fighting zone with respect to the construction of buildings and trenches, as well as the draining of trenches. However, the typical function of members of the Preußische Geologische Landesanstalt in occupied territories was to render political advice, resulting in a more efficient exploitation of the occupied territories' raw material deposits, which were important for the war economy (Krusch, 1919, clvii–clxii; Bärtling, 1916, 70–85).[13] The zoologist and paleontologist Eberhard Stechow (born 1883), working as a conservator at the Bayerische Zoologische Staatssammlung in Munich, can be seen more obviously as acting in a *Grauzone* ("gray area") between scientific research under the protection of the occupation force and the looting of cultural possessions. During the war years 1915, 1916, and 1918, Stechow made several "naturwiss.[enschaftliche] Sammelreis.[en] i.[n] d.[ie] besetzt.[en] Geb.[iete] i.[m] Ost.[en]" with such substantial yields that he was afterward considered to be the discoverer of the Lithuanian Mesolithikum (Middle Stone Age). In the aftermath of Stechow's war activities, various single studies were published in his publication series Beiträge zur Natur- und Kulturgeschichte Litauens und angrenzender Gebiete.[14]

At the end of 1918, when he came back to his institute and museum of geology and paleontology, Huene was glad to resume the prewar correspondence with North American and British paleontologists (on their initiative) in the first year after the war (Huene, 1944, 27). Then, in 1920, the American paleontologist William Diller Matthew (1871–1930), a member of the world-renowned American Museum of Natural History in New York, initiated a common American–German research project on his journey to European natural history museums during a visit Huene made to Tübingen—the so-called second excavations of Trossingen, held in the small town of Trossingen in southern Wuerttemberg. Financed by the American Museum of Natural History (and thus securing half of the later findings for himself) and carried out by Huene and his colleagues and students in Tübingen, the excavations on Germany's most famous and profitable dinosaur excavation site brought to light a total of 14 single findings (among them two complete *Plateosaurus* skeletons and 12 skeleton components of the lizard pelvis dinosaur) between 1921 and 1923.[15] In the eyes of German paleontology, the American commitment in Trossingen was at the same time a major step (not to be underestimated)—tantamount to a *Satz zurück aus dem Dunklen* ("leap back out of the dark")—out of the international isolation of German science since 1914. In particular, the justification of German warfare, the destruction of the university library of Leuven, Belgium, in August 1914, and Prussian militarism in general by well-known German scientists, authors, and intellectuals (in the notorious

Aufruf der '93 an die Kulturwelt from October 1914) had led to the exclusion of German research institutions and scientists from foreign academies and international science institutions. The initiative of the American Museum of Natural History can be seen as an expression of the (successful) general American efforts for a reorganization of the international science relations in the postwar period. They had aimed—in opposition to the unyielding position of France and Belgium—for Germany's speedy return to the international scientific community.[16] The growing importance of the United States as a science power after 1918 also became apparent from the fact that the Musée Royal d'Histoire Naturelle de Belgique and the American Museum of Natural History in New York applied for financial support of an American foundation—for the planned (and finally unrealized) continuation (Gosselin, 2000, 67–70) of the new excavation at Bernissart, which had begun during the German occupation of Belgium.[17]

If and to what extent German paleontologists and natural history museums and collections again initiated—besides the renewed attempt of a German excavation in Bernissart during the second occupation of Belgium in 1940–1944[18]—research and excavation projects in European countries occupied by Germany between 1939 and 1945 and took part in the national socialist looting of cultural possessions (Heuss, 2000) in my view remains a question for further research. The same applies to a systematic history of German paleontology during national socialism.

At the end of World War II, French paleontology began to show interest—in view of the forthcoming French occupation of the German southwest—in several natural history museums and collections in southern Wuerttemberg. It is true that the *Durchsichts- und Ausbeutungsphase* ("examination and exploitation policy") (Heinemann, 1990, 417) of the French science policy in the southern occupation zone in Germany only lasted until June 1945. Several institutes of the Kaiser-Wilhelm-Gesellschaft (moved to southern Wuerttemberg because of the air raids)[19] were in the center of French interest. In May 1945, KWI für Metallforschung in Stuttgart was completely dismantled, and some of the instruments of KWI für Chemie were transported to Tailfingen (Defrance, 1994, 86ff.; Heinemann, 1990, 417; Krafft, 1981, 340–342; and in general Ludmann-Obier, 1988a, 397–414, esp. 403ff.). Only a few laboratory instruments were transported from the natural science institutes of Tübingen University to France (Zauner, 1994, 203; Fassnacht, 2000, 63). For a while during the summer of 1945, the plan to transfer laboratory instruments from Freiburg University to Tübingen was discussed by the French occupation force; it was finally given up in September 1945. There also was a *Wunsch interessierter Kreise in Paris* ("desire of interested circles in Paris") behind all this (according to Zauner), "wissenschaftliches Material und insbesondere Forschungslabors aus dem grenznahen Freiburg nach Frankreich abzutransportieren" ("to transport scientific materials and especially research labs from Freiburg, near the border, to France") (Zauner, 1994, 201ff.; see also Fassnacht, 2000, 60).

But the French culture and science policy in the French occupation zone—which until 1947 oscillated between a claim for superiority and understanding, characterized by improvisation and managed by a considerable number of involved persons and institutions—facilitated great variety in culture policy and parallel developments (Hudemann, 1983, 237–240, and 1987, 15–33; Ruge-Schatz, 1983, 91–110, esp. 91ff., 94ff., 110; Vaillant, 1989, 203–217, esp. 203–207; Henke, 1983, 49–89; Wolfrum, 1991, 34–36, 215–218) and apparently also helped to enforce the activities of French paleontology in southern Wuerttemberg until 1946.

The beginning was the destruction and withdrawal of parts of the Württembergische Naturaliensammlung Stuttgart (today called Staatliches Museum für Naturkunde Stuttgart)—scattered in the course of evacuations over the territories of the later French and American zones—by French occupation soldiers in the last months of the war (Maier, 2003, 275; details in Adam, 1991, 81–84). In September 1945, a French commission led by Camille Arambourg (1885–1969),[20] who had worked at the Muséum National d'Histoire Naturelle de Paris since 1936, arranged a confiscation of some pieces of the Stuttgart stocks of the slate plate fossils from Holzmaden, located in the French occupation zone. They were only returned to the museum in 1947 after the intervention of the Stuttgart collection conservator, Fritz Berckhemer (Maier, 2003, 275; Adam, 1991, 84f; Ludmann-Obier, 1988b, 78). Likewise, probably in the fall of 1945, it is said that Arambourg came to the Hauff-Museum in Holzmaden (in the middle Swabian Alb) with a group of French soldiers and confiscated a large ichthyosaur fossil for the Paris natural history museum.[21] Similar to the case of Bernissart during World War I, the interests of the Parisian paleontologist Arambourg and of the French military government in Germany were by no means identical in 1945–1946—on the contrary. The department of culture of the French military government, which was already counting on communication with the former enemy, was so alienated by Arambourg's activities that for a time, it even made efforts (which turned out to be futile in the end) to have the paleontologists expelled from the French occupation zone (Ludmann-Obier, 1988b, 78).

Probably in the same period, perhaps slightly later in 1946, a French commando unit (not described in detail) tried to confiscate fossils from the Institut und Museum für Geologie und Paläontologie der Universität Tübingen. This was only prevented because of the presence of the noted *Hauptkonservator*, Friedrich Freiherr von Huene (Turner, 2009, 234). Whether these confiscation activities (and possibly those in other collections) were in fact retaliatory actions against all natural history institutions in the French occupation zone that had collaborated or sympathized[22] with the national socialists (as Turner, 2009, 234, plausibly argues from the personal comments of the French paleontologist, paleontology historian, and previous director of the Muséum National d'Histoire Naturelle de Paris, Philippe Taquet), as well as how far-reaching these activities were, require broader investigation.

In the summer of 1946, Arambourg finally led an even bigger excavation in the occupied southern Wuerttemberg—namely in the fossil-rich plate limestone quarry of Nusplingen in the southwest Swabian Alb. It was reopened by the initiative of the Paris universities. Some inhabitants of the town of Nusplingen were enlisted by force to execute the excavation works. However, it is not clear how successful the excavation of fossils (subsequently transported to France) finally was; sources speak of only a few bigger specimens to two trucks full of findings packed in crates.[23]

Conclusion

Without a doubt, further research will be necessary to describe and define the specific role of paleontology in the history of international science relations. In this context, it would be interesting and promising to investigate the causal connection between the prevailing processes and events of cooperation, competition, conflict, and rapprochement of the persons and institutions involved.

Acknowledgments

I thank Heinz Krölls and Ingo Juknat for their careful reading and corrections. I am indebted to Eric Buffetaut for the information on the (realized and tried) French confiscations and excavations of fossils 1945 and 1946 in Holzmaden, Tübingen, and Nusplingen in the French occupation zone in southern Wuerttemberg. I thank Philippe Taquet for his explanations and for mailing me literature about the transportation of fossils out of Lisbon's natural history museum to Paris during the French occupation of Portugal since 1807. Jean-Paul Caulier (living in Bernissart, a great-grandson of Jules Caulier, 1867–1941, who was among the workers of the Bernissart coal mine between 1916 and 1918, and who carried out the new German excavation at the dinosaur excavation site, facing the premises of the mine) kindly made available a photograph (taken during the new excavation) from his family archive, showing a group of excavation workers, among them Jules Caulier. I thank Angela and René Delcourt for helping me procure the photograph. Susan Turner and Pierre-Yves Lacour helpfully made available their papers about German paleontologist Friedrich von Huene. I thank Susan Turner and Sofie De Schaepdrijver for carefully reading the chapter's final version and commenting on it. Christoph Bartz, Nicolas Beaupré, and Christoph Jahr, who were spontaneously ready to help and who sent me useful documents, read a 1975 biography about the French paleontologist Camille Arambourg published in a French biographical dictionary, and suggested that I write a longer biographical contribution.

Notes

1. The current state of research in this field is outlined in Roolf (2009, 137–139).

2. This contribution is mainly based—for those sections that deal with the case of the German new excavation in Bernissart during World War I—on my detailed investigation in Roolf (2004, 5–26, and references therein). The investigation was at that time also submitted as a French translation prepared by René Delcourt (Roolf, 2006, 7–34) and in a shorter version written in English (Roolf, 2005, 271–281).

3. Here, only the contribution of Förster (1999–2000, 12–29) is to be referred to, which is programmatic for the research conception of "total warfare."

4. The *Kulturtransfer* research conception has, with varying emphases, also been given the names "*Verflechtungsgeschichte*," "*histoire croisée*," "shared history," and "entangled history." About the model of the *Kulturtransfer* and the research history of the conception (signed on a transnational historiography), see the summary of Struck and Gantet (2008, 12–13, 193–198, 206–207, the quotes here). The suitability of the research concepts, especially for the investigation of international science relations, is also emphasized by Rödder (2006, 660–669, esp. 666–669).

5. See for this the definitive investigation by Maier (2003).

6. Jaekel had deliberately consulted the Brussels Museum in 1901 because of an investigation of the 3,000 fish fossils that were also found in Bernissart during the 1878–1881 excavations. See about this in retrospect the letter of Gustave Gilson (director of the Musée Royal d'Histoire Naturelle de Belgique, Brussels) to the Conseil d'Administration of the Société anonyme des Charbonnages de Bernissart, December 23, 1915, p. 5, Royal Belgian

Institute of Natural Sciences (Brussels), Archives, dossier "Bernissart 1915 à 1921."

7. For the German *Rückforderungsaktion* during World War I, see Roolf (2007, 433–477); Kott (2006, especially about the question of the museum's confiscation in occupied northern France in the context of the German museum and art policy); Savoy (2003, 293–307, 476–484); and Heuss (2000, 251–259).

8. See Lacour (2009, 101–122) and Lacour (2007, 21–40, esp. 22–26, 38–39). For the importance of the *Mosasaurus* skull fossil and its confiscation in Maastricht in 1795, see Desmond (1978, 7–14). For the biographies of Faujas de Saint-Fond (1741–1819) and Thouin (1747–1824), see Jassaud and Brygoo (2004, 210–211, entry Barthélémy Faujas de Saint-Fond, and 491–493, entry André Thouin).

9. For a general overview of this period of Portugal's history, see Bernecker and Pietschmann (2001, 71–77).

10. Large parts of the Lisbon natural history museum collections were destroyed by a fire in March 1978, including a part of the Parisian donation of 1855—the collection of birds. See Antunes and Taquet (2002, 646).

11. For the notions of the persons thus or similarly involved, and in general regarding the significance of the activities in occupied territories in the prevailing scientists' biographies, see Roolf (2009, 137–154).

12. About his biography, see Turner (2009, 223–243, esp. 225–227, 229–231, 233–234); Hölder (1977, 147–152); Probst and Windolf (1993, 70–71); Colbert (1968, 107–112); and Huene's memoirs (1944).

13. In his memoirs, Huene (1944) himself considered the importance of his two-year-long activities as a *Kriegsgeologe* to be an earlier slight: "Wissenschaftlich kam dabei nichts heraus." See Huene (1944, 26). But in 1918 he had still written within the frame of a longer contribution for the semiofficial compendium *Bilder aus der Dobrudscha*, edited by the German occupation administration: "Die Geologie der Dobrudscha ist sogar eine ungewöhnlich mannigfaltige und interessante" (Huene, 1918, 1). For the German occupation policy in Romania since 1916, see Mayerhofer (2008, 119–149).

14. See Degener (1935, 1537–1538, entry Eberhard R. Th. W. Stechow), quotation 1,538. About the German cultural and educational policy in occupied East Europe, see Liulevicius (2002, 143–188, esp. 166–171) and Häpke (1919, 17–33).

15. The three excavations at Trossingen (1911–1912, 1921–1923, and 1932) resulted in a total of 95 *Plateosaurus* findings, among them 35 complete or nearly complete skeletons of dinosaurs with a length of just under 6 m (Probst and Windolf, 1993, 64–78). For Matthew's visit to Huene in Tübingen and the American–German excavation cooperation during the second excavations of Trossingen, see Colbert (1992, 187–190); Turner (2009, 226, 231); Hölder (1977, 141–142); Colbert (1968, 126–127); and Huene (1944, 28).

16. For the rapid isolation of German science since 1914 and the Aufruf der '93, see Schroeder-Gudehus (1990, 858–885); Ungern-Sternberg and Ungern-Sternberg (1996); vom Brocke (1985, 649–719); and Krumeich (2008, 29–38). The United States, being neutral during the war, was supported by the smaller countries in their endeavors for a speedy end to the international isolation of German science; see especially Fuchs (2002, 263–284, esp. 283–284) as well as Düwell (1990, 747–777). It was perhaps therefore not a coincidence that the visit of New York paleontologist Matthew to the Tübingen dinosaur researcher Huene, "als erster Ausländer [. . .] nach dem Kriege," was shortly followed by that of Swedish paleontologist Gustaf T. Troedsson (1891–1954) from the University of Lund (as the second foreigner) (Huene, 1944, 28).

17. For this matter, see the letter of Gilson (director of the Musée Royal d'Histoire Naturelle de Belgique, Brussels) to Herbert Hoover (president of the Commission for Relief in Belgium, Educational Foundation, Stanford University, California, and later president of the United States), December 28 1921, pp. 1–8, Royal Belgian Institute of Natural Sciences (Brussels), Archives, dossier "Bernissart—Dommages de Guerre 1919–1923." Gilson had mentioned to the American foundation president the name of Henry Fairfield Osborn, president of the American Museum of Natural History in New York, as a person to contact for further information about the importance of the dinosaur excavation site of Bernissart. For a short overview of the history of the American Museum of Natural History, see Dingus (1997, 14–16). For the foundation, established in the summer of 1919 with the rest of the funds of the American Commission for Relief in Belgium (which during the war supported the people in occupied Belgium with food donations), see De Schaepdrijver (2004, 107–116, 217–219) and Schivelbusch (1993, 129–130).

18. See Roolf (2004, 19), which is based on the scant information of Gosselin (2000, 71). I reserve to myself the investigation of the continuation of the Bernissart case during World War II.

19. For moving institutes into regions less threatened by air raids, see Hachtmann (2007, 1022–1034).

20. For the biography of Arambourg and his scientific work (without mentioning his activities in Wuerttemberg 1945 and 1946), see Coppens (1975, 30–39) and Jassaud and Brygoo (2004, 42–43, entry Camille Louis Joseph Arambourg).

21. From a personal written comment by the French paleontologist Eric Buffetaut (based on his research stay at the Hauff-Museum in Holzmaden in the early 1980s) to me, April 1, 2009. For the slate fossils of Holzmaden in general, see Hauff and Hauff (1981).

22. The director of the Tübingen institute and museum, paleontologist Edwin Hennig (1882–1977), had lost his professorship in October 1945 because he was regarded by the French occupation force as a supporter of national socialism, even if he was no party member (Maier, 2003, 275; Turner, 2009, 234). For the French denazification policy in Wuerttemberg and Hohenzollern, see Henke (1981, 20–53) and Wolfrum (1991, 205–215).

23. See Dietl and Schweigert (2001, 21). The excavation in Nusplingen is also mentioned, although without an explanation of the specific circumstances, by Hölder (1977, 146: "Die nächste Grabung daselbst fand nach dem Krieg von französischer Seite [. . .] statt."

References

Archives of the Department of Paleontology, Institut royal des Sciences naturelles de Belgique, Brussels.

Adam, K. D. 1991. Die Württembergische Naturaliensammlung zu Stuttgart im Zweiten Weltkrieg. Stuttgarter Beiträge zur Naturkunde (ser. C) 30: 81–97.

Antunes, M. T., and P. Taquet. 2002. Le roi Dom Pedro V et le paléontologue Alcide d'Orbigny: un episode des relations scientifiques entre le Portugal et la France. Comptes Rendus Palevol 1: 639–647.

Bärtling, R. 1916. Grundzüge der Kriegsgeologie. Zeitschrift der Deutschen Geologischen Gesellschaft— B. Monatsberichte 68: 70–85.

Bernecker, W. L., and H. Pietschmann. 2001. Geschichte Portugals. Vom Spätmittelalter bis zur Gegenwart. C. H. Beck, Munich, 135 pp.

Brocke, B. vom. 1985. Wissenschaft und Militarismus. Der Aufruf der 93 "An die Kulturwelt"! und der Zusammenbruch der internationalen Gelehrtenrepublik im Ersten Weltkrieg; pp. 649–719 in W. M. Calder III, H. Flashar, and T. Lindken (eds.), Wilamowitz nach 50 Jahren. Wissenschaftliche Buchgesellschaft, Darmstadt.

Colbert, E. H. 1968. Men and dinosaurs. The search in field and laboratory. Evans Brothers, London, 283 pp.

———. 1992. William Diller Matthew, paleontologist: the splendid drama observed. Columbia University Press, New York, 275 pp.

Coppens, Y. 1975. Camille Arambourg (1885–1969); pp. 1:30–39 in Hommes et destins. Dictionnaire biographique d'outre-mer. Académie des sciences d'Outre-Mer, Paris.

Defrance, C. 1994. La politique culturelle de la France sur la rive gauche du Rhin 1945–1955. Presses universitaires de Strasbourg, Strasbourg, 363 pp.

Degener, H. A. L. (ed.) 1935. Degeners Wer ist's? Eine Sammlung von rund 18000 Biographien mit Angaben über Herkunft, Familie, Lebenslauf, Veröffentlichungen und Werke, Lieblingsbeschäftigung, Mitgliedschaft bei Gesellschaften, Anschrift und andere Mitteilungen von allgemeinem Interesse; Auflösung von ca. 50000 Pseudonymen. Degener, Berlin, 1833 pp.

De Schaepdrijver, S. 2004. La Belgique et la Première Guerre mondiale. Peter Lang, Brussels, 334 pp.

Desmond, A. J. 1978. Das Rätsel der Dinosaurier. Kiepenheuer & Witsch, Cologne, 378 pp.

Dietl, G., and G. Schweigert. 2001. Im Reich der Meerengel. Der Nusplinger Plattenkalk und seine Fossilien. Verlag Dr. Friedrich Pfeil, Munich, 144 pp.

Dingus, L. 1997. American Museum of Natural History; pp. 14–16 in P. J. Currie and K. Padian (eds.), Encyclopedia of dinosaurs. Academic Press, San Diego.

Düwell, K. 1990. Die deutsch-amerikanischen Wissenschaftsbeziehungen im Spiegel der Kaiser-Wilhelm- und der Max-Planck-Gesellschaft; pp. 147–777 in R. Vierhaus and B. vom Brocke (eds.), Forschung im Spannungsfeld von Politik und Gesellschaft.

Geschichte und Struktur der Kaiser-Wilhelm-/Max-Planck-Gesellschaft. Deutsche Verlags-Anstalt, Stuttgart.

Fassnacht, W. 2000. Universitäten am Wendepunkt? Die Hochschulpolitik in der französischen Besatzungszone (1945–1949). Verlag Karl Alber, Freiburg, 279 pp.

Förster, S. 1999–2000. Das Zeitalter des totalen Kriegs, 1861–1945. Konzeptionelle Überlegungen für einen historischen Strukturvergleich. Mittelweg 36. Zeitschrift des Hamburger Instituts für Sozialforschung 8: 12–29.

Fuchs, E. 2002. Wissenschaftsinternationalismus in Kriegs- und Krisenzeiten. Zur Rolle der USA bei der Reorganisation der internationalen "scientific community," 1914–1925; pp. 263–284 in R. Jessen and J. Vogel (eds.), Wissenschaft und Nation in der europäischen Geschichte. Campus Verlag, Frankfurt.

Gosselin, R. 2000. L'Iguanodon de Bernissart. Histoire d'une gigantesque découverte. Administration communale de Bernissart, Bernissart, 118 pp.

Hachtmann, R. 2007. Wissenschaftsmanagement im "Dritten Reich." Geschichte der Generalverwaltung der Kaiser-Wilhelm-Gesellschaft. Zweiter Band. Wallstein Verlag, Göttingen, 723 pp.

Häpke, R. 1919. Die geschichtliche und landeskundliche Forschung in Litauen und Baltenland 1915–1918. Hansische Geschichtsblätter 45: 17–33.

Hauff, B., and R. B. Hauff. 1981. Das Holzmaden-Buch. Holzmaden, 136 pp.

Heinemann, M. 1990. Der Wiederaufbau der Kaiser-Wilhelm-Gesellschaft und die Neugründungen der Max-Planck-Gesellschaft (1945–1949); pp. 407–470 in R. Vierhaus and B. vom Brocke (eds.), Forschung im Spannungsfeld von Politik und Gesellschaft. Geschichte und Struktur der Kaiser-Wilhelm-/Max-Planck-Gesellschaft. Deutsche Verlags-Anstalt, Stuttgart.

Henke, K.-D. 1981. Politische Säuberung unter französischer Besatzung. Die Entnazifizierung in Württemberg-Hohenzollern. Deutsche Verlags-Anstalt, Stuttgart, 205 pp.

———. 1983. Politik der Widersprüche. Zur Charakteristik der französischen Militärregierung in Deutschland nach dem Zweiten Weltkrieg; pp. 49–89 in C. Scharf and H.-J. Schröder (eds.), Die Deutschlandpolitik Frankreichs und die französische Zone 1945–1949. Franz Steiner Verlag, Wiesbaden.

Heuss, A. 2000. Kunst- und Kulturgutraub. Eine vergleichende Studie zur Besatzungspolitik der Nationalsozialisten in Frankreich und der Sowjetunion. Winter, Heidelberg, 385 pp.

Hölder, H. 1977. Geschichte der Geologie und Paläontologie an der Universität Tübingen; pp. 87–270 in W. F. von Engelhardt and H. Hölder, Mineralogie, Geologie und Paläontologie an der Universität Tübingen von den Anfängen bis zur Gegenwart. J. C. B. Mohr (Paul Siebeck), Tübingen.

Hudemann, R. 1983. Französische Besatzungszone 1945–1952; pp. 205–248 in C. Scharf and H.-J. Schröder (eds.), Die Deutschlandpolitik Frankreichs und die französische Zone 1945–1949. Franz Steiner Verlag, Wiesbaden.

———. 1987. Kulturpolitik im Spannungsfeld der Deutschlandpolitik. Frühe Direktiven für die französische Besatzung in Deutschland; pp. 15–33 in F. Knipping and J. Le Rider (eds.), Frankreichs Kulturpolitik in Deutschland, 1945–1950. Tübinger Symposium, September 19 and 20, 1985. Attempto Verlag, Tübingen.

Huene, F. von. 1918. Überblick über die Geologie der Dobrudscha; pp. 1–32 in Bilder aus der Dobrudscha. Deutsche Etappen-Verwaltung in der Dobrudscha, Constanza.

———. 1944. Arbeitserinnerungen. Kaiserliche Leopoldinisch-Carolinische Deutsche Akademie der Naturforscher, Halle (Saale), 52 pp.

Jassaud, P., and É.-R. Brygoo (eds.) 2004. Du jardin au Muséum en 516 biographies. Muséum national d'Histoire naturelle, Paris, 630 pp.

Kott, C. 2006. Préserver l'art de l'ennemi? Le patrimoine artistique en Belgique et en France occupées, 1914–1918. Peter Lang, Brussels, 441 pp.

Krafft, F. 1981. Im Schatten der Sensation. Leben und Wirken von Fritz Straßmann. Dargestellt nach Dokumenten und Aufzeichnungen. Verlag Chemie, Weinheim, 541 pp.

Krumeich, G. 2008. Bruch der Wissenschaftsbeziehungen im Ersten Weltkrieg und die Schwierigkeiten seiner Überwindung; pp. 29–38 in D. Breuer and G. Cepl-Kaufmann (eds.), Das Rheinland und die europäische Moderne. Kulturelle Austauschprozesse in Westeuropa 1900–1950. Klartext-Verlag, Essen.

Krusch, P. 1919. Die Kriegsaufgaben der Geologischen Landesanstalt. Jahrbuch der Preußischen Geologischen Landesanstalt 40: 125–162.

Lacour, P.-Y. 2007. Les Commissions pour la recherche des objets d'art et de Sciences en Belgique, Allemagne, Hollande et Italie. 1794–1797: des voyages naturalistes?; pp. 21–40 in N. Bourguinat and S. Venayre (eds.), Voyager en Europe de Humboldt à Stendhal. Contraintes nationales et tentations cosmopolites. 1790–1840. Nouveau Monde Éditions, Paris.

———. 2009. Les Amours de Mars et Flore aux cabinets. Les confiscations naturalistes en Europe septentrionale, 1794–1795. Annales historiques de la Révolution française 4: 101–122.

Liulevicius, V. G. 2002. Kriegsland im Osten. Eroberung, Kolonisierung und Militärherrschaft im Ersten Weltkrieg. Hamburger Edition, Hamburg (original edition 2000), 373 pp.

Ludmann-Obier, M.-F. 1988a. Le contrôle de la recherche scientifique en zone française d'occupation en Allemagne, 1945–1949. Revue d'Allemagne 4: 397–414.

———. 1988b. La mission du CNRS en Allemagne (1945–1949). Cahiers pour l'histoire du CNRS, 1939–1989, 1: 73–84.

Maier, G. 2003. African dinosaurs unearthed. The Tendaguru expeditions. Indiana University Press, Bloomington, 380 pp.

Mayerhofer, L. 2008. Making friends and foes: occupiers and occupied in First World War Romania, 1916–1918; pp. 119–149 in H. Jones, J. O'Brien, and C. Schmidt-Supprian (eds.), Untold war: new perspectives in First World War studies. Brill, Leiden.

Probst, E., and R. Windolf. 1993. Dinosaurier in Deutschland. C. Bertelsmann, Munich, 315 pp.

Rödder, A. 2006. Klios neue Kleider. Theoriedebatten um eine Kulturgeschichte der Politik in der Moderne. Historische Zeitschrift 283: 657–688.

Roolf, C. 2004. Dinosaurier-Skelette als Kriegsziel: Kulturgutraubplanungen, Besatzungspolitik und die deutsche Paläontologie in Belgien im Ersten Weltkrieg. Berichte zur Wissenschaftsgeschichte 27: 5–26.

———. 2005. German scientists in Belgium at the First World War between occupation policy and planning of plundering cultural assets—the case example of palaeontology; pp. 271–281 in S. Jaumain et al. (eds.), Une guerre totale? La Belgique dans la Première Guerre mondiale. Nouvelles tendances de la recherche historique. Actes du Colloque international organisé à l'Université Libre de Bruxelles du 15 au 17 janvier 2003. Archives générales du Royaume/Algemeen Rijksarchief, Brussels.

———. 2006. Des squelettes de dinosaures comme objectif de guerre: projets de pillage de biens culturels, politique d'occupation et paléontologie allemande en Belgique pendant la première guerre mondiale; pp. 7–34 in Les iguanodons de Bernissart comme objectif de guerre. Mercuriale—Cercle d'Histoire et d'Archéologie Louis Sarot, Hors série.

———. 2007. Die Forschungen des Kunsthistorikers Ernst Steinmann zum Napoleonischen Kunstraub im Ersten Weltkrieg zwischen Kulturgeschichtsschreibung, Auslandspropaganda und Kulturgutraub; pp. 433–477 in Y. Dohna (ed.), E. Steinmann. Der Kunstraub Napoleons. Bibliotheca Hertziana—Max-Planck-Institut für Kunstgeschichte Rom, Rom. http://edoc.biblhertz.it/editionen/steinmann/kunstraub/.

———. 2009. Eine "günstige Gelegenheit"? Deutsche Wissenschaftler im besetzten Belgien während des Ersten Weltkrieges (1914–1918); pp. 137–154 in M. Berg, J. Thiel, and P. T. Walther (eds.), Mit Feder und Schwert. Militär und Wissenschaft—Wissenschaftler und Krieg. Franz Steiner Verlag, Stuttgart.

Ruge-Schatz, A. 1983. Grundprobleme der Kulturpolitik in der französischen Besatzungszone; pp. 91–110 in C. Scharf and H.-J. Schröder (eds.), Die Deutschlandpolitik Frankreichs und die französische Zone 1945–1949. Franz Steiner Verlag, Wiesbaden.

Savoy, B. 2003. Patrimoine annexé. Les biens culturels saisis par la France en Allemagne autour de 1800, vol. 1. Éditions de la Maison des Sciences de l'Homme, Paris, 494 pp.

Schivelbusch, W. 1993. Eine Ruine im Krieg der Geister. Die Bibliothek von Löwen August 1914 bis Mai 1940. Fischer Taschenbuch Verlag, Frankfurt am Main, 243 pp.

Schroeder-Gudehus, B. 1990. Internationale Wissenschaftsbeziehungen und auswärtige Kulturpolitik 1919–1933. Vom Boykott und Gegen-Boykott zu ihrer Wiederaufnahme; pp. 858–885 in R. Vierhaus and B. vom Brocke (eds.), Forschung im Spannungsfeld von Politik und Gesellschaft. Geschichte und Struktur der Kaiser-Wilhelm-/Max-Planck-Gesellschaft. Deutsche Verlags-Anstalt, Stuttgart.

Struck, B., and C. Gantet. 2008. Revolution, Krieg und Verflechtung 1789–1815. Wissenschaftliche Buchgesellschaft, Darmstadt, 272 pp.

Turner, S. 2009. Reverent and exemplary: "dinosaur man" Friedrich von Huene (1875–1969); pp. 223–243 in M. Kölbl-Ebert (ed.), Geology and religion: a history of harmony and hostility. Geological Society, London.

Ungern-Sternberg, J. von, and W. von Ungern-Sternberg. 1996. Der Aufruf "An die Kulturwelt!." Das Manifest der 93 und die Anfänge der Kriegspropaganda im Ersten Weltkrieg. Franz Steiner Verlag, Stuttgart, 247 pp.

Vaillant, J. 1989. Frankreichs Kulturpolitik in Deutschland 1945–1949; pp. 203–217 in P. Hüttenberger and H. Molitor (eds.), Franzosen und Deutsche am Rhein 1789–1918–1945. Klartext-Verlag, Essen.

Wolfrum, E. 1991. Französische Besatzungspolitik und deutsche Sozialdemokratie. Politische Neuansätze in der "vergessenen Zone" bis zur Bildung des Südweststaates 1945–1952. Droste Verlag, Düsseldorf, 366 pp.

Zauner, S. 1994. Erziehung und Kulturmission. Frankreichs Bildungspolitik in Deutschland 1945–1949. R. Oldenbourg Verlag, Munich, 351 pp.

A Short Introduction to the Geology of the Mons Basin and the Iguanodon Sinkhole, Belgium

3

**Jean-Marc Baele*, Pascal Godefroit,
Paul Spagna, and Christian Dupuis**

Bernissart is located in the northern part of the Mons Basin, which consists of a 300-m-thick pile of Meso-Cenozoic sediments that accumulated in a small but actively subsiding area. Sedimentation initiated in the Lower Cretaceous with continental siliciclastics, from which the iguanodons were recovered at Bernissart, and continued under marine conditions during the Cretaceous and more changing environments during the Tertiary. Subsidence in the Mons Basin was mainly controlled by intrastratal dissolution of deep evaporite beds in the Mississippian basement. Localized collapse structures, such as sinkholes or natural pits, developed throughout the basin and trapped the Barremian lacustrine clay with dinosaurs and other taxa at Bernissart.

Introduction

Bernissart is located in the northwestern part of the Mons Basin, western Belgium, just next to the French border. The Mons Basin is a small but peculiar subsiding zone predominantly originating from deep karstification processes. Here we provide the essentials of the geological context and processes in the Bernissart area for understanding the geological environment of the deposits that have yielded the *Iguanodon* skeletons.

General Structure of the Mons Basin

The Mons Basin is traditionally defined by the extension area of Meso-Cenozoic, mainly Cretaceous, sediments that accumulated within an east–west elongate subsiding zone in southwestern Belgium (Marlière, 1970; Fig. 3.1). The basin developed uncomfortably on Pennsylvanian coal measures and is bounded by Mississippian carbonate in the north and by overthrusted Devonian siliciclastics in the south (Fig. 3.2). The subsiding area is rather small, less than 40 by 15 km in dimension, and the maximum depth of the basin is only 300 m. However, the Mons Basin has attracted many geologists because its sedimentary record is significantly different from that of other nearby basins, such as the Paris Basin, to which it is connected westward. In addition, the structure of the basin is uncommon: the maximum thickness for each sedimentary unit is observed in different region of the basin (Cornet, 1921a). There is therefore no single perennial depocenter for the basin but rather several depocenters that moved over time (Fig. 3.2). A sigmoid or clinoform-like sedimentary architecture developed, especially in the northern part of the basin. However, this is not the result of sediment

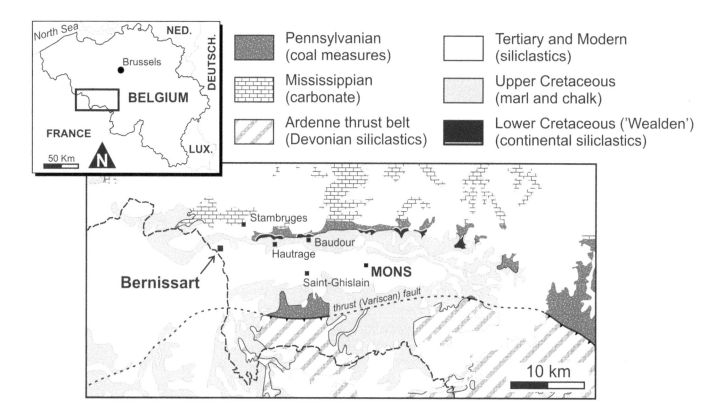

3.1. Location of Bernissart and other localities of interest within the simplified geological framework of the Mons Basin.

Sedimentary Record

prograadation, as it is the case for actual clinoforms, but of a southward migration of the depocenters through time.

Sediment accumulation in the Mons Basin started in Early Cretaceous times (Fig. 3.3). The Wealden facies (including the Sainte-Barbe Clays Formation, the Baudour Clays Formation, and the Hautrage Clays Formation), as defined by Allen (1955), appears principally as the first sediments trapped and conserved in the Mons Basin. They outcrop exclusively on the northern border of this structure, trapped either in kilomter-wide deposits (Marlière, 1946), including the Hautrage Clays Formation and the Baudour Clays Formation or infilling of sinkholes, known as resulting from deep dissolution processes (see Quinif and Licour, Chapter 5 in this book). Successive depocenter migration and erosion account for the unusual location of these oldest sediments, which would be otherwise expected to lie deeply buried in the middle of the basin. In the whole Mons Basin, the Wealden facies is clearly diachronous (Fig. 3.3), with ages extending from middle (to upper) Barremian in the western part of the basin to upper Turonian in its eastern part (Yans et al., 2006; Yans, 2007; Dejax et al., 2007, 2008). The Baudour Clays Formation and the Hautrage Clays Formation consist of lignitic clays and sands that deposited in fluvial, deltaic, and lacustrine environments (Yans, 2007; Spagna et al., Chapter 9 in this book; Godefroit et al., Chapter 13 in this book).

Eustatic transgressive pulses during the Albian and Cenomanian left mixed siliciclastic–carbonate formations known as *meule* that again are found mainly in the north of the basin but that extend deeper and farther southward than the Wealden formations. Maximum flooding of the basin was initiated with the Turonian transgression, during which marls (or

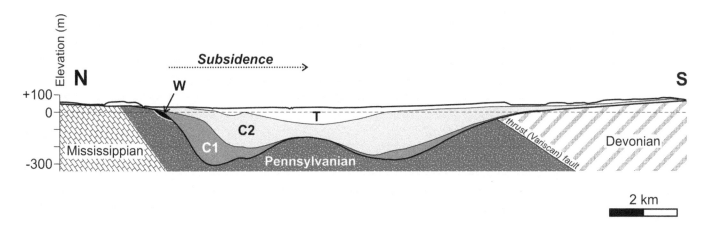

3.2. Geological cross section of the Mons basin in the Bernissart area. The arrow shows the direction of depocenter migration in the northern part of the basin since the Barremian. *Abbreviations:* W, Barremian (Wealden); C1, Albian–Cenomanian; C2, Turonian–Campanian; T, Tertiary and Quaternary.

dièves) were deposited. After a short fall in the sea level, an important transgressive phase began, and carbonate calcilutite (chalk) accumulated during the Coniacian, Santonian, and Campanian. Receding sea then resulted in an increase in detrital and phosphate input in the Maastrichtian chalk as well as a sedimentary hiatus that lasted until the Early Paleogene. Various shallow marine to continental environments were subsequently induced throughout the Tertiary by a multitude of transgressive–regressive phases. Sustained lowland conditions with frequent swamp environment, occurrence of decametric-thick Quaternary peat beds, and microseismic activity in the Mons area suggest that subsidence was active in recent times and is still active today.

Subsidence in the Mons Basin: Heritage from Deep Evaporite

The main control of the subsidence in the Mons Basin was not satisfactorily unraveled until deep anhydrite layers were discovered by drilling exploration in the 1970s (Delmer, 1972). The Saint-Ghislain borehole revealed massive anhydrite layers and associated brecciated/karstified horizons producing large quantities of sulfate-rich geothermal water (see Quinif and Licour, Chapter 5 in this book). Progressive dissolution of deep (>1,500 m) evaporite in underlying Mississippian carbonate is now considered as a major subsidence process in the Mons Basin, although tectonic activity may have also played a significant role (Dupuis and Vandycke, 1989; Vandycke and Spagna, Chapter 6 in this book). As a result of intrastratal karstification, collapse structures developed at different scales depending on factors that are not yet well understood. The highly irregular surface contact between the Paleozoic basement and overlying Cretaceous formations, formerly interpreted as a fluvial erosional surface by Cornet (1921b), now receives a better explanation through karstic-induced deformations. Among the karstic-induced collapse structures produced by deep evaporite dissolution, sinkholes, or natural pits, are the smallest in horizontal extension but perhaps the most spectacular, as they can reach more than 1,000 m in vertical extension (see Quinif and Licour, Chapter 5 in this book). The term *sinkhole* will be used in the following, although it is usually restricted to collapse structures that form at the surface. Sinkholes in the Mons Basin consist in decametric- to hectometric-wide pipes filled with downdropped and often brecciated geological formations that may originate from more than 150 m above (Delmer, 2004; Fig. 3.5). Mining

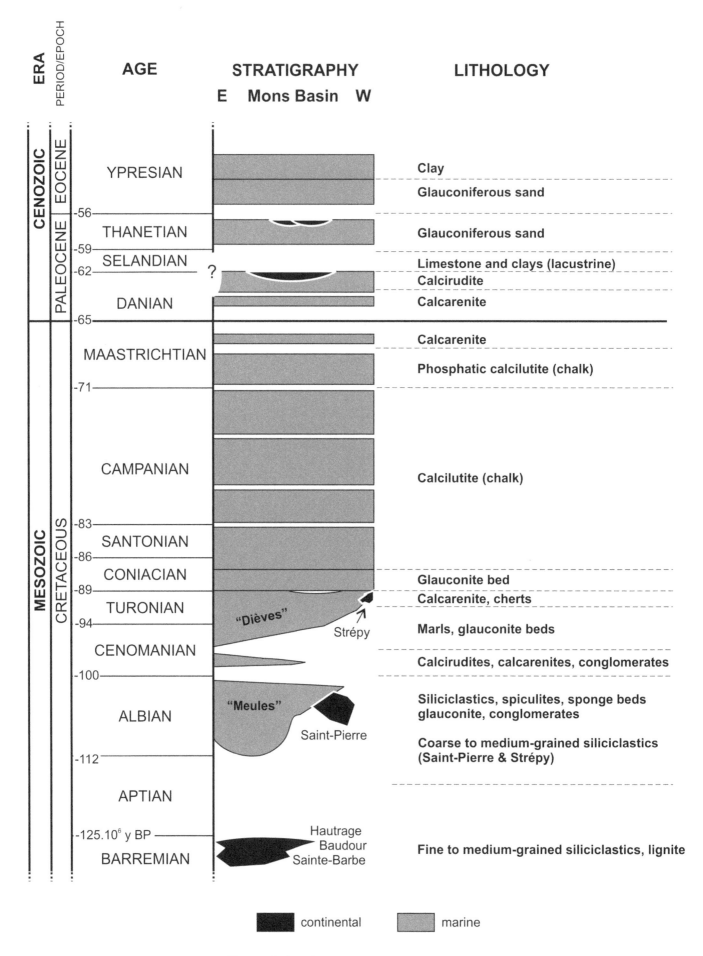

ERA	PERIOD/EPOCH	AGE	STRATIGRAPHY E Mons Basin W	LITHOLOGY

records have reported a large number of these sinkholes in the Mons area (Delmer and Van Wichelen, 1980). They are often found concentrated within larger-scale subsiding regions (Bernissart area, for example; Fig. 3.4). In these areas, geological formations are heavily fractured in decametric-wide corridors, termed *brouillages* by miners. These corridors often radiate from the sinkholes and may represent the boundaries of large blocks that have collapsed.

3.3. Cretaceous lithostratigraphic scale of the Mons Basin (modified after Marlière, 1970, and Robaszynski et al., 2001).

A Closer Look at Bernissart: The Iguanodon Sinkhole

Three sinkholes were recognized by coal miners in the Bernissart area (Fig. 3.4): the North, the South, and the Iguanodon sinkholes. Wealden facies sediments were recognized in the North Sinkhole (at –160 m) and in the Iguanodon Sinkhole (Cornet and Schmitz, 1898; Cornet, 1927). The South Sinkhole was never explored. The main filling in the North Sinkhole consists of Pennsylvanian coal strata that have just downdropped with very little deformation (to the point that they were still mineable in the past).

Figure 3.5 is a north–south cross section passing through the Iguanodon Sinkhole, adapted from Delmer and Van Wichelen (1980) and including new data from the 2003 drilling program (Tshibangu et al., 2004; Yans et al., 2005). It shows south-dipping Cretaceous strata lying with slight unconformity on the Pennsylvanian basement.

Several observations indicate sustained but fading karstic subsidence since Barremian times in the geological formation overlying the Iguanodon Sinkhole: (1) marine Cretaceous formations are downdropped and thicker than in surrounding areas, (2) Tertiary rocks also show this trend, although to a lesser extent (Van den Broeck, 1899), and (3) today a small swampy circular area is noticeable at the surface right above the sinkhole (already mentioned by De Pauw, 1898).

Coring of the BER 3 borehole drilled in 2003 yielded about 50 m of Lower Cretaceous clay (Sainte-Barbe Formation) out of the Iguanodon Sinkhole (Fig. 3.5). A middle Barremian to earliest Aptian age was obtained by palynologic dating (Yans et al., 2006; Yans et al., Chapter 8 in this book). The environment at Bernissart was formerly interpreted as lacustrine on the basis of grain size and varvelike laminar stratification (Van Den Broeck, 1898). This was confirmed by recent studies (Yans, 2007; Schnyder et al., 2009; Spagna, 2010; Spagna et al., Chapter 9 in this book).

Conclusions and Further Developments

The Bernissart iguanodons were discovered in a unique and particularly complex geological context. In this chapter, we have only presented basic information about the geology of the Mons Basin and of the Iguanodon Sinkhole. Further aspects of the geology and paleontology of the Mons Basin and of the Wealden facies at and around Bernissart will be developed in the following chapters in this book:

· Information about the geometry of the Iguanodon Sinkhole and its integration in a 3D model of the top surface of the Paleozoic basement of the Mons Basin are presented in Chapter 4.

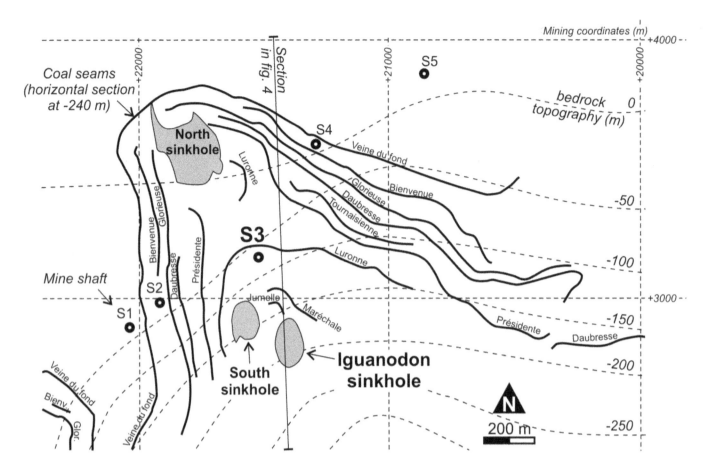

3.4. Plan view of the horizontal section at −240 m in the Bernissart area showing the relation between the sinkhole distribution and the larger-scale subsiding zone revealed by the coal seam deformation pattern (modified from Delmer, 2000). *Abbreviations:* S1, Négresse pit; S2, Moulin pit; S3, Sainte-Barbe pit; S4, Sainte-Catherine pit; S5, unnamed pit.

· In Chapter 5, the sinkholes in the Mons Basin are considered from a karstologic point of view, and their genesis and evolution are discussed.

· The interactions between tectonic and karstic processes that contributed to the trapping and the conservation of the Wealden fossil-rich deposits on the northern part of the Mons Basin are described in Chapter 6.

· Chapter 7 focuses on the stratigraphy of the Cretaceous sediments overlying the dinosaur-bearing Wealden facies cut by the BER 3 borehole in the Bernissart Sinkhole.

· The age of the *Iguanodon*-bearing Wealden facies at Bernissart is refined in Chapter 8 according to recent works based on palynology and chemostratigraphy.

· In Chapter 9, Wealden facies at Bernissart and Hautrage are investigated following different sedimentological parameters, including lithofacies evolutions, mineralogical and granulometric data, and organic matter properties. A schematic east–west paleovalley map of the Mons Basin, integrating all the new paleoenvironmental results, is proposed.

· Mesofossil plant remains, sampled from the Hautrage Clays Formation and described in Chapter 10, provide important data for the reconstruction of the paleoenvironment of the Mons Basin during the Early Cretaceous.

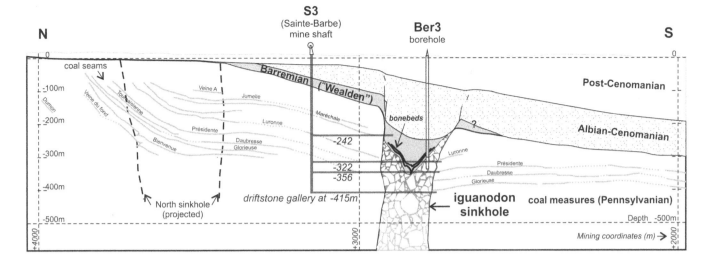

3.5. Cross section of the Bernissart area showing the geological setting of the Iguanodon Sinkhole (adapted from Delmer and Van Wichelen, 1980). In this hypothesis, downwarping of the bonebed at −322 m would explain the fossils that were found at −356 m. Therefore, the bonebeds in both galleries would be stratigraphically equivalent (see Fig. 14.4 for a second hypothesis, based on the occurrence of additional, deeper bonebeds).

· The bone diagenesis (postmortem modification of their chemical composition after their burial) of the *Iguanodon* skeletons discovered at Bernissart between 1878 and 1881 are investigated in Chapter 11.

· The bone fragments discovered at the occasion of the coring of the BER 3 borehole, drilled in 2003, are described and tentatively identified in Chapter 12. Comparison between "fresh" (from the borehole) and "old" (kept in the museum for more than 130 years under ordinary conditions) *Iguanodon* bones allowed researchers to check at the tissue level the degradation process experienced by pyritized bones.

· The rare dinosaur bones discovered in Wealden formations of the Mons Basin, outside the Iguanodon Sinkhole, are described in Chapter 13.

· As a synthesis, an integrated geological model is proposed in Chapter 14 to explain the exceptional mass accumulation of articulated skeletons in the Iguanodon Sinkhole. This model is then used as a framework for discussing different taphonomic scenarios. The role of site-specific geological factors, such as subsidence due to solution collapse deep underground and possible upwelling of sulfate-rich brines, is emphasized.

Allen, P. 1955. Age of the Wealden in north-western Europe. Geological Magazine 92: 265–281.

Cornet, J. 1921a. Études sur la structure du bassin crétacique du Hainaut. Annales de la Société géologique de Belgique 45: 43–122.

———. 1921b. Sur les détails du relief du terrain houiller recouvert par le Crétacique. Annales de la Société géologique de Belgique 45: 166–169.

———. 1927. L'époque wealdienne dans le Hainaut. Bulletin de la Société belge de Géologie, Paléontologie et Hydrologie 50: 89–104.

Cornet, J., and G. Schmitz. 1898. Note sur les puits naturels du terrain houiller du Hainaut et le gisement des Iguanodons de Bernissart. Bulletin de la Société belge de Géologie, Paléontologie et Hydrologie 12 : 196–206, 301–318.

Dejax, J., D. Pons, and J. Yans. 2007. Palynology of the dinosaur-bearing Wealden facies sediments in the natural pit of Bernissart (Belgium). Review of Palaeobotany and Palynology 144: 25–38.

———. 2008. Palynology of the Wealden facies from Hautrage quarry (Mons Basin, Belgium). Memoirs of the Geological Survey 55: 45–52.

References

Delmer, A. 1972. Origine du Bassin crétacique de la Vallée de la Haine. Professional Paper, Service géologique de Belgique
5: 1–13.
———. 2000. Les gisements houillers du
Hainaut. Vol. 1. Le Couchant de Mons,
Bruxelles, 22 pp.
———. 2004. Tectonique du front varisque
en Hainaut et dans le Namurois. Memoirs
of the Geological Society of Belgium 50:
1–61.
Delmer, A., and P. Van Wichelen. 1980.
Répertoire des puits naturels connus en
terrain houiller du Hainaut. Professional
Paper, Publication du Service Géologique
de Belgique 172: 1–79.
De Pauw, L. 1898. Observations sur le gisement de Bernissart. Bulletin de la Société
belge de Géologie, Paléontologie et
Hydrologie 12: 206–216.
Dupuis, C., and S. Vandycke. 1989. Tectonique et karstification profonde: un
modèle de subsidence original pour le
Bassin de Mons. Annales de la Société
géologique de Belgique 112: 479–487.
Marlière, R. 1946. Deltas wealdiens du Hainaut; sables et graviers de Thieu; argiles
réfractaires d'Hautrage. Bulletin de la
Société belge de Géologie 55: 69–100.
———. 1970. Géologie du bassin de Mons
et du Hainaut: un siècle d'histoire.
Annales de la Société géologique du Nord
4: 171–189.

Robaszynski, F., A. V. Dhondt, and J. W. M.
Jagt. 2001. Cretaceous lithostratigraphic
units (Belgium); pp. 121–134 in P. Bul-
tynck and L. Dejonghe (eds.), Guide
to a revised lithostratigraphic scale of
Belgium. Geologica Belgica 4.
Schnyder, J., J. Dejax, E. Keppens, T. T.
Nguyen Tu, P. Spagna, S. Boulila, B. Galbrun, A. Riboulleau, J.-P. Tshibangu, and
J. Yans. 2009. An Early Cretaceous lacustrine record: organic matter and organic
carbon isotopes at Bernissart (Mons Basin,
Belgium). Palaeogeography, Palaeoclimatology, Palaeoecology 281: 79–91.
Spagna, P. 2010. Les faciès wealdiens du
Bassin de Mons (Belgique): paléoenvironnements, géodynamique et valorisation
industrielle. Ph.D. thesis, Faculté Polytechnique de l'Université de Mons, 138
pp.
Tshibangu, J.-P., F. Dagrain, H. Legrain,
and B. Deschamps. 2004. Coring
performance to characterize the geology
in the "Cran au iguanodons" of Bernissart, Belgium; pp. 359–367 in R. Hack,
R. Azzam, and R. Charlier (eds.), Engineering geology for infrastructure planning in Europe: a European perspective.
Lecture Notes in Earth Sciences 104,
Springer-Verlag, Berlin/Heidelberg.
Van den Broeck, E. 1898. Les coupes du
gisement de Bernissart. Caractères et

dispositions sédimentaires de l'argile ossifère du Cran aux Iguanodons. Bulletin de
la Société belge de Géologie, Paléontologie et Hydrologie 12: 216–243.
———. 1899. Nouvelles observations
relatives au gisement des iguanodons de
Bernissart. Bulletin de la Société belge de
Géologie, Paléontologie et Hydrologie 13:
6–13, 175–181.
Yans, J. 2007. Lithostratigraphie, minéralogie et diagenèse des sédiments à faciès
wealdiens du Bassin de Mons (Belgique).
Mémoires de la Classe des Sciences,
Académie royale de Belgique (ser. 3) 9:
1–179.
Yans, J., P. Spagna, C. Vanneste, M. Hennebert, S. Vandycke, J. M. Baele, J.-P.
Tshibangu, P. Bultynck, M. Steel, and
C. Dupuis. 2005. Description et impli-
cations géologiques préliminaires d'un
forage carotté dans le "Cran aux Iguanodons" de Bernissart. Geologica Belgica 8:
43–49.
Yans, J., J. Dejax, J. Pons, L. Taverne, and
P. Bultynck. 2006. The iguanodons of
Bernissart are middle Barremian to ear-
liest Aptian in age. Bulletin de l'Institut
royal des Sciences naturelles de Belgique,
Sciences de la Terre 76: 91–95.

3D Modeling of the Paleozoic Top Surface in the Bernissart Area and Integration of Data from Boreholes Drilled in the Iguanodon Sinkhole

Thierry Martin*, Johan Yans, Christian Dupuis, Paul Spagna, and Olivier Kaufmann

Since 1878–1881 and the discovery of numerous complete skeletons of dinosaurs in Bernissart (Belgium), many studies have been dedicated to the paleontological content of the Iguanodon Sinkhole. However, little is known about the geometry of the sinkhole and its integration within the Paleozoic basement of the Mons Basin. In 2002–2003, three new boreholes (BER 2, BER 3, and BER 4) provided us with the opportunity to improve our understanding of the geometry of the sinkhole by integrating the new data into a 3D model of the top surface of the Paleozoic basement. To achieve this, the 3D model of the top surface of the Paleozoic was created at a regional scale (an area of 340 km^2), in the western part of the Mons Basin, near the French border. This area was delimited in order to contain enough data to outline the overall geometry. The methodology used in this study was previously developed for modeling the Meso-Cenozoic cover in the eastern part of the Mons Basin. Both the BER 2 ($Z = 33$ m) and BER 3 ($Z = 24$ m) cores cut the Wealden facies before reaching the Carboniferous basement, respectively at −291 m and −315 m below ground level. The BER 4 did not reach the natural pit, with the clayey sediments found at −246 m being attributed to the weathered Namurian basement.

Introduction

Despite specialized software that allows the modeling of complex and irregular geological bodies in 3D on the basis of geological maps, databases from geologic surveys, and structural information, it remains a challenge to build a correct model. However, a primary source of information is often abundant. Local geological data available at low cost usually consist of cross sections, geological maps, and punctual data (e.g., borehole logs and outcrop descriptions).

One of the main difficulties in using such information in 3D geological modeling lies in the heterogeneity of the descriptions and interpretations. Among the variety of recent or older data available for modeling the subsurface geology, only a small proportion is easily accessible, accurate enough, and representative at the scale of interest (Galerini et al., 2009; Jonesa et al., 2009; Kaufmann and Martin, 2008).

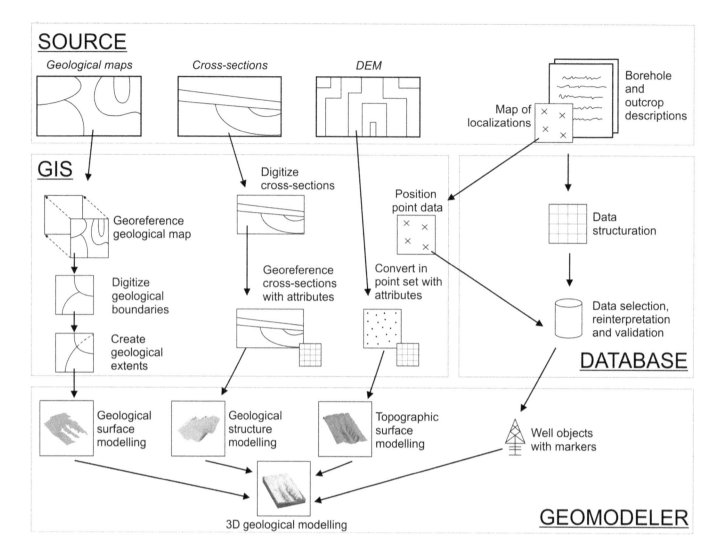

SOURCE

Geological maps *Cross-sections* *DEM*

Borehole and outcrop descriptions

Map of localizations

GIS

Georeference geological map

Digitize geological boundaries

Create geological extents

Digitize cross-sections

Georeference cross-sections with attributes

Convert in point set with attributes

Position point data

Data structuration

Data selection, reinterpretation and validation

DATABASE

Geological surface modelling

Geological structure modelling

Topographic surface modelling

Well objects with markers

GEOMODELER

3D geological modelling

Methodology

Sources of Information

In most cases, the major source of geological information is the geological survey record. Punctual data consist of borehole and outcrop descriptions collected during geological mapping or other ground investigations—for instance, mining. Interpretative information like geological maps is often available as well. These data can be used to constrain a 3D reconstruction of the subsurface geology.

Because elevations in these data sets are often unknown or wrong, or are based on former topography maps with poor relief contours, a digital elevation model (DEM) and a topographic map were used as spatial references. Special attention must be paid to areas where the relief was modified, such as embankments and slag heaps, between the original observation time and the DEM.

Developed Methodology

To build accurate 3D geological models, a methodology has been developed taking into account the variety of available data. In the methodology that we adopted herein, special attention has been paid to data structures and processing flows. The overall aim is to achieve a comprehensive data description, an effective data validation, and easier model updates.

The developed methodology includes several steps to process data depending on their type. Some steps were automated to speed up data processing and to allow easier model updates. Other steps, such as data reinterpretation or validation, require user interaction.

Older information usually exists only in a nondigital or unstructured format. This latter format must be structured, encoded, and digitized, then positioned in a referenced spatial coordinate system. Finally, the original descriptions must be reinterpreted within a consistent geological framework.

As discussed above, elevations are often not reliable, and therefore a spatial reference surface is needed to position observations in a consistent altitudinal reference system. A DEM may be used to model the topographic surface and assign new elevations to all data.

The developed methodology has been implemented within a system based on ArcGIS as a geographic information system, GOCAD as a geomodeler, and a simple database (Fig. 4.1). The data transfers between these software components were simply made through file exchanges.

4.1. Organigram of the methodology developed. Data are collected in order to structure and store them in a GIS and a database. They are later integrated into a geomodeler for the 3D modeling process (Kaufmann and Martin, 2008).

Geological Information

A 3D model of the Paleozoic top surface was created from geological boundaries and punctual data such as outcrops and wells. The top surface of the Paleozoic was built with geological information from the Geological Surveys of Belgium and France. To refine the model around the Iguanodon Sinkhole at Bernissart, information was collected from coal mine maps (e.g., pit descriptions) and articles published in mining and/or geological journals.

These data were stored in a structured database. At the same time, the punctual data were positioned in the geographic information system (GIS) on a georeferenced topographic map that created a set of points; then their projected coordinates were computed and imported into the database. In the GIS, the Paleozoic boundaries were digitized as line objects from the 1:25,000 geological maps drawn by Marlière (1969, 1977).

3D Modeling of the Paleozoic Top Surface

Before modeling the top surface of the Paleozoic, the relief, extracted from a DEM, was modeled in order to position the data in a same altitudinal reference system. The Belgian Lambert 72-coordinate system, which is the present projection system used in Belgium, was adopted. The relief was modeled from a DEM consisting of a grid of 30-m by 30-m cells, with a vertical resolution of 1 m (with precision from 3 to 6 m). This DEM was processed in a GIS to be transformed as a point set with elevation attributes. Then it was imported into GOCAD, where the Z property was calculated from the elevation values. A homogenous triangulated surface was then created from these points and interpolated in order to be used as the ground elevation reference surface. This interpolation was used to gently smooth out the surface where there is strong relief. This operation was conducted in such a way that altitudes were kept within the precision range of the DEM.

The next step consisted of data validation and selection. In the validation process, every piece of information deemed reliable enough—that is, accurate enough, well located, and consistent with surrounding information—was validated and used in the 3D modeling process.

The validation process included analyses and judgment about the available data, based on information such as the author, the period, or the objective when acquiring these data. Each information type was validated manually and separately; depth of the top basement in the logs of neighboring data, position, and interpretation were checked. Accurate and reliable data were kept, whereas questionable and imprecise data were rejected.

In the selection process, information relevant to the 3D modeling of a specific area (close to the area of interest and relevant to the modeling) was extracted from the database (i.e., geological formation limits in boreholes of the area of interest). In this way, new data may be added or existing data may be reinterpreted to produce specific models and update them.

Only validated data in the studied area were selected for modeling the top surface. Among the 1,549 punctual data, only 165 boreholes were regarded as reliable and were hence validated for 3D modeling.

To give a rough estimate of the amount and the spatial distribution of validated information, a proximity map was computed for the Paleozoic top surface (Fig. 4.2). This proximity map shows the distance to the nearest Paleozoic boundary or to the nearest borehole that encountered the top surface of the Paleozoic basement. Figure 4.2 shows that the quantity of information is sufficient, except in the southwestern part of the modeled area. In the northern and southeastern parts, the geological boundaries have a strong influence on the modeling of the Paleozoic surface during the interpolation. In the Bernissart area, the density of punctual data is particularly high.

Punctual data were extracted from the database to be imported in GO-CAD as points. Geological boundaries were digitized from georeferenced geological maps, corrected when necessary using newer information, and managed in the GIS. These lines were imported like curve objects into GOCAD.

Once these steps were completed, the data were imported in the GO-CAD geomodeler to be integrated into the 3D modeling process (Fig. 4.3). Paleozoic boundaries and punctual data cutting the Paleozoic top were projected vertically onto the DEM (Fig. 4.3 A). Punctual data were likewise moved vertically at the depth where the Paleozoic was reached (Fig. 4.3 B). Once the data were positioned in the spatial referential (Fig. 4.3 C), the top surface of the Paleozoic was modeled (Fig. 4.3 D) using the discrete smooth interpolation (DSI) method (Mallet, 1997). DSI equations minimize roughness criteria while taking into account geometrical constraints.

Results and Discussion

In the studied area, the basement of the Mons Basin is characterized by a relatively vertical northern flank, whereas the southern flank is relatively flat, with many depressions. Figure 4.4 shows the resulting 3D model of the top surface of the Paleozoic basement at a regional scale, with the position of the limit between the Namurian schists and the overlying discordant Wealden facies (Early Cretaceous) in the borehole BER 3 ($Z = -290$ m).

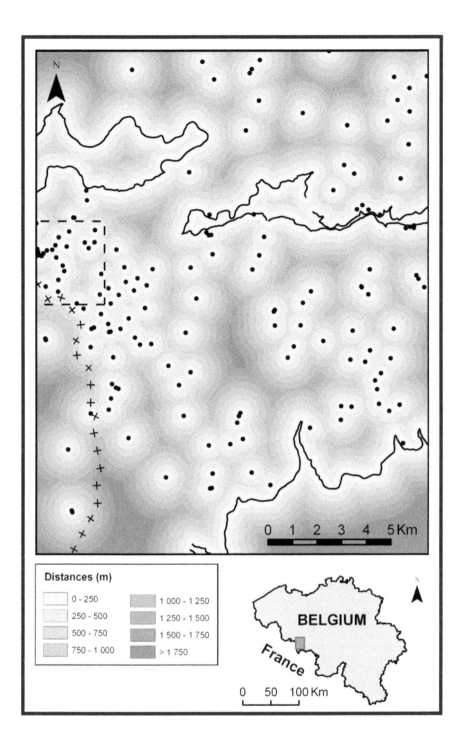

4.2. Data distribution in the studied area (point, borehole; fine line, geological boundary) and in the Bernissart area (dashed line). Direct evidence proximity map was computed for Paleozoic top surface.

Wealden facies are only represented in the northern part of the Mons Basin, in weakly buried sediments, or as infill in sinkholes developed on Namurian sediments, for example in the Bernissart pit (Sainte-Barbe Clays Formation; see Chapter 3 in this book). Wealden facies are absent in the southern part of the Mons Basin (see Fig. 3.2 in Chapter 3 in this book), where the oldest sediments are Turonian in age, indicating that the subsidence was clearly diachronic. The relative vertical northern flank of the Mons Basin probably reflects rapid karstic–tectonic activity at large scale, leading to the deposition and conservation of Wealden facies (see Chapter 5 in this book). The southern flank of the Mons Basin opened later, during a less rapid subsidence episode.

4.3. 3D modeling of the Paleozoic top surface. A, Paleozoic boundaries and punctual data cutting the Paleozoic top are projected vertically on the relief. B, Punctual data are moved vertically at the depth where the Paleozoic is reached. C, Spatial distribution of data in the studied area. D, 3D modeling of the Paleozoic top surface.

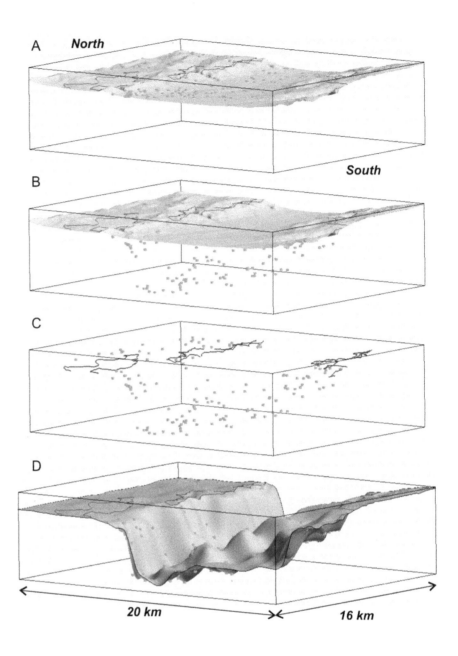

A **North**

South

B

C

D

← 20 km → ← 16 km →

The BER 3 borehole does not perfectly fit with the top of the Paleozoic basement in the 3D model presented in Figure 4.4. This is due to the local (around 50 m) spatial impact of the natural pit, characterized by subvertically oriented walls, as shown in Figure 14.3 (Chapter 14 in this book). On the basis of depth data from both the BER 2 and BER 3 boreholes, together with older mining information, the model represents the natural pit as a subcylindrical hole filled, from −230 to around −360 m, with Wealden sediments from the Sainte-Barbe Clays Formation. This geometrical configuration (reduced diameter size and vertical walls) easily explains the position of the natural pit under the Paleozoic basement top surface on the regional-scale 3D modeling that uses a grid of 30-m by 30-m cells.

Acknowledgment

We thank Katherine R. Royse, who reviewed an earlier version of this chapter and made many useful comments.

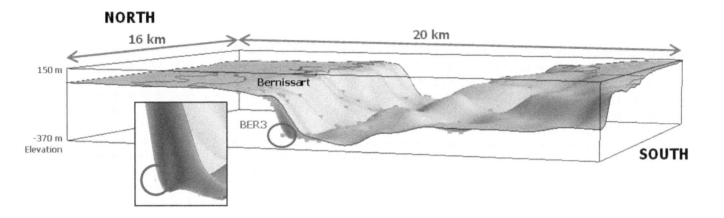

NORTH

16 km   20 km

150 m

Bernissart

BER3

-370 m
Elevation

SOUTH

Galerini, G., and M. De Donatis. 2009.
3D modelling using geognostic data:
the case of the low valley of Foglia
River (Italy). 3D Modeling in Geology. Computers and Geosciences 35:
146–164.

Jonesa, R., K. McCaffrey, P. Clegg,
R. Wilson, N. Holliman, R. Holdworth,
J. Imber, and S. Waggo. 2009. Integration of regional to outcrop digital data:
3D visualisation of multi-scale geological models. 3D Modeling in Geology.
Computers and Geosciences 35: 4–18.

Kaufmann, O., and Th. Martin. 2008. 3D
geological modelling from boreholes,

cross-sections and geological maps,
application over former natural gas
storages in coal mines. Computers and
Geosciences 34: 278–290.

Mallet, J.-L. 1997. Discrete modelling for
natural objects. Mathematical Geology
29 (2): 199–219.

Marlière, R. 1969. Carte géologique
1:25,000 de Quièvrain-Saint-Ghislain N°
150. Service Géologique de Belgique.

———. 1977. Carte géologique 1: 25,000
de Beloeil-Baudour N°139. Service
Géologique de Belgique.

References

4.4. 3D model of the top surface of the Paleozoic
basement at a regional scale, with position of the
limit between the Namurian and Wealden facies in
the borehole BER 3.

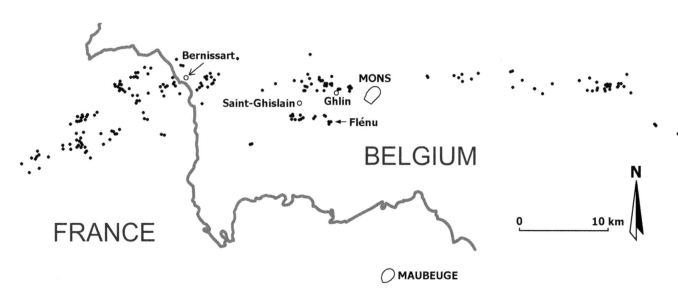

5.1. Distribution of the sinkholes in Hainaut province (Belgium) and northeastern France.

The Karstic Phenomenon of the Iguanodon Sinkhole and the Geomorphological Situation of the Mons Basin during the Early Cretaceous

5

Yves Quinif* and Luciane Licour

During the Late Jurassic and the Early Cretaceous, an extensional tectonic regime induced fracturation in carbonated Mississippian formations, notably enhanced their permeability, and initiated karstification. The low hydraulic potential that prevailed during the Cretaceous gave birth to the ghost rock karstification of the outcropping Mississippian limestone north of the Mons Basin. Deep water circulation also set in carbonated and sulfated strata, following convection induced by thermal contrast effects on water density, with the outcrop acting both as recharge and discharge area. Karstification resulting from these circulations left traces in the breccia pipes locally called "natural pits," including the famous Iguanodon Sinkhole at Bernissart.

The Iguanodon Sinkhole is one of the numerous collapse features crossing Pennsylvanian formations that are known from the French Nord-Pas-de-Calais coal basin (Puits de Dièves) to the region of Charleroi (Belgium; see Fig. 5.1). These geological structures are locally named "natural pits."

These sinkholes are roughly vertical structures filled with breccia, with various diameters and fillings. The aim of this chapter is to consider these morphological features from a karstologic point of view and to discuss their genesis and evolution.

Introduction

The sinkholes are located in the Paleozoic substratum of the Meso-Cenozoic Mons Basin. The 5,400-m-deep Saint-Ghislain borehole (Fig. 5.2), described by Groessens et al. (1979), provided precious information about the geology of the Paleozoic substratum and particularly of the Viséan series. It revealed the presence of thick anhydrite layers between −1,900 m and −2,500 m. Highly permeable karstic breccias were discovered under the anhydrite. This borehole was turned into a geothermal well, exploiting the deep Dinantian reservoir as an energy source. Two other wells were drilled in Douvrain and Ghlin. These two wells were stopped in the Upper Viséan and did not meet any anhydrite, except as pseudomorphs in breccia.

Paleozoic Substratum of the Mons Basin

Delmer and Van Wichelen (1980) localized and described the sinkholes in Hainaut province (Belgium). Most of them are located on the northern and center part of the Mons Basin, as shown in Figure 5.3A, where

Description of the Sinkholes

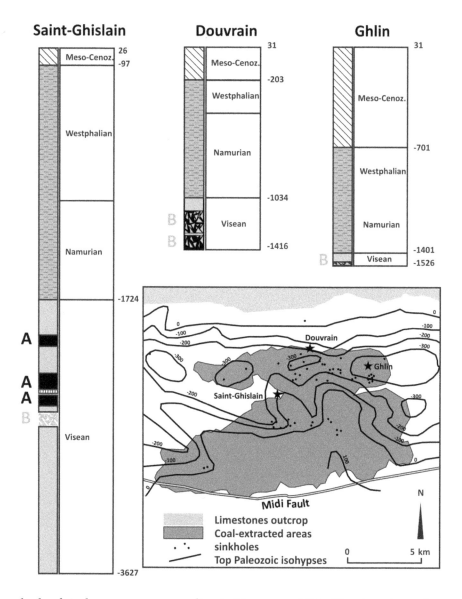

5.2. Anhydrite (A) and breccia (B) distribution in three geothermal wells in the Mons Basin.

bedrock isohypse curves according to Stevens and Marlière (1944) are also represented.

The sinkholes are irregular cylindrical features, with a diameter ranging from several meters to hundreds of meters for the greatest ones. Figure 5.3B represents the disparity of the sinkhole diameter in the Mons Basin.

The sinkholes are mostly filled by Westphalian rocks in breccia. Some of them apparently stopped their progression to the surface and found their stability within the Upper Carboniferous strata, because voids were sometimes met during coal mining works (Fig. 5.4), or, more frequently, sinkholes crossed by exploitation were not met at lower depth (e.g., the pit of Flénu; see below).

The filling of the sinkholes that emerged at the surface during the Mesozoic times is partly formed by younger sediments topping the Pennsylvanian breccia. It is thus possible to determine the period when the sinkhole reached the surface by examining its filling. A good example is the Iguanodon Sinkhole, partially filled by Wealden facies sediments.

With a known height of 1,200 m and a diameter approaching 100 m, the Flénu Sinkhole is one of the largest in the Mons Basin. This sinkhole is

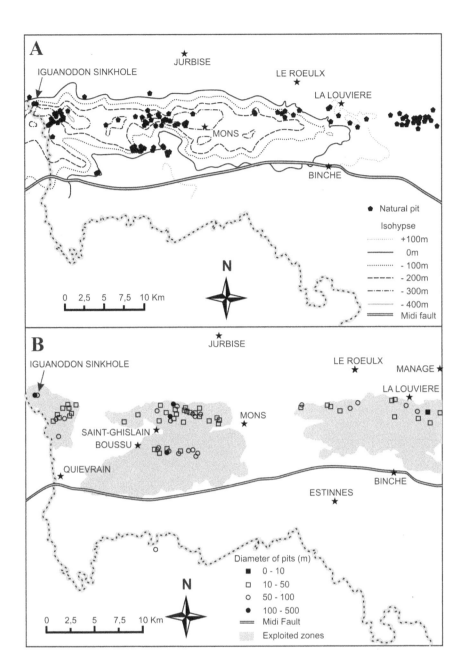

not vertical but has a southeast tilt angle of a few degrees (Fig. 5.5 A). The tilt angle can reach 20 degrees, toward the south in most cases, in other inclined sinkholes from the Mons Basin (Delmer and Van Wichelen, 1980). As in other sinkholes from the Mons Basin, the Flénu Sinkhole (Quinif, 1994) is not cylindrical, but its diameter progressively decreases toward its top, from more than 150 m at −1,300 m to less than 80 m at −150 m deep (Fig. 5.5 B).

The Ghlin geothermal well, drilled through the Ghlin Sinkhole, provided for the first time a complete stratigraphic section within the filling of a sinkhole in the Mons Basin (Fig. 5.6). From the bottom to the top, the stratigraphic series include Westphalian breccias, Wealden facies, marine Albian *meule,* and lower Turonian. It seems that Upper Turonian sediments seal the pit filling, indicating the period when it stopped functioning (Delmer et al., 1982).

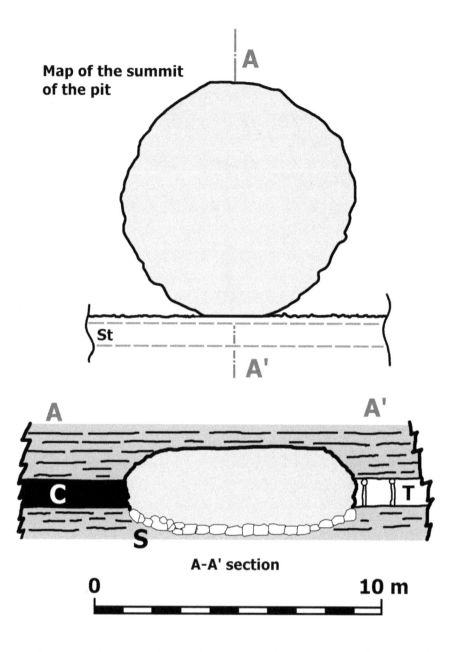

5.4. Map and section of the end of a gallery in the Quaregnon coal mine of (W. Bourgeois, pers. comm.). *Abbreviations:* T, gallery; C, coal; S, shale; st, roof support.

Map of the summit of the pit

St

A-A' section

0   10 m

In 2002 and 2003, two boreholes (BER 2 and BER 3) were drilled through the Iguanodon Sinkhole at Bernissart, providing precious stratigraphic and geological information about the Wealden facies in this sinkhole.

Genesis of the Sinkholes in the Mons Basin

In the nineteenth century, coal miners already knew about the strange pocketlike formations that interrupted the continuity of their coal seams. They called them crans or *failles circulaires*. Cornet and Briart (1870) published the first morphological study of the sinkholes in the Mons Basin. After the discovery of the Bernissart iguanodons, Dupont (1878), imagined the existence of a *Vallée bernissartienne*, supposed to be a great incision through the Paleozoic formations (Fig. 5.7).

Cornet and Schmidt (1898) were the first to hypothesize that the sinkholes resulted from collapses in cavities located in underlying Dinantian limestone and formed by underground rivers. The presence of overburden

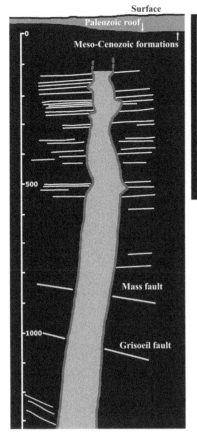

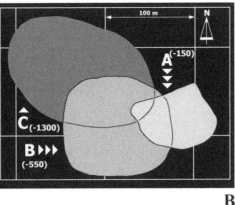

A

5.5. The Flénu Sinkhole. A, Vertical section (north-west–southeast). B, Three horizontal sections, at −150 m, −550 m, and −1,300 m, showing inclination of the sinkhole.

Mesozoic terranes in the sinkholes and the occurrence of Paleozoic breccia within the Paleozoic bedrock indicate a karstic piping.

Figure 5.8 is the synthetic model of a typical sinkhole explaining its genesis. The deep Mississippian limestone formation (8) is karstified. The voids (5) initiate breakdowns and piping, which progress across the upper formations: Namurian shales and sandstones (7), and coal-bearing formations (6) with breccia in the sinkhole (4). The sinkhole reaches the top of the Paleozoic formations. Overburden formations (2–3) are gradually decanted into the pit. The first horizontal formation (1) above the sinkhole indicates when the karstic activity leading to the formation of the sinkhole came to an end.

Karstification of the Mississippian Formations on the Northern Border of the Mons Basin

The occurrence of deep karstification in Mississippian formations reflects the geodynamical context of the Mons Basin during the Mesozoic. The study of outcropping Mississippian strata in Tournai and Soignies explains the development of deep karstification during the Late Jurassic and the Early Cretaceous.

The karstification observed in these formations is essentially a ghost rock karstification (Quinif, 1999; Quinif et al., 1993; Vergari and Quinif, 1997). The ghost rock weathering is characterized by a transformation of the massive rock into a powderlike texture (Fig. 5.9). The porosity of the rock consequently increases. Scanning electron microscope images reveal

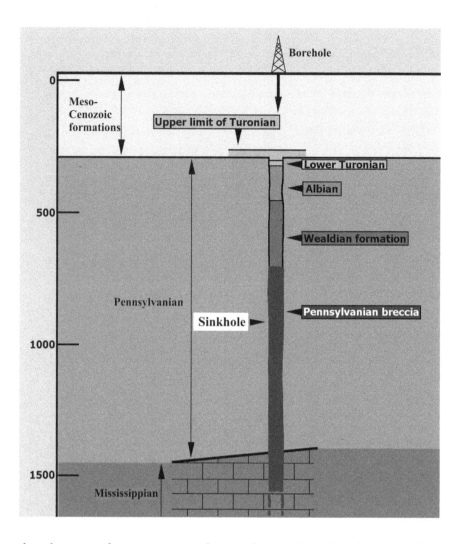

that the intact limestone is made up of grains brought closer together. The grains' surfaces are the cleavage planes. In a weathered sample, the grains are separated by interconnected voids. The sparitic crystals are isolated from the micritic cement and corrosion gulfs develop. In silica-rich limestones (near Tournai), only the silica skeleton is preserved.

The morphological characteristics of the ghost rock karstification are the vertical channels and the pseudoendokarsts (Fig. 5.10).

The formation of the ghost rock depends on the very low hydraulic potential and on the presence of aggressive water. The water must be able to penetrate the formations, which implies opened joints and therefore an extensional tectonic regime. In the Tournai area, the ghost rock structures are encountered more than 250 m below the Cretaceous paleosurface. This karstification was only possible because of the existence of an extensional tectonic phase during Cretaceous times.

It is usually difficult to date karstic features. The ghost rocks of the Mississippian carbonate of the north border are sealed by Cenomanian formations (carbonated conglomerate; Tourtia facies) and Turonian marls (Dièves facies), indicating that the karstification ended at the beginning of the Late Cretaceous. It is more hazardous to evaluate the beginning of the karstification. Tectonic arguments may be used to hypothesize that it started at the end of the Jurassic or at the beginning of the Cretaceous (Quinif, 1999; Quinif et al., 1993; Vergari and Quinif, 1997).

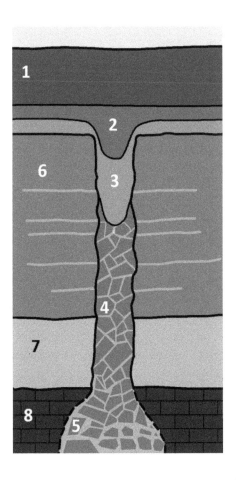

5.7. Artistic representation of Dupont's (1878) *Vallée bernissartienne*. Archives IRSNB/KBIN.

5.8. Theoretical vertical section in a sinkhole. 1, Covering of the sinkhole; 2 and 3, overburden formations; 4, sinkhole; 5, voids; 6, coal-bearing formations; 7, Namurian shales and sandstones; 8, karstified deep limestone formation.

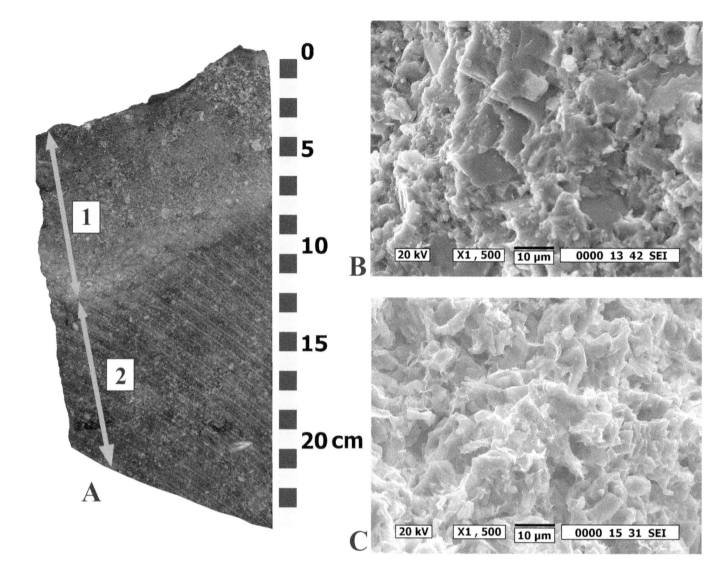

5.9. The ghost rock. A, Weathering of a limestone called *petit-granit*. B, Scanning electron micrograph of the bedrock; the grains are soldered together, without weathering. C, Scanning electron micrograph of the ghost rock; the grains are separated by voids and the porosity is interconnected. 1, Weathered zone; 2, intact limestone

Deep Karstification under the Mons Basin

The Saint-Ghislain borehole revealed that Mississippian formations contained—and still contain in some places—highly soluble rocks. Conversely, the Douvrain and Ghlin geothermal wells crossed only breccias, and no information exists about the hypothetic presence of anhydrite under the investigated depth (Fig. 5.2).

It may be hypothesized that karstification initiated in the north of the Mons Basin where Mississippian limestones outcropped, and that the dissolution phenomenon started during extensional tectonic regime in the Late Jurassic/Early Cretaceous, as shown by the earliest post-Paleozoic fillings of the sinkholes. The karstification could then propagate as a rough east–west front toward the south through the Mississippian massif, using opened joints that allowed intense fluid flows. The localization of Wealden facies deposits and of the sinkhole disposition on the northern border of the Mons Basin, and the tilt angle of some sinkholes toward the south, which might be caused by later regional subsidence influence on the whole massif, are arguments supporting this model.

Several conditions had to be fulfilled in order to allow deep karstification. Water had to be renewed, which includes recharge of fresh water and discharge of saturated solutions. In this case, recharge and discharge zones

5.10. A ghost rock feature in the Gaurain-Ramecroix quarry. Residual weathered filling remains inside this pseudocave.

are both located in the outcropping carbonates, and the movements are induced by density contrasts that are due to thermal heterogeneity and flow organization in convection cells.

The sinkholes can thus be regarded as early manifestations of deep karstification, occurring at a local scale in favorable spots, like crossing fractures, and then acting like seeds from which larger-scale dissolution

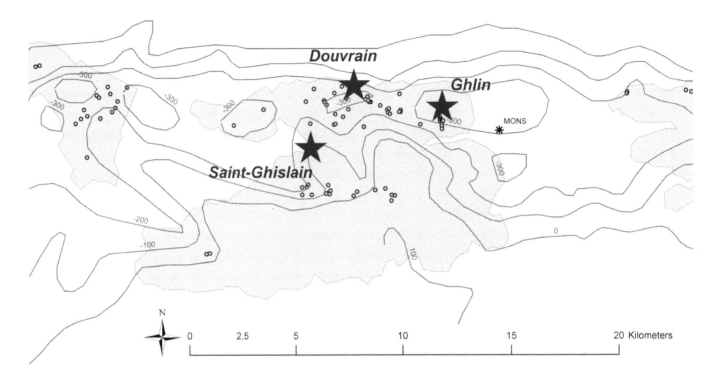

5.11. The Saint-Ghislain area and its singularities.

could propagate at the interfaces between the impermeable anhydrite and the permeable fractured carbonate.

The distribution of the sinkholes in the Saint-Ghislain area show some particularities (Fig. 5.11): none of them has been discovered in the Saint-Ghislain area itself, but many of them are located southward, quite a long way from the main basin axis. Another singularity is the fact that thick anhydrite layers still remain at Saint-Ghislain, whereas it seems to have been dissolved away in both northern and southern adjacent areas. Tectonic influence may explain both the abnormal thickness of the anhydrites and their intense deformation. Rouchy et al. (1984), among others, recognized Variscan orientations in the deformation of the anhydrite layer in the Saint-Ghislain area. The above-mentioned distribution of sinkholes south of the Saint-Ghislain area can thus be related to the same tectonic disorder that enhanced fracturing and multiplication of carbonate–sulfate interfaces.

Acknowledgments

We thank R. Maire, who reviewed a previous version of this chapter. We thank H. De Potter for redrawing Figure 5.3.

References

Cornet, J., and A. Briart. 1870. Notice sur les puits naturels du terrain houiller. Bulletin de l'Académie Royale des Sciences, des Lettres et des Beaux-Arts de Belgique 29: 477–490.

Cornet, J., and Schmitz, S. 1898. Note sur les puits naturels du terrain houiller du Hainaut et le gisement des iguanodons de Bernissart. Bulletin de la Société Belge de Géologie 12: 301–318.

Delmer, A., and P. Van Wichelen. 1980. Répertoire des puits naturels connus en terrain houiller du Hainaut. Professional Paper, Publication du Service Géologique de Belgique 172: 1–79.

Delmer, A., V. Leclerq, R. Marlière, and F. Robaszynsky. 1982. La géothermie en Hainaut et le sondage de Ghlin (Mons, Belgique). Annales de la Société Géologique du Nord 101: 189–206.

Dupont, E. 1878. Sur la découverte d'ossements d'*Iguanodon*, de poissons et de végétaux dans la fosse Sainte-Barbe du Charbonnage de Bernissart. Bulletin de l'Académie royale de Belgique 46: 387.

Groessens, E., R. Conil, and M. Hennebert. 1979. Le Dinantien du Sondage de Saint-Ghislain. Stratigraphie et Paléontologie. Mémoires pour servir à l'Explication des Cartes géologiques et minières de la Belgique 22: 1–137.

Quinif, Y. 1994. Le puits de Flénu: la plus grande structure endokarstique du monde (1200 m) et la problématique des puits du Houiller (Belgique). Karstologia 24: 29–36.

————. 1999. Fantômisation, cryptoaltération et altération sur roche nue, le triptyque de la karstification. Études de géographie physique, Travaux 1999—Supplément 18. Cagep, Université de Provence, 159–164.

Quinif, Y., A. Vergari, P. Doremus, M. Hennebert, and J.-M. Charlet. 1993. Phénomènes karstiques affectant le calcaire du Hainaut. Bulletin de la Société belge de Géologie 102: 379–394.

Rouchy, J.-M., E. Groessens, and A. Laumondais. 1984. Sédimentologie de la formation anhydritique viséenne de Saint-Ghislain (Hainaut, Belgique). Implications paléogéographiques et structurales. Bulletin de la Société belge de Géologie 93: 105–145.

Stevens, C., and R. Marlière. 1944. Révision de la carte du relief du socle paléozoïque du Bassin de Mons. Annales de la Société belge de Géologie 67: 145–175.

Vergari, A., and Y. Quinif. 1997. Les paléokarsts du Hainaut. Geodinamica Acta 10: 175–187.

6.1. Situation of the Mons Basin within the Meso-Cenozoic tectonic pattern of northwestern Europe. The Mons Basin is influenced by the tectonic activities of the Nord Artois Shear Zone (NASZ) and the Lower Rhine Graben. *Abbreviations:* L, London; B, Brussels. In gray is the extension of the Cretaceous formations in the European platform.

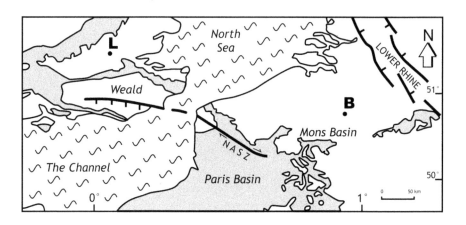

Geodynamic and Tectonic Context of Early Cretaceous *Iguanodon*-Bearing Deposits in the Mons Basin

Sara Vandycke* and Paul Spagna

The Wealden facies sediments of the Mons Basin, where the Bernissart iguanodons were discovered, are affected by multiple tectonic features. Different systems of faulting and fracturing are observed in terms of type, orientation, movements, and dating: reverse and normal faults, and strike-slip faults and joints. The synsedimentary character of the deformation is particularly clear in the clayey sediments of the Early Cretaceous Hautrage Clays Formation. The local interaction between tectonic and karstic phenomena discussed here contributes to the trapping and the conservation of the Wealden fossil-rich deposits in a floodplain depositional environment confined to the northern flank of the future Mons Basin. In this area, Lower Cretaceous sediments also recorded several younger tectonic phases due to the proximal influences of regional crustal zones, such as inversion tectonics at the end of the Cretaceous and extensional regimes linked to the dynamics of the lower Rhine Graben Embayment.

Introduction

The Mons Basin is situated on the Northwest Europe platform (Fig. 6.1). Today it is surrounded by the Paris Basin to the west, the North Sea to the north(west), and the Rhine Graben to the east. Its tectonic history has been influenced by several crustal active zones such as the North Artois Shear Zone and the Rhine Graben. But at the dawn of its activity, karstic phenomena linked to the presence of particular deep anhydrous levels (Delmer, 1972; Dupuis and Vandycke, 1989) played a major role (see Chapter 5 in this book).

The Wealden facies are mainly localized on the northern edge of the Mons Basin (Spagna et al., 2008). In the village of Hautrage, about 10 km from Bernissart, the Early Cretaceous Hautrage Clays Formation (middle Barremian to earliest Aptian; Spagna et al., 2008; Yans et al., 2002; Yans, 2007) is composed of continental clays, silts, and sands, and it contains lignite remains, pyrite, and siderite nodules in various proportions. The depositional environment of those sediments is currently interpreted as an east–west-oriented floodplain, settled by swamp and/or lacustrine environment, and crossed by numerous meandering channels (see Chapter 9 in this book). This paleovalley was connected westward to the main gutter of the Wealden sediments, ranging from the Weald (southeast England) to the Paris Basin (Thiry et al., 2006).

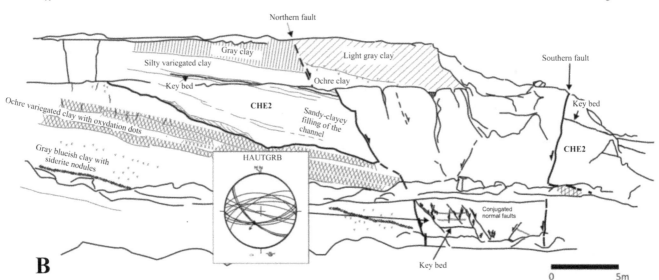

Fractures, Faults, and Other Deformations in Wealden Deposits of the Mons Basin

Different kinds of faulting and fracturing have been observed and studied in the Wealden facies sediments of the Mons Basin. They are described hereafter in terms of type and orientation.

1. Hydroplastic faults are characterized by their east–west orientation and by the presence of clay injections along their fault planes, in association with complex features of plastic deformations. This "soft" deformation suggests the synsedimentary character of the deformation. It reflects the presence of a very plastic clay layer in the Hautrage Clays Formation succession, localized just under a channel bed, suggesting a deflocculation process, when the faulting acted (Fig. 6.2A).

6.2. Two north–south outcrops in the Danube-Bouchon quarry at Hautrage. A, Graben structure with clay diapirs and injections along fault plane, upturning and normal brittle faulting in its filling. B, Same structure observed 20 m eastward, showing a total vertical throw of about 10 m. Inset, Lower Schmidt projection of fault planes.

2. Brittle normal faults are organized in grabens (Fig. 6.2 B) or found isolated in the Hautrage Clays Formation (Spagna et al., 2007). The synsedimentary aspect of this faulting event is underlined by the presence of the unfaulted covering Wealden strata (Fig. 6.2 A).

   In the complex graben structure observed at Hautrage, the combination of both those types of faults induced a total vertical throw of about 10 m. The faults are all preferentially oriented, with a measured strike around N100E° (Fig. 6.2). Striations are also observed on some fault planes, indicating the main north–south extension direction. Actually, this north–south extensional direction is coherent with the major regional movements observed in other Early Cretaceous clay sediments from western Europe (Dupuis and Vandycke, 1989), as in the Boulonnais (northern France; Bergerat and Vandycke, 1994).

   In the Sainte-Barbe Clays Formation succession, both normal and reverse small faults were observed (Fig. 6.3). These faults can be related to the accommodation of metric to decametric blocs of sediments in the karstic subsidence context of the natural pit (Spagna, 2010).

3. Some northeast–southwest-trending strike-slip faults, with lateral displacement, and northwest–southeast regularly spaced joints were also observed (Fig. 6.4). Both may be related to younger dynamics, widely recognized in the Mons Basin but also present on the European platform (Vandycke, 2002). The northeast–southwest strike-slip faults are associated with inversion tectonic events, well recognized in the Mons Basin during the lower Maastrichtian (Upper Cretaceous) and related to the North Artois Shear Zone activity (Vandycke et al., 1991). Cretaceous sediments, chalks in particular, in northwest Europe are generally affected by the northwest–southeast system of jointing. Inherited from the Late Hercynian framework, this northwest–southeast orientation is periodically remobilized until the present time. Morphological studies and tectonic analysis have highlighted these recent tectonics, generally linked to Neogene–Present rifting of the Lower Rhine

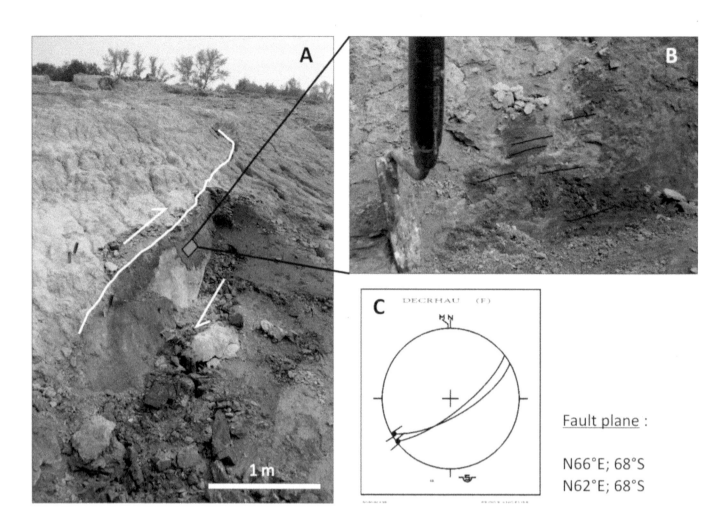

C DECRHAU (F)

Fault plane :

N66°E; 68°S
N62°E; 68°S

Embayment. In the Mons Basin, northwest–southeast orientations are found since the Early Cretaceous and are still sometimes registered morphologically in Quaternary alluvial plains (Vandycke, 2007).

Geodynamic Context

6.4. Example of northeast–southwest strike-slip fault with striation in the Wealden facies sediments (Hautrage Clays Formation) at Hautrage, in the Mons Basin (Belgium). The analysis of fault plane striation suggests a dextral movement of the fault. It is related to inversion tectonics active during the Upper Cretaceous in the Mons Basin.

The deformations registered in the Hautrage Clays Formation and Sainte-Barbe Clays Formation sediments provide pieces of information about the geodynamic context during the Early Cretaceous, when the Wealden alluvial sediments were deposited in the Mons Basin (Fig. 6.5). It has already been noted that the Wealden facies are located along the northern flank of the Mons Basin (Dupuis and Vandycke, 1989; Delmer, 1972). From the middle Barremian to the earliest Aptian, the space required for the trapping of the sediments was created by the subsidence of the basement along the northern side of the Mons Basin due to the dissolution of east–west-oriented Viséan deep anhydrite layers. This karstic dynamic may probably be related to the Mesozoic uplift of Caledonian highs by erosion of the Paleozoic cover (Vandycke et al., 1991).

A geometric model for the subsidence dynamic of both Hautrage and Bernissart deposits has been developed, in which all the deformations previously presented (except the strike-slip fault and the joints) are interpreted as accommodations of the Wealden sediments in response to the evolution of the karstic subsidence (Spagna, 2010). In this model, the Hautrage complex

Chronostratigraphy			Age M.a.	Lithostratigraphy Group	Formation	Registered deformations in the HCF serie	Geodynam.
CENOZOIC · PALEOGENE · PALEOCENE		THANETIAN	53 / 58	LANDEN	TIENEN / HANNUT / BERTAIMONT	NW-SE joints	Regional crustal activity / Local deep dissolution
		SELANDIAN	58 / 61	HASPENGOUW	HEERS / OPGLABBEEK		
		DANIAN	61 / 65	HAINE	HAININ / MONS / CIPLY / HOUTEM		
MESOZOIC · CRETACEOUS · UPPER CRETACEOUS		MAASTRICHTIAN		CHALK GROUP	MAASTRICHT / ST-SYMPHORIEN / CIPLY-MALOGNE	NE-SW strike-slip fault	
		CAMPANIAN	72		SPIENNES / GULPEN / NOUVELLES / OBOURG / Sm de Herve / VAALS / TRIVIERES	Tectonic tilting	
		SANTONIAN	83		ST-VAAST / AACHEN		
		CONIACIAN	87		MAISIERES / HAUTRAGE / ST-DENIS		
		TURONIAN	88 / 92		VILLE-POMMEROEUL / THULIN / STREPY / THIVENCELLES		
		CENOMANIAN	92 / 96		BERNISSART / BRACQUEGNIES		
MESOZOIC · CRETACEOUS · LOWER CRETACEOUS		ALBIAN (VRAC, U, M, L)		HAINE	CATILLON / HARCHIES / ST-PIERRE / POMMEROEUL	Complex graben (brittle and ductile)	
		APTIAN	108 / 113	HAINAUT	HAUTRAGE/BAUD./ST-B.	Karstic tilting (+ flexion)	? ?
		BARREMIAN	113 / 117				
		HAUTERIVIAN	117 / 123				
		VALANGINIAN	123 / 131				
		BERRIASIAN	131 / 135				

graben is understood as a graben, localized at the breakpoint of the deposits' bending, induced by a north–south gradient in the subsidence rate. The well-oriented and well-structured fault system inside the graben filling is then explained as the result of the stress-field constraint (north–south extensional), which is known as being required for the deep dissolution phenomena to occur (Quinif et al., 1997; Chapter 5 in this book).

After this first karstic subsidence phase, the subsidence of the Mons Basin increased during the Late Cretaceous, mainly because of the extensional crustal tectonic (Vandycke et al., 1991).

6.5. Synthesis of the main observed deformations in the Wealden facies sediments of the Danube-Bouchon quarry at Hautrage, in the Mons Basin. The deformations are positioned following their ages and their regional tectonic and/or local dissolution causes.

Conclusion

This work shows that sands and clays are good markers of tectonic and geodynamic evolution in a synsedimentary context. The Mons Basin is unique in the northwest European platform because of the presence of the Viséan deep layers in its underground. The interaction between tectonic and karstic phenomena resulting from the dissolution of east–west-oriented Viséan deep anhydrite layers led to the trapping and conservation of the Wealden fossil-rich deposits at the dawn of the formation of the Mons Basin.

Acknowledgments

S.V. is a research associate at the National Research Foundation of Belgium (FNRS). This work was part of the Ph.D. thesis of P.S. under the supervision of C. Dupuis. Special thanks are extended to him and to J. Yans for their help during fieldwork. Financial support was provided by university and industrial collaboration between the Faculté Polytechnique de Mons (University of Mons) and the CBR-Heidelbergcement cement manufactory of Harmignies.

References

Bergerat, F., and S. Vandycke. 1994. Cretaceous and Tertiary fault systems in the Boulonnais and Kent areas: paleostress analysis and geodynamical implications. Journal of the Geological Society of London 151: 439–448.

Delmer, A. 1972. Le Bassin du Hainaut et le sondage de Saint-Ghislain. Professional Paper, Geological Survey of Belgium, 1–143.

Dupuis, C., and S. Vandycke. 1989. Tectonique et karstification profonde: un modèle de subsidence original pour le Bassin de Mons. Annales de la Société Géologique de Belgique 112: 479–487.

Quinif, Y., S. Vandycke, and A. Vergari. 1997. Chronologie et causalité entre tectonique et karstification. L'exemple des paléokarsts crétacés du Hainaut (Belgique). Bulletin de la Société Géologique de France 168: 463–472.

Spagna, P. 2010. Les faciès wealdiens du Bassin de Mons (Belgique): paléoenvironnements, géodynamique et valorisation industrielle. Ph.D. thesis, University of Mons, 181 pp.

Spagna, P., S. Vandycke, J. Yans, and C. Dupuis. 2007. Hydraulic and brittle extensional faulting in the wealdien facies of Hautrage (Mons Basin, Belgium). Geologica Belgica 10: 158–161.

Spagna, P., C. Dupuis, and J. Yans. 2008. Sedimentological study of the wealden clays in the Hautrage quarry. Memoirs of the Geological Survey of Belgium 55: 35–44.

Thiry, M., F. Quesnel, J. Yans, R. Wyns, A. Vergari, H. Théveniaut, R. Simon-Coinçon, M.-G. Moreau, D. Giot, C. Dupuis, L. Bruxelles, J. Barbarand, and J.-M. Baele. 2006. La France et la Belgique continentaux au Crétacé inférieur: paléoaltérations et paléotopographies. Bulletin de la Société Géologique de France 177: 155–175.

Vandycke, S. 2002. Paleostress records in Cretaceous formations in NW Europe: synsedimentary strike-slip and extensional tectonics events. Relationships with Cretaceous–Tertiary inversion tectonics. Tectonophysics 357: 119–136.

———. 2007. Déformations cassantes du nord-ouest européen. Thèse d'Agrégation de l'Enseignement Supérieur, Faculté Polytechnique de Mons (Belgium), 462 pp.

Vandycke, S., F. Bergerat, and C. Dupuis. 1991. Meso-Cenozoic faulting and inferred paleostresses of the Mons Basin (Belgium). Tectonophysics 192: 261–271.

Yans, J. 2007. Lithostratigraphie, minéralogie et diagenèse des sédiments à faciès wealdien du Bassin de Mons (Belgique). Mémoires de la Classe des Sciences, Académie royale de Belgique (ser. 3) 9: 1–179.

Yans, J., P. Spagna, J.-C. Foucher, A. Perruchot, M. Streel, P. Beaunier, F. Robaszynski, and C. Dupuis. 2002. Multidisciplinary study of the Wealden deposits in the Mons Basin (Belgium): a progress report. Aardkundige Mededelingen 12: 39–42.

Yans, J., P. Spagna, C. Vanneste, M. Hennebert, S. Vandycke, J.-M. Baele, J.-P. Tshibangu, P. Bultynck, M. Streel, and C. Dupuis. 2005. Description et implications géologiques préliminaires d'un forage carotté dans le "Cran aux Iguanodons" de Bernissart. Geologica Belgica 8: 43–49.

Biostratigraphy of the Cretaceous Sediments Overlying the Wealden Facies in the Iguanodon Sinkhole at Bernissart

7

Johan Yans*, Francis Robaszynski, and Edwige Masure

The stratigraphy of the Cretaceous sediments overlying the dinosaur-bearing Wealden facies intersected by the BER 3 borehole in the Iguanodon Sinkhole at Bernissart (Mons Basin, Belgium) is assessed. These Cretaceous strata are Late Albian to Coniacian in age, according to the foraminifera and dinocyst assemblages. The beds directly overlying the Wealden facies (Harchies Formation) are early Late Albian in age. The Catillon Formation is late Late Albian. The Bracquegnies Formation is "Vraconnian" (latest Albian, Dispar Zone). The lowermost part of the Bernissart Calcirudites Formation would be Early Cenomanian in age. The overlying Thivencelles Marls Formation is Early Turonian. The uppermost Cenomanian is lacking in the BER 3 borehole as it has been observed elsewhere in Bernissart and along the northern margin of the Mons Basin. The overlying Thulin Marls, Ville-Pommeroeul Chert, and Hautrage Flints formations are Turonian. The uppermost chalks are Coniacian. All of these Cretaceous strata overlay the Wealden facies of late Late Barremian to earliest Aptian age and are unconformably overlain by the Hannut Formation of Thanetian age. These results refine the stratigraphic framework of the Mons Basin.

Introduction

Three boreholes have been drilled within and around the Iguanodon Sinkhole at Bernissart in 2002–2003 (Tshibangu et al., 2004; Yans et al., 2005). One of these (BER 3) provided exceptional material to improve our knowledge of the *Iguanodon*-bearing Wealden facies and overlying strata. Previous investigations focused on the Wealden facies from the BER 3 borehole to document the paleoenvironments of the iguanodons, such as palynology and paleobotany of wood and plant mesofossils (Gerards et al., 2007; Gomez et al., 2008), characterization of organic matter (Schnyder et al., 2009), diagenesis and paleohistology of bone fragments (Ricqlès and Yans, 2003), sedimentology of the lacustrine Wealden facies, clay mineralogy, grain-size analysis, and magnetic susceptibility measurements (Spagna et al., 2008). Moreover, samples of Wealden facies from the Iguanodon Sinkhole housed in the Royal Belgian Institute of Natural Sciences collection (historical searches of 1878–1881) and other sites in the Mons Basin (Hautrage, Thieu, Baudour) were also investigated (Dejax et al., 2007a, 2007b; Yans et al., 2007, 2010). By contrast, the sediments overlying the Wealden facies in Bernissart are poorly characterized. Yans et al. (2005) provided a preliminary lithostratigraphy of the entire BER 3 borehole, correlated with the lithostratigraphic chart of Belgium (Robaszynski et al., 2001).

7.1. Stratigraphical framework of the Cretaceous of the Mons Basin (Wealden facies and overlying strata).

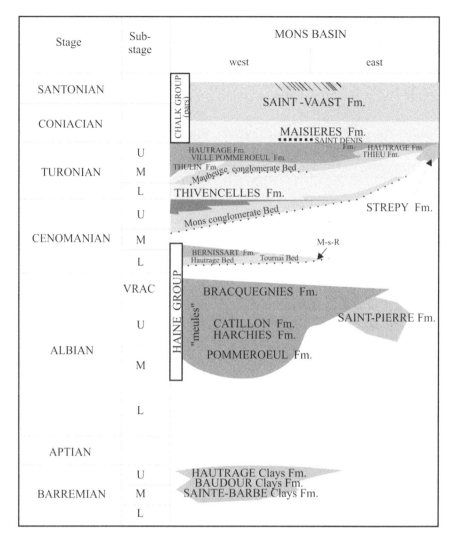

Here we present a detailed study of the dinocysts and the foraminifera in the Cretaceous sediments overlying the Wealden facies in the BER 3 borehole in Bernissart to refine the biostratigraphy and document the different steps of the development of the Mons Basin.

Methodology

Twenty-three preparations for palynological observations (dinocysts) were carried out by mechanical and chemical methods in the sediments of the Haine Green Sandstone Group (locally called *meule*). Each sample was tested for the presence of carbonate by the addition of a few drops of dilute hydrochloric acid; 20 g was then crushed in a mortar. Seventy percent hydrofluoric acid was added for 24 hours. The insoluble fluorides were eliminated with boiling 10% hydrochloric acid; the residue was then washed three times with water. The organic residue was poured in a tube with 10% nitric acid for oxidation, heated in a double boiler for one or two minutes, then washed, centrifuged, and filtered through a 100-μm mesh sieve to remove large debris. The filtrate was sieved through a 5-μm micromesh nylon sieve, and then the 5–100-μm fraction was washed with water and concentrated by centrifuging. The final residue was strewn on a coverslip using hydroxyethyl cellulose (Cellosize), dried, and mounted upside down with Canada balsam on a microscope slide.

Fifty-two samples were collected from 7 to 245.5 m and processed to release the foraminiferal content, 36 of which yielded benthic and/or planktonic forms. The 22 chalky to marly samples from 7 to 73.10 m were first dried, then soaked overnight in water to which 5 g/L of polyphosphate powder had been added. The next day, the samples were washed under water on a 55-μm mesh sieve. The residues were dried and separated into four fractions (<149 μm, 149–210 μm, 210–297 μm, and >297 μm) for observation under a binocular microscope. The 30 samples below the base of the Thivencelles Marls Formation were processed differently because the rocks are harder, like the sediments of the Haine Green Sandstones Group. Samples were gently broken with a hammer to prevent crushing, then soaked in water with polyphosphate during two days, and finally washed under water and separated into four fractions as above. Glauconite was removed from fractions containing too many of these grains with a Frantz isodynamic magnetic separator. Only 14 samples yielded some foraminifera, and these only in small to very small numbers.

Geological Setting

The oldest sediments of the Mons Basin are the Wealden facies ("middle"–Late Barremian to earliest Aptian in age). The Wealden facies of the Iguanodon Sinkhole at Bernissart is attributed to the Sainte-Barbe Clays Formation (Hainaut Group; Robaszynski et al., 2001; Yans et al., 2005; Yans, 2007; Fig. 7.1). In the BER 3 borehole at Bernissart, these sediments are encountered from 265 to 315 m. They consist of laminated dark pyritic clays with millimeter-thick brown and white silty levels.

The overlying sediments are attributed to the Haine Green Sandstone Group, which can be divided into five formations (from base to top): Pommeroeul Greensand Formation, Harchies Formation, Catillon Formation, Bracquegnies Formation, and Bernissart Calcirudites Formation (Robaszynski et al., 2001, fig. 7.2).

The Harchies Formation (−265−−170.7 m in the BER 3 borehole) is composed of argillaceous and calcareous glauconiferous sandstone and calcarenite, locally silicified, rich in sponges. A conglomerate is located at its base between −265 and −258.4 m.

The Catillon Formation (−170.7−−135.4 m) consists of locally indurated glauconiferous calcarenite with trigonia shells. Green conglomeratic levels are locally observed.

The Bracquegnies Formation (−135.4−−105.3 m) displays alternating gray fossiliferous, glauconiferous sandy calcarenite and conglomeratic calcarenite.

The Bernissart Calcirudites Formation (−105.3−−73.1 m) consists of sandy, glauconiferous calcarenite, locally indurated, cellular, and geodic, with bioclast levels and dispersed flints and pebbles.

The Haine Green Sandstone Group is overlain by the following sedimentary units (from base to top):

The Thivencelles Marls Formation (−73.1−−55 m in BER 3 borehole; locally named *partie inférieure des dièves*) is composed of greenish marls and clays; a basal conglomeratic level between −73.1 and −71 m (locally named *Tourtia de Mons*) is present.

The Thulin Marls Formation (−55−−45.6 m; locally called *partie supérieure des dièves*) consists of gray marls.

The Ville-Pommeroeul Chert Formation (−45.6−−33.6 m; locally named *fortes-toises*) is chalky, with a siliceous cementation and numerous cherts and marl levels.

The Hautrage Flints Formation (−33.6−−27.5 m; locally named *rabots*) consists of white chalk with black flints.

The Maisières Chalk Formation (−26.3 to −24.45 m of depth) consists of glauconiferous gray chalk with many bioturbations and phosphatic pellets.

From −24.45 to −8 m, cuttings of white to gray fragments of chalk with some glauconiferous and phosphatic levels can be observed. Because the fragments are superposed to the Maisières Chalk Formation, these white to grayish chalks were interpreted as the Saint-Vaast Formation.

The Hannut Formation (−8 to −4 m; Landen Group) consists of green micaceous fine clayey sand attributed to the Grandglise Member, Thanetian in age (Kaaschieter 1961; Maréchal and Laga, 1988).

From −4 to 0 m, the BER 3 borehole encountered a Pleistocene loam with humus and arable layer.

Results and Discussion

Dinocysts

The biostratigraphy of dinocysts is based on first appearances data (FAD) of boreal index species that are correlated with the ammonite and foraminiferal zonal schemes (Williams and Bujak, 1985; Costa and Davey, 1992; Foucher and Monteil, 1998). The Appendix lists the dinocysts cited in this chapter.

Harchies Formation. The occurrence of *Surculosphaeridium longifurcatum* (Fig. 7.2L) at 261.6 m in the conglomerate indicates that the sample cannot be older than the Albian: *S. longifurcatum* ranges from Early Albian to Late Campanian (Williams and Bujak, 1985). *Litosphaeridium conispinum* (Fig. 7.2E) and *Stephodinium coronatum* (Fig. 7.2J) were recorded in sample 252.6m. The range of *L. conispinum* is from Middle Albian to Cenomanian. *Xiphophoridium alatum* is recorded at 257.5 m. The FAD of *X. alatum* is known to be located in the Late Albian and the last appearance datum (LAD) in the Santonian (Williams and Bujak, 1985, Costa and Davey, 1992; Foucher and Monteil, 1998). The FAD of *Stephodinium coronatum* characterizes the Middle Albian; the LAD is located in the Turonian. Sample 231.8m contains *Atopodinium chleuh* and *Tehamadinium mazaganense*. These species have been reported in Late Albian from Morocco (Below, 1984; Masure, 1988) and Italy (*T. mazaganense*, Fiet and Masure, 2001; Torricelli, 2006) and are not reported from Cenomanian sediments.

Catillon Formation. Dinoflagellate cyst assemblages are diversified. *Xiphophoridium alatum* (Fig. 7.2K), recorded at 155.2 m, is a stratigraphical index species (Late Albian–Santonian). Species of *Achomosphaera* (Fig. 7.2A), *Spiniferites*, and *Pervosphaeridium* are frequent in Harchies and Catillon formations. The occurrence of *Atopodinium chleuh* in samples B3-231.8m and B3-148m, and *A. haromense* in sample B3-138.1m, with *E. dettmaniae*, suggests Tethyan influences in the Mons Basin during the Late Albian.

Bracquegnies Formation. Species of biostratigraphic interest are *Epelidosphaeridia spinosa* (Fig. 7.2C) and *Litosphaeridium siphoniphorum* (Fig.

7.2G), recorded in sample 129.2m. Their FADs are located at the base of the Dispar Zone ("Vraconnian") (Williams and Bujak, 1985; Costa and Davey, 1992; Foucher and Monteil, 1998), and their LADs are in the Cenomanian. The FAD of *Endoceratium dettmaniae* (Fig. 7.2B), recorded in sample 112.4m, is well known in the "Vraconnian" (Cookson and Hughes, 1964); its LAD is recorded in the Cenomanian. *Oligosphaeridium complex* (Fig. 7.2H) and *Surculosphaeridium longifurcatum* are dominant species during the Late Albian (Harchies and Catillon formations); they are scarcer during the "Vraconnian" (Bracquegnies Formation).

Bernissart Calcirudites Formation. Epelidosphaeridia spinosa is more frequent in sample 102.7m. A peak of abundance is recorded in the Early Cenomanian (Fauconnier *in* Juignet et al., 1983; Foucher, 1979). Hence, an Early Cenomanian age is suggested for sample 102.7m. A Cenomanian age is suggested for sample 80.8m, dominated by *Ovoidinium* (Fig. 7.2I) species.

The age assignment of the lithostratigraphic units based on the dinocyst zonation is as follows:

1. Late Albian: ?261.6 m–138.1 m (Harchies Formation and Catillon Formation).
2. Latest Albian ("Vraconnian"): 129.2–112.4 m (Bracquegnies Formation).
3. ?Early Cenomanian: 102.7–80.8 m (Bernissart Formation).

Foraminifera

On the basis of the foraminifera assemblages (Fig. 7.3), the following biostratigraphy is proposed:

–245.50––189.40 m—Late Albian sensu stricto, characterized by the association of *Arenobulimina chapmani* Cushman, *Orithostella jarzevae* (Vassilenko), *A*. cf. *sabulosa* (Chapman), and *Gavelinella cenomanica* (Brotzen).
–109.60––103.10 m—"Vraconnian," based on the association of *G. baltica* (Brotzen), *Epistomina* cf. *cretosa/spinulifera* (Reuss), and *Vaginulina strigillata bettenstaedti* Albers.
–77.50––70.60 m—Probably latest Cenomanian by the association of *G. cenomanica*, *G. baltica*, and *Lingulogavelinella globosa* (Brotzen) as the last occurrence of *G. cenomanica* found in the sample at –70.60 m is not reworked (the last occurrence of *G. cenomanica* is situated in the latest Cenomanian, and *L. globosa* already starts in the Late Cenomanian before fully developing in the Early Turonian). If the microfauna in the –70.60-m sample is reworked, the conglomerate (or Tourtia) would be Early Turonian, which is in agreement with earlier interpretations of numerous boreholes in the Mons Basin.
–69.50––26.20 m—Turonian, because of the association of *Whitenella archaeocretacea* Pessagno, *Praeglobotruncana stephani* (Gandolfi), *Dicarinella hagni* (Scheibnerova), *Bdelloidina cribrosa*

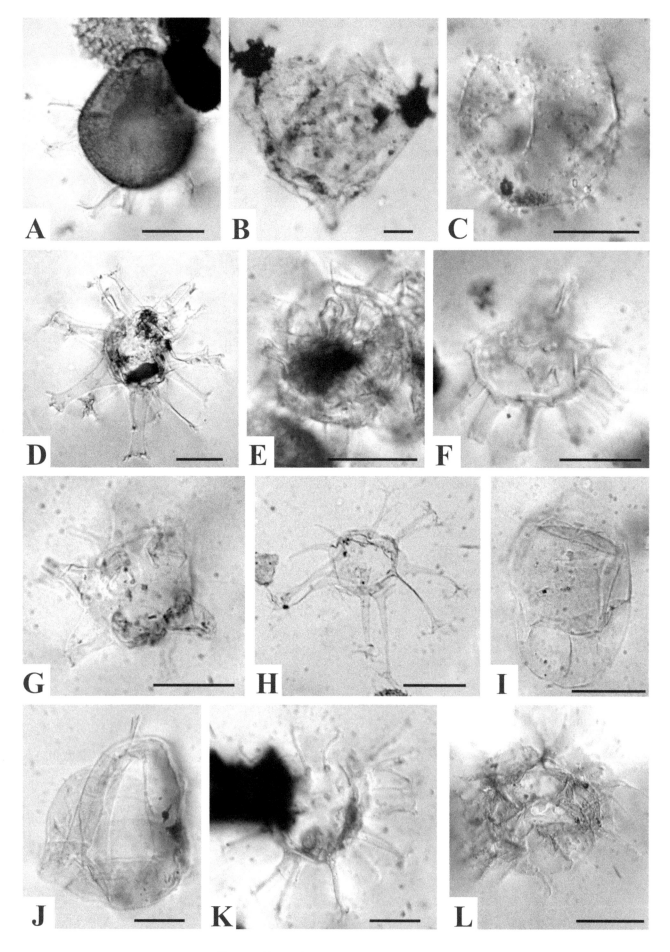

(Reuss), *Helvetoglobotruncana helvetica* (Bolli), *Marginotruncana sigali* (Reichel), *M. pseudolinneina* Pessagno, *Globorotalites micheliniana* (d'Orbigny), and the mesofossil *Terebratulina rigida.* −24.80−−8 m—Coniacian, because of the association of *Gavelinella arnagerensis* Solakius, *Reussella kelleri* Vassilenko, *Stensioeina granulata granulata* (Olbertz), and *Globotruncana linneiana* (d'Orbigny).

Integrated Biostratigraphy of the Haine Green Sandstone Group

On the basis of the dinocyst and foraminifera content, the stratigraphy of the Haine Green Sandstone Group in the BER 3 borehole could be refined. In the Harchies Formation (−265−−170.7 m), the foraminifera confirm the Late Albian sensu stricto age. In the *locus typicus* of Harchies, the Harchies Formation contains *Actinoceramus concentricus* and *Actinoceramus sulcatus*, suggesting an early Late Albian sensu stricto age (Robaszynski and Amédro, 1986).

In the Catillon Formation (−170.7−−135.4 m), no foraminifera could be recognized; only dinocysts suggest, according to their range charts, a Late Albian age. In Baudour, the ammonite *Mortoniceras inflatum* was observed in the Catillon Formation, suggesting a late Late Albian sensu stricto age (Robaszynski and Amédro, 1986).

In the Bracquegnies Formation (135.4–105.3 m), the association of both dinocysts and foraminifera confirms a "Vraconnian" age (latest Albian; Dispar Zone), which is in good agreement with previous studies dedicated to ammonites in this formation (Amédro, 2002, 2008).

The lowermost part (sample at −102.7 m) of the Bernissart Calcirudites Formation (−105.3−−75 m) would be Early Cenomanian in age. This new zonation refines the stratigraphic chart of the Cretaceous of Belgium, which indicated an Early to Middle Cenomanian age for this formation. In Bettrechies, the Bernissart Calcirudites Formation is covered by sediments containing the late Middle Cenomanian ammonite *Acanthoceras jukesbrownei*.

In the BER 3 borehole, the typical facies of the Middle Albian (Amédro, 1984) Pommeroeul Greensand Formation is lacking. In the *locus typicus* of the Haine Green Sandstone Group in Harchies (2.9 km from Bernissart), the Pommeroeul Greensand Formation is 23.5 m thick. The complete Haine Green Sandstone Group in Harchies is, however, thinner than the incomplete Haine Green Sandstone Group in BER 3. Moreover, at Harchies, Marlière (1939) observed an angular unconformity between the Pommeroeul Greensand Formation and the overlying Harchies Formation These features confirm the relative complexity of tectonics and related subsidence in the Mons Basin during the Early Cretaceous (e.g., Dupuis and Vandycke, 1989).

Above the Haine Greensand Group, the Thivencelles Marls Formation (−73.1−−56 m) is Early Turonian in age. The uppermost Cenomanian is lacking between the Haine Greensand Group and the Thivencelles Formation because no *Rotalipora cushmani* was found below the basal Turonian conglomerate (Tourtia de Mons from −73.1 to −70 m). The same feature was observed at the borehole Bernissart 41 (Robaszynski, 1972a). Both the

7.2. Dinocysts from BER 3 borehole. Coordinates: England finder Graticule, scale bar = 20 μm. A, *Achomosphaera sagena*, sample B3–257.5m, Y42, 55 μm. B, *Endoceratium dettmanniae*, sample B3–112.4m, Q44, 120 μm. C, *Epelidosphaeridia spinosa*, sample B3–102.7m, T31, 38 μm. D, *Kleithriasphaeridium loffrense*, sample B3–257.5m, X49, 76 μm. E, *Litosphaeridium conispinum*, sample B3–257.5m, R48–49, 45 μm. F, *Litosphaeridium fucosum*, sample B3–257.5m, Q28, 40 μm. G, *Litosphaeridium siphoniphorum*, sample B3–129.2m, L47, 47 μm. H, *Oligosphaeridium complex*, sample B3–245.2m, ST 39, 63 μm. I, *Ovoidinium scabrosum*, principal sutures of archeopyle open, sample B3–80.8m, H23, 52 μm. J, *Stephodinium coronatum*, sample B3–245.2m, S31, 75 μm. K, *Surculosphaeridium longifurcatum*, sample B3–261.6m, V36, 78 μm. L, *Xiphophoridium alatum*, sample B3–155.2m, W41, 76 μm.

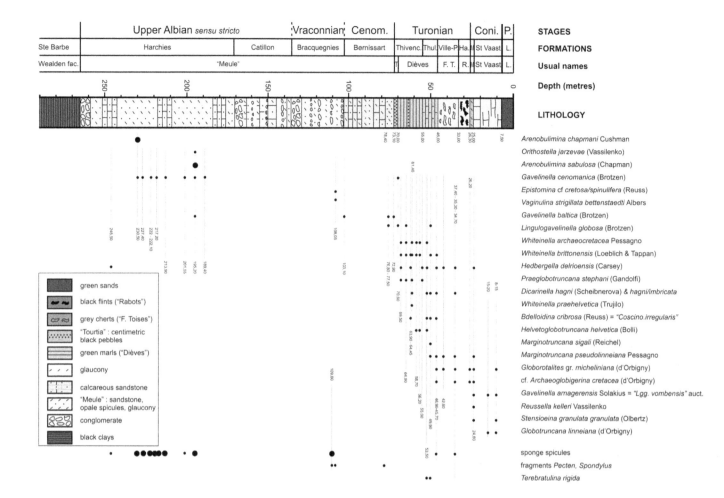

7.3. Vertical distribution of foraminifera in the BER 3 borehole *i*, focusing on the strata overlying the Wealden facies. *Abbreviations:* Coni., Coniacian; F.T., *fortes-toises;* H., Hannut; Ha., Hautrage Flints; M., Maisières Chalk; P., Paleocene; R., *rabots;* T., Tourtia; Thivenc., Thivencelles Marls; Thul, Thulin Marls; Ville-P., Ville-Pommeroeul Chert.

boreholes BER 3 and Bernissart 41, where the uppermost Cenomanian marls are lacking, are situated at the marginal northern part of the Mons Basin. By contrast, in the central part of the basin, the Cenomanian transgression corresponds to several meters of marls containing *R. cushmani* (Robaszynski, 1972b; Leplat and Robaszynski, 1972). This highlights once again the diachronous sedimentation in and around the Mons Basin during the Cretaceous.

The overlying Thulin Marls Formation (−56−−45.6 m), Ville-Pommeroeul Chert Formation (−45.6−−33.6 m), and Hautrage Flints Formation (−33.6−−27.5 m) are Turonian in age. Robaszynski et al. (2001) suggest a Middle Turonian age for the Thulin Marls Formation and a Late Turonian age for the two overlying formations.

Finally, the overlying chalks (the Maisières Chalk and Saint-Vaast Chalk formations) are Coniacian. The upper part of the Saint-Vaast Formation is probably not preserved in the BER 3 borehole because this should be Santonian in age (Robaszynski et al., 2001).

Conclusion

We have refined the Cretaceous stratigraphic framework of the Mons Basin, especially around the Bernissart area. The dinosaur-bearing Wealden facies are overlain by early Late Albian to Coniacian strata. Despite the lack of the Pommeroeul Greensand Formation in BER 3 borehole of Bernissart, the Haine Green Sandstone Group is thicker here than in the *locus typicus* of Harchies. As observed elsewhere in the northern margin of the Mons Basin, the uppermost Cenomanian is lacking.

List of Cited Dinocysts

Achomosphaera sagena Davey and Williams, 1966, Fig. 4.1
Atopodinium chleuh (Below, 1981) Masure, 1991
Atopodinium haromense Thomas and Cox, 1988
Endoceratium dettmanniae (Cookson and Hughes, 1964) Stover and Evitt, 1978; *emend.* Harding and Hughes, 1990, Fig. 4.2
Epelidosphaeridia spinosa Cookson and Hughes, 1964 ex Davey, 1969, Fig. 4.3
Kleithriasphaeridium loffrense Davey and Verdier, 1976, Fig. 4.4
Litosphaeridium conispinum Davey et Verdier, 1973; *emend.* Lucas-Clark, 1984, Fig. 4.5
Litosphaeridium fucosum (Valensi, 1955) Masure *in* Fauconnier and Masure, 2004, Fig. 4.6
Litosphaeridium siphoniphorum (Cookson and Eisenack, 1958), Davey and Verdier, 1966; *emend.* Lucas-Clark, 1984, Fig. 4.7
Oligosphaeridium complex (White, 1842) Davey and Williams, 1966, Fig. 4.8
Ovoidinium scabrosum (Cookson and Hughes, 1964) Davey, 1970, Fig. 4.9
Pervosphaeridium spp.
Spiniferites spp.
Stephodinium coronatum Deflandre, 1936, Fig. 4.10
Surculosphaeridium longifurcatum (Firtion, 1952) Davey et al., 1966, Fig. 4.11
Tehamadinium mazaganense (Below, 1984) Jan du Chêne et al., 1986
Xiphophoridium alatum (Cookson and Eisenack, 1962) Sarjeant, 1966, Fig. 4.12

D. Batten, M. Dusar, and N. Vandenberghe reviewed an earlier version of this chapter and made many valuable comments.

Acknowledgments

References

Amédro, F. 1984. L'Albien de la bordure septentrionale du bassin de Paris. Mise en évidence d'un contrôle tectonique de la sédimentation. Bulletin du Bureau de Recherches Géologiques et Minières (section IV) 3: 179–192.
———. 2002. Plaidoyer pour un étage Vraconnien entre l'Albien sensu stricto et le Cénomanien (système Crétacé). Académie Royale de Belgique. Mémoires in 4° de la Classe des Sciences (ser. 3) 4: 1–128.

———. 2008. Support for a Vraconnian stage between the Albian sensu stricto and the Cenomanian (Cretaceous System). Carnets de Géologie/Notebooks on Geology, Brest, Memoir 2008/02, 83 pp.
Below, R. 1984. Aptian to Cenomanian dinoflagellate cysts from the Mazagan plateau, Northwest Africa (sites 545 and 547, DSDP leg 79); pp. 621–649 in K. Hinz et al. (eds.), Initial Reports DSDP 79, Sites 544–547, Washington, D.C.

Cookson, I. C., and N. F. Hughes. 1964. Microplankton from the Cambridge Greensand (mid-Cretaceous). Palaeontology 7: 37–59.

Costa, L. I., and R. J. Davey. 1992. Dinoflagellate cysts of the Cretaceous System; pp. 99–153 in A. J. Powell (ed.), A stratigraphic index of dinoflagellate cysts. Chapman and Hall, London.

Dejax, J., E. Dumax, F. Damblon, and J. Yans. 2007a. Palynology of Baudour Clays Formation (Mons Basin, Belgium): correlation within the "stratotypic" Wealden. Notebooks on Geology e-Journal, http://paleopolis.rediris.es/cg/CG2007_M01/CG2007_M01.pdf, 16–28.

Dejax, J., D. Pons, and J. Yans. 2007b. Palynology of the dinosaur-bearing Wealden facies sediments in the natural pit of Bernissart (Belgium). Review of Palaeobotany and Palynology 144: 25–38.

Dupuis, C., and S. Vandycke. 1989. Tectonique et karstification profonde: un modèle de subsidence original pour le Bassin de Mons. Annales de la Société géologique de Belgique 112: 479–487.

Fiet, N., and E. Masure. 2001. Les dinoflagellés albiens du bassin de Marches-Ombrie (Italie); proposition d'une biozonation pour le domaine téthysien. Cretaceous Research 22: 63–77.

Foucher, J.-P. 1979. Distribution des kystes de Dinoflagellés dans le Crétacé moyen et supérieur du Bassin de Paris. Palaeontographica B 169: 78–105.

Foucher, J.-C., and E. Monteil. 1998. Cretaceous biochronostratigraphy, chart 5: Mesozoic and Cenozoic sequence chronostratigraphic framework of European basins, in P. C. de Graciansky, J. Hardenbol, T. Jacquin, and P.-R. Vail (eds.), Mesozoic and Cenozoic sequence stratigraphy of European basins. SEPM Special Publication 60. Society for Sedimentary Geology.

Gerards, T., J. Yans, and P. Gerrienne. 2007. Growth rings of Lower Cretaceous softwoods: some plaeoclimatic implications. Notebooks on Geology e-Journal. http://paleopolis.rediris.es/cg/CG2007_M01/CG2007_M01.pdf, 29–33.

Gomez, B., T. Gillot, V. Daviero-Gomez, P. Spagna, and J. Yans. 2008. Paleoflora from the Wealden facies strata of Belgium: mega- and meso-fossils of Hautrage (Mons Basin). Memoirs of the Geological Survey 55: 53–60.

Juignet, P., R. Damotte, D. Fauconnier, W. J. Kennedy, F. Magniez-Jannin, C. Monciardini, and G.-S. Odin. 1983. Étude de trois sondages dans la région-type du Cénomanien. La limite Albien-Cénomanien dans la Sarthe (France). Géologie de la France 3: 193–234.

Kaaschieter, J. 1961. Foraminifera of the Eocene of Belgium. Mémoire de l'Institut royal des Sciences naturelles de Belgique 147: 1–271.

Leplat, J., and F. Robaszynski. 1972. Une couche à Rotalipores dans les "Dièves" (Crétacé supérieur) dans un sondage à Trith (Nord). Annales de la Société géologique du Nord 91: 199–202.

Maréchal, R., and P. Laga. 1988. Voorstel lithostratigraphische indeling van het Paleogeen, Commissie Tertiair. Nationale Commissie voor Stratigrafie, Brussels, 207 pp.

Marlière, R. 1939. La transgression albienne et cénomanienne dans le Hainaut. Mémoire du Musée royal d'Histoire naturelle de Belgique 8: 1–440.

Masure, E. 1988. Albian–Cenomanian dinoflagellate cysts from Sites 627 and 635, Leg 101, Bahamas; pp. 121–138 in J. A. Austin et al., Ocean Drilling Project Scientific Results, Proceedings 101, Washington, D.C.

Ricqlès, A. de., and J. Yans. 2003. Bernissart's iguanodons: the case for "fresh" versus "old" dinosaur bone. Journal of Vertebrate Paleontology 23 (supplement to 3): 45A.

Robaszynski, F. 1972a. Les foraminifères pélagiques des "Dièves" aux abords du golfe de Mons (Belgique). Annales de la Société géologique du Nord 91: 31–38.

———. 1972b. Les "Dièves" de Maubeuge et leurs deux Tourtias (Crétacé supérieur). Annales de la Société géologique du Nord 91: 193–197.

Robaszynski, F., and F. Amédro. 1986. The Cretaceous of the Boulonnais (France) and a comparison with the Cretaceous of Kent (United Kingdom). Proceedings of the Geological Association 97: 171–208.

Robaszynski, F., A. V. Dhondt, and J. W. M. Jagt. 2001. Cretaceous lithostratigraphic units (Belgium); pp. 121–134 in P. Bultynck and L. Dejonghe (eds.), Guide to a revised lithostratigraphic scale of Belgium. Geologica Belgica 4.

Schnyder, J., J. Dejax, E. Keppens, T. Nguyen Tu, P. Spagna, S. Boulila, B. Galbrun, A. Riboulleau, J.-P. Tshibangu, and J. Yans. 2009. An Early Cretaceous lacustrine record: Organic matter and organic carbon isotopes at Bernissart (Mons Basin, Belgium). Palaeogeography, Palaeoclimatology, Palaeoecology 281: 79–91.

Spagna, P., C. Dupuis, and J. Yans. 2008. Sedimentology of the Wealden clays in the Hautrage quarry. Memoirs of the Geological Survey of Belgium 55: 35–44.

Torricelli, S. 2006. Dinoflagellate cyst stratigraphy of the Scisti a Fucoidi Formation (Early Cretaceous) from Piobbico, Central Italy: calibrated events for the Albian of the Tethyan realm. Rivista Italiana di Paleontologia e Stratigrafia 112: 95–112.

Tshibangu, J.-P, F. Dagrain, B. Deschamps, and H. Legrain. 2004. Nouvelles recherches dans le Cran aux Iguanodons de Bernissart. Bulletin de l'Académie Royale de Belgique, Classe des Sciences 7: 219–236.

Williams, G. L., and J. Bujak. 1985. Mesozoic and Cenozoic dinoflagellates; pp. 847–964 in H. M. Bolli, J. B. Saunders, and K. Perch-Nielsen (eds.), Plankton stratigraphy. Cambridge Earth Science series. Cambridge University Press, Cambridge.

Yans, J. 2007. Lithostratigraphie, minéralogie et diagenèse des sédiments à faciès wealdien du Bassin de Mons (Belgique). Mémoire de la Classe des Sciences, Académie Royale de Belgique (ser. 3) 2046: 1–179.

Yans, J., P. Spagna, C. Vanneste, M. Hennebert, S. Vandycke, J.-M. Baele, J.-P. Tshibangu, P. Bultynck, M. Streel, and C. Dupuis. 2005. Description et implications géologiques préliminaires d'un forage carotté dans le "Cran aux Iguanodons" de Bernissart. Geologica Belgica 8: 43–49.

Yans, J., E. Masure, J. Dejax, D. Pons, and F. Amedro, F. 2007. Influences boréales dans le bassin de Mons (Belgique) à l'Albien; in L. G. Bulot, S. Ferry, and D. Groshény (eds.), Relations entre les marges septentrionale et méridionale de la Téthys au Crétacé. Notebooks on Geology, Mémoire 2007/02, Résumé 06 (CG2007_M02/06).

Yans, J., T. Gerards, P. Gerrienne, P. Spagna, J. Dejax, J. Schnyder, J.-Y. Storme, and E. Keppens. 2010. Carbon-isotope of fossil wood and dispersed organic matter from the terrestrial Wealden facies of Hautrage (Mons Basin, Belgium). Palaeogeography, Palaeoclimatology, Palaeoecology 291: 85–105.

On the Age of the Bernissart Iguanodons

Johan Yans*, Jean Dejax, and Johann Schnyder

We summarize the studies dealing with the dating of the Bernissart iguanodons. Both palynology (especially pollens of angiosperm affinities, such as the biorecord Superret-*croton* and probably the paleotaxon Superret-*subcrot*) and chemostratigraphy (carbon isotope composition of dispersed organic matter and fossil wood) have recently been applied to refine the age of the *Iguanodon*-bearing Wealden facies trapped in the Iguanodon Sinkhole at Bernissart (Sainte-Barbe Clays Formation). These studies suggest that this formation is late Late Barremian to earliest Aptian in age.

Johan Yans*, Jean Dejax, and Johann Schnyder

Introduction

Although numerous studies are dedicated to the faunal and floral content of the Iguanodon Sinkhole at Bernissart, the age of the Sainte-Barbe Clays Formation (Wealden facies) remained poorly constrained until recently. In the most recent synthesis on the Cretaceous of Belgium, Robaszynski et al. (2001) concluded that this formation was Late Jurassic to Early Cretaceous in age (161.2 to 99.6 Ma, according to Gradstein et al., 2004). Here we summarize the previous attempts to date the Bernissart iguanodons and discuss new datings on the basis of both palynology of the angiosperm pollen content and carbon isotope chemostratigraphy on organic matter.

Material and Methods

Palynology

The present study is based on, first, samples collected at −322 m in the Iguanodon Sinkhole at Bernissart during the 1878–1881 excavations and stored in the collections of the Royal Belgian Institute of Natural Sciences of Brussels (Dejax et al., 2007a), and second, on the preliminary results of samples from the BER 3 borehole, drilled in 2002–2003 (Yans et al., 2005).

Standard procedures, as described in Dejax et al. (2007a), were used to eliminate the carbonate and the insoluble fluorides in each sample and to concentrate the organic matter. The final residue, obtained through sieving on a 5-µm micromesh nylon sieve and centrifuging, was strewn on a coverslip by means of hydroxyethyl cellulose (Cellosize), dried, and mounted upside down with Canada balsam on a microscope slide. The observations and determinations noted herein are based on light microscopic examination, mainly using an interferential–differential contrast objective. The preparations were stored in the paleontological collections of the Royal Belgian Institute of Natural Sciences in Brussels.

8.1. Stratigraphic distribution of the Wealden facies from the Sainte-Barbe Clays Formation in the Iguanodon Sinkhole at Bernissart (in gray), correlated with the succession of the MCT phases defined in the Wealden facies from the reference sections of the Weald and Wessex subbasins (England). Because of the presence of biorecord Superret-*croton* and paleotaxon Superret-*subcrot*, the studied sediments belong to MCT phase 4, which extends from the late Late Barremian to the earliest Aptian. Modified from Yans et al. (2005) and Schnyder et al. (2009).

The carbon isotope ratios of bulk organic matter and isolated wood fragments were measured on 141 samples from the BER 3 borehole, collected between −315 and −265 m.

Samples were prepared according to standard procedures, as described in Yans et al. (2010). Isotopic measurements were carried out on bulk samples with a Finnigan MAT Delta^{plus} mass spectrometer connected to a Thermo-Finnigan Flash EA1112 series microanalyzer at the Free University of Brussels by means of standard techniques. Up to three distinct measurements have been made for each sample. Reproducibility of standards is within ~0.2 ‰.

Previous Works

Numerous various ages, based on different approaches, have been proposed for the dating of the Bernissart iguanodons (Yans et al., 2005). In 1900, Vandenbroeck compared the paleofloras from Bernissart and southeastern England and suggested a Late Jurassic age for the Bernissart locality. In 1954, Marlière proposed an age ranging from the Late Jurassic to the "Neocomian," also on the basis of the paleoflora. However, the same author subsequently revised this hypothesis and argued for an age more recent than the Valanginian–Hauterivian (Marlière, 1970). In 1975, Taquet compared the morphologies of different taxa within the Iguanodontidae and concluded that the Bernissart iguanodons are Barremian in age. On the basis of a comparative study of the lithology and the paleontological content of Bernissart and England, Allen (1955) suggested a Late Aptian–Early Albian age. Later, Allen and Wimbledon (1991) proposed a Hauterivian–Barremian age, and Pelzer and Wilde (1987) and Vakhrameev (1991) a "Neocomian" age. In a recent synthesis (Norman and Weishampel, 1990), the species *Iguanodon bernissartensis* (including *I. orientalis*, according to the recommendations of Norman, 1996) was suggested to range from the Valanginian to the Albian and *Mantellisaurus atherfieldensis*, from the Berriasian, to the Aptian. Taking into account all these previous studies, Robaszynski et al. (2001) concluded that the *Iguanodon*-bearing Wealden facies (Sainte-Barbe Clays Formation) were Late Jurassic to Early Cretaceous in age.

Results and Discussion

Several previous studies have dealt with the palynology of the Wealden facies from the northern part of Europe, including the pioneering palynological works of Delcourt and Sprumont (1955; see references in Dejax et al., 2007a). The Wealden facies, encountered at −322 m depth in the Iguanodon Sinkhole at Bernissart (the level where the majority of the iguanodons were found in 1878–1881), contains the biorecord Superret-*croton*, a pollen grain of angiospermous affinity (Yans et al., 2005, 2006; Yans, 2007; Dejax et al., 2007a). This taxon, defined following the binominal nomenclature of Hughes et al. (1979), is monosulcate and shows a crotonoid sculpture, which is quite similar to those exhibited by Liliaceae, Euphorbiaceae, and Buxaceae. Penny (1986) considered *Stellatopollis hughesii*, from the Upper Barremian (?) of Egypt, to be equivalent to the biorecord Superret-*croton*. In the "stratotypic" Wealden facies of the Weald and Wessex subbasins, well dated by the occurrence of interstratified levels with ammonites and dino-flagellate cysts (Harding, 1986, 1990), the biorecord Superret-*croton* ranges from the middle Barremian to the earliest Aptian (Hughes, 1994). This corresponds to monosulcate columellate tectate (MCT) biostratigraphic phases 3 to 5 (Fig. 8.1).

Here we also figure specimens of the biorecord Superret-*croton* and, for the first time, specimens probably belonging to the paleotaxon Superret-*subcrot*. The latter, encountered at −265 m in the BER 3 borehole (Fig. 8.2), has been previously recognized within the Baudour Clays Formation at Baudour (Wealden facies, Mons Basin; Dejax et al., 2007b). This unit also provided a left coracoid confidently attributed to the ornithopod *Iguanodon bernissartensis*, and a left tibia belonging to an indeterminate sauropod (Chapter 13 in this book). The paleotaxon Superret-*subcrot* was defined by Hughes et al. (1979) as SUPERRET-(CAND)SUBCROT and reexamined by Hughes in 1994 with a scanning electron microscope. In the English Wealden facies, this paleotaxon is restricted to MCT phase 4 (Fig. 8.1). According to the stratigraphic distribution of its pollen of angiospermous affinity, the Sainte-Barbe Clays Formation should be assigned to the MCT phases 3 and 4 (probably 4), evidencing its late Late Barremian–earliest Aptian age.

Carbon isotope ratios are a powerful tool that may reflect worldwide geological events. The isotopic composition of dispersed organic matter in a sediment sample may represent the isotopic composition of the atmosphere in which plants and other organisms lived, at or near the time of death and burial. Dispersed organic carbon and fossil wood are reliable materials for carbon-isotope chemostratigraphy (e.g. Yans et al., 2010). Schnyder et al. (2009) observed that the 50-m-thick succession of the Sainte-Barbe Clays Formation in the BER 3 borehole presents a negative $\delta^{13}C$ trend in both dispersed organic carbon (DOC; $\delta^{13}C_{DOC}$) and fossil wood ($\delta^{13}C_{WOOD}$). In the "stratotypical" Isle of Wight succession, Robinson and Hesselbo (2004) deciphered such a negative $\delta^{13}C_{WOOD}$ trend in the Upper Barremian to Lower Aptian Wealden facies. Erba et al. (1999) and Sprovieri et al. (2006) observed a similar $\delta^{13}C$ negative trend in Upper Barremian to Lower Aptian pelagic carbonates of the Tethyan realm (Cismon and Umbria-March composite sections, respectively; Fig. 8.3). Together with the palynological results, the $\delta^{13}C$ negative trend would confirm the late Late Barremian to earliest Aptian age for the Wealden facies of Bernissart (Sainte-Barbe Clays Formation), as encountered at −322 m in the Iguanodon Sinkhole and at −265 m in the BER 3 borehole. This age is slightly younger than the age (late Early to early Late Barremian) of the Hautrage Clays Formation at Hautrage (Mons Basin; Yans et al., 2010).

Moreover, on the basis of spectral analysis of the total gamma ray variations in the BER 3 borehole core, the total duration of the sedimentation of the 50-m-thick Sainte-Barbe Clays Formation in the Iguanodon Sinkhole at Bernissart is estimated to range between 0.55 to 2.2 myr (probably closer to 0.55 myr; Schnyder et al., 2009).

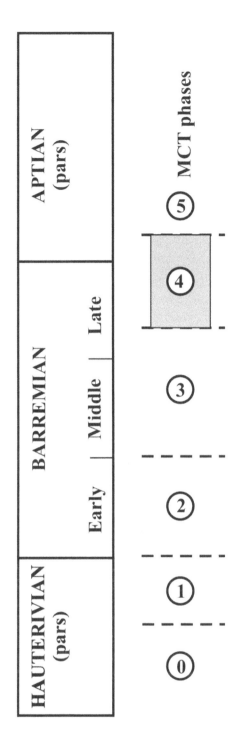

Conclusion

The Sainte-Barbe Clays Formation in the Iguanodon Sinkhole at Bernissart contains the biorecord Superret-*croton* and probably the paleotaxon Superret-*subcrot*, suggesting a late Late Barremian to earliest Aptian age

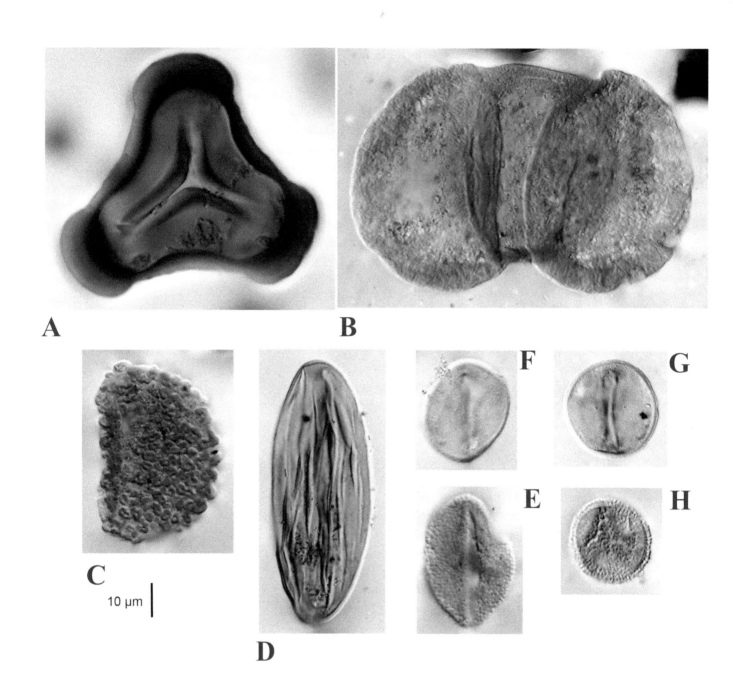

A

B

C

10 μm

D

F

G

E

H

(MCT phase 4). Correlation of the negative δ¹³C trend in both dispersed organic matter and fossil wood between the Bernissart borehole and a series of well-dated reference sections in Wessex (U.K.), Cismon (Italy), and Umbria-March, confirms its age.

References

Allen, P. 1955. Age of the Wealden in north-western Europe. Geological Magazine 92: 265–281.

Allen, P., and W. A.Wimbledon. 1991. Correlation of NW European Purbeck–Wealden (nonmarine Lower Cretaceous) as seen from the English type-areas. Cretaceous Research 12: 511–526.

Dejax, J., D. Pons, and J. Yans. 2007a. Palynology of the dinosaur-bearing Wealden facies sediments in the natural pit of Bernissart (Belgium). Review of Palaeobotany and Palynology 144: 25–38.

Dejax, J., E. Dumax, F. Damblon, and J. Yans. 2007b. Palynology of Baudour Clays Formation (Mons Basin, Belgium): correlation within the "stratotypic" Wealden. Notebooks on Geology e-Journal. http://paleopolis.rediris.es/cg/CG2007_M01/CG2007_M01.pdf, 16–28.

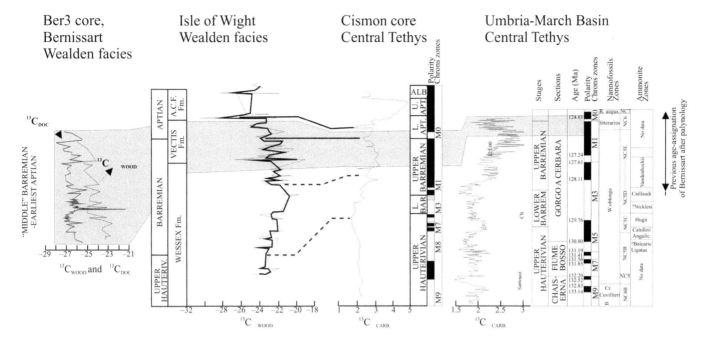

Ber3 core,
Bernissart
Wealden facies

Isle of Wight
Wealden facies

Cismon core
Central Tethys

Umbria-March Basin
Central Tethys

Delcourt, A., and G. Sprumont. 1955. Les spores et grains de pollen du Wealdien du Hainaut. Mémoires de la Société belge de Géologie, de Paléontologie et d'Hydrologie 4: 1–73.

Erba, E., J. E. T. Channell, M. Claps, C. Jones, R. Larson, B. Opdyke, I. Premoli Silva, A. Riva, G. Salvini, and S. Torricelli. 1999. Integrated stratigraphy of the Cismon Apticore (southern Alps, Italy); a "reference section" for the Barremian–Aptian interval at low latitudes. Journal of Foraminferal Research 29: 371–391.

Gradstein, F., J. Ogg, and A. Smith. 2004. A geological time scale, Cambridge University Press, Cambridge, 589 pp.

Harding, I. C. 1986. An Early Cretaceous dinocyst assemblage from the Wealden of Southern England. Special Papers in Palaeontology 35: 95–109.

———. 1990. A dinocyst calibration of the European Boreal Barremian. Palaeontographica B 218: 1–76.

Hughes, N. F. 1994. The enigma of angiosperm origins. Cambridge University Press, Cambridge, 303 pp.

Hughes, N. F., and A. B. McDougall. 1987. Records of angiosperm pollen entry into English Early Cretaceous succession. Review of Palaeobotany and Palynology 50: 255–272.

Hughes, N. F., G. E. Drewry, and J. F. Laing. 1979. Barremian earliest angiosperm pollen. Palaeontology 22: 513–535.

Marlière, R. 1954. Le Crétacé; pp. 417–443 in P. Fourmarier (ed.), Prodrôme d'une description géologique de la Belgique. Université de Liège, Liège.

———. 1970. Géologie du Bassin de Mons: un siècle d'histoire. Annales de la Société géologique du Nord 110: 171–189.

Norman, D. B. 1996. On Mongolian ornithopods (Dinosauria: Ornithischia). 1. Iguanodon orientalis Rozhdestvenskii, 1952. Zoological Journal of the Linnean Society 116: 303–315.

Norman, D. B., and D. B. Weishampel. 1990. Iguanodontidae and related ornithopods; pp. 510–533 in D. B. Weishampel, P. Dodson, and H. Osmólska (eds.), The Dinosauria, 2nd ed. University of California Press, Berkeley.

Pelzer, G., and V. Wilde. 1987. Klimatische Tendezen während der Ablagerung der Wealden-Fazies in Nordwesteuropa. Geologisches Jahrbuch A 96: 239–263.

Penny, J. H. 1986. An Early Cretaceous angiosperm pollen assemblage from Egypt. Special Papers in Palaeontology 35: 119–132.

Potonié, R. 1956. Synopsis des Gattungen der Sporae dispersae I. Teil: Sporites. Beihefthe zum Geologischen Jahrbuch 23: 1–103.

Robaszynski, F., A. V. Dhondt, and J W. M. Jagt. 2001. Cretaceous lithostratigraphic units (Belgium), pp. 121–134 in P. Bultynck and L. Dejonghe (eds.), Guide to a revised lithostratigraphic scale of Belgium. Geologica Belgica 4.

Robinson, S. A., and S. P. Hesselbo. 2004. Fossil-wood carbon-isotope stratigraphy of the non-marine Wealden Group (Lower Cretaceous, Southern

8.2. (facing) Selected palynomorphs from a sample at −265 m in the BER 3 borehole, slide Ber 3/265A/#58950. The position of each illustrated palynomorph on this slide is provided in square brackets after the England Finder. A, *Trilobosporites hannonicus* (Delcourt and Sprumont, 1955) Potonié, 1956 [O49–2]; B, *Parvisaccites radiatus* Couper, 1958 [P39–4]; C, biorecord Hauterivian-*cactisulc* (in Hughes and McDougall, 1987), alias *Cerebropollenites* sp. [J41–1]; D, *Ephedripites montanaensis* Brenner, 1968 [R44–2]; E, biorecord Superret-*croton* (in Hughes et al., 1979) [N49]; F–G, probable paleotaxon Superret-*subcrot* (in Hughes et al., 1979) [F, H35/H36] [G, P43–4/Q43–2]; H, trichotomosulcate semitectate columellate pollen grain (probable biorecord Retichot-*baccat* in Hughes et al., 1979) [X49–4].

8.3. Comparison between the C-isotopes curves of Bernissart, Isle of Wight (Robinson and Hesselbo, 2004), the Cismon section (Erba et al., 1999), and an Umbria-March composite (Sprovieri et al., 2006) in the Barremian–Aptian interval. Solid line indicates independently verified correlation with the Cismon core. Dashed line indicates tentative correlation. Data from the Umbria-March Basin are modified from Sprovieri et al. (2006). The magnetostratigraphic ages for the magnetic chron boundaries are from Gradstein et al. (2004). Gray surface indicates tentative correlation between Bernissart, Cismon, and the Umbria-March composite, based on palynological age assignment for Bernissart and the shape of carbon isotope curves. From Schnyder et al. (2009).

England). Journal of the Geological Society London 161: 133–145.

Schnyder, J., J. Dejax, E. Keppens, T. Nguyen Tu, P. Spagna, S. Boulila, B. Galbrun, A. Riboulleau, J.-P. Tshibangu, and J. Yans. 2009. An Early Cretaceous lacustrine record: organic matter and organic carbon isotopes at Bernissart (Mons Basin, Belgium). Palaeogeography, Palaeoclimatology, Palaeoecology 281: 79–91.

Sprovieri, M., R. Coccioni, F. Lirer, N. Pelosi, and F. Lozar. 2006. Orbital tuning of a lower Cretaceous composite record (Maiolica Formation, central Italy). Palaeoceanography 21: PA4212.

Taquet, P. 1975. Remarques sur l'évolution des Iguanodontidés et l'origine des Hadrosauridés. Colloque international C.N.R.S. n° 218: Problèmes actuels de paléontologie-évolution des vertébrés, Paris, 503–511.

Vakhrameev, V. A. 1991. Jurassic and Cretaceous floras and climates of the Earth. Cambridge University Press, Cambridge, 318 pp.

Vandenbroeck, E. 1900. La question de l'âge des dépôts wealdiens et bernissartiens. Pourquoi, dans la nouvelle édition de la légende de la Carte géologique de la Belgique, les dépôts à Iguanodons de Bernissart viennent d'être classés dans le Jurassique supérieur. Bulletin de la Société belge de Géologie 14: 70–73.

Yans, J. 2007. Lithostratigraphie, minéralogie et diagenèse des sédiments à faciès wealdien du Bassin de Mons (Belgique). Mémoire de la Classe des Sciences, Académie Royale de Belgique (ser. 3) 2046: 1–179.

Yans, J., P. Spagna, C. Vanneste, M. Hennebert, S. Vandycke, J.-M. Baele, J.-P. Tshibangu, P. Bultynck, M. Streel, and C. Dupuis. 2005. Description et implications géologiques préliminaires d'un forage carotté dans le "Cran aux Iguanodons" de Bernissart. Geologica Belgica 8: 43–49.

Yans, J., J. Dejax, D. Pons, L. Taverne, and P. Bultynck. 2006. The iguanodons of Bernissart are middle Barremian to earliest Aptian in age. Bulletin de l'Institut royal des Sciences naturelles de Belgique 76: 91–95.

Yans, J., T. Gerards, P. Gerrienne, P. Spagna, J. Dejax, J. Schnyder, J.-Y. Storme, and E. Keppens. 2010. Carbonisotope of fossil wood and dispersed organic matter from the terrestrial Wealden facies of Hautrage (Mons Basin, Belgium). Palaeogeography, Palaeoclimatology, Palaeoecology 291: 85–105.

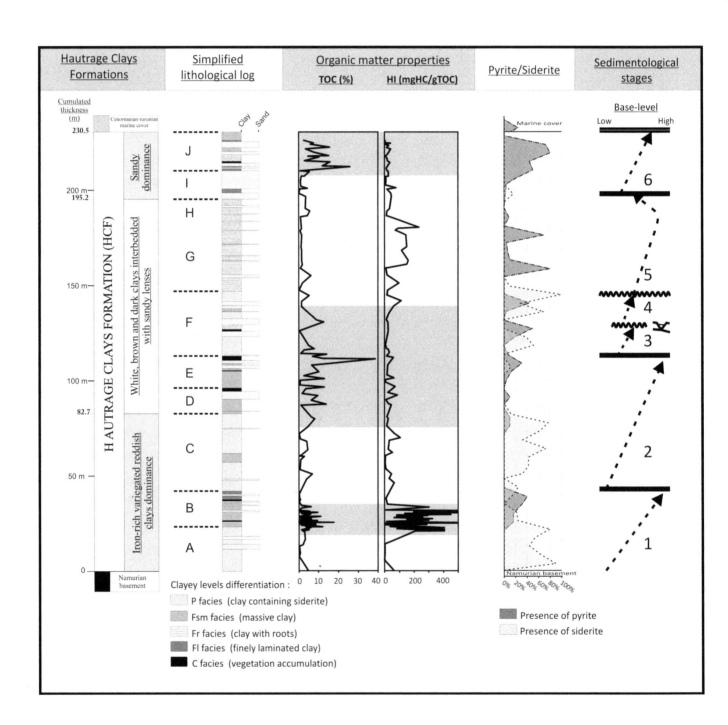

Hautrage Clays Formations

Simplified lithological log

Organic matter properties
TOC (%)   HI (mgHC/gTOC)

Pyrite/Siderite

Sedimentological stages

Cumulated thickness (m)

230.5   Cenomanian-turonian marine cover

J

200 m
195.2   I

H

G

150 m

F

100 m
82.7   E

D

C

50 m

B

A

0   Namurian basement

Clayey levels differentiation :

P facies (clay containing siderite)

Fsm facies (massive clay)

Fr facies (clay with roots)

Fl facies (finely laminated clay)

C facies (vegetation accumulation)

HAUTRAGE CLAYS FORMATION (HCF)

White, brown and dark clays interbedded with sandy lenses

Iron-rich variegated reddish clays dominance

Sandy dominance

Clay — Sand

0 10 20 30 40 0   200   400

Marine cover

Namurian basement

0% 20% 40% 60% 80% 100%

Base-level
Low   High

6

5

4

3

2

1

Presence of pyrite

Presence of siderite

86

The Paleoenvironment of the Bernissart Iguanodons: Sedimentological Analysis of the Lower Cretaceous Wealden Facies in the Bernissart Area

Paul Spagna*, Johan Yans, Johann Schnyder, and
Christian Dupuis

The Wealden facies at Hautrage and Bernissart (Mons Basin, Belgium)
have been investigated following different sedimentological parameters,
including lithofacies evolution, mineralogical and granulometric data,
and organic matter properties. A six-step paleoenvironmental evolution
can be observed in the Hautrage Clays Formation at Hautrage (10 km
from Bernissart), in relation with the variation of the base level (deepening upward) inside a floodplain. Three sedimentological units can be
recognized in the Sainte-Barbe Clays Formation of the Iguanodon Sinkhole at Bernissart, leading to a 2D modeling of the Bernissart paleolake.
A schematic east–west paleovalley map is finally proposed, integrating all
the new paleoenvironmental information collected in the Wealden facies
from the Mons Basin.

The sedimentological study of the Wealden facies presented in this chapter
has been held in two different sites localized on the northern border of
the Mons Basin: the Hautrage Danube-Bouchon quarry, prospected and
exploited since a few decades for its Al-rich Wealden clays (Hautrage Clays
Formation), and the Iguanodon Sinkhole at Bernissart, sampled by a drilling in 2002–2003 (Sainte-Barbe Clays Formation; see Fig. 3.1 in this book).
The lithological and sedimentological comparisons of these two Wealden
formations, deposited and conserved in different environments (pluriki-
lometrical pockets trapping parts of an alluvial floodplain at Hautrage,
and plurimetrical natural pit developing lacustrine to swampy environment at Bernissart), allow the integration of several paleoenvironmental
components.

The Hautrage Site: An Active Quarry Digging Clays in a Flooding Plain

The numerous geological surveys and studies led in the Hautrage Clays
Formation revealed an accessible succession with a cumulated thickness of
more than 230 m (e.g., Yans, 2007), with a wide range of different lithologies: variegated reddish clays, black and brown to white silty and sandy
clays, silt and sands of different colors and clayey contents, and even conglomerates. These sediments contain various quantities of pyrite, siderite,

Introduction

9.1. Log and synthesis of different sequences recognized in the Hautrage Clays Formation in the Wealden facies of the Hautrage pocket. From Spagna (2010).

and organic matter (charcoal, coalified fragments of wood, dispersed organic matter; see Gerards et al., 2007, 2008; Gomez et al., 2008; Dejax et al., 2008; Yans et al., 2010). On the basis of lithological characterization, the succession can be divided into nine units (Fig. 9.1), which can be further regrouped in three major sets: a dominance of reddish (Fe-rich) clays at the bottom (Units A to C); black and brown to white silty clays set interbedded with numerous sandy lenses (Units D to G); and a sandy (to conglomerate) dominance at the top (Units H and I). This wide range of lithologies and their arrangement (widespread clayey layers cut by sandy lenses) already suggest a floodplain environment crossed by numerous meandering channels, a hypothesis confirmed by sedimentological study (see below).

The Bernissart Site: A Buried Paleolake Developed in a Local Subsidence Area (Sinkhole) that Trapped Numerous Basal Iguanodontia Dinosaurs

A total of approximately 70 m of core belonging to the Sainte-Barbe Clays Formation were collected from the two new boreholes, BER 2 (20 m) and BER 3 (50 m), within the Iguanodon Sinkhole at Bernissart (Yans et al., 2005). Unlike the Hautrage Wealden facies, the Wealden sediments from the Iguanodon Sinkhole are homogeneous, composed of brown to black silty clays laminated by thin (millimetric) silty (to sandy) layers, rich in organic matter. The homogeneity of the sediments, together with the omnipresence of thin laminations and the freshwater fauna found during the nineteenth-century excavations, suggest a lacustrine paleoenvironment. The Sainte-Barbe Clays Formation overlies a basal breccia, with a contact defined around −315 m in BER 3, containing basement clasts (centimetric up to half-metric in size) that can be mixed in a Wealden clayey matrix. The Wealden facies from the Iguanodon Sinkhole is covered by Cretaceous marine sediments, composed, among others, of chalks, calcarenites, and conglomerate sediments (from −8 to −265 m; see Chapter 7 in this book), then by Cenozoic fine clayey sands (from 0 to −8 m; Fig. 9.2).

Sedimentological Study

Many parameters of the Wealden deposits from both the Hautrage Clays Formation and Sainte-Barbe Clays Formation have been analyzed, including texture and structure of the sediments; mineralogy of the clayey fraction; granulometry of lacustrine, flooding plain, and channel deposits; and organic matter contents and properties. These parameters have been combined with the detailed lithological descriptions and correlations made in the field and on the borehole cores.

Hautrage Clays Formation

We focus on parameters that present a clear evolution in the Hautrage Clays Formation series: lithofacies evolution (sensu Miall, 1996), organic matter parameters (Yans et al., 2010), for example, total organic content (TOC in %) and hydrogen index (HI in mgHC/gTOC), and the presence or absence of pyrite and oxidized siderite (Fig. 9.1). The integration of these parameters leads to the definition of a six-step sequential evolution of the

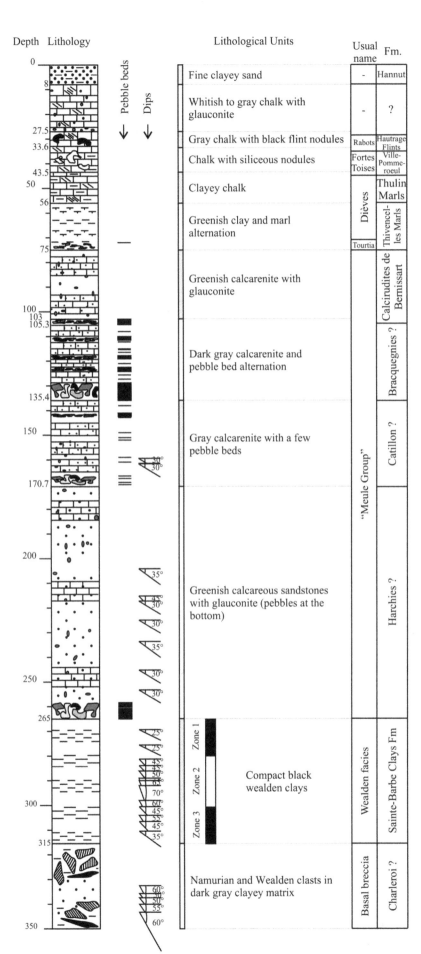

Depth	Lithology			Lithological Units	Usual name	Fm.
0		Pebble beds	Dips	Fine clayey sand	-	Hannut
8				Whitish to gray chalk with glauconite	-	?
27.5				Gray chalk with black flint nodules	Rabots	Hautrage Flints
33.6				Chalk with siliceous nodules	Fortes Toises	Ville-Pomme-roeul
43.5				Clayey chalk	Dièves	Thulin Marls
50						
56				Greenish clay and marl alternation		Thivencel-les Marls
75					Tourtia	
				Greenish calcarenite with glauconite		Calcirudites de Bernissart
100						
103						
105.3				Dark gray calcarenite and pebble bed alternation	Bracquegnies ?	
135.4						
150				Gray calcarenite with a few pebble beds	Catillon ?	"Meule Group"
170.7			30° 30°			
200			35°			
			45° 30°	Greenish calcareous sandstones with glauconite (pebbles at the bottom)	Harchies ?	
			30°			
			35°			
250			30°			
265			30°			
			25° 25°	Zone 1		
			45° 45° 50° 60° 70°	Zone 2 Compact black wealden clays	Wealden facies	Sainte-Barbe Clays Fm.
300			60° 45° 55° 45° 35°	Zone 3		
315						
			60° 55° 50° 55°	Namurian and Wealden clasts in dark gray clayey matrix	Basal breccia	Charleroi ?
350			60°			

9.2. Preliminary description of the BER 3 borehole (from Yans et al., 2005). For explanations of the zones in the Wealden facies, see Figure 9.4 and text.

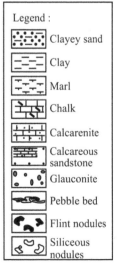

Legend :

- Clayey sand
- Clay
- Marl
- Chalk
- Calcarenite
- Calcareous sandstone
- Glauconite
- Pebble bed
- Flint nodules
- Siliceous nodules

paleoenvironment in the Hautrage Clays Formation series. Interpretation keys suggest that this sequential evolution is linked to variations of the base level in the floodplain. The notion of base level, introduced by Wheeler (1964) to transpose the sequential stratigraphy to continental environments, delineates areas of erosion and deposition associated to a detrital flow. Although the role of the water level seems predominant in the evolution of the succession, this interpretation is based on a large-scale evolution (more than 230 m) of the different parameters; therefore, it must be considered a first attempt to define the sequences.

The interpretation keys used in this chapter are briefly described below.

· P facies are characterized by variegated siderite-rich clayey layers, indicating soils developed in variable oxidation-reduction conditions (frequently exposed soils), whereas C and Fr facies, respectively representing organic matter accumulation beds and clayey levels including root fossils, are more likely to occur in more hydromorphous environments;
· The diagram indicating the presence of siderite/pyrite reflects exactly the same idea of exposed versus hydromorphous paleosoils. It is based on a counting of the beds in 5-m-thick sets containing each of those minerals before expressing their accumulated thickness as a percentage.
· TOC reflects the accumulation of the organic carbon within the sediments, which may be linked to the abundance of the terrestrial vegetation cover, algal–bacterial productivity within ponds, or local accumulation of organic particles due to the destabilization of soils.
· HI is an indicator of the sources of the organic matter and the alteration state of the organic particles.

The six sequences always start with a low base level and progress to higher ones (deepening upward). This evolution may be explained as a combination of subsidence effects and sedimentary input variations, possibly reinforced by climatic changes.

Sequences 1 and 2 characterize a border-condition environment, starting with frequently exposed soils (pseudogleys; Fig. 9.3A) that are progressively replaced by hydromorphous soils, richer in organic matter and pyrite. Small swamps and lake or ponds, where hydrogen-rich organic matter accumulates, are then installed in the alluvial plain.

Sequences 3 and 4 are characterized by the same evolution that took place in a fully developed alluvial plain, composed of numerous channels cutting floodplain deposits. In the field, those levels (top of Unit E and all of Unit F) present synsedimentary deformations (Fig. 9.3B) attributed to karstic subsidence activity (Spagna et al., 2007; Spagna, 2010; Chapter 6 in this book) that perturbed the sedimentary dynamics.

Sequence 5 took place just after the intense subsidence activity period of sequences 3 and 4. It characterizes a more frequently immersed (swampy to lacustrine) environment, developed in the alluvial plain and marked by the sedimentation of the black thinly laminated clays of Unit G in a relatively quiet environment (Fig. 9.3C).

Finally, sequence 6 represents a higher-energy environment, occupied by sandy channels containing variable quantities of pebble lenses, plurimetrical wood fragments, pyrite, and so on (Fig. 9.3D).

Because of the southward dipping of the layers in the Danube-Bouchon quarry, this "vertical" recorded evolution is geographically positioned from north (sequence 1) to south (sequence 6). Even if the current configuration is also probably partially linked to the late tectonic history of the Mons Basin (tilting since the mid to Late Cretaceous combined with later erosional phases; Vandycke, 2002), the synsedimentary character of the deformations described in sequences 3 and 4 clearly reflects the mobile aspect (southward moving) of the subsidence acting at the dawn of the development of the Mons Basin, contributing to the north-to-south succession of the sequences (Spagna, 2010).

Sainte-Barbe Clays Formation

Lithological descriptions alone were not sufficient to correlate the BER 2 and BER 3 boreholes. We therefore used the granulometric parameter "means of the D(v,0.9)" (the size of a mesh sieve that would let pass 90% in volume of the sample), to define at least three units in the Wealden

9.3. Core and exposures showing the sequences. A, Reddish variegated clays (pseudogleys) of the sequences 1 and 2. B, Deformations affecting a channel of the sequences 3 and 4. C, Dark, finely laminated clays of the sequence 5. D, Sandy dominance cover of the sequence 6 (south front of the quarry).

9.4. 2D modeling of the Iguanodon Sinkhole. From Spagna (2010).

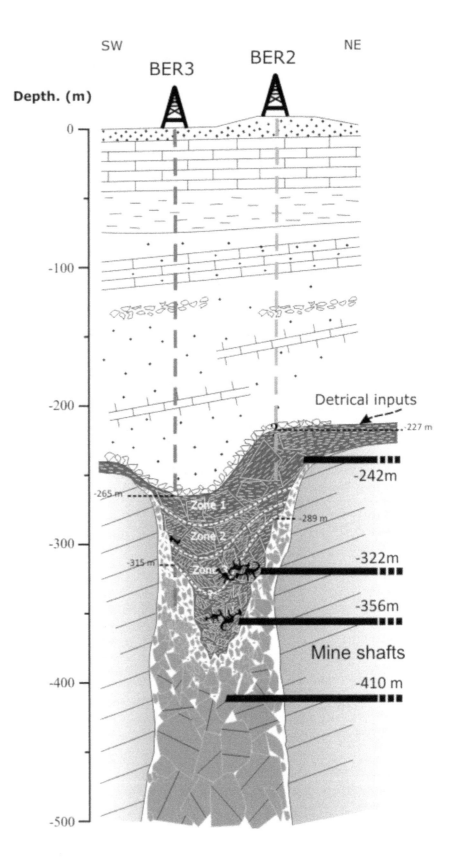

SW   NE

BER3   BER2

Depth. (m)

0

-100

-200   Detrical inputs
                                  ·-227 m

                                  -242m

-265 m ···· Zone 1
                    ···· 289 m.

-300   Zone 2

-315 m ···   Zone ?   -322m

                                  -356m

                                  Mine shafts

-400   -410 m

-500

cores, recorded in both boreholes (Figure 9.4). Those limits, pointed respectively at −272 m and −283 m in BER 2 and at −280 m and −302 m in BER 3, fit perfectly with the limits of the organic matter, defined by Schnyder et al. (2009) on the basis of evolution of TOC, HI, and surface percentage of amorphous organic matter within palynofacies slides. Together, those parameters provide indications of the history of the water level in the Bernissart paleolake and on the related fluctuations through time of fine sandy/silty detrital inputs, which have been interpreted to be partly controlled by orbitally induced climate cycles (Schnyder et al., 2009).

Dinosaur bone fragments found at −296.5 and −309 m (see Chapter 12 in this book) in the BER 3 borehole highlight the fantastic accumulation of dinosaur bones at Bernissart. The iguanodons moved in a wetland landscape marked by strong seasonality with significant dry periods. The sedimentological data contribute to understanding the trapping conditions of the Bernissart natural pit, which at the time probably appeared at the surface as an inoffensive lake. This lake masked the muddy quicksand-like area developed at the top of the natural pit, the result of the nature of its continuous local subsidence. The quicksand acted as a final collector for the dinosaur carcasses.

Regional Integration

Figure 9.5 tentatively integrates at a local scale the sedimentological data collected from the Hautrage Clays Formation and the Bernissart Clays Formation, together with extrapolations from similar deposits in the Mons Basin, such as the Baudour Clays Formation (Dejax et al., 2007) and the sinkhole belt localized a few hundred meters to the south of this area. The schematic east–west paleovalley drawn here was occupied by rivers flowing westward, probably more braided on the upstream part and more meandering on the downstream part. The bordering relief was more or less deeply eroded by many tributaries that supplied sediment inputs into the valley. The northern border of the valley was occupied by stabilized sediments from sequences 1 and 2 (frequently exposed with abundant pseudogley formations), whereas the alluvial plain (sequences 3 to 6) expanded southward (following the displacement of the main subsidence vector). Consequently, the southern edge of the valley was probably more irregular than the northern one because it was going for collapse, following the karstic subsidence movement (induced by deep dissolution phenomena) in this direction. In this context, the sinkholes, which formed an east–west-oriented belt, are interpreted as a local expression of the same phenomena of deep dissolution, localized right over its (heterogeneous) progression front. In the landscape, those natural pits probably appeared as swampy and/or lacustrine restricted areas (closed environment) that could trap and preserve many traces of life.

Conversely, traces of the macrofauna in the open environment of the alluvial plain in the Hautrage area are rare. However, a few pebbles found in situ might be interpreted as gastroliths of large herbivorous dinosaurs, reflecting their presence in the valley (see Chapter 13 in this book).

Acknowledgments

The results presented here are part of the Ph.D. thesis of P.S., funded principally by collaboration between the Faculté Polytechnique de Mons (UMons) and the CBR-Heidelbergcement cementery of Harmignies (Belgium). This work was also supported by grant FRFC-FNRS of Belgium, n° 2.4.568.04.F, "Age, palaeoenvironments and formation processes of the Cran aux Iguanodons of Bernissart: integration of the results in the global stratigraphic and palaeogeographic context." The authors are grateful to P. Bultynck for his constructive suggestions to improve this chapter.

Brabant Massif
Quenast
Tournai
"Paratochtone brabançon"
Mons
Mons Basin
Fleurus
Namur
Charleroi
"Allochtone ardennais"
0
16km
Albo-Cenomanian "Meule Group"
Wealden facies
Variscan front and Dinant Basin
"Houiller" (mainly)
Dinantian (Limestone)
Mid. and Up. Devonian
Brabant Massif Lower Paleozoic
HA
BA
BE
Channel
Abandonned channel
Point-bar
Pseudogleys
Lacustrine environment
Marshy environment
Alluvial cone
Namurian basement
N
0
~2 km

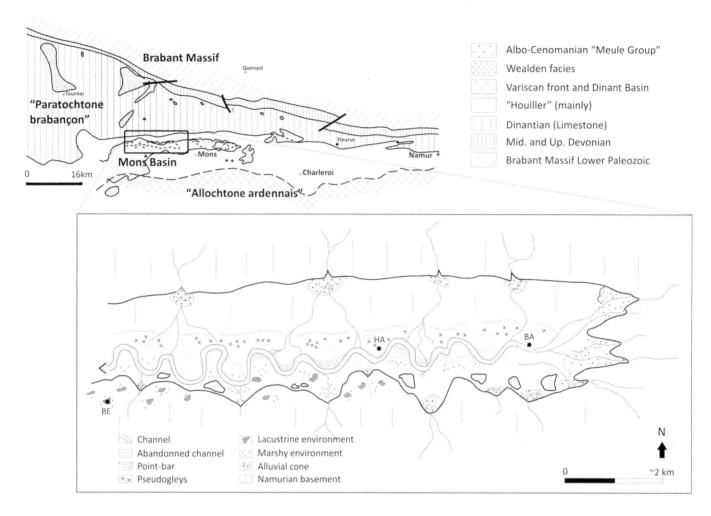

9.5. Schematic regional map of the Wealden paleovalley in the western part of the Mons Basin northern edge. *Abbreviations:* BA, Baudour; HA, Hautrage; BE, Bernissart. From Spagna (2010).

References

Dejax, J., D. Pons, and J. Yans. 2007. Palynology of the dinosaur-bearing Wealden facies sediments in the natural pit of Bernissart (Belgium). Review of Palaeobotany and Palynology 144: 25–38.

Dejax, J., D. Pons, and J. Yans. 2008. Palynology of the Wealden facies from Hautrage quarry (Mons Basin, Belgium). Memoirs of the Geological Survey 55: 45–52.

Gerards, T., J. Yans, and P. Gerrienne. 2007. Growth rings of Lower Cretaceous softwoods: some plaeoclimatic implications. Notebooks on Geology e-Journal. http://paleopolis.rediris.es/cg/CG2007_M01/CG2007_M01.pdf, 29–33.

Gerards, T., J. Yans, P. Spagna, and P. Gerrienne. 2008. Wood remains and sporomorphs from the Wealden facies of Hautrage (Mons Basin, Belgium): paleoclimatic and paleoenvironmental implications. Memoirs of the Geological Survey 55: 61–70.

Gomez B., T. Gillot, V. Daviero-Gomez, P. Spagna, and J. Yans. 2008. Paleoflora from the Wealden facies strata of Belgium: mega- and meso-fossils of Hautrage (Mons Basin). Memoirs of the Geological Survey 55: 53–60.

Miall, A. D. 1996. The geology of fluvial deposits: sedimentary facies, basin analysis and petroleum geology. Springer-Verlag, Berlin, 582 pp.

Schnyder, J., J. Dejax, E. Keppens, T. Nguyen Tu, P. Spagna, S. Boulila, B. Galbrun, A. Riboulleau, J.-P. Tshibangu, and J. Yans. 2009. An Early Cretaceous lacustrine record: organic matter and organic carbon isotopes at Bernissart (Mons Basin, Belgium). Palaeogeography, Palaeoclimatology, Palaeoecology 281: 79–91.

Spagna, P. 2010. Les faciès wealdiens du Bassin de Mons (Belgique): paléoenvironnements, géodynamique et valorisation industrielle. Ph.D. thesis, Faculté Polytechnique de l'Umons, 138 pp.

Spagna, P., S. Vandycke, J. Yans, and C. Dupuis. 2007. Hydraulic and brittle extensional faulting in the Wealden facies of Hautrage (Mons Basin, Belgium). Geologica Belgica 10: 158–161.

Vandycke, S. 2002. Palaeostress records in Cretaceous formations in NW Europe: extensional and strike-slip events in

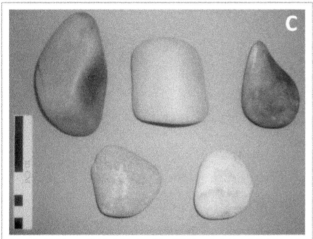

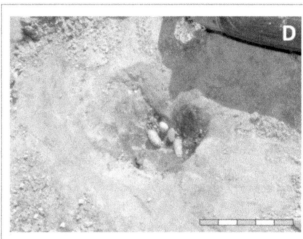

relationship with Cretaceous–Tertiary inversion tectonics. Tectonophysics 357: 119–136.

Wheeler, H. E. 1964. Base-level, lithosphere surface and time stratigraphy. Geological Society of America Bulletin 75: 599–610.

Yans, J. 2007. Lithostratigraphie, minéralogie et diagenèse des sédiments à faciès wealdien du Bassin de Mons (Belgique). Mémoire de la Classe des Sciences, Académie Royale de Belgique (ser. 3) 2046: 1–179.

Yans, J., P. Spagna, C. Vanneste, M. Hennebert, S. Vandycke, J.-M. Baele, J.-P. Tshibangu, P. Bultynck, M. Streel, and C. Dupuis. 2005. Description et implications géologiques préliminaires d'un

forage carotté dans le "Cran aux Iguanodons" de Bernissart. Geologica Belgica 8: 43–49.

Yans, J., J. Dejax, D. Pons, L. Taverne, and P. Bultynck. 2006. The iguanodons of Bernissart are middle Barremian to earliest Aptian in age. Bulletin Institut Sciences naturelles Belgique 76: 91–95.

Yans, J., T. Gerards, P. Gerrienne, P. Spagna, J. Dejax, J. Schnyder, J.-Y. Storme, and E. Keppens. 2010. Carbon-isotope of fossil wood and dispersed organic matter from the terrestrial Wealden facies of Hautrage (Mons Basin, Belgium). Palaeogeography, Palaeoclimatology, Palaeoecology 291: 85–105.

9.6. Specimens of the presumed "gastroliths" from the Hautrage Clays Formation. A–B, In situ isolated element, with size ranging from 5 to 10 cm. C, Isolated elements extracted from the sediment. D, Pocket of smaller grouped elements (about half a centimeter in size). Scale = 5 cm, except for C, 10 cm. From Spagna (2010).

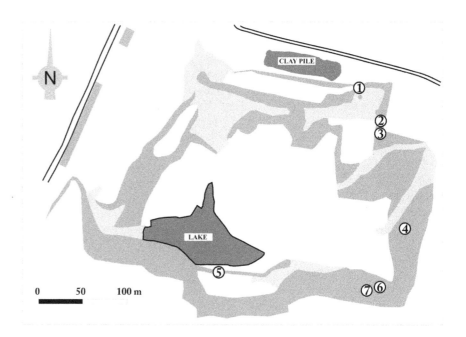

10.1. Topographic map of the Danube-Bouchon quarry at Hautrage showing the sampling locations (Gomez et al., 2008).

Mesofossil Plant Remains from the Barremian of Hautrage (Mons Basin, Belgium), with Taphonomy, Paleoecology, and Paleoenvironment Insights

10

Bernard Gomez*, Thomas Gillot, Véronique Daviero-Gomez, Clément Coiffard, Paul Spagna, and Johan Yans

Seven beds bearing mesofossil plant remains have been sampled from the late Early to early Late Barremian Hautrage Clays Formation in the Danube-Bouchon quarry at Hautrage (Mons Basin, Belgium). They include various fertile and sterile parts of ferns (*Weichselia reticulata* (Stokes et Webb) Fontaine, *Phlebopteris dunkeri* Schenk, *Gleichenites nordenskioeldii* (Heer) Seward), Cheirolepidiaceae (*Alvinia* Kvaček, *Frenelopsis* (Schenk) Watson), Miroviaceae (*Arctopitys* Bose et Manum), Taxodiaceae (*Sphenolepis* Schenk), other conifers (*Brachyphyllum* Brongniart and *Pagiophyllum* Heer), and Ginkgoales (*Pseudotorellia* Florin). Although the plant assemblages vary from one bed to another, the taxa remain globally unchanged, suggesting repeated vegetation changes that may be related to lateral divagations of stream channels in a continental freshwater floodplain. Integration of taphonomic and sedimentological data suggest that fires may have played a role in the production, transport, and preservation of the mesofossil plant remains that may mostly represent the local vegetation.

Introduction

Megafossil plant remains were first reported from the Mons Basin in Belgium by Coemans (1867). However, the paleontological interest for this area really began from 1878, with the discovery of the Bernissart iguanodons in the Iguanodon Sinkhole at Bernissart. Megafossil plant specimens were also collected from different localities in the Mons Basin, and all are housed in the paleontological collections of the Royal Belgian Institute of Natural Sciences (RBINS) in Brussels. The taxonomy and systematics of the megafossil plant specimens from Bernissart were studied by Seward (1900), Harris (1953), and Alvin (1953, 1957, 1960, 1968, 1971). Only three species of conifer female cones were described from Hautrage (Alvin, 1953, 1957, 1960).

The Danube-Bouchon quarry is located about 20 km northwest of Mons (Fig. 10.1). The quarry cuts the Hautrage Clays Formation (Robaszynski et al., 2001). It consists of continental clays, silts, and sands, containing lignite debris, pyrite, and siderite nodules in variable proportions (Spagna et al., 2007). The sedimentological study suggests that it was a floodplain with east–west-oriented meandering channels and a westward flow, and that clayey sediments were deposited during major floods (Spagna et al., 2008; see Chapter 9 in this book). The Wealden facies of Hautrage are

10.2. Log of the Hautrage Formation with the stratigraphic location of the seven samples collected in the quarry.

10.3. Mesofossil plant remains from bed 1 in the Danube-Bouchon's Hautrage clay pit quarry (Mons Basin, Belgium). A, Wood preserved as fusains (charcoals; RBINS B5760); B, yellow to orange pieces of amber (RBINS B5761); C–I, sterile fern pinnules (*Gleichenites nordenskioeldii* [RBINS B5762] and *Weichselia reticulata* [RBINS B5763]); J–K, circinate fern fronds (RBIN B5764); L–L', fertile fern pinnule (*Phlebopteris dunkeri,* RBINS B5765) showing rounded sori; M–M', fragment of pinna showing insertion scars of sori (*W. reticulata,* RBINS B5766); N–S, conifer leafy twigs showing more or less adpressed, tiny leaves (*Sphenolepis;* RBINS B5767); T–V, leaf fragments of *Arctopitys* (T, apex; U, base; V, several middle laminas; RBINS B5768); W–Y, cone and scales of uncertain affinities (RBINS B5769); Z, seeds of uncertain affinities (RBINS B5770). Scale bars = 10 mm (A, B, and N) and 1 mm (others).

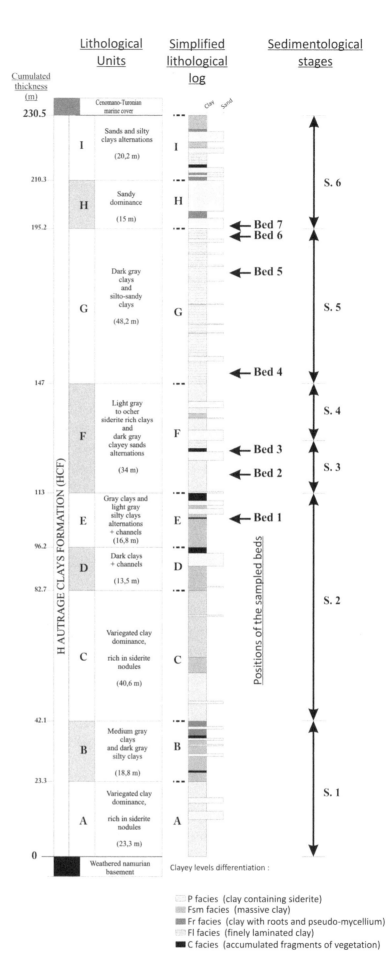

Lithological Units | Simplified lithological log | Sedimentological stages

Cumulated thickness (m)

230.5 — Cenomano-Turonian marine cover

I — Sands and silty clays alternations (20,2 m)

210.3

H — Sandy dominance (15 m) — Bed 7

195.2 — Bed 6

G — Dark gray clays and silto-sandy clays (48,2 m) — Bed 5

— Bed 4

147

F — Light gray to ocher siderite rich clays and dark gray clayey sands alternations (34 m) — Bed 3 — Bed 2

113

E — Gray clays and light gray silty clays alternations + channels (16,8 m) — Bed 1

96.2

D — Dark clays + channels (13,5 m)

82.7

C — Variegated clay dominance, rich in siderite nodules (40,6 m)

42.1

B — Medium gray clays and dark gray silty clays (18,8 m)

23.3

A — Variegated clay dominance, rich in siderite nodules (23,3 m)

0 — Weathered namurian basement

H AUTRAGE CLAYS FORMATION (HCF)

Positions of the sampled beds

S. 6
S. 5
S. 4
S. 3
S. 2
S. 1

Clayey levels differentiation :

▫ P facies (clay containing siderite)
▨ Fsm facies (massive clay)
▓ Fr facies (clay with roots and pseudo-mycellium)
▤ Fl facies (finely laminated clay)
■ C facies (accumulated fragments of vegetation)

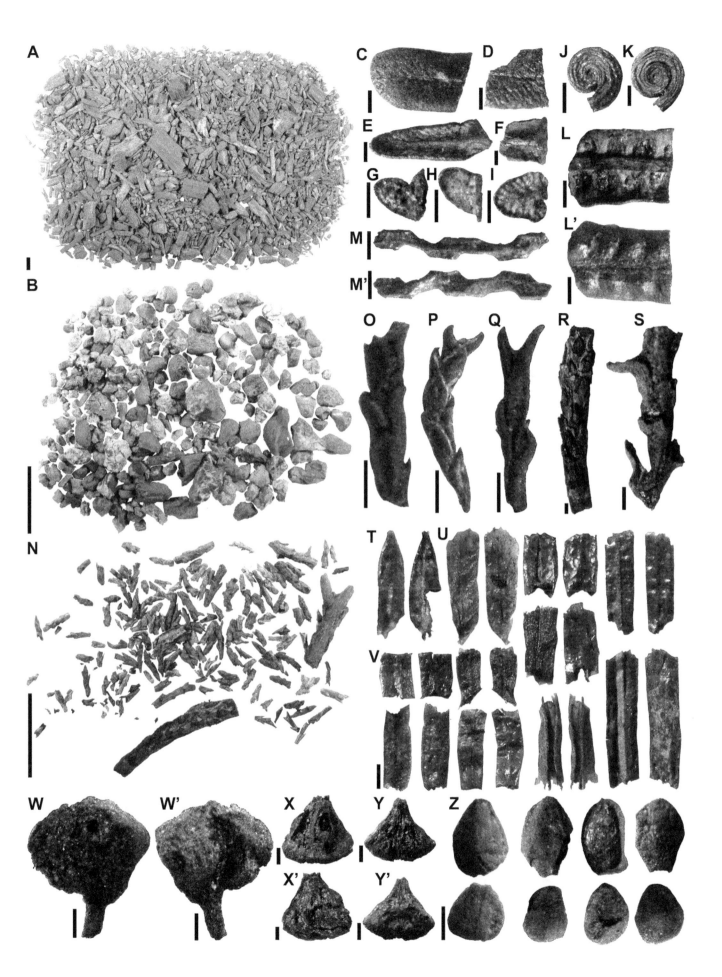

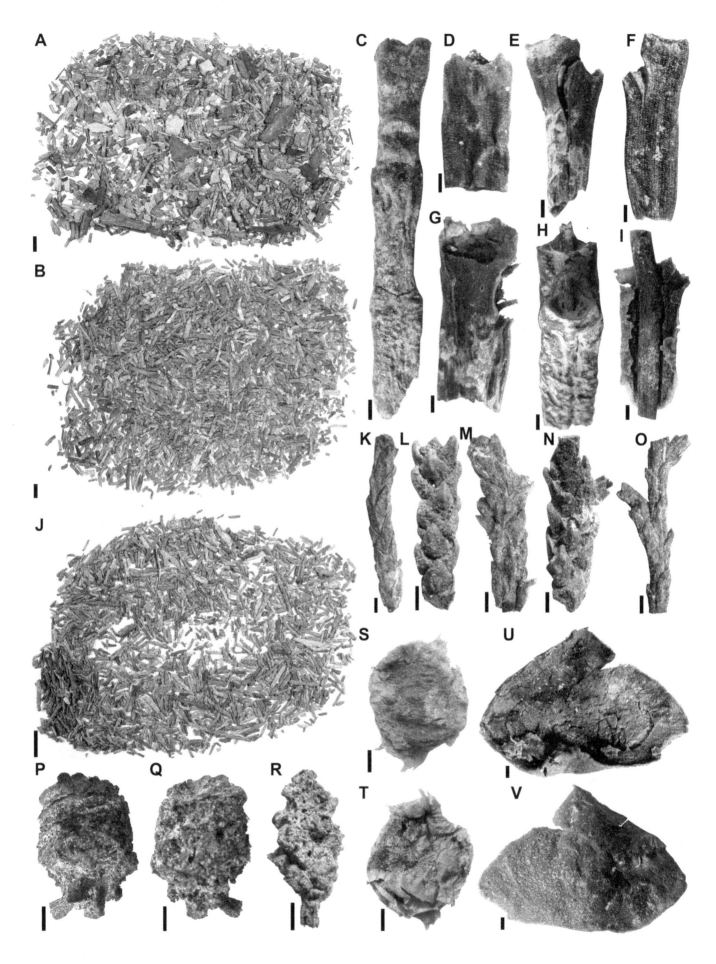

late Early to early Late Barremian in age, based on the occurrence of rare angiosperm pollen grains and chemostratigraphy (Dejax et al., 2008; Yans et al., 2006, 2010).

The present work shows that the mesofossil plants from the late Early to early Late Barremian Hautrage Clays Formation are abundant and more diversified than previously reported, and that important changes in plant composition occur from one bed to another. We finally propose a paleoenvironmental reconstruction based on the taphonomic analysis of the mesofossil plant remains and integration of the sedimentological data.

10.4. Mesofossil plant remains from bed 2 in the Danube-Bouchon's Hautrage clay pit quarry (Mons Basin, Belgium). A, Wood preserved as fusains (charcoals; RBINS B5771); B–I, conifer leafy axes showing whorls of three leaves and a central solid vascular tissue (*Frenelopsis;* RBINS B5772); J–O, conifer leafy twigs showing more or less adpressed, tiny leaves (*Sphenolepis;* RBINS B5773); P–R, conelike structure of uncertain affinities (RBINS B5774); S–T, conifer cheirolepidiaceous female cone scale (*Alvinia;* RBINS B5775); U–V, conifer female cone scale of uncertain affinity (RBINS B5776). Scale bars = 10 mm (A, B, and J) and 1 mm (others).

Materials and Methods

Seven clayey beds, apparently rich in mesofossil plant remains, were sampled in the Danube-Bouchon quarry of Hautrage in 2007 by three of us (B.G., P.S., J.Y.; Fig. 10.2). The sediments were bulk macerated in hydrogen peroxide (H$_2$O$_2$, 35%) or simply dissolved with tap water (Gomez et al., 2008). They were washed under tap water through a column of sieves with meshes of 1 mm, 500 µm, and 100 µm. The residues were air dried. Only fractions more than 1 mm in size were sorted with the naked eye or under a stereomicroscope.

Results

Bed 1 is lignitic. Woody fragments dominate; most are charcoalified, and a few are vitrified (Fig. 10.3A). Twigs of *Sphenolepis*, with small leaves arranged spirally along the axes, are also present (Figs. 10.3O–S). Orange to yellow pieces of amber occur and are quite abundant (Fig. 10.3B). Proximal, middle, and a few distal parts of leaves of the miroviaceous conifer genus *Arctopitys* Bose et Manum (see Nosova and Wcislo-Luraniec, 2007, for a review) are preserved as cuticle compressions (Figs. 10.3T–V). They can be easily recognized by clear longitudinal band of sediments at the center of the leaf lamina, which corresponds to a depressed groove, in which stomatal apparatus are crowded together. Among ferns, fertile and sterile pinnules, circinate fronds, and isolated rachis of *Weichselia reticulata* (Stokes et Webb) Fontaine, *Phlebopteris dunkeri* Schenk, and *Gleichenites nordenskioeldii* (Heer) Seward have been identified (Figs. 10.3C–L). They are preserved as charcoals and are very fragmented. Such ferns were similarly reported from the Wealden facies of England by Harris (1981), who suggested a production by wildfires. Other retrieved mesofossil plants are represented by scales, cones, and reproductive organs of uncertain affinities (Figs. 10.3W–Z).

Bed 2 corresponds to bluish clays with siderite nodules. It mainly bears millimetric to centimetric woody fragments that are less numerous than in bed 1 (Fig. 10.4A). Leafy axes of conifers are also present. They include cuticle compressions of the cheirolepidiaceous conifer genus *Frenelopsis* (Schenk) Watson, with typical whorls of three leaves and a central vascular charcoalified bundle (Figs. 10.4B–I), and twigs bearing small spirally arranged leaves that resemble the taxodiaceous conifer genus *Sphenolepis* Schenk (Figs. 10.4J–O). Some ovulate scales are related to the cheirolepidiaceous genus *Alvinia* Kvaček (Figs. 10.4S–T), while other scales (Figs. 10.4U–V) and reproductive organs (Figs. 10.4P–R) are of uncertain affinities. They are preserved as cuticles or as charcoals.

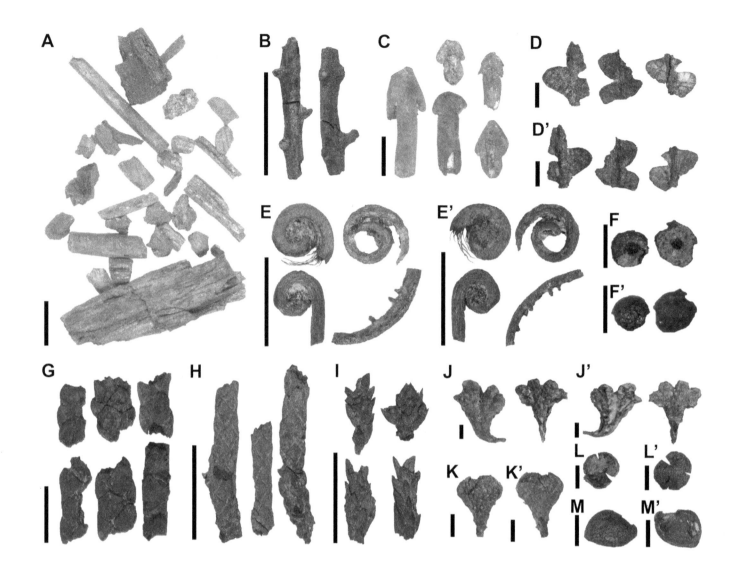

10.5. Mesofossil plant remains from bed 3 in the Danube-Bouchon's Hautrage clay pit quarry (Mons Basin, Belgium). A, Wood preserved as fusains (charcoals; RBINS B5777); B–C, rachis of ferns (RBINS B5778); D–D', pinnules of *Weichselia reticulata* (RBINS B5779); E–E', circinate fronds (RBINS B5780); F–F', indusia of ferns (RBINS B5781); G–I, leafy axes of conifers (G–H) *Sphenolepis* or *Brachyphyllum* [RBINS B5782] and (I) *Pagiophyllum* [RBINS B5783]); J–K, reproductive organs of uncertain affinities (RBINS B5784); L–M, seeds of uncertain affinities (RBINS B5785). Scale bars = 5 mm (A–C, E, G–I) and 1 mm (others).

Bed 3 consists of dark sandy clays. It contains charcoalified wood (Fig. 10.5A). Ferns include pinnules of *W. reticulata* and unidentified rachis and circinate fronds (Figs. 10.5B–F). Leafy conifer twigs (Fig. 10.5G–I), seeds (Figs. 10.5L–M), and diverse reproductive organs (Figs. 10.5J–K) are also found.

Bed 4 consists of white sands in lenses interbedded in black clays. Charcoalified woody fragments are also found (Fig. 10.6A). Different fragments of ferns have been sorted: rachis (Fig. 10.6B), circinate fronds (Fig. 10.6C), pinnules of *G. nordenskioeldii* (Figs. 10.6D–D') and of *P. dunkeri* (Figs. 10.6E–E') showing distinct venation pattern, and even isolated peltate indusia (Figs. 10.6G–G'). Conifer leafy twigs include the genus *Sphenolepis*, but also, depending on the lengths of the free leaf part, the morphogenera *Brachyphyllum* Brongniart and *Pagiophyllum* Heer (Figs. 10. 6 H–M). Fragments of *Frenelopsis* have cuticles with typical longitudinal parallel stomatal rows (Figs. 10.6M–M'). Many scales, cones, and reproductive organs are present. Among them, three cones in connection with leafy twigs, observed in situ, have been identified as *Sphenolepis* (Gomez et al., 2008). A winged seed probably belongs to a conifer (Figs. 10.6W–W'), and cuticles of fruits or ovules may belong to some kind of ginkgo (Figs. 10.6Y–Y').

Bed 5 consists of sandy clays. It is the most diversified mesofossil plant assemblage. It includes centimetric wood fragments (Fig. 10.7A). It also contains numerous and well-preserved remains of ferns: *G. nordenskioeldii* (Figs. 10.7B–C) and *P. dunkeri* (Figs. 10.7D–E). Peltate indusia are found isolated (Fig. 10.7F). The size of the indusia and the central position of the peltate attachment are similar to the circular scars shown by the fertile pinnules on Figure 10.7E, which suggest that some, if not all, indusia may belong to *P. dunkeri*. There are also fern rachis (Figs. 10.7G,J–M), circinate fronds (Fig. 10.7I), and unidentified pinnules (Fig. 10.7H). The leafy twigs are identified as *Brachyphyllum* type (Figs. 10.7O–P) and *Pagiophyllum* type (Fig. 10.7Q). Conifers also include leaves of *Arctopitys* (Fig. 10.7R) and leafy internodes of *Frenelopsis* (Fig. 10.7S). Scales (Figs. 10.7U–V), seeds (Figs. 10.7W–X), reproductive organs (Fig. 10.7Y), and cuticular envelopes (Fig. 10.7T) are also present, but are of uncertain affinities.

Bed 6 consists of dark clays. It contains mainly charcoalified and few vitrified wood fragments (Fig. 10.8A). Ferns are very scarce fragments, being circinate fronds (Figs. 10.8B–B') and indusia (Figs. 10.8E–E'). Leafy axes of *Sphenolepis* (or of *Brachyphyllum* type) are present, but rarer than in beds 1, 2, and 4 (Figs. 10.8C–D). One of the main characteristics of bed 6 is a higher abundance of cuticle debris. The mesofossil plant remains include *Frenelopsis* (Figs. 10.8F–G), *Arctopitys* (Fig. 10.8L), and probably the ginkgoalean *Pseudotorellia* Florin, characterized by leaf laminas with three (Figs. 10.8H–J) or five (Fig. 10.8K) longitudinal stomatal bands. Cupules (Fig. 10.8M) and ovules (Fig. 10.8N) of Ginkgoales are quite frequent, as well as are cuticle envelopes of fruits or ovules (Fig. 10.8P). A few millimetric pieces of amber occur.

Bed 7 consists of clayey sands deposited in a large channel. It contains very few and poorly diversified mesofossil plant remains. Apart from small-size, but abundant, woody fragments (Fig. 10.9A), there are some charcoalified ferns of *W. reticulata* (Fig. 10.9C) and *G. nordenskioeldii* (Fig. 10.9D). The leafy twigs of *Brachyphyllum* type (Fig. 10.9F) and of *Pagiophyllum* type (Fig. 10.9G), an axis of a disarticulated cone (Fig. 10.9E), cuticles of a probable fruit (Fig. 10.9H), and millimetric amber pieces (Fig. 10.9I) are also present. It is noteworthy that some elongate amber pieces are surrounded by wood, which could belong to the tree producing the amber.

The beds 1–7 are replaced in stratigraphic order on the log in Figure 10.2 to interpret the paleoenvironmental changes in time and space. A synthesis of the composition and preservation types is proposed in Table 10.1.

Taxonomy and Plant Assemblages

Discussion

Up to now, only conifer fossil plants were described from the Lower Cretaceous of Hautrage. These include three pinaceous cones: *Pityostrobus corneti* (Coemans) Alvin (Alvin, 1953), *Pityostrobus villerotensis* Alvin (Alvin, 1957), and *Pityostrobus hautrageanus* Alvin (Alvin, 1960). Gérards (2007) and Gérards et al. (2007, 2008) identified *Brachyoxylon* Hollick et Jeffrey, *Podocarpoxylon* Gothan, cf. *Sequoioxylon* Torrey, *Taxodioxylon* Hartig, and *Thujoxylon* (Unger) Hartig *emend.* Süss et Velitzelos. They also recognized three different unidentified Pinaceae and two new genera. The present

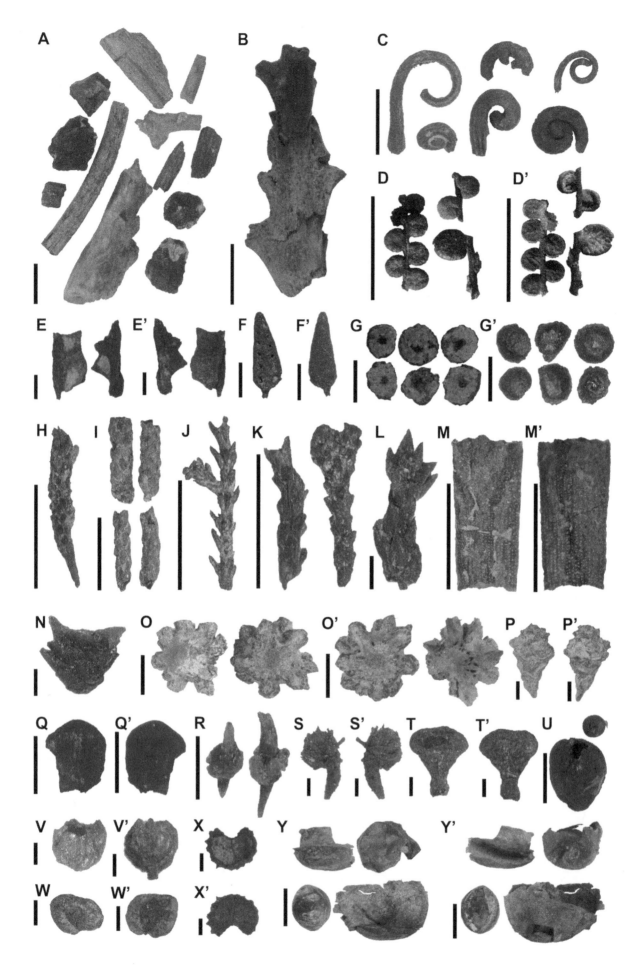

study, based on the morphology of leaves and reproductive organs, reveals much more diversified mesofossil plant assemblages. The associations of ferns and gymnosperms—mainly conifers—are similar to those previously described in coeval Wealden localities from the Mons Basin (Bommer, 1892; Alvin, 1953; Dejax et al., 2007; Pirson et al., 2008). Among the ferns, *G. nordenskioeldii*, *P. dunkeri*, and *W. reticulata* are common in several beds. As far as conifers are concerned, the spirally arranged, leafy axes are diverse. On the basis of their gross morphology, they can be provisionally identified as *Brachyphyllum*, *Pagiophyllum*, and *Sphenolepis*. The whorled leafy axes of *Frenelopsis* and leaf fragments of *Arctopitys* and *Pseudotorellia* are also recognized. In all the beds, isolated scales and entire cones of conifers, as well as seeds and reproductive organs, have also been collected. The wide range of the gross morphologies and the apparent excellent preservations of these organs as cuticle or charcoalified mommies are consistent with their belonging to diverse families, as previously suggested both by palynology (Dejax et al., 2008) and xylology (Gérards, 2007; Gérards et al., 2007, 2008).

Taphonomy

From a necrobiosis point of view, the mesofossil plant remains were probably produced by two means: first, traumatic loss linked to the environment, and second, physiologic loss linked to natural processes as seed dispersion (Martín-Closas and Gomez, 2004). Fragments of spirally arranged, leafy axes of conifers may have been produced by simple fall of dried twigs, as can be observed in living Cupressaceae. One may also argue that wildfires might produce similar fragmentation, especially when considering that many specimens are charcoalified (Harris, 1981; Yans et al., 2010). Future studies could allow confirmation of the occurrence and role of fires in the production of the mesofossil plant remains of Hautrage, or even to characterize them. Fires may explain the fragmentations of ferns. Thus, although these nonsegmented plants do not break easily, only fragments of ferns have been sorted out. Amber is also produced during trauma and ensures the cicatrization of tissues. Several authors suggested that amber production increases during fires (e.g., Martínez-Delclòs et al., 2004; Najarro et al., 2010).

From a biostratinomy point of view, the mesofossil plant remains probably have allochthonous and parautochthonous origins. The parautochthonous elements may have been shortly transported by gravity, wind, or water erosion of the soil. Fires heat air masses that rise in the atmosphere and may move up the lightest charcoalified plant particles. Fires are also often associated with strong winds that may transport even large and heavy plant fragments far away. The abrasion and degree of fragmentation of some debris are probably related to hydraulic transport. Agitation and suspension load in sediments are factors that play an important role in buoyancy, fragmentation, and preservation. The plant remains were finally deposited in the stream bottoms in sandy sediments or in the surrounding floodplain in clayey sediments of overbank floods.

From a diagenetic point of view, the mesofossil plant remains from Hautrage are usually well preserved, probably because the organisms grew in the same environment in which they were quickly buried, and the

10.6. Mesofossil plant remains from bed 4 in the Danube-Bouchon's Hautrage clay pit quarry (Mons Basin, Belgium). A, Wood preserved as fusains (charcoals; RBINS B5786); B, rachis of fern (RBINS B5787); C, circinate frond of ferns (RBINS B5788); D–D', pinnules of *Gleichenites nordenskioeldii* (RBINS B5789); E–E', fragment of pinnules of *Phlebopteris dunkeri* (RBINS B5790); F–F', pinnule or reproductive organ of unknown affinity (RBINS B5791); G–G', indusia of ferns (probably *P. dunkeri;* RBINS B5792); H, leafy axis of conifer or disarticulated cone (RBINS B5793); I–L, leafy axes of conifer of *Sphenolepis* (or I, K, and L, *Brachyphyllum* [RBINS B5794]); J, *Pagiophyllum* [RBINS B5795]); M–M', middle part of internode of *Frenelopsis* (RBINS B5796); N–P, conifer cones of uncertain affinities (RBINS B5797); Q–Q', scale of conifer cone of uncertain affinity (RBINS B5798); R, axes of disarticulated conifer cones (RBINS B5799); S–T, V, reproductive organs of uncertain affinities (RBINS B5800); U, seeds of uncertain affinities (RBINS B5801); W–W', winged seed (RBINS B5802); X–X', unidentified envelope (RBINS B5803); Y–Y', cuticles of fruits or berries (RBINS B5804). Scale bars = 5 mm (A–D, H–K, M, O, Q–R, U, Y) and 1 mm (others).

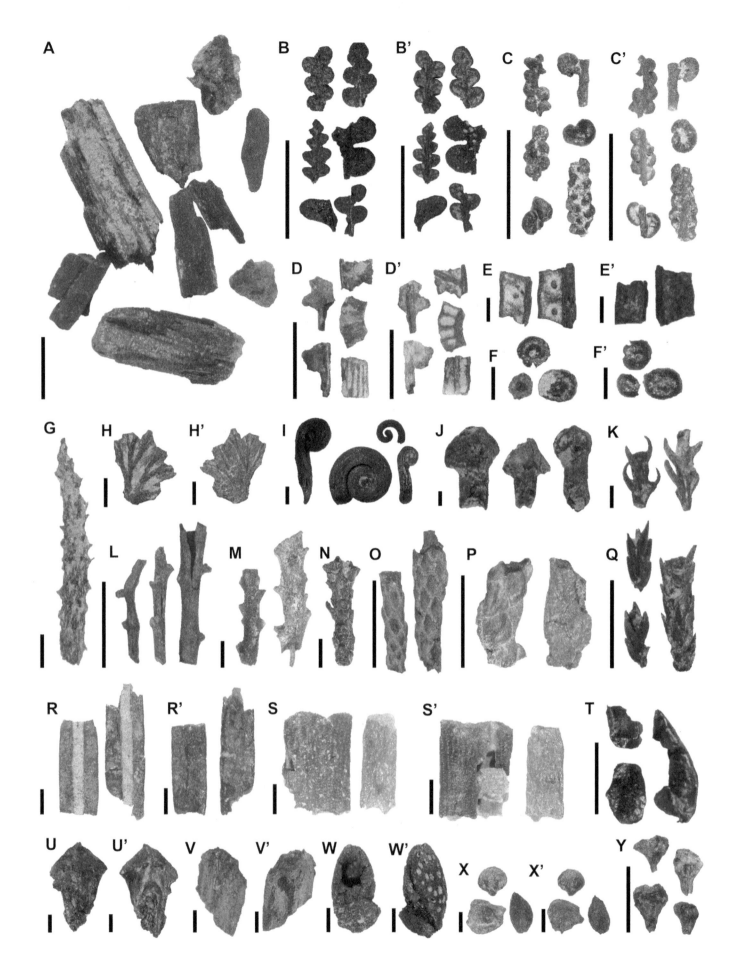

diagenesis was consequently low (Dejax et al., 2008). Different preservation types are observed. A large percentage is charcoalified, but a few woods are coalified. The transformation of living organic matter into charcoal may have diagenetic or fire origins (Harris, 1981). The presence of jet may be also explained by the fall of wood fragments in acid water (Gérards, 2007). The rest of the plant debris are preserved as cuticles and are mostly unburned parts. The fact that several *Frenelopsis* axes show one charcoalified central cylindrical vascular bundle surrounded by cuticle is difficult to explain; preferential combustion of some of the tissues of the same plant during fires is improbable.

Comparisons with Sedimentological Data

Sedimentological studies interpreted the Hautrage Clays Formation deposits as fluvial sediments that deposited in a floodplain crossed by east–west-oriented stream channels (Spagna et al., 2008; see Chapter 9 in this book). The complete Hautrage Clays Formation succession is more than 230 m thick and is divided into six stages (Chapter 9 in this book). Among these, stages 3 to 5 (in which beds 2 to 6 were sampled) characterize a well-developed alluvial floodplain crossed by east–west meandering channels, with a clear rise in the water level in a calm environment during stage 5 (swampy to lacustrine environments in beds 4, 5, and 6). In contrast, stage 6, which contains bed 7, is characterized by higher-energy conditions and is marked by the presence of large channels, sometimes carrying conglomeratic lenses. In this context, the clays are supposed to have been deposited during major floods of the alluvial plain channels, feeding the surrounding marshes in which plants grew (Spagna et al., 2008). Gérards (2007) suggested that Taxodiaceae were particularly well adapted to this environment and climate. The clay beds contain the greatest diversity, especially the beds 4 and 5 (i.e., beds of stage 5), indicating the good preservation conditions of these layers (limited oxygen content resulting from stagnation of water and/or consummation by decomposition of the accumulated plants). On the contrary, the beds representing channel deposits show a relatively low diversity. For example, bed 7 was sampled in one of the largest channels of the series, with many pebbles and other indices of high water energy (Spagna et al., 2008); mesofossil plant remains, and especially cuticles, are less abundant in this bed than in the others. The size of the wood fragments (millimetric to metric) is consistent with the energy of the environment. Finally, although the depositional environment shows an evolution from intermittent-flooded paludal soils (stages 1 and 2), then a fully developed alluvial floodplain (stages 3 to 5), and finally sandy channel deposits (stage 6) along the series (see Chapter 9 in this book), the vegetations remained globally unchanged; the abundance and diversity of the mesofossil plant assemblages appear directly correlated with their position (clayey flooding plain deposits or sandy channel beds) in the paleoplain.

Paleoecological and Paleoenvironmental Reconstructions

The paleoclimatic data obtained from growth rings of Hautrage wood and the plant environmental affinities suggest a warm, temperate to tropical

10.7. Mesofossil plant remains from bed 5 in the Danube-Bouchon's Hautrage clay pit quarry (Mons Basin, Belgium). A, Wood preserved as fusains (charcoals; RBINS B5805); B–C, pinnules of *Gleichenites nordenskioeldii* (RBINS B5806); C–C' revolute pinnules; D–D', fragment of sterile pinnules of *Phlebopteris dunkeri* (RBINS B5807); E–E', fragments of fertile pinnules of *P. dunkeri* (RBINS B5808); F–F', indusia of ferns (probably of *P. dunkeri;* RBINS B5809); H–H', pinnule fragment of fern of uncertain affinity (RBINS B5810); I, circinate fronds of ferns (RBINS B5811); J–M, rachis of ferns (RBINS B5812); N–Q, leafy axes of conifers (RBINS B5813; *Sphenolepis* or (N–P) *Brachyphyllum* and (Q) *Pagiophyllum*); R–R', leafy fragments of *Arctopitys* (RBINS B5814); S–S') leafy fragments of *Frenelopsis* (RBINS B5815); G, axis of disarticulated cone of conifer (RBINS B5816); U–V, scales of conifer cones (RBINS B5817); W–X, diverse seeds (RBINS B5818); Y, reproductive organs of uncertain affinity (RBINS B5819); T, fragments of cuticular envelopes of fruits or ovule (RBINS B5820). Scale bars = 5 mm (A–D, L, O–Q, T, Y) and 1 mm (others).

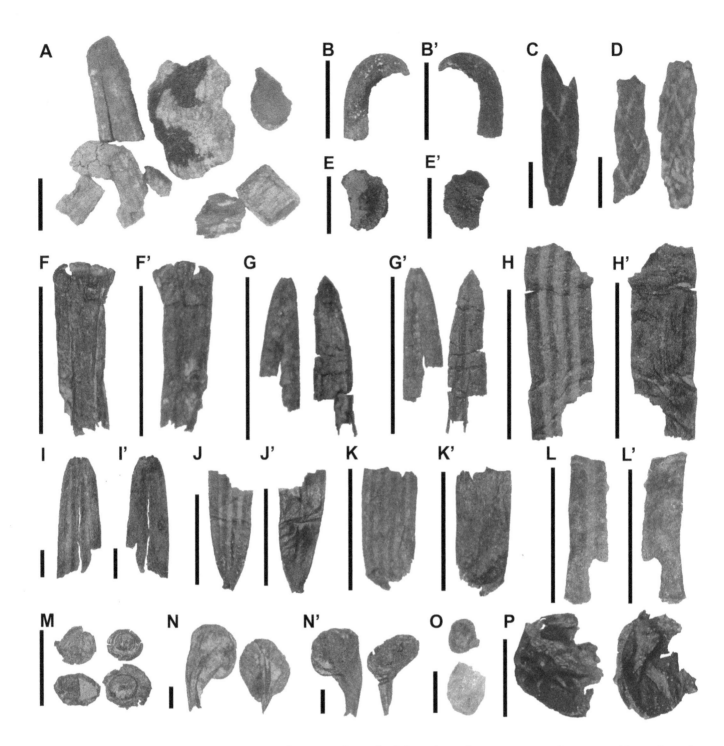

10.8. Mesofossil plant remains from bed 6 in the Danube-Bouchon's Hautrage clay pit quarry (Mons Basin, Belgium). A, Wood preserved as fusains (charcoals; RBINS B5821); B–B', circinate frond of fern (RBINS B5822); C–D, fragments of *Sphenolepis* (or *Brachyphyllum;* RBINS B5823); F–G, fragments of leafy internodes of *Frenelopsis* (RBINS B5824; F–F', middle part; G–G', apex); H–J, leaf fragments of *Pseudotorellia* (RBINS B5825; H–H', middle lamina, I–I', apex with mucron; J–J', base); K–K', leaf fragment with five stomatal bands (RBINS B5826); L–L', leaf fragment of *Arctopitys* (RBINS B5827); M, ovules of Ginkgoales (RBINS B5828); N–N', cupules of Ginkgoales (RBINS B5829); O, amber (RBINS B5830); P, cuticular envelope of fruit or ovule (RBINS B5831). Scale bars - 5 mm (A–B, F–H, J–M, P) and 1 mm (others).

climate with marked drought and rainy seasons alternating (Gérards et al., 2007). In this context, fires may have occurred at the end of the drought season and before the rains, when storms and winds were intense (Harris, 1981). The vegetation from the late Early to early Late Barremian of Hautrage includes arborescent and herbaceous ferns (*W. reticulata, P. dunkeri* and *G. nordenskioeldii*), numerous arborescent conifers (Cheirolepidiaceae, Miroviaceae, Taxodiaceae, Pinaceae), and Ginkgoales (*Pseudotorellia*) that likely grew close to water, being deposited and fossilized in marshes, ponds, and channels. It is noteworthy that *Frenelopsis*, collected in the typical continental Hautrage Clays Formation, lived not only in saline, coastal, or mangrove environments (Gérards, 2007; Uličný et al.,

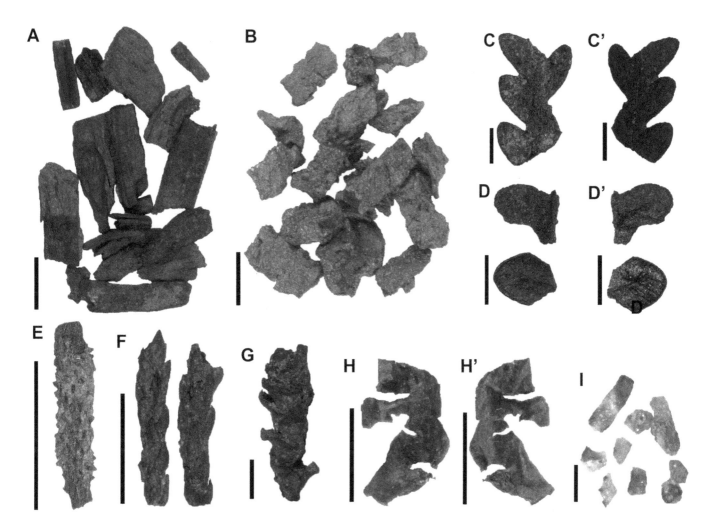

1997), but also in freshwater continental environments (Gomez et al., 2001, 2002; Mendes et al., 2010). Forests probably mainly composed of the conifer families Pinaceae and Taxodiaceae underwent seasonal wildfires. The burned remains found downstream are the witnesses of these recurrent wildfires. Moreover, the evolution of the flora in the Hautrage area during the Barremian appears particularly dynamic (Table 10.1).

The present study provides important data for the reconstruction of the paleoenvironment of the Mons Basin during the Barremian. The Bernissart iguanodons and other animals collected some kilometers away in the Iguanodon Sinkhole walked around in and fed in this wet, freshwater, continental landscape.

10.9. Mesofossil plant remains from bed 7 in the Danube-Bouchon's Hautrage clay pit quarry (Mons Basin, Belgium). A, Wood preserved as fusains (charcoals; RBINS B5832); B, plant debris preserved as cuticles of uncertain affinity (RBINS B5833); C–C', pinnules of *Weichselia reticulata* (RBINS B5834); D–D', pinnules of *Gleichenites nordenskioeldii* (RBINS B5835); E, axis of disarticulated conifer cone (RBINS B5836); F–G, leafy axes of conifers of *Sphenolepis* (RBINS B5837; or F, *Brachyphyllum;* G, *Pagiophyllum*); H–H', cuticular envelope of fruit or ovule (RBINS B5838); I, yellow amber pieces (RBINS B5839). Scale bars = 5 mm (A–B, E–F, H) and 1 mm (others).

Acknowledgments

We thank the editor and the reviewers for constructive suggestions for improvement of the chapter. This research received support from the FP-VI European-funded Integrated Infrastructure Initiative grant SYNTHESIS Project (http://www.synthesys.info/), which is financed by European Community Research Infrastructure Action under the FP6 Structuring the European Research Area Project. This article is a contribution to projects ANR AMBRACE (No. BLAN07-1-184190) of the French "Agence Nationale de la Recherche." The research of B.G. and V.D.-G. was supported by CNRS-UMR 5125 PEPS, projects CGL2008-00809/BTE, CGL2008-00550/BTE

Table 10.1. Abundance and preservation of the mesofossil plant remains in the seven beds sampled[a]

Fossils beds	Wood	Ferns				Conifers			Reproductive Organs	Seeds	Cuticles (fruits, ovules)	Amber
		Pinnules	Fronds	Rachis	Indusia	Leaf or Leafy Axes	Cones	Scales				
⑦ clayey sands	+++ mm	+ *W. reticulata* *G. nordenskioeldii*				+(+) *Brachyphyllum* +(+) *Pagiophyllum*	+ disarticulated				+	+ one in wood
⑥ dark clays	+++ mm dominant		+		+	+ **Arctopitys** + *Brachyphyllum* ++ **Frenelopsis** ++ **Pseudotorellia**			++ **Ginkgoales** (cupules + ovules)		+	+
⑤ sandy clays	+++ mm to cm some cylindrical	++(+) *G. nordenskioeldii* *P. dunkeri*	+(+)	++	++(+)	+ **Arctopitys** ++(+) *Brachyphyllum* + **Frenelopsis** + *Pagiophyllum*	+ disarticulated	+	+(+)	+(+)	+	
④ white sands	+++ mm to cm	+ *G. nordenskioeldii* *P. dunkeri*	++	+	++	++ *Brachyphyllum* + **Frenelopsis** + *Pagiophyllum*	+ some disarticulated	+	+	+	+(+)	
③ dark sandy clays	+++ cm some large	++ *W. reticulata*	++	++	++	++ *Brachyphyllum* + *Pagiophyllum*	+		+	+		
② blue clays with siderite nodules	+++ cm some large				++	+++ **Frenelopsis** ++ **Sphenolepis** (*Brachyphyllum* + *Pagiophyllum*)	+	+ **several cm**				
① lignite sandy clays	+++ mm to cm	+ *W. reticulata* *G. nordenskioeldii*	+	++	+	+ **Arctopitys** ++ **Sphenolepis** (*Brachyphyllum* + *Pagiophyllum*)			+	+	+	++

[a]. +++, very abundant; ++, abundant; +, rare; (blank), absent; ( ) uncertain; **bold**, cuticles (others are preserved as charcoal).

and CGL2009-11838/BTE of the "Ministerio de Educación y Ciencia" of the Spanish government and project 2009SGR1451 of the Catalan government.

References

Alvin, K. L. 1953. Three abietaceous cones from the Wealden of Belgium. Mémoires de l'Institut royal des Sciences naturelles de Belgique 125: 3–42.

———. 1957. On the two cones *Pseudoaraucaria heeri* (Coemans) n. comb. and *Pityostrobus villerotensis* n. sp. from the Wealden of Belgium. Mémoires de l'Institut royal des Sciences naturelles de Belgique 135: 3–27.

———. 1960. Further conifers of the Pinaceae from the Wealden formation of Belgium. Mémoires de l'Institut royal des Sciences naturelles de Belgique 146: 3–39.

———. 1968. The spore-bearing organs of the Cretaceous fern *Weichselia* Stiehler. Journal of the Linnean Society (Botany) 61: 87–92.

———. 1971. *Weichselia reticulata* (Stokes et Webb) Fontaine from the Wealden of Belgium. Mémoires de l'Institut royal des Sciences naturelles de Belgique 166: 1–33.

Bommer, C. 1892. Un nouveau gîte à végétaux découvert dans l'argile wealdienne de Bracquegnies. Bulletin de la Société Belge de Géologie 6: 160–161.

Coemans, E. 1867. Description de la flore fossile du premier étage du terrain Crétacé du Hainaut. Mémoires de l'Académie Royale des Sciences, des Lettres et des Beaux-Arts de Belgique 36: 1–21.

Dejax, J., E. Dumax, F. Damblon, and J. Yans. 2007. Palynology of Baudour Clays Formation (Mons Basin, Belgium): correlation within the "stratotypic" Wealden. Carnets de Géologie M01/03: 16–26.

Dejax, J., D. Pons, and J. Yans. 2008. Palynology of the Wealden facies from Hautrage Quarry (Mons Basin, Belgium). Memoirs of the Geological Survey of Belgium 55: 45–51.

Martínez-Delclòs, X., D. E. G. Briggs, and E. Peñalver. 2004. Taphonomy of insects in carbonates and amber. Palaeogeography, Palaeoclimatology, Palaeoecology 203: 19–64.

Gérards, T. 2007. Étude des végétaux des sédiments à faciès Wealdien du Bassin de Mons. Ph.D. dissertation, University of Liège, Belgium, 2 vols., 221 pp.

Gérards, T., J. Yans, and P. Gerrienne. 2007. Quelques implications paléoclimatiques de l'observation de bois fossiles du Wealdien du Bassin de Mons (Belgique)—Résultats préliminaires. Carnets de Géologie 01/04: 29–34.

Gérards, T., J. Yans, P. Spagna, and P. Gerienne. 2008. Wood remains and sporomorphs from the Wealden facies of Hautrage (Mons Basin, Belgium): palaeoclimatic and palaeoenvironmental implications. Memoirs of the Geological Survey of Belgium 55: 61–70.

Gomez, B., C. Martín-Closas, H. Méon, F. Thévenard, and G. Barale. 2001. Plant taphonomy and palaeoecology in the lacustrine delta of Uña (Upper Barremian, Iberian Ranges, Spain). Palaeogeography, Palaeoclimatology, Palaeoecology 170: 133–148.

Gomez, B., C. Martín-Closas, G. Barale, N. Solé de Porta, F. Thévenard, and G. Guignard. 2002. *Frenelopsis* (Coniferales: Cheirolepidiaceae) and related male organ genera from the Lower Cretaceous of Spain. Palaeontology 45: 997–1036.

Gomez, B., T. Gillot, V. Daviero-Gomez, P. Spagna, and J. Yans. 2008. Paleoflora from the Wealden facies strata of Belgium: mega- and meso-fossils of Hautrage. Memoirs of the Geological Survey of Belgium 55: 53–59.

Harris, T. M. 1953. Conifers of the Taxodiaceae from the Wealden Formation of Belgium. Mémoires de l'Institut royal des Sciences naturelles de Belgique 126: 3–43.

———. 1981. Burnt ferns from the English Wealden. Proceedings of the Geologists' Association 92: 47–58.

Martín-Closas, C., and B. Gomez. 2004. Taphonomie des plantes et interprétations paléoécologiques. Une synthèse. Geobios 37: 65–88.

Mendes, M. M., J. L. Dinis, B. Gomez, and J. Pais. 2010. The cheirolepidiaceous conifer *Frenelopsis teixeirae* Alvin et Pais from the lower Hauterivian of Vale Cortiço (Torres Vedras, western Portugal). Review of Palaeobotany and Palynology 161: 30–42.

Najarro, M., E. Peñalver, R. Pérez-de la Fuente, J. Ortega-Blanco, C. Menor-Salván, E. Barrón, C. Soriano, I. Rosales, R. López del Valle, F. Velasco, F., Tornos, V. Daviero-Gomez, B. Gomez, and X. Delclòs. 2010. A review of the El Soplao amber outcrop, Early Cretaceous of Cantabria (Spain). Acta Geologica Sinica 84: 959–976.

Nosova, N., and E. Wcislo-Luraniec. 2007. A reinterpretation of *Mirovia* Reymanówna (Coniferales) based on the reconsideration of the type species *Mirovia szaferi* Reymanówna from the Polish Jurassic. Acta Palaeobotanica 47: 359–377.

Pirson, S., P. Spagna, J.-M. Baele, F. Damblon, P. Gerrienne, Y. Vanbrabant, and J. Yans. 2008. An overview of the geology of Belgium. Memoirs of the Geological Survey of Belgium 55: 5–25.

Robaszynski, F., A. V. Dhondt, and J. W. M. Jagt. 2001. Cretaceous lithostratigraphic units (Belgium); pp. 121–134 in P. Bultynck and L. Dejonghe (eds.), Guide to a revised lithostratigraphic scale of Belgium. Geologica Belgica 4.

Seward, A. C. 1900. La flore wealdienne de Bernissart. Mémoires du Musée royal d'Histoire naturelle de Belgique 1: 1–39.

Spagna, P., S. Vandycke, J. Yans, and C. Dupuis. 2007. Hydraulic and brittle extensional faulting in the Wealden facies of Hautrage (Mons Basin, Belgium). Geologica Belgica 10: 158–161.

Spagna, P., C. Dupuis, and J. Yans. 2008. Sedimentology of the Wealden clays in the Hautrage quarry. Memoirs of the Geological Survey of Belgium 55: 35–44.

Uličný, D., J. Kvaček, M. Svobodová, and L. Špičáková. 1997. High-frequency sea-level fluctuations and plant habitats in Cenomanian fluvial to estuarine succession: Pecínov quarry, Bohemia. Palaeogeography, Palaeoclimatology, Palaeoecology 136: 165–197.

Yans, J., J. Dejax, D. Pons, L. Taverne, and P. Bultynck. 2006. The iguanodons of Bernissart are middle Barremian to earliest Aptian in age. Bulletin de l'Institut royal des Sciences naturelles de Belgique Sciences de la Terre 76: 91–95.

Yans J., T. Gerards, P. Gerrienne, P. Spagna, J. Dejax, J. Schnyder, J.-Y. Storme, and E. Keppens. 2010. Carbon-isotope of fossil wood and dispersed organic matter from the terrestrial Wealden facies of Hautrage (Mons Basin, Belgium). Palaeogeography, Palaeoclimatology, Palaeoecology 291: 85–105.

Diagenesis of the Fossil Bones of *Iguanodon bernissartensis* from the Iguanodon Sinkhole

11

Thierry Leduc

We investigate the bone diagenesis of the *Iguanodon* skeletons discovered in the Iguanodon Sinkhole in 1878–1881. By means of x-ray diffraction and energy-dispersive spectrometry analysis, about 30 mineral phases were identified in the fossil bones of the Bernissart iguanodons. During burial, recrystallization took place: the slightly crystallized carbonated hydroxylapatite (the mineral phase of fresh bone tissue) was replaced by well-crystallized carbonated fluorapatite, currently present in the *Iguanodon* bones. Whereas some minerals infiltrated the bone during cavity filling by the sediment (detrital quartz, "argillaceous" phyllosilicates, anatase, and rutile), others are authigenic and precipitated in the cavities during burial (pyrite, barite, sphalerite, celestine, and to a lesser extent quartz) or after the exhumation (other sulfates, oxides, and/or hydroxides). Pyrite is the most abundant authigenic mineral in the *Iguanodon* fossil bones, making them brittle. Different pyrite morphologies can be observed, such as crystals, framboids, thin coatings, and fibroradial structures. Barite is the second most abundant phase. When both minerals are present, pyrite was formed first. Their distribution throughout the bones seems to be random. Since their discovery, the fossil bones have been treated for conservation in several ways. However, this treatment did not prevent the alteration of pyrite into an assemblage of 13 different secondary sulfate minerals, more particularly szomolnokite ($FeSO_4 \cdot H_2O$) and rozenite ($FeSO_4 \cdot 4 H_2O$) and in a less way roemerite ($Fe^{2+}Fe^{3+}_2SO_4 \cdot 14 H_2O$), jarosite [$KFe_3(SO_4)_2(OH)_6$], natrojarosite [$NaFe_3(SO_4)_2(OH)_6$], halotrichite [$Fe^{2+}Al_2(SO_4)_4 \cdot 22 H_2O$], tschermigite [$(NH_4)Al(SO_4)_2 \cdot 12 H_2O$)], melanterite ($FeSO_4 \cdot 7 H_2O$), coquimbite [$Fe^{3+}_2(SO_4)_3 \cdot 9 H_2O$], metavoltine [$(K,Na)_8Fe^{2+}Fe^{3+}_6(SO_4)_{12}O_2 \cdot 18 H_2O$], gypsum ($CaSO_4 \cdot H_2O$), anhydrite ($CaSO_4$), and pure sulfur (S). Szomolnokite and rozenite are the most abundant of these minerals and can be found in nearly all samples. These two minerals differ only by their degree of hydration and can easily transform into each other. This mechanism depends on the degree of humidity in the environment. Barite and anhydrite are the result of the neutralization of the acid by phosphate or carbonate.

Introduction

Taphonomy studies the passage of organisms from the biosphere to the lithosphere. These processes include diagenesis. The diagenesis of animal or vegetal remains reflects the postmortem modification of their chemical composition after their burial (Sandford, 1992). Specifically, bone diagenesis is the result of cumulative physical, chemical, and biological processes that modify the original chemical and/or structural properties of an organic object and that determine its ultimate fate in terms of preservation or destruction. Existing studies mainly report discrete aspects of bone diagenesis, including the permineralization (Briggs et al., 1996; Pfretzschner, 2000, 2001; Parker and Toots, 1970), the recrystallization of bone apatite (Bonar et al., 1983; Chipera and Bish, 1991; Gillette, 1994; Jackson et al., 1978; Kolodny et al., 1995; Lees and Prostak, 1988), and the enrichment of the bone in trace elements (Parker and Toots, 1972; Sillen and Sealy, 1995; Tuross et al., 1989; Williams and Marlow, 1987). Most researches have been conducted so far on archaeological bone material or on a limited quantity of samples.

From this perspective, the complete skeletons of Bernissart iguanodons constitute a rich source of information. Although the presence of pyrite and the influence of its oxidation on the conservation of the bones are well known, no detailed mineralogical study of this material has been undertaken so far. The present study was thus undertaken to obtain a better knowledge of the mechanisms of the diagenetic processes so that better conservation techniques could be developed for fossilized bones.

Bone diagenesis is a complex process that includes the degradation of organic matter, the recrystallization of bone apatite, the enrichment in trace elements, the precipitation of new minerals in the bone cavities (permineralization), and the compaction process—and on top of that, all the mechanisms that took place more recently during the storage in natural history museums. Only some of these processes will be discussed in this article.

Institutional abbreviation. RBINS, Royal Belgian Institute of Natural Sciences, Brussels, Belgium.

Materials and Methods

Material

When the iguanodons were discovered in the Iguanodon Sinkhole, pyrite was abundantly present in the fossil bones. This made them brittle. The clay blocks that contained the fossils remains were covered with plaster to preserve the cohesion of the bones. Afterward, in the museum, all the visible pyrite was curetted and the voids were filled with plaster or a mixture of carpenter's glue and chalk (De Pauw, 1902). Throughout time, the fossil bones of iguanodons have been treated in several ways (carpenter's glue, shellac, and mowilite) to keep their cohesion and reduce oxidation. Those treatments, however, did not prevent the alteration of the remaining pyrite and further increased the fragility of the bones.

A recent restoration of the skeletons presented in an *en gisement* display allowed us to take about 300 bone samples. They only come from bone areas that were cracked, fissured, or pyritized, or that are rich in sulfates. The undamaged bones were not investigated. Approximately 180 samples were selected for this study. These samples mainly come from the long bones, vertebrae, and ribs of the 13 individuals currently displayed in their original *en gisement* display. Some bone fragments were also taken from an individual displayed in life position (RBINS Vert-05144-1714, individual "G").

One bone fragment from BER 3 (see Chapter 12 in this book) was also studied. Contrary to the bones preserved in the RBINS, this new sample did not receive any preservation treatment and is consequently strongly disaggregated.

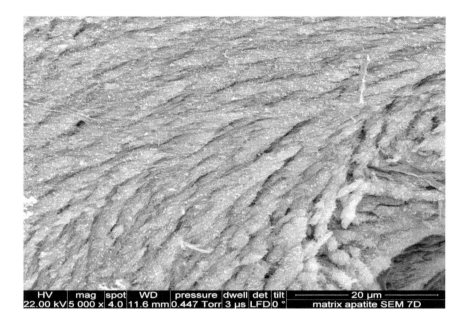

11.1. Matrix apatite of a cancellous bone [sample S16(3)].

Methodology

The selection of the samples was done macroscopically with a binocular microscope WILD M5 (maximum magnification 50×). The photographs were taken with a Nikon Coolpix 4500 (4.0 Mp) camera.

The identification of the mineral phase mixtures was carried out with a Philips diffractometer PW 3710 mpd control, Bragg-Brentano geometry, $Fe_{k\alpha}$ radiation ($\lambda = 1.09373$ Å) supplied with a PANalytical PW 3830 x-ray generator (40 kV and 30 mA). The spectra covered the range 5–75° 2θ. Samples were crushed and pulverized in an agate mortar. Samples that could be isolated were crushed to powder and analyzed with an x-ray powder camera (Debye-Sherrer or Gandolfi) with a Philips PW 1729 x-ray generator (40 kV and 20 mA), using $Cu_{k\alpha}$ ($\lambda = 1.5404$ Å) radiation.

Some samples were investigated with a scanning electron microscope (SEM; FEI XL 30) using two kinds of detectors: secondary electron and backscattered electron. The secondary electron detector was used to observe various morphologies, whereas the backscattered electron detector made it possible to distinguish mineral phases on the basis of the difference in electron density.

Qualitative and semiquantitative energy-dispersive spectrometry (EDS) analyses were performed with an EDAX Apollo 10 SDD silicon drift detector between 10 and 30 kV.

Infrared spectroscopy was carried out with a spectrometer Nicolet Nexus in the range 400–4,000 cm^{-1} with a spectral resolution of 1 cm^{-1}.

Crystallographic cell parameter determination of mineral phases was performed on samples mixed with lead nitrate as internal standard using the program LCLSQ 8.4 (least squares refinement of crystallographic lattice parameters; Burnham, 1991).

The clay composition was identified on the basis of major and some trace elements by x-ray fluorescence on an ARL 9400XP spectrometer.

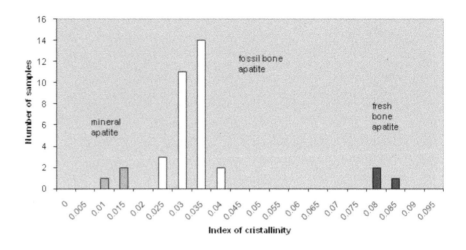

11.2. Crystallinity of carbonated fluorapatite from bones of *Iguanodon bernissartensis*, compared with mineral apatite and carbonated hydroxylapatite from fresh bones.

Observations

Recrystallization of the Apatite Matrix

The x-ray powder diffractograms of 37 bone apatite samples identified fluorapatite [$Ca_5(PO_4)_3F$] as the mineral phase constituting the *Iguanodon* bones. All EDS analyses carried out on rough matrix apatite samples and thin sections cut from certain bones confirmed the presence of fluorine. In addition, these analyses show that the major elements are Ca, P, O, and F; elements such as Fe, S, Sr, Na, and K are also present, but to a lesser extent.

Recrystallized apatite generally forms the bulk of the bone when the structure is preserved. The individual crystals could not be discriminated by our observation methods (Fig. 11.1).

The powder diffractograms also allowed us to calculate Klugg and Alexander's (1962) crystallinity index of the bone apatite. This index is based on the 2θ value of the width of the peak corresponding to the (002) reflexion measured at half height on the x-ray diffractogram. The index is inversely proportional to the crystallinity of the mineral. The crystallinity index was measured on 30 apatite samples taken from *Iguanodon* bones (Fig. 11.2).

For comparison, the crystallinity index of three apatite samples of mineralogical origin (mineralogy collection of RBINS R.V. 7/700, R.C. 4927 and R.C. 6047) and of three carbonated hydroxylapatite from fresh bones (fox, bird, and rat) are also provided (Fig. 11.2). The values obtained for the *Iguanodon* bones vary between 0.25 and 0.43; it is therefore closer to the values of mineral apatite (0.13–0.17) than to the apatite of fresh bones (0.813–0.863).

Examination of the x-ray powder diffractograms of two bone samples [S15(3) and L14 (1bis)], mixed with lead nitrate, allowed calculation of the values of the cell parameters of the matrix apatite. The values are included in Table 11.1.

The infrared spectra of seven apatite samples from bones of *I. bernissartensis* reveal all the characteristic peaks of the CO_3^{2-} group (Fig. 11.3).

Nevertheless, it seems that the presence of this group, which replaces PO_4^{3-} in apatite, is not sufficient to distinguish a carbonated fluorapatite from a fluorapatite on the x-ray powder diffractograms. The values of the absorption bands of the infrared spectra are reported in Table 11.2.

EDS analysis revealed no significant traces of heavy metals or of radioactive elements incorporated into the structure of the apatite or adsorbed on

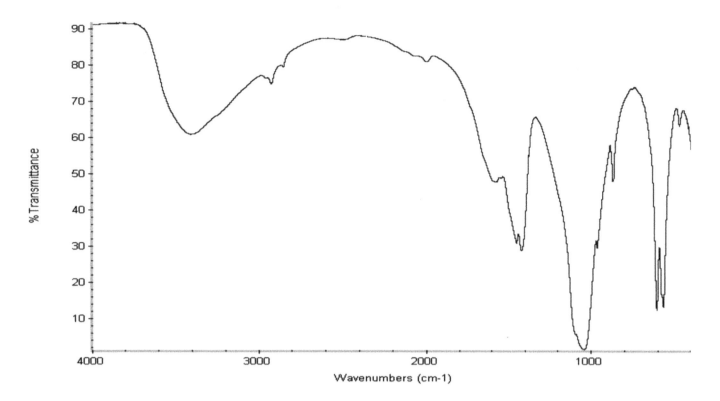

11.3. Infrared spectra of a selected sample of bone fluorapatite from bones of *Iguanodon bernissartensis.*

the surface of crystals. The use of a Geiger-Muller counter (CB1C counter, Nardeux, sensitive to the α, ß, and γ radiations) confirms the absence of radioactivity. Fluorescence X and microprobe analyses identified trace elements in the bone apatite.

Attempts of collagen extraction were made on the *Iguanodon* bones by a technique developed by J.-F. Godart (2005) and used in the Anthropological Service of the Royal Institute of Natural Sciences of Belgium. This technique has already given good results on archaeological material. But in the case of the Bernissart iguanodons, the organic material collected in this way is widely contaminated by fish collagen that was used in the manufacturing of the glue used to strengthen the bones at the end of the nineteenth century. The presence of collagen in plaster of Paris samples (purely inorganic gypsum) below the *Iguanodon* specimens' *en gisement* display irrefutably proves this contamination.

Permineralization

X-ray powder diffraction SEM and EDS were used to identify 26 mineral phases in the bones of the Bernissart iguanodons (Table 11.3).

Pyrite (FeS₂). Pyrite is the most common mineral within the bone voids of all sizes. Pyrite can also be found in the surrounding clay as isolated grains and nodules.

Pyritization is observed, in different forms, everywhere in the bones. Euhedral and subhedral crystals can be isolated (Fig. 11.4A), accumulated (Fig. 11.4B), or aggregated (Fig. 11.4C). Octahedra, cubes, pyritohedra, and mixed forms are scattered or form a continuous crust on the walls of bone voids and cracks (Fig. 11.4D). Many bone voids are also partially or completely filled by fine-grained pyrite. It even replaces the bone matrix

Table 11.1. Unit cell data for two selected samples of carbonated fluorapatite (hexagonal, P6₃/m) from the bone matrix of *Iguanodon*

Parameter	L14(1bis)	S15(3)	Bhatnagar (1)	Calderin (2)	Corbridge (3)
a (Å)	9,373(6)	9,378(7)	9,37(1)	9.3642(3)	9,367
b (Å)	9,373(6)	9,378(7)	9,37(1)	9.3642(3)	9,367
c (Å)	6,878(5)	6,897(6)	6,88(1)	6,8811(2)	6,884

(1) Bhatnagar (1967), (2) Calderin et al. (2003), (3) Corbridge (1974)

Table 11.2. Position and assignment of the infrared absorption bands recorded for selected samples of bone fluorapatite from bones of *Iguanodon bernissartensis*

Structural group	Vibrational mode	Wave number (cm⁻¹)			
		Ref. (1)	Ref. (2)	Ref. (3)	*Iguanodon* samples
$(OH)^-$	Stretching	3566	3578	3576	3573(1)
H_2O	H-O-H stretching			3407	3404
H_2O	H-O-H "scissor" bending			1633	—
$(CO_3)^{2-}$	v_3 antisymmetric stretching			1457	1455
$(CO_3)^{2-}$	v_3' antisymmetric stretching			1420	1423
$(PO_4)^{3-}$	v_3 antisymmetric stretching	1095	1098	1093	1094
$(PO_4)^{3-}$	v_3' antisymmetric stretching	1075	1060	1064	—
$(PO_4)^{3-}$	v_3'' antisymmetric stretching	1040	1045	1037	1042
$(PO_4)^{3-}$	v_1 symmetric stretching	960	968	961	963
$(CO_3)^{2-}$	v_2 out-of-plane bending (O-C-O)	870		873	872 and 880
$(OH)^-$	OH libration	633	635	635	—
$(PO_4)^{3-}$	v_4 in-plane bending (O-P-O)	603	608	602	604
$(PO_4)^{3-}$	v_4' in-plane bending (O-P-O)	574	572	574	574
$(PO_4)^{3-}$	v_4'' in-plane bending (O-P-O)	566		564	566
$(PO_4)^{3-}$	v_2 out-of-plane bending (O-P-O)	471	475	472	471

(1) Baddiel and Berry (1966), (2) Bhatnagar (1968), (3) Dumitras et al. (2004). The position of the bands is established as the mean of the wave numbers recorded for similar bands in seven different spectra of representative samples. Analyzed samples: RBINS E17(1), L7, L9(2), L14(1bis), M17(12), N9(8), and teeth of G/1714.
(1) That band is only present in four samples.

in some areas or is present as a core surrounded by ferrous sulfates. Fibro-radial pyrite (Fig. 11.4E) appears as isolated nodules in the bone structure (e.g., in neurapophysis) or as crust mammillated on the external surface of bones. Framboidal pyrites (Fig. 11.4F) are usually scattered on the walls of cavities in some cancellous bones. Pyrite often coats the walls of the bone voids as a thin film.

EDS analyses of pyrite do not reveal any detectable substitution either of iron or sulfur.

Most of the analyzed samples (76%) contain pyrite. In bone fragments, pyrite is present in the apatite matrix, with or without traces of sulfates (14%). In sulfate-rich samples, pyrite is associated with szomolnokite and/or rozenite (62%).

The quantity of pyrite found in bone voids is quite variable, even within a limited portion of the bone.

Barite $[Ba(SO_4)]$ *and Celestine* $[Sr(SO_4)]$. The presence of small quantities of barite is difficult to distinguish because of overlapping d-spacings with those of other major phases. Barite is often present in samples, but it is much less abundant than pyrite. It fills some bone voids and cracks. Usually barite forms clusters with a subparallel crystal growth (Fig. 11.5A). Sometimes long and slender crystals with well-marked ends stick out from the clusters (Fig. 11.5B). Barite also fills the voids within the bone.

Table 11.3. Minerals identified in the fossil bones of *Iguanodon bernissartensis*

Name	Formula
Sulfur	S
Pyrite	FeS_2
Sphalerite	ZnS
Rutile-anatase	TiO_2
Goethite	$Fe^{3+}O(OH)$
Fluorapatite	$Ca_5(PO_4)_3F$
Szomolnokite	$Fe^{2+}(SO_4) \cdot H_2O$
Rozenite	$Fe^{2+}(SO_4) \cdot 4\,H_2O$
Roemerite	$Fe^{2+}Fe^{3+}_2(SO_4)_4 \cdot 14\,H_2O$
Melanterite	$Fe^{2+}(SO_4) \cdot 7\,H_2O$
Coquimbite	$Fe^{3+}_2(SO_4)_3 \cdot (6+3)\,H_2O$
Halotrichite	$Fe^{2+}Al_2(SO_4)_4 \cdot 22\,H_2O$
Jarosite	$KFe^{3+}_3(SO_4)_2(OH)_6$
Natrojarosite	$NaFe^{3+}_3(SO_4)_2(OH)_6$
Metavoltine	$(K,Na)_8Fe^{2+}Fe^{3+}_6(SO_4)_{12}O_2 \cdot 18\,H_2O$
Tschermigite	$(NH_4)Al(SO_4)_2 \cdot 12\,H_2O$
Gypse	$Ca(SO_4) \cdot 2\,H_2O$
Anhydrite	$Ca(SO_4)$
Barite	$Ba(SO_4)$
Celestine	$Sr(SO_4)$
Hemimorphite	$Zn_4(Si_2O_7)(OH)_2 \cdot H_2O$
Clinochlore	$(Mg,Fe^{2+})_5Al(Si_3Al)O_{10}(OH)_8$
Illite	$KAl_2(Si_3Al)O_{10}(OH)_2$
Kaolinite	$Al_2Si_2O_5(OH)_4$
Montmorillonite	$(Na,Ca)_{0.3}(Al,Mg)_2Si_4O_{10}(OH)_2 \cdot n\,H_2O$
Quartz	SiO_2

Barite occurs isolated or in association with pyrite. Pyrite seems to form first; while pyrite coats the walls of the voids, barite fills the remaining space.

Concomitant use of SEM images obtained with the backscattered electron detector together with the EDS revealed the presence of celestine within the samples. This mineral is frequently present in the samples, although always in small quantities. It usually forms sheaves of elongated slender crystals (Fig. 11.5C) or more prismatic crystal aggregates.

EDS analyses show partial substitution of Ba by Sr in barite (Fig 11.5D) and Ba inversely replacing Sr in celestine.

Sphalerite (ZnS). It is difficult to identify small quantities of sphalerite by x-ray powder diffraction because the most intensive d-spacings coincide with those of pyrite. Sphalerite could be identified only when Debye-Sherrer or Gandolfi cameras were used on small, pure crystals.

However, observations under binocular microscope, SEM, and EDS highlighted the presence of sphalerite in the form of aggregates of brown plates. These plates consist of an assemblage of tetrahedra with subparallel growth (Fig. 11.6).

Sphalerite is locally present in some cancellous bones, with or without pyrite, but only where no sulfates are present.

In one sample (O5), white micrograins proved to be a sphalerite core surrounded by hemimorphite.

Iron oxides and hydroxides. Goethite [α-FeO(OH)] is the only mineral of this category to be identified with certainty in five samples, one of them

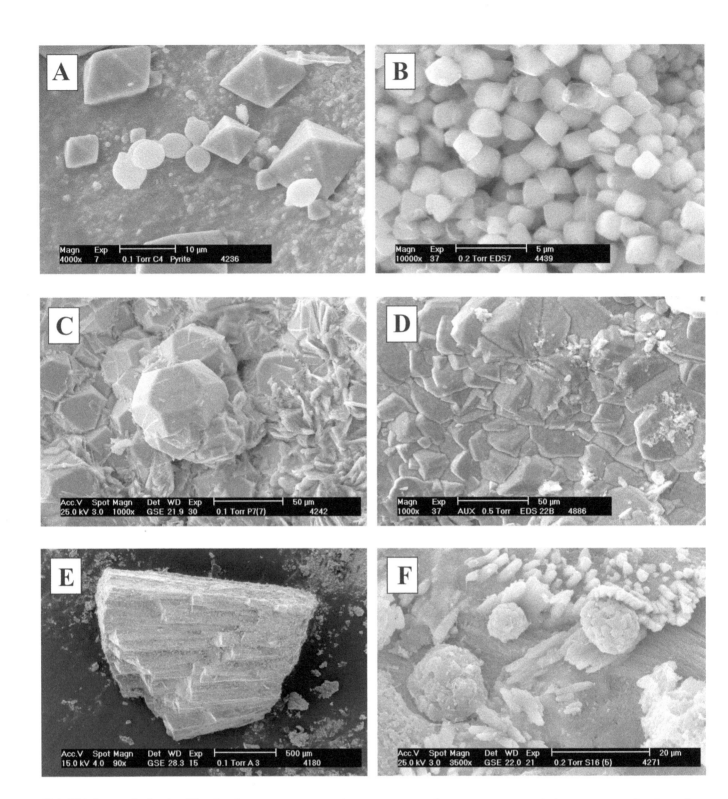

11.4. SEM photographs showing different pyrite morphologies. A, Isolated octahedral crystals (sample C4); B, accumulated octahedral crystals (sample from the BER 3 core hole); C, pyrite aggregates [sample P7(7)]; D, crust of pyrite [sample N19(8bis)]; E, fragment of radiated pyrite (sample A3); F, framboidal pyrite [sample S16(5)].

being a clay fragment. Clay in contact with bones or inside them has sometimes an ochre color caused by enrichment in goethite. EDS analyses show that the iron enrichment is also clearly visible in some framboïdal pyrites (Fig. 11.7A–B), in some compact masses within the bone, and in red reticular pyrite films that cover some walls of osseous cavities (Fig. 11.7C). However, goethites could not be identified with absolute precision in these cases.

Szomolnokite [Fe(SO$_4$)·H$_2$O] and rozenite [Fe(SO$_4$)·4 H$_2$O]. Szomolnokite and rozenite are the most abundant secondary minerals. The

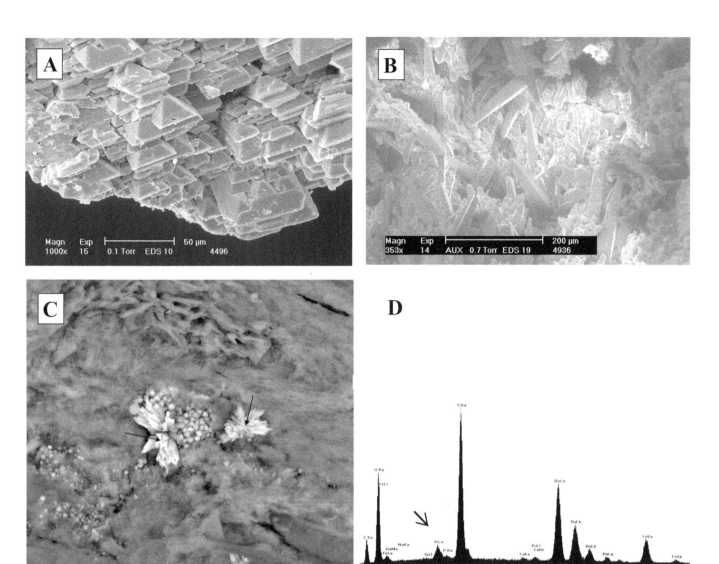

11.5. SEM photographs. A, Barite forms clusters with parallel crystal growth; B, long and slender crystals of barite; C, sheaves of elongated slender crystals of celestine; D, EDS analysis shows partial substitution of Ba by Sr in barite.

efflorescences of these two minerals can be observed in nearly all the areas where pyrite is oxidized. They only differ by their degree of hydration and can transform into each other when the environmental moisture varies. X-ray powder diffraction differentiates them because they have different structures.

They primarily form microglobular cauliflower-like efflorescences (Fig. 11.8A), and also white compact masses. On some thin pyrite films covering the walls of bone cavities, they form iridescent aureoles. These structures are formed by a radiate assemblage of small ferrous sulfate rods (Fig. 11.8B). Unfortunately, the size of these aureoles (approximately 100 µm) prevented their sampling for x-ray identification.

Szomolnokite and rozenite are usually mixed together, but they can also be observed separately. Out of 137 sulfate-rich samples, 131 contained at least one of these phases. In 55% of the analyzed samples, both minerals occurred together, in 30% only szomolnokite was present, and in 15% only rozenite.

The crystallographic parameters of szomolnokite and rozenite were calculated on the basis of the diffractograms of four samples with white

11.6. SEM photograph showing an isolated crystal of sphalerite in sample P1.

Acc.V	Spot	Magn	Det	WD	Exp		200 μm
20.0 kV	4.0	160x	GSE	30.3	1	0.1 Torr P1	4242

efflorescences [S16(6), K10, E16, and E4] mixed with lead nitrate. Table 11.4 shows the values of the calculated parameters.

Roemerite [$Fe^{2+}Fe^{3+}_2(SO_4)_4 \cdot 14\ H_2O$], melanterite [$Fe^{2+}(SO_4) \cdot 7\ H_2O$] and coquimbite [$Fe^{3+}_2(SO4)_3 \cdot (6+3)H_2O$]. Roemerite is often present, whereas melanterite and coquimbite were only identified in one sample, respectively, in S15(7) and in the lower jaw of the specimen RBINS Vert-05144-1714. In several samples, roemerite is present in the form of crystals (thick tablets or pseudocubic, Fig. 11.9), always together with szomolnokite but with or without rozenite. Conversely, roemerite is associated with metavoltine in sample F17(3).

Tschermigite [$(NH_4)Al(SO_4)_2 \cdot 12\ H_2O$]. Based on x-ray diffraction, tschermigite was found in about 30% of the sulfate-rich samples. It is mainly associated with szomolnokite and/or rozenite. Nevertheless, observations under a binocular microscope showed that tschermigite grew directly on the clay surface in the absence of any other sulfate.

Tshermigite can be observed as octahedral (Fig. 11.10A) or thick tabular crystals of maximum 80 μm, and as fibrous columnar and twisted aggregates (Fig. 11.10B).

EDS analyses revealed the presence of aluminium, sulfur and nitrogen, which allowed distinguishing tschermigite from other isotypic minerals of the alum group.

Natrojarosite [$NaFe^{3+}_3(SO_4)_2(OH)_6$] and jarosite [$KFe^{3+}_3(SO_4)_2(OH)_6$]. Natrojarosite and jarosite are yellow minerals typically present as an earthy crust, pulverous masses, or as a powder, often in small amounts and in association with rozenite and szomolnokite. These two phases were identified in 18% of the sulfate-rich samples. Observation of the samples under binocular microscope or SEM indicates the presence of this mineral in most of the samples, but in quantities too small to be detected by x-ray diffraction. Natrojarosite occurs primarily in two forms: aggregates of pseudohexagonal tablets (Fig. 11.11B), and aggregates of flattened rhombohedral crystals (Fig. 11.11A).

EDS analyses indicate the preponderance of natrojarosite to jarosite, the latter being identified in only two sulfate-rich samples. However, there is a partial replacement of Na by K in many cases.

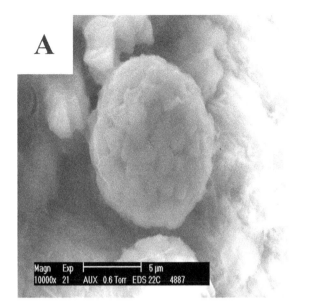

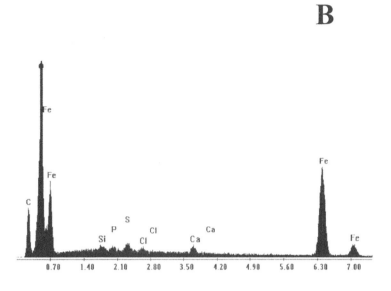

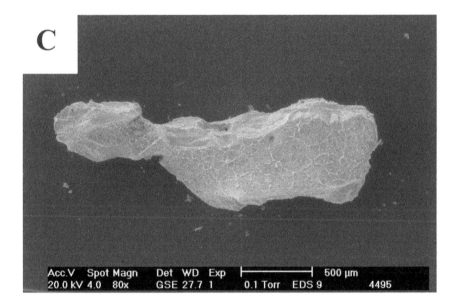

11.7. SEM photographs (A) and EDS analysis (B) of the pseudomorphose of pyrite in iron oxide and/or hydroxide in a framboid. C, Reticule of iron oxide and/or hydroxide in a thin coating of pyrite in cancellous bone.

11.8. SEM photographs. A, Microglobular cauliflower-like efflorescences of ferrous sulfates (sample A3); B, "aureoles" of ferrous sulfates on a thin coating of pyrite in a cancellous bone [sample S16(3)].

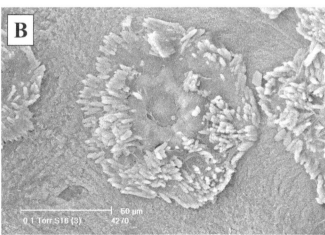

Table 11.4. Unit cell data for two selected samples of szomolnokite (monoclinic C2/c) and two of rozenite (monoclinic P21/n) in sulfatized areas from *Iguanodon bernissartensis* bones

Szomolnokite				
Parameter	S16 (6)	K10	Giester (1)	Pistorius (2)
a (Å)	7,078(3)	7,091(3)	7,078(3)	7,123(9)
b (Å)	7,543(3)	7,533(4)	7,549(3)	7,469(9)
c (Å)	7,610(4)	7,622(4)	7,773(3)	7,837(9)
β (°)	116,17(5)	115,98(5)	118,65(2)	118,94(9)
Rozenite				
Parameter	E16	E4	Baur (3)	Jambor (4)
a (Å)	5,841(5)	5,966(1)	5,97	5,945
b (Å)	13,643(8)	13,615(4)	13,64	13,59
c (Å)	7,964(5)	7,965(2)	7,98	7,94
β (°)	90,87(9)	90,47(2)	90,26	90,3

(1) Giester (1988), (2) Pistorius (1960), (3) Baur (1960), (4) Jambor and Traill (1963).

11.9. SEM photograph showing an aggregate of thick tablets of roemerite (arrows) with ferrous iron [sample S17(9)].

Metavoltine [(K,Na)$_8$Fe^{2+}Fe$^{3+}_6$(SO$_4$)$_{12}$O$_2$·18 H$_2$O]. Metavoltine was identified in only one sample [F17(3)]. Optically, this mineral forms a crust or a yellow powder. The crystals are hexagonal flattened tablets, similar to those of natrojarosite (Fig. 11.11C).

Halotrichite [Fe^{2+}Al$_2$(SO$_4$)$_4$·22 H$_2$O]. Halotrichite is present as acicular crystals in diverging (Fig. 11.12A) or parallel aggregates (bundles). Generally it develops at the surface of clays and more rarely on bones in sulfated zones. Halotrichite often occurs disseminated among other sulfates and is present in small quantities. It is therefore not easily identifiable by x-ray diffraction.

Sulfur (S). Observation under optical and electronic microscope shows that sulfur is commonly present but often in quantities too small to be highlighted by x-ray diffraction. The crystals have a rounded shape (Fig. 11.12B), are honey colored, and regularly show strong traces of dissolution (Fig. 11.12C). Sulfur is localized either in the sulfated efflorescences or directly on the surface of the clay, where it is accompanied by tschermigite. It is rarely found on the surface of the bones.

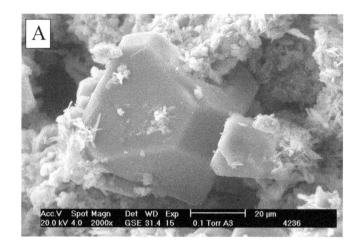

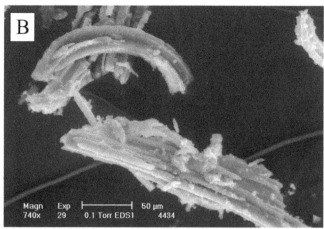

11.10. SEM photographs of tschermigite. A, Octahedral crystal with ferrous sulfate and gypsum (sample A3); B, fibrous columnar and twisted aggregates [sample S16(7)].

Gypsum [Ca(SO₄)·2 H₂O] and anhydrite [Ca(SO₄)]. Gypsum and/or anhydrite are present in 55% of the analyzed samples. Gypsum and anhydrite are primarily associated with the two major ferrous iron sulfates. Gypsum is sometimes associated with the pyrite and more closely with apatite in the samples without sulfates. EDS analyses performed on thin sections in bone fragments confirm that apatite and gypsum (or anhydrite) are closely associated.

Whereas the anhydrite crystals usually occur as crusts (Fig. 11.13E) or pinkish clusters of squat crystals, gypsum displays different morphologies (i.e., Fig. 11.13A,B). It occurs as white grains disseminated on the surface of the bones and clays, as white saccharoidal masses, or as crystals of various forms. The white, brown, or colorless crystals are acicular and prismatic; they also occur isolated, in sheaves, or in aggregates.

The growth of gypsum crystals is locally responsible for the fissuring of the bones (Fig. 11.13C). Gypsum crystals also grow in the surrounding clays (Fig. 11.13D).

Quartz (SiO₂). Quartz is present in nearly all the samples. It is the result of the filling of bone voids and cracks by clay and of the precipitation of authigenic rounded grains.

Minerals present in shale. Analyses of Wealden gray clays in and outside the bones show equivalent mineralogical compositions. In addition to the prevalent quartz, clinochlore $[(Mg,Fe^{2+})_5Al(Si_3Al)O_{10}(OH)_8]$, illite $[KAl_2(Si_3Al)O_{10}(OH)_2]$, montmorillonite $[(Na,Ca)_{0.3}(Al,Mg)_2Si_4O_{10}(OH)_2·nH_2O]$, and kaolinite $[Al_2Si_2O_5(OH)_4]$ are the major components of the clays. The accessory minerals include pyrite, rutile, and/or anatase.

Recrystallization of the Apatite Matrix

Discussion

The study of the apatite matrix of the *Iguanodon* bones by x-ray diffraction, combined with the qualitative chemical EDS analysis, confirms that fluorapatite is the mineral phase of the bone, as previously observed by Barker et al. (1997) and Wings (2004) on other fossil bones.

This result, combined with a relative low crystallinity index, confirms without any doubt the recrystallization of the slightly crystallized carbonated hydroxylapatite mineral phase of the fresh bones into the well-crystallized phase currently present in the fossil bones. The index is indeed

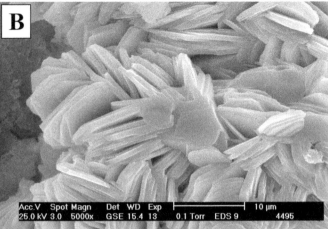

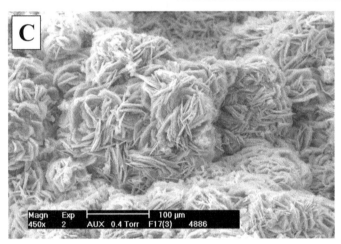

11.11. SEM photographs of natrojarosite (A–B) and metavoltine (C). A, Aggregates of flattened rhombohedral crystals (sample from the BER 3 core hole); B, aggregates of pseudohexagonal tablets [sample S15(2)]; C, hexagonal flattened tablets of metavoltine very similar to those of natrojarosite [sample F17(3)].

closer to the values of apatite of mineral origin than to those of the apatite present in fresh bones.

The introduction of fluorine replacing the $(OH)^-$ groups in the crystal lattice of apatite increases the stability of the structure. Fluorine increases the size of the crystallites and their thermal stability, whereas it reduces the tensions inside the crystal and decreases their solubility (LeGeros and LeGeros, 1984).

The increase of the crystallinity may be partly explained by a loss of CO_3^{2-} in the carbonated hydroxylapatite during the recrystallization (Person et al., 1995, 1996; Shemesh, 1990). It is now generally accepted that the CO_3^{2-} absorption bands in infrared spectra of carbonated apatite are due to structural substitutions and not to admixed carbonate phases, CO_3^{2-} adsorption on the grain surface, or bicarbonate ion substitutions (LeGeros et al., 1970). The ion group CO_3^{2-} is usually regarded as a substitute for PO_4^{3-}. The splitting into two fundamental vibrations of the antisymmetric stretching (v_3) of the carbonate group (Table 11.2) agrees with the hypothesis that carbonate groups in our samples occupy two structural positions (cf. Elliott, 2002). The corresponding out-of-plane (v_2) bending must also be doubly degenerate because of the presence of carbonate groups in two different structural environments. This second band recorded at 880 cm^{-1} is only present as a little shoulder on the side of the main peak.

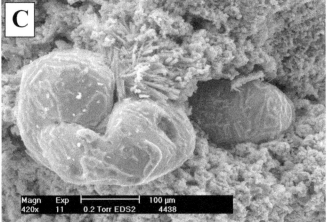

A substitution of hydroxyl group by fluorine can be observed during fossilization. This is reflected in the infrared spectrum, where similarities with fluorapatite become apparent (Brophy and Nash, 1968).

Absorption bands related to the vibrations of OH^- groups (633 cm^{-1}, OH liberation; 3,573 cm^{-1}, OH stretch) are usually absent. However, a weak shoulder in the range of 3,573 cm^{-1} is present in the spectrum of four out of seven samples, which should indicate the presence of small quantities of hydroxyl groups. The band at 633 cm^{-1} is always absent.

Although both the mineral powder and the KBr were previously stored in desiccators, infrared spectra clearly show an absorption band due to molecular H_2O (3,404 cm^{-1}). This may be due to the adsorption of H_2O molecules or to the presence of structural water in the lattice of the analyzed samples if they are Ca-deficient. In the case of Ca-deficient hydroxylapatite, the protonation of some of the phosphate groups respects a charge-balance scheme $[Ca^{2+} + OH^- + PO_4^{3-} \longleftrightarrow (HPO_4)^{2-}]$, resulting in vacancies in both Ca and OH sites. It is likely that molecules of H_2O occupy some of the vacated hydroxyl sites present in the c-axis anion channel (Elliott, 2002). This mechanism has already been described in hydroxylapatite but not in fluorapatite. Therefore, in our case, adsorption should be considered as the most important mechanism.

The presence of sulfur, often associated with iron in EDS analyses of bone apatite, is most probably related to the presence of microscopic pyrite finely disseminated in the structure of the bone rather than to a substitution

11.12. SEM photographs of halotrichite (A) and pure sulfur (B–C). A, Acicular crystals in diverging aggregates on clay [sample S16(1)]; B, rounded crystal of sulfur with barite and ferrous sulfates [sample E10(1)]; C, crystals of sulfur show strong traces of dissolution together with gypsum and ferrous sulfates [sample S16(7)].

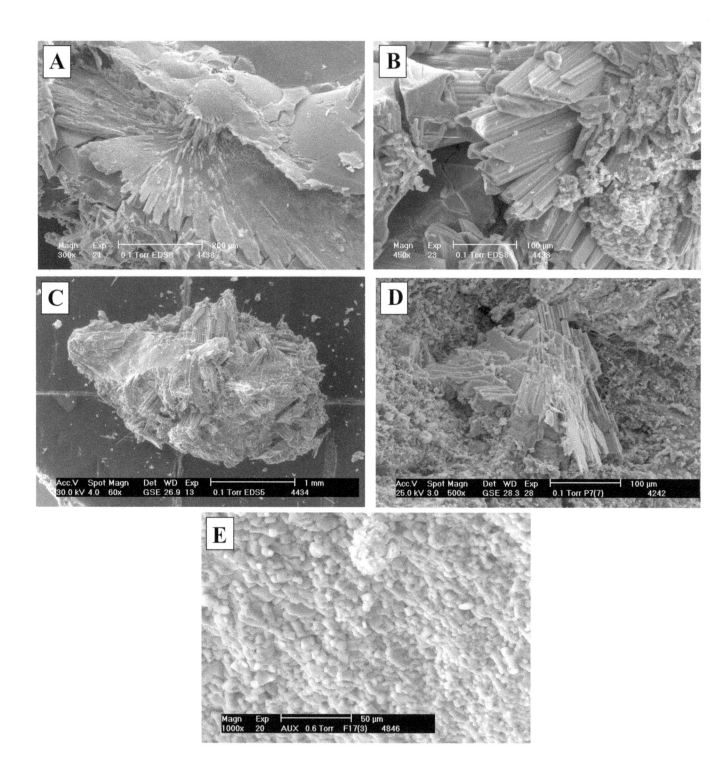

11.13. SEM photographs of gypsum (A–D) and anhydrite (E). A, Fan-shaped crystals in bone [sample F17(9)]; B, aggregate of elongated crystals in bone (sample A1); C, growth of gypsum crystals breaks the bone (sample from BER 3 core hole); D, growth of gypsum crystals fissures the clay surface [sample P7(7)]; E, crust of prismatic crystals of anhydrite [sample F17(9)].

of PO_4^{3-} group by SO_4^{2-}. The absence of a shoulder at 1,170–1,130 cm^{-1} and of a platform near 645 cm^{-1} in the same spectral range of PO_4^{3-} (Baumer et al., 1990) indicates the absence of a SO_4^{2-} group in the structure of apatite. Other elements identified during EDS analyses of bone apatite are Sr, Na, K, and Fe. These elements could partially replace Ca in the lattice.

Permineralization

Permineralization is the filling of the fractures and cavities (marrow and cancellous bone cavities, Haversian canals, osteocyte lacunae, and canaliculi) of the bone by foreign minerals. Although some minerals infiltrated the bone during the filling of the cavities by surrounding sediments (detrital quartz, "argillaceous" phyllosilicates, anatase, and rutile), others are authigenic and precipitated in the cavities during burial (pyrite, barite, sphalerite, celestine, and to a lesser extent quartz) or after the exhumation (other sulfates, oxides, and/or hydroxides).

The precipitation of minerals in the bone cavities and their replacement are parts of a complex diagenetic process that began during the fossilizations and continued after the exhumation. The nature and the importance of these processes depend on the quantity and composition of seepage waters, the local concentration in ions, the solubility of the mineral, and the local conditions of pH and Eh. The localization and succession of minerals are indicators of the diagenesis phase during which they formed and also of the physicochemical conditions inside the bone during this period.

Pyrite was by far the most abundant mineral in the *Iguanodon* bones at the moment of their discovery. The growth of pyrite crystals caused a partial destruction of the osseous structures, making the bones very brittle.

Thermodynamic calculations (Berner, 1964; Garrels and Christ, 1965) indicate that pyrite is stable only in the absence of air and in the presence of sulfides in solution. In sediments, these conditions are encountered during the accumulation of organic matter and its metabolization by bacteria. This results in the replacement of the oxygen in pore water by sulfide ions in solution (Berner, 1970). The Wealden clays represent an excellent example of environment favorable for intense pyritization. The main stages in the formation of pyrite are the bacterial reduction of sulfates, the reaction of H_2S with ferrous minerals, and the transformation of black ferrous monosulfures (intermediate phases) into pyrite. These stages can be synthesized by the total reaction (Allen et al., 1912):

$$Fe^{2+} + H_2S + S^0 \rightarrow FeS_2 + 2 H^+$$

The two independent sources for H_2S generation in the sediments are the dissolved sulfates of bacterial origin, dominating in marine environment, and the decomposition of organic compounds rich in sulfurs (collagen, etc.), dominating in nonmarine environment (Berner, 1970). In the Iguanodon Sinkhole, the Sainte-Barbe Clays Formation deposited in a freshwater paleolake (see Chapter 9 in this book), and iron was primarily of sedimentary origin. If the sediment contained a limited quantity of organic matter, iron remained dissolved in pore water and was thus mobile in the sediment. The degradation of the organic matter surrounding the bones provided a local source of sulfurs, allowing the formation of pyrite (Briggs et al., 1996; Canfield and Raiswell, 1991). The analysis of the clays indicates that the pyrite is present in the form of scattered micrograins or, in contrast, as isolated centimetric nodules. The pyritization of the bones could subsequently be accentuated by the external diffusion of sulfates derived from seawater covering the clays from the Late Albian (see Chapter 7 in this book).

Pfretzschner (2000) identified a first generation of endogenous pyrite formation localized in the osseous cavities close to the marrow cavity of the bone. This pyrite would be formed from sulfur and iron generated, respectively, by the decomposition of the collagen and the hydrolysis of hemoglobin. In the case of the Bernissart iguanodons, no observation confirms the existence of such a process.

However, all observations are distorted by the fact that the bones were curetted at the time of their original preparation. It is no longer possible to determine the quantity and the localization of the extracted pyrite. The pyrite still present in the bones is nevertheless sufficient to testify to the abundant and widespread pyritization of the entire *Iguanodon* skeletons. Macroscopic observations in situ, as well as the study of thin sections of bones (long bones, ribs), indicates a highly variable pyritization. These bones show that pyrite is present as fibroradial pyrite in mammillated crusts on the external surfaces; in the osseous cavities or replacing osseous matrix material located under the bone surface; as a thin coating on some internal walls of the bone; and as important masses of pyrite located in the interior zone of the bone. However, this in-depth pyritization is too abundant to be only the result of the mechanism proposed by Pfretzschner (2000).

These various localizations speak for a primary exogenic pyritization. The elements necessary for the formation of pyrite diffused gradually into the bone. The process was accentuated by the appearance of fractures. The filling of these fractures by pyrite confirms the formation of this mineral after the fissuring of the bone. The degree of pyritization varies not only from one individual skeleton to another, but also from one bone to another, and even within an area of the same bone: in 1 cm^3 of cancellous bone in a vertebra, some cavities are completely empty, whereas others contain pyrite or barite or even both.

Barite is the second most common mineral formed in the bones. It can precipitate starting from a Ba-rich fluid in anaerobic sediments that are rich in organic matter (Torres et al., 2003). A chemical analysis of the shale shows that among the trace elements, Ba is the most abundant. Strontium is also present in the sediment, but in a lower concentration. This can explain the partial substitution of Ba by Sr in barite as well as the presence of celestine in the bones. Barite represents a later generation than pyrite. Indeed, when they are both present in a cavity, pyrite generally coats the walls, whereas barite partially or completely fills the remaining space.

Celestine and sphalerite are regularly present in small quantities and are often much localized.

The exhumation of the Bernissart iguanodons induced a radical environmental change for the bones and for the pyrite inside the bones. Contact with atmospheric water and oxygen induced the oxidation of the pyrite with a release of acid, according to the following reaction (Canfield and Raiswell, 1991):

$$2\,FeS_2 + 7\,O_2 + 2\,H_2O \rightarrow 2\,Fe^{2+} + 4\,SO_4^{2-} + 4\,H^+$$

This reaction is dominating when the pH is higher than 4. In the presence of oxygen, ferrous iron oxidizes into ferric iron (Moses et al., 1987). When the pH becomes lower than 4, pyrite is oxidized by the ferric iron according to the following reaction (Moses et al., 1987):

$$FeS_2 + 14\,Fe^{3+} + 8\,H_2O \rightarrow 15\,Fe^{2+} + 2\,SO_4^{2-} + 16\,H^+$$

When the pH is higher than 4, ferric iron can also precipitate to form iron oxide hydroxides following the reaction (Nicholson et al., 1988)

$$FeS_2 + 15/4\,O_2 + 7/2\,H_2O \rightarrow Fe(OH)_3 + 2\,SO_4^{2-} + 4\,H^+$$

In the *Iguanodon* bones, iron oxides and/or hydroxides were locally identified. These minerals are soluble when the pH is lower than 4. The release of acid during these reactions notably induced an increased the acidification of the environment, which could explain the absence of oxides and/or iron hydroxides in significant quantity.

When the pH is acid, ferrous iron can react with sulfate in solution and precipitate to form ferrous sulfates of the $FeSO_4.n\ H_2O$ type according to the reaction proposed by Bylina et al. (2009):

$$FeS_2 + 2\ O_2 + n\ H_2O \rightarrow FeSO_4.n\ H_2O + S^0$$

Szomolnokite and rozenite, the two most abundant sulfates in the bones, belong to this category, like melanterite. However the latter, identified only in one sample, is regarded as an early sulfate formed during the oxidation of pyrite (Jambor et al., 2000). Melanterite, rozenite, and szomolnokite differ only by their degree of hydration. Melanterite is readily soluble in water; when exposed to air, it tends to lose hydration water and change into a whitish powder (generally rozenite or szomolnokite) with increasing temperature and/or decreasing relative humidity (Alpers et al., 1994).

Pyrite oxidation is not a constant mechanism. Indeed, oxidation rate varies in particular with pH, temperature, oxygen content, and degree of saturation in water. Another important criterion is the specific surface of pyrite (Bellaloui et al., 2002; Lowson, 1982; Nordstrom, 1982; Van Breemen, 1972). In the *Iguanodon* bones, the aggregates of micrograined pyrite and the framboidal pyrite oxidize more quickly than the pyrite with a fibroradial structure. Isolated crystals of pyrite or in aggregates show only small traces of oxidation in the form of a slight iridescence. Thin coatings of pyrite on the surface of some bone cavities also have a slightly grayish color. Moreover, they are regularly punctuated by iridescent encrusting aureoles (100 μm in diameter) composed of needles of a ferrous sulfate that could not be identified.

Roemerite is a mixed ferrous–ferric iron sulfate, which is formed according to the reaction (Bayless and Olyphant, 1993)

$$3\ FeS_2 + 23/2\ O_2 + 15\ H_2O \rightarrow Fe^{2+}Fe^{3+}_2(SO_4)_4 \cdot 14\ H_2O + 2\ SO_4^{2-} + 2\ H^+$$

In comparison with complete oxidation by oxygen, only one third of H^+ is released, while the remaining acidity is stored as unhydrolized partly oxidized iron in the solid phase (Bayless and Olyphant, 1993).

A release of acids caused the degradation of the clays near the bones. Cations like Al, Na, and K could then react with iron and sulfate to form another kind of sulfate salt. The most common one is natrojarosite. It is a yellow mineral, almost always present in small amounts in association with szomolnokite and/or rozenite.

The semiquantitative analysis of natrojarosite shows a partial replacement of Na by K and sometimes a deficit of these two cations (report/ratio (Na+K)/Fe < 1/3). This deficit could be explained by the occupation of this crystallographic site by H_3O^+ or NH_4^+, not detected by EDS analyses.

Tschermigite is regularly observed in the white sulfate efflorescences, although it is not easily detectable when in small quantities by optical observations via x-ray powder diffraction.

Metavoltine was first identified by x-ray powder diffraction in one sample. Metavoltine and natrojarosite have the same color and very close morphologies; both are composed of the same chemical elements: Na, K,

Fe, and S. Nevertheless, these two minerals have different theoretical Na:Fe ratios (1/1 for metavoltine and 1/3 for natrojarosite). A semiquantification of EDS analyses identified metavoltine in one other sample.

Halotrichite is rare and can only be found in very small quantities. Generally, it is in direct contact with the clays, but it is also present in the white sulfate efflorescences. Optically, the clusters of halotrichite are not easily discernible from those of gypsum. EDS analyses distinguish halotrichite from other isotypic minerals of the of pickeringite group [$MgAl_2(SO_4)_4 \cdot 22 H_2O$].

Sulfur is regularly present and is the last secondary mineral resulting from the pyrite oxidation observed in the bones. It could be formed following the reaction proposed by Bylina et al. (2009) (see above). Most of the crystals show traces of dissolution probably related to the acid release linked to the mechanisms of pyrite oxidation.

The oxidation mechanism of pyrite forms a series of secondary minerals and releases acid, as discussed above. However, the acid can be neutralized by the presence of carbonates or phosphates according to a reaction similar to the one given by Evangelou (1995):

$$CaCO_3 + 2 H^+ + SO_4^{2-} + 2 H_2O \rightarrow CaSO_4 \cdot 2 H_2O + H_2CO_3$$

The dissolution of the apatite matrix in the *Iguanodon* bones releases calcium and phosphate. Calcium can react with sulfates in solution, and according to the temperature and the moisture of the environment, they can precipitate as gypsum or anhydrite. These two minerals differ by their degree of hydration and can easily transform into each other. Calcite is the major mineral of chalk that was mixed with the fish glue. This mixture was initially used to fill the cavities in the curetted or repaired bones. This is the only calcium carbonate identified during this study. It seems that calcite did not intervene significantly in the neutralization of the acid, and when it did so, it was always close to these repaired zones. The study of an untreated bone fragment from BER 3 core indicated a significant presence of gypsum in spite of the absence of carbonate. This confirms the role of bone apatite as a buffer.

Although the presence of gypsum is relatively easy to determine, this is not the case for anhydrite, which in general is only identifiable by its most intense peak at $d_{002} = 3.498$ Å. In sulfate-rich samples, this peak closely corresponds to a shoulder at the base of the principal szomolnokite peak at 3.44 Å or is completely lost in the background noise when matrix apatite is present. Nevertheless this mineral appears quite widespread throughout the samples.

Conclusions

The diagenesis of the *Iguanodon* bones was a complex phenomenon, as shown by the recrystallization of the matrix apatite and the identification of about 30 mineral phases. Some of them are authigenic, whereas others are the result of sediment infiltrations in the bones.

Paradoxically, the burying of the bones in the Wealden clays is responsible for their excellent conservation, but also for their degradation. Indeed, the fast burial preserved the bones from decomposing under the atmospheric constraints, but it kept them in an environment favorable for the pyritization, responsible for making the bones brittle during the burial.

Since the exhumation of the iguanodons, pyrite oxidation has had a disastrous effect on the conservation of the bones. The growth of pyrite crystals had already weakened the bones during their burial; the degradation of the Bernissart iguanodon bones worsened in contact with the fresh air because of the oxidation and transformation of the pyrite into sulfate minerals, and also because of the growth of anhydrite and gypsum crystals. The mechanisms directly or indirectly linked to the pyrite oxidation are often responsible for the bursting of the bone structure. So far, the safeguarding techniques proved to be inefficient to prevent these destructive processes, but in some cases they have partially slowed them down.

In this work, we present the results of qualitative or semiquantitative analyses from bony samples. It was regrettably impossible to obtain quantitative data for the mineral paragenesis for several reasons, including the original curettage of the bones that extracted the main part of the pyrite, as well as the lack of information about the quantities taken in this period and their localization in the bones. There was also important variability of primary and secondary minerals (nature, quantities, and localization). Finally, the samples were preferentially taken from bones that urgently needed restoration.

Quantitative data would have revealed osseous zones more favorable to the pyritization or to the development of the other mineral phases. Nevertheless, studies of the bones indicate the absence of a relation between the nature of the skeleton elements and their rate of pyritization or degradation. Even if spongy bone contains vaster cavities than compact bone, it is not necessarily more pyritized. Pyritization seldom appears within the whole skeleton.

Acknowledgments

The author acknowledges financial supports from the departments of Palaeontology and Geology of the Royal Institute of Natural Sciences, Belgium, and from the Laboratory of Mineralogy, University of Liège, Belgium. The FRS-FNRS is acknowledged for research grant 1.5.112.02. Many thanks are due to A. M. Fransolet, H. Goethals, Tom Thijs, and A. de Ricqlès for their constructive comments; J. Cillis, J. Jedwab, and C. Henrist for EDS analyses and SEM images; M. Rondeux for her help in using the program LCLSQ; F. Hatert for the infrared spectroscopic investigation; and J. Vander Auwera for x-ray fluorescence measurements.

References

Allen, E. T., J. L. Crenshaw, J. Johnston, and E. S. Larsen. 1912. Mineral sulfides of iron. American Journal of Science (ser. 4) 169–236.

Alpers, C. N., D. W. Blowes, D. K. Nordstrom, and J. L. Jambor. 1994. Secondary minerals and acid mine-water chemistry; pp. 247–270 in D. W. Blowes and J. L. Jambor (eds.), The environmental geochemistry of sulfide mine-wastes, 22. MAC Shortcourse Handbook. Mineralogical Association of Canada, Waterloo, Ontario.

Baddiel, C. B., and E. E. Berry. 1966. Spectra structure correlations in hydroxyl and fluorapatite. Spectrochimica Acta 22: 1407–1416.

Barker, M. J., J. B. Clarke, and D. M. Martill. 1997. Mesozoic reptile bones as diagenetic windows. Bulletin de la Société Géologique de France 168: 535–545.

Baumer, A., R. Caruba, and M. Ganteaume. 1990. Carbonate–fluorpapatite: mise en évidence de la substitution 2 $PO_4^{3-} \rightarrow SiO_4^{4-} + SO_4^{2-}$ par spectrométrie infrarouge. European Journal of Mineralogy 2: 297–304.

Baur, W. H. 1960. Die kristallstruktur von FeSO$_4$·4 H$_2$O. Naturwissenschaften 47: 467.

Bayless, E. R., and G. A. Olyphant. 1993. Acid-generating salts and their relationship to the chemistry of groundwater and storm runoff at an abandoned mine site in southwestern Indiana, USA. Journal of Contaminant Hydrology 12: 313–328.

Bellaloui, A., G. Nkurunziza, and G. Ballivy. 2002. Caractérisation en laboratoire du potentiel expansif de granulats de remblais de fondation. Canadian Geotechnical Journal 39: 141–148.

Berner, R. A. 1964. Stability fields of iron minerals in anaerobic marine sediments. Journal of Geology 72: 826–834.

———. 1970. Sedimentary pyrite formation. American Journal of Science 268: 1–23.

Bhatnagar, V. M. 1967. IR-spectra of fluorapatite and fluor-chlorapatite. Cellular and Molecular Life Sciences 23: 10–12.

———. 1968. Infrared spectra of hydroxyapatite and fluorapatite. Bulletin de la Société Chimique de France 1968: 1771–1773.

Bonar, L. C., A. H. Roufosse, W. K. Sabine, M. D. Grynpas, and M. J. Glimcher. 1983. X-ray diffraction studies of the cristallinity of bone minerals in newly synthesised and density fractionated bone. Calcified Tissue International 35: 202–209.

Briggs, D. E. G., R. Raiswell, S. H. Bottrel, D. Hatfield, and C. Bartels. 1996. Controls on the pyritization of exceptionally preserved fossils: an analysis of the Lower Devonian Hunsbrück Slate of Germany. American Journal of Science 296: 633–663.

Brophy, G. P., and J. T. Nash. 1968. Compositional, infrared, and x-ray analysis of fossil bone. American Mineralogist 53: 445–454.

Burnham, C. W. 1991. LCLSQ: lattice parameter refinement using correction terms for systematic errors. American Mineralogist 76: 663–664.

Bylina, I., L. Trevani, S. C. Mojumdar, P. Tremaine, and U. G. Papangelakis. 2009. Measurement of reaction enthalpy during pressure oxidation of sulfide minerals. Journal of Thermal Analysis and Calorimetry 96: 117–124.

Calderin, L., M. J. Stott and A. Rubio. 2003. Electronic and crystallographic structure of apatites. Physical Revue B 67: 134106 (7 pp.).

Canfield, D. E., and R. Raiswell. 1991. Pyrite formation and fossil preservation; pp.

337–387 in P. A. Allison and D. E. Briggs (eds.), Taphonomy. Plenum Press, New York.

Chipera, S. J., and D. L. Bish. 1991. Application of x-ray diffraction crystallite size/strain analysis to *Seismosaurus* dinosaur bone. Advance in X-ray Analysis 34: 481–482.

Corbridge, D. E. C. 1974. The structural chemistry of phosphorus. Elsevier Scientific Publication, Amsterdam, 542 pp.

De Pauw, L. F. 1902. Notes sur les fouilles du charbonnage de Bernissart. Découverte, solidification et montage des Iguanodons. Imprimerie photo-litho, J. H. & P. Jumpertz, Brussels, Belgium, 56 pp.

Dumitras, D. G., S. Marincea, and A. M. Fransolet. 2004. Brushite in the guano deposit from the "dry" Cioclovina Cave (Sureanu Mountains, Romania). Neues Jahrbuch für Mineralogie, Monatshefte: 45–64.

Elliott, J. C. 2002. Calcium phophate biomineral. Reviews in Mineralogy and Geochemistry 48. Phosphate: Geochemical, Geobiological and Materials Importance 48: 427–454.

Evangelou, V. P. B. 1995. Pyrite oxidation and its control. CRC Press, Boca Raton, Fla, 295 pp.

Garrels, R. M., and C. L. Christ. 1965. Solutions, minerals and equilibria. Harper & Row, New York, 450 pp.

Giester, G. 1988. The crystal structure of CuSO$_4$·H$_2$O and CuSeO$_4$·H$_2$O and their relationship to kieserite. Mineral Petrology 38: 277–284.

Gillette, D. D. 1994. *Seismosaurus*, the earth shaker. Columbia University Press, New York, 205 pp.

Godart, J.-F. 2005. Collaboration à la création d'un laboratoire d'extraction du collagène osseux à l'Institut Royal des Sciences Naturelles de Belgique. Application aux ossements de l'île de Pâques. B.A. thesis, Haute École Provinciale du Hainaut Occidental, Ath, 85 pp.

Jackson, S. A., A. G. Cartwright, and D. Lewis. 1978. The morphology of bone mineral crystals. Calcified Tissue Research 25: 217–222.

Jambor, J. L., and R. J. Traill. 1963. On rozenite and siderotil. Canadian Mineralogist 7: 751–763.

Jambor, J. L., D. K. Nordstrom, and C. N. Alpers. 2000. Metal–sulfate salts from sulphide mineral oxidation; pp. 303–350 in C. N. Alpers et al. (eds.), Sulfate minerals—crystallography, geochemistry, and environmental significance. Reviews in Mineralogy and Geochemistry

40. Mineralogical Society of America, Geochemistry Society, Washington, D.C.

Klugg, H. P., and L. E. Alexander. 1962. X-ray diffraction procedures for polycrystalline and amorphous materials. J. Wiley & Sons, New York, 716 pp.

Kolodny, Y., B. Luz, M. Sander, and W. A. Clemens. 1995. Dinosaur bones: fossil or pseudomorphs? The pitfalls of physiology reconstruction from apatitic fossils. Palaeogeography, Palaeoclimatology, Palaeoecology 126: 161–171.

Lees, S., and K. S. Prostak. 1988. The locus of mineral crystallites in bone. Connective Tissue Research 18: 41–54.

LeGeros, R. Z., and J. P. LeGeros. 1984. Phosphate minerals in human tissues; pp. 351–385 in J. O. Nriagu and P. B. Moore (eds.), Phosphate minerals. Springer-Verlag, New York.

LeGeros, R. Z., J. P. LeGeros, O. R. Trautz, and E. Klein. 1970. Spectral properties of carbonate in carbonate-containing apatites. Developments in Applied Spectroscopy 7B: 3–12.

Lowson, R. T. 1982. Aqueous oxidation of pyrite by molecular oxygen. Chemical Revue 82: 461–497.

Moses, C. O., D. K. Nordstrom, J. S. Herman, and A. L. Mills. 1987. Aqueous pyrite oxidation by dissolved oxygen and by ferric iron. Geochimica et Cosmochimica Acta 51: 1561–1571.

Nicholson, R. V., R. W. Gillham, and E. J. Reardon. 1988. Pyrite oxidation in carbonate-buffered solution: 1. Experimental kinetics. Geochimica et Cosmochimica Acta 52: 1077–1085.

Nordstrom, D. K. 1982. Aqueous pyrite oxidation and the consequent formation of secondary iron minerals; pp. 37–56 in D. K. Nordstrom (ed.), Acid sulfate weathering. Special Publication 10. Soil Science Society of America, Madison, Wis.

Parker, R. B., and H. Toots. 1970. Minor elements in fossil bone. Geological Society of America Bulletin 81: 925–932.

———. 1972. Hollandite–coronadite in fossil bone. American Mineralogist 57: 1527–1530.

Person, A., H. Bocherens, J. F. Saliège, F. Paris, V. Zeitoun, and M. Gérard. 1995. Early diagenetic evolution of bone phosphate: an x-ray diffratometry analysis. Journal of Archaeological Science 22: 211–221.

Person, A., H. Bocherens, A. Mariotti, and M. Renard. 1996. Diagenetic evolution and experimental heating of bone phosphate. Palaeogeography, Palaeoclimatology, Palaeoecology 126: 135–149.

Pfretzschner, H. U. 2000. Pyrite formation in Pleistocene bones—a case of very early formation during diagenesis. Neue Jahrbuch für Geologie und Paläontologie Abhandlungen 217: 143–160.

———. 2001. Pyrite in fossil bone. Neue Jahrbuch für Geologie und Paläontologie Abhandlungen 220: 1–23.

Pistorius, C. W. F. T. 1960. Lattice constants of $FeSO_4 \cdot H_2O$ (artifical szomolnokite) and $NiSO_4 \cdot H_2O$. Bulletin de la Société Chimique de Belgique 69: 570–574.

Sandford, M. K. 1992. A reconsideration of trace element analysis in prehistoric bone; pp. 79–103 in S. R. Saunders and M. R. Katzenberg (eds.), Skeletal biology of past peoples: research methods. J. Wiley & Sons, New York.

Shemesh, A. 1990. Crystallinity and diagenesis of sedimentary apatites. Geochimica et Cosmochimica Acta 54: 2433–2438.

Sillen, A., and J. C. Sealy. 1995. Diagenesis of strontium in fossil bone: A reconsideration of Nelson et al. (1986). Journal of Archaeological Science 22: 313–320.

Torres, M. E., G. Bohrmann, T. E. Dubé, and F. G. Poole. 2003. Formation of modern and Paleozoic stratiform barite at cold methane seeps on continental margins. Geology 31: 897–900.

Tuross, N., A. K. Behrensmeyer, E. D. Eanes, L. W. Fisher, and P. E. Hare. 1989. Molecular preservation and crystallographic alterations in a weathering sequence of wildebeest bones. Applied Geochemistry 4: 261–270.

Van Breemen, N. 1972. Soil forming processes in acid sulfate soils; pp. 66–130 in H. Wageningen (ed.), Acid sulfate soils. Proceedings of the International Symposium on Acid Sulfate Soils; I, Introductory papers and bibliography. International Institute for Land Reclamation and Improvement.

Williams, C. T., and C. A. Marlow. 1987. Uranium and thorium distributions in fossil bones from Olduvai Gorge, Tanzania and Kanam, Kenya. Journal of Archaeological Science 14: 297–309.

Wings, O. 2004. Authigenic minerals in fossil bone from the Mesozoic of England: poor correlation with depositional environments. Palaeogeography, Palaeoclimatology, Palaeoecology 204: 15–32.

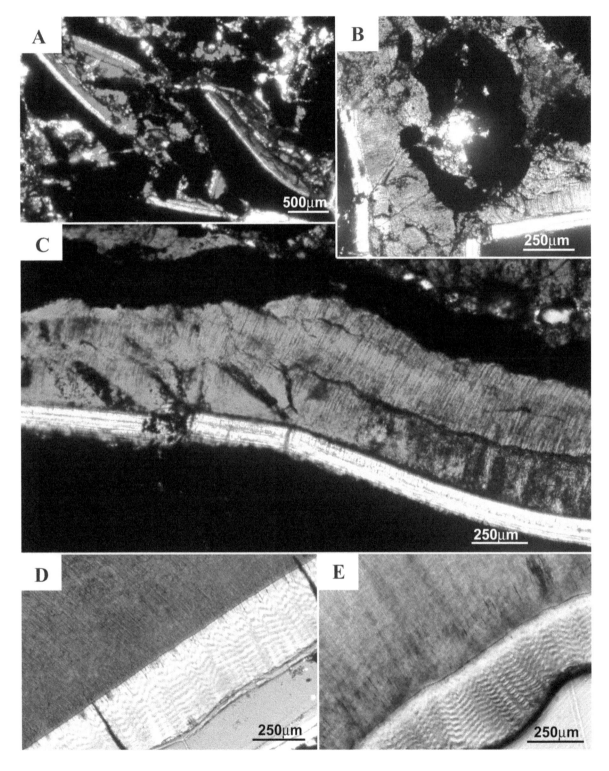

Histological Assessment of Vertebrate Remains in the 2003 Bernissart Drill

12

Armand de Ricqlès*, Pascal Godefroit, and Johan Yans

After the 2003 drilling in the Iguanodon Sinkhole at Bernissart (BER 3 borehole), examination of the column revealed stony dark grayish remains at levels −296.5 m and −309 m, and hence within the Wealden clays levels. Microscopic examinations of the remains (paleohistology) conclusively demonstrate the occurrence of bone and tooth tissues. Whether the histological characteristics of the tissues allow a precise identification, especially whether the remains belong to *Iguanodon*, is quite another matter. The small teeth material clearly does not belong to ornithopod dinosaurs; however, their precise taxonomic origin cannot be assessed. However, the bony material shows structures compatible with a dinosaurian origin. A survey of the literature devoted to *Iguanodon* bone and tooth histology, as well as comparisons with *Iguanodon bernissartensis* bone and tooth material and with *Bactrosaurus johnsoni* teeth, do not demonstrate that the material definitely belongs to *Iguanodon*, although the possibility is likely for several reasons, detailed herewith. Comparison between "fresh" (from the borehole) and "old" (kept in the RBINS for more than 130 years under ordinary conditions) *Iguanodon* bones also allows checking the degradation process experienced by pyritized bones at the tissue level.

In 2002–2003, three new boreholes were drilled within and around the Iguanodon Sinkhole at Bernissart. They provided exceptional material used for a multidisciplinary research to improve our knowledge of the *Iguanodon*-bearing Wealden facies (see Chapter 1 in this book). Detailed examination of the BER 3 column revealed stony dark grayish remains at levels −296.5 m and −309 m, and hence within the Wealden clays levels. The likeliness that those remains could be vertebrate skeletal fragments was high because of their phosphatic nature and because the borehole was drilled at the presumed site where the Bernissart iguanodons were discovered in 1878. Microscopic examinations of these fragments (paleohistology) conclusively demonstrate that these fragments are actually bone and tooth remains. Here, we describe the histology of the skeletal fragments discovered in the BER 3 borehole. Histological comparisons are attempted with data compiled from the literature and also with bone and teeth fragments taken from *Iguanodon bernissartensis* and *Bactrosaurus johnsoni* specimens. The "fresh" material from the borehole is also compared with "old" bones discovered at Bernissart between 1878 and 1881 in order to check at the tissue level the degradation process experienced by pyritized bone.

Institutional abbreviation. RBINS, Royal Belgian Institute of Natural Sciences, Brussels, Belgium.

12.1. A, BER 296.5, section 1, general view of the apical region (tip) of two small teeth, plus tooth fragment. B, BER 296.5, section 3, detail of possibly the proximal region of a small tooth crown (broken). Irregular large, rounded bays in the dentine may be evidence of biological dentinoclasy linked to tooth replacement. C, BER 296.5, section 3, detail of dentine and enamel along a tooth crown. The thin, highly birefringent enamel is nonprismatic and appears to be divided in four to five superimposed sheets; the numerous dentine canaliculi are obvious. D, *Iguanodon bernissartensis* from Bernissart (RBINS unregistered specimen "H"), detail of the enamel–dentine junction (EDJ) in a maxillary tooth. The thick enamel shows extensive superimposed zigzagging bright and dark bands, typical of advanced ornithopod enamel; the thickness of the dentine toward the pulp cavity would extend upward over the full height of the plate. E, *Bactrosaurus johnsoni* maxillary tooth; detail of the EDJ for comparison. Superposition of the bandings in the enamel suggests the fake occurrence of juxtaposed vertical pillars forming the tissue.

Introduction

Material and Methods

Four fragments of a few cubic centimeters each were carefully extracted from the BER 3 column for examination by histological technics. The fragments, numbered Bernissart 3 296.5, Bernissart 3 309 A, Bernissart 3 309 B, and Bernissart 3 309 C, were dried and embedded in resin under gentle vacuum, with the resin temperature monitored to secure a slow polymerization. The resulting blocks were trimmed and sawed with a thin diamond/copper circular blade and further processed to obtain thin sections following routine paleohistological techniques (e.g., Wilson, 1994).

For comparative purposes, a fragment of rib and one maxillary tooth from *Iguanodon bernissartensis* (RBINS unregistered specimen "H"; see Norman, 1986, appendix 1), and maxillary and dentary teeth from *Bactrosaurus johnsoni*, a basal hadrosaurid from the Iren Dabasu Formation in Inner Mongolia (P.R. China), were also histologically processed. The resulting thin sections were examined under dissecting and compound microscopes, in ordinary and polarized lights. Some preliminary analyses by x-ray diffraction and scanning electron microscopy were also conducted to check differences between fresh material from the borehole and the old bones discovered at Bernissart between 1878 and 1881.

Histological Description

The Bernissart 3 296.5 specimen. All the sections show compact bone fragments around a small cavity filled with a black material containing some tooth remains. The bone fragments have a complex structure of compacted secondary endosteal trabeculae and Haversian systems. The absence of Sharpey fibers indicates that the observed bone tissues are not of periosteal or dermal origin. The bone tissue is entirely secondary (reconstructed in vivo) and varies from region to region. Typical mature secondary osteons (Haversian systems) are locally superimposed on each other (Fig. 12.2E). Other structures suggest large secondary endosteal trabeculae collapsed on each other.

The tooth remains (Fig. 12.1A–C) suggest numerous thin elongate small teeth. The crown may have been cylindrical with a pointed arch-shaped apex ending in a rather acute tip (Fig. 12.1A). The enamel is thin, highly anisotropic under crossed Nicols, and nonprismatic. It is divided into four to five superimposed sheets (Fig. 12.1C). The dentine shows the traditional radially oriented canaliculi and some evidence of a clastic activity locally, perhaps linked to tooth replacement (Fig. 12.1B). No root system or ankylosis on dentigerous bone could be observed.

The Bernissart 3 309A, B, C specimens. These specimens show a more or less dense Haversian bone tissue intimately associated with massive pyrite deposition. Some regions apparently preserve the natural free surface of the bone (Fig. 12.2A), and it is possible to observe there some primary (periosteal) bone tissue, more or less invaded by secondary osteons (Haversian systems) (Fig. 12.2B). The secondary osteons are numerous, forming a dense Haversian bone with superposition of osteon generations. It is nevertheless still possible to observe remains of primary (periosteal) bone tissues forming some of the interstitial systems between the secondary osteons (Fig. 12.2D). The primary bone tissue appears to be a poorly defined modulation of the fibrolamellar complex, where small longitudinal primary osteons are the prevailing vascular component (Fig. 12.2A–D). The circular and especially

12.2. A, BER 309 C, section 3, low-power view of a subperiosteal surface in cross section; scattered Haversian reconstruction into the primary cortex; some large, unfinished secondary osteons almost reach the bone surface, which does not show an external fundamental system (EFS). B, BER 309 C, section 3, detail of the primary bone tissue forming the superficial cortex. A few poorly developed small primary osteons oriented longitudinally permeate the tissue. C, BER 309 B, section 1, detail of the primary cortex with a LAG (=line of arrested growth; arrow) parallel to the bone-free surface. D, BER 309 C, section 1. The deep cortex is formed by large secondary osteons (=Haversian systems) with extensive evidence of periosteal bone tissue still forming the interstitial systems between them. E, BER 296.5, section 3; detail of dense Haversian bone in superimposed generations. F, Rib of IRSNB unregistered specimen "H," section 4; general view of the cortex at low magnification. After more than a century of pyrite degradation, the whole structure is fragmented by multiple larger, smaller, and minute cracks. The numerous whitish spots (in the secondary osteons) are artifacts caused by the resin monomer having differentially permeated the tissue along the minute cracks before polymerization.

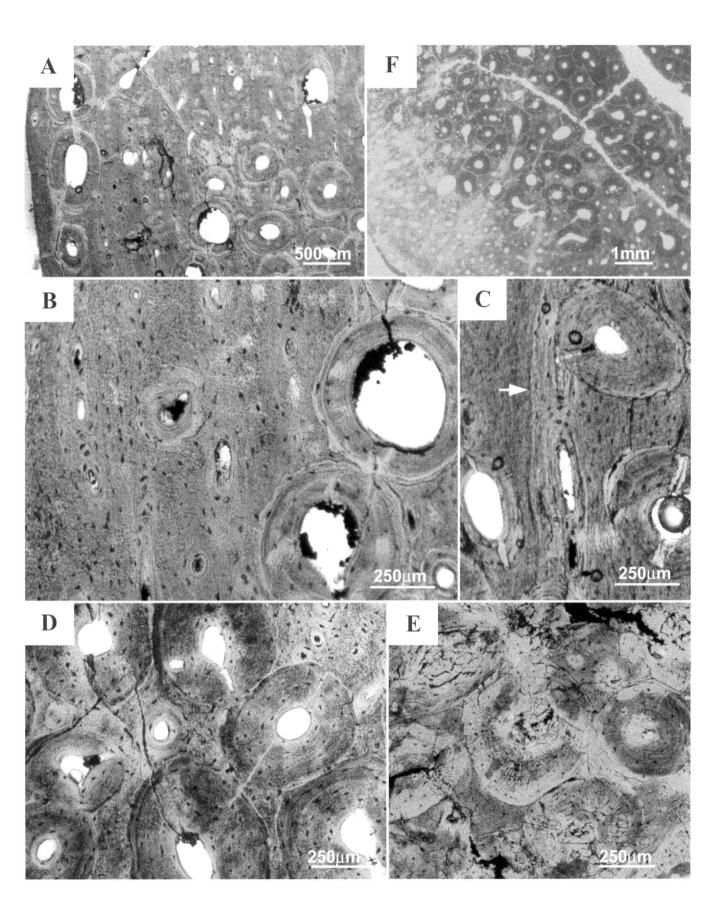

radial vascular canals are almost lacking. There is some evidence of lines of arrested growth (LAGs) parallel to the free surface of the bone (Fig. 12.2C).

Iguanodon Bone and Tooth Histology: A Bibliographical Survey

Iguanodon Mantell, 1825, appears to be one of the first dinosaurs submitted to histological analyses. In Queckett's catalog (1855), one can easily recognize dense Haversian tissue as depicting *Iguanodon's* compact bone structure. The rediscovery of old thin sections of fossil bone created in the 1870s at the request of Professor Paul Gervais at the Paris Museum was recently published (Ricqlès et al., 2009b). It unveils previously unknown early thin sections from *Iguanodon*. Some of the thin sections were made in England, while others appear to have been processed in Paris. The material predates (1875–1876) the Bernissart discovery (1878) and would thus come from England, although its precise origin is unknown. One section (made in England) depicts coarse cancellous bone of secondary origin, and the two others show primary bone of the laminar type, with little Haversian reconstruction. The structures of the latter sections (Ricqlès et al., 2009b, fig. 2C) of compact bone suggest an immature, actively growing individual.

Most later histological descriptions of *Iguanodon* are based on the Bernissart material. Seitz (1907) meticulously described (1907, 325–330) and figured (1907, pl. 10, fig 58; pl. 11, figs. 59–61) the bone structures of *Iguanodon* from a left femur. To summarize his findings in modern terms, he observed (1907, fig. 58) a primary bone cortex formed by a fibrolamellar complex dominated by longitudinal primary osteons, with evidence of growth cycles and of rather discrete, scattered reconstruction by secondary osteons of larger diameters than the primary osteons. At higher magnification (1907, fig. 59) the bone tissue shows a good structural preservation, with a precise morphology of the bone cells lacunae and of their canaliculi. The primary bone tissue is permeated by numerous longitudinal primary osteons and shows evidence of lines of arrested growth, and perhaps also of Sharpey fibers locally. The larger, well-finished secondary osteons interrupt the primary structures and clearly show the reversion line at their periphery. Another region (1907, fig. 60) experienced a more intensive process of bone substitution, as evidenced by the higher density of secondary osteons and their partial superposition. A very peculiar image at high magnification (1907, fig. 61) is provided by Seitz, showing a vascular canal cut longitudinally and filled up by what is tentatively interpreted as mass of blood cells (*blutkörperchen*). His discussion of this observation (1907, 329–330) interestingly predates the current findings and discussions (e.g., Martill and Unwin, 1997; Schweitzer and Horner, 1999; Schweitzer et al., 2005) on pyrite framboids versus original organic remains in fossil bones.

Nopcsa and Heidsieck (1933) and Gross (1934) both used *Iguanodon* in their paleohistological studies. The first one mostly dealt with the histological differences likely to be observed following the ontogeny of ornithopods, suggesting that several recognized ornithopod taxa were merely ontogenetic growth stages, a situation further analyzed by more recent researches (Chinsamy, 1995; Horner et al., 2000, 2009; Knoll et al., 2010). Gross (1934) described dinosaur bone tissues following the then-recent understanding of bone fibrillar organization brought by, for example, Weidenreich (1930), noting important distinctions among types of primary bone tissues and dense Haversian bone among dinosaurs. Regarding *Iguanodon*, Gross referred to Seitz's material and described (1934, 759, fig. 21) dense Haversian tissue in polarized light, noticing that the lamellar organization of the secondary osteons is identical to the situation observed in mammals and that the two groups cannot be distinguished on this basis. Enlow and Brown's (1957, 203) descriptions of ornithischian dinosaur bone derived in part from Seitz (1907) and Gross (1934). Indeed, *Iguanodon* tissues in Seitz (1907, pl. 10, fig. 58, and pl. 11, fig. 60) appear in Enlow and Brown (1957, respectively pl. 23, fig. 4, and pl. 22, fig. 9). Among his important studies of dinosaur bone tissues, Reid (1985) described primary bone in an *Iguanodon* femur (1985, plate 1, figs. 5 and 6). The tissue appears as the familiar modulation of the fibrolamellar complex described as laminar, with the characteristic development of superimposed rows of circular vascular canals uniting the longitudinal primary osteons. However, as noted by Reid, the structural differences between the fibrous (woven) and lamellar (osteonal) components of the complex are almost indistinguishable under ordinary light (1985, fig. 5), but become obvious only under crossed Nicols (1985, fig. 6). Interestingly, this tissue closely resembles the one forming the thin sections of *Iguanodon* processed at the request of Professor Paul Gervais in about 1875 (see above). From another point of view, Reid (1984, 1997) also used *Iguanodon* to analyze and illustrate the growth dynamics in the length of the long bones in the epiphyses of dinosaurs.

More recently, focus has turned toward the detection of proteins and other organic components in *Iguanodon* bones. Embery et al. (2000) reported extraction of noncollagenous proteins from a rib cortex with a dense Haversian structure. The bone comes from the collections of the British Natural History Museum in London, and hence its origin is presumably from the United Kingdom and not from Bernissart. Later works on the same material (Embery et al., 2003) provided evidence for the partial preservation of biomolecules from both the compact and the cancellous bone tissues. A glycoprotein akin to osteocalcin, phosphoproteins, and mucopolysaccharides was recovered, all fractions of the extracellular bone matrices whose high interactions with the phosphatic mineral phases may be linked to their amazing preservation potential.

The tooth histology of *Iguanodon* and other advanced ornithopods is peculiar and diagnostic. As observed in thin sections in *Iguanodon*, *Rhabdodon*, and various more advanced ornithopods including *Bactrosaurus* and neonate, juvenile, and adult *Maiasaura* (pers. obs.), the enamel has a characteristic structure, autapomorphic for the group. The highly anisotropic enamel shows zigzag structures, from the enamel–dentine junction to almost its outer free surface (Fig. 12.1D,E). The zigzags are superposed in an ordered way, causing the subjective appearance of vertical pillars set side by side. Analysis by scanning electron microscopy allows us to interpret this enamel as a "coarse wavy enamel" for most of the enamel thickness and as a "fine wavy enamel" forming a very thin layer at the surface (Sander, 1999, pl. 15, figs. 1–3), according to this Sander's terminology. The appearance under crossed Nicols probably derives from what Sander describes as the "staggered" or "whorled" arrangement of the enamel crystallites.

The dentine is a thick, regular orthodentine, showing very numerous faint growth cycles parallel to the enamel–dentine junction. The dentine canaliculi are long and gently curling toward the pulp cavity (Fig. 12.1D,E). Irregularities in the shape of the enamel–dentine junction and/or local differences in dentine centripetal growth create a complex situation at the level of the front where the dentine reaches the pulp cavity. There, the dentine does not completely fill up various extensions of the pulp cavity. In this way, many elongate canals are formed, and the dentine can be described as a vasodentine in the circumpulpar region.

| Discussion | The bone tissues observed at level −309 m in the Bernissart borehole closely resemble previous descriptions of adult dinosaurian bone tissues (dense Haversian bone) in general, and particularly of *Iguanodon*. Direct comparison with a small Bernissart *Iguanodon* rib (RBINS, unregistered specimen "H," diameter 12×20 mm; Fig. 2F) even suggests that the material collected at −309 m might pertain to an early adult animal because of the low number of superposed generations of secondary osteons, leaving some primary tissues between them; and the incomplete Haversian replacement in the superficial region, leaving a region of primary bone of periosteal origin (Fig. 12.2A–D). The occurrence of LAGs in this primary bone (Fig. 12.2C) and its poor vascularity suggest that this individual was close to the adult condition, although a clear external fundamental system was not observed. What can be observed in bone fragments from level −296.5 m concurs with the ones from −309 m. In both cases, dense Haversian tissues are observed, with a moderate amount of substitution cycles among the secondary osteons (Fig. 12.2E). Lack of radial cracks at the periphery of the secondary osteons is not characteristic for an early aquatic taphonomic episode (Pfretzschner, 2000). |

The primary periosteal bone does not show all the tissue variability already observed in *Iguanodon*. The primary bone tissue appears moderately vascularized by longitudinally oriented primary osteons, with a grossly pseudolamellar organization and some evidence of LAGs (Fig. 12.2C). There is no evidence of the laminar pattern of the fibrolamellar complex, as described by, for example, Reid (1985), suggesting active growth (perhaps among grossly immature individuals), nor of an external fundamental system suggesting a mature adult condition with almost no further growth (e.g., Horner et al., 2009). Instead, a moderately active radial growth with some cyclicity seems indicated, again suggesting an almost mature or subadult condition.

The teeth observed at level −296.5 m (Fig. 12.1A–C) clearly differ in size and structure from the ones of *Iguanodon* and other Iguanodontia (e.g., *Bactrosaurus johnsoni*, Fig. 12.1D,E). Their small size, slenderness, and thin enamel do not fit with crushing functions. Histological comparisons with teeth of the Bernissart actinopterygians (15 species), urodeles (*Hylaeobatrachus*), and crocodiles (*Goniopholis*, *Bernissartia*) have not been attempted. The very small size probably exclude a crocodilian origin (apart from tiny neonates or juveniles); the lack of a bicuspidate apex and a pedicellate structure does not support a lissamphibian origin, and both size and statistics would favor an actinopterygian origin, although no teeth structures peculiar to them (acrodine, etc.) could be observed. Unfortunately, the lack of information on the mode of ankylosis of the teeth and on the occurrence of a root precludes further diagnosis. There is also no clear evidence of actual (anatomical) relationships with the surrounding bone fragments.

The tooth structure of *Iguanodon* does not show significant histological differences with those of *Bactrosaurus* (Fig. 12.1D,E). In both cases, the thick enamel is highly structured, and as explained by Sander (1999, 74), this enamel structure may indeed be regarded as a diagnostic feature or autapomorphy for advanced ornithopods (iguanodontids and hadrosaurids).

Comparisons between fresh (from the borehole) and old (from the RBINS) bony material are interesting because they clearly reflect at the histological level the physicochemical changes induced in the fossils by the oxidation of pyrite under standard museum conditions (Ricqlès and Yans, 2003; see also Chapter 11 in this book). RBINS unregistered specimen "H" (rib) was extracted during the 1878–1881 period and probably received preparation standard at that time (Godefroit, 2009) before thereafter being kept in a wood drawer under standard museum conditions of temperature and humidity. Histological observations compared to the fresh specimen show an advanced process of fragmentation of the museum material at both organ and tissue levels (Fig. 12.2E). The intimate tissue structure is hardly changed, if at all, but the ground color of the tissue has changed (with localized whitish spots), perhaps indicting chemical changes (see Chapter 11 in this book), and above all, the tissue shows multiple cracks that weaken its structure. They are well underlined by the embedding plastic medium; under gentle vacuum, the fluid plastic monomer percolated into the bone, following the multiple cracks, and ultimately permeated the bone before hardening by polymerization. Without this process, it would have been impossible to obtain thin sections from the material.

Preliminary chemical analyzes by x-ray diffraction show a spectacular decrease of the pyrite spikes in the museum-kept specimen, compared to the fresh specimens taken from the borehole. This is in agreement with the more detailed analytical results of Leduc (Chapter 11 in this book).

Concluding Remarks

This histological description of the fragments from BER 3 borehole definitely brings evidence of the occurrence of bone and tooth materials at the −296.5 m and −309 m levels. The tooth material definitely does not belong to *Iguanodon*; its most likely origin is from one of the 15 species from seven actinopterygian (bony fishes) orders described from Bernissart (Godefroit, 2009, 131). It is likely that histological comparisons with the teeth of the actinopterygian taxa known from Bernissart will allow precise determination.

What agrees with a dinosaurian origin for the bone fragments at the −309 m level are the general structure suggesting an origin from large- to very large-size bones; the prevalence of dense Haversian bone tissue, known to be common among large mature dinosaurs; and the occurrence of primary bone tissues with primary osteons also known from the external cortex of submature dinosaurian bones.

The bone tissue structures observed from the borehole match reasonably well with the ones already described for *Iguanodon*, but for all that, no diagnostic structures (as would have been the case for teeth) remain to compel us to ascribe the finds to this taxon. We are left only with the statistical likelihood argument to consider that we are dealing with *Iguanodon*—a tentative conclusion that is neither disproved or enforced by any available data.

Whatever it may be, the lucky occurrence of bone and tooth material at two superposed levels in a small-diameter drill again emphasize the fossil abundance and value of the Wealden sediments within the Iguanodon Sinkhole.

Acknowledgments

We thank T. Coradin (Chimie de la matière condensée, UMR 7574 CNRS-Université Paris 6) for his x-ray diffraction analyses and Louise Zylberberg (ISTEP/ UMR 7093 CNRS-Université Paris 6) for her interest and practical help in setting of the figures. J. R. Horner and K. Padian reviewed this chapter and made helpful comments.

References

Chinsamy, A. 1995. Ontogenetic changes in the bone histology of the late Jurassic ornithopod *Dryosaurus lettowvorbecki*. Journal of Vertebrate Paleontology 15: 96–104.

Embery, G., A. C. Milner, R. J. Waddington, R. C. Hall, M. S. Langley, and A. M. Milan. 2000. The isolation and detection of non-collagenous proteins from the compact bone of the dinosaur *Iguanodon*. Connective Tissue Research 41: 249–259.

Embery, G., A. C. Milner, R. J. Waddington, R. C. Hall, M. S. Langley, and A. M. Milan. 2003. Identification of proteinaceous material in the bone of the dinosaur *Iguanodon*. Connective Tissue Research 44 (supplement to 1): 41–46.

Enlow, D. H., and S. O. Brown. 1957. A comparative histological study of fossil and recent bone tissue. Part 2. Texas Journal of Science 9: 186–214.

Godefroit, P. 2009. 130 years ago: the discovery of the Bernissart iguanodons; pp. 129–135 in P. Godefroit and O. Lambert (eds.), Tribute to Charles Darwin and the Bernissart iguanodons: new perspectives on vertebrate evolution and Early Cretaceous ecosystems. Darwin-Bernissart Symposium, Brussels, February 9–13, 2009.

Gross, W. 1934. Die Typen des mikrokospischen Knochenbaues bei fossilen Stegocephalen und Reptilien. Zeitschrift für Anatomie 103: 731–764.

Horner, J. R., A. de Ricqlès, and K. Padian. 2000. The bone histology of the Hadrosaurid dinosaur *Maiasaura peeblesorum*: growth dynamics and physiology based on an ontogenetic series of skeletal elements. Journal of Vertebrate Paleontology 20: 109–123.

Horner J. R., A. de Ricqlès, K. Padian, and R. D. Scheetz. 2009. Comparative long bone histology and growth of the "Hypsilophodontid" dinosaurs *Orodromeus makelai*, *Dryosaurus altus*, and *Tenontosaurus tilletti* (Ornithischia: Euornithopoda). Journal of Vertebrate Paleontology 29 (3): 734–747.

Knoll, F., K. Padian, and A. de Ricqlès. 2010. Ontogenetic change and adult body size of the early ornithischian dinosaur *Lesothosaurus diagnosticus*: implications for basal ornithischian taxonomy. Gondwana Research 17: 171–179.

Martill, D. M., and D. M. Unwin. 1997. Small spheres in fossil bones: blood corpuscles or diagenetic products? Paleontology 40: 619–624.

Nopcsa, F. von, and E. Heidsieck. 1933. On the bone histology of the ribs of immature and half-grown trachodont dinosaurs. Proceedings of the Royal Zoological Society 1933: 221–226.

Norman, D. B. 1986. On the anatomy of *Iguanodon atherfieldensis* (Ornithischia: Ornithopoda). Bulletin de l'Institut Royal des Sciences Naturelles de Belgique Sciences de la Terre 56: 281–372.

Pfretzschner, H.-U. 2000. Microcracks and fossilization of Haversian bone. Neues Jahrbuch für Geologie und Paläontologie, Abhandlungen 216: 413–432.

Queckett, J. 1855. Descriptive and illustrated catalogue of the histological series contained in the Museum of the Royal College of Surgeons of England, Vol. 2. London.

Reid, R. E. H. 1984. The histology of dinosaurian bone, and its possible bearing on dinosaurian physiology; pp. 629–663 in M. W. J. Fergusson (ed.), The structure, development and evolution of reptiles. Academic Press, Orlando, Fla.

———. 1985. On supposed haversian bone from the Hadrosaur *Anatosaurus*, and the nature of compact bone in dinosaurs. Journal of Paleontology 59: 140–148.

———. 1997. How dinosaurs grew; pp. 403–413 in J. O. Farlow and M. K. Brett-Surman (eds.), The complete dinosaur. Indiana University Press, Bloomington.

Ricqlès, A. de, and J. Yans. 2003. Bernissart's *Iguanodon*: the case for "fresh" versus "old" dinosaur bone. Journal of Paleontology 23 (supplement to 3): 45A.

Ricqlès, A. de, P. Godefroit, and J. Yans J. 2009a. Vertebrate remains in the 2003 Bernissart drill: histological assessement; p. 33 in P. Godefroit and O. Lambert (eds.), Tribute to Charles Darwin and the Bernissart iguanodons: new perspectives on vertebrate evolution and Early Cretaceous ecosystems. Darwin-Bernissart Symposium, Brussels, February 9–13, 2009.

Ricqlès, A. de, P. Taquet, and V. de Buffrenil. 2009b. "Rediscovery" of Paul Gervais' paleohistological collection. Geodiversitas 31: 943–971.

Sander, P. M. 1999. The microstructure of reptilian tooth enamel: terminology, function, and phylogeny. Münchner Geowissenschaftliche Abhandlungen A 38: 1–102.

Schweitzer, M. H., and J. R. Horner. 1999. Intravascular microstructures in the trabecular bone tissues of *Tyrannosaurus rex*. Annales de Paléontologie 85: 179–192.

Schweitzer, M. H., J. F. Wittmeyer, J. R. Horner, and J. K. Toporski. 2005. Soft tissue vessels and cellular preservation in *Tyrannosaurus rex*. Science 307: 1952–1955.

Seitz, A. L. 1907. Vergleichenden Studien über den mikroskopischen Knochenbau fossiler und rezenter Reptilien. Nova Acta Leopoldina: Abhandlungen der Kaiserlich Leopoldinisch-Carolinisch Deutschen Akademie der Naturforscher 37: 230–370.

Weidenreich, F. 1930. Das Knochengewebe; pp. 391–520 in V. Mollendorff (ed.), Handbuch der mikroskopischen Anatomie des Menschen, Vol. 2. Springer Verlag, Berlin.

Wilson, J. W. 1994. Histological techniques, pp. 205–234 in P. Leiggi and P. May (eds.), Vertebrate paleontological techniques, Vol. 1. Cambridge University Press, Cambridge.

13.1. Location of the "Bois de Baudour" site (modified from Spagna, 2010).

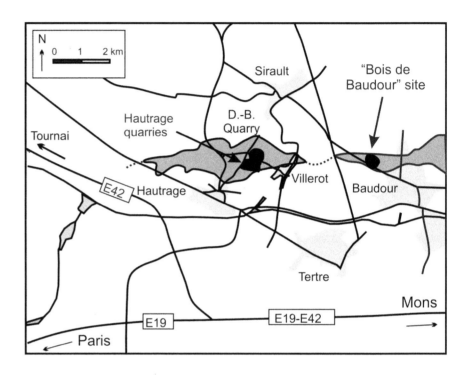

Early Cretaceous Dinosaur Remains from Baudour (Belgium)

13

Pascal Godefroit*, Jean Le Loeuff, Patrick Carlier, Stéphane Pirson, Johan Yans, Suravech Suteethorn, and Paul Spagna

We describe two dinosaur bones found in the Bois de Baudour clay quarries (Mons Basin, Belgium) of the Baudour Clays Formation (middle Barremian to earliest Aptian) during their exploitation period. Apart from the numerous skeletons found in the Sainte-Barbe pit at Bernissart, these are the only dinosaur fossils discovered in Wealden deposits in the Mons Basin. The first bone is a left coracoid that can confidently be attributed to the ornithopod *Iguanodon bernissartensis*. The second bone is a left tibia belonging to an indeterminate sauropod. This is the first sauropod bone from Cretaceous deposits in Belgium. Recent drillings in the Baudour Clays Formation at Bois de Baudour suggest that the Baudour and Hautrage areas were probably parts of the same floodplain environment. A bone fragment, probably a fragment of a vertebral centrum, was found in the drilling core about 14 m below the ground surface. Paleoecological conditions in the Baudour and Hautrage formations were apparently not favorable at all for the preservation of in situ complete skeletons, like those from the Bernissart Sinkhole. However, drilling through a bone does suggest a high concentration of fossils.

Introduction

Although vertebrate fossils are extraordinarily abundant in the Sainte-Barbe Clays Formation from the Iguanodon Sinkhole at Bernissart, they have never been described from other Wealden deposits in the Mons Basin, although clay quarries were—and are still—intensively exploited in this area from the nineteenth century.

Here we describe two dinosaur bones discovered at Bois de Baudour (Fig. 13.1). Small clay quarries were actively exploited in this area during the twentieth century for their aluminium-rich clay content, mainly intended for the manufacture of bricks, ceramics, and refractories (Fig. 13.2). Nowadays the exploitation is totally abandoned, and the site, covered by forest and lakes that mark the flooded former clay pits, is protected by regional authorities as a zone of biological interest.

In 1952, a worker discovered a fossil bone about 10 m below the ground surface in the Degand-Dutalys clay quarry in Bois de Baudour. Victor Degand, the owner of the quarry, was aware of the importance of the paleontological discoveries at Bernissart and kept this precious bone in a shed of his house at Sirault, believing that it belonged to an *Iguanodon*. After his death in 1983, the bone was kept at Jemeppes-sur-Sambre in the garage of his daughter, Anne-Marie Dufrane-Degand, where it was used for blocking

13.2. Exploitation of the Baudour Clays Formation in the Dutalys quarries at "Bois-de-Baudour" (Belgium). From Casier (1978).

a window. From 1995 to 1999, Patrick Carlier, the son-in-law of Anne-Marie Dufrane-Degand, exhibited the fossil in the local museum of Sainte-Croix and Notre-Dame College in Hannut. Since 1999, this specimen is housed in the Paleontological collections of the Royal Belgian Institute of Natural Sciences (RBINS), under the catalog number RBINS R267.

A second dinosaur bone (RBINS R268), with the label "Wealdien de Baudour (Hainaut)," was found in the Coupatez-Wouters collections, also housed in the paleontological department of the RBINS. Unfortunately, the circumstances of the discovery of this fossil remain unknown. However, the preservation and patina of this bone are nearly identical to those of RBINS R267, so it is likely that this fossil also comes from a clay quarry at Bois de Baudour.

Geological Context

Geological information from the active exploitation period of the clay quarries at Bois de Baudour is limited. Some indications in the publications of Cornet (1927) and Casier (1978) simply confirm the presence of black, gray, white, and reddish clays (this latter Fe-rich clay was named *bolus* by the exploiters) in the exploited raw materials.

In their recent stratigraphic synthesis, Robaszynski et al. (2001) define the Baudour Clays Formation, formerly outcropping in the Bois de Baudour quarries, as one of the three Wealden formations in the western part of the Mons Basin, the two other being the Hautrage Clays Formation (Hautrage) and the Sainte-Barbe Clays Formation (Bernissart). These formations have been recently dated by matching palynological and geochemical

Unit	Depth (m)	Lithological description
1	0 to 7.4	Packing
2	7.4 to 10	Dark brown clay, becoming lighter and silty at the bottom
3	10 to 11.7	Whitish fine sands and thin brown clayey layers alternation
4	11.7 to 13.5	Light brown and dark gray variegated clays, with pyritized wood fragments; indurated centimetric level at the bottom
5	13.5 to 14.7	Dark gray sandy clays, rich in lignite and wood fragments, with one bone fragment (Fig. 13.4)
6	14.7 to 19.8	Light gray silty clay
7	19.8 to 21	Light gray clay rich in millimetric nodules of siderite
8	21 to 23.5	Brownish clay to light gray silty clay

Table 13.1. Lithological content of the BAUD1 borehole

Note: From Spagna (2010).

studies as middle Barremian to earliest Aptian (Dejax et al., 2007a, 2007b, 2008), the Sainte-Barbe Clays Formation being most probably younger than the Hautrage Clays Formation (Schnyder et al., 2009; see Chapter 8 in this book).

In 2003, the Faculté Polytechnique de Mons (Belgium) and the Bureau de Recherches Géologiques et Minières (France) dug an auger drilling (called BAUD1) at Bois de Baudour. More than 20 m of the Baudour Clays Formation succession were crossed and studied on this occasion (Table 13.1 and Fig. 13.3).

The few lithological data collected from the Baudour Clays Formation series were compared with the more extensive observations on the Hautrage Clay Formation (see Chapter 9 in this book) and suggest that the Baudour area was part of the same floodplain environment. Should a lithological continuity between both sites be attested, the BAUD1 sediments could probably be compared with those at the transition between units A and B or those from the units C to E interval at Hautrage.

A fragmentary bone was found in the drilling core, about 14 m below the ground's surface (Fig. 13.4). The important expansion of cancellous bone and the thin cortical bone suggest that this fragment belongs to the vertebral centrum of a large vertebrate. The sandy composition of the sediments around this bone seems typical for a channel environment deposit, and it is therefore likely that this fossil was transported. The chances of having touched an in situ complete skeleton during this drilling are consequently quite small. However, this discovery might reveal the presence of a bonebed accumulation in the Baudour Clays Formation.

Iguanodon bernissartensis Boulenger in Van Beneden, 1881

Dinosaur Remains from Baudour

RBINS R268 is a massive and dish-shaped element. On the dorsal edge of the coracoid, the suture with the scapula is straight, thick, and rugose. The coracoid foramen forms a notch positioned on the caudal portion of the scapulocoracoid suture in lateral view. The lower half of the glenoid forms a wide and cup-shaped depression at the caudal end of the scapulocoracoid suture. Beneath the glenoid, the caudal edge of the coracoid is embayed, and at the junction with the cranial edge, it forms a blunt hooklike process. The lateral side of the coracoid is marked by a large muscle scar for Musculus costo-coracoideus along the caudoventral embayment. The cranial

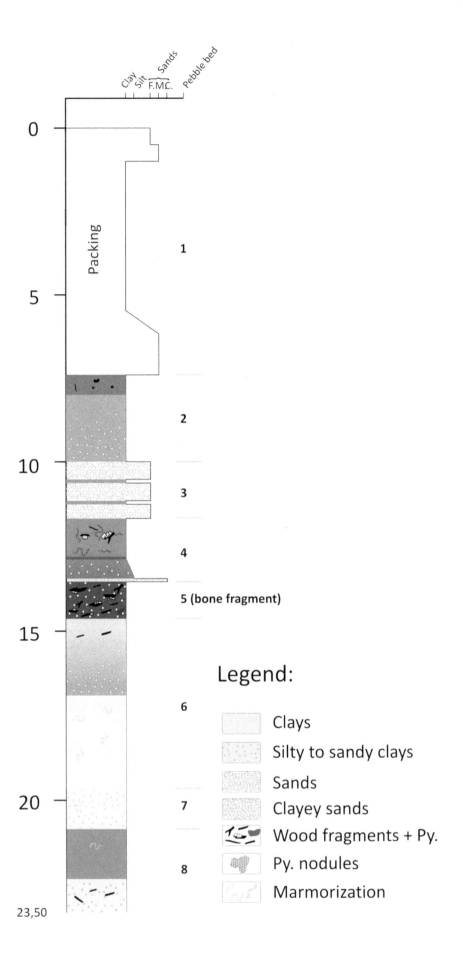

13.3. BAUD1 lithological log (modified from Spagna, 2010).

edge of the coracoid is convex and rugose. At midheight, it is notably thickened by a raised ridge on the lateral surface of the bone, which probably articulated against an ossification in the sternal cartilage (Norman, 1980).

RBINS R268 is identical in size and morphology to the left coracoid of the ornithopod *Iguanodon bernissartensis* Boulenger *in* Van Beneden, 1881 (Fig. 13.5). Because *I. bernissartensis* is particularly abundant in the neighboring and contemporaneous Iguanodon Sinkhole at Bernissart (24 more or less complete skeletons, as well as several partly preserved individuals), it can confidently be referred to this taxon.

Sauropoda indet.

RBINS R267 is a left tibia with incompletely preserved proximal and distal ends (Fig. 13.6). It is a straight bone with a cranially well-developed cnemial crest (C/D = 0.9: cf. Royo Torres, 2009). The bone is rather slender with a tibial robustness index of 0.21 and a distal tibia robustness index of 0.14. In lateral view, the cnemial crest is located in the upper third of the bone. It is triangular in shape, but its proximal end is not preserved. In cranial view, a sharp crest runs distally from the basis of the cnemial crest and stops some centimeters before the distal end. Distally, the craniolateral process for the articulation of the fibula is preserved, but the caudolateral process is broken off. Thus, the caudal part of the distal extremity is missing. However, the articular surface for the ascending process of the astragalus can be observed.

In Europe, Barremian–Aptian sauropods are mostly known from fragmentary material, with the exception of *Tastavinsaurus sanzi*, a sauropod from the Early Aptian of Spain recently referred by Royo Torres (2009) to the new clade Laurasiformes. Laurasiformes include *Aragosaurus* (Aptian of Spain) as well as the Early Cretaceous American genera *Cedarosaurus* and *Venenosaurus*. Other sauropods from the same time interval in Spain include a rebbachisaurid (Pereda-Suberbiola et al., 2003) and Titanosauriformes. The Wessex Formation of England (Barremian) has also yielded an abundant sauropod assemblage with various Titanosauriformes, putative diplodocids and camarasaurids as well as rebbachisaurids (see Mannion, 2009). Although the general aspect of the tibia is reminiscent of some Titanosauriformes more than of diplodocids and camarasaurids, the available material from Baudour is too fragmentary to warrant a precise identification, and we regard it as an indeterminate Sauropoda.

13.4. Bone fragment found during the drilling in Bois de Baudour.

Conclusions

Two dinosaur bones were unearthed during the exploitation period of the Bois de Baudour clay quarries. The first one is a left coracoid that undoubtedly belongs to the ornithopod *Iguanodon bernissartensis*, known by more than 24 complete skeletons in the neighboring and contemporaneous Iguanodon Sinkhole at Bernissart. The second one is a left tibia that belongs to an indeterminate sauropod. This is the first sauropod specimen from Cretaceous deposits in Belgium, and the second sauropod fossil from Belgium (one isolated tooth was described from the Upper Triassic of Habay-la-Vieille; Godefroit and Knoll, 2003). A fragment of bone, probably the vertebral centrum of a large vertebrate, was found about 14 m below

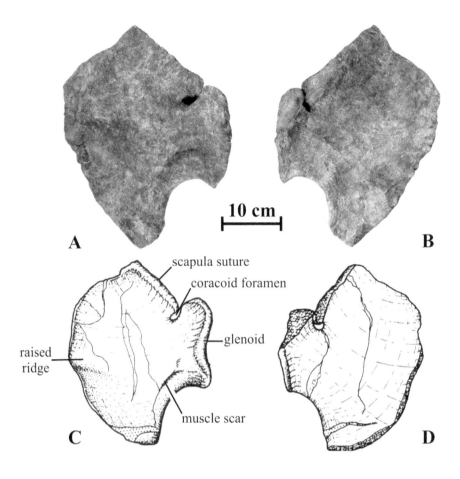

13.5. Left coracoids of *Iguanodon bernissartensis* in lateral (A, C) and medial (B, D) views. A, B, Photographs of RBINS R268, from the Wealden of Bois de Baudour. C, D, Interpretative drawings of RBINS R51 (holotype of *I. bernissartensis*) from the Iguanodon Sinkhole at Bernissart (modified after Norman, 1980, fig. 53).

the ground surface on the occasion of a recent drilling campaign. During Wealden times, the Baudour and Hautrage areas were probably parts of the same floodplain environment. These paleoecological conditions were apparently not at all favorable for the preservation of complete in situ skeletons, as in the Iguanodon Sinkhole. Indeed, vertebrate fossils are extremely rare in the Baudour and Hautrage clay quarries, although they were intensively exploited during the twentieth century and the quarrymen were aware of the importance of the paleontological discoveries at Bernissart.

Acknowledgments

The authors are grateful to E. Buffetaut for reviewing this chapter and T. Hubin for the photographs of the specimens.

References

Casier, E. 1978. Les Iguanodons de Bernissart. L'Institut Royal des Sciences Naturelles de Belgique, Brussels, 166 pp.

Cornet, J. 1927. L'époque wealdienne dans le Hainaut. Annales de la Société géologiques de Belgique 50: 89–103.

Dejax, J., E. Dumax, F. Damblon, and J. Yans. 2007a. Palynology of Baudour Clays Formation (Mons Basin, Belgium): correlation within the "stratotypic" Wealden; pp. 16–28 in P. Steemans and E. Javaux (eds.), Recent advances in palynology. Notebooks on Geology Memoir, Abstract 3.

Dejax, J., D. Pons, and J. Yans. 2007b. Palynology of the dinosaur-bearing Wealden facies in the natural pit of Bernissart (Belgium). Review of Palaeobotany and Palynology 144: 25–38.

Dejax, J., D. Pons, and J. Yans. 2008. Palynology of the Wealden facies from Hautrage quarry (Mons Basin, Belgium). Memoirs of the Geological Suvey of Belgium 55: 45–51.

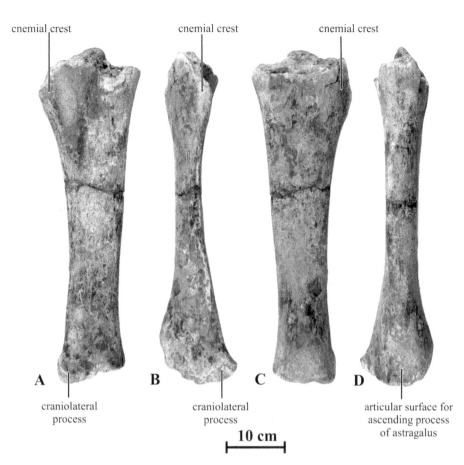

cnemial crest   cnemial crest   cnemial crest

A   B   C   D

craniolateral process   craniolateral process   articular surface for ascending process of astragalus

10 cm

13.6. Left tibia of Sauropoda indet. from the Wealden of Bois de Baudour in lateral (A), cranial (B), medial (C), and caudal (D) views.

Godefroit, P., and F. Knoll. 2003. Late Triassic dinosaur teeth from southern Belgium. Comptes Rendus Palévol 2: 3–11.

Mannion, P. D. 2009. A rebbachisaurid sauropod from the Lower Cretaceous of the Isle of Wight, England. Cretaceous Research 30: 521–526.

Norman, D. B. 1980. On the ornithischian dinosaur *Iguanodon bernissartensis* of Bernissart (Belgium). Mémoires de l'Institut royal des Sciences naturelles de Belgique 178: 1–103.

Pereda-Suberbiola, X., F. Torcida, L. A. Izquierdo, P. Huerta, D. Montero, and G. Perez. 2003. First rebbachisaurid dinosaur (Sauropoda, Diplodocoidea) from the Early Cretaceous of Spain: palaeobiogeographical implications. Bulletin de la Société Géologique de France 174: 471–479.

Robaszynski, F., A. Dhondt, and J. W. M. Jagt. 2001. Cretaceous lithostratigraphic units (Belgium); pp. 121–134 in P. Bultynck P. and L. Dejonghe (eds.), Guide to a revised lithostratigraphic scale of Belgium. Geologica Belgica 4.

Royo Torres, R. 2009. El sauropodo de Penarroya de Tastavins. Monografias Turolenses 6: 1–548.

Schnyder, J., J. Dejax, E. Keppens, T. T. Nguyen Tu, P. Spagna, A. Riboulleau, and J. Yans. 2009. An Early Cretaceous lacustrine record: organic matter and organic carbon isotopes at Bernissart (Mons Basin, Belgium). Palaeogeography, Palaeoclimatology, Palaeoecology 281: 79–91.

Spagna, P. 2010. Les faciès wealdiens du Bassin de Mons (Belgique): paléoenvironnements, géodynamique et valorisation industrielle. Ph.D. thesis, Faculté Polytechnique de l'Umons, Mons, 138 pp.

14.1. Plan view of the driftstone galleries that penetrated the Iguanodon Sinkhole (after Cornet and Schmitz, 1898). Inferred locations for the BER 2 and BER 3 boreholes are shown.

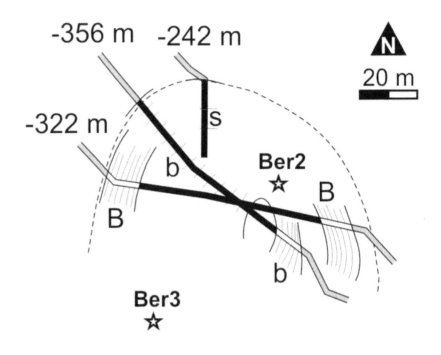

Geological Model and Cyclic Mass Mortality Scenarios for the Lower Cretaceous Bernissart *Iguanodon* Bonebeds

14

**Jean-Marc Baele*, Pascal Godefroit,
Paul Spagna, and Christian Dupuis**

The Iguanodon Sinkhole at Bernissart (Belgium) is an exceptional fossil deposit as a result of the quantity and preservation quality of Cretaceous basal Iguanodontia found by coal mine workers in 1878. Efforts to unravel the processes that caused the accumulation and preservation of many dinosaurs, along with other taxa, are here based on a new geological model that relies on several discrete, continuous bonebeds. Several taphonomic scenarios are proposed and discussed within the specific geological and environmental specificities of the so-called Lower Cretaceous Bernissart paleolake. On the basis of sedimentological and taphonomic evidence, attrition and obrution processes appear less likely than mass death by drowning and/or intoxication. Contamination of the aquatic environment by sulfate-rich brines related to deep solution–collapse processes could support the hypothesis of intoxication by H_2S or biological toxins as a direct or indirect lethal agent in a context of seasonally shrinking water.

Introduction

The discovery, from 1878 until 1881, of about 40 skeletons of iguanodontid dinosaurs in Lower Cretaceous ("Wealden") deposits at Bernissart (Belgium) is exceptional in the history of paleontology, and immediately it intrigued geologists. Indeed, 117 similar sinkholes are known within Mississippian deposits of the Mons Basin, and only one has yielded dinosaur remains. Outside the Iguanodon Sinkhole, only two dinosaur bones have been identified to date in Wealden outcrops from the Mons Basin (see Chapter 13 in this book). Since 1878, geologists have tried to explain the processes leading to the formation of this unique accumulation. On the basis of geological sections of the site, Dupont (1878, 1897) hypothesized that the Bernissart environment back to Lower Cretaceous times was a narrow gorge (*crevasse*) in which iguanodontids lived, died, and were periodically buried during flooding episodes.Soon, Cornet and Schmitz (1898) and Cornet (1927) proposed an alternative explanation. They believed that the accumulation of numerous iguanodontids skeleton at Bernissart was clearly a slow, attritional process, resulting from the sliding or stacking of carcasses of dead animals in a subsiding lake. Because the fossiliferous layers were trapped within a sinkhole, they could escape the erosion that removed the coeval surrounding layers in other places in the Mons Basin (Bultynck, 1989).

Louis Dollo, the original describer of all the terrestrial vertebrates from Bernissart in articles published between 1882 and 1923, proposed several conflicting hypotheses to explain the accumulation of iguanodontid skeletons at Bernissart (see reviews in Casier, 1960, and Bultynck, 1989). After noting that most of the dinosaurs were old-age individuals, Dollo proposed that Bernissart was some kind of dinosaur graveyard by Early Cretaceous times, or that flash floods selectively killed older and less agile animals. He also suggested that some iguanodontid specimens from Bernissart showed evidence of a violent death, maybe through combat.

Casier (1960) provided two further explanations for the mass burial at Bernissart. On the basis of the supposition that iguanodontids usually retreated into water to escape from predators or other startling events, he first supposed that some dinosaurs might have inadvertently slipped into the steep-sided marshy depression at Bernissart. In a second hypothesis (the "*Hippopotamus*" hypothesis), Casier assumed that iguanodontids were amphibious and therefore dependent on a permanent body of water; a period of low rainfall may have led to these animals becoming mired in the muddy ooze around shrinking water holes.

More recently, Norman (1987) compared Bernissart with another Early Cretaceous iguanodontid bonebed in Nehden, Germany, and refuted a mass kill scenarios on the basis of a more detailed taphonomic analysis. Bultynck (1989) also spoke for an attritional scenario, agreeing with the sinkhole environment hypothesis previously developed by Cornet and Schmitz (1898) and Cornet (1927).

Here we present a refined geological model for the Bernissart *Iguanodon* deposit. This model is then used as a framework for evaluating different taphonomic scenarios. We placed emphasis on the role of site-specific geological factors such as subsidence due to solution collapse deep underground and possible upwelling of sulfate-rich brines.

Institutional abbreviation. RBINS, Royal Belgian Institute of Natural Sciences, Brussels, Belgium.

Inside the Iguanodon Grave: Geological Model

Historical Discovery and Fossil Assemblage

The iguanodon bonebeds were discovered in April 1878 by coal miners excavating a driftstone gallery at level −322 m approximately 250 m south–southeast from the Sainte-Barbe shaft (see Chapter 1 in this book for details on the discovery). Earlier in March, the gallery went through 10 m of faulted and brecciated Pennsylvanian rocks with minor streaks of white sand, pyrite veinlets, and lignite fragments. Then it encountered 60 m of well-stratified Barremian lignitic clay and reentered a zone of heavily fractured rocks (Figs. 14.1 and 14.2). Most fossils were found in the first 14 m of clay, where 14 more or less complete and four partial skeletons were excavated (first series in Fig. 14.2). Eight other iguanodons were found in the second half of the gallery (second series) and three more in a second, deeper gallery (−356 m) that cut through 8 m of Barremian clay (third series). In this latter gallery, a small exploration pit reached the bottom of the clay infill 3–4 m below the ground. A third gallery entered the sinkhole at −242 m and breached a sandy aquifer beyond a 9-m-thick wall of Barremian clay. Exploration then continued by a horizontal borehole that showed 5–6 m of this sand and was aborted after having reentered 8 m of clay formation. These are the shallowest Barremian sediments ever recognized in the Iguanodon Sinkhole, and there were no fossils in it (as is seemingly the case for the 26 m of Barremian clay observed in the Sainte-Barbe shaft).

The recovered fossil assemblage includes freshwater animals that lived in the lake, and terrestrial animals that lived around the lake. The freshwater component mainly consists of more than 3,000 fishes belonging to 15 different taxa. Amiiformes are particularly diversified, suggesting swampy and poorly oxygenated waters. One amphibian, six turtles, and four crocodilian specimens complete the freshwater fauna. The terrestrial fauna is largely dominated by iguanodontid dinosaurs: at least 43 specimens were unearthed from 1878 until 1881, including 25 complete to moderately complete (>60% of the skeletal elements) individuals (see Norman, 1986, for a detailed catalog of the iguanodontid specimens in the collections of the Royal Belgian Institute of Natural Sciences) and 18 partial skeletons or fragmentary material. At least 33 specimens are referred to as *Iguanodon bernissartensis*; six other incomplete individuals probably also belong to this taxon (Norman, 1986). The smaller iguanodontid *Mantellisaurus atherfieldensis* is only represented by one complete specimen (skeleton labeled "T" in Fig. 14.2) and probably by one incomplete skeleton. A small tooth, caudal vertebrae, and ossified tendons collected from the coal tips possibly represent a third *M. atherfieldensis* specimen (Norman, 1986). Many aspects of their anatomy indicate that *I. bernissartensis* and *M. atherfieldensis* were highly active on land. *I. bernissartensis* probably spent most of its time in a quadrupedal posture, whereas *M. atherfieldensis* spent considerable periods of time walking or running bipedally (Norman, 1980, 1986). Noniguanodontid dinosaurs are represented by only one theropod phalanx. One hemipteran wing completes the terrestrial fauna from Bernissart. Numerous coprolites belonging to carnivorous reptiles were discovered in Bernissart pit (Bertrand, 1903), but their producer (perhaps a crocodile or theropod) remains unknown.

Numerous fossil plants were also collected, with abundant remains of the fern *Weichselia* (Seward, 1900), which also thrived in swamps.

Stratigraphy

Cornet and Schmitz (1898) published the only detailed geological section of the Iguanodon Sinkhole, according to the original drawings made by the mining engineer Sohier, who was assisted by De Pauw and Sonnet in his work (Fig. 14.2). Unfortunately, data for the first 14 m, which yielded the greatest number of fossils, were lacking as a result of missing documents. Although Arnould made a general description of this missing zone and collected a few structural measurements, Cornet and Schmitz (1898) preferred not to use this information to fill the gap in their original figure. However, De Pauw (1898) published a geological section from his own observations, which is consistent with the description from Arnould (in Cornet and Schmitz, 1898). De Pauw (1898) wrote that he reported the small-scale details of folds and fractures, but only the general structure was reproduced in his article, and each clay bed was measured (location and thickness).

The clay sediments in the Iguanodon Sinkhole are well stratified, as was subsequently confirmed by the 2003 coring program (Yans et al., 2005). Both flat and lenticular layerings were identified in the gallery section. Sand and lignite are the major sediments forming lenses, and several lignite beds up to 8 cm thick were observed. This may reflect a more variable (and perhaps shallower) depositional environment than is commonly suggested by varvelike stratification. Hard coal fragments derived from the Pennsylvanian rocks occur along the banks of the lake and are also found concentrated in discrete strata. Continuous millimeter-thick sandy layers clearly separate the clay in decimeter-thick beds (see De Pauw, 1898, fig. 2). Inside each bed, small-scale lamination (a millimeter thick or less) is clearly visible, although the detailed description of the BER 3 and BER 2 cores showed several nonlaminated intervals (Spagna and Van Itterbeek, 2006). Bone fragments were found in these nonlaminated intervals (see Chapter 12 in this book); therefore, they could be significant from a taphonomic point of view, such as homogeneization of the clay material might be due to bioturbation or trampling. However, it is not possible to differentiate between accidental drilling-induced homogenization and the actual absence of original lamination in the clay. These nonlaminated intervals were not described in the nineteenth-century galleries, but whether have they been overlooked or are actually missing remains a pending question.

Sedimentological structures include small-scale faulting, slumping, and erosional surfaces. Small synsedimentary faults are particularly abundant and often show pinching-out structures (*boudinage*) suggestive of soft sediment deformation due to dewatering or degassing.

Structure

The clay beds bounding the sinkhole are steeply (60–70 degrees) inclined (Fig. 14.2). Dip angles rapidly decrease toward the center of the collapse structure, with the strata lying almost horizontally within about 10 m from the walls. This overall flat-bottomed basin structure induced by sinkhole subsidence is slightly more complex as a small anticline appears at approximately 40 m from the western wall. In addition, deformation in the Barremian clay did not proceed merely in folding, but rather by disruption and displacement of a multitude of clay blocks in which parallel stratification is preserved. This is clearly evidenced by plotting the strata dip angle as a function of depth in the BER 3 borehole (Spagna, 2010). The blocky structure of the clay is more pronounced in the eastern part of the sinkhole filling (Fig. 14.2), where subsidence-induced deformation and downdrop amplitude are highest. The structure of the Barremian clay in the iguanodon sinkhole may therefore be idealized as a north–south elongated, asymmetrical funnel, as suggested by the geological model in Fig. 14.3.

Bonebeds

One critical observation made by De Pauw (1898) is that fossils in the first 14 m of Barremian clay, with the exception of plant remains, were found in discrete 35- to 55-cm-thick layers that are clustered within the first 14 m of Barremian clay (Fig. 14.2). The section of Cornet and Schmitz (1898) is

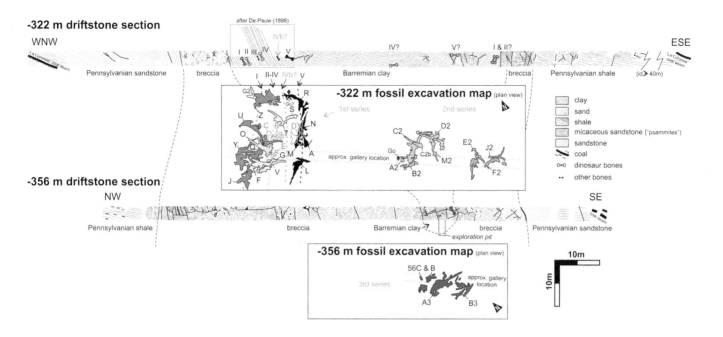

14.2. Detailed geological section and excavation plan of the −322 and −356 m galleries in the Iguanodon Sinkhole at Bernissart (adapted from Cornet and Schmitz, 1898, and Norman, 1986, respectively). The −322 m section, for which a hiatus resulting from missing documents was originally mentioned for the first 14 m, was completed after Van den Broeck (1898). Color correspondence between specimens and bonebeds in the first series of excavation was made only for the lowermost and uppermost beds. The dashed line shows the extension of possible additional bonebed IVb that yielded fossil turtles, crocodiles, and fishes (after Gosselin, 1997).

unfortunately the only geological record for the second and third series of fossils. Fourteen more or less complete and four partial iguanodon skeletons were excavated from beds I, II, IV, and V, which are stratigraphically distant from each other by approximately 1, 2, and 5 m, respectively. Because there is no obvious change in bed thickness, this could indicate that the cyclic conditions leading to dinosaur accumulation were met at a rate decreasing with time. Skeletons from the excavation plan (adapted from Norman, 1986) are tentatively assigned to fossiliferous beds, although no conclusive attribution was found for the specimens in the central zone (in gray in Fig. 14.2). Another uncertainty arises from De Pauw's comments about the fish bonebeds. In his figure, only bed III is indicated as an accessory fish bonebed observed at the footwall of the cavity created for excavating an iguanodon from bed IV (Bultynck, 1989, fig. 63). However, he claimed that the principal fish bonebed lay 4 m below and also contained chelonians and crocodiles. This bonebed is not indicated in the original figure, and the comment on its location is puzzling because "4 m below" would stratigraphically correspond to bed I or II. Perhaps De Pauw did not use stratigraphical but local references because of the excavation chamber's complex geometry and the ground's irregular surface. It is therefore impossible to locate this bonebed more precisely. Fortunately, on the basis of archive analysis, Gosselin (1997) mentioned a chelonian- and crocodile-bearing bonebed between beds IV and V (bed IVb in fig. 2). Although this point necessitates further verification, we shall consider Gosselin's solution, which we deem consistent with stratigraphic continuity. Indeed, a crocodile seemingly corresponding to bed IVb was found at 38.4 m from the west wall of the sinkhole, and iguanodons from bed IV (second series) were then recovered a little farther eastward. It is worth noting that as a result of its continuity, one of these fish-rich beds was used as a stratigraphical marker for drawing the geological section.

Data for the western region of the iguanodon sinkhole at −322 m thus suggest four dinosaur bonebeds, with one or two additional bonebeds depending on whether bonebeds III and/or IVb are taken into account (De Pauw, 1898, considered five main beds). Although fishes were

14.3. Proposed model section for the Iguanodon Sinkhole based on the continuity of the bonebeds recognized in the −322 m gallery and the BER 3 borehole. Only the lowermost and uppermost bonebeds are indicated (dotted lines). Others lie within the bonebed interval (light gray shading). The findings at −356 m could relate to either the existence of an additional, deeper bonebed (suggested here) or to downwarping of one of the beds above (as depicted in Fig. 3.2).

found concentrated in bonebeds, isolated fishes apparently also occurred within dinosaur bonebeds. To illustrate this, a fossil fish was observed sandwiched between the left foreleg and the head of an *Iguanodon* (RBINS R51, specimen Q).

Correlation between the western and eastern bonebeds at −322 m is straightforward, although beds III, IV, and IVb were not explicitly reported on the east. This is likely related to the more intense deformation in that region of the sinkhole filling where fewer fossils were recovered. However, it is not clear whether this is due to fossil scarcity or lack of exploration. Indeed, the first western fossil series received more time, attention, and money than limited in-depth exploration in other regions of the deposit.

Correlation with the single (?) bonebed at −356 m is more difficult because of extreme deformation and the lack of stratigraphic markers. As a first explanation, additional, deeper bonebeds would be lying near the bottom of the clay deposit (Fig. 14.3). Excessive subsidence-induced deformation and downdropping of the −322 m bonebeds would provide a second explanation (see Fig. 3.2 in Chapter 3 in this book).

Figure 14.3 shows a nice correspondence between the bonebed occurrences in the nineteenth-century galleries and the 2003 coring program, especially in BER 3, where several bone fragments were found between −396.5 and −309 m in the lower part of the clay formation.

The Crime Scene: Taphonomy

Taphonomic Elements

No detailed taphonomic study is available for the Bernissart bonebeds so far. However, we can summarize the taphonomically relevant elements as follows.

1. The terrestrial component (animals that lived around the Bernissart lake) of the vertebrate assemblage is nearly monotaxic (in this case, dominated by multiple individuals of one taxon; Fiorillo and Eberth, 2004; Eberth and Currie, 2005). *Iguanodon bernissartensis* is the dominant species, representing about 90% of the terrestrial specimens discovered at Bernissart.
2. Most iguanodontid specimens discovered at Bernissart can be regarded as fully ossified adults. In all specimens, the transverse processes and neural arches are fused to the centra of the vertebrae, and the sacral centra are fused together. Open cranial sutures are absent, even in the smaller *Mantellisaurus atherfieldensis* specimen (Norman, 1986). In some *Iguanodon bernissartensis* specimens (RBINS R55, specimen N, for example), the scapula and coracoid are completely fused together, and the tarsus appears co-ossified to the tibia. Intersternal ossification, which presumably originated within the cartilage of the sternal plate, occurs in most *I. bernissartensis* specimens (Norman, 1980). Norman (1986) recognized three "subadult" specimens of *I. bernissartensis* from smaller individuals, although these are all rather poorly preserved.
3. Most iguanodontid skeletons discovered at Bernissart are complete or subcomplete and are articulated. This is also the case for most

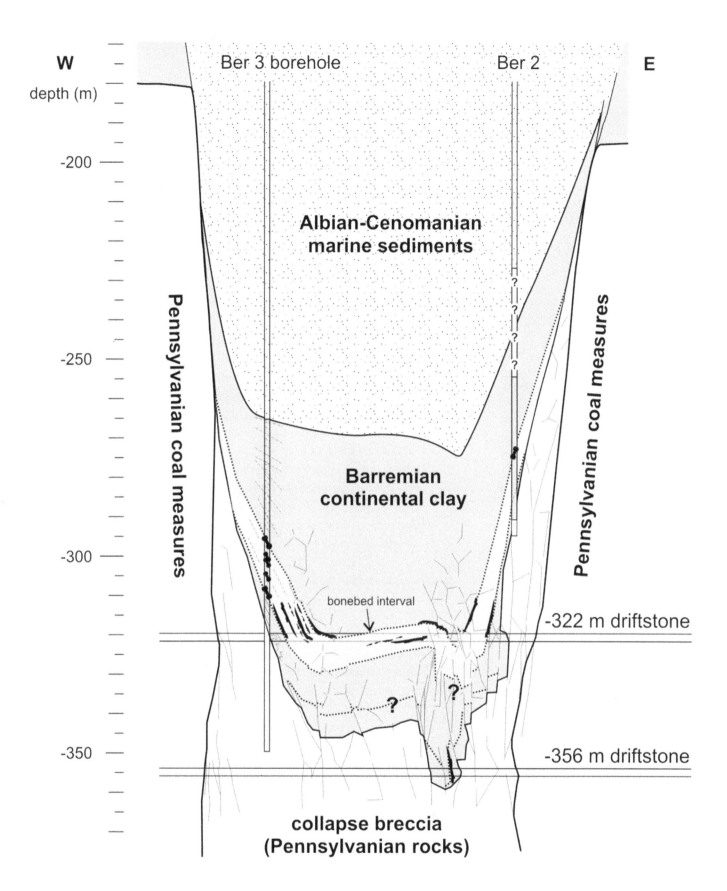

other vertebrates found in this locality, with the notable exception of the theropod isolated phalanx. Some iguanodontid specimens (RBINS R56, specimen L) were found slightly disarticulated. Even incomplete skeletons were found in connection with this locality. The extremely difficult excavation conditions, in dark galleries more than 300 m below ground level, could explain why at least some of the specimens were probably incompletely unearthed and why skeletons might have been partially destroyed by the excavation team. For example, RBINS VERT-5144-1716 (specimen A/B) is a partial skeleton comprising the articulated tail, the posterior portion of the pelvis, and the distal portions of the femur and hind limbs. This was the first skeleton to be systematically excavated and then mounted. The anterior portion of the body was undoubtedly destroyed during initial gallery construction in April 1878 (Norman, 1986).

The presence of subcomplete and articulated skeletons indicates that the iguanodontid carcasses were quickly buried in the sediments of the lake just after the death of the animals. Scavengers (theropods and crocodiles are represented at Bernissart) had limited opportunities for dismembering the carcasses. Moreover, the carcasses were buried in the sediments before they had decayed too much and were kept floating by gases inside the body cavity. Schäfer's (1962) studies on the gradual decay of floating mammal carcasses in the sea (notably those of seals) are especially interesting in this respect because they show that bloated carcasses can drift for more than a month before what is left of them finally settles to the bottom, after having lost a large part of their bony elements. Observations of dead marine mammals cannot be directly extrapolated to carcasses of dinosaurs in a freshwater environment, but they suggest that the gradual decay of floating carcasses certainly leads to the burial in bottom sediments of isolated bones instead of subcomplete skeletons, as observed in Bernissart. In confined lakes, this process is accelerated by the proximity of scavengers.

4. The animals are exceptionally preserved. Ossified tendons are not displaced in most cases, and skin impressions can be observed on many specimens. De Pauw (1898) even described the presence of potential flesh relics. Dinosaur skeletons preserved with skin impressions are common in regions of the Western Interior of North America; however, there are only a few detailed taphonomic studies of these occurrences (Anderson et al., 1999). Exceptional soft tissue preservation is evidently the result of critical timing of diagenetic phosphatization due to the rapid postmortem morphological changes that occur in delicate animal tissue. Such mineralization must occur soon after the death of the animal and its burial. Fossilization of soft tissues most usually occurs in anoxic fossilization conditions (Allison and Briggs, 1991).

5. Iguanodontid skeletons from Bernissart are usually lying on their side in a rather passive position, and many specimens have their necks bent sharply backward. A notable exception is specimen RBINS VERT-5144-1716 (specimen O), which was discovered lying on its back with its skeleton dorsoventrally flattened. Recurving of the neck over the

back, or an opisthotonic posture, is observed in many well-preserved amniote skeletons (Faux and Padian, 2007). Numerous postmortem processes have been proposed to account for this posture, including rigor mortis, differential contraction, desiccation, and water currents. However, in their recent experiments and a review that included the clinical literature, Faux and Padian (2007) showed that perimortem processes related to the central nervous system are more likely to produce an opisthotonic posture. The posture is achieved in the final moment of life by muscle spasms and is temporarily fixed by rigor mortis. Preservation in the fossil record of the opisthotonic posture would therefore indicate rapid burial as well as no significant transport or scavenging of the carcasses.

6. Bone modification can be defined as features on bones that were the result of any postmortem, prediagenetic process (e.g., trampling, scavenging, weathering), which alters the morphology of a once-living bone (Fiorillo, 1991a). At the occasion of the renovation of the Janlet Aisle of the museum of the RBINS , from 2004, the iguanodontid skeletons were entirely dismounted, and all the bones were treated against pyrite. It was a unique opportunity for the technical team to look systematically at bone modification features.

Postmortem fractures other than those resulting from block faulting in the clay mass or excavation incidents could not be observed with certainty. According to Ryan et al. (2001) and Eberth and Getty (2005), the observation of a large number of broken limb bones indicate a destructive history, such as the breakdown of trabecular bone and collagen, before and/or during final transportation. Behrensmeyer (1988, 1991) observed that fresh limb bones from large mammals often show no evidence of breakage during vigorous hydraulic transport. Thus, the absence of broken limb bones in the Bernissart sample suggests that the carcasses were not transported over a long distance and/or they did not experience an earlier taphonomic episode that weakened the specimens and increased their susceptibility to hydraulically induced breakage before final burial.

Perthotaxic features (bone modification processes that are active on the land surface; Clark et al., 1967) could not be positively observed in the iguanodontid bones from Bernissart. Weathering features (flaking and cracking) are apparently absent, indicating that the bones were not exposed subaerially for any significant length of time before or after transportation. Nor have we recognized trample marks, indicative of the trampling activity of another animal whose feet pressed the bone against a sandy surface, which are characterized by shallow, subparallel scratch marks left on the surfaces of bones (see, e.g., Behrensmeyer et al., 1986; Fiorillo, 1984, 1987), in the studied sample. Therefore, the absence of perthotaxic features is additional supporting evidence that the dinosaur bones discovered at Bernissart were quickly buried.

Carnivorous tooth marks on bone surfaces can be identified either as grooves, often several millimeters deep, with a V-shaped cross section or isolated punctures (Fiorillo, 1991b). Tooth marks on mammal and dinosaur bones are usually attributed to scavenging and prey carcass utilization, not to the killing process (Fiorillo, 1991a, 1991b). Tooth marks could not be identified with certainty on the bones of the Bernissart iguanodontoids. This observation can be related to the absence of theropod shed teeth at Bernissart and reflects the low incidence of scavenging on the iguanodontid carcasses. Alternatively, it may also be hypothesized that scavengers ate only the fleshy parts of the dead iguanodontid specimens because prey was abundant (Eberth and Getty, 2005). In any case, Fiorillo (1991b) observed that tooth-marked bones are uncommon in dinosaur localities. Theropod dinosaurs did not routinely chew bones during prey carcass utilization. They may have used prey bones more like modern Komodo monitors and crocodiles than mammalian carnivores—that is, by passive consumption rather than by actively seeking out the bones for nutrient intake.

It can therefore be concluded that bone modification is virtually absent for the iguanodontid skeletons from Bernissart. Again, it suggests that the dinosaur carcasses were quickly buried in the sediments after the animals' death.

7. Fossil concentration is high, but skeletons are neither systematically piled up nor excessively concentrated in certain regions for a given stratigraphic interval. Rather, they are homogeneously scattered over the strata surface. A simple calculation suggests that each dinosaur bonebed could originate from the burial of a single layer of adjacent, almost side-by-side carcasses: 300 m³ of fossiliferous clay were excavated from the first series from which 17 specimens were collected. Volume by specimen is thus 300/17 = ~17 m³, or 10 × 3 × 0.57 m (length × width × thickness), which is a slight overestimation of the volume that could be filled by a typical iguanodon skeleton in the deposit (bonebed thickness ranges from 0.35 to 0.55 m). Overestimation occurs because some unfossiliferous clay is taken into account. Stretching of strata surface caused by the sinkhole subsidence may also have caused slight dispersion of the skeletons. The stratigraphic correlations established here suggest the presence of a great number of fossil dinosaurs in the Iguanodon Sinkhole, probably exceeding at least 100 specimens. Extrapolation of the first series excavation—that is, 17 iguanodons found within an area of 20 × 15 m, to the inner 15-m-wide band of a 60 × 80 m ellipse representing a simplified section of the sinkhole at −322 m— yields 102 specimens. This estimation assumes no change in fossil concentration, which may be questionable (see below), but it does not consider additional fossils in other areas of the main bonebed group, both in the center and outside, and in possible additional bonebeds in the deepest regions of the sinkhole fill.

8. As suggested by Norman (1987), there are three groups of skeletons showing common orientation. However, two of these groups share specimens from different layers (I, II, and IV). Therefore, the common orientation of these skeletons cannot reasonably be explained by hydraulic processes producing exactly the same orientation of the carcasses at different times. Nevertheless, alignment

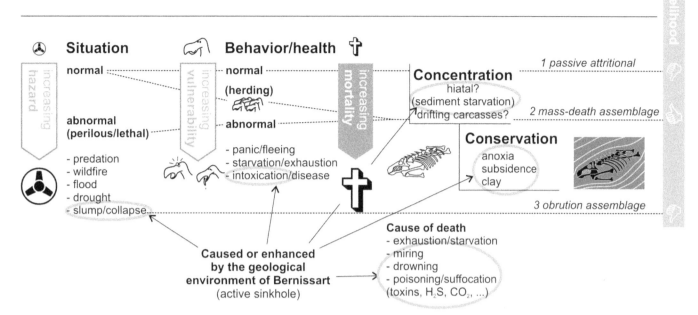

living environment + biocenosis = thanatocenosis + sedimentary environment = taphocenosis

Situation

normal

abnormal
(perilous/lethal)

- predation
- wildfire
- flood
- drought
- slump/collapse

Behavior/health

normal

(herding)

abnormal

- panic/fleeing
- starvation/exhaustion
- intoxication/disease

increasing mortality

Concentration
hiatal?
(sediment starvation)
drifting carcasses?

1 passive attritional

2 mass-death assemblage

Conservation
anoxia
subsidence
clay

3 obrution assemblage

Caused or enhanced
by the geological
environment of Bernissart
(active sinkhole)

Cause of death
- exhaustion/starvation
- miring
- drowning
- poisoning/suffocation
(toxins, H_2S, CO_2, ...)

is obvious for skeletons in bed V, but they were excavated in a zone that corresponds to a northeast–southwest-oriented syncline hinge showing fractured (and likely displaced) blocks. As an alternative explanation to hydraulic processes (albeit one not completely ruled out here), early subsidence processes in the still-hydroplastic mud could have affected the original orientation of the buried carcasses.

Discussion of elements 7 and 8 stresses the need for better assessment of postdepositional mechanical processes, which could have significantly modified the original orientation and attitude of the original iguanodon carcasses.

9. Although it is not yet firmly established, the absence of lamination in the bonebed sediment could result from bioturbation by burrowing or miring animals when the environment episodically changed to shallower, shrinking water conditions in the lake. However, no observation of contorted bed was made to support the miring hypothesis. No desiccation cracks were described either, and it is unlikely that past researchers could have overlooked these sedimentological structures.

Obrution Scenario

Obrution assemblages are formed by catastrophic sedimentological events. A typical situation is represented by herds buried by bank collapse or landslides triggered by heavy rains or earthquakes (Rogers and Kidwell, 2007). It is worth mentioning this taphonomic category here because sudden ground or bank collapses are valid processes that could have occurred when the sinkhole reached the surface, or later as a consequence of sinkhole margin instability (Fig. 14.4). Although most taphonomic elements do not contradict this hypothesis, sedimentological evidence that could support it, such as medium- to large-scale slumpings, are completely lacking.

14.4. Synopsis of the taphonomic parameters identified for the Bernissart bonebeds, listed as mortality and preservation parameters from left to right, respectively. Mass death caused either by herding or abnormal behavior under perilous situation is the most likely scenario. Passive attrition, which results from normal biological activity under normal circumstances but requires efficient processes of concentration and conservation, is unlikely. Quick burying resulting from catastrophic sedimentary events (obrution) seems unlikely as well. The possible influence of the specific geological environment of Bernissart by Lower Cretaceous times is shown. See text for details.

Attrition Scenario

Attritional (or passive attritional) assemblages result from death under normal circumstances but require efficient processes that concentrate as bonebeds the carcasses of animals dead at different times and in different places; hydraulic transportation usually plays an important role in the formation of vertebrate attritional assemblages (Lauters et al., 2008; Behrensmeyer, 2007a; Rogers and Kidwell, 2007). Lignite lenses within the Iguanodon Sinkhole most probably derived from floods that occurred after wildfires. As discussed above (elements 3, 6, and 8), there is no indication that the iguanodontid skeletons were transported over a significant distance, and the carcasses probably did not float for long before being buried. The high skeleton concentration at Bernissart (element 7) could have been achieved by a very low sedimentation rate, such as in some oxbow lakes (hiatal concentration), but this hypothesis is not consistent with sedimentological evidence of cyclic but sustained arenitic sediment influx in likely shallow conditions (see above). Moreover, the exceptional state of preservation of the iguanodontid skeletons (elements 3 and 4) and the absence of bone modification (element 6) indicate that the carcasses were quickly buried just after the death of the animals; these elements also point to high sedimentation rates.

Moreover, the age profile of the *Iguanodon bernissartensis* population (element 2) is not compatible with an attritional mortality scenario. In an ideal attritional profile, age class abundances reflect the number of animals dying from one class to the next (Lyman, 1994), showing peaks corresponding to ages where mortality rates are the highest: among the very young and, to a lesser extent, the very old. Therefore, in this case, the observed death profile of the fossil assemblage is completely different from the age profile of the living population: younger individuals are overrepresented. In attritional dinosaur assemblages, late juveniles and small subadult individuals represent over 90% of the recovered fossils (Lauters et al., 2008). This is not the case for the *I. bernissartensis* assemblage discovered at Bernissart, which is largely dominated by adult specimens. This age profile is more consistent with a catastrophic, nonselective scenario: in a catastrophic profile, the age class abundance of the assemblage corresponds to the age profile of the living population when the catastrophic event happened (Lyman, 1994).

Mass Mortality Scenario

Mass death assemblages may arise from a variety of causes or combinations of them (Behrensmeyer, 2007a; Rogers and Kidwell, 2007). The living, normal environment can become a lethal trap under certain panic/fleeing or starvation/exhaustion situations caused, for example, by predation or herding (either natural or forced by extreme circumstances). Perilous situations due to environmental or biological hazards, which usually induce abnormal behavior, also increase vulnerability and hence mortality. To date, mass death assemblages of ceratopsid (horned) dinosaurs have been found in Late Cretaceous deposits of western North America (see Dodson et al., 2004 and references therein).

Wildfires, floods, and drought were likely perilous situations in the Lower Cretaceous environment of Bernissart. Wildfires probably occurred, as shown by lignite beds and typical postfire colluvia in the Lower Cretaceous of Hautrage (Spagna, 2010).

In western North America, large ceratopsid bonebeds in channel deposits indicate that flooding may have been responsible for mass drowning in some cases (Currie and Dodson, 1984). Floods were naturally pacing the sedimentary environment at Bernissart, but there is no sedimentological evidence for extreme or flash floods that could have drowned and buried animals en masse.

Drought is the primary cause of mass death accumulations today (e.g., Corfield, 1973; Haynes, 1988), with carcasses of large herbivores often concentrated around water holes. Evidence of strong seasonality and semiarid conditions in association with ceratopsid bonebeds indicates that drought may also have been a major killing agent for dinosaurs during the Cretaceous (Rogers, 1990). Drought was also probably recurrent in the Bernissart environment, but again, there is no convincing evidence of severe situations, such as mud cracks, evaporitic deposits, or typical paleosols. Nevertheless, drought and concomitant shrinking water could have created circumstances under which the vulnerability of iguanodons was increased—for example, by forcing them to overcrowd around water holes. In this respect, the Iguanodon Sinkhole most probably controlled the location of the residual water body by sustained low ground conditions due to subsidence.

The dominantly clayey composition of sediments in swampy areas such as Bernissart indeed suggests miring as a highly probable cause of death. (Death would actually occur as a result of exhaustion or starvation because of being stuck in the mud.) If sedimentological data could possibly agree with that hypothesis (see discussion of element 9), other observations do not. First, it is surprising that most iguanodons are not lying on their ventral side with their legs fully extended (element 5) if we consider a miring scenario (Rogers and Kidwell, 2007; Weigelt, 1989). The above-mentioned postdepositional processes due to enhanced karstic subsidence are not thought to have changed such a typical miring posture for virtually all the specimens. Second, mired animals cannot sink deep into the mud because of buoyancy (it is far easier to sink into water). Therefore, most mired carcasses have their upper side exposed over a sufficient period of time to permit scavenging (Weigelt, 1989). Contrary to similar trap situations such as tar pits (Behrensmeyer, 2007b), remains of scavengers and/or predators that should have been attracted by mired animals are rare at Bernissart, comprising one theropod phalanx and four crocodile skeletons (with the possible exception of coprolites). Tooth marks cannot be recognized on the iguanodontid bones (element 6), and scavenging damages, if proven, would be obviously be minimal, as indicated by the full articulation and the high degree of preservation of most iguanodontid skeletons (element 4).

Mass drowning remains another possible scenario to explain the high concentration of complete iguanodontid skeletons in lake deposits at Bernissart. However, iguanodontids were probably used to living close to swamp and bog environments, and it is thus difficult to conceive that they could have easily drowned in normal situations. Abnormal, high-vulnerability situations such as herding when crossing water bodies or fleeing are therefore possible scenarios to explain mass drowning, but this hypothesis cannot be substantiated by any concrete elements. The opisthotonic posture observed in many skeletons could support the hypothesis of mass drowning because it can be caused by several brain afflictions resulting from drowning but also by asphyxiation, lack of nourishment, environmental toxins, or viral infections (Faux and Padian, 2007).

Possible Role of Chemical or Biological Intoxication

Death by gas inhalation is often related to volcanic activity. However, suffocating (CO_2) or poisonous (H_2S) gases are common products of organic matter decay and are frequently released in the sedimentary environment. An example of a recent gas-related fatal accident is provided by the H_2S emanation from a 1-m-thick stand of rotting seaweed that killed a horse and left its rider unconscious on a beach in western France on July 28, 2009. H_2S-based death scenarios at Bernissart may tentatively be hypothesized on the basis of the following observations.

1. Pyrite (FeS_2), a common product of diagenetic H_2S fixation in sediments, is an abundant mineral in the Barremian clays of Bernissart, including the iguanodontid bones (see Chapter 11 in this book).
2. H_2S is formed by sulfate-reducing bacteria decomposing organic matter in anaerobic conditions. Barremian clays are indeed rich in organic matter, and there are abundant sedimentological structures possibly attributable to gas escape. Moreover, the preservation of soft tissues also suggests anaerobic conditions for the fossilization of iguanodontids fossils (element 4).
3. Artesian, sulfate-rich brines are known in the underlying Mississippian strata. This geothermal (70°C) water is in fact the result of the dissolution of anhydrite layers that induced collapse and subsidence in the Mons Basin, including the Iguanodon Sinkhole. Although most sinkhole fillings, which are composed of predominantly shaley Pennsylvanian rocks, are now rather impervious, uprising flow of deep brines could have existed in the earliest stages of sinkhole formation. Brines could also have arisen from adjacent fractures or faults because the whole Bernissart area was subjected to karstic subsidence processes (Fig. 3.3 in Chapter 3 in this book). Delmer (2000) suggested that sulfate-rich water from underlying Mississippian limestone can locally flow up through Pennsylvanian rocks on the basis of mine records reporting on the technical efforts that were necessary to reduce H_2S concentration in the Bernissart coal mines. Today, deep sulfate-rich geothermal water forms several springs, such as the bubbling spring of Stambruges, the temperature of which is 18°C throughout the year (whereas aquifers are typically 11–12°C). In contrast to other nearby low-temperature conventional springs, life is extremely scarce and monospecific in the warm spring, forming a decametric pool of crystal-clear water in which strange cyanobacterial mats develop.

Biological intoxication such as cyanobacterial toxicosis (Varrichio, 1995; Koenigswald et al., 2004 and references therein) could also be invoked as a cause of vertebrate death or disease. Increasing temperature caused by sulfate-rich water upwelling along the sinkhole breccia column or some nearby fissure could also have created unusual chemical and biological conditions in the environment, for example by boosting cyanobacterial activity in water and/or sulfate-reducing bacteria in sediments. Of particular significance for both chemical and biological intoxication scenarios, Schnyder et al. (2009) reported a strong positive $\delta^{13}C$ anomaly in sedimentary organic matter of the Sainte-Barbe clays excluding fossil wood. They interpreted this anomaly as an enhanced algal–bacterial or macrophytal productivity.

These intoxication scenarios of drinking poisonous water or breathing toxic gas are consistent with the opisthotonic posture and are totally compatible with—and perhaps best explained by—abnormal situations such as droughts, in which iguanodons were forced to seek food or water in biologically hazardous or H_2S-emanating areas. Trampling would have had no other effect but to increase the emanation rate of lethal gas. Animals could have been killed directly or been overcome and then drowned. A context of shrinking water during drought periods is particularly interesting because it could have amplified the effects of springing geothermal brines. It is also consistent with a massive die-off of fishes, which is often reported as anoxia driven but can also be due to H_2S intoxication (Weigelt, 1989). In such situation, mass death accumulation of local freshwater fauna would have logically preceded that of terrestrial animals, and it is therefore normal that fish and iguanodontids bonebeds are separated in the sedimentological record. However, this hypothesis does not explain why the dinosaur bonebeds at Bernissart are near monotypic (element 1); indeed, *Iguanodon bernissartensis* was not the only large vertebrate to live around the Bernissart lake, and other animal should also have been strongly affected by poisonous waters or toxic gases.

Conclusions

Several discrete and laterally continuous bonebeds occur in the Iguanodon Sinkhole. They are clustered within 8 m of Barremian clay in the lower part of the Sainte-Barbe Clays Formation. Four dinosaur beds plus one or two fish- and freshwater reptile-bearing beds were recognized, although the possible occurrence of earlier-formed beds cannot be ruled out. Cyclic mass death under as yet unresolved conditions should receive more attention than obrution and attrition as formation processes for the Bernissart fossil deposit. Intoxication by H_2S or biological toxins in a context of seasonally shrinking water is an interesting scenario that remains to be assessed; it fits within the specific karstic-induced environment of Bernissart. This includes solution–collapse subsidence processes and possible sulfate-rich brine contamination that could have boosted bacterial/cyanobacterial activity.

Acknowledgments

We thank J.-P. Tshibangu for access to the 2003 drilling program documents, especially the excavation plan redrawn by R. Gosselin in 1997. We are also grateful to David A. Eberth and to P. Lauters for fruitful comments and discussions.

References

Allison, P. A., and D. E. G. Briggs. 1991. The taphonomy of soft-bodied animals; pp. 120–140 in S. K. Donovan (ed.), The process of fossilization. Belhaven Press, London.

Anderson, B. G., R. E. Barrick, M. L. Droser, and K. L. Stadtman. 1999. Hadrosaur skin impressions from the Upper Cretaceous Neslen Formation, Book Cliffs, Utah: morphology and paleoenvironmental context. Utah Geological Survey Miscellaneous Publications 99: 295–301.

Behrensmeyer, A. K. 1988. Vertebrate preservation in fluvial channels. Palaeogeography, Palaeoclimatology, Palaeoecology 63: 183–199.

———. 1991. Terrestrial vertebrate accumulations; pp. 291–335 in P. A. Allison and D. E. G. Briggs (eds.), Taphonomy: releasing the data locked in the fossil record. Plenum, New York.

———. 2007a. Bonebeds through time; pp. 65–97 in R. R. Rogers, D. A. Eberth, and A .R. Fiorillo (eds.), Bonebeds: genesis, analysis and paleobiological significance. University of Chicago Press, Chicago.

———. 2007b. Terrestrial vertebrates; pp. 318–321 in D. E. G. Briggs and P. R. Crowther (eds.), Paleobiology II. Blackwell Science, Oxford.

Behrensmeyer, A. K., K. D. Gordon, and G. T. Yanagi. 1986. Trampling as a cause of bone surface damage and pseudo-cutmarks. Nature 319: 768–771.

Bertrand, C. E. 1903. Les coprolithes de Bernissart. 1. Les coprolithes qui ont été attribués aux iguanodons. Mémoires du Musée Royal d'Histoire naturelle de Belgique 4: 1–154.

Bultynck, P. 1989. Bernissart et les iguanodons. Institut Royal des Sciences naturelles de Belgique, Bruxelles, 115 pp.

Casier, E. 1960. Les iguanodons de Bernissart. Institut Royal des Sciences Naturelles de Belgique, Bruxelles, 134 pp.

Clark, J. R., J. R. Beerbower, and K. K. Kietzke. 1967. Oligocene sedimentation, stratigraphy, paleoecology, and paleoclimatology in the Big Badlands of South Dakota. Fieldiana Geology 5: 1–158.

Corfield, T. F. 1973. Elephant mortality in Tsavo National Park, Kenya. East Africa Wildlife Journal 11: 339–368.

Cornet, J. 1927. L'époque wealdienne dans le Hainaut. Bulletin de la Société géologique de Belgique 50(4): B89–B104.

Cornet, J., and G. Schmitz. 1898. Note sur les puits naturels du terrain houiller du Hainaut et le gisement des Iguanodons de Bernissart. Bulletin de la Société belge de Géologie, Paléontologie et Hydrologie 12: 196–206 and 301–318.

Currie, P. J., and P. Dodson. 1984. Mass death of a herd of ceratopsian dinosaurs; pp. 52–60 in W.-E. Reif and F. Westphal, Third Symposium of Mesozoic Terrestrial Ecosystems. Attempto Verlag, Tübingen.

Delmer, A. 2000. Les gisements houillers du Hainaut. Vol. 1, Le Couchant de Mons. Geological Survey of Belgium, Brussels, 22 pp.

De Pauw, L. 1898. Observations sur le gisement de Bernissart. Bulletin de la Société belge de Géologie, Paléontologie et Hydrologie 12: 206–216.

Dodson, P., C. A. Forster, and S. D. Sampson. 2004. Ceratopsidae; pp. 494–513 in D. B. Weishampel, P. Dodson, and H. Osmólska (eds.), The Dinosauria. 2nd ed. University of California Press, Berkeley.

Dupont, E. 1878. Sur la découverte d'ossements d'*Iguanodon*, de poissons et de végétaux dans la fosse Sainte-Barbe du Charbonnage de Bernissart. Bulletin de l'Académie royale de Belgique 46: 387.

———. 1897. Musée Royal d'Histoire naturelle de Belgique. Guide dans les collections. Bernissart et les Iguanodons. Editions Polleunis and Centerick, Brussels.

Eberth, D. A., and P. J. Currie. 2005. Vertebrate taphonomy and taphonomic modes; pp. 453–477 in P. J. Currie and E. B. Koppelhus (eds.), Dinosaur Provincial Park, a spectacular ancient ecosystem revealed. Indiana University Press, Bloomington.

Eberth, D. A., and M. A. Getty. 2005. Ceratopsian bonebeds: occurrence, origins, and significance; pp. 501–506 in P. J. Currie and E. B. Koppelhus (eds.), Dinosaur Provincial Park, a spectacular ancient ecosystem revealed. Indiana University Press, Bloomington.

Faux, C. M., and K. Padian. 2007. The optisthotonic posture of vertebrate skeletons: postmortem contraction or death throes? Paleobiology 33: 201–226.

Fiorillo, A. R. 1984. An introduction to the identification of trample marks. Current Research, University of Maine 1: 47–48.

———. 1987. Trample marks: caution from the Cretaceous. Current Research in the Pleistocene 4: 73–75.

———. 1991a. Taphonomy and depositional settings of Careless Creek Quarry (Judith River Formation), Wheatland County, Montana, USA. Palaeogeography, Palaeoclimatology, Palaeoecology 81: 281–311.

———. 1991b. Prey bone utilization by predatory dinosaur. Palaeogeography, Palaeoclimatology, Palaeoecology 88: 157–166.

Fiorillo, A. R., and D. A. Eberth. 2004. Dinosaur taphonomy; pp. 607–613 in D. B. Weishampel, P. Dodson, and H. Osmólska (eds.), The Dinosauria. 2nd ed. University of California Press, Berkeley.

Gosselin, R. 1997. Summary excavation map of the Bernissart iguanodons (first series) from IRScNB archives. Unpublished files of the 2002–2003 drilling program in the Bernissart sinkhole.

Haynes, G. 1988. Longitudinal studies of African elephant death and bone deposits. Journal of Archaeological Sciences 15: 131–157.

Koenigswald, W. V., A. Braun, and T. Pfeiffer. 2004. Cyanobacteria and seasonal death: a new taphonomic model for the Eocene Messel Lake. Paläontologische Zeitschrift 78 (2): 417–424.

Lauters, P., Y. Bolotsky, J. Van Itterbeeck, and P. Godefroit. 2008. Taphonomy and age profile of a latest Cretaceous dinosaur bonebed in Far Eastern Russia. Palaios 23: 153–162.

Lyman, R. L. 1994. Vertebrate taphonomy. Cambridge University Press, Cambridge, 550 pp.

Norman, D. B. 1980. On the ornithischians dinosaur Iguanodon bernissartensis of Bernissart (Belgium). Mémoires de l'Institut royal des Sciences naturelles de Belgique 178: 1–103.

———. 1986. On the anatomy of Iguanodon atherfieldensis (Ornithischia: Ornithopoda). Bulletin de l'Institut royal des Sciences naturelles de Belgique, Sciences de la Terre 56: 281–372.

———. 1987. A mass-accumulation of vertebrates from the Lower Cretaceous of Nehden (Sauerland), West Germany. Proceedings of the Royal Society of London, Biological Sciences 230: 215–255.

Rogers, R. R. 1990. Taphonomy of three dinosaur bone beds in the Upper Cretaceous Two Medicine Formation of Northwestern Montana: evidence for drought-related mortality. Palaios 5: 394–413.

Rogers, R. R., and S. M. Kidwell. 2007. A conceptual framework for the genesis and analysis of vertebrate skeletal concentrations; pp. 1–64 in R. R. Rogers, D. A. Eberth, and A. R. Fiorillo (eds.), Bonebeds: genesis, analysis and paleobiological significance. University of Chicago Press, Chicago.

Ryan, M. J., A. P. Russell, D. A. Eberth, and P. J. Currie. 2001. The taphonomy of a Centrosaurus (Ornithischia: Ceratopsidae) bone bed from the Dinosaur Park Formation (Upper Campanian), Alberta, Canada, with comments on cranial ontogeny. Palaios 16: 482–506.

Schäfer, W. 1962. Aktuo-Paläontologie nach Studien in der Nordsee. Verlag Waldemar Kramer, Frankfurt am Main, 666 pp.

Schnyder, J., J. Dejax, E. Keppens, T. T. Nguyen Tu, P. Spagna, S. Boulila, B. Galbrun, A. Riboulleau, J.-P. Tshibangu, and J. Yans. 2009. An Early Cretaceous lacustrine record: organic matter and organic carbon isotopes at Bernissart (Mons Basin, Belgium). Palaeogeography,

Palaeoclimatology, Palaeoecology 281: 79–91.

Seward, A. C. 1900. La flore wealdienne de Bernissart. Mémoires du Musée Royal d'Histoire naturelle de Belgique 1: 1–39.

Spagna, P. 2010. Les faciès wealdiens du Bassin de Mons (Belgique): paléoenvironnements, géodynamique et valorisation industrielle. Ph.D. thesis, Faculté Polytechnique de l'Université de Mons, 138 pp.

Spagna, P., and J. Van Itterbeeck. 2006. Lithological description and granulometric study of the Wealden facies of two borehole core drilled in the "Cran aux Iguanodons de Bernissart." Second Geologica Belgica Meeting, September 7–8, Liège, Belgium.

Van den Broeck, E. 1898. Les coupes du gisement de Bernissart. Caractères et dispositions sédimentaires de l'argile ossifère du Cran aux Iguanodons. Bulletin de la Société belge de Géologie, Paléontologie et Hydrologie 12: 216–243.

Varrichio, D. J. 1995. Taphonomy of Jack's Birthday Site, a diverse dinosaur bonebed from the Upper Cretaceous Two Medicine Formation of Montana. Palaeogeography, Palaeoclimatology, Palaeoecology 114: 297–323.

Yans, J. P., Spagna, C. Vanneste, M. Hennebert, S. Vandycke, J. M. Baele, J. P. Tshibangu, P. Bultynck, M. Streel, and C. Dupuis. 2005. Description et implications géologiques préliminaires d'un forage carotté dans le "Cran aux Iguanodons" de Bernissart. Geologica Belgica 8: 43–49.

Weigelt, J. 1989. Recent vertebrate carcasses and their paleobiological implications. University of Chicago Press, Chicago, 188 pp.

2

15.1. NHMUK R526. Ornithopod (iguanodontian) tibia of moderate size (distal end only) collected by (or for) William Smith in the area of Cuckfield (from one of the quarries at Whiteman's Green on the outskirts of Cuckfield) during the survey work that led to the first geological map of England and Wales. The specimen was collected in 1809 or slightly earlier.

10 cm

Iguanodontian Taxa (Dinosauria: Ornithischia) from the Lower Cretaceous of England and Belgium

15

David B. Norman

This review summarizes current understanding of the history, anatomy, and taxonomy of British and Belgian iguanodontian dinosaurs. The earliest iguanodontian from this circumscribed region is Berriasian in age and represented by a well-preserved but crushed dentary with many teeth in situ; originally named *Iguanodon hoggii* Owen, 1874, this specimen has been studied and reassessed several times, and decisions concerning its taxonomic status and systematic position have proved to be consistently inconclusive. *I. hoggii* has recently been renamed *Owenodon hoggii*; however, the diagnostic anatomical characters that form the foundation for this new name are few and not taxonomically or systematically robust. It is considered appropriate to regard this undoubtedly important taxonomic entity as indicative of a basal (ankylopollexian) iguanodontian and to encourage new exploration for additional skeletal remains from Berriasian-aged deposits in England. Wealden iguanodontian taxonomy in England has also begun to be scrutinized more thoroughly. Difficulties encountered when trying to diagnose the original (Valanginian) type genus (*Iguanodon* Mantell, 1825) and species (*Iguanodon anglicus* Holl, 1829) created problems that were resolved using a rather unfortunate workaround that involved the use of a Barremian–Lower Aptian species: *I. bernissartensis* Boulenger *in* Van Beneden 1881. With regard to remains collected from numerous Wealden localities in southern England, it was recognized that known iguanodontians can be subdivided into anatomically and chronologically distinct groupings: an earlier (Valanginian) "fauna" represented by *Barilium dawsoni* (Lydekker, 1888) and *Hypselospinus fittoni* (Lydekker, 1889), and a later (Barremian–Lower Aptian) "fauna" comprising *Iguanodon bernissartensis* Boulenger *in* Van Beneden, 1881, and *Mantellisaurus atherfieldensis* (Hooley, 1925). The Belgian locality at Bernissart, assigned to the Sainte-Barbe Clays Formation (late Barremian–Lower Aptian) has yielded two taxa that have been recognized as anatomically similar to those identified in the contemporaneous Wealden deposits of southern England (the Weald Clay Group of the Wealden District and the Wealden Group of the Isle of Wight). Recent suggestions that further taxa can be diagnosed within the English and Belgian Wealden sequences are assessed (and rejected) on the basis of the evidence presented.

Historical Perspective

The geological surveys undertaken by William Smith (1769–1839) that led to the first geological map of England and Wales during the early years of the nineteenth century involved the collection of rock samples and (crucially) fossil objects from a range of outcrops, whether natural or man made. During one survey across southeast England in 1809, Smith collected samples of large and unidentified fossil vertebrate bones from working quarries in Tilgate Forest near the village of Cuckfield, Sussex. Among these remains, which are mostly fragments of vertebral centra that, with the benefit of hindsight, can be assigned to a medium-size ornithopod (iguanodontian) dinosaur, there is a reasonably distinctive iguanodontian distal left tibia (NHMUK R526; Fig. 15.1), which I identified in 1976 for Alan Charig (1979).

The locality from which this material was collected and its stratigraphic setting have become as important as details of comparative osteology. The quarries (there were several that produced both building stone and aggregate for road construction and repair) bordering Whiteman's Green, on the outskirts of Cuckfield, also produced abundant fossil remains of plants, invertebrates, and vertebrates (Mantell, 1833; Norman, in press a). Given the fashion at this time for gentlemen of means to create personal cabinets of curiosities, quarrymen were able to supplement their earnings by selling fossils that they found. Gideon Algernon Mantell (1790–1852) had qualified as a physician and set up his medical practice at the village of Lewes (10 miles [16 km] southeast of Cuckfield) during the latter half of the second decade of the nineteenth century. As an enthusiastic natural philosopher, collector, and amateur geologist, Mantell was keen to document the geology of the countryside that he visited while on his medical rounds, and to assemble a comprehensive fossil collection in preparation for a detailed publication on the fossils and geology of this area of England (Mantell, 1822).

Mantell was able to obtain several notable ridged and channeled fossil teeth, as well as fragments of rib of unusually large size, from the Cuckfield quarries. The teeth he described in some detail (Mantell, 1822, 54–55), and the rib he (very presciently) observed bore "a greater resemblance to the rib of a quadruped [=large mammal], than to those of the lacertae [=reptile]" (Mantell, 1822, 55). Shortly afterward (with advice from Georges Cuvier, William Clift, and William Daniel Conybeare), the distinctive teeth were used to establish the iconic dinosaurian genus *Iguanodon* (Mantell, 1825; Fig. 15.2). The William Smith specimens collected from near Cuckfield, although mostly battered and broken, represent the earliest preserved remains of fossil material that is genuinely attributable to Mantell's famous genus.

Some of the isolated teeth and disarticulated bones recovered by William Smith and Mantell (as well as others since) are embedded in a medium-grained sandstone, while others are associated with a well-cemented but poorly sorted coarse sandstone (known locally as Tilgate Grit). Clearly more than one facies was exposed in the Cuckfield quarries (Gallois and Worssam, 1993), and their precise lithostratigraphic relationships are uncertain. As a consequence of the combination of the disarticulated nature of individual fossils and the presence of more than one bone-bearing lithology and horizon within the quarries in the vicinity of Whiteman's Green, it is unwise to refer individual elements to the same taxon in the absence of other compelling evidence (Norman, in press a).

The strata exposed in the quarries at Whiteman's Green (which were abandoned and filled in during the middle decades of the nineteenth

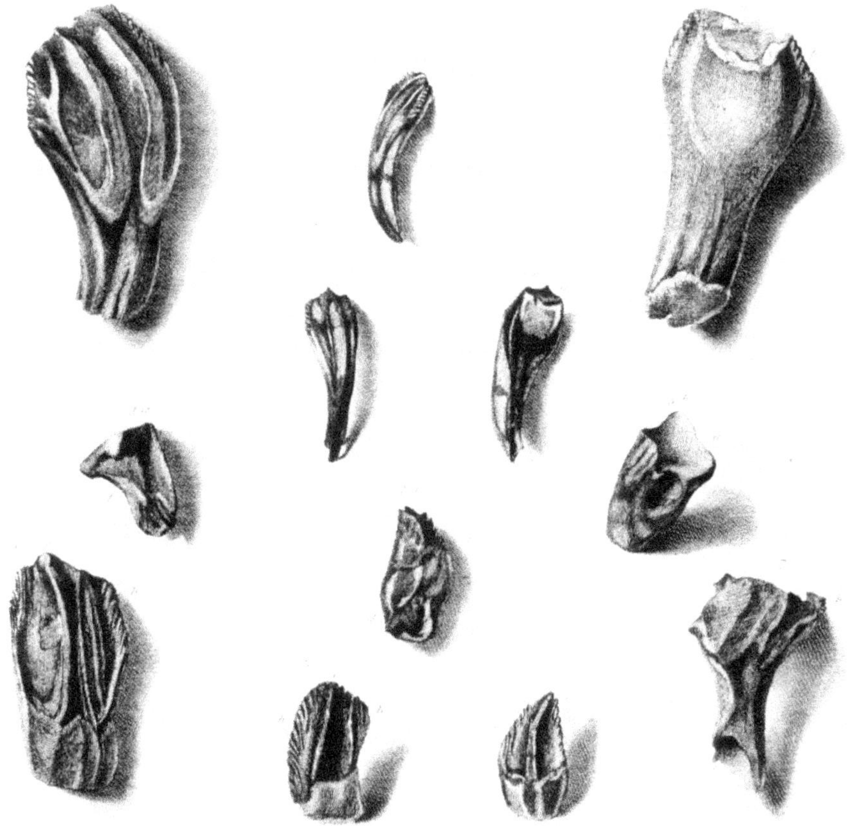

15.2. The original teeth collected from quarries in the vicinity of Whiteman's Green to the west of the village of Cuckfield; illustrated in Mantell (1825, fig. XIV) and named *Iguanodon*.

century) appear to belong to the Grinstead Clay Formation (a variable lenticular lithological unit that occurs between the Lower and Upper Tunbridge Wells Sand formations). These latter formations lie above the Ashdown Beds and Wadhurst Clay formations, and all of these formations are combined to form what is currently recognized as the Hastings Group (D.J. Batten, pers. comm., March 2011; Rawson, 2006, fig. 15.5; see also Fig. 15.3). The Hastings Group spans the entire Valanginian stage (141–137 Ma; Rawson, 2006), and the Grinstead Clay Member is therefore regarded as being of upper Valanginian age (Fig. 15.3).

The posterity that attaches itself to such iconic material cannot disguise the fact that, as with much historically important fossil material, these early described remains, though clearly unusual and distinctive at the time, would prove themselves to be inadequate for the purposes of defining species objectively (Charig and Chapman, 1998).

The generic name *Iguanodon* has been taxonomically problematic since Mantell coined the name in 1825 without designating either the holotype or a specific name.[1] Holl (1829) suggested the species name *I. anglicum* shortly thereafter on the basis of the range of disassociated teeth that had been described by Mantell (the species name ending was corrected to *anglicus* by Norman, 1986, following Norman, 1977). The publication by Friedrich Holl was (almost) universally ignored, and throughout much of the nineteenth century, material collected from the Weald that appeared to be referable to the dinosaurian genus *Iguanodon* was by default attributed to Hermann von Meyer's (1832) suggestion of what was in many respects a far more appropriate binomial: *Iguanodon mantelli* von Meyer, 1832. *I. mantelli* was occasionally attributed to Georges Cuvier (e.g., Owen, 1842).

15.3. The stratigraphy and nomenclature of the English Wealden-aged beds and contemporary beds from Belgium. (Modified from Rawson, 2006, fig. 15.5).

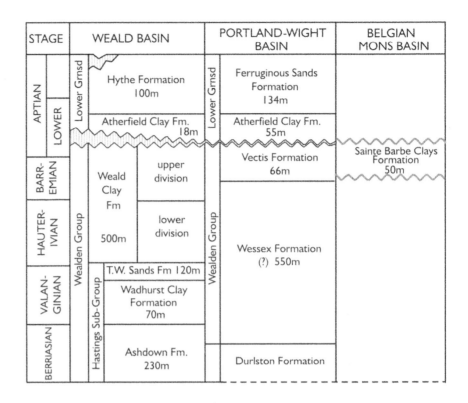

STAGE	WEALD BASIN			PORTLAND-WIGHT BASIN		BELGIAN MONS BASIN
APTIAN — LOWER	Lower Grnsd		Hythe Formation 100m	Lower Grnsd	Ferruginous Sands Formation 134m	
			Atherfield Clay Fm. 18m		Atherfield Clay Fm. 55m	
BARR-EMIAN	Wealden Group	Weald Clay Fm 500m	upper division	Wealden Group	Vectis Formation 66m	Sainte Barbe Clays Formation 50m
HAUTER-IVIAN			lower division		Wessex Formation (?) 550m	
VALAN-GINIAN		Hastings Sub-Group	T.W. Sands Fm 120m			
			Wadhurst Clay Formation 70m			
BERRIASIAN			Ashdown Fm. 230m		Durlston Formation	

Owen's magnificently illustrated Palaeontographical Society Monographs on fossil reptiles (notably *Iguanodon* and related Wealden dinosaurs; Owen, 1851, 1855, 1858, 1864, 1872, 1874) dominated the understanding of this dinosaur through the middle decades of the nineteenth century. Significant taxonomic reassessment of the Wealden material and new suggestions followed only in the wake of Owen's intellectual retirement from the active study of dinosaurs from the mid-1870s onward (Norman, 2000).

New observations and insights (complicated by the addition of further layers of taxonomic confusion and recognition of some of the inherent uncertainties,,notably by Harry Govier Seeley) emerged as a result of the work of John Whitaker Hulke (1871, 1879 [*Vectisaurus valdensis*], 1880 [*Iguanodon prestwichii*], 1882 [*Iguanodon seelyi*; see also the reported discussion with Seeley], 1885); Seeley (1875, 1883 [*Sphenospondylus*], 1887b [doubting the validity of the generic status of species assigned to the genus *Iguanodon*], 1888); Richard Lydekker (1888a [*Iguanodon dawsoni*], 1889a [*Iguanodon fittoni, I. hollintoniensis*], 1889b, 1990a, 1990b); George Albert Boulenger [*Iguanodon bernissartensis*] (*in* Pierre-Joseph Van Beneden, 1881); Louis Antoine Marie Joseph Dollo (1882a, 1882b, 1883a, 1883b, 1883c, 1884, 1885a, 1885b, 1888, 1906, 1923); and Reginald Walter Hooley (1912a, 1912b, 1917, 1925 [*Iguanodon atherfieldensis*]).

With the few exceptions of more widely geographically spread occurrences of material, *Iguanodon orientalis* Rozhdestvensky, 1952 — based exclusively on Mongolian material (see Norman, 1996); and the nomen dubium *I. ottingeri* Galton and Jensen, 1979; and *I.* [=*Dakotadon*] *lakotaensis* Weishampel and Bjork, 1989 — from the United States, no generic additions to English Wealden iguanodontians were made through the remainder of the twentieth century.

John Harold Ostrom described new iguanodontian material (*Tenonto-saurus tilletti*) from the Cloverly Formation (Albian) of Montana, and in drawing comparison and distinction with English Wealden taxa observed (with a kind of heroic understatement) that "the nomenclature of Wealden ornithopods is in a chaotic state to say the very least" (Ostrom, 1970, 132). The closing decades of the twentieth century and first decade of the twenty-first century document attempts to resolve this chaotic situation.

The status of the genus *Iguanodon* and its binomial stability was discussed by Norman (1986, 1987a, 1990) and Charig and Chapman (1998). Paul (2007) unconsciously revived the opinions expressed by Seeley (1887b) by informally proposing a new generic name for the gracile species (*Iguanodon atherfieldensis*) of upper Wealden (and its Belgian equivalent) as *Mantellisaurus atherfieldensis* (Hooley, 1925). Paul (2008) further expanded this taxonomic proposal by separating the English upper Wealden gracile iguanodontian (*M. atherfieldensis*) from the gracile Bernissart iguanodontian: proposing for the latter an entirely new binomial (*Dollodon bampingi*) and suggested that this newly created species was also to be found in English upper Wealden outcrops, Paul supported this new taxonomic proposal with several diagnoses (and also dismissed his earlier paper because it contained many errors). The taxonomic status and diversity of lower Wealden iguanodontians was also called into question (Paul, 2008), although no specific suggestions were made. Galton (2009) proposed a new generic name (*Owenodon*) for the material assigned to *I. hoggii* and supported Paul's (2008) doubts concerning lower Wealden iguanodontian taxonomy; however, he made no mention of the new upper Wealden Group genus *Dollodon*. Norman (2010), in contrast, appeared to reduce the overall diversity of lower Wealden (Hastings Group: Valanginian) iguanodontians by proposing two new generic names: *Barilium dawsoni* (Lydekker, 1888a) and *Hypselospinus fittoni* (Lydekker, 1889a), using as their basis material that had previously been assigned to three taxa: *Iguanodon dawsoni*, *I. fittoni*, and *I. hollingtoniensis*.

This contribution reviews the taxonomy of the type material and describes some of the referred skeletal remains upon which various English Wealden Group and Belgian-equivalent taxa have been based. More detailed descriptions of the individual taxa based on the holotypes and referred specimens are in preparation.

Institutional abbreviations. BBM, Booth [Bird] Museum, Brighton, U.K.; RBINS, Royal Belgian Institute of Natural Sciences; NHMUK, Natural History Museum, London (formerly the British Museum [Natural History]), U.K.

The Genus *Iguanodon* (Mantell): A Nominal Species?

As described by Norman (1993), the identity of the taxon *Iguanodon* and its nominal species (commonly assumed to have been *I. mantelli*) proved problematic. Mantell (1839) described and illustrated the first partial associated skeleton (NHMUK 3741) collected on a large slab of Kentish Rag (Worssam and Tatton-Brown, 1993) from Bensted's quarry (Maidstone, Kent; Hythe Formation [Lower Aptian]) as pertaining to the genus *Iguanodon*. Some teeth (subsequently lost) and crown impressions in the slab (Norman,

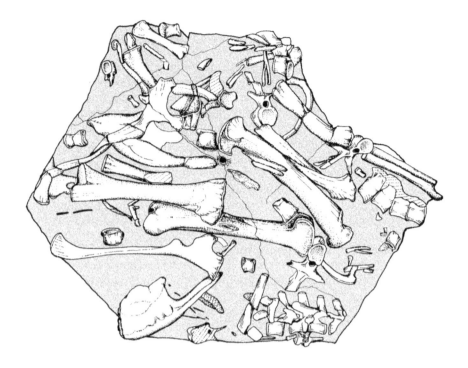

15.4. Cf. *Mantellisaurus atherfieldensis*. The "Mantel-piece" or "Maidstone Specimen" (NHMUK 3741) from the Lower Aptian (Kentish Rag) of Maidstone in Kent (Norman, 1993).

2 cm

15.5. *Iguanodon anglicus* Holl, 1829. NHMUK 2392. The lectotype of *Iguanodon anglicum* Holl, 1829 (Norman, 1986).

1993) were comparable, in general form, to those first described by Mantell (1825). This important specimen (Fig. 15.4) was redescribed and assigned formally to the species *I. mantelli* by Owen (1851).

As new taxa began to be recognized from the late 1870s onward, the need to establish some form of anatomical identity for *Iguanodon* and its nominal species became increasingly pressing. John Whitaker Hulke was the first to make an explicit statement on the matter when describing the new and extremely robust iguanodontian species: *I. seelyi*. Hulke proposed "the Iguanodon, indicated by the remains in the well-known slab figured in the Foss. Rept. Of the Cretaceous formations, pls. xxiii, xxiv, is taken as the type of *I. Mantelli*" (1882, 144). This pragmatic approach was immediately repeated by Dollo (1882a, 70): "nous croyons qu'il convient d'adopter comme type le bloc de Maidstone," and was followed by authors from that date onward, without particular regard to the earlier work of Mantell and Owen.

Norman (1986) attempted to stabilize the nomenclature of the historically important genus *Iguanodon* by focusing on the original illustrations. A lectotype was designated for *I. anglicus* (NHMUK R2392; Fig. 15.5) from the paratype series of teeth illustrated in the original description by Mantell (1825) (Fig. 15.2). Subsequent assessment of the type specimen (Charig and Chapman, 1998) led to the declaration that the lectotype and paratypes possess no unique diagnostic characters or character combinations that allow them to be distinguished from those displayed by other known Wealden iguanodontian taxa. *I. anglicus* Holl, 1829, was therefore proposed to be a *nomen dubium*, and *I. bernissartensis* Boulenger *in* Van Beneden, 1881 (by virtue of its essentially complete holotype skeleton RBINS R51 [1534] from the Sainte-Barbe Clays Formation [late Late Barremian–earliest Aptian; Yans et al., 2006], of Belgium) was proposed and subsequently recommended as the type species of the genus *Iguanodon* (ICZN 2000). The status of Mantell's original type teeth is currently being reassessed (Norman, in prep.).

A

B

5 cm

15.6. *Iguanodon hoggii* Owen, 1874. NHMUK R2998 (the holotype). An almost-complete, crushed and distorted dentary with teeth, collected from Durlston Bay, Swanage, Dorset. A, Lingual (medial) aspect; B, labial (lateral) aspect. More recently, Carpenter and Wilson (2008) referred to this specimen as an "unidentifiable euornithopodan," while Galton (2009) used it to establish a new genus: *Owenodon hoggii* (Owen, 1874) (Norman and Barrett, 2002).

Systematic Paleontology

Dinosauria Owen, 1842
Ornithischia Seeley, 1887a
Ornithopoda Marsh, 1881
Iguanodontia Dollo, 1888 [sensu Sereno, 1986]

Owenodon hoggii (Owen, 1874)

Nomenclatural Summary

Iguanodon hoggii Owen, 1874, 4.
Camptosaurus hoggii (Owen, 1874) *in* Norman and Barrett, 2002, 162.
Owenodon hoggii (Owen, 1874) *in* Galton, 2009, 241.

Holotype. NHMUK R2998, a nearly complete right dentary (Fig. 15.6) containing 15 alveoli, but missing several teeth. The dentary is crushed transversely and its anterior tip is broken; the dentary ramus shows some signs of having been abraded medially (the latter is probably the result of clumsy cold-chisel preparation of the medial surface during or after its original discovery in the 1870s).

Locality and horizon. Durlston Bay—the Durlston Formation (Lower Cretaceous; Fig. 15.3). Collected from the "Cherty freshwater beds" (middle Berriasian; Calloman and Cope, 1995).

DIAGNOSIS (*POTENTIAL AUTAPOMORPHIES)

Dentary crowns equally subdivided lingually by primary and secondary ridges that are equal in prominence, well defined, narrow, raised, and parallel*.

Tertiary ridges that originate from the bases of the apical denticles and extend thecally over the lingual surface of the crown are not present*.

Additional characters (shared notably with camptosaurian iguanodontians) that, in combination, assist in establishing the taxon's unique attributes: denticles present along the mesial, apical, and distal edges of the crowns are simple, tongue-shaped elements that lack mammillations. Denticulate margin at the widest part of the distal (posterior) corner of the crown shows no sign of inrolling to form an oblique ledge (cingulum) that borders a vertical channel along the distal edge of the root that is recessed to accommodate the immediate posterior tooth. Dentary ramus straight. (N.B. There is no obvious ventral deflection near the symphysis [Fig. 15.6]; however, postmortem crushing obscures the natural form of this area of the dentary). Fifteen alveoli. Main dentition has alveoli that form an essentially linear array, but the most posterior alveoli curve laterally into the base of the coronoid process. Coronoid process elevated, but obliquely inclined posterodorsally and not expanded anteroposteriorly near its apex. No obvious gap (usually termed a *diastema*) appears to exist between the presumed area for attachment of the predentary and the tooth-bearing margin of the dentary.

DISCUSSION

The holotype represents part of a Berriasian-aged (146–141 Ma) iguanodontian ornithopod of 2 m (or more) body length (dependent on the unknown ontogenetic stage of the holotype). The holotype, and some additional (approximately contemporaneous—Berriasian) remains, were illustrated and described in detail by Norman and Barrett (2002). The holotype, following acid preparation by the author in 1975, was reillustrated, more comprehensively described, and tentatively assigned to the genus *Camptosaurus* on the basis of an overall similarity in dentary structure and dental morphology and an apparent lack of autapomorphies (Norman and Barrett, 2002, 172). As such, the assignment then represented the youngest known "camptosaur-grade" iguanodontian.

Carpenter and Wilson (2008, 234) reassessed *I. hoggii* and stated that the dentary of the holotype "is deeper relative to its length, its dental margin and tooth row is arched, and the symphyseal ramus longer and lower. We consider these differences too great to refer NHMUK R2998 to *Camptosaurus*, contrary to Norman and Barrett (2002). At present we consider it an unnamed euornithopod." A little later in their discussion, Carpenter and Wilson (2008, 257) note that the differences (primarily the arching of the dentition) "cannot be explained as due to crushing of *C. hoggii* as suggested by Norman and Barrett (2002) because in the approximately dozen crushed dentaries of the iguanodontoid *Eolambia* Kirkland, 1998 (Lower Cretaceous, Utah), none shows arching of the dental margin or tooth row (K.C. pers. obs.)."

Galton (2009) further considered *I. hoggii* (noting the description by Norman and Barrett, as well as the commentary on it by Carpenter and Wilson, 2008). Galton concluded that this specimen merited a new generic name (*Owenodon*) on the basis of comparison with *Camptosaurus dispar*.

It should be clear from the new illustrations (Fig. 15.6; see also Norman and Barrett, 2002) that there has been a significant degree of postmortem crushing of the dentary under discussion, to the extent that the dentition has been pushed laterally (labially) and turned somewhat ventrally such that the occlusal surfaces of the functional dentary teeth face ventrolaterally, rather than dorsolaterally as they would have in life position. What the relevance of observations taken from the dentaries of an at best distantly related taxon (*Eolambia*) have to do with the interpretation of the structure of the dentition in NHMUK R2998 is quite unclear. *Eolambia* is significantly larger and considerably more robustly constructed than *Owenodon*, so postmortem effects will differ, as will those of the substantially different environmental conditions that prevailed in the two depositional environments. The length of the dentary symphysis (another claimed character) can neither be described nor measured in the holotype (as is clear from the illustrations in Norman and Barrett, 2002).

Galton (2009) followed the suggestions contained within the analysis of Carpenter and Wilson (2008) and assigned NHMUK R2998 to the new genus *Owenodon*. Galton (2009) repeated, almost verbatim, the descriptive comments and the illustrations from Norman and Barrett (2002), as well as the discussion from Carpenter and Wilson (2008), which he misquoted (241). In conclusion, Galton observed that, "given [. . .] the differences from the dentary of *Camptosaurus dispar*, and the definable nature of the teeth, *Iguanodon hoggii* Owen, 1874 is made the type species of the new genus *Owenodon*. Compared to the dentary of *Camptosaurus*, that of the holotype (NHMUK R2998) of *Owenodon hoggii* (Owen, 1874) is deeper with the dental margin and tooth row arched and a longer symphysis; the lingual surface of crowns equally subdivided by a simple, well-defined, primary ridge and a parallel and equal-size secondary ridge, tertiary ridges essentially absent, no abbreviated cingulum formed by inrolling of the denticulate distal edge of the lower distal corner of the crown" (Galton, 2009, 241).

As stated above, in refutation of the claims of Carpenter and Wilson (2008), lateral crushing of the dentary of NHMUK R2998 cannot be discounted as the causative agent for either the comparative appearance of depth of the lower jaw relative to its overall length, or to the arching of the tooth row. Equally, the length of the dentary symphysis is unknowable because this area of the lower jaw (as illustrated by Norman and Barrett, 2002, fig. 4A) is abraded and cannot be described. What remains by way of taxonomically distinct characteristics are fine details of the individual tooth crowns, as originally described and discussed by Norman and Barrett (2002). Assigning taxonomic names on the basis of the structure of reptilian teeth is a matter that needs to be approached with considerable caution.

The systematic and taxonomic status of this specimen is, as expressed earlier (Norman and Barrett, 2002, 172), "considered to be at best provisional." The dentary of the holotype (NHMUK R2998) is of a size, shape (diastema absent, alveoli that merge into the base of the coronoid process, oblique and relatively short and apically unexpanded coronoid process), and overall proportion that most closely resemble those seen in "camptosaur-grade" ankylopollexians; the teeth also share similarities in size, shape (structure of marginal denticles and pattern of principal crown ridges), and proportions with known camptosaur taxa.

Whether the differences noted in tooth morphology merit the erection of a new generic name, *Owenodon hoggii* (Owen, 1874), seems a matter of opinion based more on stratigraphic position (Lower Cretaceous rather than Upper Jurassic) combined with geographic location (Britain rather than the United States) than on the weight of anatomical differences. It is clear that this specimen records the existence of a basal (ankylopollexian) iguanodontian in the Berriasian of southern England; more skeletal material is needed to clarify the taxonomic and systematic position of this iguanodontian.

Iguanodontia Dollo, 1888 [sensu Sereno, 1986]
Ankylopollexia Sereno, 1986
Styracosterna Sereno, 1986

Barilium dawsoni (Lydekker, 1888)

Nomenclatural Summary

Iguanodon dawsoni Lydekker, 1888a, 51.
Iguanodon dawsoni [Lydekker, 1888b, 196].
Iguanodon dawsoni [Lydekker, 1890a, 38].
Iguanodon dawsoni [Lydekker, 1890b, 259].
Iguanodon dawsoni Lydekker, 1888a [*in* Steel, 1969, 18].
Iguanodon dawsoni Lydekker, 1888a [*in* Norman, 1987a, 168].
Iguanodon dawsoni Lydekker, 1888a [*in* Norman and Weishampel, 1990, 530].
Iguanodon dawsoni Lydekker, 1888a [*in* Blows, 1998, 31].
Iguanodon dawsoni Lydekker, 1888a [*in* Norman, 2004, 416].
Iguanodon dawsoni Lydekker, 1888a [*in* Galton, 2009, 245].
Barilium dawsoni (Lydekker, 1888a) Norman, 2010, 47.

Holotype. NHMUK: R798, R798b, R799, R800–R806, R4771/2, R4742 (Figs. 15.7, 15.8; illustrative examples). Despite the variety of registered numbers, these all refer to a single individual obtained from one quarry at Shornden, near Hastings.

Referred material. Includes NHMUK: R3788, R3789, R2357, R2358, R2848, R2849, R2850, R.4743, R4746. N.B. NHMUK 28660 may also be referable to this taxon, but this needs to be verified by work that is currently in preparation; NHMUK R2357 and R2358 were erroneously assigned to *H. fittoni* by Norman, (2010).

Locality and horizon. Collected from exposures of the Hastings Beds Group, Wadhurst Clay Formation (Valanginian) in quarries in the Hastings area of East Sussex, England.

DIAGNOSIS

Iguanodontian with the following unique character combination (*indicates potential autapomorphies; all other characters are apomorphic within

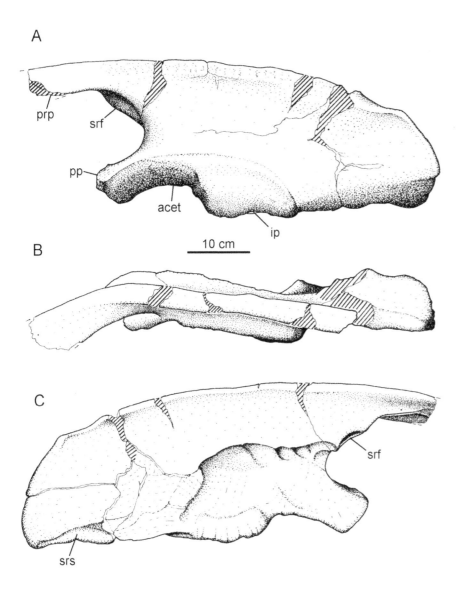

A

prp

srf

pp

acet

ip

10 cm

B

C

srf

srs

15.7. *Barilium dawsoni* (Lydekker, 1888a), NHMUK R802 (holotype left ilium). There are associated acetabular portions of the left ischium and pubis (Norman, in press b). A, Lateral; B, dorsal; and C, medial views. *Abbreviations:* acet, acetabulum; ip, ischiadic peduncle; pp, pubic peduncle; prp, preacetabular process; srf, sacral rib facet; srs, posterior sacral rib scar. Broken surfaces indicated by cross-hatching (Norman, 2010).

basal iguanodontians even if they occur sporadically within ornithopods more generally), summarized by element:

Pelvis (Fig. 15.7): preacetabular process (prp) with a prominent medial flange that arises from a large sacrodorsal rib/transverse process facet near the base of the process; rim of the sacrodorsal rib facet (srf) is visible, in lateral view, in the upper part of the recess between the preacetabular process and the pubic peduncle (pp)*; lateral surface of the preacetabular process twists axially to face dorsally toward its anterior end*; the principal body of the iliac blade is flat (Fig. 15.7); in profile, almost the entire dorsal margin of the iliac blade is gently convex, and there is a shallow indentation in the upper postacetabular margin; the small dorsal indentation is in an area of the dorsal iliac blade that shows modest lateral flaring of the (obscured by the fracturing and displacement seen across this part in the holotype; Fig. 15.7); the dorsal edge of the acetabular blade is flattened and transversely thick* (Fig. 15.7B); the postacetabular portion of the iliac blade is bluntly truncated and its lower half curves gently medially to contact the sacral yoke; very restricted brevis shelf area that is not bounded by a prominent lateral ridge*; the ventral border of the postacetabular blade lies horizontally and at the same general level as the ischiadic peduncle (ip)*;

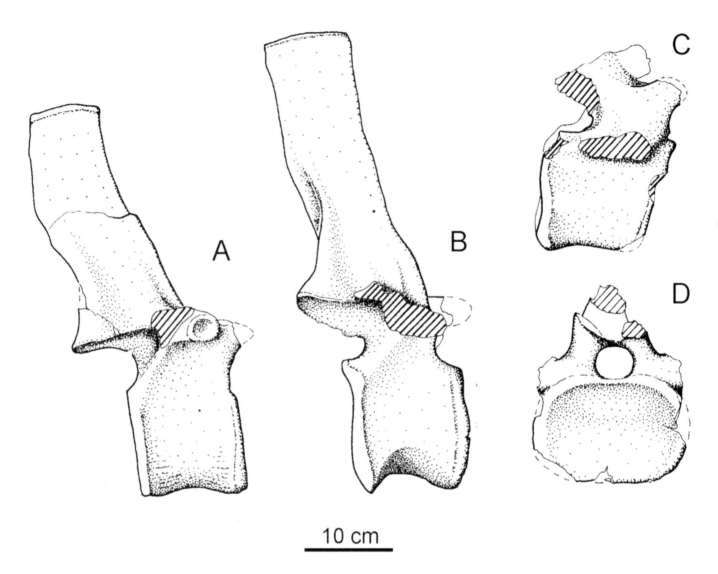

10 cm

15.8. *Barilium dawsoni* (Lydekker, 1888a). Additional examples from the holotype associated skeleton. A, NHMUK R798, middle dorsal vertebra; B, NHMUK R805, posterior dorsal vertebra; C, D, NHMUK R805, proximal caudal vertebra (in lateral and anterior views, respectively. Broken surfaces indicated by cross-hatching (Norman, 2010).

the medial surface of the ilium is prominently scarred across its ventral half where it anchors the sacral yoke but, uniquely, bears yoke scars low down along the ventral margin of the postacetabular blade* (Fig. 15.7C, srs); the entire acetabulum (formed by articulating the pubis and ischium) is remarkably large, bordering an exceptionally broad, laterally oriented, cup-shaped depression*.

Vertebrae (Fig. 15.8): middle dorsal vertebrae have spool-shaped, cylindrical centra with moderate to thin everted articular margins and transversely rounded hemal surfaces*; posterior dorsal centra have strongly everted and thickened margins; neural spines are tall, inclined and anteroposteriorly broad blades with transversely thickened apices; proximal caudal vertebrae (Fig. 15.8C,D) have broad, dorsoventrally compressed centra* (subrectangular in anterior view; Fig. 15.8D); distal caudal centra are strongly angular sided, and their articular surfaces have thin rims and are deeply amphicoelous* (see Norman, in press b, for a detailed description of this taxon).

DISCUSSION

Although Norman (1977, 1987a) retained this species within the genus *Iguanodon* while admitting, following Ostrom (1970), that the nomenclature

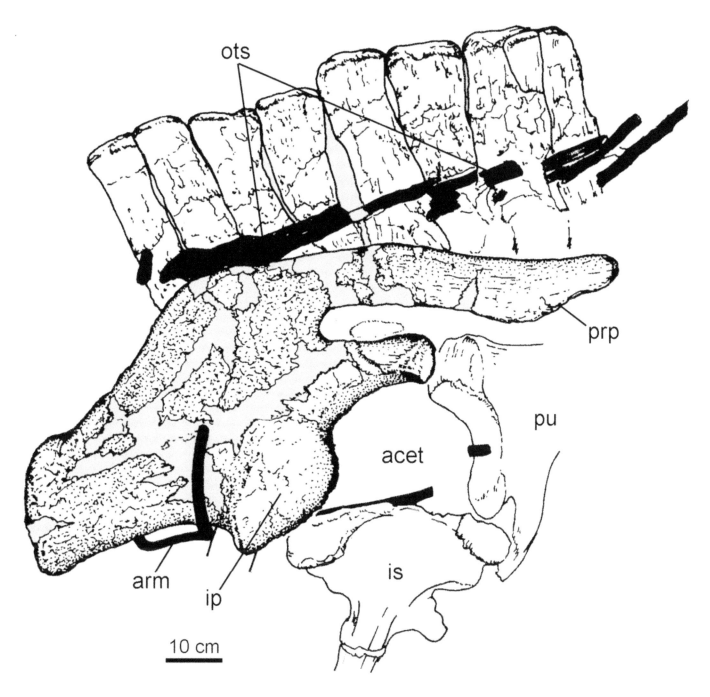

of Wealden ornithopods was in a chaotic state, the diagnosis establishes that the anatomy of this taxon, as well as its stratigraphic horizon within the Wadhurst Clay Formation (Valanginian; Fig. 15.3), separates it osteologically and chronologically from the far better described taxa of the Barremian–Lower Aptian (Weald Clay Group and Wealden Group of the United Kingdom, and the Sainte-Barbe Clays Formation of Belgium; Norman, 1980, 1986; Yans et al., 2006). For comparative observations that justify the generic distinction, see Norman (2010). The massively constructed ilium, dorsal, and sacral vertebrae with prominent deep and planklike neural spines, distinctive shapes of proximal and more distal caudal vertebral centra, and unusually large and open acetabulum combine to indicate the existence of an extremely robust/heavy-bodied/basal iguanodontian during lower Wealden Group (Valanginian) times.

15.9. Cf. *Barilium dawsoni* (Lydekker, 1888a). NHMUK R3788. Portion of a partial articulated skeleton comprising a series of dorsal, sacral vertebrae and associated pelvic elements viewed from the right side. *Abbreviations:* acet, acetabulum; arm, armature; ip, ischiadic peduncle; is, proximal end of ischium; ots, ossified tendons preserved on sides of neural spines; prp, preacetabular process of ilium; pu, acetabular portion of pubis. Even tone indicates plaster infilling.

Additional material that can be referred to this taxon also has a bearing on the understanding of its osteology, and hence of its systematic placement.

Old Roar Quarry Specimen 1 (NHMUK R3788)

Blows (1998) referred a partial articulated skeleton collected by Charles Dawson from Old Roar Quarry, near Hastings (NHMUK R3788), to *Iguanodon dawsoni* (Norman, 1977); this specimen comprises the anteroposteriorly compressed posterior dorsal vertebral column, sacrum, both ilia, and parts of the right pubis and ischium. This assignment is supported here, despite the unusual shape of the right ilium (Fig. 15.9), for reasons that are explained below. The dorsal vertebrae are massive and have expanded, cylindrical centra with expanded (but not grossly thickened — as in *H. fittoni*; see below) articular margins that conform closely to the holotype posterior dorsal vertebra (Fig. 15.8B). The neural spines are elongate and have moderately inclined blades that are slightly anteriorly curved toward the apex, and the apex of each spine is truncated and thickened transversely, as in the holotype. The acetabular region is notably large and open in construction (allowing for distortion). A composite restoration of this specimen by Blows (1998, fig. 3A) combined an outline of the holotype ilium with the dorsal and sacral series of NHMUK R3788 and some unassociated lower pelvic bones. The outline and general form of the right ilium of NHMUK R3788 as shown in Fig. 15.9, displays a sinuous dorsal margin (reminiscent of that seen in some hadrosaurids) and a very constricted preacetabular embayment beneath a notably elongate and comparatively slender preacetabular process. These features are markedly different from the general form and profile of the holotype ilium (Fig. 15.7) and require explanation if the specimen is to be considered (as proposed here) to be correctly assigned to this species and genus.

The bone texture of the right ilium exhibits evidence that its morphology has been significantly altered by preburial desiccation and weathering of the right side of the carcass and by postmortem compaction-related distortion. The surface of the ilium is extensively cracked, filled, and crazed in appearance (Fig. 15.9). The body of the ilium has also been distorted and subjected to significant shearing forces imposed upon it by the counterclockwise rotation of the sacrum, and in particular the sacral yoke that was sutured to the ventromedial surface of the iliac blade; this relative motion has sheared the lower half of the ilium dorsoposteriorly against the upper half. Confirmation that the right ilium has been significantly distorted comes from a comparison that can be made with the left ilium, preserved in place on the opposite side of the sacral block (Fig. 15.10); this side of the specimen needs to be prepared mechanically to remove much of the adhering sediment and some of the cement and plaster infill. Judging by the difference between the preservational condition of right and left sides of this partial skeleton, it seems plausible to deduce that the carcass of this animal came to rest on its left side, which became relatively quickly embedded in sediment while the right side of the carcass remained exposed to the atmosphere.

The left ilium (Fig. 15.10) does not exhibit the obvious surface crazing seen on the right side; it is robustly constructed and more closely resembles

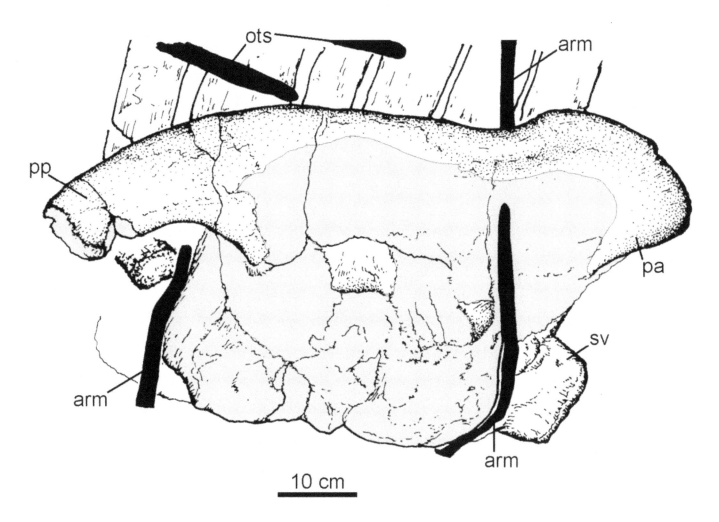

10 cm

the general shape of the holotype (Fig. 15.7). The preacetabular process is extremely robust, and its base is thick and subtriangular in cross section, as in the holotype, but its anterior tip is broken and incomplete (as is also the case in the holotype). The dorsal edge of the iliac blade describes a gently convex curve, and there is a small indentation along the postacetabular dorsal margin, accompanied by a modest lateral flaring of the upper part of the blade. The medial surface of the postacetabular blade swings smoothly ventromedially so that its lower edge contacts the adjacent sacral rib, and there is a very abbreviated brevis shelf, but there is no ridge forming the lateral border to a dorsally arched brevis fossa as seen in *H. fittoni*; the lower portion of the left ilium is broken and also obscured by matrix and a cement/plaster mix that was (presumably) used to consolidate the specimen.

Old Roar Specimen 2 (NHMUK R3789)

This almost complete sacrum (Fig. 15.11) was collected by Charles Dawson from the same quarry that yielded the partial skeleton NHMUK R3788 and may also prove to be referable to *B. dawsoni*. The sacrum of NHMUK R3788 is only properly visible dorsally where the sacral ribs and neural spines are clear and well exposed; the ventral portion is partly obscured by armature, but this area is clearly poorly preserved and infilled. The referred sacrum (Fig. 15.11) indicates the presence of 7 co-ossified vertebrae, including the sacrodorsal (sd). All the sacrals are keeled and the sacral ribs fuse laterally to

15.10. Cf. *Barilium dawsoni* (Lydekker, 1888a). NHMUK R3788. An articulated series of dorsal and sacral vertebrae and ilium viewed from the left side. *Abbreviations:* arm, portions of the supporting armature; ots, ossified tendons; pa, postacetabular blade of ilium; pp, preacetabular process; sv, portion of a posterior sacral vertebra. Even tone indicates broken bone surfaces, matrix, and some cement/plaster infill.

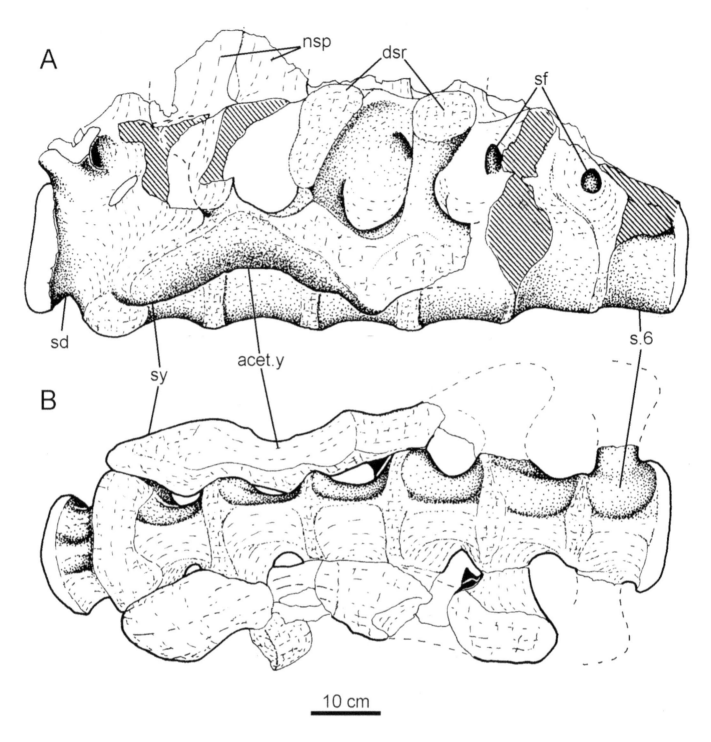

A

nsp

dsr

sf

sd

sy

acet.y

s.6

B

10 cm

15.11. Cf. *Barilium dawsoni* (Lydekker, 1888a). NHMUK R3789. A nearly complete sacrum, only missing the neural spines and some parts of the sacral ribs, collected from the same quarry that yielded the articulated remains of NHMUK R3788 (Figs. 15.9 and 15.10). A, Left lateral view; B, ventral view. *Abbreviations:* acet.y, acetabular portion of sacral yoke; dsr, dorsal portion of sacral rib; nsp, sacral neural spines (broken); s.6, sacral vertebra six; sd, sacrodorsal vertebra; sf, foramina for spinal nerves; sy, sacral yoke (fused ventrolateral ends of sacral ribs). Cross-hatching indicates broken bone surface; dashed lines suggest reconstruction outlines.

create a massive sacral yoke (sy) that forms a robust, undulating bar of bone that was sutured to the ventromedial surface of the iliac blade; the arch in the yoke that is visible anteriorly (Fig. 15.11A, acet.y) marks the sacral support for the ilium around the inner and dorsal portion of the acetabulum. The posterior part of the sacral yoke is broken away on the left side of the sacrum but may be complete on the right; it appears to have projected more or less horizontally (rather than being inclined gently dorsally), as might be expected given the relative positioning of the lower border of the postacetabular blade of the ilium seen in the holotype (Fig. 15.7C, srs). The transverse processes of each sacral vertebra form horizontal shelves that cap the dorsal portion of the sacral ribs. The anterior sacral ribs are massive and project laterally from the intercentral suture, which they span and structurally reinforce; their ventral portions expand anteroposteriorly as they extend laterally to form, when fused to their neighbors, the sacral yoke; dorsally, the sacral rib is anteroposteriorly compressed and curves medially briefly before reexpanding (dsr) to contact the midsection of the medial iliac blade. The posterior sacral ribs are substantially dorsoventrally compressed (see also Norman et al. 1987), forming just the ventral portion of the yoke. The sixth sacral (s.6) may have had a free rib, rather than contributing its rib to the yoke. Only the bases of the neural spines (nsp) are preserved and can be visibly distinguished, and so were not completely fused.

The St. Leonard's Specimen: NHMUK R2357/R2358

This material, collected by Charles Dawson from West Marina, Hastings, is also of importance in relation to our understanding of the anatomy and relationships of *B. dawsoni* (Norman, in press b). The specimen represents a partial associated skeleton of a large and robustly constructed iguanodontian comprising three articulated dentary teeth; an almost complete right shoulder girdle and forelimb: scapula, coracoid, sternal bone, humerus, radius, and ulna, the co-ossified carpus and base of a fused pollex spine; parts of the hind limb; and a number of vertebrae, including heavily constructed cylindrical dorsal centra and autapomorphically squat proximal caudals (Norman, 2010, fig. 4C,D) and angular-sided and amphicoelous distal caudals (Norman, in press b).

NHMUK R2357/8 demonstrates that this taxon was unquestionably ankylopollexian and styracosternan (*contra* Norman, 2010); as a supplementary and rather tantalizing observation, its teeth appear to be similar (if not identical) to the specimen that was designated as the lectotype (NHMUK 2392) of *Iguanodon anglicus* (Fig. 15.5; Norman, 1986, and in prep.).

COMMENT

Barilium dawsoni is of extremely robust build and contrasts (partially) with its contemporary (*H. fittoni*), which though clearly robust (certainly in the construction of its forelimb and manus) is still relatively smaller and of lighter overall construction compared with *B. dawsoni*. A similar degree of disparity in form has been demonstrated among contemporary iguanodontians of the upper Wealden Group (*I. bernissartensis* [robust] and *M. atherfieldensis* [gracile] as well as in the similarly contemporary Niger species (*Lurdusaurus arenatus* [robust] and *Ouranosaurus nigeriensis* [gracile]). The biological significance of this apparent pattern of disparity among contemporary pairs of ornithopods is not yet understood.

NOTE ADDED IN PROOF

McDonald et al. (2010) erected a new taxon, *Kukufeldia tilgatensis*, on the basis of a large, near-complete, dentary (NHMUK 28660) and have also referred a partial skeleton (NHMUK R1834) to the taxon *Barilium dawsoni*. It is considered here (see also Norman, in press b) that the type specimen of *Kukufeldia* is more probably referable to *B. dawsoni* (on the basis of its size, robust build and similarities between the teeth preserved in this specimen and in a specimen; NHMUK R2358). While there is a single autapomorphy that supports the new taxon (the arrangement of vascular openings found near the anterolateral tip of the dentary), this is not considered to be of great value, given the rarity of occasions when such details of jaw anatomy are preserved in these Wealden taxa. It is also the case that the referred specimen (NHMUK R1834) was mistakenly referred to *B. dawsoni* by McDonald et al. (2010); the ilium of NHMUK R1834, which formed the basis of the referral, does not resemble that of the holotype of *Barilium* (NHMUK R802) and is practically indistinguishable from that of the holotype and other referred specimens associated with *Hypselospinus fittoni* (Norman, 2010, and in prep.). The importance of the latter point is that a very poorly preserved dentary fragment associated with NHMUK R1834, which displays shallow arching of the dentary ramus, clearly does not represent the form of the dentary in *Barilium*. The robust and straight form of the dentary ramus of NHMUK 28660 and its distinction from that of the misidentified dentary of NHMUK R1834 was one of the primary reasons for erecting the new taxon; these taxonomic points will be dealt with more thoroughly elsewhere (Norman, in press b, and in prep.).

Carpenter and Ishida (2010) published a further attempt to review iguanodontian taxonomy; this work was published after the papers by Norman (2010) and McDonald et al. (2010). Several new taxa were created, which this article describes; these prove to be synonyms of previously created taxa or unfounded suggestions. This appears to reflect a lack of detailed knowledge of the original material and is representative of a style of work that is not appropriate within the discipline of taxonomy. The issues raised by the content of this paper will be considered elsewhere (see Norman, in press b, and in prep.).

Iguanodontia Dollo, 1888 [sensu Sereno, 1986]
Ankylopollexia Sereno, 1986
Styracosterna Sereno, 1986

Hypselospinus fittoni (Lydekker, 1889a)

Nomenclatural Summary

Iguanodon fittoni Lydekker, 1889a, 354.
Iguanodon hollingtoniensis Lydekker, 1889a, 355.
Iguanodon fittoni [Lydekker, 1890a, 38].
Iguanodon hollingtoniensis [Lydekker, 1890a, 40].
Iguanodon fittoni Lydekker, 1889a [Lydekker, 1890b].
Iguanodon hollingtoniensis Lydekker, 1889a [Lydekker, 1890b].
Iguanodon fittoni Lydekker, 1889a [*in* Steel, 1969, 18].
Iguanodon hollingtoniensis Lydekker, 1889a [*in* Steel, 1969, 19].
Iguanodon fittoni Lydekker, 1889a [*in* Norman, 1987a, 164].
Iguanodon fittoni Lydekker, 1889a [*in* Norman and Weishampel, 1990, 530].
Iguanodon fittoni Lydekker, 1889a [*in* Blows, 1998, 31].
Iguanodon fittoni Lydekker, 1889a [*in* Norman, 2004, 416].
"*Iguanodon*" *hollingtoniensis* [Paul, 2008, 214].
"*Iguanodon*" *hollingtoniensis* [Galton, 2009, 245].

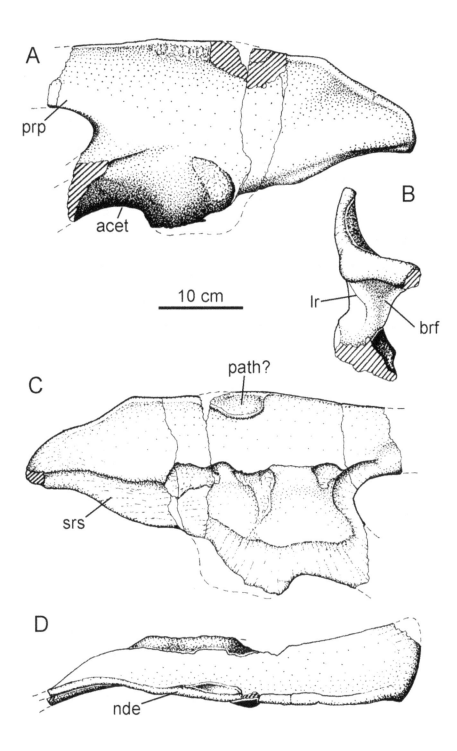

15.12. *Hypselospinus fittoni* (Lydekker, 1889a). NHMUK R1635 (holotype). Right ilium. A, Lateral view; B, posterior view showing the inflected shelf, lateral ridge, and brevis fossa; C, medial view showing the form of the sacral rib scars; D, dorsal view showing the transversely narrow dorsal margin of the ilium compared with that of *B. dawsoni*. The mediodorsal edge of the ilium shows an abscess-like excavation that may be pathological in origin. Even though the dorsal margin is eroded laterally posterior to the ischial peduncle (cross-hatched area), there is relatively little indication of pronounced thickening or lateral eversion. *Abbreviations:* acet, acetabulum; brf, brevis fossa; lr, lateral ridge that bounds the brevis fossa; nde, narrow dorsal edge to the iliac blade; path?, probable pathological structure; prp, preacetabular process; srs, sacral rib scar on medial surface of iliac blade. Broken surfaces indicated by cross-hatching (Norman, 2010).

Holotype. NHMUK R1635 (ilium; Fig. 15.12). Loosely associated (paratype series) elements include a partial posterior sacrum, anterior caudal centrum, and (more dubiously) a water-worn proximal ischial fragment (see Norman, 2010).

Referred material. NHMUK R1148 (an associated skeleton that includes registered numbers R1629, R1632); this may also include some of the specimens numbered R604 and R811, R811a,b; R33, R966, R1627, R1636, R1831 (R1831 is an associated skeleton that also includes material with the registered numbers R1832, R1833 and R1835) and an entirely separate partial skeleton numbered R1834. A more complete listing and full osteological description of this taxon is in preparation.

Locality and horizon. Quarry site at Shornden, near Hastings, East Sussex. Hastings Beds Group, Wadhurst Clay Formation (Valanginian).

DIAGNOSIS (BASED SOLELY ON THE HOLOTYPE)

Iguanodontian with the following unique character combination (*indicates potential autapomorphies—all other characters are apomorphic within basal iguanodontians, even if they occur sporadically within ornithopods more generally), summarized by element:

Ilium (Fig. 15.12): preacetabular process (prp) vertically orientated and concave laterally, laterally compressed, and comparatively deep proximally*; ventral edge of the preacetabular process is slightly thicker than the dorsal edge*; a low-relief, curved ridge is present on the medial surface of the preacetabular process; the main portion of the iliac blade is flat; the dorsal edge of the iliac blade is narrow and transversely compressed* (Fig. 15.12D, nde); the dorsal edge of the central portion of the iliac blade, in profile view (Fig. 15.12A), is nearly straight (rather than being bowed dorsally)*; the postacetabular blade tapers to a blunt, transversely thickened bar (Fig. 15.12A,B); strong medial inturning of the ventral half of the postacetabular blade forms a vaultlike roof to a well-defined brevis fossa (brf); the brevis fossa is bordered laterally by a horizontal ridge (lr); the scarring for the posterior sacral ribs (Fig. 15.12C, srs) follows the ventral margin of the postacetabular blade as it rises obliquely toward the posterior end of the blade to merge with the more dorsally positioned facets along the lower border of the postacetabular blade.

Vertebrae (Norman, 2010, fig. 6): the ventral surfaces of the posterior centra of the sacrum bear narrow midline keels and the intervertebral suture are comparatively little expanded; midanterior caudal vertebral centrum cylindrical*.

ANATOMICAL OBSERVATIONS

Comparisons that confirmed the establishment of this genus, and summarized some aspects of its unique anatomy, were presented in Norman (2010). This Valanginian-aged styracosternan iguanodontian (~6-m body length) is known from a substantial amount of material representing several skeletons collected from quarries and foreshore localities in the Hastings area by two resourceful collectors: Samuel Husbands Beckles and Charles Dawson; their discoveries resulted in a number of publications (Owen, 1872, 1874; Lydekker, 1888–1890). As indicated (Norman, 2010), *Hypselospinus* differs consistently from *Barilium* in the structure of its jaws and teeth, the axial skeleton, and pelvis; a full description of *H. fittoni* is in preparation.

The neural spines of the dorsal, sacral, and proximal caudal series of vertebrae are striking because of their oblique, narrow, and elongate form (Fig. 15.13). Cervicodorsal vertebral centra have the shape of low cylinders, with markedly thickened articular margins; viewed laterally, the cervico-anterodorsal vertebral centra are tilted posteriorly; proximal caudals also exhibit very elongate, narrow neural spines (Norman, 2010, fig. 9C), and the lower portion of the anterior and posterior edges of the spines have a prominent midline ridge, on either side of which is a deep vertical recess; proximal caudal centra appear to be deep and strongly

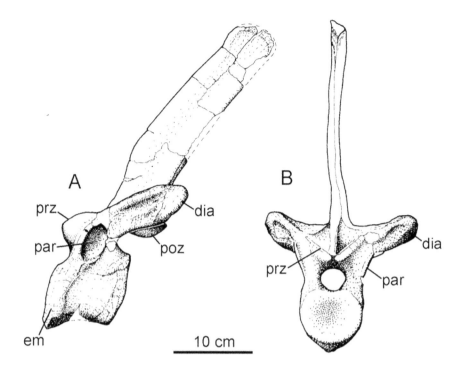

15.13. Cf. *Hypselospinus fittoni.* NHMUK R604. Anterior–middorsal vertebra. A, Lateral view; B, anterior view. *Abbreviations:* dia, diapophysis; em, everted articular margin of centrum; par, parapophysis; poz, postzygapophysis; prz, prezygapophysis.

anteroposteriorly compressed, and they exhibit a vertical rather than tilted orientation (Norman, 2010, fig. 9C).

The postacetabular blade of the ilium (Fig. 15.12) tapers to a point, with a well-developed brevis shelf marked off by a horizontal lateral ridge on the postacetabular blade; the dorsal margin of the postacetabular blade is marked by a rugose lateral flare (eroded away in the holotype) of the dorsal margin, positioned marginally posterodorsal to the ischiadic peduncle; the anterior iliac process is elongate and little twisted along its length, laterally compressed and bears a low-relief, curved, medial ridge (Norman, 2010, fig. 8D, mr). Forelimb and shoulder girdle are robust and show evidence of partial or complete fusion of elements such as the sternals (Hulke, 1885, pl. XIV); the carpals display co-ossification, and there is a large, transversely compressed (triangular in lateral aspect) pollex ungual (Owen, 1872, pls. I–III). The hind limb was clearly robust: the femur (Norman, 2010, fig. 7) is relatively stout and bowed, and has a thick, pillarlike anterior (lesser) trochanter projecting from the anterolateral corner of the greater trochanter; there is also a narrow anterior (slotlike) exposure of the anterior intercondylar groove.

COMMENT

The type material of *I. fittoni* and *I. hollingtoniensis* is indistinguishable. The type material attributed to *I. fittoni* establishes the species and takes priority over *I. hollingtoniensis* by virtue of the sequence in which these two taxa appear in print; the latter species has been declared a junior subjective synonym of *I. fittoni* (Norman, 2010). *H. fittoni* was evidently a relatively robustly constructed styracosternan iguanodontian, judging by its forelimb structure. Nevertheless, it has a comparatively slender and slightly ventrally deflected anterior portion of its dentary (NHMUK R1831, R1834) which also displays a modest diastema (*contra* Paul 2008). *Barilium* appears to have had a deep, essentially straight dentary ramus with little or no room for a diastema (NHMUK 28660). *H. fittoni* also had a much more lightly built pelvis and a vertebral column that combined comparative small vertebral centra with unusually expanded centrum articular surfaces and unusually elongate, slender neural spines throughout the dorsal, sacral (probably), and anterior caudal series.

Iguanodontia Dollo, 1888 [sensu Sereno, 1986]
Ankylopollexia Sereno, 1986
Styracosterna Sereno, 1986

Iguanodon bernissartensis Boulenger *in* Van Beneden, 1881

Nomenclatural Summary

Iguanodon bernissartensis Boulenger *in* Van Beneden, 1881, 606.
Iguanodon seelyi Hulke, 1882, 135.
Iguanodon bernissartensis Boulenger, 1881 [*in* Dollo, 1882a, 55].
Iguanodon bernissartensis Boulenger, 1881 [*in* Steel, 1969, 17].
Iguanodon bernissartensis Boulenger, 1881 [*in* Ostrom, 1970, 132].
Iguanodon bernissartensis Van Beneden, 1881 [*in* Norman, 1980, 10].
Iguanodon bernissartensis Boulenger, 1881 [*in* Norman, 1986, 282].
Iguanodon bernissartensis Boulenger, 1881 [*in* Norman and Weishampel, 1990, 530].
Iguanodon bernissartensis Boulenger, 1881 [*in* Norman, 2004, 416].
Iguanodon bernissartensis Boulenger, 1881, *in* Van Beneden [*in* Charig and Chapman, 1998, 61]
N.B. *I. seelyi* was, by general consent, relegated to a position of junior subjective synonym of *I. bernissartensis* following the review of Dollo (1882), summarized by Lydekker (1888a, 46), Ostrom (1970, 133), and Norman (1980, 10).

Holotype I. bernissartensis. RBINS R51 [1534], an almost complete articulated skeleton (Norman, 1980).

Holotype I. seelyi. NHMUK R2501–2509. Partial remains of the pelvis, hind limb (see Figs. 15.14, 15.15), forelimb, and vertebral column.

Locality and horizon of the holotypes. I. bernissartensis: Bernissart, Hainaut, Belgium. Sainte-Barbe Clays Formation (Late Barremian–earliest Aptian; Yans et al., 2006). *I. seelyi*: Brook Chine, Isle of Wight (U.K.). Wealden Group, Wessex Formation, (Late Barremian–Early Aptian).

Referred specimens. See Norman (1986, appendix 1) for a complete listing of specimens from the Sainte-Barbe Clays Formation considered referable to *Iguanodon bernissartensis*. Hulke (1882) listed the holotype material of *I. seelyi*. Additional referable material is also present in the collections at the Natural History Museum (Lydekker, 1888b, 1890b), notably as part of the Rev. W. D. Fox collection and more recent acquisitions (e.g., the Rivett and Chase collections; Norman, 1977, manuscript catalog) and a substantial discovery of remains of similar type were reported from Germany (Norman, 1987c; Norman et al., 1987).

Iguanodontian with the following unique character combination (*indicates potential autapomorphies—all other characters are apomorphic within basal styracosternan iguanodontians, even if they occur sporadically within ornithopods more generally), summarized by element:

Cranial: two palpebral bones* (the larger is elongate and bowed dorsally and has a blunt termination, and articulates against the lateral surface of the prefrontal while its medial portion was anchored in connective tissue to the dorsal orbital margin); the accessory palpebral is a small irregular nubbin found immediately behind the distal tip of the larger palpebral and lying against the surface of the postorbital*. A short (~1 tooth width) diastema is present between the posterior end of the predentary and the first tooth position. Massive, robust, and straight, or very slightly arched, dentary ramus* (significant ventral deflection of the anterior portion of the dentary ramus is invariably associated with breakage and distortion in the Bernissart specimens; see Norman, 1980, pls. 1–4). Coronoid process elevated and perpendicular to the long axis of the dentary, and its apical region is anteroposteriorly expanded. The antorbital fossa forms an oblique channel between the anteromedial orbit margin and the lateral surface of the snout, which opens out on to the maxilla via a small oval fenestra. Quadrate shaft is pillarlike and straight and has an extensive vertical buttress that extends down the posterior edge of the shaft of the quadrate from its dorsal (articular) head. Quadrate (paraquadratic) foramen is formed by the quadratojugal bridging an embayment, positioned just beneath the midheight of the quadrate*, in the jugal/quadratojugal wing of the quadrate.

Dental: vertical tooth column numbers are variable (depending up individual size and presumed ontogenetic stage), reaching an apparent general maximum of 29 (maxilla) and 25 (dentary). Dentary crowns broad and slightly recurved, with a blunt apex; the lingual surface exhibits a prominent distally offset primary ridge and well-developed (but broader) mesially offset secondary ridge, tertiary ridges seem to be comparatively rare. Marginal denticles on the mesial and distal edges of the crown for ledges that bear small, irregular mammillae. Maxillary crowns are narrower than dentary crowns and display a very prominent primary ridge that is slightly offset distally: the labial surface of the crown is bisected by the primary ridge, thereby creating two subequal channels delimited by the thickened denticulate edges of the crown; these surfaces are essentially flat, and the mesial surface is commonly adorned by tertiary ridges that run thecally (toward the base of the crown) roughly parallel to the primary ridge. (These ridges are not consistently expressed on all teeth in any one dentition.) Thickening and partial inrolling of the distal and mesial denticulate margins of the crown, just thecal to its widest point leads to the formation of oblique, thickened ridges (loosely referred to as cingula) that bound the labial edges of grooves to accommodate adjacent replacement crowns.

Axial skeleton: dorsal, sacral, and caudal vertebrae have large, parallel-sided neural spines that are comparatively closely packed with little space between successive spine blades (resembling those seen in B. dawsoni). Dorsal centra become increasingly opisthocoelous as the sacrum is approached*; centra are generally tall and narrow in lateral and axial aspects, with a transversely rounded ventral surface; posterior dorsals remain tall sided, but broaden and produce an almost circular (indented dorsally by the neural canal) producing a broadly heart-shaped articular face when viewed axially*. Ossified tendons form a complex, multilayered trellis on either side of the neural spines. Ossified tendons exhibit fusion to the lateral faces of the sacral neural spines. Sacrum comprises eight fused vertebrae with a broad ventral sulcus in the midline of the more posterior sacral centra*.

Appendicular skeleton: Scapular blade tends to be comparatively slightly expanded distally and modestly recurved in lateral view (though there is some variability in these proportions, some scapulae have a more flared and curved distal blade), with the proximal portion of the shaft expanded and massive adjacent to the scapulocoracoid suture and humeral glenoid*. Coracoid massive and dished, bearing a coracoid foramen that forms a pronounced notch positioned on the scapulocoracoid suture when viewed externally*. An "intersternal ossification" (Norman, 1980, 47) is commonly preserved in the midline of the upper chest in the space between the coracoids and sternal bones* (the sternals have never been observed to co-ossify as observed in the specimen referred to *H. fittoni*; NHMUK R1835; Hulke, 1885, Norman, in prep.). Forelimb at least 70% of the length of the hind limb, and robustly constructed*. Carpals heavily co-ossified and supported by ossified ligaments*. Phalangeal count of the manus: 2:3:3:2:4*. First phalanx of digit 1 of the manus is preserved as a flattened and warped plate that lies in a recess in the base of the pollex ungual*. Pollex is long and conical with a slight medial curvature along its length*; and the united phalanx 1–pollex can articulate freely (and transversely) against the rollerlike articular surface of metacarpal 1*. Ungual II of the manus is elongate but flattened and twisted along its length; ungual III is shorter and broader, and resembles a foreshortened pedal ungual. Precetabular process has a robust base that is triangular in cross section, the ramus projects anteriorly and is deflected ventrally, while its lateral surface becomes twisted dorsolaterally toward its tip (the latter is flattened and slightly expanded; the dorsal margin of the ilium is primarily convex in lateral view and is notably thickened transversely and appears to flare laterally over the ischiadic peduncle; the ischiadic peduncle appears not to be strongly expanded laterally to form a prominent stepped boss as seen in *Mantellisaurus*, *Barilium*, and *Hypselospinus*; the postacetabular blade of ilium narrows distally in profile view and bears a broad, elongate brevis shelf and fossa that is shallowly arched and bounded laterally by a prominent ridge*. Anterior pubic ramus is stout, having a comparatively narrow, relatively elongate, and transversely thickened proximal portion, which becomes laterally compressed distally and expands both dorsally and ventrally*; the posterior pubic ramus slender and tapering and less than half of the length of the ischial shaft. Ischium shaft stout, J-shaped in lateral view, and terminates in an anteriorly expanded boot. Three distal tarsals are present*. Metatarsal I reduced to a transversely flattened splint that lies obliquely against the shaft of metatarsal II*.

Despite numerous short (preliminary) notes by Dollo (1882–1923; see the bibliography in Norman, 1980) on the aspects of the anatomy of *I. bernissartensis*, full details of the anatomy of this taxon were only described for the first time by Norman (1980). At the time of its original description (by Dollo), the issue of greatest concern was the status of the taxon: it had been named on the basis of a published review of a manuscript by G. A. Boulenger that had been rejected by the Royal Belgian Academy of Sciences. The review (Van Beneden, 1881) quoted information from Boulenger's manuscript and thereby reproduced the substance of his original proposal; this was that a new name (*Iguanodon bernissartensis*) should be created for the large and robust Bernissart skeletons on the basis of one anatomical difference: the larger number of sacral vertebrae (six), compared to five described in a sacrum (NHMUK 37685) attributed to *I. mantelli* by Owen (1842, 1855, tables III to VI). Van Beneden's review criticized Boulenger's proposal by observing that the proposed anatomical difference was most probably attributable to sexual dimorphism. Despite the critical nature of the review and the fact that no illustrations were provided, the taxonomic name became current within the scientific community at large, as it became obvious that two distinct morphs (large and robust versus small and gracile) were preserved in the material being retrieved from Bernissart. Boulenger left the Royal Museum of Natural History in Brussels shortly after this episode to take up a position at the Natural History Museum, London; the task of studying the material from Bernissart was then given to Louis Dollo.

In 1882, J. W. Hulke described some large iguanodontian remains that he collected from Brook [Brooke] Chine on the Isle of Wight in 1870 (Figs. 15.14 and 15.15). Having visited the Belgian collections in August 1879 and ascertained that a detailed description of the remains was not "to be expected for several years" (Hulke, 1882, 135), he decided to describe and illustrate his material in detail and compared it with other known material (using the by then widely accepted convention of regarding the type of *Iguanodon mantelli* for reference purposes, as the Maidstone Block or "Mantel-piece"—NHMUK 3741; Norman, 1993). Clear anatomical differences between the supposed type specimen and the 1870 discoveries prompted Hulke to suggest a new name for this distinctive material: *Iguanodon seelyi*.[2] In the formal discussion that was reported at the end of the scientific paper (as part of the published proceedings of the Geological Society of London meetings), it should be noted that H. G. Seeley "doubted whether a form with an ilium so different from that of *Iguanodon Mantelli* ought to be referred to the same genus. The same doubts were suggested by the different proportions of the limb-bones, and by peculiarities of the vertebrae, as compared to those of *Iguanodon*" (Seeley *in* Hulke, 1882, 144); Seeley's somewhat contrary views were published at greater length later (Seeley, 1887b).

Later that same year (1882), Louis Dollo published the first of a series of comparatively brief papers covering topics concerned with the taxonomy, osteology, functional morphology, systematics, and paleobiology of these Bernissart fossil reptiles; these articles appeared regularly and became

15.14. *Iguanodon seelyi* Hulke, 1882. NHMUK R2502. Holotype left ilium. A, Lateral view; B, medial view. *Abbreviations:* acet, acetabulum; bs, brevis shelf/fossa bounded by lateral ridge; ca, healed fracture at base of preacetabular process of ilium; edm, everted and thickened dorsal margin of postacetabular iliac blade; ip, ischiadic peduncle; pp, pubic peduncle; prp, preacetabular process (distal tip broken off).

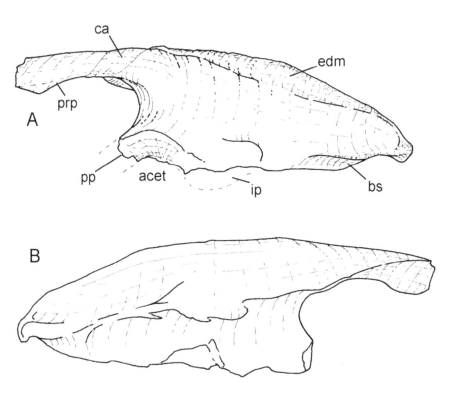

increasing telegraphic in style until his last paper on the topic (Dollo, 1923), which was, in essence, a numbered list of observations. The first paper (Dollo, 1882a) examined the basis for the creation of the new species *I. bernissartensis* as distinct from *I. mantelli*. Dollo established an overall similarity in anatomy between the smaller and more gracile species from Bernissart RBINS R57 [1551] and the remains of the "Mantel-piece" (NHMUK 3741) and therefore by convention identified RBINS R57 [1551] as *Iguanodon mantelli* (Norman, 1993).

With respect to the larger species, Dollo (1882a) circumvented the problems of sexual dimorphism in the sacral count by demonstrating a wider range of additional anatomical differences: skull proportions, size of narial openings, shape of the orbit, size and shape of the infratemporal openings, shape of scapular blade, completeness of external coracoid foramen and overall shape of the coracoid, size of the humerus, proportions of the manus and pollex, shape of anterior pubic blade. Dollo proceeded to discuss (and dismiss) the validity of Van Beneden's assumption that these were sexual morphs; he finally concluded (given what was then known of the range and variation within and between species and genera of dinosaurs) that in view of their overall similarity in anatomy, they merited being considered as separate species. Seeley (1887b) used precisely the same evidence to argue that these two forms were generically separate. Despite the interventions by Seeley, a vaguely worded consensus over these vexed taxonomic issues seems to be indicated by Lydekker (1888a, 46): "I think we ought to adopt the former and earlier name for the species which has been so well described by the Belgian naturalist."

This short and essentially historical review is not an appropriate forum within which to discuss the relative merits of *Iguanodon seelyi* and *I. bernissartensis* in detail; a complete description of *I. seelyi* is currently

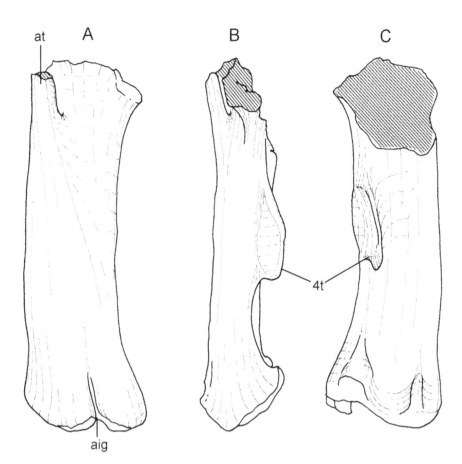

15.15. *Iguanodon seelyi* Hulke, 1882. NHMUK R2503. Holotype right femur. A, Anterior view; B, medial view; C, posterior view. *Abbreviations:* aig, anterior intercondylar groove; 4t, fourth trochanter; at, anterior (lesser) trochanter. Broken surfaces indicated by cross-hatching.

in preparation, and work by C. L. Marquart (in prep.) will address the taxonomical, and by implication biological, significance of morphometric differences between these iguanodontian taxa, as well as the morphological variation displayed by several Wealden ornithopod taxa. For the present, it is worth simply recording at this stage that *I. seelyi* is larger and more robust than the most robust morphs of *I. bernissartensis* collected from Bernissart.

Iguanodontia Dollo, 1888 [sensu Sereno, 1986]
Ankylopollexia Sereno, 1986
Styracosterna Sereno, 1986

Mantellisaurus atherfieldensis (Hooley, 1925)

Nomenclatural Summary

Iguanodon atherfieldensis Hooley, 1925, 1.
Iguanodon atherfieldensis Hooley, 1924 [*sic*] [*in* Steel, 1969, 19].
Iguanodon atherfieldensis Hooley, 1925 [*in* Ostrom, 1970, 132].
Iguanodon atherfieldensis Hooley, 1925 [*in* Norman, 1986, 283].
Iguanodon atherfieldensis Hooley, 1925 [*in* Norman and Weishampel, 1990, 530].
Iguanodon atherfieldensis Hooley, 1925 [*in* Norman, 2004, 416].
Mantellisaurus atherfieldensis (Hooley, 1925) Paul, 2007, 69.
Dollodon bampingi Paul, 2008, 192. (*partim*)

Holotype. NHMUK R5764 (major proportion of an associated skull and partially articulated skeleton).

Referred specimens. RBINS R57 [1551], an almost complete skeleton from the Sainte-Barbe Clays Formation at Bernissart, Belgium, along with some associated remains in the Conservatoire of the Royal Belgian Institute of Natural Sciences that was listed in Norman (1986); there are also important collections of referable material, notably those made by Fox, Rivett, and Chase in the Natural History Museum (London), as well as important material preserved in the collections of the Museum of Isle of Wight Geology (Norman catalog MS, in prep.).

Locality and horizon. Holotype: Atherfield Point, Brook Bay, Isle of Wight (U.K.). Wealden Group, Wessex Formation (upper Barremian).

N.B. Similar remains have been recovered from the Weald Clay Group (Barremian–Lower Aptian) of Sussex and Surrey (U.K.); Kentish Rag (Barremian–Lower Aptian) of Kent (U.K.); Norman (1986) consolidated into this species the material assigned to the nominal taxon *I. mantelli* by Dollo (1882)—notably RBINS R57 [1551] and one smaller articulated skeleton, which was later assigned to the new *Dollodon bampingi* (Paul, 2008).

DIAGNOSIS

Styracosternan iguanodontian with the following unique character combination (*indicates potential autapomorphies—all other characters are apomorphic within basal iguanodontians, even if they occur sporadically within ornithopods more generally), summarized by element:

Cranial and dental: long and low with an extended, ventrally deflected snout* and a slender lower jaw with a well-developed (~3–4 crown widths) diastema separating the presumed position of the posterior tip of the predentary and the first dentary tooth position*. Palpebral (single element) articulates with the prefrontal and is long, dorsally bowed to follow the dorsal orbital margin, and tapers to a point* (the medial edge of the palpebral was bound to the orbital margin by connective tissue). Shaft of quadrate bowed anteriorly*, short posterior buttress to the articular head*. Tooth column counts: maxilla 23+, dentary 20+, dependent largely on ontogenetic stage of the individual.

Axial skeleton: dorsal vertebrae have centra that have the form of low, rectangular cylinders (when viewed in lateral aspect) with a tendency to form a small, ventral midline keel, posterior dorsals are broader, deeper, and more circular in axial aspect; neural spines are well developed across the dorsal–sacral–anterior caudal series and form thick, parallel-sided blades that are inclined posteriorly and proportionally taller (compared to the height of the dorsal centrum) than the spines of *I. bernissartensis*; on average, spines of *M. atherfieldensis* are 2.5 times centrum height, compared to 2 times centrum height in *I. bernissartensis*; ossified tendons form a multilayered lattice and shows evidence of anchorage by fusion to the sacral plate of neural spines (also seen in *I. bernissartensis*), and there is some evidence to suggest that the sacral neural spines are partially fused in both species. Ventral surfaces of the sacrum, especially posteriorly, are either rounded transversely or slightly keeled*; they do not show the broad ventral sulcus seen in *I. bernissartensis*. Sacrum comprises seven sutured vertebrae (including one sacrodorsal).

Appendicular skeleton: scapular blade contracted above the articular region and flares modestly distally, the whole blade being slightly recurved (several individuals of *I. bernissartensis* exhibit modest flaring of the scapular blade and recurvature, but they differ primarily in overall mediolateral thickness); coracoid has a fully enclosed coracoid foramen (laterally) that is separated by a short distance from the scapulocoracoid suture*; phalangeal formula (2?:3:3:3:3); manus is narrow and lightly constructed; carpus appears to be lightly co-ossified and does not show evidence of reinforcement by means of ossified ligaments; metacarpals II–IV comparatively slender*; pollex ungual is notably short, conical, and symmetrical*; proximal phalanx of digit II is slender and twisted medially along its length*; sternal bones have a shaft that is twisted laterally* (medially in *I. bernissartensis*). Ilium: preacetabular process is long, slender, laterally compressed with a modestly developed medial ridge; the entire process curves gently ventrally (unlike that of *I. bernissartensis*, there is relatively little axial twisting of the lateral surface into a more dorsolateral orientation*); the main body of the dorsal iliac blade is relatively straight and narrow, with a slight lateral overhang to its dorsolateral edge*, and beyond the region above the ischiadic peduncle, the dorsolateral surface is beveled and more expanded by a roughened surface beyond which the dorsal edge turns ventrally toward the posterior tip; this region is also expanded medially to contact the sacral rib medially, and generates a small posteriorly positioned, arched brevis fossa; the ventral margin of the postacetabular iliac blade rises to meet the upper edge of the iliac blade* and has a small medioventral flange that is visible in lateral aspect* (see Norman, 1986, fig. 54); the ischiadic peduncle is expanded laterally and stepped; the anterior pubic blade is short, deep, and laterally compressed, forming a thin blade*, and flares distally more gradually than that of *I. bernissartensis*; ischial shaft appears straight or only slightly decurved along its length and appears to have a relatively small distal expansion to form a boot*; metatarsal I appears to form a narrow, pencil-like, splint that lies parallel to and sutured against the medial surface of metatarsal II* (metatarsal I was not recovered in either NHMUK R5764 or RBINS R57 [1551], but this is not yet regarded as evidence of absence). One specimen of *M.* cf. *atherfieldensis*—NHMUK R1829 (Norman, 1986, fig. 62)—shows the existence of a thin splintlike metatarsal I that is partially displaced, suggesting that this bone was not firmly sutured in place against metatarsal II (and may well be more readily lost during normal biostratinomic processes).

DISCUSSION: THE TAXONOMIC STATUS OF GRACILE BARREMIAN–LOWER APTIAN IGUANODONTIANS

Paul (2007) proposed a new generic name *Mantellisaurus atherfieldensis* for the holotype material of *Iguanodon atherfieldensis* (following a similar line of reasoning to that of Seeley in the 1880s). There are a number of clear anatomical differences summarized within the composite diagnoses for *I. bernissartensis* and *M. atherfieldensis* (above) that allow these taxa to be distinguished, which prompted Norman to identify them as "osteological species" (Norman 1980, 81; 1986, 362). When analyzed systematically (Norman, 2002, 2004), the taxa can appear as discrete operational taxonomic units that form a clade but, under some circumstances, form successive sister taxa; this hints at their anatomical differences being sufficient to merit generic distinction. The basis for this distinction will be reviewed in detail shortly, but there is without doubt a *prima facie* osteological case for the separation of the two upper Wealden taxa as *Iguanodon* and *Mantellisaurus*, respectively, and the proposal articulated by Paul (2007) is tacitly accepted, notwithstanding serious concerns about the presentation and argument in favor of this proposal.

Paul (2008) attempted to reformulate the taxonomy outlined in Paul (2007) and extended the scope of taxonomic revision of iguanodontians more generally. In relation to this particular article, he offered diagnoses of the Barremian–Lower Aptian iguanodontians from England and Belgium. An additional proposal (with respect to these specific taxa) was made: that the material from Belgium, attributed originally to *I. mantelli* by Dollo (1882a) and to *I. atherfieldensis* (Norman, 1986)—primarily the articulated skeleton RBINS R57 [1551]—merits its own distinct generic and specific identity: *Dollodon bampingi* Paul, 2008; the justification for this taxonomic revision will be considered in more detail below. The diagnoses of the two taxa distinguished by Paul (*M. atherfieldensis* and *D. bampingi*) above are quoted verbatim (below) to facilitate an assessment of the merit of the taxonomic revision.

Mantellisaurus atherfieldensis—Holotype: NHMUK R5764.

Diagnosis (*verbatim* Paul, 2008, 200):

> "Probably modest sized as adults. Overall lightly constructed. Premaxillary tip to anterior orbital rim/latter to paraoccipital process tip length ratio ~1.25; dentary pre-coronoid process length/minimum depth ratio under 5. Rostrum subtriangular in lateral view. Maxillary process of premaxilla shallow. Dorsal apex of maxilla set posteriorly. Antorbital fossa and fenestra reduced. Lacrimal short, does not contact nasal. Lateral temporal fenestra moderate in size. Posterior portion of jugal short. Quadratojugal short. Quadrate tall, transversely narrow, shaft curved, lateral foramen set high, dorso-posterior buttress small. Diastema absent. Tooth positions 23 in maxilla, 22 in dentary. Dorso-sacral/hindlimb length ratio ~1. Posterior dorsal centra not compressed antero-posteriorly. 7 fused sacrals. Scapula blade narrow and constricted at middle of blade, base rather narrow, acromion process placed rather dorsally and directed anteriorly. Forelimb ~50% of hindlimb length. Deltopectoral crest of humerus distally placed, fairly large and incipiently hatchet shaped. Manual phalanx 1 of digit I absent, pollex spike and other unguals moderate in size. Pelvis *relatively large. Main body of ilium deep. Prepubic process of pubis deep. Femoral shaft curved. Metatarsal I present, II short. Distal phalanges of toes *not strongly abbreviated." [N.B. An asterisk precedes unambiguous autapomorphies; Paul, 2008, 198.]

Dollodon bampingi—Holotype: RBINS R57 [1551].

Diagnosis (*verbatim* Paul, 2008, 200):

> Probably modest sized as adults at 6? m and 1? tonnes. Overall lightly constructed. Premaxillary tip to anterior orbital rim/latter to paraoccipital process tip length ratio ~1.6 via elongation of posterior nasal, maxillary process of premaxilla and maxilla; dentary pre-coronoid process length/minimum depth ratio over 5. Maxillary process of premaxilla shallow. Dorsal apex of maxilla set posteriorly. Antorbital fossa and fenestra reduced. Lacrimal short, does not contact nasal. Accessory palpebral absent. Posterior border of occiput straight. Lateral temporal

fenestra small. Posterior portion of jugal short. Quadratojugal short. Quadrate short, transversely narrow, shaft curved, lateral foramen set high, dorso-posterior buttress small. Diastema short. Dorso-sacral/hindlimb length ratio 1.2. Posterior dorsal centra not compressed antero-posteriorly. Neural spines of dorsals, sacrals and caudals form moderately tall sail. 7 fused sacrals. Scapula blade narrow and constricted at middle of blade, base rather narrow, acromion process placed ventrally and directed distally. Forelimb 55% of hindlimb length. Deltopectoral crest of humerus distally placed, fairly large and incipiently hatchet shaped. Manus narrow. Manual phalanx 1 of digit I present, pollex spike and other unguals moderate in size. Main body of ilium shallow. Prepubic process of pubis deep. Femoral shaft moderately curved. Metatarsal I present, II long. [N.B. An asterisk precedes unambiguous autapomorphies; Paul, 2008, 198].

Discussion of the principal features identified by the diagnoses. Although many of the characters used are imprecise or ambiguously phrased, each is discussed, and its status assessed below.

1. *General attributes.* General attributes such as overall length, stature, and weight have no specific taxonomic value; however, if the two diagnoses above are compared, these characters are in fact exactly the same and therefore do not discriminate between the two taxa. Assessment: invalid.

2. *Relative elongation of the preorbital region of the skull (M. atherfieldensis ~1.25, D. bampingi ~1.6).* Elongation is crucially dependent on the reconstruction of the skull and the ability to allow for, or estimate accurately, distortion. Given that Paul is comparing the two holotypes, simple measurement of these proportions (Norman, 1986, figs. 3, 6) yields ratios that are similar for *M. atherfieldensis* of 1.5 and *D. bampingi* of 1.6; these cannot be regarded as legitimate evidence of difference. Assessment: invalid.

3. *Precoronoid process dentary length relative to depth (<5 M. atherfieldensis, >5 D. bampingi).* So far as it can be accurately estimated, the ratio in both specimens is similar (~5). Assessment: invalid.

4. *Rostrum subtriangular in lateral view (in M. atherfieldensis).* This character (and precisely what is being described or compared is not at all clear) is not commented on in the diagnosis of *D. bampingi.* Assessment: invalid.

5. *Maxilllary process of the premaxilla shallow.* This character is the same for both taxa. Assessment: invalid.

6. *Dorsal apex of the maxilla set posteriorly.* This character is the same for both taxa. Assessment: invalid.

7. *Antorbital fossa and fenestra reduced.* This character is the same for both taxa. Assessment: invalid.

8. *Lacrimal short, does not contact nasal.* This character is the same for both species. Assessment: invalid.

9. *Lateral temporal fenestra moderate in size (M. atherfieldensis) and small in size (D. bampingi).* An ill-defined character and therefore this is an entirely subjective consideration. No significant difference in the size of the lateral temporal fenestra is discernible (Norman, 1986, figs. 3 and 6). Assessment: invalid.

10. *Posterior portion of the jugal short.* This character is described as the same for both taxa. Assessment: invalid.

11. *Quadratojugal short.* This character is described as the same for both taxa. Assessment: invalid.

12. *Quadrate tall, transversely narrow, shaft curved, lateral foramen set high, dorsoposterior buttress small (in M. atherfieldensis); quadrate differs in M. bambingi by being short.* This is a character-complex, but upon careful examination there is no discernible difference in length of the quadrates when they are compared directly. Assessment: invalid.

13. *Diastema absent (M. atherfieldensis), diastema short (M. bampingi).* The holotype of *I. atherfieldensis* (NHMUK R5764) displays a clear elongate narrow region between the first alveolus and the area for attachment of the predentary (Hooley, 1925, fig. 4; see also Norman, 1986, fig. 19), a similar anatomical configuration is observed in RBINS R57 [1551]. Assessment: invalid.

14. *Tooth positions 23 in maxilla, 22 in dentary (M. atherfieldensis), more than 25 tooth positions in the dentary (diagnostic of the genus Dollodon; see Paul, 2008, 200).* The tooth number cannot be ascertained with any certainty in RBINS R57 [1551] because the skull is laterally compressed and the whole tooth row in both dentaries cannot be counted. Assessment: probably invalid.

15. *Dorsosacral/hindlimb length ratio ~1 (M. atherfieldensis), ~1.2 (D. bampingi).* This is an entirely subjective assessment based (I suspect) on the use of "technical skeletal restoration" advocated by Paul (2008, 202) derived from photographs of mounted museum specimens. Assessment: invalid. [N.B. A "technical skeletal restoration" based on a photograph of a mounted specimen does not, and cannot, account for postmortem distortion in the original skeletal material, missing elements, restoration undertaken in preparation for the mounting for display, or indeed structural compromises made by technicians in order to position and arrange skeletal elements in an aesthetically pleasing and biologically plausible way, during the installation of a skeletal mount. While such comments might not seem even worthy of mention, such considerations are actually paramount when material such as that from Bernissart, in which postmortem distortion and breakage are apparent (Norman, 1980, 1986, 1987b). In this context, these factors render the concept associated with the phrase "technical skeletal restoration" singularly inappropriate. In this instance, there really is no substitute for closely studying the original material.]

16. *Posterior dorsal vertebrae not compressed anteroposteriorly.* This is stated to be the same for both taxa. Assessment: invalid.

17. *Seven fused sacrals.* This character is stated to be the same for both taxa. Assessment: invalid.

18. *Scapula[r] blade narrow and constricted at the middle of the blade, base rather narrow, acromion process placed rather dorsally and directed anteriorly (M. atherfieldensis); the scapula of D. bampingi is claimed to differ in having a ventrally positioned acromion that is directed distally.* The form of the scapula in RBINS R57 [1551] (Norman, 1986, fig. 37) contradicts the description that is offered by

Paul, and it differs in no significant way from that of NHMUK 5764. Assessment: invalid.

19. *Forelimb/hind limb length 50% in M. atherfieldensis, 55% in D. bampingi.* As shown in Norman (1986, appendix 2, and illustrations throughout), the individual forelimb and hind limb bones are significantly crushed and distorted, making any estimates of overall limb length best estimates; such minor differences in proportion represent error inherent in using the "technical skeletal restoration" approach to fossil skeleton anatomy. Assessment: invalid.

20. *Deltopectoral crest of humerus distally placed, fairly large and incipiently hatchet shaped.* This character is described as being the same for both taxa. Assessment: invalid.

21. *Manual phalanx 1 of digit I absent (M. atherfieldensis), present (D. bampingi).* As stated by Norman (1986, 316), there is no evidence for the presence of this proximal phalanx in either the disarticulated remains of NHMUK 5764 or in the mounted articulated manus of RBINS R57. Assessment: invalid.

22. *Pollex spike and other unguals moderate in size.* This is described as being the same for both taxa. Assessment: invalid.

23. *Pelvis *relatively large.* This applies to M. atherfieldensis, but no comparative comments relating to this character are made in the other diagnosis, and the character status is impossible to evaluate. Assessment: invalid.

24. *Main body of ilium deep (M. atherfieldensis), main body of ilium shallow (D. bampingi).* The ilia of both specimens NHMUK R5764 and RBINS R57 are of similar form, after allowing for the significant distortion suffered by the latter (see Norman, 1986, fig. 53). Assessment: invalid.

25. *Prepubic process of pubis deep.* This character is described as being the same for both taxa. Assessment: invalid.

26. *Femoral shaft curved (M. atherfieldensis), moderately curved (D. bampingi).* The form of the femur is well shown in NHMUK R5764 (Norman, 1986, fig. 57), and allowing for the clear distortion and infilling seen in RBINS R57 [1551] that is illustrated in the same figure, the two are similar in shape. Assessment: invalid.

27. *Metatarsal I present (M. atherfieldensis), absent (D. bampingi).* Norman (1986) did not find metatarsal in RBINS R57 [1551], and as stated clearly by Norman (1986, 324), this bone was not preserved in NHMUK R5764. If the bone was present (as is strongly suspected), it may have been relatively easily detached during or before burial (see comments in the formal Diagnosis section for *Mantellisaurus atherfieldensis* above). Assessment: invalid.

28. *Metatarsal II short (M. atherfieldensis), long (D. bampingi).* Such a difference simply does not exist in the known material of NHMUK R5764 and RBINS R57 [1551]. Assessment: invalid.

Additional characters mentioned in the listings above, but not compared across taxa, including the following: (1) a palpebral bone was found in RBINS R57 [1551] but was not preserved in the skeleton of NHMUK R5764 (Norman, 1986, fig. 16), so claiming that an "accessory palpebral" was absent in *D. bampingi* is not only inexplicable, but its relevance as a diagnostic character in relation to these two taxa is also unclear. (2) Neural spines of dorsals, sacrals, and caudals form a moderately tall sail (*D. bampingi*); none of the vertebrae in this region of the skeleton of NHMUK R5764 has preserved neural spines, so no valid comparison can be made. (3) Manus narrow (*D. bampingi*); the manus bones of NHMUK R5764 are slender and lightly constructed by comparison with *I. bernissartensis*, and are closely similar to those seen in the crushed and distorted manuses preserved in RBINS R57 [1551] (Norman, 1986, figs. 50–52).

Summary. The proposal of Paul (2007)—that *Mantellisaurus* be adopted as a new generic name for *I. atherfieldensis* in order to emphasize the differences in anatomy that can be demonstrated when comparing the contemporary large and robustly constructed taxon (*Iguanodon bernissartensis*) with the smaller and gracile taxon previously referred to as *I. atherfieldensis*—is tacitly accepted (the foresight of Harry Seeley may be vindicated!).

On the basis of the published diagnoses of Paul (2008) that have been evaluated above, there are no valid osteological differences that distinguish the skeletons NHMUK R5764 and RBINS R57 [1551]. The two specimens are approximate contemporaries, differ but little in size, and were sufficiently close geographically to be considered sympatric, and there appears to be no justification for the erection of the new binomial. Accordingly, *Dollodon bampingi* Paul, 2008, should be relegated in synonymy as a junior subjective synonym of *Mantellisaurus atherfieldensis* (Hooley, 1925).

Conclusions

1. The Berriasian ornithopod jaw that was named *Iguanodon hoggii* by Richard Owen (1874) is regarded as a camptosaur-grade iguanodontian. The proposal to rename this species *Owenodon hoggii* by Galton (2009) can only be justified on the basis of minor differences in tooth morphology when compared to other similar basal iguanodontians. More material that can be attributed to this taxon is urgently needed to enable its taxonomic position to be stabilized. Other anatomical features that have been claimed in the recent past to diagnose this taxon lack credibility or appear to be simple, repeated factual errors.

2. Two Lower Wealden (Valanginian) taxa of iguanodontian ornithopod can be recognized: *Barilium dawsoni* (Lydekker, 1888a), based on the type material of *Iguanodon dawsoni*, a massively built styracosternan iguanodontian; and *Hypselospinus fittoni* (Lydekker, 1889a), based on the type material of *I. fittoni* and incorporating (by synonymy) material also referred to the taxon *I. hollingtonensis*.

3. *Barilium dawsoni* is considerably more robust and of larger maximum body length when compared to the skeletal form of the contemporary taxon *Hypselosaurus fittoni*. *H. fittoni* is notable for the possession of unusually slender and elongate neural spines (compared to the long, but much deeper and more plank-like, neural spines seen in *B. dawsoni*), but *H. fittoni* does appear to possess a rather robust forelimb and ankylopollexian manus (as also seen in material that is clearly referable to *B. dawsoni*). Some of the material attributable to *H. fittoni* shows significant additional (possibly

pathological) bone growth in the areas that surround normal articular and sutural surfaces; it is also evident—and this is probably linked to the latter's anatomy—in the midline fusion between the sternal bones, as described by Hulke (1885) in at least one example (Norman, in prep.).

4. Upper Wealden iguanodontian taxa have been the subject of taxonomic revision in recent years. Brief consideration of these new proposals has shown that there is some justification for the creation of a new generic name (*Mantellisaurus*) for the taxon previously described as *Iguanodon atherfieldensis*.

5. The attempt to designate an entirely new taxonomic name for the gracile iguanodontian from Bernissart (RBINS R57 [1551]) as *Dollodon bampingi* cannot be justified on the basis of any of the evidence that has been suggested and this name should be relegated in synonymy with *Mantellisaurus atherfieldensis* (Hooley, 1925).

6. Presently, the Wealden District of Britain and deposits of similar age in Belgium contain at least two distinct chronofaunas of styracosternan iguanodontians: lower Wealden (Valanginian), *Barilium dawsoni* and *Hyselospinus fittoni*; and upper Wealden/ Sainte-Barbe Clays Formation (Barremian–Lower Aptian), *Iguanodon bernissartensis* and *Mantellisaurus atherfieldensis*.

Acknowledgments

This chapter is offered as an affectionate tribute in memory of Dr. Annie V. Dhondt (1942–2006), who, as a research scientist at the Royal Belgian Institute of Natural Sciences, will be forever associated with my time spent living and working in Brussels (principally during the period 1975–1978, but also on several occasions since then). Annie was not just a wonderfully generous host and colleague, but—far more importantly—a sincere and kind friend who is much missed.

This contribution formed the substance of a plenary presentation at the New Perspectives on Vertebrate Evolution and Early Cretaceous Ecosystems meeting held at the Royal Belgian Institute of Natural Sciences in Brussels (and also at the town of Harchies, near the village of Bernissart, southwest Belgium), February 9–13, 2009. I thank Pascal Godefroit and Olivier Lambert for their invitation to attend the meeting, and for their seemingly endless and effortless geniality and hospitality. The staff of the Natural History Museum, London (Angela Milner, Paul Barrett, and Sandra Chapman), provided assistance with, and access to, the iguanodontian collections necessary for preparation of this chapter. More detailed accounts of several of the dinosaurs discussed in this article are either in press (Norman in press a, b) or are nearing completion (notably a detailed account of the osteology of *Hypselospinus* that will complement the monograph on *Barilium*, which forms the focus of Norman, in press b).

This work was made possible through financial support from the Royal Belgian Institute of Natural Sciences in Brussels, the Royal Society of London, the Department of Earth Sciences at the University of Cambridge, and the Master and Fellows of Christ's College, Cambridge. I would also like to acknowledge Xabier Pereda-Suberbiola and Pascal Godefroit; their insightful comments have done much to improve this chapter.

Notes

1. Mantell's reticence to offer a species name might be interpreted to represent nomenclatural inexperience (perhaps apeing William Buckland's 1824 naming of *Megalosaurus*); however, it may equally have its origin in his geologically based appreciation of the disassociated nature of the remains (a consideration that applies with equal force to the Stonesfield material).

2. Given that Hulke had visited and examined the Belgian collections before publication of this paper, he must have been aware of the similarities that existed between the new material he had collected from Brook Chine and the large and robustly constructed Bernissart species.

References

Blows, W. T. 1998. A review of Lower and Middle Cretaceous dinosaurs of England. New Mexico Museum of Natural History and Science Bulletin 14: 29–38.

Buckland, W. 1824. Notice on the *Megalosaurus* or great fossil lizard of Stonesfield. Transactions of the Geological Society of London (ser. 2) 1: 390–396.

Calloman, J. H., and J. C. W. Cope. 1995. The Jurassic geology of Dorset. Field Geology of the British Jurassic. P. D. Taylor. London, The Geological Society: 51–104.

Carpenter, K., and Y. Wilson. 2008. A new species of *Camptosaurus* (Ornithopoda: Dinosauria) from the Morrison Formation (Upper Jurassic) of Dinosaur National Monument, Utah, and a biomechanical analysis of its forelimb. Annals of Carnegie Museum 76: 227–263.

Carpenter, K, and Y. Ishida. 2010. Early and "Middle" Cretaceous Iguanodonts in time and space. Journal of Iberian Geology 36: 145–164.

Charig, A. J. 1979. A new look at the dinosaurs. British Museum (Natural History)/Heinemann, London, 175 pp.

Charig, A. J., and S. D. Chapman. 1998. Case 3037. *Iguanodon* Mantell, 1825 (Reptilia, Ornithischia): proposed designation of *Iguanodon bernissartensis* Boulenger *in* Beneden, 1881 as type species, and proposed designation of a lectotype. Bulletin of Zoological Nomenclature 55: 99–104.

Dollo, L. 1882a. Première note sur les dinosauriens de Bernissart. Bulletin du Musée Royal d'Histoire Naturelle de Belgique 1: 55–80.

———. 1882b. Deuxième note sur les dinosauriens de Bernissart. Bulletin du Musée Royal d'Histoire Naturelle de Belgique 1: 205–211.

———. 1883a. Note sur la présence chez les oiseaux du "troisième trochanter" des dinosauriens et sur la fonction de celui-ci. Bulletin du Musée Royal d'Histoire Naturelle de Belgique 2: 13–20.

———. 1883b. Troisième note sur les dinosauriens de Bernissart. Bulletin du Musée Royal d'Histoire Naturelle de Belgique 2: 85–126.

———. 1883c. Quatrième note sur les dinosauriens de Bernissart. Bulletin du Musée Royal d'Histoire Naturelle de Belgique 2: 223–252.

———. 1884. Cinquième note sur les dinosauriens de Bernissart. Bulletin du Musée Royal d'Histoire Naturelle de Belgique 3: 129–146.

———. 1885a. L'appareil sternal de l'Iguanodon. Revue des Questions Scientifiques 18: 664–673.

———. 1885b. Les Iguanodons de Bernissart. Revue des Questions Scientifiques 18: 5–55.

———. 1888. Iguanodontidae et Camptonotidae. Comptes Rendus hebdomadaires de l'Académie des Sciences, Paris 106: 775–777.

———. 1906. Les allures des Iguanodons, d'après les empreintes des pieds et de la queue. Bulletin Scientifique de la France et de la Belgique 40: 1–12.

———. 1923. Le centenaire des Iguanodons (1822–1922). Philosophical Transactions of the Royal Society of London Series B 212: 67–78.

Gallois, R. W., and B. C. Worssam (eds.). 1993. Geology of the country around Horsham: memoir for 50,000 geological sheet 302 (England and Wales). Her Majesty's Stationary Office, London, 130 pp.

Galton, P. M. 2009. Notes on Neocomian (Lower Cretaceous) ornithopod dinosaurs from England—*Hypsilophodon*, *Valdosaurus*, "*Camptosaurus*," "*Iguanodon*"—and referred specimens from Romania and elsewhere. Revue de Paléobiologie, Genève 28: 211–273.

Galton, P. M., and J. A. Jensen. 1979. Remains of ornithopod dinosaurs from the Lower Cretaceous of North America. Brigham Young University Geology Studies 25: 1–10.

Holl, F. 1829. Handbuch der Petrefactenkunde Pt 1. Leipzig, Quedlinberg, 232 pp.

Hooley, R. W. 1912a. On the discovery of remains of *Iguanodon mantelli* in the Wealden beds of Brighstone Bay, Isle of Wight. Geological Magazine 5: 444–449.

———. 1912b. On the discovery of remains of *Iguanodon mantelli* in the Wealden beds of Brightstone bay, I.W., and the adaptation of the pelvic girdle in relation to an erect position and bipedal progression. Geological Magazine 5: 520–521.

———. 1917. On the integument of *Iguanodon bernissartensis*, Boulenger, and of *Morosaurus becklesii*, Mantell. Geological Magazine 6: 148–150.

———. 1925. On the skeleton of *Iguanodon atherfieldensis* sp. nov., from the Wealden shales of Atherfield (Isle of Wight). Quarterly Journal of the Geological Society of London 81: 1–61.

Hulke, J. W. 1871. Note on a large reptilian skull from Brooke, Isle of Wight, probably dinosaurian, and referable to the genus *Iguanodon*. Quarterly Journal of the Geological Society of London 27: 199–206.

———. 1879. *Vectisaurus valdensis*, a new Wealden dinosaur. Quarterly Journal of the Geological Society of London 35: 421–424.

———. 1880. *Iguanodon prestwichii*, a new species from the Kimmeridge Clay founded on numerous fossil remains lately discovered at Cumnor, near Oxford. Quarterly Journal of the Geological Society of London 36: 433–456.

———. 1882. Description of some *Iguanodon* remains indicating a new species, *I. seelyi*. Quarterly Journal of the Geological Society of London 38: 135–144.

———. 1885. Note on the sternal apparatus in *Iguanodon*. Quarterly Journal of the Geological Society of London 41: 473–475.

Kirkland, J. I. 1998. A new hadrosaurid from the Upper Cedar Mountain Formation (Albian–Cenomanian: Cretaceous) of Eastern Utah—the oldest known hadrosaurid (lambeosaurine?). New Mexico Museum of Natural History and Science Bulletin 14: 283–295.

Lydekker, R. 1888a. Note on a new Wealden iguanodont and other dinosaurs. Quarterly Journal of the Geological Society of London 44: 46–61.

———. 1888b. Catalogue of the fossil Reptilia and Amphibia in the British Museum (Natural History), London (Part 1). Trustees of the British Museum (Natural History), London, 309 pp.

———. 1889a. Notes on new and other dinosaur remains. Geological Magazine 6: 352–356.

———. 1889b. On the remains of five genera of Mesozoic reptiles. Quarterly Journal of the Geological Society 45: 41–59.

———. 1890a. Contributions to our knowledge of the dinosaurs of the Wealden and the sauropterygians of the Purbeck and Oxford Clay. Quarterly Journal of the Geological Society of London 46: 36–53.

———. 1890b. Catalogue of the fossil Reptilia and Amphibia in the British Museum (Natural History). Pt IV. Trustees of the British Museum (Natural History), London, 295 pp.

Mantell, G. A. 1822. The Fossils of the South Downs; or, Illustrations of the Geology of Sussex. Lupton Relfe, London.

———. 1825. Notice on the *Iguanodon*, a newly discovered fossil reptile, from the sandstone of Tilgate forest, in Sussex. Philosophical Transactions of the Royal Society of London 115: 179–186.

———. 1833. The geology of the south east of England. Longman, Rees, Orme, Brown, Green and Longman, London, 415 pp.

———. 1839. The Wonders of Geology; of a familiar Exposition of Geological Phenomena; being the substance of a course of lectures delivered at Brighton. 3rd ed., 2 vols. Relfe and Fletcher, London, 804 pp.

Marsh, O. C. 1881. Principal characters of American Jurassic dinosaurs. Part V. American Journal of Science (ser. 3) 21: 417–423.

McDonald, A. T., P. M. Barrett, and S. D. Chapman. 2010. A new basal iguanodont (Dinosauria: Ornithischia) from the Wealden (Lower Cretaceous) of England. Zootaxa 2569: 1–43.

Norman, D. B. 1977. On the anatomy of the ornithischian dinosaur *Iguanodon*. Ph.D. thesis, King's College, London. 631 pp.

———. 1980. On the ornithischian dinosaur *Iguanodon bernissartensis* from Belgium. Mémoires de l'Institut Royal des Sciences Naturelles de Belgique 178: 1–105.

———. 1986. On the anatomy of *Iguanodon atherfieldensis* (Ornithischia: Ornithopoda). Bulletin de l'Institut royal des Sciences naturelles de Belgique, Sciences de la Terre 56: 281–372.

———. 1987a. Wealden dinosaur biostratigraphy; pp. 161–166 in P. J. Currie and E. H. Koster (eds.), Fourth Symposium on Mesozoic Terrestrial Ecosystems, short papers. Royal Tyrrell Museum of Palaeontology, Drumheller.

———. 1987b. On the discovery of fossils at Bernissart (1878–1921) Belgium. Archives of Natural History 13: 131–147.

———. 1987c. A mass-accumulation of vertebrates from the Lower Cretaceous of Nehden (Sauerland), West Germany. Proceedings of the Royal Society of London B230: 215–255.

———. 1990. A review of *Vectisaurus valdensis*, with comments on the family Iguanodontidae; pp. 147–162 in K. Carpenter and P. J. Currie (eds.), Dinosaur systematics: approaches and perspectives. Cambridge University Press, Cambridge.

———. 1993. Gideon Mantell's Mantelpiece: the earliest well-preserved ornithischian dinosaur. Modern Geology 18: 225–245.

———. 1996. On Mongolian ornithopods (Dinosauria: Ornithischia). 1. *Iguanodon orientalis* Rozhdestvensky, 1952. Zoological Journal of the Linnean Society (London) 116: 303–315.

———. 2000. Professor Richard Owen and the important, but neglected, dinosaur *Scelidosaurus harrisonii*. Historical Biology 14: 235–253.

———. 2002. On Asian ornithopods (Dinosauria: Ornithischia). 4. Redescription of *Probactrosaurus gobiensis* Rozhdestvensky, 1966. Zoological Journal of the Linnean Society (London) 136: 113–144.

———. 2004. Basal Iguanodontia; pp. 413–437 in D. B. Weishampel, P. Dodson, and H. Osmólska (eds.), The Dinosauria. 2nd ed. University of California Press, Berkeley.

———. 2010. A taxonomy of iguanodontians (Dinosauria: Ornithopoda) from the lower Wealden Group (Cretaceous: Valanginian) of southern England. Zootaxa 2489: 47–66.

———. In press a. Ornithopod dinosaurs. In D. J. Batten (ed.), Field guide to the Wealden of England. Oxford/Blackwell, The Palaeontological Association.

———. In press b. On the osteology of the lower Wealden (Valanginian) iguanodontian *Barilium dawsoni* (Iguanodontia: Styracosterna). Special Papers in Palaeontology, Palaeontological Association 86: 1–28.

Norman, D. B., and P. M. Barrett. 2002. Ornithischian dinosaurs from the Lower Cretaceous (Berriasian) of England; pp.161–189 in A. R. Milner and D. J. Batten (eds.), Life and environments in Purbeck times. Special Papers in Palaeontology 8. Blackwell, Oxford.

Norman, D. B., and D. B. Weishampel. 1990. Iguanodontidae and related ornithopods; pp. 510–533 in D. B. Weishampel, P. Dodson, and H. Osmólska (eds.), The Dinosauria. University of California Press, Berkeley.

Norman, D. B., K.-H. Hilpert, and H. Hölder. 1987. Die Wirbeltierfauna von Nehden (Sauerland), Westdeutschland. Geologie und Paläontologie in Westfalen 8: 1–77.

Ostrom, J. H. 1970. Stratigraphy and paleontology of the Cloverly Formation (Lower Cretaceous) of the Big Horn Basin area, Wyoming and Montana. Peabody Museum of Natural History 35: 1–234.

Owen, R. 1842. Report on British Fossil Reptiles. Part 2. Report of the British Association for the Advancement of Science (Plymouth) 11: 60–204.

———. 1851. Monograph of the fossil reptilia of the Cretaceous formations. Part 1. Chelonia (Lacertilia, &c.). Palaeontographical Society Monographs 5: 1–118.

———. 1855. Monograph of the fossil reptilia of the Wealden and Purbeck formations. Part II. Dinosauria (*Iguanodon*). Palaeontographical Society Monographs 8: 1–54.

———. 1858. Monograph of the fossil reptilia of the Wealden and Purbeck formations. Supplement I. Dinosauria (*Iguanodon*). Palaeontographical Society Monographs 10: 1–7.

———. 1864. Monograph of the fossil reptilia of the Wealden and Purbeck formations. Supplement III. Dinosauria (*Iguanodon*). Palaeontographical Society Monographs 16: 19–21.

———. 1872. Monograph of the fossil reptilia of the Wealden and Purbeck formations. Supplement IV. Dinosauria (*Iguanodon*). Palaeontographical Society Monographs 25: 1–15.

———. 1874. Monograph of the fossil reptilia of the Wealden and Purbeck formations. Supplement V. Dinosauria (*Iguanodon*). Palaeontographical Society Monographs 27: 1–18.

Paul, G. S. 2007. Turning the old into the new: a separate genus for the gracile iguanodont from the Wealden of England; pp. 69–77 in K. Carpenter (ed.), Horns and beaks: ceratopsian and ornithopod dinosaurs. Indiana University Press, Bloomington.

———. 2008. A revised taxonomy of the iguanodont dinosaur genera and species. Cretaceous Research 29: 192–216.

Rawson, P. F. 2006. Cretaceous: sea levels peak as the North Atlantic opens; pp. 365–393 in J. Brenchley and P. F. Rawson (eds.), The geology of England and Wales. Geological Society, London.

Rozhdestvensky, A. K. 1952. Otkritiye iguanodonta v Mongolii. Doklady Akademiya Nauk CCCP 84: 1243–1246.

Seeley, H. G. 1875. On the axis of a dinosaur from the Wealden of Brook in the Isle of Wight, probably referable to *Iguanodon*. Quarterly Journal of the Geological Society of London 31: 461–464.

———. 1883. On the dorsal region of the vertebral column of a new dinosaur (indicating a new genus *Sphenospondylus*) from the Wealden of Brook in the Isle of Wight, preserved in the Woodwardian Museum of the University of Cambridge. Quarterly Journal of the Geological Society of London 39: 55–61.

———. 1887a. On the classification of the fossil animals commonly named Dinosauria. Proceedings of the Royal Society of London 43: 165–171.

———. 1887b. Mr Dollo's notes on the dinosaurian fauna of Bernissart (parts 1 and 2). Geological Magazine 3: 80–87, 124–130.

———. 1888. On the reputed clavicles and interclavicles of *Iguanodon*. Reports to the British Association for the Advancement of Science 1888 (57th Meeting): 698.

Sereno, P. C. 1986. Phylogeny of the bird-hipped dinosaurs (Order Ornithischia). National Geographic Society Research 2: 234–256.

Steel, R. 1969. Ornithischia. Handbuch der Paläoherpetologie 15: 1–84.

Van Beneden, P. J. 1881. Sur l'arc pelvien chez les dinosauriens de Bernissart [Review of G. A. Boulenger]. Bulletin de l'Académie Royale des Sciences, des Lettres et des Beaux-Arts de Belgique (3rd ser.) 1: 600–608.

Von Meyer, H. 1832. Paleologica zur Geschichte der Erde und ihrer Geshöpfe. S. Schmerber, Frankfurt-am-Main.

Weishampel, D. B., and P. R. Bjork. 1989. The first indisputable remains of *Iguanodon* (Ornithischia: Ornithopoda) from North America: *Iguanodon lakotaensis*, sp. nov. Journal of Vertebrate Paleontology 9: 56–66.

Worssam, B. C., and T. Tatton-Brown. 1993. Kentish rag and other Kent building stones. Archaeologia Cantiana 112: 93–125.

Yans, J., J. Dejax, D. Pons, L. Taverne, and P. Bultynck. 2006. The iguanodons of Bernissart are middle Barremian to earliest Aptian in age. Bulletin de l'Institut royal des Sciences naturelles de Belgique, Sciences de la Terre 76: 91–95.

The Brain of *Iguanodon* and *Mantellisaurus:* Perspectives on Ornithopod Evolution

16

Pascaline Lauters*, Walter Coudyzer,
Martine Vercauteren, and Pascal Godefroit

Information on the structure of the brain of the basal iguanodontian dinosaurs *Iguanodon bernissartensis* and *Mantellisaurus atherfieldensis,* from the Early Cretaceous of Bernissart, is presented on the basis of computed tomographic scanning and 3D reconstruction of three braincases. The resulting digital cranial endocasts are compared with physical and digital endocasts of other dinosaurs. The orientation of the brain is more horizontal than in lambeosaurine hadrosaurids. The large olfactory tracts indicate that the sense of smell was better developed than in hadrosaurids. The primitive flexures of the midbrain are virtually absent in *I. bernissartensis* but appear to be better developed in *M. atherfieldensis,* which might be explained by the smaller body size of the latter. The brain of *Iguanodon* was relatively larger than in most extant nonavian reptiles, sauropods, and ceratopsians. However, it was apparently smaller than in lambeosaurines and most theropods. The relative size of the cerebrum was low in *Iguanodon.* In *Mantellisaurus,* the cerebrum was proportionally larger than in *Iguanodon* and compares favorably with lambeosaurines. The behavioral repertoire and/or complexity were therefore probably different in the two iguanodontoids from Bernissart, *Iguanodon* and *Mantellisaurus.* The enlargement of the cerebrum appeared independently, together with possible capabilities for more complex behaviors, at least two times during the evolution of Iguanodontoidea.

Two natural endocasts of basal Iguanodontia were discovered in England during the eighteenth century. The first one, from the Wealden of Brooke (Isle of Wright), was described by Hulke (1871) and Andrews (1897). In his Ph.D. thesis, Norman (1977) studied a second nicely preserved endocast from the Wealden of Sussex, England. This specimen was subsequently illustrated and briefly discussed in both editions of *The Dinosauria* (Norman and Weishampel, 1990; Norman, 2004) and in the book *Dinosaurs: A Very Short Introduction* (Norman, 2005). As noted by Norman (1980), the skulls of the Bernissart iguanodons are damaged because of oxidation of pyrite, and the matrix inside the endocranial cavities has never been completely removed. Consequently, the skulls of the Bernissart iguanodons are particularly fragile, and for that reason, to date, nobody has tried to study their neurocranial anatomy and reconstruct their endocranial space. However, new medical imaging technologies now allow studying inaccessible areas inside the skull without damaging these fragile fossils.

Introduction

During the renovation of the Janlet Wing of the Royal Belgian Institute of Natural Sciences, from 2003 to 2007, all the skeletons of the Bernissart iguanodons were completely dismantled and treated against pyrite deterioration. On this occasion, the best-preserved skulls were scanned by computed tomography (CT) at Gasthuisberg Hospital in Leuven (Belgium). CT scanning generates a three-dimensional image of an object from a large series of x-ray images taken around a single axis of rotation. CT scanning was recently used to reconstruct the endocranial space and inner ear of pterosaurs (Witmer et al., 2003), ankylosaurians (Witmer and Ridgely, 2008a), sauropods (Sereno et al., 2007; Witmer et al., 2008), theropods (Rogers, 1998, 1999, 2005; Brochu, 2000, 2003; Larsson, 2001; Franzosa, 2004; Franzosa and Rowe, 2005; Sanders and Smith, 2005; Norell et al., 2009; Sampson and Witmer, 2007; Witmer and Ridgely, 2009), ceratopsians (Zhou et al., 2007; Witmer and Ridgely, 2008a), and hadrosaurids (Evans et al., 2009).

Here we present the first reconstructions of the endocranial space of *Iguanodon bernissartensis* Boulenger *in* Van Beneden, 1881, and *Mantellisaurus atherfieldensis* (Hooley, 1925) from the Early Cretaceous of Bernissart, based on CT scanning of skulls from the RBINS collection.

Institutional abbreviations. AEHM, Palaeontological Museum of the Institute of Geology and Nature Management, Far East Branch, Russian Academy of Sciences, Blagoveschensk, Russia; CMN, Canadian Museum of Nature, Ottawa, Ontario, Canada; MNHN, Muséum national d'Histoire naturelle, Paris, France; RBINS, Royal Belgian Institute of Natural Sciences, Brussels, Belgium; ROM, Royal Ontario Museum, Toronto, Ontario, Canada.

Materials and Methods

Ten skulls were regarded as sufficiently well preserved to be CT scanned:

- RBINS R51 (="IRSNB 1534[Q]" in Norman (1986, appendix 1). Holotype of *Iguanodon bernissartensis* Boulenger *in* Van Beneden, 1881.
- RBINS R52 (="IRSNB 1536[A3]"). Paratype of *I. bernissartensis.*
- RBINS R54 (="IRSNB 1731[F]"). Paratype of *I. bernissartensis.*
- RBINS R55 (="IRSNB 1535[N]"). Paratype of *I. bernissartensis.*
- RBINS R56 (="IRSNB 1561[L]"). Paratype of *I. bernissartensis.*
- RBINS R57 (="IRSNB 1551[T]"). *Mantellisaurus atherfieldensis* (Hooley, 1925).
- RBINS VERT-5144-1562 (="IRSNB 1562[E2]"). *I. bernissartensis.*
- RBINS VERT-5144-1657 (="IRSNB 1657[D2]"). *I. bernissartensis.*
- RBINS VERT-5144-1680 (= »IRSNB 1680[J] »). *I. bernissartensis.*
- RBINS VERT-5144-1715 (= »IRSNB 1715[C2] »). *I. bernissartensis.*

These specimens were CT scanned at the Gasthuisberg Hospital, Leuven, Belgium, with a Siemens Sensation 64 device. All specimens were scanned helically in the coronal plane with a slice thickness of 1 mm and a 0.5-mm overlap. Data were output in DICOM format and imported into Amira 5.1 or Artecore 1.0-rc3 (Nespos, VisiCore Suite) for viewing, reconstruction/analysis, visualization, and measurement. Anatomical features of

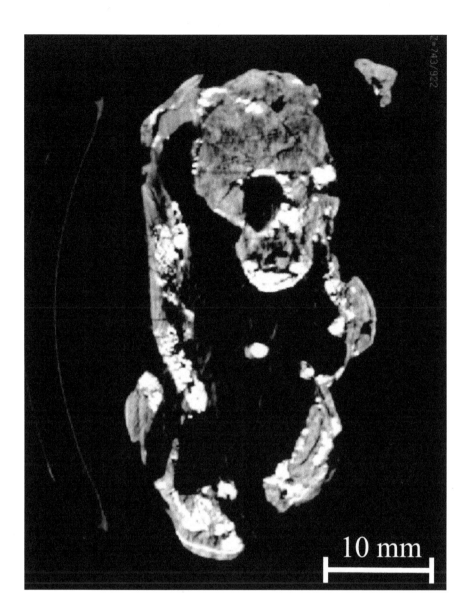

16.1. Coronal CT slice through the skull of *Iguanodon bernissartensis* (RBINS R51, holotype). The white spots are caused by pyrite, which extensively reflects x-rays.

interest (endocranial space) were highlighted and digitally extracted with Artecore's segmentation tools.

Several specimens were highly pyritized (Fig. 16.1) and/or squashed after taphonomic processes. High levels of pyritization cause artifacts, typically long-axis lines that blurred the pictures and prevented the reconstruction of the endocranial space. After close examination, only three data sets were suitable for study: RBINS R51 and RBINS R54 (*Iguanodon bernissartensis*), and RBINS R57 (*Mantellisaurus atherfieldensis*).

The measurements taken on the endocranial endocasts are illustrated in Figure 16.2.

Brain Cavity Endocasts of *Iguanodon* and *Mantellisaurus*

It has often been hypothesized that the brain of sauropsids does not fill the endocranial cavity, and the reasonable assumption has been that the endocast is essentially a cast of the dura mater (Osborn, 1912; Jerison, 1973; Hopson, 1979). In extant reptiles, the endocranium reflects forebrain surface morphology, and thus the endocast is likely a relatively accurate representation of the shape of the telencephalon (Hopson, 1979). In crocodilians,

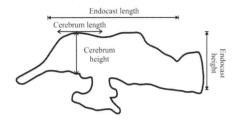

16.2. Explanation of the measurements taken on the cranial endocasts.

the proportion of the endocranial cavity occupied by the brain varies with body size (Hopson, 1979; Rogers, 1999). Traditionally nonavian dinosaurs have been regarded as reptilian in that their brains were thought to have filled a relatively small portion of the endocranial cavity, in contrast to the condition in mammals and birds (Jerison, 1969; Hopson, 1979; Rogers, 1999; Larsson, 2001).

However, previous observations led to the discovery of vascular impressions on the internal surfaces of the braincase of dinosaurs (Hopson, 1979; Osmólska, 2004). These vascular imprints (valleculae) show that the brain was closely appressed to the wall of the braincase. Evans (2005) and Evans et al. (2009) demonstrated for hadrosaurids and pachycephalosaurs that the portion of the endocast corresponding to the telencephalon faithfully represents the contours of the underlying brain. Moreover, valleculae were also observed on endocasts of Russian hadrosaurids (Lauters et al., in prep.) and on the cast of the basal hadrosauroid *Batyrosaurus rozhdestvenskyi* from Kazakhstan (see Chapter 20 in this book). Therefore, we assume that the endocast generally reflects the shape of the rostral and ventral regions of the brain. Evans (2005) and Evans et al. (2009) showed that much of the hindbrain of hadrosaurids was not in close relationship to the endocranial wall. It is also probably the case in *Iguanodon* and *Mantellisaurus* because the dorsal region of the endocast appears largely undefined in these basal iguanodontians.

The digital endocasts of the three studied specimens do not show considerable detail for several reasons. With a slice thickness of 1 mm, the resolution of the CT scan does not allow an accurate reconstruction of smaller anatomical features such as the endosseous labyrinth and the smaller cranial nerves. Moreover, CT scanning revealed that the deformation and the deterioration of the braincases, after taphonomic processes and oxidation of the pyrite, were more important than expected. Finally, the high concentration of pyrite within both the bone and the matrix creates artifacts that obscure smaller details of the endocranial cavities, such as the valleculae.

Iguanodon bernissartensis (RBINS R51 and RBINS R54)

Only two braincases were sufficiently well preserved to allow a suitable reconstruction of the endocranial cavity in *Iguanodon bernissartensis*. The reconstruction of the endocranial cavity was made by Artecore 1.0-rc3. In *Iguanodon*, the major axis of the cerebrum is oriented ~15 degrees to the horizontal plane (as seen in lateral view; Figure 16.3A). This is the usual orientation encountered in hadrosaurines and other ornithopods (Hopson, 1979), as opposed to the more oblique (~45 degrees) orientation observed in lambeosaurine hadrosaurids. According to Evans et al. (2009), the rotation of the cerebrum is certainly related to the caudodorsal expansion of the nasal cavity and crest in the evolution of lambeosaurines.

The endocranial cavity of the holotype RBINS R51 is fairly complete (Fig. 16.3B). The brain endocast measures 183 mm from the foramen magnum to the rostral margin of the cerebrum and has a total volume of 357 cm³. The olfactory tracts and olfactory bulbs are not included in this measurement. The endocranial cavity of RBINS R54 is less complete (Fig. 16.4). Its estimated volume of 194 cm³ is probably largely underestimated.

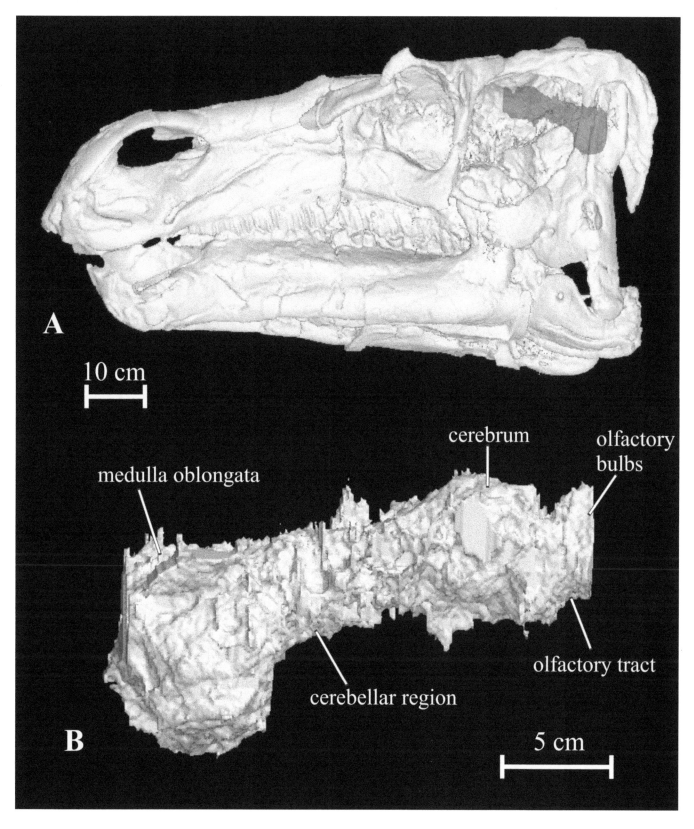

A

10 cm

medulla oblongata   cerebrum   olfactory bulbs

cerebellar region   olfactory tract

B   5 cm

Although their volume cannot be adequately estimated, the olfactory tracts and bulbs appear to be particularly large, reflecting a well-developed sense of smell in *Iguanodon bernissartensis*, as also observed in most dinosaurs, including the ceratopsians *Psittacosaurus* (Zhou et al., 2007) and *Pachyrhinosaurus* (Witmer and Ridgely, 2008b), and virtually all theropods (Hopson, 1979; Zelenitsky et al., 2008). The great proportions of the visible

16.3. Cranial endocast of *Iguanodon bernissartensis* (RBINS R51, holotype). A, Position and orientation of the endocranial cavity within the skull in left lateral view. B, Cranial endocast reconstructed from CT scans in right lateral view.

16.4. Cranial endocast of *Iguanodon bernissartensis* (RBINS R54) reconstructed from CT scans in right lateral view.

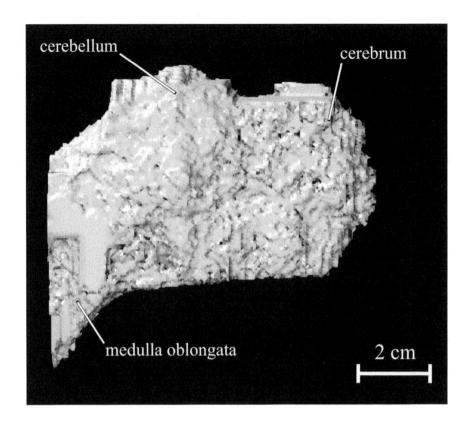

parts of the olfactory system (tracts and bulbs) in *Iguanodon* contrast with the situation observed in adult hadrosaurines and lambeosaurines, in which the bulbs usually form ~5% of the total endocast volume. Evans (2006) and Evans et al. (2009) consider that the small size of the olfactory system in hadrosaurids is a plesiomorphic character. However, it is more plausible that the reduction of the olfactory bulbs and tracts is in fact synapomorphic for hadrosaurids because the olfactory system is well developed in the basal Iguanodontia *Iguanodon* (and also in *Mantellisaurus*; see below) and in ceratopsians.

The cerebrum is enlarged and round, but not proportionally as large as in hadrosaurids (see below). The shape of the cerebrum prefigures the condition observed in later hadrosaurids, in which it is more expanded.

In RBINS R51, the cerebellar region is constricted and elongated. The cranial and pontine flexures are virtually absent, and the endocranial cavity is consequently nearly straight. According to Giffin (1989), these characters are derived for "iguanodontids" and hadrosaurids. The cerebellar region is less constricted in RBINS R54, and a large triangular peak is developed above the midbrain. Similar peaks are observed in a wide variety of dinosaur brains, including the theropods *Allosaurus*, *Tyrannosaurus* (Hopson, 1979), and *Majungasaurus* (Sampson and Witmer, 2007), and to a lesser extent in the lambeosaurines *Hypacrosaurus*, *Corythosaurus*, and *Lambeosaurus* (Evans et al., 2009). It is possible that the dural space housed a well-developed pineal apparatus (epiphysis). Pineal glands are present in extant birds (Breazile and Hartwig, 1989), and evidence for pineal-like tissue in alligators was presented by Sedlmayr et al. (2004).

Table 16.1. Measurements of brain cavity endocasts in three basal Iguanodontia from the Early Cretaceous of Bernissart

Specimen	Taxon	Skull Length (mm)	Endocast Length (mm)	Endocast Height (mm)	Endocast Max. Width (mm)	Endocast Volume (cm³)
RBINS R51	*Iguanodon bernissartensis*	840	183	105	110	357
RBINS R54	*Iguanodon bernissartensis*	794	>107	100	62	>194
RBINS R57	*Mantellisaurus atherfieldensis*	>540	141	166	89	>131

Note: Measurements were taken on digital endocasts by means of digital segmentation by the Artecore and Amira programs.

Mantellisaurus atherfieldensis (RBINS R57)

Although the skull appears to be superficially well preserved, the endocranium is partially squashed, and the midbrain is consequently lost. The reconstruction of the endocranial cavity was made by Amira 5.1 (Fig. 16.5). The endocranial cavity is 14.1 cm long from the foramen magnum to the rostral margin of the telencephalon. Its maximal height and width are, respectively, 16.6 cm and 8.9 cm. The volume of the reconstructed endocranial cavity is about 131 cm³. The brain volume of the living animal is, underestimated, however, because a large portion of the midbrain is missing. The cerebrum is only slightly larger than the medulla oblongata. As in *Iguanodon*, the olfactory tract is particularly large, suggesting large olfactory bulbs.

The endocranial cavity is proportionally higher in *Mantellisaurus* than in *Iguanodon* (Table 16.1). Moreover, although a large portion of the midbrain is missing in RBINS R57, the major axis of the hindbrain appears out of line with that of the cerebrum. Therefore, the cerebral cavity does not appear as straight as in *Iguanodon*. It can therefore be hypothesized that the primitive flexures in the midbrain region were better developed in *Mantellisaurus* than in *Iguanodon*. Hopson (1979) and Giffin (1989) hypothesized that the most likely causes of variation in the angles of the primitive flexure pattern are absolute skull size and relative eye size. Smaller genera and individuals tend to have more highly flexed endocasts than do large genera and individuals because of the negative allometry of the brain and eye sizes in reptiles (Hopson, 1979). Brains are therefore less constrained by space limitation in larger reptiles like *Iguanodon*.

Encephalization Quotient

Relative Brain Size in *Iguanodon* and *Mantellisaurus*

The encephalization quotient (EQ) is an estimation of the relative brain size and represents the actual brain size of an individual divided by the expected brain size for its particular body size, calculated by using an allometric relationship derived from a large extant sample (Jerison, 1969, 1973). According to Jerison (1969) and Hopson (1979), there is a negative allometry in vertebrates between brain size and body size. On the basis of EQ, Jerison (1969) noted that living vertebrates cluster into two groups, endotherms and ectotherms. Hopson (1979) concluded that the EQs of dinosaurs usually fall between those of modern ectotherms and endotherms. Hurlburt

16.5. Cranial endocast of *Mantellisaurus atherfieldensis* (RBINS R57) reconstructed from CT scans in right lateral view.

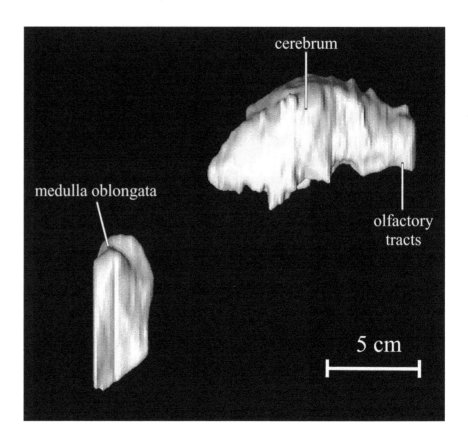

(1996) adapted Jerison's "lower" vertebrate equation for nonavian reptiles and defined a reptile encephalization quotient (REQ) as follows:

$$REQ = M_{Br}/(0.0155 \times M_{Br}^{0.553})$$

where M_{Br} is the mass of the brain (in grams) and M_{Bd} is the mass of the body (in grams). The mass of the brain is obtained by multiplying the volume of the brain by 1.036 g/mL (Stephan, 1960).

In extinct taxa, both the brain and body masses must be estimated, leading to many uncertainties in the calculation of the REQ. REQ calculations for dinosaurs usually estimated the volume of the brain under the assumption that the brain occupied 50% of the endocranial volume (Jerison, 1973; Hopson, 1977; Hurlburt, 1996). According to Evans (2005) and Evans et al. (2009), the extensive valleculae in hadrosaurids imply that the brain occupied a relatively larger portion of the endocranial cavity than in other ornithischians, and they calculated the REQ on the basis of a brain size estimate of 60% of the endocast volume. Because it cannot be decided whether valleculae are present in *Iguanodon* and *Mantellisaurus*, we calculated the REQ for these basal iguanodontians using both the 50% and the 60% estimates (Table 16.2).

The body weight of the specimens was estimated using both the bipedal and quadrupedal regression formulae of Anderson et al. (1985). Indeed, although Norman (1980) suggested that quadrupedality was the predominant posture for adults of *Iguanodon*, subsequent discussions (Alexander, 1985; Norman, 1986, 2004; Norman and Weishampel, 1990; Bultynck, 1993) suggest a semibipedal (sensu Bultynck, 1993) gait and locomotion for *Iguanodon* and *Mantellisaurus*. Therefore, the actual weight for these basal

Lauters et al.

Specimen	Taxon	Brain Weight	Body Weight	REQ Bipedal	REQ Quadrupedal
RBINS R51	*Iguanodon bernissartensis*	185 g (50%) 222 g (60%)	3,880 kg (bipedal), 7,727 kg (quadrupedal)	2.71 (50%) 3.25 (60%)	1.85 (50%) 2.22 (60%)
RBINS R57	*Mantellisaurus atherfieldensis*	>68 g (50%) >81 g (60%)	926 kg (bipedal), 1,547 kg (quadrupedal)	>2.2 (50%) >2.6 (60%)	>1.66 (50%) >1.97 (60%)

Table 16.2. Brain weight, body weight, and nonavian reptile encephalization quotient (REQ) for two basal Iguanodontia from the Early Cretaceous of Bernissart

Note: Brain weight was calculated based on assumptions that the brain occupied either 50% (Jerison, 1973; Hurlburt, 1996) or 60% (Evans, 2005; Evans et al., 2009) of the endocranial cavity. Body weight was estimated by both the bipedal and quadrupedal regression formulae of Anderson et al. (1985).

iguanodontians probably fell somewhere between the mass estimates for the bipedal and quadrupedal formulae of Anderson et al. (1985).

Depending on the chosen model (50% versus 60% relative brain size; bipedal versus quadrupedal gait), our REQ estimates for *I. bernissartensis* range between 1.88 and 3.3 (Table 16.2). The lowest of our REQ estimates is marginally higher than most extant nonavian reptiles (Hurlburt, 1996), sauropods (*Diplodocus*, 0.53–0.69; *Nigersaurus*, 0.4–0.8; Franzosa, 2004; Witmer et al., 2008), and ceratopsians (*Psittacosaurus*, 1.7; *Triceratops*, 0.7; Zhou et al., 2007; Witmer et al., 2008). Both the lowest and highest REQ estimates are lower than those calculated for the lambeosaurine hadrosaurid *Hypacrosaurus altispinus* (2.3–3.7; Evans et al., 2009). Estimated REQ values for *Iguanodon* also appear significantly lower than most nonavian theropods (*Carcharodontosaurus*, 2.3–3.23; *Ceratosaurus*, 3.31–5.07; *Allosaurus*, 2.4–5.24; *Acrocanthosaurus*, 2.75–5.92; *Citipati* 3.6; *Tyrannosaurus*, 5.44–7.63; *Troodon*, 7.76; Franzosa, 2004).

Cerebrum

According to Evans et al. (2009), the most striking aspect of the brain endocast of lambeosaurine hadrosaurids is the relatively large size of the cerebrum. Its estimated relative volume (CRV = cerebrum volume/endocast volume) calculated for four late Campanian lambeosaurines from North America varies between 35% and 42% (Table 16.3). The cerebrum of lambeosaurines is therefore larger than that of large theropods such as *Carcharodontosaurus* (24%) and *Tyrannosaurus rex* (33%), but compares favorably with the maniraptoran theropod *Conchoraptor* (43%) and even with the basal bird *Archaeopteryx* (45%).

We estimated the volume of the cerebrum in *Iguanodon* and *Mantellisaurus*, in *Lurdusaurus arenatus* Taquet and Russell, 1999 (a basal styracosternan Iguanodontia, according to Norman, 2004), in the basal hadrosauroid *Batyrosaurus rozhdestvenskyi* (see Chapter 20 in this book), and in the basal lambeosaurine *Amurosaurus riabinini* Bolotsky and Kurzanov, 1991 (Table 16.3). Because the cerebrum has roughly the shape of an ellipsoid, its volume can be deduced from the following formula: $V = 4/3\pi \times L \times W \times H$, where L, W, and H are, respectively, half of the maximal length, width, and height of the cerebrum. In RBINS R51, the holotype of *Iguanodon bernissartensis*, the CRV is 19%, as in the basal styracosternan *Lurdusaurus* (Table 16.3). Figure 16.6A shows that the cerebrum is proportionally less developed

Table 16.3. Comparisons of the estimated endocast and cerebrum volume in styracosternan ornithopods

Specimen	Taxon	Endocast Length (mm)	Endocast Volume (cm³)	Cerebrum Volume (cm³)	CRV (%)
RBINS R51	*Iguanodon bernissartensis*	183	357	68	19
RBINS R54	*Iguanodon bernissartensis*	>107	>194	58	—
RBINS R57	*Mantellisaurus atherfieldensis*	141	>131	90	—
MNHN GDF 1700	*Lurdusaurus arenatus*	137	167	32	19
AEHM 4/1	*Batyrosaurus rozhdestvenskyi*	100	—	27	—
AEHM 1/232	*Amurosaurus riabinini*	145	290	88	30
ROM 758a	*Lambeosaurus* sp.	113	94.1	35.1	37
ROM 759a	*Corythosaurus* sp.	110	97.9	41.6	42
CMN 34825a	*Corythosaurus* sp.	142	145.4	51.1	35
ROM 702a	*Hypacrosaurus altispinus*	204	289.9	117.5	41

Note: CRV = cerebrum relative volume (cerebrum volume/endocast volume).
[a] Data from Evans et al. (2009).

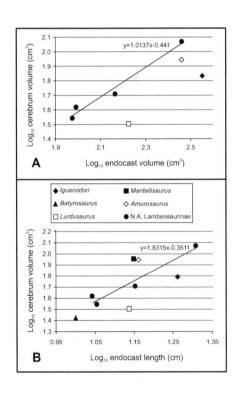

16.6. Relative volume of the cerebrum in selected Iguanodontia compared with endocast volume (A) and length (B). The best-fitting lines are calculated for North American Lambeosaurinae by the least squares regression method.

in *Iguanodon* and *Lurdusaurus* than in North American lambeosaurines and in the Russian basal lambeosaurine *Amurosaurus* (CRV 30%).

Although the skull of RBINS R57 (referred specimen of *Mantellisaurus atherfieldensis*) is much smaller than that of RBINS R51 (holotype of *Iguanodon bernissartensis*), the volume of the cerebrum is larger in RBINS R57 (90 cm³) than in RBINS R51 (68 cm³). Unfortunately, the CRV of *Mantellisaurus* cannot be adequately estimated because the endocast is squashed at the level of the midbrain region in RBINS R57. However, it is still possible to compare the estimated volume of the cerebrum with the length of the endocast (Fig. 16.6B); the latter measurement is not affected by the squashing of the midbrain region in RBINS R57. The cerebrum is similarly enlarged in *M. atherfieldensis* and in the basal lambeosaurine *Amurosaurus*, and it compares favorably with the North American lambeosaurines. It is proportionally larger than in *Iguanodon* and *Lurdusaurus*.

Enlarged brain and cerebrum relative to body size are usually equated with increased behavioral complexity in vertebrates (Jerison, 1969, 1973; Hopson, 1977; Hurlburt, 1996). According to Evans et al. (2009), the relatively large size of the brain and the cerebrum in lambeosaurines is consistent with the range and complexity of social behaviors inferred from the hypothesis that the supracranial crest was an intraspecific signaling structure for visual and vocal communication. However, a similar increase in the relative size of the cerebellum can be observed in *Mantellisaurus atherfieldensis*. On the other hand, calculated REQ and CRV values remain relatively low in *Iguanodon*. This observation suggests that the behavioral repertoire and/or complexity were different in *Iguanodon* and *Mantellisaurus*, two taxa that most likely occupied the same territories in western Europe during the late Barremian and lower Aptian (see Chapter 15 in this book).

According to Wu and Godefroit (Chapter 19 in this book) and Godefroit et al. (Chapter 20 in this book), *Mantellisaurus* and *Iguanodon* belonged, with *Ouranosaurus*, to a monophyletic clade named Iguanodontidae. Iguanodontidae is the sister taxon of Hadrosauroidea. If this phylogeny is correct, it means that the enlargement of the cerebrum appeared independently, together with possible capacities for more complex behaviors, at least two times during the evolution of Iguanodontoidea.

We thank the technical team of the Department of Palaeontology of the RBINS who helped in the restoration and the transport of *Iguanodon* skulls. P. L. thanks P. Semal, who helped her with the Amira and Artecore programs. E. G. is thanked for fruitful discussions and review of a previous version of this paper. CT scanning was supported by Fonds de la Recherche Scientifique grant 1.5.102.07-C.C. to P. G. This paper constitutes part of a Ph.D. thesis by P.L. at the Free University of Brussels. P. L.'s research was supported by the Fonds pour la Formation à la Recherche dans l'Industrie et dans l'Agriculture.

Acknowledgments

Alexander, R. McN. 1985. Mechanics of posture and gait of some large dinosaurs. Zoological Journal of the Linnean Society 83: 1–25.

Anderson, J. F., A. Hall-Martin, and D. A. Russell. 1985. Long-bone circumference and weight in mammals, birds and dinosaurs. Journal of Zoology A 207: 53–61.

Andrews, C. W. 1897. Note on a cast of the brain-cavity of *Iguanodon.* Annals and Magazine of Natural History (ser. 6) 19: 585–591.

Bolotsky, Y. L., and S. K. Kurzanov. 1991. The hadrosaurs of the Amur Region; pp. 94–103 in Glubinnoe Tihookeanskogo Obranleniâ. Amur KNII, Blagoveschensk.

Breazile, J. E., and H.-G. Hartwig. 1989. Central nervous system; pp. 485–566 in A. S. King and J. McLelland (eds.), Form and function in birds, vol. 4. Academic Press, New York.

Brochu, C. A. 2000. A digitally rendered endocast for *Tyrannosaurus rex.* Journal of Vertebrate Paleontology 20: 1–6.

———. 2003. Osteology of *Tyrannosaurus rex*: insights from a nearly complete skeleton and high-resolution computed tomographic analysis of the skull. Society of Vertebrate Paleontology Memoir 7. Journal of Vertebrate Paleontology 22 (supplement to 2): 1–140.

Bultynck, P. 1993. An assessment of the posture and gait in *Iguanodon bernissartensis* Boulenger, 1981. Bulletin de l'Institut royal des Sciences naturelles de Belgique, Sciences de la Terre 63: 5–11.

Evans, D. C. 2005. New evidence on brain–endocranial cavity relationships in ornithischian dinosaurs. Acta Palaeontologica Polonica 50: 617–622.

———. 2006. Nasal cavity homologies and cranial crest function in lambeosaurine dinosaurs. Paleobiology 32: 109–125.

Evans, D. C., R. Ridgely, and L. M. Witmer. 2009. Endocranial anatomy of lambeosaurine hadrosaurids (Dinosauria:

Ornithischia): a sensorineural perspective on cranial crest function. Anatomical Record 292: 1315–1337.

Franzosa, J. W. 2004. Evolution of the brain in Theropoda (Dinosauria). Ph.D. dissertation, Department of Geological Sciences, University of Texas, Austin, Texas, 357 pp.

Franzosa, J. W., and T. Rowe. 2005. Cranial endocast of the Cretaceous theropod dinosaur *Acrocanthosaurus atokensis.* Journal of Vertebrate Paleontology 25: 859–864.

Giffin, E. B. 1989. Pachycephalosaur paleoneurology (Archosauria: Ornithischia). Journal of Vertebrate Paleontology 9: 67–77.

Hooley, R. W. 1925. On the skeleton of *Iguanodon atherfieldensis* from the Wealden shales of Atherfield (Isle of Wight). Quarterly Journal of the Geological Society of London 81: 1–61.

Hopson, J. A. 1977. Relative brain size and behavior in archosaurian reptiles. Annual Review of Ecology, Evolution, and Systematics 8: 429: 448.

———. 1979. Paleoneurology; pp. 39–146 in C. Gans, R. Glenn Northcutt, and P. Ulinski (eds.), Biology of the Reptilia. Academic Press, London.

Hulke, J. W. 1871. Note on a large reptilian skull from Brooke, Isle of Wight, probably dinosaurian, and referable to the genus *Iguanodon.* Quarterly Journal of the Geological Society 27: 199–206.

Hurlburt, G. 1996. Relative brain size in recent and fossil amniotes: determination and interpretation. Ph.D. thesis, University of Toronto, Department of Zoology, 250 pp.

Jerison, H. J. 1969. Brain evolution and dinosaur brains. American Naturalist 103 (934): 575–588.

———. 1973. Evolution of the brain and intelligence. Academic Press, New York, 482 pp.

Larsson, H. C. E. 2001. Endocranial anatomy of *Carcharodontosaurus saharicus* (Theropoda: Allosauroidea) and its

References

implications for theropod brain evolution; pp. 19–33 in D. H. Tanke and K. Carpenter (eds.), Mesozoic vertebrate life: new research inspired by the paleontology of Philip J. Currie. Indiana University Press, Bloomington.

Norell, M. A., P. J. Makovicky, G. S. Bever, A. M. Balanoff, J. M. Clark, E. Barsbold, and T. Rowe. 2009. A review of the Mongolian Cretaceous dinosaur Saurornithoides (Troodontidae: Theropoda). American Museum Novitates 3654: 1–63.

Norman, D. B. 1977. On the anatomy of the ornithischian dinosaur *Iguanodon*. Ph.D. dissertation, King's College, London, 331 pp.

———. 1980. On the ornithischian dinosaur *Iguanodon bernissartensis* of Bernissart (Belgium). Mémoires de l'Institut Royal des Sciences Naturelles de Belgique 178: 1–105.

———. 1986. On the anatomy of *Iguanodon atherfieldensis* (Ornithischia: Ornithopoda). Bulletin de l'Institut royal des Sciences naturelles de Belgique, Sciences de la Terre 56: 281–372.

———. 2004. Basal Iguanodontia; pp. 413–437 in D. B. Weishampel, P. Dodson, and H. Osmólska (eds.), The Dinosauria. 2nd ed. University of California Press, Berkeley.

———. 2005. Dinosaurs: a very short introduction. Oxford University Press, Oxford, 176 pp.

Norman, D. B., and D. B. Weishampel. 1990. Iguanodontidae and related ornithopods; pp. 510–533 in D. B. Weishampel, P. Dodson, and H. Osmólska (eds.), The Dinosauria. University of California Press, Berkeley.

Osborn, H. F. 1912. Crania of *Tyrannosaurus* and *Allosaurus*. Memoirs of the American Museum of Natural History 1: 1–30.

Osmólska, H. 2004. Evidence on relation of brain to endocranial cavity in oviraptorid dinosaurs. Acta Palaeontologica Polonica 49: 321–324.

Rogers S. W. 1998. Exploring dinosaur neuropaleobiology: viewpoint computed tomography scanning and analysis of an *Allosaurus fragilis* endocast. Neuron 21: 673–679.

———. 1999. *Allosaurus*, crocodiles, and birds: evolutionary clues from spiral computed tomography of an endocast. Anatomical Record 257: 162–173.

———. 2005. Reconstructing the behaviors of extinct species: an excursion into comparative paleoneurology. American Journal of Medical Genetics 134A: 349–356.

Sampson, S. D., and L. M. Witmer. 2007. Craniofacial anatomy of *Majungasaurus crenatissimus* (Theropoda: Abelisauridae) from the Late Cretaceous of Madagascar. Society of Vertebrate Paleontology Memoir 8, Journal of Vertebrate Paleontology 27 (supplement to 2): 32–102.

Sanders, R. K., and D. K. Smith. 2005. The endocranium of the theropod dinosaur *Ceratosaurus* studied with computed tomography. Acta Palaeontologica Polonica 50: 601–616.

Sedlmayr, J. C., S. J Rehorek, E. J. Legenzoff., and J. Sanjur 2004. Anatomy of the circadian clock in avesuchian archosaurs. Journal of Morphology 260: 327.

Sereno, P. C., J. A. Wilson, L. M. Witmer, J. A. Whitlock, A. Maga, O. Ide, and T. Rowe. 2007. Structural extremes in a Cretaceous dinosaur. PloS One 2: e1230.

Stephan, H. 1960. Methodische Studien über den quantitativen Vergleich architektonischer Struktureinheiten des Gehirns. Zeitschrift für wissenschaftliche Zoologie 164: 143–172

Taquet, P., and D. E. Russell. 1999. A massively-constructed Iguanodont from Gadoufaoua, Lower Cretaceous of Niger. Annales de paléontologie 85: 85–96.

Van Beneden, P. J. 1881. Sur l'arc pelvien chez les dinosauriens de Bernissart. Bulletin de l'Académie Royale des Sciences, des Lettres et des Beaux-Arts de Belgique (ser. 3) 1: 600–608.

Witmer, L. M., and R. C. Ridgely. 2008a. The paranasal air sinuses of predatory and armored dinosaurs (Archosauria: Theropoda and Ankylosauria) and their contribution to cephalic architecture. Anatomical Record 291: 1362–1388.

———. 2008b. Structure of the brain cavity and inner ear of the centrosaurine ceratopsid *Pachyrhinosaurus* based on CT scanning and 3D visualization; pp. 117–144 in P. J. Currie, W. Langston Jr., and D.H. Tanke (eds.), A new horned dinosaur from an Upper Cretaceous bone bed in Alberta. National Research Council Research Press, Ottawa.

———. 2009. New insights into the brain, braincase, and ear region of tyrannosaurs (Dinosauria, Theropoda), with implications for sensory organization and behavior. Anatomical Record 292: 1266–1296.

Witmer, L. M., S. Chatterjee, J. W. Franzosa, and T. Rowe. 2003. Neuroanatomy of flying reptiles and implications for flight, posture and behavior. Nature 425: 950–953.

Witmer, L. M., R. C. Ridgely, D. L. Dufeau, and M. C. Semones. 2008. Using CT to peer into the past: 3D visualization of the brain and ear regions of birds, crocodiles, and nonavian dinosaurs; pp. 67–88 in H. Endo and R. Frey (eds.), Anatomical imaging: towards a new morphology. Springer-Verlag, Tokyo.

Zelenitsky, D., F. Therrien, and Y. Kobayashi. 2008. Olfactory acuity in theropods: palaeobiological and evolutionary implications. Proceedings of the Royal Society B 276: 667–673.

Zhou, C.-F., K.-Q. Gao, R. C. Fox, and X.-K. Du. 2007. Endocranial morphology of psittacosaurs (Dinosauria: Ceratopsia) based on CT scans of new fossils from the Lower Cretaceous, China. Palaeoworld 16: 285–293.

Hypsilophodon foxii and Other Smaller Bipedal Ornithischian Dinosaurs from the Lower Cretaceous of Southern England

17

Peter M. Galton

The cranial and postcranial anatomy and biology is detailed for *Hypsilophodon foxii*, a small cursorial (not arboreal) basal euornithopod with no confirmed record outside of the Isle of Wight (Late Barremian) of southern England. Three large distal femora with a wide distal extensor groove are reidentified. Two with a narrow more medially situated lateral flexor condyle are basal Ornithopoda. The other distal femur from Potton, which also has a lateral expansion of the medial flexor condyle partly overlaping the flexor groove, is identified as Iguanodontia. The record of the dryosaurid *Valdosaurus* is restricted to England. "*Camptosaurus*" *valdensis* Lydekker, 1889a, a femur, is a nomen dubium and basal Ornithopoda. *Iguanodon hoggii* Owen, 1874, the type species of *Owenodon* Galton, 2009, is represented by the small holotype dentary with tooth row from Dorset. *Owenodon*, which is also present in the Lower Cretaceous of Spain and Romania, is a basal member of the Styracosterna. A small dentary with three replacement teeth from the Isle of Wight (Late Barremian) is basal Iguanodontoidea indet., as are a small distal half femur from the Isle of Wight and a small femur from Dorset (Middle Berriasian). However, histological studies are needed to determine whether they represent juveniles of an already known taxon or whether they are adults of a new taxon. The taxonomic and size diversity of ornithischians of the Lower Cretaceous of southern England are calculated and listed for the different horizons

Introduction

Terrestrial deposits yielding remains of dinosaurs are rare in the Lower Cretaceous and are listed with faunas in Weishampel et al. (2004, 558–570). Some of the best-known dinosaur faunas are from the Lower Cretaceous of southern England. These yield abundant remains, especially those of herbivorous ornithischian dinosaurs (for faunas see Ostrom, 1970, 129–141; updated Pereda-Suberbiola, 1993; Benton and Spencer, 1995; for dinosaurs of the Isle of Wight, see Martill and Naish, 2001a).

The large Wealden iguanodontid ornithopod *Iguanodon* Mantell, 1825, from West Sussex was erected on the basis of isolated teeth. It was one of the three genera upon which the Dinosauria Owen, 1842, was founded. Since then, numerous articulated remains have been described from the Wealden of southern England and Belgium (Chapter 1 in this book) as *Iguanodon* (see Naish and Martill, 2001a; Norman, 1980, 1986). However, several of these specimens are now the basis for different genera and species (Paul, 2006, 2008; Norman, 2010; Chapter 15 in this book).

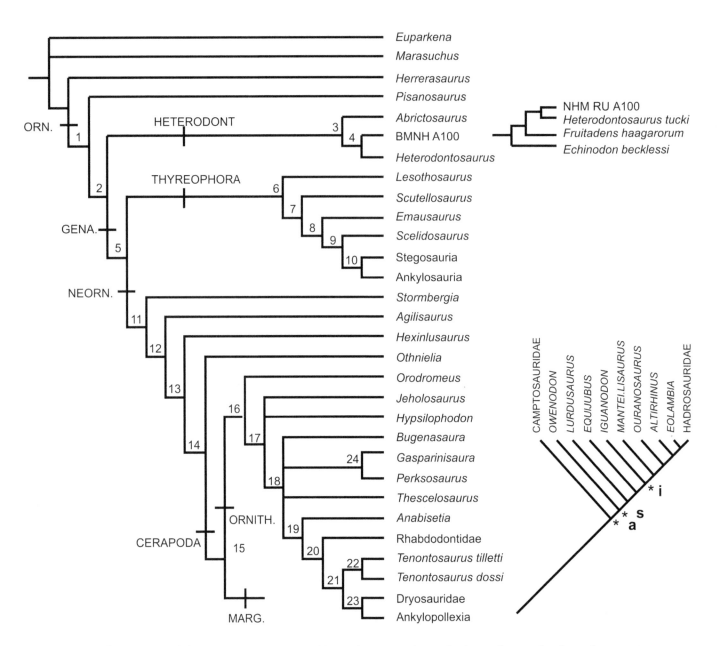

17.1. Cladogram of nonmarginocephalian Ornithischia modified from Butler et al. (2008; see appendix 4 for unambiguous synapomorphies for nodes 1–23), with details of Heterodontosauridae updated from Butler et al. (2010, ESM) and expansion of the Ankyllopolexia, with tentative position shown for *Owenodon,* modified from Galton (2009; after Norman 2004). *Abbreviations:* a, Ankylopollexia; GENA, Genasauria; HETERODONT, Heterodontosauridae; i, Iguanodontoidea; ia, Iguanodontia; MARG, Marginocephalia; NEORN, Neornithischia; ORN, Ornithischia; ORNITH, Ornithopoda; s, Styracosterna; 10, Eurypoda; 11, Neornithischia; 16, Ornithopoda.

Small jaw bones with teeth from the Purbeck Beds near Swanage, Dorset, were described as those of a piscivorous lizard, *Echinodon becklesii* Owen, 1861a. It was subsequently recognized as a dinosaur and is currently referred to the basal ornithischian family Heterodontosauridae (Butler et al., 2010) (Fig. 17.1).

Two small articulated specimens, supposedly juvenile individuals of *Iguanodon mantelli,* were described from the Wealden (Wessex Formation) of the southwest coast of the Isle of Wight, a partial skeleton (Mantell, 1849; Owen, 1855) and a skull (Owen, 1874) (Fig. 17.2). Fox (1868) discussed the skull and other articulated specimens from the same bed in his collection as a new unnamed small species of *Iguanodon.* However, these specimens represent a basal ornithopod, *Hypsilophodon foxii,* which was erected and described by Huxley (1869, 1870). This was the first skull (Fig. 17.2C) described for an ornithopod and the second for an ornithischian dinosaur, the first being that of the basal armored dinosaur *Scelidosaurus* (Lower Jurassic, southern England; Owen, 1861b). Huxley (1870) reinterpreted the tibia and

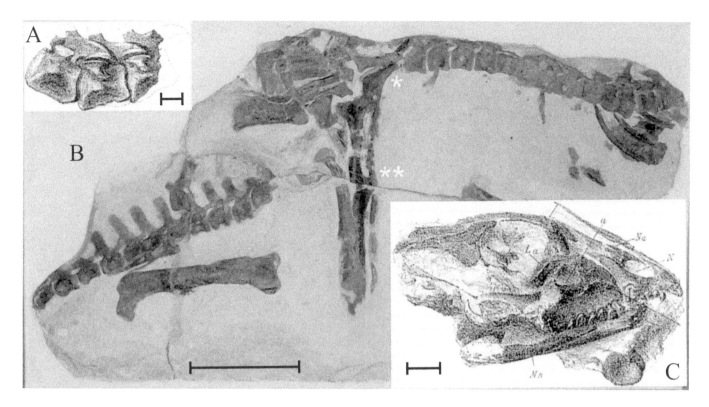

17.2. Basal ornithopod *Hypsilophodon foxii* from *Hypsilophodon* Bed, at top of Wessex Formation (Lower Cretaceous, late Barremian), of southwest coast of the Isle of Wight, southern England (same data for other figures of *H. foxii*), paratype (A, B, NHMUK 28707, 39560–1) and holotype specimens (C, R197). A, Cervical vertebrae 7–9 in left lateral view, from Mantell (1849). B, Sandstone block originally figured by Owen (1855), with posterior cervicals and rest of column to proximal part of tail plus pelvic and hind limb bones in right lateral view; *, anterior process of pubis; **, posterior process of pubis. C, Holotype skull in right lateral view with dorsal vertebral centrum in end on view, from Huxley (1870). Scale bars = 10 (A, C) and 100 mm.

fibula of Owen (1855) as parts of the pelvic girdle, the ischium, and pubis (Fig. 17.2B). In so doing, Huxley showed that the pubis of *Hypsilophodon* has an anterior process and, in addition, a posterior process consisting of a long, slender rod lying alongside the length of the ischium. This is the opisthopubic pelvic girdle, which also occurs in birds. On the basis of the form of the pelvic girdle, Seeley (1887) divided the known Dinosauria into two groups. These were the Saurischia (reptile hipped, triradiate pelvis; for Theropoda and Sauropoda) and the Ornithischia (bird hipped, quadriradiate pelvis; for Ornithopoda and Stegosauria).

A small basal pachycephalosaurid or dome-headed dinosaur, *Yaverlandia bitholos* Galton, 1971a, is based on a markedly thickened skull roof (IWCMS 1530, Wessex Formation, Yaverlandia, southeast shore of Isle of Wight). As discussed by Naish and Martill (2001b, photographs pl. 26), *Yaverlandia* has been referred to various ornithischian groups, but Naish, who identifies it as the skull roof of a theropod or carnivorous dinosaur, lists four of its theropod characters in Sullivan (2006, 361).

A pair of small femora from the Wealden of the Isle of Wight, incorrectly referred to *Hypsilophodon foxii* by Lydekker (1888a, 194), were made the holotype of the dryosaurid ornithopod *Dryosaurus? canaliculatus* Galton, 1975. This is the type species of *Valdosaurus* Galton, 1977a.

Owen (1864) described a small dentary, from the Wealden of the Isle of Wight, as a juvenile individual of *Iguanodon mantelli*. After being referred to several other genera, it was identified as Iguanodontoidea indet. by Galton (2009).

The small holotype dentary of *Iguanodon hoggii* Owen, 1874, came from the Purbeck Beds near Swanage. It included a complete tooth row, the basis for the first description of the pattern of tooth replacement in any herbivorous dinosaur. After being redescribed as *Camptosaurus hoggii*

17.3. Small basal heterodontosaurid dinosaur *Echinodon becklesii* Owen, 1861a, from Purbeck Limestone Formation (Berriasian) of Durlston Bay, Swanage, Dorset. A–K, Lectotype, split slab with jaws of a single individual, NHMUK 48209 (A, B, H–J) and 48210 (C–G, K). A, B, Premaxillae and anterior left maxilla in lateral view with teeth and impression of medial surface; C, D, posterior left maxilla in medial view and partial palatine; E, F, sectioned maxilla (lateral edge to left) in anterior view showing middle tooth (E, * in D) and crown (F, G). H, Right premaxilla and left premaxillary teeth after removal of left premaxilla (cf. A, B); I–K, left teeth in labial (I, J) and lingual views (K). L, Premaxillary teeth 1 and 3; J, maxillary teeth 1–4; K, four posterior maxillary teeth in lingual view (cf. C, D). L, M, Anterior right maxilla NHMUK 48211 in lateral view: L, bone (with palatine, cf. Fig. 17.5C); M, teeth (cf. Fig. 5G). N, Posterior maxillary teeth NHMUK 48212 in lingual view (cf. Fig. 17.5F). Teeth inverted in K, M, N; B, D, F from Owen (1861a). Scale bars = 5 (A–D, H) or 1 mm.

by Norman and Barrett (2002), it was made the type species of *Owenodon* Galton, 2009.

Basal iguanodontians (Fig. 17.1) were first recognized from the Lower Cretaceous of England on the basis of large distal femora from the Wealden of the Isle of Wight and the Potton Sandstone of Potton, Bedfordshire (Galton, 2009).

Unless indicated to the contrary, comparisons with other ornithischians are based on the references listed below (see Fig. 17.1).

Heterodontosauridae (see Norman et al., 2004), now regarded as the most basal of the well-known groups of ornithischians (Fig. 17.1; Butler et al., 2008, 2010), includes *Heterodontosaurus tucki* (Lower Jurassic, South Africa; Norman et al., 2004), *Fruitadens haagarorum* (Upper Jurassic, Colorado, USA; Butler et al., 2010), and *Tianyulong confuciusi* (Middle to Upper Jurassic; see Lü et al., 2011; China; Zheng et al., 2009).

Basal Neoornithischia: *Lesothosaurus diagnosticus* (Lower Jurassic, southern Africa; Sereno, 1991).

The "Hypsilophodontidae" is not monophyletic and mostly represents a grade of basal ornithopods (Fig. 17.1; see Norman et al., 2004; Butler et al., 2008). Cited taxa include *Jeholosaurus shangyuanensis* (Lower Cretaceous, Barremian, China; Barrett and Han, 2009) and *Anabisetia saldiviai* (Upper Cretaceous, Argentina; Coria and Calvo, 2002).

The Iguanodontia comprise the rest of a paraphyletic array of ornithopod taxa (Fig. 17.1; see Norman, 2004; Paul, 2008).

Basal Iguanodontia include *Tenontosaurus* (Lower Cretaceous, Aptian, Wyoming, *T. tilletti* Ostrom, 1970; Forster, 1990; Texas, *T. dossi* Winkler et al., 1997) and *Muttaburrasaurus langdoni* (Lower Cretaceous, Albian, Australia; Bartholomai and Molnar, 1981).

Dryosauria for Dryosauridae: *Dryosaurus*: Upper Jurassic, western USA, *D. altus* (Galton, 1983, 1989); Tanzania, *D. lettowvorbecki* (Janensch, 1955; Galton, 1983, 1989); *Elrhazosaurus* Galton, 2009 (Lower Cretaceous, Aptian, Niger, type species *Valdosaurus nigeriensis* Galton and Taquet, 1982), and *Kangnasaurus coetzeei* (Lower Cretaceous, South Africa; Cooper, 1985; in family as a nomen dubium, Sues and Norman, 1990).

Ankylopollexia for Camptosauridae: Upper Jurassic: *Camptosaurus*, western USA, *C. dispar* (Gilmore, 1909, 1912); *C. aphanoectes* (Carpenter and Wilson, 2008) and England: *C. prestwichii* (Galton and Powell, 1980); and *Draconyx loureiroi* (Portugal; Mateus and Antunes, 2001).

Styracosterna: for *Lurdusaurus arenatus* (Lower Cretaceous, Aptian, Niger; Taquet and Russell, 1999) and *Equijubus normani* (Barremian–Albian, China; You et al., 2003).

Iguanodontoidea: consists of increasingly derived taxa that include *Iguanodon bernissartensis* (Lower Cretaceous, Valanginian–Albian, Bernissart, Belgium and England; Norman, 1980); *Mantellisaurus* Paul, 2006 ("2007") (type species *Iguanodon atherfieldensis* Hooley, 1925; color photograph of mounted skeleton in Gardom and Milner, 1993, 27; Lower Cretaceous, Valanginian–Albian, Bernissart and England; Norman, 1986); *Ouranosaurus nigeriensis* (Lower Cretaceous, Aptian, Niger; Taquet, 1976); *Altirhinus kurzanovi* (Lower Cretaceous, Aptian–Albian, Mongolia; Norman, 1998), and *Eolambia caroljonesa* (Lower Cretaceous, Utah, USA; Kirkland, 1998).

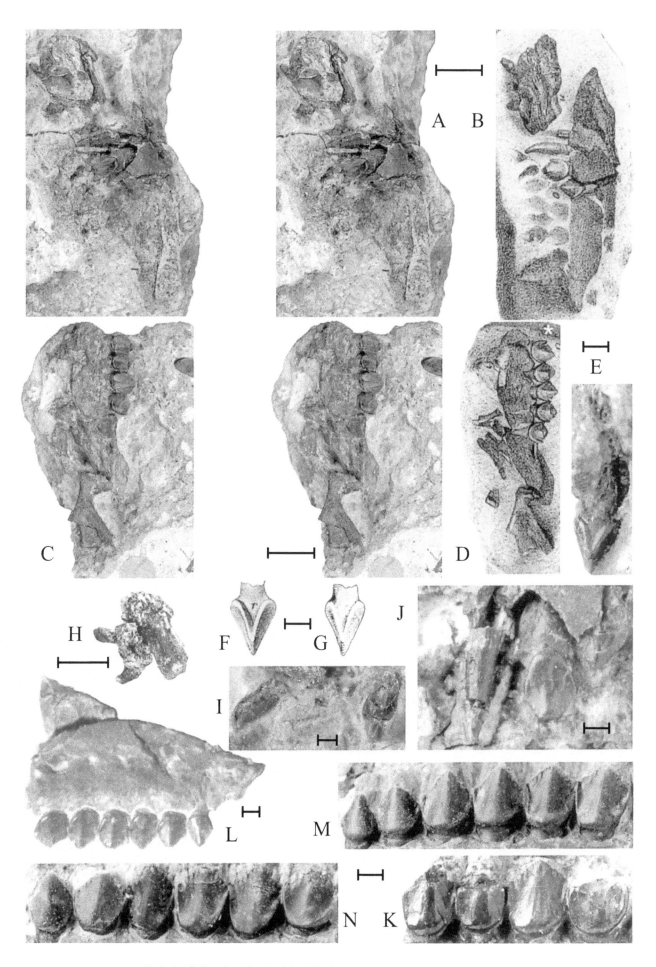

17.4. A–C, Right mandible NHMUK 48214, previously referred to *Echinodon becklesii*, in (A) lateral, (B) dorsal, and (C) medial views (cf. Fig. 17.5J–L). D–P, Basal heterodontosaurid *Echinodon becklesii*, right (D–F) and left dentaries (G–N) with details of tooth rows (D, G, H, J, K, O, P) in medial or lingual (D, F, G, I, J, N, P), lateral or buccal (D, G, H, L, O), and dorsal views (M). D, NHMUK 40723. E–H, NHMUK 48215a. I–K, NHMUK 46213. L–P, NHMUK 48215b. Scale bars = 10 mm (A–C, E, F, I, L–N) and 1 mm (D, G, H, J, K, O, P).

The holotypes and referred specimens of the smaller bipedal ornithischian dinosaurs from the Lower Cretaceous of England, with particular emphasis on *Hypsilophodon foxii*, are the subject of this chapter.

Institutional abbreviations: AMNH, American Museum of Natural History, New York, USA; BELUM, Ulster Museum (National Museums of Northern Ireland), Belfast, U.K.; BMB, Booth Museum of Natural History, Brighton, Sussex, U.K.; CAMSM, Sedgwick Museum, University of Cambridge, U.K.; CM, Carnegie Museum of Natural History, Pittsburgh, Pennsylvania, USA; GPIT, Geologcal und Paläontological Institut, Universitat der Tübingen, Germany; RBINS, Royal Belgian Institute of Natural Sciences, Brussels, Belgium; IWCMS, Museum of Isle of Wight Geology (abbreviation formerly MIWG), Sandown, Isle of Wight, U.K.; KINUA, Hull Museums and Art Gallery, Kingston upon Hull, UK; MFCPTD, Museo de la Fundación Conjunto Paleontológico de Teruel-Dinópolis; Teruel, Spain; MFNB, Museum für Naturkunde — Leibniz Institute for Research on Evolution and Biodiversity at the Humboldt University, Berlin, Germany; MTCO, Muzeul Țării Crișurilor, Oradea, Romania; NHMUK, The Natural History Museum (abbreviation formerly NHM and also BMNH for British Museum [Natural History]), London, U.K., with R indicating NHMUK R, the museum housing most of the cited specimens; SAM-PK, South African Museum (Iziko Museums of Cape Town), South Africa; YPM, Yale University, Peabody Museum of Natural History, New Haven, Connecticut, USA.

Heterodontosauridae: *Echinodon becklesii* Owen, 1861a

Echinodon becklesii Owen, 1861a, is based on small jaw bones with teeth (Figs. 17.3, 17.4, and 17.5C–K) from the Purbeck Limestone Formation (=Durlston Formation, Upper Purbeck, mid-Berriasian, Allen and Wimbledon, 1991) of Durlston Bay near Swanage, Dorset. Although originally clasified under Lacertilia (lizards), Owen (1861b, 7) noted similarities to the teeth of *Scelidosaurus* (now a basal thyreophoran or armor-bearing dinosaur, Fig. 17.1) and that *Echinodon* "may prove to be a small kind, or young, of a Dinosaur." Later he included *Echinodon* in the Dinosauria (Owen, 1874, 15; also Lydekker, 1888a, 247), but it was not formally referred to the Ornithischia until Hennig (1915) listed it in the Stegosauria/Scelidosauridae. Galton (1978) described most of the material, much of which was refigured by Norman and Barrett (2002), who critically reviewed the basis for all previous systematic assignments. The "granicones," first referred to *Echinodon* by Galton (1981a), are the limb osteoderms of solemyid turtles (see Barrett et al., 2002).

The lectotype individual of *Echinodon becklesii* Owen, 1861a, consists of a split slab. One part contains the premaxillae and the left maxilla, with the anterior part in lateral view (Fig. 17.3A,B). The other part has the rest of maxilla in medial view, along with the posterior portion of the left pterygoid (Fig. 17.3C,D). The fragmentary left premaxilla (Fig. 17.3A,B) is removable to reveal the palatal ramus of the right premaxilla (Fig. 17.3H,I; only part shown by Norman and Barrett, 2002, fig. 7A, pl. 1, fig. 1). The long anterior process of the maxilla is visible dorsal to the premaxillae (Fig. 17.3A,B). A right maxilla, originally more complete posteriorly, is exposed in lateral view along with the palatine (Figs. 17.3L,M and 17.5C,E). The lateral sheet, which is pierced by several nutrient foramina and contacts the lacrimal and

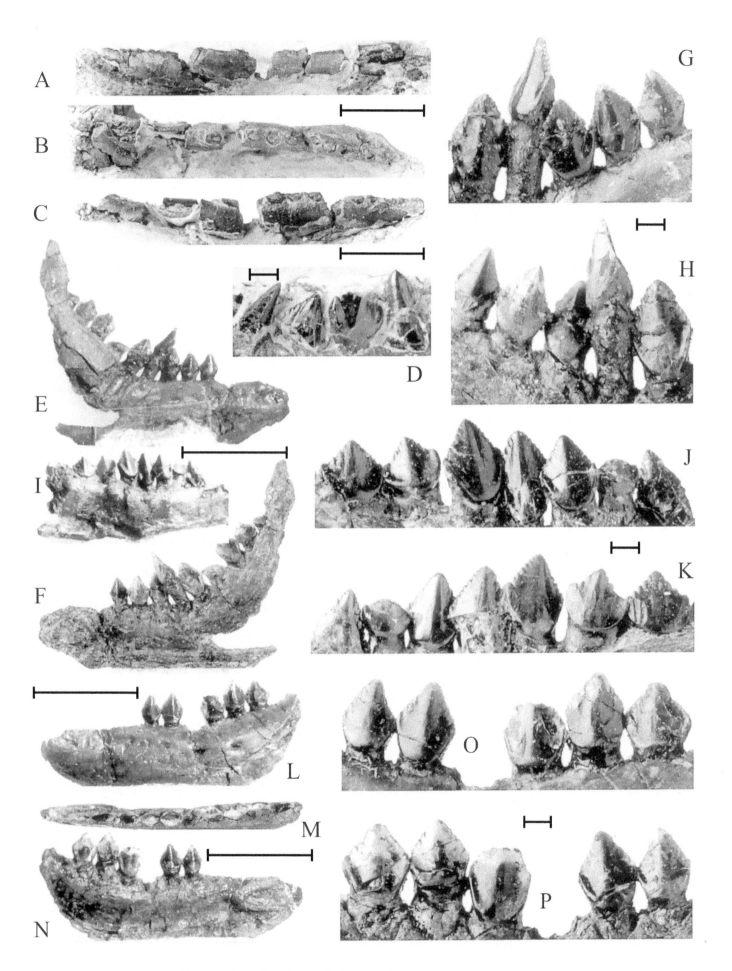

17.5. Heterodontosauridae. A, B, Skeletal reconstructions of *Heterodontosaurus tucki* (Lower Jurassic, South Africa), SAM-PK-K1332 in (A) bipedal and (B) quadrupedal walking poses. Given the close similarities of the postcranial bones of *Fruitadens* (Upper Jurassic, USA) to those of *Heterodontosaurus* (see Butler et al., 2010), this scaled reconstruction should give a fairly accurate idea of the skeleton of *Echinodon*. C–H, *Echinodon becklesii*. C, Right maxilla NHMUK 48211 as originally preserved, in lateral view (cf. Fig. 17.3L). D–E, Partial right maxillary tooth rows: D, NHMUK 48211, posterior part in medial (lingual) view; E, NHMUK 48212, anterior part in lateral (labial) view. F–G, Dentary tooth row, NHMUK 48215a, in (F) lateral and (G) medial views. H, Reconstruction of partial skull, based on NHMUK 48209–48212, 48215. *Abbreviations:* c, caniniform tooth (there is another one immediately anterior to this one in the lectotype maxilla); cor, coronoid process; d, dentary; j, jugal; l, lacrimal; mx, maxilla; pd, predentary; pmx, premaxilla. I–K, Right mandible NHMUK 48214, previously referred to *Echinodon becklesii*. I, Right mandible in lateral view with outline drawing and detail of replacement tooth crown; J–K, mandible in (J) dorsal and (K) medial views (cf. Fig. 17.4A–C). * Empty alveolus. A, B kindly provided by G. S. Paul, who retains the copyright; C, I–K from Owen (1861a); D–H modified from Galton (1978). Scale bars = 100 mm (A, B) and 10 mm (C–K).

jugal posteriorly, is much deeper than in other heterodontosaurids (Figs. 17.3L, 17.5C,H, and cf. 17.5A). Medially, the posterior part of the lectotype has a sutural area that was probably for the palatine (Fig. 17.3C,D); this area is shown by Norman and Barrett (2002, fig. 7B, pl. 1, fig. 2), who identified the bone posterolateral to the last alveolus as part of either the ectopterygoid or the jugal, but it is probably the latter.

The anterior part of the dentary is slender (Figs. 17.4E,F,L and 17.5H), with several nutrient foramina opening laterally; there is quite a prominent rounded longitudinal ridge (along with a corresponding ridge on the maxilla; Fig. 17.3L) that was probably for the attachment of cheeks (see below under *Hypsilophodon*), and it deepens posteriorly so the coronoid eminence was prominent (Fig. 17.5C). Norman and Barrett (2002, 182) noted that a deep coronoid eminence "cannot be assessed confidently from the known material." In this connection, a slightly incomplete syntypic mandible (Figs. 17.4A–C and 17.5I–K) appears to have a low coronoid eminence (Figs. 17.4A,C and 17.5H–I,K). However, the jaw is now broken and in need of repair and further preparation, including the exposure of the complete crown of a replacement tooth, to ascertain whether the referral to *Echinodon becklesii* is correct or not. Medially, there are no special foramina, and the Meckelian groove is shallow (Fig. 17.4E,L). The blunt-ended anterior end bears the prominently ridged symphysis that is set at a slight angle to the rest of the surface (Figs. 17.4M, 17.5J), as in *Fruitadens* and *Heterodontosaurus*, and it is not borne on the medial edge of a spoutlike anteromedially directed process as in *Lesothosaurus* and other ornithopods such as *Hypsilophodon* (Fig. 17.10F). A left dentary has a slight depression along the anteroventral edge (Fig. 17.4L). Galton (1978) thought that this was for the ventral process of a forked predentary (as in *Hypsilophodon*, Fig. 17.10B), but it is not present on the other dentary (Fig. 17.4E; Norman and Barrett, 2002) or on those of *Fruitadens*. In *Echinodon*, a short lateral process of the predentary probably slightly overlapped the anterodorsally facing edge (Fig. 17.4E,L), as in *Lesothosaurus*, to give a wedge-shaped bone (Fig. 17.5H), rather than extending beyond the angle as reconstructed by Galton (1978).

There are two premaxillary teeth, separated by a gap with a fragment of a root, so there were three teeth originally. Although the teeth are not well preserved, the crowns are subequal in size and expanded relative to the root, as in *Fruitadens*, rather than being conical and unexpanded above the root with the last one greatly enlarged and caniniform as in other heterodontosaurids (Fig. 17.5A) except *Tianyulong*, in which the single tooth is large and caniniform. The second tooth lacks denticles and is caniniform and long, so the crown height is greater than that of the other teeth. Anterior to this is the impression of another similar tooth (Fig. 17.3A; Norman and Upchurch, 2002, fig. 7A), the base of which was originally preserved (Fig. 17.3B). However, more distally, there is a thin anteroposteriorly concave layer that represents the inside of part of the enamel layer from the lingual surface of the anterior tooth (Fig. 17.3J). The length of this crown was probably intermediate between that of the better-preserved caniniform and the other teeth. Another maxilla has a large and empty alveolus anteriorly that would have housed a caniform tooth, bordered by an arched diastema (Figs. 17.3L and 17.5C,E). The third tooth of the

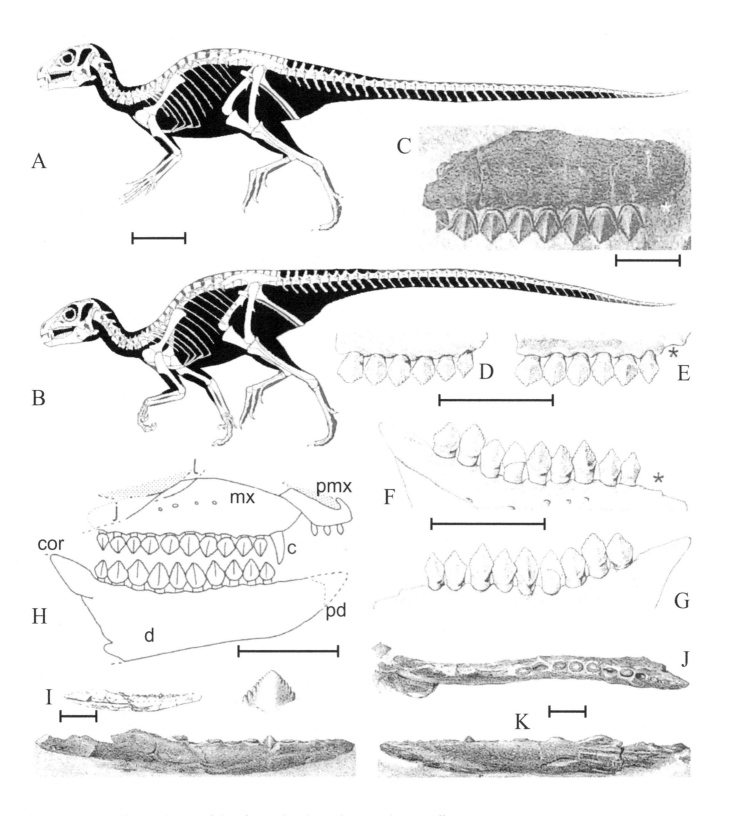

lectotypic maxilla is of normal basal ornithischian form with a small, highly inclined apical wear facet that was probably formed by tooth–food contact (Fig. 17.3A,B,J). Immediately posterior to this functional tooth is the unerupted distal crown of a replacement tooth. The counterpart slab contains five tooth crowns, the last of which is only partially erupted, followed by an empty alveolus (Fig. 17.3C,D,K). On the basis of the two parts, the maxilla bore 12 teeth (Fig. 17.5H), or 13 counting the incomplete caniniform (Fig. 17.3J).

The tooth count for the associated dentaries is 10 (Fig. 17.4E,F,L–N), and there are no indications anteriorly of a caniniform tooth or of an enlarged alveolus (Fig. 17.4M; also 10 teeth and no caniniform in the *?Echinodon* dentary, Figs. 17.4A–C and 17.5J).

The crowns of maxillary and dentary teeth are mesiodistally expanded with a prominent central ridge, on either side of which the distal part bears five or six coarse denticles (Figs. 17.3A–D, J–N, 17.4D–P, 17.5C–H). In midrow teeth, this involves the distal third as in other heterodontosaurids except *Fruitadens*, in which the denticles occupy over half of the crown height. Except in a few teeth (Fig. 17.4D,J,K), the nonapical denticles are not supported by secondary ridges. The most basal denticle is supported on each side by a prominent ridge that borders the rest of the crown and merges with a prominent swelling or cingulum across the base of the crown (Figs. 17.3A–D,K and 17.5C–H). In mesial view, the root curves dorsomedially (Fig. 17.3E), as in *Hypsilophodon* (Fig. 17.10D), but the crown is uniformally enameled. The bordering ridges form a V-shaped border for a depression that lacks enamel and is continuous with the root (Fig. 17.3E–G; inverted V shape for dentary teeth, Fig. 17.4G,H,K). Two most posterior teeth show prominent wear surfaces buccally near the base of the crown (Fig. 17.4K). One is planar whereas the other shows prominent vertical striae, but neither was caused by tooth-to-tooth wear. Instead they were the result of interdental pressure during development by contact within the alveolus as reported in other ornithischians, in which they are usually sharply defined, small, and circular; have flat or irregular surfaces; and occur toward the base of the crown, often on the mesial or distal surface but also on the lingual or labial surface (Thulborn, 1974).

Norman and Barrett (2002) referred *Echinodon* to the Heterodontosauridae on the basis of the presumedly wedge-shaped predentary, the restriction of the denticles of the midcheek teeth to the apicalmost third of the crown (Figs. 17.3J–N, 17.4G–P, and 17.5C–H), and the absence medially of any special foramina for replacement teeth (Figs. 17.3C,D and 17.4F,N). The Heterodontosauridae, a group whose previous most recent record was the Early Jurassic (Fig. 17.5A,B; Norman et al., 2004), have since been described from the Upper Jurassic of Colorado, USA (*Fruitadens haagarorum* Butler et al., 2010), and the Middle to Upper Jurassic of China (*Tianyulong confuciusi* Zheng et al., 2009). The Heterodontosauridae, long regarded as basal ornithopods, are now regarded as the most basal of the well-known ornithischians. This position, first suggested by the noncladistic work of Bakker and Galton (1974), is based on recent cladistic analyses (Butler et al., 2008, 2010; Zheng et al., 2009), with *Echinodon* as the most basal member of the family (Butler et al., 2010) (Fig. 17.1).

The jaws of *Echinodon* are from animals of unknown age that are extremely small, with an estimated body length of ~56 cm and a body mass of ~0.33 kg (see the Appendix). In contrast to the adults of the other Early Cretaceous ornithischians, this taxon was probably an ecological generalist with an omnivorous diet, as were the other more recent survivors of the Heterodontosauridae, *Fruitadens* and *Tianyulong*. These three genera were smaller and lacked the cranial specializations for herbivory present in the Early Jurassic members of the group (Fig. 17.5A; Butler et al., 2010).

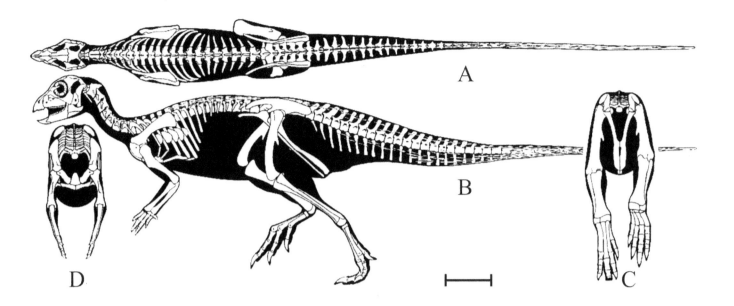

Introduction

Hypsilophodon foxii comes from the *Hypsilophodon* Bed in the cliffs close to Cowleaze Chine (near Barnes High) on the southwest coast of the Isle of Wight, southern England. It occurs at top of Wessex Formation (=old Wealden Marls, Late Barremian). Nearly all the material comes from a 1-m-thick variable bed that forms the highest part of the Wessex Formation, which underlies the Vectis Formation (=old Wealden Shales; Insole and Hutt, 1994; Benton and Spencer, 1995; Martill and Naish, 2001b). The *Hypsilophodon* Bed forms the floor of Cowleaze Chine, gradually rising northwesterly to the top of the sea cliff just beyond Barnes High about 1.2 km away, and the productive region is at the base of the cliff about 90 m west of the chine (Benton and Spencer, 1985, fig. 8.13; Galton, 2001a, fig. 1). About 160 m west of the chine are well-developed dessication cracks in the underlying marls. These cracks, which are about 450 mm deep and 40 mm wide, are filled with sandstone continuous with that of the overlying rocky band that represents the base of the *Hypsilophodon* Bed (Galton, 1967, pl. 3, fig. D). This bed is unusual in the preservation of numerous articulated or slightly disarticulated remains of this small dinosaur with an estimated body length of ~1.95 m and a mass of ~21 kg. The remains are suggestive of catastrophic burial, possibly by a flash flood (Galton, 1974). However, soft sediment structures within the unit show that liquefication of the sediment occurred, so the smaller animals of a herd may have been trapped in quicksand (Stewart, 1978; Insole and Hutt, 1994).

Supposed records of *Hypsilophodon foxii* from England outside of the Isle of Wight are incorrect. They were based on distal femora, one of *Valdosaurus* sp. from West Sussex (Lydekker, 1888a; Swinton, 1936a; Weishampel, 1990; NHMUK 36509, Galton, 1975, 2009, fig. 7T–U). The other (R12555), from East Sussex, was figured as a large individual of *Hypsilophodon foxii* (Galton and Taquet, 1982, fig. 1Q; Molnar and Galton, 1986, fig. 4C). However, it is either that of a nodosaurid (armored dinosaur *Hylaeosaurus armatus*, Barrett, 1996, fig. 1a,b) or of a basal Ornithopoda indet. (Galton,

17.7. Basal ornithopod *Hypsilophodon foxii*, flesh reconstruction, kindly supplied by R. T. Bakker, who retains the copyright.

2009, fig. 16A–F). Supposed records from outside England (Portugal, Spain, Romania, South Dakota, Texas) are discussed by Galton (2009), who concluded on the basis of the available descriptions that all are incorrect, with most representing basal Ornithopoda indet. and a few as different genera.

"An attempt at a complete osteology of *Hypsilophodon foxii*, a British Wealden dinosaur," a paper by Hulke (1882a), was the first description of most of the anatomy of an ornithopod dinosaur. Dollo (1882) erected the family Hypsilophodontidae for *Hypsilophodon*. This recognized the distinctive form of *H. foxii* that was clearly quite different from *Iguanodon*, which by then was known from numerous complete articulated skeletons from the Wealden of Bernissart, Belgium (RBINS; see Norman, 1980, 1986; Martin and Bultynck, 1990).

The articulated specimens of *Hypsilophodon foxii* described in the 1800s were mostly visible only on the surface of hand-prepared blocks of matrix (Figs. 17.2B,C and 17.9A,B), several of which I prepared chemically (using acetic acid) and mechanically. Hulke (1874) described a block of wave-weathered sandstone that included two partial vertebral columns and a slightly disarticulated but almost complete skull (Figs. 17.8B, 17.10A–C, 17.11, 17.1E–G, and 17.14, R2477; Galton, 1967, pl. 22, 2001a, fig. 5) that was larger than the type skull (Figs. 17.3C, 17.9A,B). Another important specimen, R196, consisted of a block split into two pieces (Galton, 1967, pls. 19–21, 2001a, figs. 17–20) that contained a practically complete articulated skeleton (Figs. 17.6, 17.8D,E,H,L, 17.13, 17.16B–E, 17.17, 17.19R–U, 17.20C,D; photographs of bones in Galton, 2001a, figs. 21–24; lacks more than half of tail, slightly disarticulated skull now around matrix core, Galton, 2001b, fig. 9). In addition, there are the overlapping distal two-thirds of

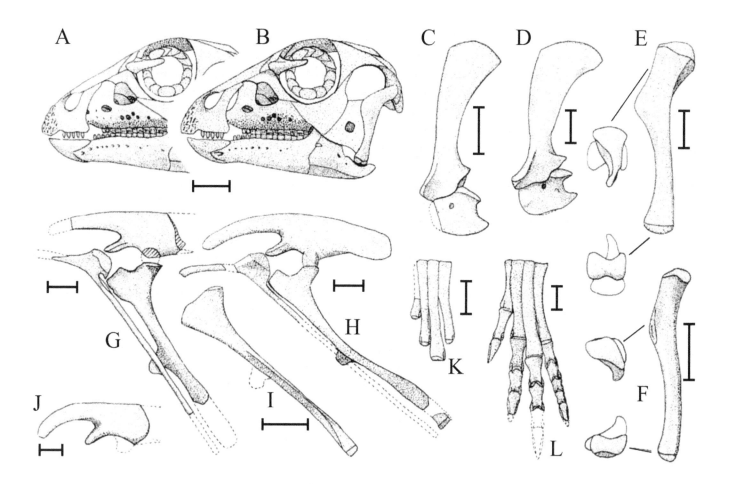

the tail from a larger individual (R196a, left in matrix, used in Fig. 17.6A,B; Galton, 1967, pls. 19, 20, 2001a, fig. 20). For earlier illustrations from the literature, along with photographs of the principal articulated specimens before (NHMUK 28707, 39560-1, R196, R2477) and after further preparation (R196, R197, R2477), see Galton (1967, 2001a, 2001b).

The holotype skull (R197) of *Hypsilophodon foxii* Huxley, 1869, has been described for the right (Huxley, 1870; Owen, 1874; Hulke, 1882a; Galton, 1974) and left sides and in ventral view (Galton, 1974). Galton (2009, fig. 2A–F) provided stereophotographs from approximately the same angles (Figs. 17.9A,A',B,B'). Most of the bones of the skull of R2477 separated slightly at the sutures before preservation so they could be completely freed from matrix and from each other. The skull was dorsoventrally compressed during preservation (Galton, 1967, pls. 23–25), but the reassembled bones give a good undistorted three-dimensional skull (Fig. 17.10A–V) that is only slightly imperfect as a result of breaks in a few of the bones (see Galton, 1974, figs. 4–8, stereophotographs pl. 1, pl. 2, figs. 1, 2, 2001b, figs. 15–17).

The anatomy of *Hypsilophodon foxii* was documented in detail by Galton (1967, 1974; including individual variation with a summary figure in Galton, 1977b, 1980a), who also discussed the pelvic musculature (Galton, 1969); the evidence for a cursorial nonarboreal mode of life (Galton, 1971b, 1971c, 1974, 2001a); the form of the skull (Galton, 1973a, figs. 6–8); the evidence for the presence of cheeks (Galton, 1973b, 2009); the reidentification of some specimens as Hypsilophodontidae indet. and *Dryosaurus? canaliculatus* Galton, 1975, the type species of *Valdosaurus* Galton, 1977a; the

17.8. Basal euornithopod *Hypsilophodon foxii*, individual variation, all bones in left lateral view except K–L (in anterior or dorsal view) and E, F (with proximal and distal views). A, B, Reconstruction of skull of R2477; B, as (A) with maxilla and quadratojugal based on R197. C, D, Scapula and coracoid. C, R5830; D, R196. E, F, Humerus. E, R196; F, R5830. G, H, Pelvic girdle. G, R195; H, R196. I, Ischium R5830. J, Ilium R2477. K, Metatarsus R5830. L, Pes R196. Modified from Galton (1977b). Scale bars = 20 mm.

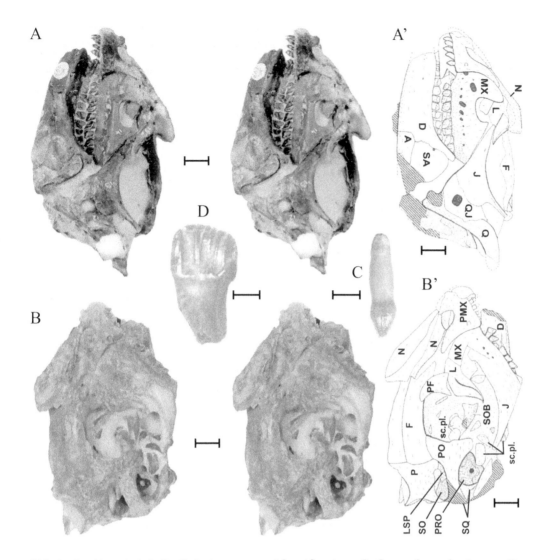

17.9. Basal ornithopods. A–C, *Hypsilophodon foxii*. A, B, Holotype skull R197, stereophotographs in (A) left lateral and (B) right lateral, views with explanatory drawings at approximately the same angle (A', B'), modified from Galton (1974). C, Right premaxillary tooth R2471 in labial view. D, Basal Ornithopoda grades 16–18 indet. of Butler et al. (2008) (Fig. 17.1), left tooth R15993 (formerly R8367) from maxilla in labial view or dentary in lingual view (see *Hypsilophodon foxii*, form of teeth). *Abbreviations:* A, angular; BO, basioccipital; BSP, basisphenoid; D, dentary; F, frontal; J, jugal; L, lacrimal; LSP, laterosphenoid; MX, maxilla; N, nasal; OP, opisthotic; P, parietal; PF, prefrontal; PMX, premaxilla; PO, postorbital; PRO, prootic; PSP, parasphenoid; PT, pterygoid; Q, quadrate; QJ, quadratojugal; SA, surangular; sc pl, sclerotic plates; SO, supraoccipital; SOB, supraorbital; SQ, squamosal. Scale bars = 10 mm (A, B) and 2 mm (C, D).

identification of a femur from the Lower Cretaceous of the western United States as *Hypsilophodon* sp. (Galton and Jensen, 1975; *H. wielandi* Galton and Jensen, 1978), and isolated ends of limb bones from the Upper Jurassic of Portugal as *Hypsilophodon* sp. (Galton, 1980b); the detailed anatomy of the braincase and endocranial cast (Galton, 1989); the form of the teeth of the holotype and dentary teeth with photographs (Galton, 2006, fig. 2.14); and reidentification of the dermal armor of Hulke (1874), Nopcsa (1905), and Swinton (1936b, 1970) as intercostal plates (Butler and Galton, 2008). The cranial joints of *Hypsilophodon foxii* were described by Weishampel (1984), as were jaw mechanics involving intracranial movements for streptostyly and pleurokinesis (see also Norman, 1984; Norman and Weishampel, 1985, 1991). Reed (1984, 642, 645) commented on the histology of the femur of *Hypsilophodon foxii*, describing fibrolamellar azonal bone, and provided a transverse section of the femur R170 to show the absence of lines of arrested growth and the absence of medullary lining bone around the medullary cavity (Reed in Chinsamy et al., 1998, fig. 3). Naish and Martill (2001a) gave a long, detailed review of the ornithopod dinosaurs of the Isle of Wight, with photographs of many specimens of *Hypsilophodon foxii* (NHMUK, IWCMS; but see Galton, 2009, 216, for corrections to some of the captions). Stereophotographs of the holotype skull (R197) and of tooth rows and teeth were provided by Galton (2009), who redescribed the teeth, tooth replacement, and wear, and confirmed the posterior–anterior sequence of closure of the neurocentral suture in R196, as suggested by Irmis (2007), and illustrated the presence of a shallow extensor depression distally in larger femora of *Hypsilophodon foxii*.

An autapomorphy of *Hypsilophodon foxii* is the large foramen in the ascending process of the maxilla, which opens medially into the antorbital fossa (Figs. 17.8A,B, 17.9A, and 17.10B) (Butler et al., 2008). There is also a prominent foramen within the quadratojugal (Figs. 17.8B and 17.9A), as also occurs in the basal euornithopod *Jeholosaurus* and the basal iguanodontian *Tenontosaurus* (Winkler, 2006).

Cranial Endocast and Special Senses

The brains of dinosaurs appear to have molded the cranial cavity to a greater extent than is usual in reptiles, and they resembled birds and mammals in this respect (Hopson, 1979). The cranial endocast of *Hypsilophodon foxii* lacks details of the structure of the brain because there were intervening meningeal membranes and spaces containing cerebrospinal fluid, blood, and some adipose tissue. In addition, anteroventrally, the walls were cartilaginous so they are not preserved. However, the main regions of the brain — the cerebrum, the cerebellum, and the medulla — are recognizable as are a few of the cranial nerves (indicated by Roman numerals) and blood vessels (Fig. 17.12E–G), some of which are shown more clearly on endocasts taken from isolated sidewall bones of the braincase (Fig. 17.12A–D; for stereophotographs of bones, see Galton, 1989, pl. 2, figs. 9–12). As is the case in most dinosaurs, the brain is proportionally larger than would be expected for the calculated body mass compared to living reptiles, but not nearly as much compared to living birds and mammals (Hopson, 1979).

The olfactory lobes are prominent (Fig. 17.12E,F), so the sense of smell was important. The optic lobes are not visible, but like most reptiles (except crocodiles and snakes) and birds, the eyeball is supported by a series of overlapping sclerotic plates (Figs. 17.8A,B, 17.9B, and 17.10B). The plates form a sclerotic ring that looks something like the adjustable diaphragm of a camera. However, these bones do not move relative to one another to alter the size of the pupil, as was long thought to occur in living forms; this is done by a soft tissue structure, the iris (Edinger, 1929). Instead, the ring forms a broad annular sulcus or groove between the transparent cornea and the rest of the eyeball, a cuplike structure supported internally by cartilage. During focusing on closer objects, the ciliary muscles press against the lens, causing its anterior surface to become more convex so the image is returned to the plane of the light-sensitive layer, the retina. The sclerotic ring prevents the resulting increased pressure inside the eyeball from evaginating this sulcus and negating the effect of the ciliary muscle (Walls, 1942; Young, 1962, figs. 218, 293). The eye was large and the frontals are narrow (Figs. 17.10A,B), so when looking anteriorly, the fields of view overlapped slightly for stereoscopic vision. On the basis of comparisons of the proportions of the orbit and of the sclerotic ring with those of living reptiles of known habits, Galton (1974) concluded that *Hypsilophodon* had good powers of accommodation (for closer vision) and was diurnal (active during the day, or photopic; see Schmitz and Motani, 2011).

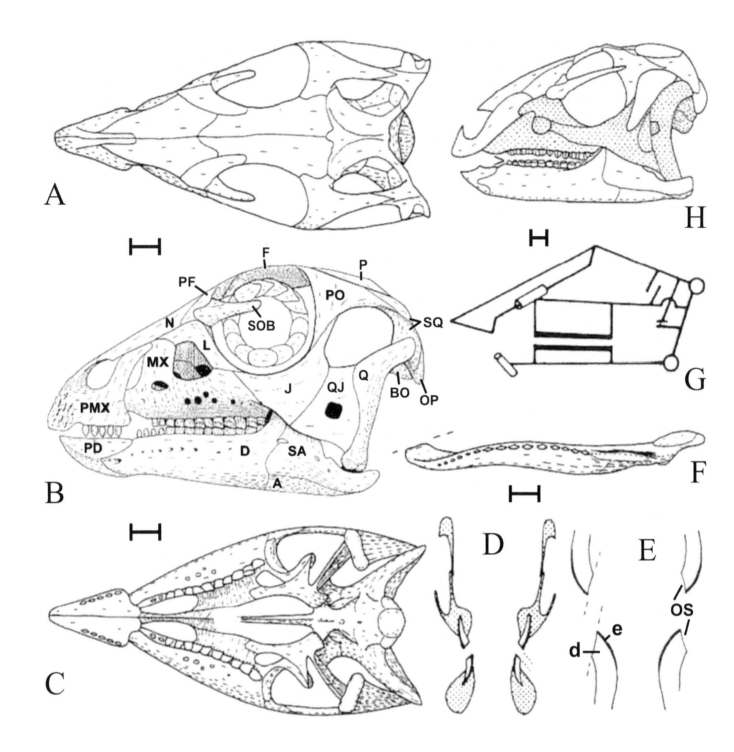

Presumably, as in *Dryosaurus altus* (Galton, 1989, pl. 2, fig. 1), the stapes was a slender straight rod connecting the inner ear to the eardrum, which was supported anteriorly by the quadrate.

Form of Teeth

The teeth of the left side of the holotype skull R197 are in perfect condition (Fig. 17.12A,B) but were only shown diagrammatically (Fig. 17.9A') by Galton (1974). The posterior five teeth of the left maxilla and eight of the dentary of R2477 are shown completely free of matrix (Fig. 17.14; shown in skull Fig. 17.10B–C; Galton, 1974). Note that the illustrations of the right dentary of IWCMS 6362 in medial view with details of tooth row in labial and lingual views (Naish and Martill, 2001b, fig. 5.5) are actually of R192 for figure 5.5A (left dentary in medial view) and *Echinodon becklessii* (Heterodontosauride, Lower Cretaceous, Dorset), left dentary NHMUK 48215 in lingual and labial views for figures 5.5 B, C (see Fig. 17.4L–P; Galton, 1978, figs. 1C,C',F,F'; Norman and Barrett, 2002, pl. 2, fig. 2).

Hulke (1873, 1882a, pl. 18, fig. 6) described a maxillary tooth (Fig. 17.15I,J; Galton, 2009, fig. 2Q,R) that differs somewhat from those of the holotype (Fig. 17.13A,C). However, this tooth is part of an associated partial skeleton (Hulke, 1873; R2471) that includes teeth (Figs. 17.9C, 17.15G,H,N; Hulke, 1882a, pl. 18, figs. 3–8; Galton, 2009, figs. 2L–W, 3M) quite similar to other premaxillary and dentary teeth of *Hypsilophodon foxii*.

An isolated tooth (Fig. 17.9D) was illustrated as a maxillary tooth (Galton, 1974, fig. 14a,b) and subsequently as a dentary tooth (Galton, 2009, fig. 2Y,Z). The mesial and distal borders of the crown resemble those of the left maxillary teeth of the holotype in labial view (Fig. 17.13A). However, the prominent apicobasally extending ridge is vertical, rather than being obliquely inclined. The prominent ridge is not nearly as prominent as in dentary teeth of *Hypsilophodon foxii* (Figs. 17.14D, 17.15A–C,E,G,K,L), and in addition, the mesial and distal edges of the crown are less similar to those of the left dentary teeth of R196 in lingual view (Fig. 17.15K). On the basis of the dental characters used by Butler et al. (2008), this tooth represents a basal ornithopod, nodes 16–18 indet. (Fig. 17.1).

Premaxillary Teeth and Beaks

The premaxillae are rugose anteriorly (Figs. 17.9A, 17.10A,B; also Naish and Martill, 2001a, pl. 4, fig. 1 for IWCMS 6362), indicating the presence of a small horny beak or rhamphotheca for the edentulous part of the skull. There are no rugosities on the predentary, but it is based on one from a juvenile individual (R2470; Nopcsa, 1905, fig. 3; Galton, 1974, fig. 11, 2001b, figs. 3, 7D; some on complete bone IWCMS 6362, Naish and Martill, 2001a, pl. 4, figs. 7, 8). It was probably also covered by a horny beak, the remains of which are preserved next to the predentary in several hadrosaurs, the common large ornithopods of the Upper Cretaceous (Morris, 1970). More posteriorly, the premaxillary teeth bit outside the predentary. The presence of five premaxillary teeth is plesiomorphic for *Hypsilophodon foxii* because in most ornithischians they are lost. Plant food was obtained initially by the cropping action of the horny beaks (and the premaxillary teeth; with lingual wear surfaces, caused by opposing predentary or the food, visible in IWCMS 6362; Naish and Martill, 2001a, pl. 4, figs. 1, 2). Krauss and Salame (2010) used models to show that *Hypsilophodon* could have fed on the seeds of cycads, using the beak to pry the seeds loose from the cones. The premaxillary teeth could have been used to slice through the seed coat, and then the cheek teeth were used to chew the pulp inside. However, in contrast to living forms with beaks, such as birds and chelonians (turtles, tortoises), the maxillary and dentary teeth were not lost in ornithopods.

17.10. A–G, Basal ornithopod *Hypsilophodon foxii*. A–F, Reconstruction of skull (mostly R2477, also R192 [not R193, Galton, 1974, 37], R194, R196, R197, R2470, R8418; for details, see Galton, 1974, 19) in (A), dorsal, (B) left lateral, and (C) ventral views. D–E, Diagramatic vertical cross section of (D), lacrimals, maxillae, and dentaries, and (E) teeth. F, Mandible in dorsal view. G, Kinematic abstraction of the jaw mechanism. H, *Dryosaurus altus* skull CM 3392 in left lateral view to show line of the pleurokinetic hinge of the bones of the tooth-bearing cheek region (stippled) from those of the central skull roof. *Abbreviations:* d, dentine; e, thick enamel layer; os, occlusal surface; for other abbreviations, see caption to Fig. 17.6. A, C, F, H from Galton (2001a); B, D, E modified from Galton (1974); G after Norman and Weishampel (1991). Scale bars = 10 mm.

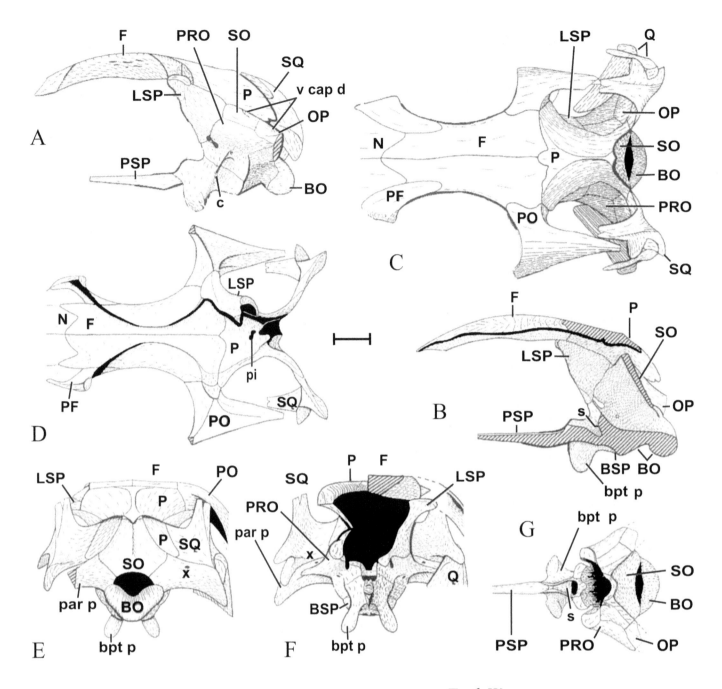

17.11. Basal ornithopod *Hypsilophodon foxii*. Cranium R2477 in (A) left lateral view and (B) medial view, as (A) with cranium sagitally sectioned; (C) dorsal and (D) ventral views of skull roof; (E) posterior view and (F) anterior view with left frontal sectioned and right frontal and laterosphenoid removed; and (G) braincase in dorsal view. *Abbreviations:* c, Vidian canal for internal carotid artery; bpt p, basipterygoid process; par p, paroccipital process; pi, pit for pineal; s, sella turcica for pituitary; v cap d, vena capitis dorsalis; x, remnant of posttemporal fenestra; for other abbreviations, see caption to Fig. 17.6. Modified from Galton (1989). Scale bar = 10 mm.

Tooth Wear

In *Hypsilophodon* the maxillary and dentary teeth have an en echelon arrangement, with each tooth slightly overlapping the one behind it (Fig. 17.14), to give an imbricate pattern in occlusal view (Fig. 17.14B,E). Isolated maxillary and dentary teeth are also shown (Fig. 17.15). The maxillary and dentary teeth are thickly enameled on one side and curve transversely in opposite directions (Fig. 17.10D–E). The vertically convex surface bears the thick enamel layer in both cases and it is five times the thickness of the thinner labial side in a section of a dentary tooth (R8419); in maxillary teeth, the labial enamel is the thickest. The thick enamel forms a sharp leading edge (Figs. 17.10E, 17.13, and 17.15) because it is more resistant than the dentine of the rest of the tooth. The cutting action of this edge is enhanced by the presence of serrations formed by the wear of the vertical ridges on the convex surface. In particular, the large apical or primary ridge on each dentary tooth forms a prominent spike on the cutting edge (Fig. 17.15A–H, K–N). The rest of the tooth forms an obliquely inclined wear or occlusal surface (Fig. 17.15B,F,I,M,N). After the initial cutting action of the leading edges, the food was further broken down when the obliquely inclined occlusal surfaces were forced past each other by further contraction of the jaw closing muscles during the transverse power stroke (see below). In this way, the food between the teeth was also crushed. The wear surfaces on the lower teeth, which face upward and outward, are exposed on the type skull (Fig. 17.13A). Initially, each wear surface obliquely truncated the crown cleanly (Fig. 17.15B,F,M–N), but as the dentary teeth become more worn, the thicker leading edge of the maxillary tooth cuts into the outer edge of the opposing dentary crown so a slight step was formed just above the cingulum (Figs. 17.13A and 17.14E). The surfaces are not just gently concave vertically but include more deeply excavated basins that correspond to individual occluding maxillary teeth. This indicates that the motion producing the wear did not involve an anterior–posterior motion of the lower jaw.

There are well-preserved occlusal wear facets on the dentary teeth of the holotype skull R197 (Fig. 17.13A), less well-preserved matching ones on the more posterior left cheek teeth of R2477 (Fig. 17.14B–C,E,F), and on isolated teeth (Fig. 17.15). The obliquely inclined wear surfaces range from 50 to 65 degrees from horizontal (mesially and distally, respectively; Weishampel, 1984), and the individual wear surfaces, which are discontinuous over adjacent teeth (Figs. 17.13A and 17.14B–C,E,F), are gently concave transversely (Galton, 1974, figs. 15b,d, 16b) and slightly sinuous mesiodistally.

Cheeks

As in other ornithischians, the skull of *Hypsilophodon* has a large space external to the upper and lower tooth rows that is roofed by the maxilla and floored by the massive dentary (Figs. 17.8B, 17.9A, and 17.10B,D). Because of the oblique orientation of the lower occlusal surface (Figs. 17.9A, 17.10E, 17.13A, and 17.14F), chewed food on it ended up external to the tooth row, and it would have been lost if there was no structure functionally analogous to the cheeks of mammals. Such a structure bordering this space would have passed from the maxilla to the prominent ridge on the outside of the massive dentary (Figs. 17.8B, 17.9A, and 17.10B,D; see flesh reconstruction, Fig. 17.7) (Galton, 1973b). Cheeks were originally suggested for the ceratopsian dinosaur *Triceratops* by Lull (1903), who later restored cheeks with a buccinator muscle for this genus (Lull, 1908). Brown and Schlaikjer (1940) argued against the presence of cheeks in ornithischian dinosaurs because the buccinator muscle, a facial muscle innervated by a branch of cranial nerve VII, is characteristic of mammals and is not present in any living reptile or bird. However, this argument does not preclude ornithischians from having a cheeklike structure that lacked a buccinator muscle (Galton, 1973b; but see Papp and Witmer, 1998). Elliptical dermal ossifications (cheek plates, Lambe, 1919) are present within the buccal emargination in the nodosaurid ankylosaurs (armored dinosaurs) *Panoplosaurus* and

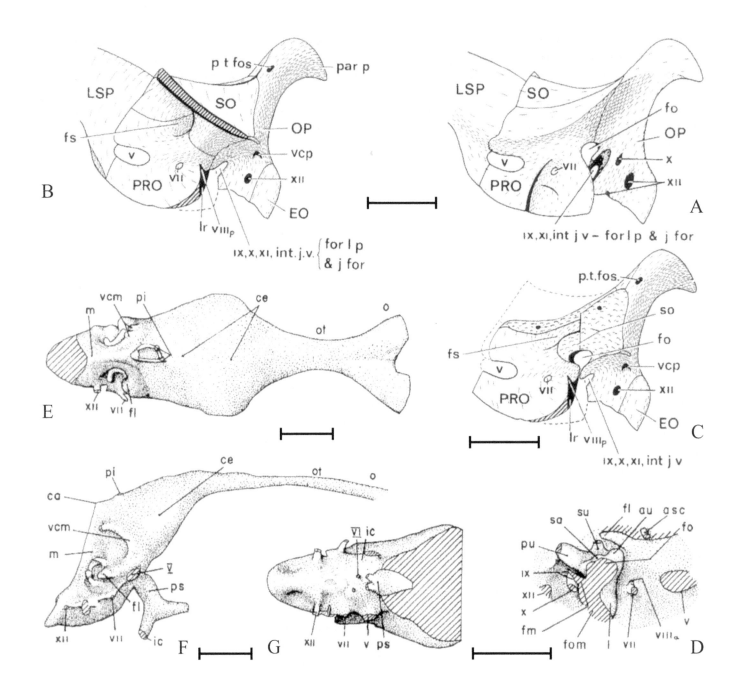

Edmontonia (Upper Cretaceous, western North America; see Carpenter, 1990; Vickaryous et al., 2004, fig. 17.9A–E; Vickaryous, 2006, fig. 2). These plates must have formed in a sheet of dermal tissue that extended between the prominent maxillary and dentary ridges, thus providing an osteological correlate for the presence of cheeks in at least some ornithischians (see Barrett, 2001, for further discussion). In addition, Paul (2010, 25–26 with figure) points out that the Andean or South American condor (*Vultur gryphus*) has a large, fleshy, elastic cheeklike structure that starts superiorly about halfway along the upper jaw and attaches at about the same level on the mandible (photograph in Hoyo et al., 1994, 37).

Jaw Mechanics

As in most other euornithopods, the skull of *Hypsilophodon foxii* has a diastema, a toothless area separating the food-cropping region (premaxillary teeth and beak; beak only in more advanced euornithopods) from the inset food-processing region of the cheek teeth, an elevated coronoid process on the lower jaw to improve the leverage of the jaw-closing muscles, and a ventrally offset jaw articulation that ensured that the opposing tooth rows engaged each other almost simultaneously in a nutcracker-like action (Figs. 17.9A and 17.10B,C). During jaw action, any anteroposterior movement of the mandibles was prevented by the curvature of the tooth rows (Figs. 17.10F and 17.12B,E), the imbricate arrangement of the teeth, and the pattern of dentary tooth wear (Figs. 17.9A and 17.14C,F). The form of the skull indicates the presence of a pleurokinetic hinge, one that permitted transverse rotation of the upper jaws laterally on both sides of the skull during isognathic (bilateral) jaw closure (see Norman, 1984; Weishampel, 1984; Norman and Weishampel, 1985, 1991). This cranial flexibility was possible because in ornithopods there is no bony brace between the two maxillae (Fig. 17.10C), so movement was possible at the pterygoid–basipterygoid joint (Figs. 17.10C and 17.11A,F–G); in mammals, the solid bony palate acts as a brace and prevents movement of the maxillae. In ornithopods, the transverse flexibility depends on the presence of movable joints between the rigid elements of the tooth-bearing bones of the cheek and the rigid median portion of the skull roof (Fig. 17.10G). The hinge, which runs along a diagonal line on the side of the skull (Fig. 17.10H), is formed by sutures between the premaxilla and maxilla, the lacrimal and prefrontal, the jugal and postorbital, and the quadrate and squamosal. This lateral rotation allowed the occlusal surfaces of the maxillary teeth to be forced across those of the dentary in forms with an isognathus skull to produce a transverse power stroke simultaneously on each side. This helped to reduce the food into smaller pieces by a combination of shearing and crushing forces. The resulting ability to grind plant fibers in a manner analogous to that used by present-day herbivorous mammals was an important factor contributing to the rise and diversification of ornithopod dinosaurs that culminated in the hadrosaurids of the Late Cretaceous (Norman and Weishampel, 1985, 1991; Weishampel and Norman, 1989).

Postcranial Anatomy

Galton (1974, figs. 18–58, 2001a, figs. 21–24) described the postcranial bones of *Hypsilophodon foxii*, summary figures of which are given for the vertebral column (Fig. 17.16; for cervical ribs 3–9, see Galton, 1974, figs. 19, 20, 2001a, fig. 22, for photograph of dorsal ribs 1–13; Galton, 1974, figs. 28, 29 for caudal vertebrae 1–12), pectoral girdle and forelimb (Figs. 17.8C–F and 17.17; stereophotographs of forelimb in Galton, 1971c, fig. 2A, and manus 1983, fig. 6J,K), and pelvic girdle and hind limb (Figs. 17.8G–L, 17.18, 17.19, 17.20A–E; stereophotograph of pubis, ischium, femur in Galton, 1969, figs. 7, 8, pes in 1971c, fig. 2B, 1974, pl. 2, fig. 3).

17.12. Basal ornithopod *Hypsilophodon foxii.* A–C, Sidewall of braincase (EO, exoccipital R8418; LSP, laterosphenoid R2477; OP, opisthotic R194, R2477; SO, supraoccipital R8418 [ex R8366 cited Galton, 1974]) in (A) left lateral view, (B) right medial view with supraoccipital sagitally sectioned, (C) as (B) with laterosphenoid and supraoccipital removed (for stereophotographs of bones, see Galton, 1989, pl. 2, figs. 9–12). D, Detailed endocast of ear region from B–C. E–G, Endocast of braincase R2477 (see Fig. 17.11B) in (E) dorsal, (F) , right lateral, and (G) ventral views. *Abbreviations:* asc, anterior semicircular canal; au, auricle; ca, cartilage; ce, cerebrum; f, foramen; fl, floccular lobe; fm, foramen metoticum (=j for + for l p); fo, fenestra ovalis; fom, fenestra ovalis plus foramen metoticum; for l p, foramen lacerum posterius; fs, fossa subarcuata; ic, internal carotid artery; int aud m, internal auditory meatus; ic, internal carotid artery; int j v, internal jugular vein; j for, jugular foramen; l, lagena; lr, lagenar recess; m, medulla oblongata; par p, paroccipital process; o, olfactory bulb; ot, olfactory tract; pi, pineal; ps, pituitary space in sella turcica; p t fos, remnant of posttemporal fenestra; pu, posterior utriculus; sa, saccule; so, surface for supraoccipital; vcm, vena capitis medius; vcp, vena capitis posterior; foramina for cranial nerves in Roman numerals; for other abbreviations, see caption to Fig. 17.8. Modified from Galton (1989). Scale bars = 10 mm for R2477.

17.13. Basal ornithopod *Hypsilophodon foxii*, holotype skull R197, stereophotographs of tooth rows in lateral (labial) view: A, left maxilla and dentary; B, left premaxilla; C, posterior part of right maxilla. Scale bar = 2 mm.

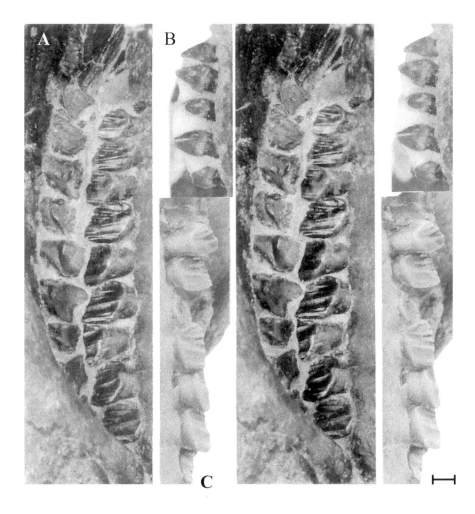

Individual Variation

Galton (1977b, 1980a; full discussion in Galton, 1974) provided a diagram summarizing the major variations in *Hypsilophodon foxii* (Fig. 17.8). However, it does not include the dimorphic variation in the six centra sacrum. In the pentapleural sacrum, the rib of the dorsosacral vertebra attaches to the end of the massive diapophysis (Fig. 17.16E, rib not preserved), as occurs in dorsal vertebrae. In the hexapleural sacrum, it arises anteroventrally from the neurocentral suture and curves back to meet the anterior surface of the first sacral rib (Fig. 17.16F–G); the diapophysis is slender and tapers distally, and the extra rib is also supported by the last dorsal vertebra. This sacral dimorphism is correlated with a few other dimorphisms in the bones of the pelvic girdle (Galton, 1974). Individuals with the hexapleural sacra may have grown larger because all the sacral sutures are closed in the pentapleural sacrum R196 (Fig. 17.16E; femur L4 65), whereas they are mostly open in the larger hexapleural sacrum R193 (Fig. 17.16G; L4 86). Unfortunately, histological studies of the femora would probably not provide any information on the age of the specimens involved because the histological section of the femur R170 showed no lines of arrested growth (Chinsamy et al., 1998, fig. 3).

In one specimen of the basal iguanodontian *Tenontosaurus dossi*, the left diapophysis of the dorsosacral vertebra is slender and tapering, with the extra sacral rib arising from the middle of the neurocentral suture (Winkler

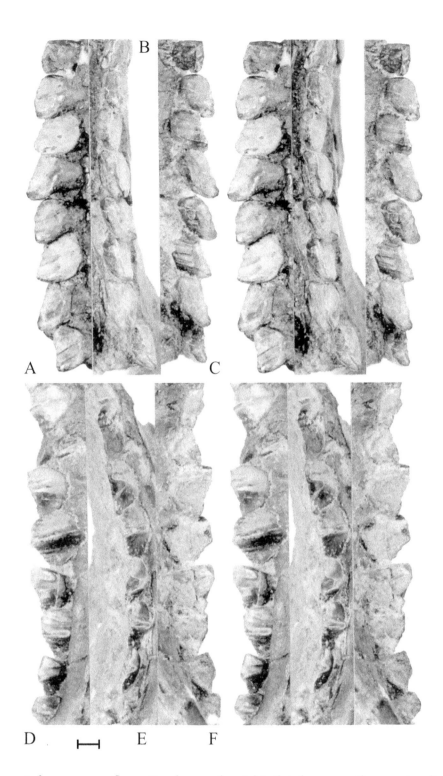

17.14. Basal ornithopod *Hypsilophodon foxii*, skull R2477, stereophotographs of right posterior eight teeth of (A–C) maxilla and (D–F) dentary in (A), (F) lateral (labial); (B, E) occlusal, and (C, D) medial (lingual) views. Scale bar = 2 mm.

et al., 1997, 337, fig. 14A), whereas the right rib is borne on the massive diapophysis, as is also the case on both sides of another dorsosacral vertebra of *T. dossi* (Winkler et al., 1997).

Although suggestive of a sexual dimorphism, with three pentapleural sacra and five hexapleural sacra, the sample for *Hypsilophodon foxii* is too small to be statistically significant. Histological studies are needed to confirm the suggestion that the former represent males (Galton, 1974) or females (Galton, 1999, 26), as in many living amniotes (especially birds; see Galton, 1999, for references). The latter reassignment was based on the assumption that the hexapleural type sacrum of *Hypsilophodon foxii*

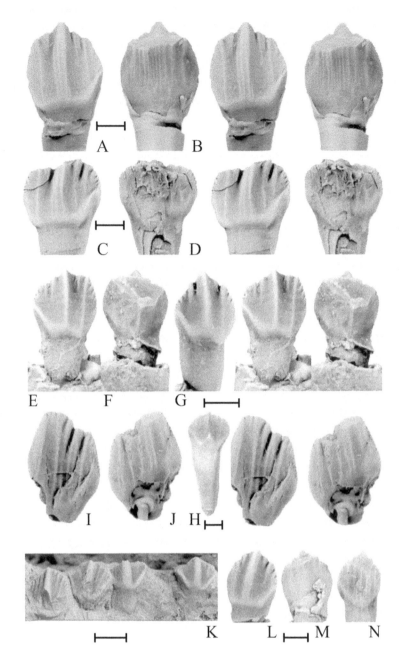

17.15. Basal ornithopod *Hypsilophodon foxii*, isolated crowns of right (A–D, L– M) and left (E–H, K, N) dentary teeth and a left maxillary tooth (I– J) (stereophotographs except G, H, K–N) in lingual (A, C, E, G, H, I, K, L) and labial (B, D, F, J, M, N) views: A, B, R8367b; C, D, R191; E, F, R5863; G–J, R2471 (H complete tooth to show length of root); K, R196 (dentary in Hulke, 1882a, pl. 72, fig. 2); L–M, R15993 (formerly R8367); N, R2471. Scale bars = 2 mm.

was equivalent to the robust morph of the theropod dinosaur *Syntarsus rhodesiensis* (Lower Jurassic, southern Africa). On the basis of anatomical (Raath, 1977, 1990) and histological evidence (Raath, 1977; Chinsamy, 1990), the robust morph probably represents the female.

The rib of the first sacral vertebra is either on its diapophysis (so four sacral ribs) or forms an anteroposteriorly thin vertical extra sacral rib (so five sacral ribs) in several species of stegosaurs or plated dinosaurs (Galton, 1982, 1991, in press; Galton and Upchurch, 2004). A geometric morphometric analysis of the femora of the stegosaur *Kentrosaurus aethiopicus* (Upper Jurassic, Tanzania) by Barden and Maidment (2011) demonstrated a statistically significant shape difference of the proximal end in anterior view that, because it can only be identified in larger femora and is then independent of size, probably represents a sexual dimorphism. The robust morph, which is twice as common (14 versus 7), proximally has a more distinct head and a pronounced and better developed greater trochanter, and the distal condyles are larger compared to the slender morph. The mounted skeleton of *Kentrosaurus aethiopicus* (MFNB) consists of numerous bones of the lectotype individual (Heinrich, 2011) that include the sacrum with four sacral ribs (Hennig, 1925, fig. 23b) and both femora, neither of which was included in the femoral analysis of Barden and Maidment (2011). However, both femora are the robust morph (right in anterior view, Galton, 1982, pl. 6, fig. 1; Mallison, pers. comm.). In the theropod dinosaur *Tyrannosaurus rex*, there is also a femoral sexual dimorphism (Larson, 2008), with the robust morph identified as female based on pelvic dimensions and fused caudal vertebrae. Analyses for the presence of medullary bone in this same series of femora showed a significant clustering with the robust forms (Schweitzer et al., 2005). This indicates that the robust morph was probably female because female birds form medullary bone as a calcium reservoir to aid in egg shell formation (Dacke et al., 1993). Consequently, in *Kentrosaurus aethiopicus*, the four sacral rib morph was probably female because it is associated with the robust femoral morph, but analysis for medullary bone in the associated femora is needed to confirm this. Thus in *Hypsilophodon foxii*, individuals with the pentapleural sacrum probably represent females, with males having the hexapleural sacrum.

Dermal Armor

One question left unresolved by Galton (1974) was the identity of the mineralized plates associated with the dorsal ribs and described by Owen (1855, pls. 1, 15, fig. 8) as skin (Fig. 17. 2B, in medial view below anterior dorsal ribs) and by Hulke (1874, fig. 1, 1882a, pl. 72, fig. 1) and Nopcsa (1905, fig. 4) as dermal armor. However, the identity of these plates was recently resolved by Butler and Galton (2008), who made comparisons with well-preserved articulated skeletons of the Upper Cretaceous *Thescelosaurus* sp. (basal Ornithopoda, western USA; Fisher et al., 2000, fig. 1, U; Boyd et al., 2009, fig. 2NCSM 15728) and *Talenkauen santacrucensis* (basal Iguanodontia, Argentina; Novas et al., 2004, figs. 1B,C, 2B). As in these taxa, the plates of *Hypsilophodon foxii* are associated with the lateral surfaces of the distal sections of the anterior dorsal ribs (Figs. 17.2B and 17.6B). In *Thescelosaurus*, the plates have a layer of calcified cartilage surrounding an inner layer of lamellar bone, so they do not represent dermal armor that is formed via an intramembraneous pathway (Boyd et al., 2008). In addition, the plates were not passively embedded in the surrounding tissue but served as an attachment site for muscles, as indicated by dense packets of obliquely oriented Sharpey's fibers on the lateral margin of the plates in *Talenkauen* (Boyd et al., 2008). These intercostals plates are not homologous to the uncinate processes of other diapsids (e.g., maniraptorans) (Boyd et al., 2008). Such intercostal plates were probably widespread in basal euornithopods but have rarely been recognized to date due to incomplete preservation or variation in the timing and degree of mineralization and ossification. In these Cretaceous taxa, the cartilaginous sternal sections of the dorsal ribs are also mineralized; the earliest record of this is in the basal neornithischian *Othnielosaurus consors* (Upper Jurassic, western USA; Galton and Jensen, 1973), but no intercostal plates are preserved, so they were probably not mineralized and/or ossified (Butler and Galton, 2008).

17.16. Basal ornithopod *Hypsilophodon foxii*, vertebrae in left lateral view. A, Proatlas, atlas, and axis, R2477. B–E, R196: B, cervicals 2–9; C, dorsals 1–8; D, dorsals 9–15; E, sacrals 1–6 (pentapleural sacral type). F, G, Hexapleural sacral type, dorsal 15 and sacrals 1–3 with extra sacral rib: F, R195; G, R193. *Abbreviations:* IC 1, intercentrum of atlas; NA, neural arch of atlas; PA, proatlas. From Galton (1974). Scale bars = 10 mm.

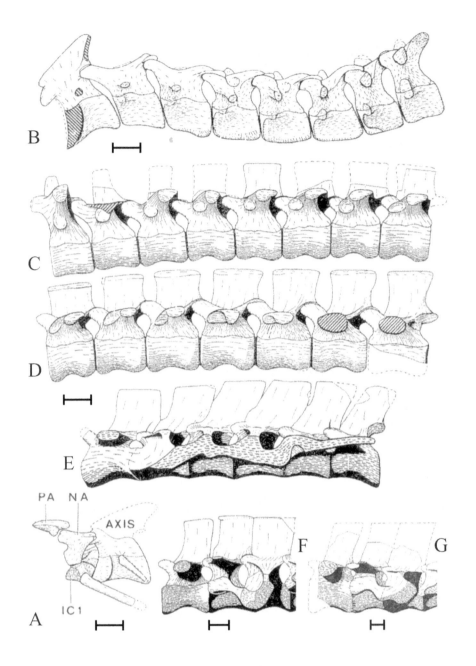

Arboreal Versus Cursorial Mode of Life

Hulke (1882a, 1062) restored the complete skeleton in a quadrupedal pose (see Galton, 2001a, fig. 6, also for copies of most of the other figures in literature cited in this section), and because of the long fingers and toes, he concluded that "*Hypsilophodon* was adapted to climbing upon rocks and trees." On the basis of a study of the NHMUK material and related forms from the Upper Jurassic of the western United States, Marsh (1895, 1896a, 1896b) reconstructed the skeleton in a bipedal pose, with the distal half of the tail trailing on the ground, following the example of Dollo (1883) for *Iguanodon bernissartensis* (but the tail was broken to achieve this pose; see Norman, 1980). Abel (1912) reconstructed the foot with an opposable first toe or hallux (Fig. 17.19U) that, with the strong flexural capabilities of the remaining toes, supposedly meant that the foot was capable of grasping branches, as in living arboreal birds. He also suggested that *Hypsilophodon* retained

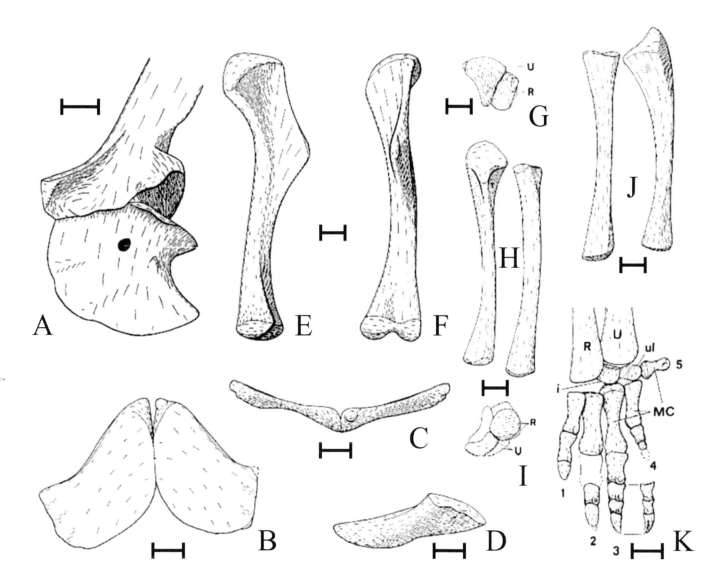

the ancestral arboreal habitat for all dinosaurs. Heilmann (1913, reprinted 1916) initially reconstructed *Hypsilophodon* in a quadrupedal arboreal pose. He regarded this arboreality as secondary because the first metatarsal is shortened as in ground-living dinosaurs, with the hallux originating at a higher level than the other digits. This secondary arboreal adaptation was analogous to that of the secondarily arboreal tree kangaroo *Dendrolagus dorianus* from New Guinea and Queensland, Australia (which was used as a model for a flesh reconstruction of *Hypsilophodon*, as it was for another one by Abel, 1922). Abel (1922, 1925, 1927) agreed and opined that metatarsal I was not further reduced because it was probably used in climbing trees. He thought that the sharp and strongly arched claws of the pes would have rendered movement on the ground difficult. However, Heilmann (1926, 163, fig. 115.3) reconsidered the evidence, noting that the reconstruction of the pes (Fig. 17.19U) by Abel (1912) does not match the figures and measurements it was supposedly based on, those of Hulke (1882a). He prepared a new reconstruction with all the toes shown forwardly directed with no opposability of the hallux, and concluded that *Hypsilophodon* was a ground-living dinosaur; unfortunately, this conclusion was largely ignored. It should be noted that the proportions of digits II to IV of *Hypsilophodon*

17.17. Basal euornithopod *Hypsilophodon foxii*, R196, left pectoral girdle and forelimb (see also Fig. 17.8C–F). A, Proximal scapula and coracoid in ventrolateral view. B–D, Sternum in (B) ventral, (C) anterior, and (D) lateral views. E, F, Humerus in (E) medial and (F) anterior views. G–J, Radius and ulna in (G) proximal, (H) anterior, (I) distal, and (J) lateral views. K, Distal radius, ulna, carpus, and manus in lateral (dorsum) view. *Abbreviations:* i, intermedium; MC, metacarpal; 1–4, digits 1–4; R, radius; U, ulna; ul, ulnare. From Galton (1974). Scale bars = 10 mm.

17.18. Basal euornithopod *Hypsilophodon foxii*. A, Reconstruction of the pelvic girdle, sacrum, proximal caudals, and femur in left lateral view to show the lines of action of the individual muscles (mainly R193; also R196, R5830, 28707). B–E, Distally incomplete left femur showing the areas of attachment of the limb muscles (mainly R193; also R196, R5830) in (B) posterior, (C) lateral, (D) anterior, and (E) medial views. *Abbreviations:* ADD, M. adductor femoralis; AMB, M. ambiens; CA-FEM BR, M. caudi-femoralis brevis; CA-FEM L, M. caudi-femoralis longus; DOR CA, M. dorsalis caudae; DOR T, M. dorsalis trunci; FEM-T 1. 2 & 3, M. femoro-tibialis 1, 2 and 3; F T E, M. flexor tibialis externus; F T, M. flexor tibialis internus; G, M. gastrocnemius; IL-CAUD, M. ilio-caudalis; IL-FEM, M. ilio-femoralis; IL-FIB, M. ilio-fibularis; IL-TIB 1 & 2, M. ilio-tibialis 1 (sartorius) and 2; IL-TROC, M. ilio-trochantericus; IS-CAUD, M. ischio-caudalis; IS-TROC, M. ischio-trochantericus; LIG, ligaments for holding head in acetabulum; O A EXT, M. obliquus abdominis externus; O A INT, M. obliquus abdominis internus; OBT, M. obturator internus; P-I-F INT 1, dorsal part of M. pubo-ischio-femoralis internus; P-I-F INT 2, ventral part of M. pubo-ischio-femoralis internus; P-TIB, M. pubo-tibialis; R ABD, M. rectus abdominis; TND, tendon inserting on fibula; TR A, M. transversus abdominis. From Galton (1969; see figs. 7, 8 for stereophotographs of pubis, ischium, and femur of R193). Scale bars = 100 mm.

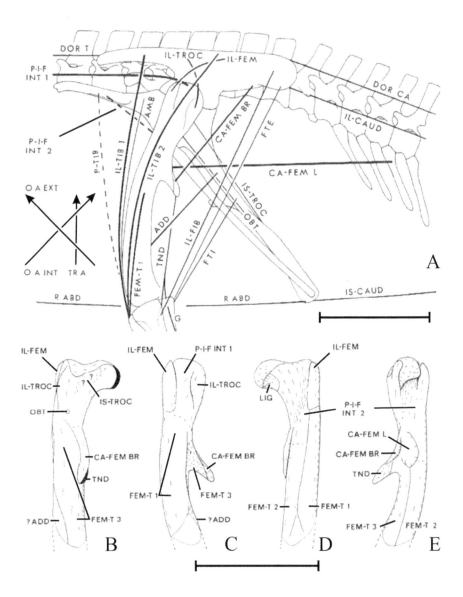

foxii closely resemble those of other basal euornithopods and small basal sauropodomorph dinosaurs.

The pes as reconstructed (Fig. 17.19U) by Abel (1912) was clearly based on the right one of R196 (Fig. 17.20D), with the missing first ungual phalanx restored after that of the other side (Fig. 17.20C; stereophotograph in dorsal view in Galton, 1971c, fig. 2B, 1974, pl. 2, fig. 3). The phalanx was incorrectly rotated through 180 degrees so it appeared to be opposable, but this is not the case (Figs. 17.8L, 17.19R,S, and 17.20C). Swinton (1936b) noted that some opposability of the human thumb is possible, even when the first metacarpal is closely approximated to the second. However, the thumb adductor muscle gets in the way so it is not right against the adjacent metapodial, as is digit I of the pes of *Hypsilophodon* (Figs. 17.8K,L, 17.19O,R,S, and 17.20D,E), and the plane of the condyle is at about 45 degrees to the plane of the other metapodials, not subparallel as in the pes of *Hypsilophodon* (Fig. 17.19O). The little finger of the human hand would be a better analogy (but with no transverse arching of the metatarsals). Consequently, the opposable hallux, the most important argument for regarding *Hypsilophodon* as arboreal, was based on misinterpretations of the material.

Swinton (1936b) cited three characters of the humerus to show that *Hypsilophodon* had a wider range of brachial movements as befits an arboreal animal. However, the lengths of the humerus and scapula are equal (Fig. 17.6B) in several specimens so the greater length of the humerus in R5829, the large mounted individual, is the result of the unnatural shortening of both scapulae due to the loss of the distal end by breakage (photograph in Glut, 1997, 488; color photograph in Gardom and Milner, 1993, 26). The more proximal position of the deltopectoral crest of the humerus (Figs. 17.8E,F and 17.17E,F) is matched in *Dryosaurus* and *Mantellisaurus*, and the medial position of the head (Figs. 17.8E,F and 17.17E,F) is matched by several other ornithopods such as *Camptosaurus* and *Iguanodon*.

The radius and ulna of *Hypsilophodon* are both slender (Fig. 17.17H; stereophotograph of articulated forearm in Galton, 1971c, fig. 2A), but the degree of development of the interforearm space is comparable to that of the ornithopods *Dryosaurus* and *Camptosaurus*; this space is also quite well developed in *Mantellisaurus*. This space is not uniquely large to *Hypsilophodon*, so contrary to the suggestion of Abel (1925, 1927), it does not differ from other dinosaurs in the same way that Carlsson (1914) demonstrated that the arboreal *Dendrolagus* differs from ground-living kangaroos. In assessing the grasping capabilities of the manus (Fig. 17.14K; stereophotographs in Galton, 1971b, fig. 2A, 1981b, fig. 6J,K), Swinton (1936b) exaggerated slightly in describing the ungual phalanges as long and thin, and in addition, the three central digits are not all long because the fourth is quite short. However, the small size of the hand would have restricted its usefulness in climbing.

The ossified tendons form a sheath distally on more than half of the tail (Fig. 17.6A,B) that would have made it more rigid, so it was more effective for balancing in trees (Swinton, 1936b, 1970). However, a similar sheath is present on the tails of two Lower Cretaceous dinosaurs from Wyoming, USA, the large basal iguanodontian *Tenontosaurus* and the theropod *Deinonychus* (Ostrom, 1969). Both taxa were obviously ground living, and the rigidity would have increased the efficiency of the tail as a dynamic stabilizer when the animal rapidly changed direction. Swinton (1936b, 1970) considered that the dermal armor shown in the flesh reconstruction by Heilmann (1913, 1916) was insufficient to protect *Hypsilophodon* from predators so it retreated to the trees. However, nearly all ornithopods lack armor, and reexamination of the material indicates that the dermal plates are actually intercostal plates (Fig. 17.6B; Butler and Galton, 2008).

Swinton (1936b) noted that although the tibia was longer than the femur, the metatarsus was not elongated as in cursorial dinosaurs (Figs. 17.6B and 17.19A,H,S). In addition, he considered that because of the low position of the fourth trochanter on the femur (Fig. 17.19A), the attaching tail–thigh leg muscles (Fig. 17.18A) would have hampered fast locomotion but were sufficient for climbing and balancing, so the animal sought refuge in the trees. However, the fourth trochanter (Fig. 17.19A–D) is more proximally placed than in most other ornithopod dinosaurs. The proportions of the hind limb, with a short femur and an elongate tibia and metatarsus that is slender (Figs. 17.6B and 19A,H,S), are comparable to those of modern rapidly running mammals such as the horse and gazelle. Consequently,

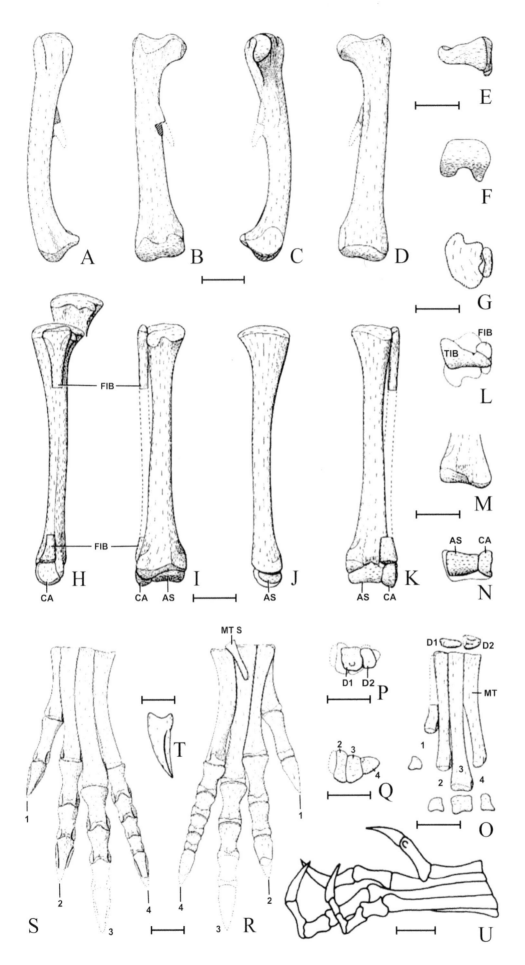

it is concluded that *Hypsilophodon* shows no specific adaptations for an arboreal dinosaur. Instead, it was clearly a cursorial or fast running ground dweller (Figs. 17.6B and 17.7; Galton, 1971b, 1971c, 1974, 2001a), the accepted mode of life for it since the early 1970s, with estimated speeds of 4.6–6.1 m/s (17–22 km/h) for R5830 and 6.7–8.9 m/s (24–32 km/h) for R5829 (Thulborn, 1982).

Iguanodontia indet.

Galton (2009) identified the distal ends of large femora (Fig. 17.21A–H) from the Lower Cretaceous of southern England as Iguanodontia indet. The anterior intercondylar groove of these femora is proportionally much deeper (Fig. 17.21C,F,H) than it is in *Hypsilophodon* (Figs. 17.19F and 17.20A; Gardom and Milner, 1993, 26), but it is proportionally wider and shallower than in *Dryosaurus, Valdosaurus, Camptosaurus,* and Wealden iguanodontoideans (Galton, 2009, figs. 12H,K, 13C,D,I,K, 16Q). The form of this groove is comparable to those of *Kangnasaurus, Tenontosaurus,* and *Muttaburrasaurus* (Galton, 2009, fig. 16M–O), iguanodontians from the Lower Cretaceous. However, the presence of this intercondylar groove, character 203 of Butler et al. (2008), is given as an unambiguous synapomorphy (i.e., it diagnoses the node under both accelerated transformation or ACCTRAN and delayed transformation or DELTRAN) for node 19, the unnamed node (Fig. 17.1; Butler et al., 2008, 40, fig. 4) for *Anabisetia* and higher ornithopods. Another unambiguous synapomorphy for this node is the form of the lateral flexor condyle, which is strongly inset medially and reduced in width relative to the medial condyle, as it is in these femora (Fig. 17.21C,F,H). In the plesiomorphic state, this condyle is positioned more laterally and is only slightly narrower than the medial condyle (Fig. 17.19F) (Butler et al., 2008, character 205). On the basis of the form of the extensor groove and lateral flexor condyle, two of these distal femora (Fig. 17.21A–C,G,H) can only be positively identified as basal Ornithopoda "Node 19" indet.

The distal end of the largest "Node 19" femur (Fig. 17.21A–C; Galton, 2009, fig. 15A–E) came from the Isle of Wight (presumably southwest shore, Wessex Formation).

The remaining three distal ends are water worn and from the Phosphate Bed, Potton Sandstone (Late Aptian–Early Albian; Rawson et al., 1978), of Potton, Bedfordshire (various localities on Old Potton–Sandy railway line; Benton and Spencer, 1995, 257), the bones of which were probably derived from the older Wealden beds. The left distal femur is basal Ornithopoda "Node 19" indet. (Fig. 17.21G,H; Galton, 2009, fig. 15J–N), and the rather incomplete right, lacking most of both flexor condyles, is probably not a femur but a distal humerus (CAMSM B29739; Galton, 2009, fig. 15O–S).

The third Potton distal end, that of a right femur, is similar to the one from the Isle of Wight (Fig. 17.21A–C), but it is smaller and the medial edges are somewhat more square (Fig. 17.21D–F; Galton, 2009, fig. 15F–I). However, one important difference from the other femora is that the posterior (flexor) groove is not fully open, the plesiomorphic condition (Figs. 17.19F and 17.21C,G), but is partly covered by a lateral inflation of the medial condyle (Fig. 17.21E). This derived condition (Butler et al., 2008, character 204) is present in Rhabdodontidae, *Tenontosaurus,* Dryosauridae (Fig. 17.22E,J), *Camptosaurus* (but not *Draconyx*), *Muttaburrasaurus,*

17.19. Basal ornithopod *Hypsilophodon foxii.* A–T, Left hind limb. A–Q, R5830; A–L, femur (A–F) and articulated tibia and ends of fibula (G–L) in (A, H) lateral, (B, L), posterior, (C, J) medial, (D, K) anterior, (E, G) proximal, and (F, L) distal views. M, Distal tibia in anterior view; N, astragalus and calcaneum in distal view; O, distal tarsals and metatarsus in anterior (dorsum) view with metatarsals 1–4 in distal view; P, distal tarsals in proximal view; Q, metatarsals in proximal view. R–U, R196; R, S, pes in (R) ventral (plantar) and (S) anterior (dorsum) views; T, ungual 1 in lateral view. U, pes as reconstructed by Abel (1912), right in ventrolateral view with ungual 1 from right but rotated 180 degrees (see Fig. 17.20C,D,). *Abbreviations:* AS, astragalus; CA, calcaneum; D, distal tarsal; FIB, fibula; MT, metatarsal; TIB, tibia. A–R, U from Galton (1974), T from Galton (1971b). Scale bars = 20 mm.

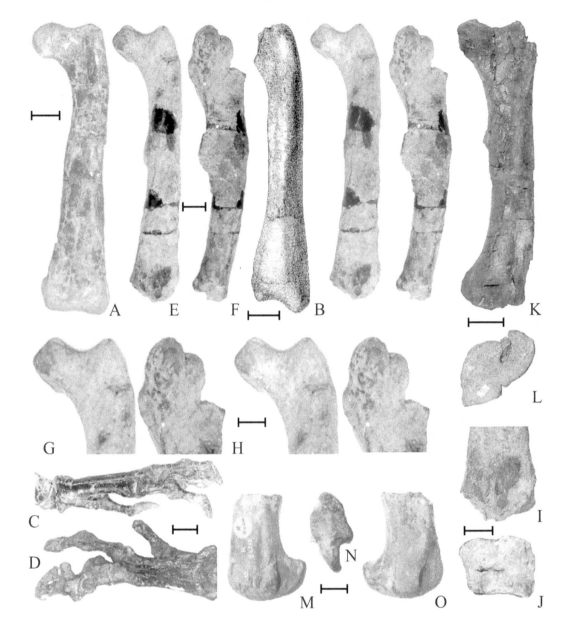

Iguanodon, and Mantellisaurus but not in more basal Ornithopoda, including Anabisetia (Butler et al., 2008; Galton, 2009, figs. 6M,N,P, 13K, 16M–P). Character 204 is an unambiguous synapomorphy for node 20, the node that represents Iguanodontia sensu Norman, 2004 (Fig. 17.1; Butler et al., 2008, 40, fig. 4). This distal femur is therefore identified as Iguanodontia ("Node 20") indet.

Dryosauria, Dryosauridae: Valdosaurus

Valdosaurus canaliculatus (Galton, 1975)

A pair of femora (R184, R185, Fig. 17.22A–E; length 140 mm), from the Hypsilophodon Bed near Cowleaze Chine, Isle of Wight, was made the holotype of Dryosaurus? canaliculatus Galton, 1975, the type species of Valdosaurus Galton, 1977a (Galton, 1975, fig. 3A–F, 2009, fig. 6A–F; color photograph as Hypsilophodon in Hamilton et al., 1974; photographs as Hypsilophodon foxii in Naish and Martill, 2001a, pl. 7, figs. 1–4 at ×0.75

magnification, not ×0.5 as in caption). R185 is readily distinguishable from the larger holotype femur of *Dryosaurus altus* (see Galton, 1975, fig. 2G–L, 1977a, fig. 2s, 1981a, fig. 13, also other femora, Galton, 1981b), but it is more similar to the smaller ones of *Dryosaurus lettowvorbecki* (Fig. 17.22F–J). The only difference is that distally, the extensor groove is more pronounced in *Valdosaurus*, in which the lip of the medial condyle slightly overlaps the extensor groove anteriorly (Fig. 17.22E cf. Fig. 17.22J). According to Butler et al. (2008), the deep pit on the shaft adjacent to the fourth trochanter (Fig. 17.22C,H) is a potential synapomorphy of Dryosauridae.

Two associated specimens of *Valdosaurus canaliculatus* (Wessex Formation, southwest coast of Isle of Wight) are currently being studied by R. Twitchett and P. M. Barrett (Barrett, pers. comm., 2011): a nearly complete leg and other bones (BELUM K17051; listed Blows, 1998, 32) and an associated nearly complete right hind limb found between Clinton Chine and Sudmoor Point. IWCMS 6879, 6438 is undescribed, but photographs are provided by Naish and Martill (2001a, 89–91, fig. 5.11, pls. 8, 9,10, figs. 1–3; femur length ~500 mm; partial pes, metatarsals, phalanges, ungual). They note that it "differs from *Hypsilophodon* in having a pes 20 percent shorter, and with only three functional toes." (Naish and Martill, 2001a, 91). The latter is also the case in *Dryosaurus* (Galton, 1981b).

Camptosaurus valdensis (Lydekker, 1889a)

A larger femur (Fig. 17.20E–J; R167, original length ~245 mm; Galton, 2009, fig. 5Q–V, stereophotographs 1974, pl. 2, fig. 4a,b), the holotype of *Camptosaurus valdensis* Lydekker, 1889a, from the *Hypsilophodon* Bed near Cowleaze Chine on the Isle of Wight, was referred to *Hypsilophodon foxii* by Galton (1974). However, Barrett (1996, 117–118, fig. 4) suggested that *Camptosaurus valdensis* Lydekker, 1889a, "may be a senior synonym of the dryosaur *Valdosaurus canaliculatus* (Galton, 1975), as the beginnings of a marked intercondylar groove can be seen on the anterior face of the femoral shaft" (Fig. 17.20E,I). In addition, the cleft between the greater and lesser trochanters was originally deep (Fig. 17.20E–H; lesser trochanter broken off across base to leave unbroken adjacent surface of greater trochanter), rather than shallow as in *Hypsilophodon foxii* (Fig. 17.19C,D). The shaft is poorly preserved, so the presence or absence of a deep pit adjacent to the fourth trochanter cannot be determined (Fig. 17.20F). The autapomorphic character of the femora of *Valdosaurus canaliculatus*, compared to those of *Dryosaurus*, is the slight anterior overlap of the extensor groove by the lip of the medial condyle (Fig. 17.22E), but because the bone is distally incomplete, this character is indeterminate for R167 (Fig. 17.20J). However, because the lesser and greater trochanters are separated by a deep cleft (Fig. 17.20E,G,H), the pendant fourth trochanter is on the proximal half of the shaft (Fig. 17.20F), and there is the beginning of a deep distal extensor groove (Fig. 17.20E,I,J). R167 is referable to *Valdosaurus canaliculatus* within the context of the Wealden of the Isle of Wight. However, *Camptosaurus valdensis* Lydekker, 1889a, is a nomen dubium because the holotype femur has no autapomorphies or unique combination of plesiomorphies, and it is basal Ornithopoda "Node 19" indet. (Fig. 17.1).

17.20. A–E, Basal ornithopod *Hypsilophodon foxii*. A, B, Left femora in anterior view: R5829 (A) and R192a (B); C, D, articulated ankle and pes of R196: C, left in medial view; D, right in ventrolateral view. E–J, Dryosauridae *Valdosaurus* sp. from *Hypsilophodon* Bed, southwest coast of Isle of Wight, R167, holotype of *Camptosaurus valdensis* Lydekker, 1889a, left femur: E, F, stereophotographs in (E) anterior and (F) medial views; G, H, proximal end in (G) anterior and (H) medial views; I, J, distal end in (I) anterior and (J) distal views. K–O, Basal Iguanodontoidea indet.: K–L, left femur CAMSM X29337 from Durlston Bay, Dorset, in (K) anteromedial and (L) distal views. M–O, Left distal femur (lateral half) R8420 from Isle of Wight in (M) lateral, (N) distal, and (O) medial views. B from Hulke (1882a). Scale bars = 20 mm.

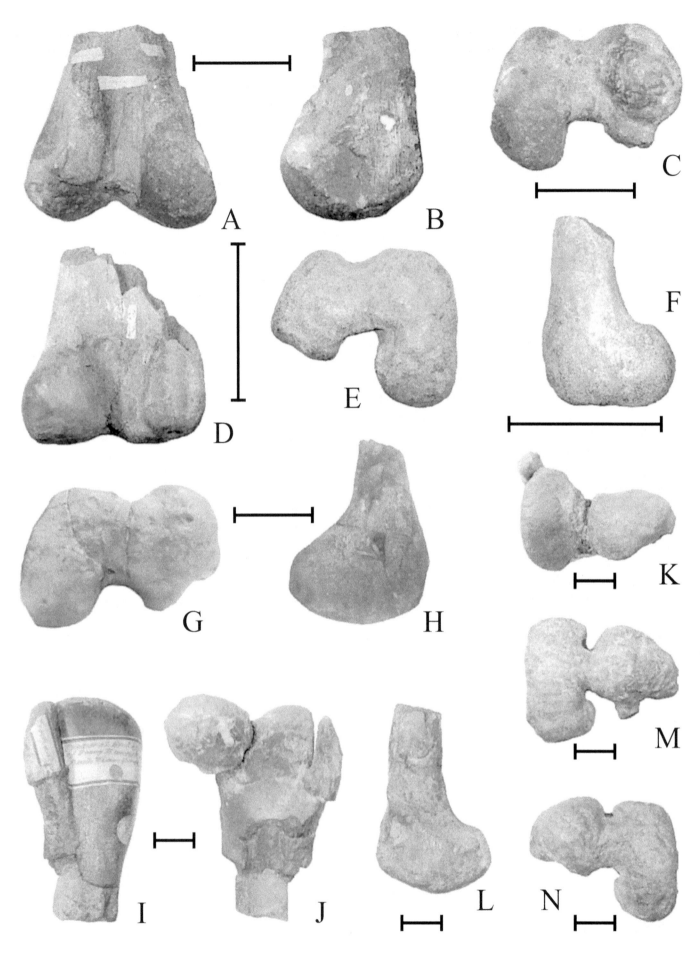

Ruiz-Omeñaca (2001; Ruiz-Omeñaca and Canudo, 2004) suggested that the dryosaurid *"Camptosaurus" valdensis* Lydekker, 1889a, is possibly a senior synonym of *"Hypsilophodon" wielandi* Galton and Jensen, 1978, citing these taxa (along with others) as evidence for a transatlantic land route, and *valdensis* is made the type species of a new unpublished genus of dryosaurid, with *wielandi* as a second species, by Ruiz-Omeñaca (2006). However, these two femora (Galton and Jensen, 1978, fig. 1A–F; Galton, 1974, pl. 2, fig. 4a,b; 2009, fig. 5K,L,Q–V) are not referable to the same genus because the cleft between the lesser and greater trochanters is deep in R167 (Fig. 17.20E,G,H), as against shallow in AMNH 2585, and the anterior intercondylar groove is much better developed in R167 (Fig. 17.19J). There is no deep pit adjacent to the fourth trochanter in AMNH 2585, so it is not a dryosaurid. The taxa based on these femora are nomina dubia, and both are listed as such in Sues and Norman (1990; *H. wielandi* by Norman et al., 2004).

Dryosaurus sp. from Sussex

Two biozones are recognized for the Wealden of England and the rest of Western Europe. These are the Upper Wealden fauna of the Isle of Wight (exposed upper part of Wessex Formation [=Wealden Marls] and overlying Vectis Formation [=Wealden Shales]) and Sussex (Upper and Lower Weald Clay) versus the Lower Wealden fauna of Sussex (Hastings Beds) with the Lower Weald Clay/Hastings Beds boundary in Sussex marking the Hauterivian–Valanginian boundary in time (Chapter 15 in this book, Fig. 15.3; Norman, 1987; Martin and Buffetaut, 1992; Pereda-Suberbiola, 1993; Benton and Spencer, 1995; Weishampel et al., 2004; Radley, 2004, 2006a, 2006b; Rawson, 2006).

The faunas of the Upper and Lower Wealden are distinguished using species of Iguanodontoidea and Nodosauridae (armored dinosaurs) as index fossils (Norman, 1987; Martin and Buffetaut, 1992; Pereda-Suberbiola, 1993), such as *Iguanodon bernissartensis*, *Mantellisaurus atherfieldensis* (see Norman, 1987; Paul, 2006) versus *Iguanodon anglicus* Holl, 1829, *Barilium* Norman, 2010, for *Iguanodon dawsoni* Lydekker, 1888b, *Hypselospinus* Norman, 2010, for *Iguanodon fittoni* Lydekker, 1889b (=*I. hollingtoniensis* Lydekker, 1889b) (see Norman, 2010; Chapter 15 in this book), and, for the Nodosauridae, *Polacanthus foxii* versus *Hylaeosaurus armatus* (Pereda-Suberbiola, 1993).

The dryosaurid specimens from the Lower Wealden of West Sussex are provisionally referred to *Valdosaurus* sp. rather than to *Valdosaurus canaliculatus*, the Upper Wealden species. Most of the finds were made during the 1800s, and they were usually misidentified as juvenile individuals of *Iguanodon*. They came from the Grinsted Clay Member, a diachronous lithological unit within the Tunbridge Wells Sand Formation (which represents the uppermost part of the Hastings Subgroup, upper Valanginian; Rawson, 2006).

Owen (1842, 139) described a complete femur of a juvenile individual of *Iguanodon mantelli* from the Hawksbourne pit near Rusper near Horsham. A pair of femora (Fig. 17.21I–N, BMB 004297–004300; original length ~300), both lacking the shaft, were identified as this femur by Blows (1998, 32),

17.21. A–H, Distal femora of basal Ornithopoda: A–H, left (A–C, G, H) and right (D–F) from Lower Cretaceous of the Isle of Wight (A–C) and Potton, Bedfordshire (D–H) in posterior (A, D), medial (B, H), and distal (C, E, G) views. A–C, Ornithopoda "Node 19" indet. (Fig. 17.1; Butler et al., 2008), NHMUK 31815 from Wessex Formation of Isle of Wight; D–F, Iguanodontia ("Node 20") indet., NHMUK 40424; and G, H, Ornithopoda "Node 19" indet., CAMSM 29738. I–N, *Valdosaurus* sp. from near Rusper, West Sussex, femora: I–M, left: I–L, proximal end BMB 004298 in (I) lateral, (J) anteromedial, (K) proximal views; L–M, distal end BMB 004299 in (L) medial and (M) distal views; and N, right distal end BMB 004300 in distal view. Scale bars = 100 mm (A–H) and 20 mm (I–N).

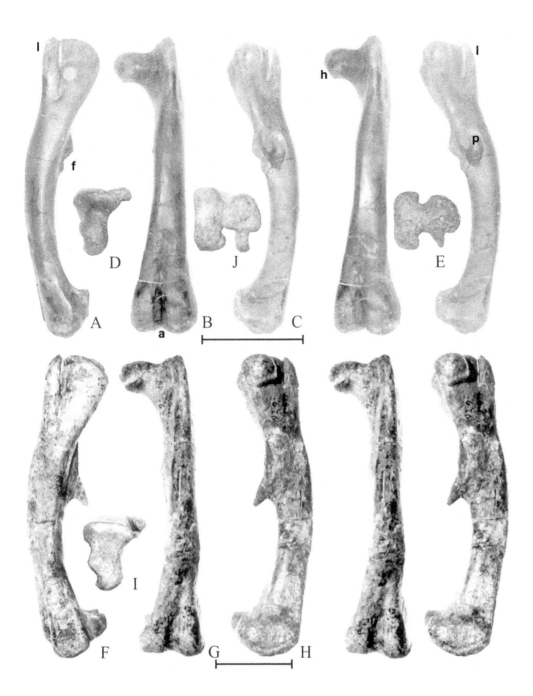

17.22. Dryosauridae, left femora (B, C, G, H stereophotographic views) in (A, F) lateral, (B, G) anterior, (C, H) medial, (D, L) proximal, and (E, J) distal views. A–F, *Valdosaurus canaliculatus* (Galton, 1975), holotype R185 from *Hypsilophodon* Bed (Wessex Formation) of the southwest coast of the Isle of Wight; G–J, *Dryosaurus lettowvorbecki*, Tendaguru Beds (Upper Jurassic) of Tanzania, East Africa, GPIT 1495/14. *Abbreviations:* a, anterior (extensor) groove; f, fourth trochanter; g, greater trochanter; h, head; l, lesser trochanter; p, pit. Scale bars = 50 mm.

who referred them to *Valdosaurus canaliculatus*. These femora, described and illustrated by Galton (2009, fig. 7A–L) as *Valdosaurus* sp., represent the first published record of *Valdosaurus*.

In ilia of *Hypsilophodon*, *Camptosaurus*, and Wealden iguanodontoideans, the brevis shelf is transversely narrow and the postacetabular process is vertically deep, so their ratio is less than 0.5 (Galton, 1974; Gilmore, 1909; Norman, 1980, 1986). Butler et al. (2008) give a wide brevis shelf of the ilium as a possible synapomorphy for the Dryosauridae. On the basis of the lowness of the posterior part of the postacetabular process, the height subequal to that of the preacetabular process, and the greater width of the incomplete brevis shelf—characters that match those of the ilia of *Dryosaurus* (Fig. 17.23H–J; see Galton, 1981b, ratio just over 2.0)—two ilia were referred to *Valdosaurus canaliculatus* by Galton and Taquet (1982, fig. 2H–K); these ilia are now referred to *Valdosaurus* sp. Lydekker (1888a, fig.

35, 1888b, fig. 3) figured a large right ilium (Fig. 17.23A,B; Galton, 2009, fig. 11A–E) from Cuckfield, previously identified as that of *Iguanodon mantelli*, as *Hylaeosaurus oweni*. A similar but smaller less complete left Cuckfield ilium (Fig. 17.23C,D; Galton, 2009, fig. 11J–L) was also referred to *H. oweni* by Lydekker (1888a). The brevis shelf is complete, with the ratio as 2.7, in an ilium referred to *Valdosaurus* sp. from Waterman's Field near Southwater, southwest of Horsham (Fig. 17.23E–G; Galton, 2009, fig. 11F–I). These ilia are less dorsoventrally compressed than are those of the dryosaurids *Planicoxa venenica* (Lower Cretaceous, Barremian–Aptian, Utah; DiCroce and Carpenter, 2001, fig. 13.5; Carpenter and Wilson, 2008, fig. 49F–H) and *Planicoxa depressus* (Gilmore, 1909) (Lower Cretaceous, Barremian, South Dakota; Gilmore, 1909, figs. 45, 46; Carpenter and Wilson, 2008, fig. 49A–E). Galton (1976, fig. 2A–D) referred a pair of ilia with a sacrum and 12 associated dorsal vertebrae (R8649; Wessex Formation, about 100 yards west of Chilton Chine, southwest coast of Isle of Wight) to *Vectisaurus valdensis*. However, contrary to Galton (1976) and Norman (1990), the dorsoventral shortness of these ilia (Fig. 17.23K–M) is not the result of compression, and these *Planicoxa*-like ilia indicate a new dryosaurid taxon, *Proplanicoxa galtoni* Carpenter and Ishida, 2010.

A medium-sized ischium (NHMUK 2183; Galton, 1975, fig. 2H, 2009, fig. 5C; Galton and Taquet, 1982, fig. 2L) and tibia (NHMUK 36506, incorrectly given as R124 by Galton, 2009, fig. 5U,L, as 36506 in 1975, fig. 1F,G) from Cuckfield, possibly referable to *Valdosaurus* sp. within the context of the English Wealden, are basal Ornithopoda indet. at best (Galton, 2009).

Valdosaurus from Outside England

The occurence of *Valdosaurus* in Spain is yet to be established (Galton, 2009). Consequently, Canudo and Salgado (2004; also Canudo et al., 2009) are incorrect in citing its occurrence as evidence for a land connection between Spain and Gondwana, especially as *V. nigeriensis* Galton and Taquet, 1982, from Niger is incorrectly referred to *Valdosaurus* on the basis of differences in age (Late Aptian versus Late Barremian) and morphology. *V. nigerensis* Galton and Taquet, 1982, is the type species of the genus *Elrhazosaurus* Galton, 2009, as *E. nigeriensis* (Galton and Taquet, 1982). This taxon is a dryosaurid, as indicated by the deep pit on the femoral shaft level with the fourth trochanter. The femur has the following combination of characters: proximal end of lesser trochanter below that of greater trochanter, transversely wide raised area separating deep pit from base of fourth trochanter, extensor groove in distal view is deep and obliquely inclined postero-medially and overlapped anteriorly by the acute-edged medial condyle (Galton, 2009, fig. 5M–P).

The cited record of *Valdosaurus* sp. from the Lower Cretaceous (Berriasian–Valanginian) of Cornet, Romania, in Benton et al. (1997, 283, 2006, 82; also Posmoşanu, 2003a, 32, but not 2009) is based on a poorly preserved distal right femur (distal width ~60 mm, estimated length ~250 mm; MTCO 33/14.295; Jurcsák and Popa, 1979, fig. 21a,b as *Valdosaurus* sp., Jurcsák, 1982, fig. 8; Jurcsák and Kessler, 1991, fig. 17a,b, as *V. canaliculatus*; Galton, 2009, fig. 6Q,R). However, in distal view, the extensor groove is displaced laterally, and because it is diagonally inclined slightly posterolaterally, the

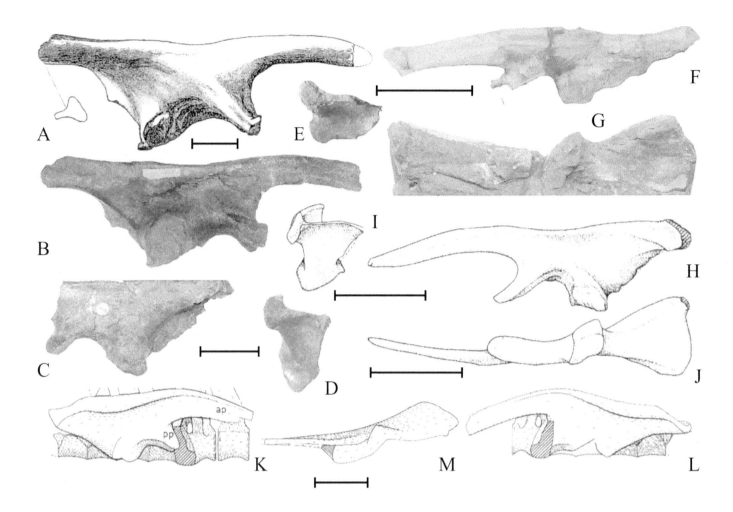

17.23. Dryosauridae, right (A, B, K) and left (C–J, L, M) ilia. A–G, *Valdosaurus* sp. from Upper Tunbridge Wells Sandstone of Cuckfield (A–D) and Waterman's Field (E–G), West Sussex (A, B, NHMUK 2150; C, D, NHMUK 2132; E–G, BMB 004274) in (A, C, F) lateral (A with posterior view of postacetabular process), (d, e) posterior and (G) ventral views. H–J, *Dryosaurus lettowvorbecki* from Upper Jurassic of Tanzania, East Africa, left ilium MFNB dy II in (H) lateral, (L) posterior, and (J) ventral views. K–M, *Proplanicoxa galtoni* Carpenter and Ishida, 2010, from Wessex Formation of the southwest coast of the Isle of Wight, part of holotype R8649; K, right ilium with sacrum in lateral view; L, M, left ilium in lateral view (with sacrum) and M, in ventral view. A from Lydekker (1889a), H–J from Galton (1981a), K–M modified from Galton (1976). Scale bars = 50 mm (A–J) and 100 mm (K–M).

groove is overlapped anteriorly by the lateral condyle, the edge of which is acute. In *Valdosaurus*, *Dryosaurus*, and *Camptosaurus*, the extensor groove is more centrally located, is subvertical, and is not overlapped by the lateral condyle, the edge of which is gently convex. Another distal femur (distal width ~34 mm, estimated length ~150 mm; MTCO 21.362/7290+9124), described by Posmoşanu (2003b, pl. 1, fig. 1a–d) as *Iguanodon* sp., is similar (Galton, 2009, fig. 6S,T). In the iguanodontoideans from the Wealden of Western Europe (*Iguanodon*, *Mantellisaurus*), the extensor groove is vertical, centrally located, deep, channellike, and almost tubular with an anterior overlap by lips from both condyles (Norman, 1980, 1986). In the dryosaurid *Elrhazosaurus nigeriensis*, the extensor groove is obliquely inclined, but as it passes posteromedially, it is the medial condyle that is acutely edged and overlaps the groove anteriorly (Galton, 2009, fig. 5N,O). The only other femur with a similar distal extensor groove, one obliquely inclined posterolaterally with an acute-edged overlapping lateral condyle like the Cornet femora, is that of the camptosaurid *Draconyx loureiroi* (Mateus and Antunes, 2001, fig. 6C), but the groove is much deeper in the Cornet femora (Galton, 2009, fig. 6Q–T). Galton (2009) tentatively referred these Cornet distal femora, which are more derived than those of *Valdosaurus*, *Camptosaurus*, and *Draconyx* but are not Iguanodontoidea, to *Owenodon* sp. However, they are reidentified as basal Styracosterna indet.

Holotype Dentary

Iguanodon hoggii Owen, 1874, is based on an almost complete small right dentary (Fig. 17.24, R2998; preserved length ~145 mm), with most of the teeth preserved in situ (Fig. 17.26), from the Purbeck Limestone Formation (=Durlston Formation, Upper Purbeck, mid-Berriasian; Allen and Wimbledon, 1991) of Durlston Bay, Swanage, Dorset. The dentary is almost complete but slightly crushed transversely, with 15 preserved alveoli, most with teeth and the last one rudimentary and on the medial side of the coronoid process (Fig. 17.24B–E, Fig. 17.26A). Norman and Barrett (2002, 164–167, figs. 1–3) redescribed R2998, which was discussed by Galton (2009, fig. 9A–F, 2006, figs. 2.17S, 2.18I). This dentary cannot be referred to the Iguanodontoidea because the most posterior teeth are slightly inset into the thin anteromedial edge of the coronoid process (Fig. 17.24A,B), as in *Camptosaurus* (Gilmore, 1909; Galton, 2009, fig. 10A–C), rather than being separated by a gap (Fig. 17.25I,J) as in Iguanodontoidea, and the teeth bear simple denticles that are not mammillated (Fig. 17.24D–E, 17.26; Norman and Barrett, 2002, 168).

Given that R2998 is definitely not referable to the Iguanodontoidea sensu Norman (2004), the differences from the dentary of *Camptosaurus dispar*, and the definable nature of the teeth (Norman and Barrett, 2002; Carpenter and Wilson, 2008), *Iguanodon hoggii* Owen, 1874, was made the type species of the genus *Owenodon* Galton, 2009. Compared to the dentary of *Camptosaurus*, that of the holotype of *Owenodon hoggii* (Owen, 1874) is deeper, with the dental margin and tooth row more arched and the symphysis longer; the lingual surface of the dentary tooth crowns are equally subdivided by a simple, well-defined, primary ridge and a parallel and equal-size secondary ridge with tertiary ridges essentially absent, and there is no abbreviated cingulum formed by inrolling of the denticulate distal edge of the lower distal corner of the crown. On the basis of the more derived form of the teeth, Galton (2009) considered that *Owenodon* could represent an unnamed clade between Ankylopollexia (for Camptosauridae: *Camptosaurus, Draconyx*) and Styracosterna Sereno, 1986, sensu Norman (2002, 2004, 434; for *Lurdusaurus, Equijubus*) (Fig. 17.1). Although the dentary of *Lurdusaurus* is unknown, the described bones look much more Iguanodontoidea-like than *Camptosaurus*-like, and in addition, the skull of *Equijubus* looks Iguanodontoidea-like with the tooth row passing medial to a subvertical coronoid process.

Styracosterna Sereno, 1986, was defined by Sereno (1998, 62) as "all Ankylopollexians closer to *Parasaurolophus* than to *Camptosaurus*," so on the basis of this stem-based definition, *Owenodon* is a basal member of the Styracosterna, the position it plots in the cladogram of McDonald et al. (2010). However, the use of more replicates yield somewhat different results with less resolution in the cladogram (McDonald, pers. comm.).

Owenodon hoggii is distinct from the taxon represented by the small dentary R180 (Fig. 17.25A–H) of about the same size but Iguanodontoidea indet.

Gasca et al. (2009) describe an isolated incomplete tooth (MFCPTD SUE-1/D1) from the El Castellar Formation (Lower Cretaceous, Upper Hauterivian–Lower Barremian) in Miravette de la Sierra, southeast Teruel province, northeast Spain. Much of the ~20-mm-wide crown is preserved, but there is practically nothing of the root. Mainly on the basis of the lack of a prominent primary ridge, it was identified and illustrated as a maxillary tooth, "Hypsilophodontidae" indet. However, this tooth crown does not resemble any in situ maxillary teeth of any basal euornithopod, although it is similar to the dentary teeth of *Owenodon hoggii* (Figs. 17.24D–E, 17.26). In particular, the crown is subdivided into three subequal areas by two subequal vertical ridges, with tertiary ridges essentially absent, and there is no abbreviated cingulum formed by inrolling of the denticulate distal edge of the lower distal corner of the crown. Comparisons with the teeth of R2998 (Fig. 17.24D–E, 17.26) indicate that the isolated tooth is a left dentary tooth and it is tentatively reidentified as *Owenodon* sp.

Referred Material from Romania

Owenodon teeth. The bauxite mine at Cornet near Oradea in northwestern Romania produced thousands of disarticulated bones of Berriasian–Valanginian age, mostly of ornithopod dinosaur, during excavations from 1978 to 1983 (see Benton et al., 1997, 2006).

Dentary teeth of *Owenodon* sp. occur (MTCO 1902, 10220a, Posmoşanu and Popa, 1997, figs. 3, 4; Posmoşanu, 2003a, as *Camptosaurus* sp., 2009, as *C. hoggii*; Galton, 2009, fig. 18G–K), as do maxillary teeth (MTCO 1908, 9903, Posmoşanu and Popa, 1997, figs. 1, 2; Galton, 2009, fig. 18A–K).

Basal Styracosterna indet. Galton (2009) tentatively referred those bones from Cornet, which were more derived than those of *Camptosaurus* but are not Styracosterna or Iguanodontoidea, to *Owenodon* sp. However, these bones are reidentified as basal Styracosterna indet. They include a maxilla and the impression of a tooth (MTCO 359, incorrectly given as MTCO 7601 and 9601 by Galton, 2009, 249, 250, correct in captions to figs. 17A–C, 18A; Benton et al., 1997, fig. 9Ba–d), cervical vertebra 6 (MTCO 1513; Jurcsák and Kessler, 1991, fig. 15a–c), fused medial carpals plus metacarpal I (MTCO 2358, 4032; Posmoşanu and Popa, 1997, figs. 5, 6), and two distal femora (see above under *Valdosaurus*) (see Galton, 2009, figs. 6Q–T, 17A–C, 18A, 20D–F, 23N–Q).

Other bones, more tentatively referred to *Owenodon* as cf. *Owenodon* sp., are probably also basal Styracosterna indet. These bones include a frontal (MTCO 7661; Benton et al., 1997, fig. 9Ca,b; Posmoşanu, 2003a, fig. 4), a braincase (MTCO 7601; Benton et al., 1997, fig. 9Aa–d; Posmoşanu, 2003a, fig. 2), a dorsosacral centrum (MTCO 5617; Jurcsák and Kessler, 1991, fig. 49), and a large humerus (MTCO 5284/21.333; Benton et al., 1997, fig. 10E; Posmoşanu, 2003a, fig. 5) (see Galton, 2009, figs. 17R–T, 19A–F, 20N, 22D–H).

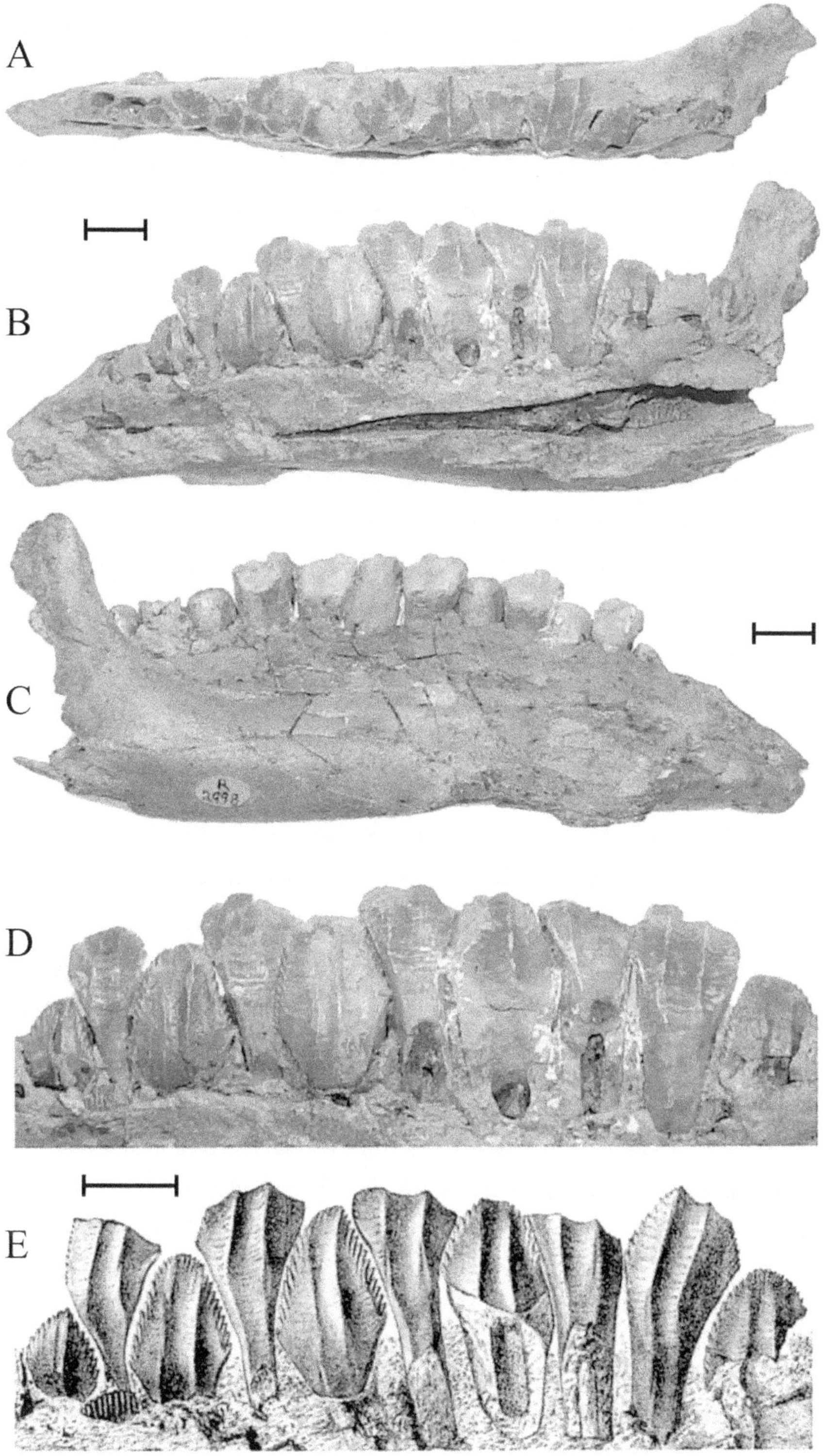

17.24. Basal styracosteran *Owenodon hoggii* (Owen, 1874), holotype right dentary R2998. A–C, Bone in (A) dorsal, (B) medial, and (C) lateral views; D–E, teeth 3–12 in lingual view. E, Modified from Owen (1874). Scale bar = 10 mm.

Dentary

Owen (1864, pl. 10) described a small dentary with 15 alveoli, three of which contain replacement teeth (Fig. 17.25A–H, R180), from the Wessex Formation near Brixton on southwest coast of the Isle of Wight. It was originally regarded as representing a juvenile *Iguanodon mantelli* and was referred to several different genera before being identified as Iguanodontoidea indet. (Galton, 2009, figs. 8A–K, 10F).

Iguanodontoidea indet.: Small Bones

The denticles of the replacement teeth of R180 bear ornamentation (mammillate processes) on the marginal denticles (Fig. 17.25D–H). Mammillate marginal denticles are absent on teeth of *Dryosaurus* (Galton, 1983, 2006), *Camptosaurus* (Galton, 2006, 2009), and *Owenodon* (Fig. 17.26A–B). Also in these taxa the posterior end of the tooth row is inset medially a little into the anterior edge of the coronoid process (Fig. 17.24A,B, 17.26A).

In Iguanodontoidea sensu Norman (2002, 2004) (Fig. 17.1), which includes *Iguanodon*, *Mantellisaurus*, *Ouranosaurus*, *Altirhinus*, and *Eolambia*, the marginal denticles form a curved, mammillated ledge (as against having a simple tongue shape) and the coronoid process forms an elevated fingerlike process that is separated from the posterior end of the tooth row by a medial shelf (Fig. 17.25I,J). Although the coronoid process is broken off at its base in R180, its beginning is visible laterally as a slight swelling (Fig. 17.25C, *), and the topography of this region (Fig. 17.25B–C) is similar to that of basal iguanodontoideans (Fig. 17.25I,J; Galton, 2009, fig. 10G–N). Norman (1986, 295) gives the adult dentary tooth count for the two iguanodontoideans occurring in the Wessex Formation: *Mantellisaurus* as 21 (based on holotype R5764, Isle of Wight) and *Iguanodon bernissartensis* as 25 (based on RBINS material, Wealden of Belgium).

R180 is Iguanodontoidea indet. on the basis of the inset nature of the posterior part of the tooth row and having teeth with mammillated denticles. Unfortunately, histological studies of the broken ends of the bone are not possible (Barrett, pers. comm., 2010), and it would probably not provide any definite evidence. Consequently, the important question regarding the age of R180 will remain unanswered. It could represent an adult (Lydekker, 1888c, 1889a, 1890) of a new small taxon or a juvenile individual (Owen, 1864; Lydekker, 1888a), as indicated by its small size (preserved length 132 mm) and the low tooth count of 15, of either a sympatric iguanodontoidean (*Iguanodon bernissartensis* or *Mantellisaurus atherfieldensis*) or a new taxon.

Femora

Histological studies are needed to determine if two small femora, described and identified by Galton (2009) as Iguanodontoidea indet., represent adult or juvenile individuals.

The complete femur (CAMSM X29337, length 300 mm) came from the Purbeck Limestone of Durlston Bay, Swanage (Fig. 17.20K; Galton, 2009, fig. 13M–P; Norman and Barrett, 2002, fig. 6A as *Camptosaurus hoggii*). The extensor groove is deep and overlapped anteriorly by lips from both condyles, so it is Iguanodontoidea indet., and as such, it is a different taxon from *Owenodon hoggii* that is represented by a dentary from a smaller-sized individual (Fig. 17.24A–C).

The lateral half of a distal femur (Fig. 17.20M–O, R8420; Galton, 2009, fig. 7X–Z; Wessex Formation, Clinton Chine, Isle of Wight) was referred to *Valdosaurus canaliculatus* by Galton and Taquet (1982, figs. 1K, 2F,G). However, in femora of *Valdosaurus* the extensor groove is proportionally not as deep, so it does not come close to the posterior intercondylar groove, and the slight anterior overlap is only by the medial condyle (Fig. 17.22E). Anterior overlap by the lateral condyle occurs in the camptosaurid *Draconyx*

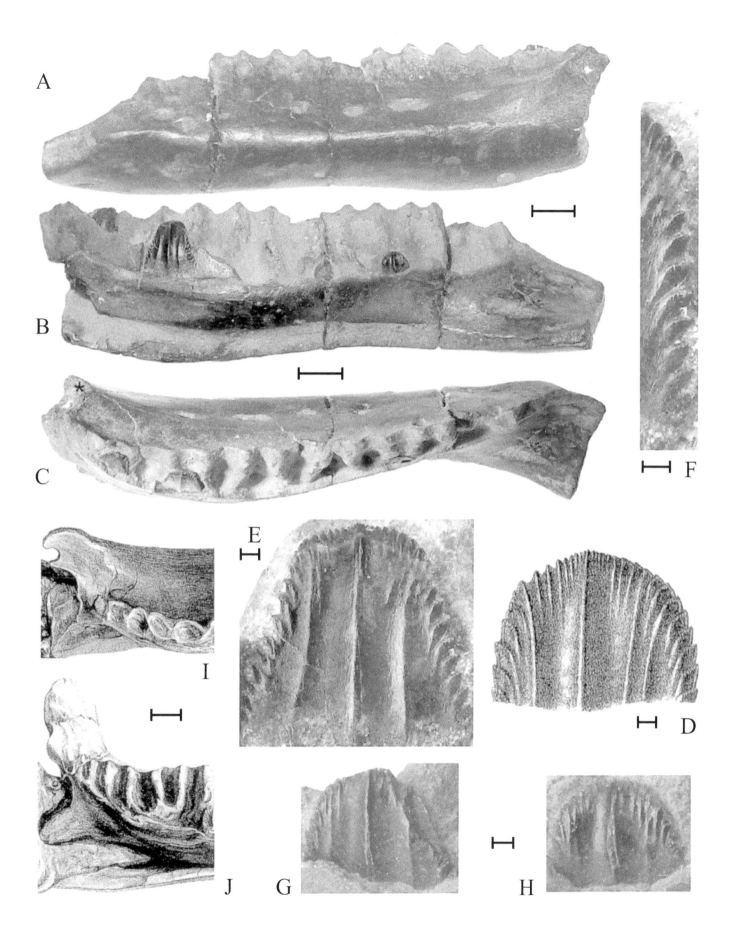

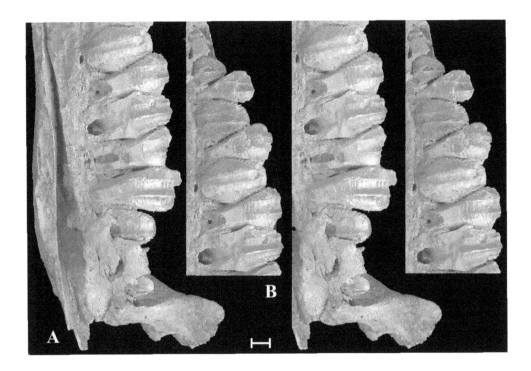

17.26. Basal styracosteran *Owenodon hoggii* (Owen, 1874), holotype right dentary R2998. Stereophotographs of tooth positions: A, 7–14 and coronoid process, and B, 1–9. Scale bar = 10 mm

(Mateus and Antunes, 2001, fig. 6; Galton, 2009, fig. 6P), but in this taxon, the extensor groove is not nearly as deep and it is obliquely inclined. Among Wealden euornithopods, extensive overlap by the lateral condyle of a deep, vertical extensor groove only occurs in iguanodontoideans (Galton, 2009, figs. 13C,D,K,L, 16Q), so R8420 is probably Iguanodontoidea indet.

English Lower Cretaceous Ornithischian Faunal Diversity

The estimated body length and mass (*AL, *M; see Appendix for details) calculated for incomplete ornithischian dinosaur remains from the Lower Cretaceous of England are minimum values, as indicated by the situation in ornithopods with a good fossil record from the Upper Jurassic (see Appendix). Thus, for *Dryosaurus*, the largest specimens did not exhibit adult histology (*D. altus*, Horner et al., 2009) or no lines of arrested growth were present (*D. lettowvorbecki*, Chinsamy, 1995). For *Camptosaurus dispar*, a specimen described by Erickson (1988) was 30% larger than the next largest specimen. The smallest and largest herbivores in the faunas, namely some mammals such as multituberculates (see Kielan-Jaworowska et al., 2004) and the sauropod dinosaurs (see Weishampel et al., 2004; Martill and Naish, 2001a, for the Isle of Wight), are not considered. The faunas are considered in descending order from the most recent to the oldest.

Aptian

Terrestrial dinosaurs are poorly represented in the Phosphate Nodule Bed of Bedfordshire because it is a remanié fossil assemblage deposited in the Upper Aptian and derived from underlying beds, with the most common terrestrial vertebrate being *Iguanodon* (Seeley, 1869, 1874; Weishampel et al., 2004, 558, list *I.* cf. *bernissartensis*, but material is undescribed).

The "Node 19" distal femur (Figs. 17.1 and 17.21G,H, CAMSM B29738, Wd 195, *L ~580; Galton, 2009, fig. 15J–N) represents a larger animal of *AL ~4.7 m and *M ~350 kg.

The Iguanodontia indet. ("Node 20") distal femur (Figs. 17.1 and 17.21D–F, NHMUK 40424, Wd 124, *L ~400; Galton, 2009, fig. 15F–I) is from an animal of *AL ~3.3 m and *M ~176 kg.

The stegosaur *Craterosaurus pottonensis* Seeley, 1874 (holotype CAMSM B.28814), an eroded neural arch of a dorsal vertebra (Galton, 1981c), represents an individual of *AL ~5 m and *M ~116 kg.

Barremian–Early Aptian: Vectis Formation, Southwest Shore, Isle of Wight

Weishampel et al. (2004, 559) list *Hypsilophodon foxii* (see next paragraph) and *Iguanodon atherfieldensis*. The holotype skeleton (R5764, L 678, *M 750 kg) of *Mantellisaurus atherfieldensis* represents a juvenile individual. An adult individual (RBINS R57, femur L 760) from Bernissart, Belgium, had an AL 6.5 m and *M 1,100 kg.

Late Berremian: Wessex Formation, Mostly Southwest Shore, Isle of Wight

The Wessex Formation of the Isle of Wight has the highest ornithischian faunal diversity, with locally abundant remains of *Hypsilophodon foxii*, the largest specimen (R5829, femur L 202, Fig. 17.20A) of which represents an animal of *AL 1.95 m and *M 21 kg (Figs. 17.6 and 17.7).

The Iguanodontoidea indet. dentary (Fig. 17.25A–C, R180 *L 145,) is from an individual of *AL ~1.95 m and *M ~ 47 kg.

The lateral half of a distal femur, Iguanodontoidea indet. (Fig. 17.20M–O, R8420, Wlc 40; Galton, 2009, fig. 7x–z) represents an animal of *AL ~3 m and *M ~134 kg.

Prolanicoxa galtoni Carpenter and Ishida, 2010 (Fig. 17.23K–M, R8649; Galton, 1976, fig. 2), represents an individual of *AL ~3.65 m and *M 164 kg (but this does not take into account the proportionally much longer neural spines compared to *Dryosaurus*).

The largest specimen of *Valdosaurus canaliculatus*, a partial skeleton (IWCMS 6879, femur *L ~500; Naish and Martill, 2001a, fig. 5.11, pls. 8, 9, 10, figs. 1–3), represents an individual of *AL ~4.34 m and *M ~277 kg.

The nodosaur or armored dinosaur *Polacanthus foxii* (R175, femur L 540; Pereda-Suberbiola, 1993, 1994) is from an animal of *AL 5 m and *M ~720 kg.

The large "Node 19" distal femur (Fig. 17.21A–C, NHMUK 31815, Wd 216; Galton, 2009, fig. 15A–E) probably had a *L of ~700 and came from an animal of *AL ~5.7 m and *M ~820 kg.

Mantellisaurus (as *Iguanodon*) *atherfieldensis* is represented by several referred specimens (NHMUK, MIWG; Martill and Naish, 2001a, for photographs). An adult of *Mantellisaurus* (RBINS R57, femur L 760) from Bernissart, Belgium, had an AL of 6.5 m and *M 1,100 kg.

Iguanodon bernissartensis is represented by the partial skeleton (R1502, 2503, 2506), near Brook Chine, southwest shore, Isle of Wight, holotype of *I. seelyi* Hulke, 1882b) with an incomplete femur (L 920, *L 1080; Hulke,

1882b, 38, fig. 1, pl. 4, fig. 2; Lydekker, 1890, fig. 46). This is slightly larger than the femur of the holotype skeleton of *Iguanodon bernissartensis* (RBINS R51, Dollo, 1883, pl. 5; femur L 1030, Wd 240, Norman, 1986) that had an AL of 8 m and *M 3,300 kg.

Upper Valanginian: Tunbridge Wells Sand Formation, Sussex

The largest femur of the dryosaurid *Valdosaurus* sp. from Sussex (Fig. 17.21I–N, BMB 004297–004300, *L 300, Wd 72) represents an individual of *AL ~2.6 m and *M 58.9 kg; for the largest ilium (Fig. 17.23A,B, NHMUK 2150, Lp 275,) the corresponding values were ~4.2 m and *M ~253 kg.

The nodosaur or armored dinosaur *Hylaeosaurus armatus* (Pereda-Suberbiola, 1994) represented an animal of *AL ~5.5 m and *M ~960 kg.

The gracile Wadhurst Clay form, *Hypselospinus fittoni* (Lydekker, 1888b) (referred specimen R1148, holotype of *Iguanodon hollingtonensis*, femur L 847; Galton, 2009, fig. 13G–I; Norman, 2010, fig. 7a,b) had an *AL of ~7.3 m and *M 1,800 kg.

The robust Wadhurst Clay form, *Barilium dawsoni* (Lydekker, 1888a) (R1627, femur L ~940) had an *AL ~7.3 m and *M ~2,430 kg.

Middle Berriasian

Most ornithischians of this age come from the Purbeck Limestone Formation of Durlston Bay, Dorset.

The small omnivorous heterodontosaurid *Echinodon becklesii* is represented only by jaw elements with teeth (Figs. 17.3, 17.4D–P, and 17.5B–G). The jaws, all of uniform size from animals of unknown age, are extremely small with an *AL ~56 and *M ~0.33 kg.

The holotype dentary of *Owenodon hoggii* (Fig. 17.24, R2998, Galton, 2009, figs. 9A–F, 10E) at L 145 represents an individual of *AL ~2.1 m and *M ~ 47 kg.

The Iguanodontoidea indet. femur (Fig. 17.20K,L 300, CAMSM X29337; Galton, 2009, fig. 13M–P) is from an animal of *AL ~3 m and *M 134 kg.

The partial hind limb tentatively referred to *"Iguanodon" hollingtoniensis* by Galton (2009, figs. 13A–F, 14; R8676, on indefinite exhibition loan at KINUA, femur L 550; Speeton Clay, Yorkshire) represents an individual of *AL ~4.4 m and *M ~413 kg. Given the difference in age, this probably represents a different genus from the iguanodontoid species of the Hastings Bed.

Appendix: Size Estimations for English Lower Cretaceous Ornithischians

Abbreviations for measurements (mostly for femora) in millimeters (unless indicated otherwise), given in the text as indicators of size, are as follows: AL, maximum length of animal; L, maximum length; Lp: maximum length from pubic process to posterior end of ilium; L4, minimum length from proximal end to distal edge of fourth trochanter; M, mass; Wd, maximum distal width; Wg, maximum width of greater trochanter; Wlc, maximum anteroposterior width of distal end across lateral condyles; Wmc, maximum anteroposterior width of distal end across medial condyles; Wp, maximum proximal width; and prefix *, calculated.

Various methods have been used to estimate the mass of dinosaurs since the pioneering work of Colbert (1962; included *Camptosaurus, Iguanodon*), which involved displacement volumes using commercially available scale models. Alexander (1985; *Iguanodon bernissartensis* at *AL 8.1 m, mass 5.4 t based on RBINS R51, Norman, 1980, fig. 84) used NHMUK models and incorporated lung volume. Anderson et al. (1985) used the correlation between femoral (or humeral + femoral for quadrupeds) cross-sectional area (calculated from circumference) and corresponding animal mass based on data from living land animals; their regressions were used by Peczkis (1994) to calculate the body mass for numerous genera of dinosaur genera on the basis of estimated cross-sectional areas from the literature. 3D mathematical slicing was used by Henderson (1999; *Iguanodon bernissartensis* at *AL 7.85 m and *M 3,7590 kg based on RBINS R51, Norman and Weishampel, 1990, fig. 25.22 by G. Paul) and the polynominal technique by Seebacher (2001; *Dryosaurus altus* *AL 3.2 m, *M 104.3 kg based on Sues and Norman, 1990, fig. 24.3B by G. Paul; *Hypsilophodon foxii* *AL 1.4 m, *M 7 kg based on R196 by G. Paul [cf. Fig. 17.6B]; *Mantellisaurus atherfieldensis* R5764 as *Iguanodon mantelli*, *AL 5.1, *M 678.4 kg based on Norman, 1980, fig. 83; *I. bernissartensis*, RBINS R51, *AL 7.9, *M 3,775.7 kg based on Norman, 1980, fig. 84, but specimen and scale not indicated).

Paul (1997) discussed the limitations of using commercial models and estimated the mass of many different species of dinosaurs. He used described museum specimens (so measurements available for different bones) to produce bone-by-bone skeletal reconstructions, with multiple views of at least one member of the group as a guide for other members of similar form (Fig. 17.6), that were the basis for reconstructing the soft tissues on scale plasticine models used for volumetric mass estimates. Mass estimates are in kilograms (1,000 g = 2.2 lb), and for consistency, the estimates of Paul (1997, appendices 1–12) are used for calculating the mass of most of the species of ornithischian dinosaurs from the Lower Cretaceous of England.

The small heterodontosaurid *Echinodon becklesi* (Purbeck Limestone Formation, Durlston Bay, Swanage, Berriasian) is represented only by jaw elements with teeth (Figs. 17.3, 17.4, 17.5C–K). However, the much better represented small heterodontosaurid *Fruitadens haagarorum* (Upper Jurassic, Colorado, USA; Butler et al., 2010), with a near adult individual in its fifth year, had *AL 70 cm and *M 0.65 kg (Butler et al., 2010, downsized to 64% from *Heterodontosaurus tucki* SAM-PK-K1332, Santa Luca et al., 1976; Santa Luca, 1980, using the method of Henderson, 1999). The dentary L of *Echinodon* at 32 mm for NHMUK 48215a (Fig. 17.4E) compares with ~40 mm for the largest preserved individual of *Fruitadens* (Butler et al., 2010), so the uniform-sized specimens of *Echinodon* represent individuals of unknown age at *AL ~56 and *M ~0.33 kg that were even smaller than the largest ones of *Fruitadens*.

The body lengths for *Hypsilophodon foxii* are based on NHMUK skeletons, the two mounts (Swinton, 1936c; Galton, 2001a), namely R5830 (femur L 101, Wd 25; Swinton, 1936c, AL 3 ft or 0.92 m) and R5829 (femur L 202, Wd 56, Fig. 17.20A; Swinton, 1936c, AL just over 5 ft or ~1.7 m; color photograph, Gardom and Milner, 1993, 26), and the paper reconstruction of R196 etc. (Fig. 17.6B; femur L ~152, AL 4.5 ft or 1.38 m; Galton, 1974, fig. 62). Paul (1997, appendix 1) gives a *M 21 kg for *AL 1.95 m (from scale line), supposedly the size of R5829, but the skeleton is scaled up from that of R196 etc. (Fig. 17.6B).

Lengths for incomplete femora of *Valdosaurus* are scaled from the proportions of complete ones of *Dryosaurus altus* and *D. lettowvorbecki* (Galton, 1981b, tables 2, 3). Paul (1997, appendix 1), for *Dryosaurus altus*, gives *AL 3.125 m and *M 103 kg for YPM 1876 (femur L 360, Wd 98), but the reconstruction is also based on CM 3392 (humerus L 190, same as YPM 1876; Galton, 1981b; Gilmore, 1925). However, it should be noted that "a histological study of the largest known individual of *Dryosaurus altus* does not display the adult histological features of other ornithopod adults, although it does suggest growth rates comparable to those of larger ornithopods, including hadrosaurs" (Horner et al., 2009, 734). The specimen involved is CM 1949 (femur L 470), a partial skeleton described by Shepherd et al. (1977; also Galton, 1981b) from Johnson County, Wyoming, contra western Colorado in Horner et al. (2009, 737). On the basis of the proportions of YPM 1876, the MFNB mounted skeleton of *D. lettowvorbecki* (femur L 268, Wd 64; Janensch, 1961) has an *AL of 2.33 m, which is close to AL 2.28 m as mounted (Janensch, 1961, largest femur a third larger so femur *L ~360). Histological studies of femora (up to *L 320) of *D. lettowvorbecki* showed rapid growth throughout ontogeny, without any pauses in bone deposition rate (Chinsamy, 1995), so again, no fully grown individuals were sampled.

Compared to the skeleton of *Dryosaurus altus* (CM 3392, YPM 1876 femur L 360, Wd 98, *AL 3.125 m, *M 103 kg; Gilmore, 1925; Galton, 1981b), the femora of *Valdosaurus* represent animals of *AL 1.22 and *M 6.11 kg [*M = 103 × (140/360),[3] R185, L 140, Wd 23, Fig. 17.22A–E], *AL ~2.13 m and *M 32.4 kg (R167, *L ~245, Fig. 17.20E–J), *AL ~2.6 m and *M 58.9 kg (BMB 004297–004300, *L 300, Wd 72, Fig. 17.21I–N), and *AL ~4.34 m and *M 277 kg (IWCMS 6879, L ~500).

Compared to the skeleton of *Dryosaurus altus* (YPM 1876, CM 3392 ilium Lp 204, *AL 3.125 m, *M 103 kg; Gilmore, 1925, fig. 6; Galton, 1981b), the ilia of *Valdosaurus* sp. represent animals of *AL 1.75 m and *M 18.1 kg (BMB 004274, Lp 114, Fig. 17.23E–G), *AL 2.42 m and *M 56.8 kg (NHMUK 2132, Lp 167, Fig. 17.23C,D), and *AL 4.2 m and *M 253 kg (NHMUK 2150, Lp 275, Fig. 17.23A,B).

The ilium of the *Proplanicoxa galtoni* Carpenter and Ishida, 2010 (Fig. 17.23K–M, R8649, Wessex Formation, southwest shore, Isle of Wight; Galton, 1976, fig. 2), has an Lp of 220 to give an *AL of ~3.17 m. However, it is probably more accurate to use the length of the articulated centra of dorsal vertebrae 4–16 and the sacrum at ~880 mm (Galton, 1976, fig. 2A), as against ~550 mm for the MFNB mounted skeleton of *Dryosaurus lettowvorbecki* (Janensch, 1961, pl. 24), to give *AL ~3.65 m and *M 164 kg for R8649. However, this *M is definitely on the low side because the neural spines are proportionally much taller in R8649 (Galton, 1976, fig. 1A).

The skeletal reconstruction of a larger individual of *Camptosaurus dispar* (holotype of *C. medius*, YPM 1877, femur L 570, Wd 180; Galton, 2009, fig. 12A–H) has an *AL of ~6 m (Marsh, 1894, 1895, 1896b, pl. 56), but as Gilmore (1909, 1912) noted, there are five too many dorsal vertebrae, so the corrected *AL would be ~5.5 m. The bones were mounted in 1938 by R. S. Lull in the classic 45-degree pose (Narendra, pers. comm., 2009;

photographs Padian, 1978, 32; Narendra, 1990, 19), and the AL (measured diagonally) is 4 m (Brinkman, pers. comm., 2009). For another larger individual (USNM 5818, femur L 592, Wd 178; Gilmore, 1909, 261, fig. 33), Paul (1997, appendix 4) gives an *AL of 4.7 m and *M 513 kg, rounded off as 500 kg by Paul (2008, fig. 1A, for USNM 5818, femur L as 580 from caption). However, these individuals are not fully grown because the humerus of a partial skeleton described by Erickson (1988) has a humerus length of 468 mm and an *AL approaching 7 m, compared to a humerus L of 360 mm for the next largest individual (USNM 4282, Gilmore, 1909, figs. 25, 26), the mounted skeleton of which has an *AL of 17 ft or 5.2 m (Gilmore, 1912, pls. 55–58). The femur of the largest individual is not preserved, but compared to USNM 4282, *Wd would probably be ~230 mm and *L ~770 mm.

Compared to USNM 5818, using the measurements in Paul (1997, 2008), the larger Isle of Wight "Node 19" distal femur (Fig. 17.21A–C, Wd 216, NHMUK 31815; Galton, 2009, fig. 15A–E) probably had a femur *L ~700, representing an individual *AL ~5.7 m and *M ~820 kg; the Potton "Node 19" distal femur (Fig. 17.21G,H, Wd 195, CAMSM B29738; Galton, 2009, fig. 15J–N) would be *L ~580, *AL ~4.7 m, and *M ~350 kg, and the Potton "Node 20" distal femur (Fig. 17.21D–F, NHMUK 40424, Wd 124; Galton, 2009, fig. 15F–I) would be *L ~400, *AL ~3.3 m, and *M ~176 kg. The partial hind limb tentatively referred to "Iguanodon" hollingtoniensis (Galton, 2009, figs. 13A–F, 14; R8676, on indefinite exhibition loan at KINUA, femur L 550) from Speeton Clay (Middle Berriasian) of Yorkshire, represents an individual of *AL ~4.4 m and *M ~413 kg.

The holotype dentary of Owenodon hoggii (Fig. 17.24A–C, R2998) at L 145 is two-thirds the length of that of Camptosaurus aphanoecetes (CM 11337, Carpenter and Wilson, 2008, fig. 5A,D; femur L 395, Gilmore, 1925), the mounted skeleton of which has an *AL of ~3.2 m (Carpenter and Wilson, 2008, figs. 4, 34A, two-thirds of tail reconstructed). This would give R2998 an *AL ~2.1 m and *M ~47 kg compared to USNM 5818 (probably same *AL and *M for Iguanodontoidea indet., dentary R180 at *L of 145, Fig. 17.25A–C).

A juvenile individual of Camptosaurus dispar, the mounted skeleton of the holotype of C. nanus (USNM 2210, femur L 258, Wd 71; Gilmore, 1909), has an *AL of 10 ft or ~3 m (Gilmore, 1912, pls. 59–61), but Paul reconstructs its *AL as 2 m (in Carpenter and Wilson, 2008, fig. 34C, from scale line). Proportional to USNM 2210, the Dorset Iguanodontoidea indet. femur (Fig. 17.20K, CAMSM X29337; Galton, 2009, fig. 13M–P) at L 300 would have an *AL of ~3.5 or ~2.5 m, for an average of ~3 m so *M ~134 kg compared to USNM 5818. In femora of Camptosaurus and Mantellisaurus, the ratio of Wd to Wlc is 1.3 and 1.2 for an average of 1.25, so the *Wd of the distal lateral half femur Iguanodontoidea indet. (Fig. 17.20M–O, R8420) was 50 so, compared to USNM 2210, the femur *L was ~182 with *AL ~1.47 m and *M ~15.3 kg compared to USNM 5818.

Several specimens (NHMUK, IWCMS), referred to Mantellisaurus (as Iguanodon) atherfieldensis, are known from the Wessex Formation (photographs in Martell and Naish, 2001a, pl. 23, femur ~L 590, NHMUK 11521). The holotype skeleton (R5764) of Mantellisaurus atherfieldensis (from overlying Vectis Formation, Barremian–Lower Aptian, Atherfield Point, southwest shore, Isle of Wight) has a femur L 678 (Hooley, 1925; Wd 144,

Norman, 1986, fig. 57H; color photograph of skeleton, Gardom and Milner, 1993, 27) and *M 750 kg (Paul, 2008, fig. 1C), but this represents a juvenile individual. A larger referred mounted skeleton (RBINS R57; femur L 760, Wd 150, Norman, 1986) from Bernissart, Belgium, originally described as *Iguanodon mantelli* by Dollo (1884, pl. 7), has an AL 6.5 m and *M 1,100 kg (Paul, 1997, appendix 7, 2008, fig. 1D, as the holotype of *Dollodon bampingi*).

Iguanodon bernissartensis is represented in the Wessex Formation by a poorly preserved humerus, ilium, and incomplete hind limb (R2502, 2503, 2506; Brook Chine, southwest shore, Isle of Wight), the holotype of *I. seelyi* Hulke 1882b. The large incomplete femur (L 920, *L 1,080; Hulke, 1882b, 38, fig. 1, pl. 4, fig. 2; Lydekker, 1890, fig. 46) is slightly larger than that of the holotype skeleton of *Iguanodon bernissartensis* (RBINS R51, Dollo, 1883, pl. 5; femur L 1,030, Wd 240, Norman, 1986) that has an AL 8 m and *M 3,300 kg (Paul, 1997, appendix 7, 2008, fig. 1B). Another Wessex Formation femur of *Mantellisaurus* has L ~1020 (Naish and Martill, 2001a, pl. 20, figs. 1–3, R120).

The ilia and dorsal vertebrae are quite distinctive in the two species of *Iguanodon* (*AL of 8 m and 6 m) recognized by Norman (1987, figs. 4, 5, 2004; also Blows, 1998, fig. 3A,B), from the Grinsted Clay Formation (later mid-Valanginian) of East Sussex: *I. dawsoni* Lydekker, 1888b, the robust form, and *I. fittoni* Lydekker, 1889b, the more lightly built form. The former is the type species of *Barilium* Norman, 2010 (=*Torilion* Carpenter and Ishida, 2010), as *B. dawsoni*, (Lydekker, 1888b) with femur of holotype (R1627, L ~940. Lydekker, 1890, 38, ~37") indicating *AL ~7.3 m and *M ~2,430 kg compared to the heavily built *I. bernissartensis* (RBINS R51, L 1025, AL 8 m, *M 3,200 kg, Paul, 1997, 2008). The latter is the type species of *Hypselospinus* Norman, 2010 (=*Wadhurstia* Carpenter and Ishida, 2010), as *H. fittoni* (Lydekker, 1889b) that includes *Iguanodon hollingtonensis* Lydekker, 1889b, the holotype femur (R1148, L 847, Galton, 2009, fig. 13G–I; Norman, 2010, fig. 7A,B) of which indicates *AL ~7.3 m and *M 1,800 kg compared to the lightly built *Mantellisaurus* (RBINS R57, L 760, AL 6.5 m, *M 1,100 kg).

The noneuornithopod ornithischians described from the Lower Cretaceous beds of southern England are both thyreophoreans, a stegosaur or plated dinosaur and an ankylosaur or armored dinosaur.

The only stegosaur from the Lower Cretaceous of England recognized by Maidment et al. (2008) is the holotype of *Craterosaurus pottonensis* Seeley, 1874 (CAMSM B.28814), an eroded neural arch of a dorsal vertebra from the Potton Sandstone of Potton, Bedfordshire (Galton, 1981c). The length at ~150 mm is slightly longer than that of the dorsal vertebrae at ~125 mm of the MFNB mounted skeleton of *Kentrosaurus aethiopicus* (Hennig, 1925, femur L 635), with AL 4.32 m (without tail spikes, with them 4.76 m; Janensch, 1925), and *M 67 kg (Paul, 1997, appendix 5); assuming the same proportions, this gives the Potton stegosaur an *M ~116 kg.

The holotype (R175) of the ankylosaur *Polacanthus foxii* (upper part of Wessex Formation, southwest coast, Isle of Wight) has a femur L 540 and *AL 5 m (Pereda-Suberbiola, 1993, 1994). Unfortunately, the specimens of *Hylaeosaurus armatus* (Grinsted Clay, Upper Valanginian, near Cuckfield, West Sussex) have no associated femora but the centrum of caudal vertebra

15 (NHMUK 3789) has a L 60 versus 54 for *Polacanthus foxii* (Pereda-Suber-biola, 1993, R175), to give an *AL ~5.5 m. The holotype of the nodosaur *Sauropelta edwardsi* (Lower Cretaceous, Western USA; AMNH 3032) has a femur L 765, *AL 4.38 m, and *M 2,100 kg (Paul, 1997, appendix 5 as AMNH 3030, but no such specimen listed by Ostrom, 1969, 104); assuming the same proportions, this gives a *M ~720 kg for *Polacanthus foxii* (R175) and ~960 kg for *Hylaeosaurus armatus*.

Acknowledgments

I thank the following people for their assistance while studying specimens at their respective institutions: J. A. Cooper (BMB, also for locality data); S. Chapman (also for arranging photographs to be taken and answering catalog questions, NHMUK); and D. Brinkman (also for measuring *Camptosaurus* skeleton YPM 1880), W. Joyce, L. K. Murray, and M. A. Turner (YPM). I thank the following people for taking or providing photographs or illustrations, for copies of papers, and/or for personal communications: R. T. Bakker (Houston, Texas, USA, Fig. 17.7), P. M. Barrett (NHMUK), H. Mallison (MFNB), A. T. McDonald (information on revised cladogram, University of Pennsylvania, Philadelphia), B. Narendra (YPM), G. S. Paul (Baltimore, Maryland, USA, Figs. 17.5A and 17.6), P. Crabb (Image Resources NHMUK; Figs. 17.3A,C,E,I–K, M–N, 17.4, 17.9A,B, 17.13–15, 17.23F–G, 17.24A–D, and 17.26), and M. Riley (CAMSM, Figs. 17.20K–L, and 17.21D–H). The remaining photographs are by the author, and all photographs are reproduced courtesy of the housing institutions, which retain the copyrights. H. De Potter (RBINS) adapted the figures for this book. A previous commitment prevented me from attending the Bernissart–Darwin Symposium, so I thank Pascal Godefroit for inviting me to contribute this chapter, the contents of which benefited greatly from the comments by R. J. Butler (Bayerische Staatssammlung für Paläontologie und Geologie, Munich, Germany) and an anonymous reviewer.

References

Abel, O. 1912. Grundzuge der Palaeobiologie der Wirbeltiere. E. Schweizerbart'sche Verlagsbuchhandlung, Stuttgart, 708 pp.

———. 1922. Lebenbilder aus der Tierwelt der Vorzeit. Fischer, Jena, 643 pp.

———. 1925. Geschichte und Methode der Rekonstruktion vorzeitlicher Wirbeltiere. Fischer, Jena, 327 pp.

———. 1927. Lebenbilder aus der Tierwelt der Vorzeit. 2nd ed. Fischer, Jena, 714 pp.

Alexander, R. McN. 1985. Mechanics of posture and gait of some large dinosaurs. Zoological Journal of the Linnean Society 83: 1–25.

Allen, P., and W. A. Wimbledon. 1991. Correlation of NW European Purbeck–Wealden (non-marine Lower Cretaceous) as seen from the English type areas. Cretaceous Research 12: 511–526.

Anderson, J. F., A. Hall-Martin, and D. A. Russell. 1985. Long-bone circumference and weight in mammals, birds and dinosaurs. Journal of Zoology A 207: 53–61.

Bakker, R. T., and P. M. Galton. 1974. Dinosaur monophyly and a new class of vertebrates. Nature 248: 168–172.

Barden, H. E., and S. C. R. Maidment. 2011. Evidence for sexual dimorphism in the stegosaurian dinosaur *Kentrosaurus aethiopicus* from the Upper Jurassic of Tanzania. Journal of Vertebrate Paleontology 31: 641–651.

Barrett, P. M. 1996. The first known femur of *Hylaeosaurus armatus* and re-identification of ornithopod material in the Natural History Museum. Bulletin of the Natural History Museum (Geology) 52: 115–118.

———. 2001. Tooth wear and possible jaw action of *Scelidosaurus harrisonii* Owen and a review of feeding mechanisms in other thyreophoran dinosaurs; pp. 25–52 in K. Carpenter (ed.), The armored dinosaurs. Indiana University Press, Bloomington.

Barrett, P. M., and F.-L. Han. 2009. Cranial anatomy of *Jeholosaurus shangyuanensis* (Dinosauria: Ornithischia) from the Early Cretaceous of China. Zootaxa 2072: 31–55.

Barrett, P. M., J. B. Clarke, D. B. Brinkman, S. D. Chapman, and P. C. Ensom. 2002. Morphology, histology and identification of the "granicones" from the Purbeck Limestone Formation (Lower Cretaceous: Berriasian) of Dorset, southern England. Cretaceous Research 23: 279–295.

Bartholomai, A., and R. E. Molnar 1981. *Muttaburrasaurus*, a new iguanodontid (Ornithischia: Ornithopoda) dinosaur from the Lower Cretaceous of Queensland. Memoirs of the Queensland Museum 20: 319–349.

Benton, M. J., and P. S. Spencer. 1995. Fossil reptiles of Great Britain. Geological Conservation Review Series 10 (4). Chapman and Hall, London, 386 pp.

Benton, M. J., E. Cook, D. Grigorescu, E. Popa, and E. Tallódi. 1997. Dinosaurs and other tetrapods in an Early Cretaceous bauxite-filled fissure, northwestern Romania. Palaeogeography, Palaeoclimatology, Palaeoecology 130: 275–292.

Benton, M. J., N. J. Minter, and E. Posmoşanu. 2006. Dwarfing in ornithopod dinosaurs from the Early Cretaceous of Romania; pp. 79–87 in Z. Csiki (ed.), Mesozoic and Cenozoic vertebrates and paleoenvironment; Tributes to the career of Prof. Dan Grigorescu. Ars Docendi, Bucharest, Hungary.

Blows, W. T. 1998. A review of Lower and Middle Cretaceous dinosaurs of England. New Mexico Museum of Natural History and Science Bulletin 14: 29–38.

Boyd, C., T. Cleland, and F. Novas. 2008. Histology, homology, and function of intercostal plates on ornithischian dinosaurs. Journal of Vertebrate Paleontology 28 (supplement to 3): 55A.

Boyd, C. A., C. M. Brown, R. D. Scheetz, and J. A. Clarke. 2009. Taxonomic revision of the basal neornithischian taxa *Thescelosaurus* and *Bugenasaura*. Journal of Vertebrate Paleontology 29: 758–770.

Brown, B., and E. M. Schlaikjer. 1940. The structure and relationships of *Protoceratops*. Annals of the New York Academy of Science 40: 133–266.

Butler, R. J., and P. M. Galton. 2008. The "dermal armour" of the ornithopod dinosaur *Hypsilophodon* from the Wealden (Early Cretaceous: Barremian) of the Isle of Wight: a reappraisal. Cretaceous Research 29: 636–642.

Butler, R. J., P. Upchurch, and D. B. Norman. 2008. The phylogeny of the ornithischian dinosaurs. Journal of Systematic Palaeontology 6: 1–40.

Butler, R. J., P. M. Galton, L. B. Porro, L. M. Chiappe, D. M. Henderson, and G. M. Erickson. 2010. Lower limits of ornithischian dinosaur body size inferred from a new Upper Jurassic heterodontosaurid from North America. Proceedings of the Royal Society B 277: 375–381.

Canudo, J. I., and L. Salgado. 2004. Los dinosaurios del Neocomiense (Cretácico Inferior) de la Península Ibérica y Gondwana occidental: implicaciones biogeográficas. Instituto de Estudios Riojanos, Ciencias de la Tierra 26: 251–268.

Canudo, J. I., J. L. Barco, X. Pereda-Suberbiola, J. I. Ruiz-Omeñaca, L. Salgado, F. Torcida Fernández-Baldor, and J. M. Gasulla. 2009. What Iberian dinosaurs reveal about the bridge said to exist between Gondwana and Laurasia in the Early Cretaceous. Bulletin de la Société de Géologie de France 180: 5–11.

Carlsson, A. 1914. Über *Dendrolagus dorianus*. Zoologische Jahrbucher (Abteilung für Systematik) 36: 547–617.

Carpenter, K. 1990. Ankylosaur systematics: example using *Panoplosaurus* and *Edmontonia* (Ankylosauria: Nodosauridae); pp. 281–298 in K. Carpenter and P. J. Currie (eds.), Dinosaur systematics: perspectives and approaches. Cambridge University Press, Cambridge.

Carpenter, K., and Y. Ishida. 2010. Early and "Middle" Cretaceous iguanodonts in time and space. Journal of Iberian Geology 36: 145–164.

Carpenter, K., and Y. Wilson. 2008. A new species of *Camptosaurus* (Ornithopoda: Dinosauria) from the Morrison Formation (Upper Jurassic) of Dinosaur National Monument, Utah, and a biomechanical analysis of its forelimb. Annals of the Carnegie Museum 76: 227–263.

Chinsamy, A. 1990. Physiological implications of bone histology of *Syntarsus rhodesiensis* (Saurischia: Theropoda). Palaeontologia Africana 27: 77–82.

———. 1995. Ontogenetic changes in the bone histology of the Late Jurassic ornithopod *Dryosaurus lettowvorbecki*. Journal of Vertebrate Paleontology 15: 96–104.

Chinsamy, A., T. Rich, and P. Vickers-Rich. 1998. Polar dinosaur bone histology. Journal of Vertebrate Paleontology 18: 385–390.

Colbert, E. H. 1962. The weight of dinosaurs. American Museum Novitates 2076: 1–16.

Cooper, M. R. 1985. A revision of the ornithischian dinosaur *Kangnasaurus coetzeei* Haughton, with a classification of the Ornithischia. Annals of the South African Museum 95: 281–317.

Coria, R. A., and J. O. Calvo. 2002. A new iguanodontian ornithopod from Neuquen Basin, Patagonia, Argentina. Journal of Vertebrate Paleontology 22: 503–509.

Dacke, C. G. S., S. Arkle, D. J. Cook, I. M. Wormstone, S. Jones, M. Zaidi, and Z. A. Bascal. 1993. Medullary bone and avian calcium regulation. Journal of Experimental Biology 184: 63–88.

DiCroce, T., and K. Carpenter. 2001. A new ornithopod from the Cedar Mountain Formation (Lower Cretaceous) of western USA; pp. 183–196 in D. H. Tanke and K. Carpenter (eds.), Mesozoic vertebrate life. Indiana University Press, Bloomington.

Dollo, L. 1882. Première note sur les dinosauriens de Bernissart. Bulletin de Musée Royal d'Histoire Naturelle de Belgique 1: 161–180. English version, pp. 380–388 in D. B. Weishampel, and N. M. White (eds.), The dinosaur papers, 1676–1906. Washington, D.C.: Smithsonian Books.

———. 1883. Troisième note sur les dinosauriens de Bernissart. Bulletin de Musée Royal d'Histoire Naturelle de Belgique 2: 85–120. English version, pp. 394–413 in D. B. Weishampel and N. M. White (eds.), The dinosaur papers, 1676–1906. Washington, D.C.: Smithsonian Books.

———. 1884. Cinquième note sur les dinosauriens de Bernissart. Bulletin de Musée Royal d'Histoire Naturelle de Belgique 3: 120–140. English version, pp. 430–442 in D. B. Weishampel and N. M. White (eds.), The dinosaur papers, 1676–1906. Washington, D.C.: Smithsonian Books.

Edinger, T. 1929. Über knochern Schleralinge. Zoologische Jahrbucher (Abteilung für Anatomie) 51: 163–226.

Erickson, B. R. 1988. Notes on the postcranium of *Camptosaurus*. Science Museum of Minnesota, Scientific Publications (n.s.) 6 (4): 1–13.

Fisher, P. E., D. A. Russell, M. K. Stoskopf, R. E Barrick, M. Hammer, and A. A. Kuzmitz. 2000. Cardiovascular evidence for an intermediate or higher metabolic rate in an ornithischian dinosaur. Science 288: 503–505.

Forster, C. A. 1990. The postcranial skeleton of the ornithopod dinosaur *Tenontosaurus tilletti*. Journal of Vertebrate Paleontology 10: 273–294.

Fox, W. 1868. On the skull and bones of an *Iguanodon*. British Association for the

Advancement of Science, Annual Report for 1867, 38: 64–65.

Galton, P. M. 1967. On the anatomy of the ornithischian dinosaur *Hypsilophodon foxii* from the Wealden (Lower Cretaceous) of the Isle of Wight, England. Ph.D. thesis, University of London, King's College, England, U.K., 513 pp.

———. 1969. The pelvic musculature of the dinosaur *Hypsilophodon* (Reptilia: Ornithischia). Postilla, Yale University 131: 1–64.

———. 1971a. A primitive dome-headed dinosaur (Ornithischia: Pachycephalosauridae) from the Lower Cretaceous of England and the function of the dome of pachycephalosaurids. Journal of Paleontology 45: 40–47.

———. 1971b. *Hypsilophodon*, the cursorial non-arboreal dinosaur. Nature 231: 159–161.

———. 1971c. The mode of life of *Hypsilophodon*, the supposedly arboreal ornithopod dinosaur. Lethaia 4: 453–465.

———. 1973a. Redescription of the skull and mandible of *Parksosaurus* from the Late Cretaceous with comments on the family Hypsilophodontidae (Ornithischia). Royal Ontario Museum, Life Sciences Contributions 89: 1–21.

———. 1973b. The cheeks of ornithischian dinosaurs. Lethaia 6: 67–89.

———. 1974. The ornithischian dinosaur *Hypsilophodon* from the Wealden of the Isle of Wight. Bulletin of the British Museum (Natural History), Geology 25: 1–152c.

———. 1975. English hypsilophodontid dinosaurs (Reptilia: Ornithischia). Palaeontology 18: 741–751.

———. 1976. The dinosaur *Vectisaurus valdensis* (Ornithischia: Iguanodontidae) from the Lower Cretaceous of England. Journal of Paleontology 50: 976–984.

———. 1977a. The ornithopod dinosaur *Dryosaurus* and a Laurasia–Gondwanaland connection in the Upper Jurassic. Nature 268: 230–232.

———. 1977b. The ornithopod dinosaur *Dryosaurus* and a Laurasia–Gondwanaland connection in the Upper Jurassic. Milwaukee Public Museum Special Publications in Biology and Geology 2: 41–54.

———. 1978. Fabrosauridae, the basal family of ornithischian dinosaurs (Reptilia: Ornithopoda). Paläontologische Zeitschrift 52: 138–159.

———. 1980a. *Dryosaurus* and *Camptosaurus*, intercontinental genera of Upper Jurassic ornithopod dinosaurs. Mémoires de Société Géologique de France (n.s.) 139: 103–108.

———. 1980b. European Jurassic ornithopod dinosaurs of the families Hypsilophodontidae and Camptosauridae. Neues Jahrbuch für Geologie und Paläontologie, Abhandlungen 160: 73–95.

———. 1981a. A juvenile stegosaurian dinosaur "*Astrodon pusillus*" from the Upper Jurassic of Portugal, with comments on Upper Jurassic and Lower Cretaceous biogeography. Journal of Vertebrate Paleontology 1: 245–256.

———. 1981b. *Dryosaurus*, a hypsilophodontid dinosaur from the Upper Jurassic of North America and Africa: Postcranial skeleton. Paläontologische Zeitschrift 55: 271–312.

———. 1981c. *Craterosaurus pottonensis* Seeley, a stegosaurian dinosaur from the Lower Cretaceous of England, and a review of Cretaceous stegosaurs. Neues Jahrbuch für Geologie und Paläontologie, Abhandlungen 161: 28–46.

———. 1982. The postcranial anatomy of stegosaurian dinosaur *Kentrosaurus* from the Upper Jurassic of Tanzania, East Africa. Geologica et Palaeontologica 15: 139–160.

———. 1983. The cranial anatomy of *Dryosaurus*, a hypsilophodontid dinosaur from the Upper Jurassic of North America and East Africa, with a review of hypsilophodontids from the Upper Jurassic of North America. Geologica et Palaeontologica 17: 207–243.

———. 1989. Crania and endocranial casts from ornithopod dinosaurs of the families Dryosauridae and Hypsilophodontidae (Reptilia: Ornithischia). Geologica et Palaeontologica 23: 217–239.

———. 1991. Postcranial remains of the stegosaurian dinosaur *Dacentrurus* from the Upper Jurassic of France and Portugal. Geologica et Palaeontologica 254: 299–327.

———. 1999. Sex, sacra and *Sellosaurus gracilis* (Saurischia, Sauropodomorpha)—or why the character "two sacral vertebrae" is plesiomorphic for Dinosauria. Neues Jahrbuch für Geologie und Paläontologie, Abhandlungen 213: 19–55.

———. 2001a. The mode of life of *Hypsilophodon*, the small supposedly arboreal ornithopod dinosaur from the Lower Cretaceous of England. Tokyo: Dino Press, Aurora Oval, 3:93–101. [In Japanese; English text 3:32–35.]

———. 2001b. The skull of *Hypsilophodon*, the small ornithopod dinosaur from the Lower Cretaceous of England. Tokyo: Dino Press, Aurora Oval, 4:118–127. [In Japanese; English text 4: 45–47.]

———. 2006. Teeth of ornithischian dinosaurs (mostly Ornithopoda) from the Morrison Formation (Upper Jurassic) of western USA; pp. 17–47 in K. Carpenter (ed.), Horns and beaks: ceratopsian and ornithopod dinosaurs. Indiana University Press, Bloomington.

———. 2009. Notes on Neocomian (Lower Cretaceous) ornithopod dinosaurs from England—*Hypsilophodon*, *Valdosaurus*, "*Camptosaurus*," "*Iguanodon*"—and referred specimens from Romania and elsewhere. Revue de Paléobiologie 28: 211–273.

———. In press. Stegosaurs. in M. K. Brett-Surman, T. R. Holtz Jr., and J. A. Farlow (eds.), The complete dinosaur 2. Indiana University Press, Bloomington.

Galton, P. M., and J. A. Jensen. 1973. Skeleton of a small hypsilophodontid dinosaur (*Nanosaurus* (?) *rex*) from the Upper Jurassic of Utah. Brigham Young University, Geology Studies 20: 137–157.

———. 1975. *Hypsilophodon* and *Iguanodon* from the Lower Cretaceous of North America. Nature 257: 668–669.

———. 1978. Remains of ornithopod dinosaurs from the Lower Cretaceous of North America. Brigham Young University Geology Studies 25: 1–10.

Galton, P. M., and P. Powell. 1980. The ornithischian dinosaur *Camptosaurus prestwichii* from the Upper Jurassic of England. Palaeontology 23: 411–443.

Galton, P. M., and P. Taquet. 1982. *Valdosaurus*, a hypsilophodontid dinosaur from the Lower Cretaceous of Europe and Africa. Géobios 13: 147–159.

Galton, P. M., and P. Upchurch. 2004. Stegosauria; pp. 343–362 in D. B. Weishampel, P. Dodson, and H. Osmólska (eds.), The Dinosauria. 2nd ed. University of California Press, Berkeley.

Gardom, T., and A. Milner. 1993. The book of dinosaurs. The Natural History Museum guide. London: Natural History Museum, 128 pp.

Gasca, J. M., J. I. Canudo, and M. Moreno-Azanza. 2009. Dientes aislados de dinosaurio de la Formacion El Castellar en Miravete de la Sierra (Cretacico Inferior, Teruel, Espana); pp. 221–234 in Colectivo Arqueológico y Paleontológico de Salas (ed.), Actas de las IV Jornadas Internacionales sobre Paleontología de Dinosaurios y su Entorno. Colectivo Arqueologico y Paleontologico de Salas, Burgos, Spain.

Gilmore, C. W. 1909. Osteology of the Jurassic reptile *Camptosaurus*, with a revision of the species of the genus, and a description of two new species. Proceedings of the United States National Museum 36: 197–332.

———. 1912. The mounted skeletons of *Camptosaurus* in the United States National Museum. Proceedings of the United States National Museum 41: 687–696.

———. 1925. Osteology of ornithopodous dinosaurs from the Dinosaur National Monument, Utah. Memoirs of the Carnegie Museum 10: 385–410.

Glut, D. F. 1997. Dinosaurs: the encyclopedia. Jefferson, N.C.: McFarland, 1,076 pp.

Hamilton, W. R., A. R. Wolley, and A. C. Bishop. 1974. The Hamlyn guide to minerals, rocks and fossils. Hamlyn, London.

Heilmann, G. 1913. Vor nuvaerende Viden om Fuglenes Afstamning. Andet Afsnit. Fugleligheder Bladt Fortidsc/lger. Dansk Ornithologisk Forenings Tidsskrift 8: 1–92.

———. 1916. Fuglenes Afstamning. Copenhagen, Denmark, 398 pp.

———. 1926. The origin of birds. Witherby, London, 210 pp. Reprint, New York: Dover, 1972.

Henderson, D. M. 1999. Estimating the masses and centers of mass of extinct animals by 3D mathematical slicing. Paleobiology 25: 88–106.

Hennig, E. 1915. Stegosauria. Fossilium Catalogus. I: Animalia 9: 1–16.

———. 1925. *Kentrurosaurus aethiopicus*, die Stegosaurier-funde von Tendaguru, Deutsch-Ostafrika. Palaeontographica, Supplement 7 (1, 1): 103–254.

Holl, F. 1829. Handbuch der Petrefactenkunde. Part 1. Hilscher, Dresden, Leipzig, Germany, 115 pp.

Hooley, R. W. 1925. On the skeleton of *Iguanodon atherfieldensis* sp. nov., from the Wealden Shales of Atherfield (Isle of Wight). Quarterly Journal of the Geological Society of London 81: 1–61.

Hopson, J. H. 1979. Paleoneurology; pp. 39–146 in C. Gans, R. G. Northcutt, and P. Ulinski (eds.), Biology of the Reptilia 9. Academic Press, New York.

Horner, J. R., A. de Ricqlès, K. Padian, and R. D. Scheetz. 2009. Comparative long bone histology and growth of the "hypsilophodontid" dinosaurs *Orodromeus makelai*, *Dryosaurus altus*, and *Tenontosaurus tillettii* (Ornithischia: Euornithopoda). Journal of Vertebrate Paleontology 29: 734–747.

Hoyo, J., A. Elliott, and J. Sargatal. 1994. Handbook of the birds of the world. Vol. 2, New World vulture to guineafowl. Lynx Edicions, Barcelona, Spain.

Hulke, J. W. 1873. Contributions to the anatomy of *Hypsilophodon foxii*. An account of recently acquired remains. Quarterly

Journal of the Geological Society 29: 522–532.

———. 1874. Supplemental note on the anatomy of *Hypsilophodon foxii*. Quarterly Journal of the Geological Society 30: 18–23.

———. 1882a. An attempt at a complete osteology of *Hypsilophodon foxii*, a British Wealden dinosaur. Philosophical Transactions of the Royal Society 172: 1035–1062.

———. 1882b. Description of some *Iguanodon*-remains indicating a new species, *I. seelyi*. Quarterly Journal of the Geological Society 38: 135–144.

Huxley, T. H. 1869. On *Hypsilophodon*, a new genus of Dinosauria. Abstracts of Proceedings of the Geological Society 204: 3–4.

———. 1870. On *Hypsilophodon foxii*, a new dinosaurian from the Wealden of the Isle of Wight. Quarterly Journal of the Geological Society 26: 3–12.

Insole, A. N., and S. Hutt. 1994. The palaeoecology of the dinosaurs of the Wessex Formation (Wealden Group, Early Cretaceous), Isle of Wight, southern England. Zoological Journal of the Linnean Society 112: 197–215.

Irmis, R. B. 2007. Axial skeletal ontogeny in the Parasuchia (Archosauria: Pseudosuchia) and its implications for ontogenetic determination in archosaurs. Journal of Vertebrate Paleontology 27: 350–361.

Janensch, W. 1925. Ein aufgestelltes Skelett des Stegosauriers *Kentrurosaurus aethiopicus* E. Hennig aus den Tendaguru-Schichten Deutsch-Ostafrikas. Palaeontographica, Supplement 7 (1, 1): 257–276.

———. 1955. Der Ornithopode *Dysalotosaurus* der Tendaguru-Schichten. Palaeontographica, Supplement 7 (1, 3): 105–176.

———. 1961. Skelettrekonstruktion von *Dysalotosaurus lettow-vorbecki*. Palaeontographica, Supplement 7 (3, 3): 237–240.

Jurcsák, T. 1982. Occurrences nouvelles des sauriens mésozoïques de Roumanie. Vertebrata Hungarica 21: 175–184.

Jurcsák, T., and E. Kessler. 1991. The Lower Cretaceous paleofauna from Cornet, Bihor County, Romania. Nymphaea 21: 5–32.

Jurcsák, T., and E. Popa. 1979. Dinozaurieni ornitopozi din bauxitele de la Cornet (Munții Pădurea, Craiului). Nymphaea 7: 37–75.

Kielan-Jaworowska, Z., R. L. Cifelli, and Z.-X. Luo. 2004. Mammals from the Age of Dinosaurs. Columbia University Press, New York, 630 pp.

Kirkland, J. I. 1998. A new hadrosaurid from the Upper Cedar Mountain Formation (Albian–Cenomanian) of Eastern Utah—the oldest known hadrosaurid (lambeosaurine?). New Mexico Museum of Natural History and Science Bulletin 14: 283–295.

Krauss, D., and I. Salame. 2010. Evidence for specialist feeding in *Hypsilophodon foxii*. Journal of Vertebrate Paleontology 30 (supplement to 3): 118A.

Lambe, L. 1919. Description of a new genus and species (*Panoplosaurus mirus*) of an armoured dinosaur from the Belly River Beds of Alberta. Transactions of the Royal Society of Canada 13: 39–50.

Larson, P. L. 2008. Variation and sexual dimorphism in *Tyrannosaurus rex*; pp. 103–128 in K. Carpenter and P. L. Larson (eds.), *Tyrannosaurus rex*: the tyrant king. Indiana University Press, Bloomington.

Lü, J, D. M. Unwin, D. C. Deeming, X. Jin, Y. Liu, and Q. Ji. 2011. An egg–adult association, gender, and reproduction in pterosaurs. Science 331: 321–324.

Lull, R. S. 1903. Skull of *Triceratops serratus*. Bulletin of the American Museum of Natural History 19: 685–695.

———. 1908. The cranial musculature and the origin of the frill in the ceratopsian dinosaurs. American Journal of Science (4) 25: 387–399.

Lydekker, R. 1888a. Catalogue of the fossil Reptilia and Amphibia in the British Museum. Part I. British Museum (Natural History), London, 309 pp.

———. 1888b. Note on a new Wealden iguanodont and other dinosaurs. Quarterly Journal of the Geological Society 44: 46–61.

———. 1888c. British Museum catalogue of fossil Reptilia, and papers on the enaliosaurians. Geological Magazine (n.s.) (3) 5: 451–453.

———. 1889a. On the remains and affinities of five genera of Mesozoic reptiles. Quarterly Journal of the Geological Society 45: 41–59.

———. 1889b. Notes on new and other dinosaurian remains. Geological Magazine (n.s.) (3) 6: 119–121.

———. 1890. Catalogue of the fossil Reptilia and Amphibia in the British Museum. Part IV. British Museum (Natural History), London, 295 pp.

Maidment, S. C. R., D. B. Norman, P. M. Barrett, and P. Upchurch. 2008. Systematics and phylogeny of Stegosauria (Dinosauria: Ornithischia). Journal of Systematic Palaeontology 6: 367–407.

Mallison, H. 2011. The real lectotype of *Kentrosaurus aethiopicus* Hennig 1915. Neues

Jahrbuch für Geologie und Pälaontologie 259: 197–206.

Mantell, G. A. 1825. Notice on the *Iguanodon*, a newly discovered fossil reptile, from the sandstone of Tilgate Forest, in Sussex. Philosophical Transactions of the Royal Society 115: 179–186.

———. 1849. Additional observations on the osteology of the *Iguanodon* and *Hylaeosaurus*. Philosophical Transactions of the Royal Society 139: 271–305.

Marsh, O. C. 1894. Restoration of *Camptosaurus*. American Journal of Science (3) 47: 245–246.

———. 1895. Restoration of some European dinosaurs, with suggestions as to their place among the Reptilia. American Journal of Science (3) 50: 407–412.

———. 1896a. Restoration of some European dinosaurs, with suggestions as to their place among the Reptilia. Geological Magazine (n.s.) (4) 3: 1–9.

———. 1896b. Dinosaurs of North America. United States Geological Survey, 16th Annual Report 1894–95: 133–244.

Martill, D. M., and D. Naish. 2001a. Dinosaurs of the Isle of Wight. Palaeontological Association Field Guides to Fossils 10: 1–433.

———. 2001b. The geology of the Isle of Wight. Palaeontological Association Field Guides to Fossils 10: 25–43.

Martin, F., and P. Bultynck. 1990. The iguanodons of Bernissart. l'Institut royal des Sciences naturelles de Belgique, Brussels, 51 pp.

Martin, V., and E. Buffetaut. 1992. Les Iguanodons (Ornithischia—Ornithopoda) du Crétacé inférieur de la région de Saint-Dizer (Haute-Marne). Revue de Paléobiologie 11: 67–96.

Mateus, O., and M. T. Antunes. 2001. *Draconyx loureiroi*, a new Camptosauridae (Dinosauria, Ornithopoda) from the Late Jurassic of Lourinhã, Portugal. Annals de Paléontologie 87: 61–73.

McDonald, A. T., P. M. Barrett, and S. D. Chapman. 2010. A new basal iguanodont (Dinosauria: Ornithischia) from the Wealden (Lower Cretaceous) of England. Zootaxa 2569: 1–43.

Molnar, R. E., and P. M. Galton. 1986. Hypsilophodontid dinosaurs from Lightning Ridge, New South Wales, Australia. Géobios 19: 231–239.

Morris, W. J. 1970. Hadrosaurian dinosaur bills—morphology and function. Los Angeles Museum of Natural Science, Contributions in Science 193: 1–14.

Naish, D., and D. M. Martill. 2001a. Ornithopod dinosaurs. Palaeontological Association Field Guides to Fossils 10: 60–132.

———. 2001b. Boneheads and horned dinosaurs. Palaeontological Association Field Guides to Fossils 10: 133–146.

Narendra, B. L. 1990. From the archives: the early years of the Great Hall, 1924–47. Discovery, New Haven 22: 19–23.

Nopcsa, F. 1905. Notes on British dinosaurs. Part I. *Hypsilophodon*. Geological Magazine (5) 2: 203–208.

Norman, D. B. 1980. On the ornithischian dinosaur *Iguanodon bernissartensis* from the Lower Cretaceous of Bernissart (Belgium). Mémoires de l'Institute Royal des Sciences Naturelles de Belgique 178: 1–103.

———. 1984. On the cranial morphology and evolution of ornithopod dinosaurs. Zoological Society of London, Symposium 52: 521–547.

———. 1986. On the anatomy of *Iguanodon atherfieldensis* (Ornithischia: Ornithopoda). Bulletin de l'Institut royal des Sciences naturelles de Belgique, Sciences de la Terre 56: 281–372.

———. 1987. Wealden dinosaur biostratigraphy; pp. 165–170 in P. Currie and E. Koster (eds.), Fourth symposium on Mesozoic terrestrial ecosystems. Tyrrell Museum of Palaeontology, Occasional Papers 3.

———. 1990. A review of *Vectisaurus valdensis*, with comments on the family Iguanodontidae; pp. 147–161 in K. Carpenter and P. J. Currie (eds.), Dinosaur systematics: approaches and perspectives. Cambridge University Press, Cambridge.

———. 1998. On Asian ornithopods (Dinosauria: Ornithischia). 3. A new species of iguanodontid dinosaur. Zoological Journal of the Linnean Society 122: 291–348.

———. 2002. On Asian ornithopods (Dinosauria: Ornithischia). 4. *Probactrosaurus* Rozhdestvensky, 1966. Zoological Journal of the Linnean Society 136: 113–144.

———. 2004. Basal Iguanodontia; pp. 413–437 in D. B. Weishampel, P. Dodson, and H. Osmólska (eds.), The Dinosauria. 2nd ed. University of California Press, Berkeley.

———. 2010. A taxonomy of iguanodontians (Dinosauria: Ornithopoda) from the lower Wealden Group (Cretaceous: Valanginian) of southern England. Zootaxa 2489: 47–66.

Norman, D. B., and P. M. Barrett. 2002. Ornithischian dinosaurs from the Lower Cretaceous (Berriasian) of England. Special Papers in Palaeontology 68: 161–189.

Norman, D. B., and D. B. Weishampel. 1985. Ornithopod feeding mechanisms: their bearing on the evolution of herbivory. American Naturalist 126: 151–164.

———. 1990. Iguanodontidae and related ornithopods; pp. 510–533 in D. B. Weishampel, P. Dodson, and H. Osmólska (eds.), The Dinosauria. University of California Press, Berkeley.

———. 1991. Feeding mechanisms in some small herbivorous dinosaurs: Processes and patterns; pp. 161–181 in J. M. V. Raynor and R. J. Wotton (eds.), Biomechanics and evolution. Cambridge University, Cambridge.

Norman, D. B., H.-D. Sues, L. M. Witmer, and R. A. Coria. 2004. Basal Ornithopoda; pp. 393–412 in D. B. Weishampel, P. Dodson, and H. Osmólska (eds.), The Dinosauria. 2nd ed. University of California Press, Berkeley.

Novas, F. E., A. V. Cambiaso, and A. Ambrosio. 2004. A new basal iguanodontian (Dinosauria, Ornithischia) from the Upper Cretaceous of Patagonia. Ameghiniana 41: 75–82.

Ostrom, J. H. 1969. Osteology of *Deinonychus antirrhopus*, an unusual theropod from the Lower Cretaceous of Montana. Bulletin of the Peabody Museum of Natural History, Yale University 30: 1–165.

———. 1970. Stratigraphy and paleontology of the Cloverly Formation (Lower Cretaceous) of the Big Horn Basin area, Wyoming and Montana. Bulletin of the Peabody Museum of Natural History, Yale University 35: 1–234.

Owen, R. 1842. Report on British fossil reptiles. Part II. British Association for the Advancement of Science, Annual Report for 1841, 9: 60–204.

———. 1855. Monograph on the fossil Reptilia of the Wealden and Purbeck formations. Part II. Dinosauria (*Iguanodon*). [Wealden.] Palaeontographical Society Monographs 7: 1–54.

———. 1861a. Monograph on the fossil Reptilia of the Wealden and Purbeck formations. Part V. Lacertilia. [Purbeck.] Palaeontographical Society Monographs 12: 31–39.

———. 1861b. The fossil Reptilia of the Liassic Formation. Part I. A monograph of a fossil dinosaur (*Scelidosaurus harrisoni*, Owen) of the Lower Lias. Palaeontographical Society Monographs 13: 1–14.

———. 1864. Monograph on the fossil Reptilia of the Wealden and Purbeck formations. Supplement 3, Dinosauria (*Iguanodon*). [Wealden.] Palaeontographical Society Monographs 16: 19–21.

———. 1874. Monograph on the fossil Reptilia of the Wealden and Purbeck formations. Supplement No. 5. Dinosauria

(*Iguanodon*) [Wealden and Purbeck.] Palaeontographical Society Monographs 27: 1–18.

Padian, K. 1978. The making of the *Brontosaurus*. Discovery, New Haven 13 (2): 32–35.

Papp, M. J., and L. Witmer, L. 1998. Cheeks, beaks, or freaks: a critical appraisal of buccal soft-tissue anatomy in ornithischian dinosaurs. Journal of Vertebrate Paleontology 18 (supplement to 3): 69A.

Paul, G. 1997. Dinosaur models: the good, the bad, and using them to estimate the mass of dinosaurs; pp.129–154 in Proceedings of Dinofest International, 1997. Dinofest International, Tempe, Ariz.

———. 2006. Turning the old into the new: a separate genus for the gracile iguanodont from the Wealden of England; pp. 69–77 in K. Carpenter (ed.), Horns and beaks: ceratopsian and ornithopod dinosaurs. Indiana University Press, Bloomington.

———. 2008. A revised taxonomy of the iguanodont dinosaur genera and species. Cretaceous Research 29: 192–216.

———. 2010. The Princeton field guide to dinosaurs. Princeton University Press, Princeton, N.J., 320 pp.

Peczkis, J. 1994. Implications of body-mass estimates for dinosaurs. Journal of Vertebrate Paleontology 14: 520–533.

Pereda-Suberbiola, J. 1993. *Hylaeosaurus*, *Polacanthus*, and the systematics and stratigraphy of armoured dinosaurs. Geological Magazine 130: 767–781.

———. 1994. *Polacanthus* (Ornithischia, Ankylosauria), a transatlantic armoured dinosaur from the Early Cretaceous of Europe and North America. Palaeontographica A 232: 133–159.

Posmoşanu, E. 2003a. Revision of the Early Cretaceous dinosaur (Ornithopoda) collection from the bauxite deposit lens 204—Cornet, Romania. Nymphaea 30: 25–38.

———. 2003b. Iguanodontian dinosaurs from the Lower Cretaceous bauxite site from Romania. Acta Paleontologica Romaniae 4: 431–439.

———. 2009. Early Cretaceous ornithopod dinosaurs from Romania [abstract 79]. Darwin-Bernissart Meeting, Brussels, Belgium, February 9–13, 2009.

Posmoşanu, E. T., and E. Popa. 1997. Notes on a camptosaurid dinosaur from the Lower Cretaceous bauxite, Cornet—Romania. Nymphaea 23–25: 35–44.

Raath, M. A. 1977. The anatomy of the Triassic theropod *Syntarsus rhodesiensis* (Saurischia: Podokesauridae) and a

consideration of its biology. Ph.D. thesis, Rhodes University, Grahamstown, Rhodesia.

———. 1990. Morphological variation in small theropods and its meaning in systematics: evidence from *Syntarsus rhodesiensis*; pp. 93–105 in K. Carpenter and P. J. Currie (eds.), Dinosaur systematics: approaches and perspectives. Cambridge University Press, Cambridge.

Radley, J. 2004. Demystifying the Wealden of the Weald (Lower Cretaceous, south-east England). Journal of the Open University Geological Society 25: 6–16.

———. 2006a. A Wealden guide I: the Weald sub-basin. Geology Today 22: 109–118.

———. 2006b. A Wealden guide II: the Wessex sub-basin. Geology Today 22: 187–193.

Rawson, P. F. 2006. Cretaceous: sea level peak as the North Atlantic opens; pp. 365–393 in P. J. Brenchley, and P. F. Rawson (eds.), The geology of England and Wales. Geological Society of London.

Rawson, P. F., D. Currey, F. C. Dilley, J. M. Hancock, W. J. Kennedy, J. W. Neale, C. J. Wood, and B. C. Worssam. 1978. A correlation of Cretaceous rocks in the British Isles. Geological Society of London, Special Report 9: 1–70.

Reed, R. E. H. 1984. The histology of dinosaurian bone, and its possible bearing on dinosaurian physiology. Symposium of the Zoological Society of London 52: 629–663.

Ruiz-Omeñaca, J. I. 2001. Dinosaurios hipsilofodóntidos (Ornithischia: Ornithopoda) en la Penénsula Iberica; pp. 175–266 in Colectivo Arqueológico y Paleontológico de Salas (ed.), Actas de las I Jornadas Internacionales sobre Paleontología de Dinosaurios y su Entorno. Colectivo Arqueologico y Paleontologico de Salas, Burgos, Spain.

———. 2006. Restos directos de dinosaurios (Saurischia, Ornithischia) en el Barremiense (Cretácico Inferior) de la Cordillera Ibérica en Aragñn (Teruel, Espana). Ph.D. thesis, Facultad de Ciencias, Universidad de Zaragoza, España, 584 pp.

Ruiz-Omeñaca, J. I., and J. I. Canudo. 2004. Dinosaurios (Saurischia, Ornithischia) en el Barremiense (Cretácico Inferior) de la Península Ibérica. Instituto de Estudios Riojanos, Ciencias de la Terre 26: 269–312.

Santa Luca, A. P. 1980. The postcranial skeleton of *Heterodontosaurus tucki* (Reptilia, Ornithischia) from the Stormberg of South Africa. Annals of the South African Museum 79: 159–211.

Santa Luca, A. P., A. W. Crompton, and A. J. Charig. 1976. A complete skeleton of the Late Triassic ornithischian *Heterodontosaurus tucki*. Nature 264: 324–328.

Schmidt, L., and R. Motani. 2011. Nocturnality in dinosaurs inferred from scleral ring and orbit morphology. *Science* 332: 705–708.

Schweitzer, M. H., J. L. Wittmeyer, and J. R. Horner. 2005. Gender-specific reproductive tissue in ratites and *Tyrannosaurus rex*. Science 308: 1456–1460.

Seebacher, F. 2001. A new method to calculate allometric length–mass relationships of dinosaurs. Journal of Vertebrate Paleontology 21: 51–60.

Seeley, H. G. 1869. Index to the fossil remains of Aves, Ornithosauria, and Reptilia from the secondary system of strata arranged in the Woodwardian Museum of the University of Cambridge. Deighton, Bell, & Co., Cambridge, 143 pp.

———. 1874. On the base of a large lacertilian cranium, presumably dinosaurian, from the Potton Sands. Quarterly Journal of the Geological Society 30: 690–692.

———. 1887. On the classification of the fossil animals commonly named Dinosauria. Proceedings of the Royal Society 43: 165–171.

Sereno, P. C. 1986. Phylogeny of the bird-hipped dinosaurs (Order Ornithischia). National Geographic Research 2: 234–256.

———. 1991. *Lesothosaurus*, "fabrosaurids," and the early evolution of Ornithischia. Journal of Vertebrate Paleontology 11: 168–197.

———. 1998. A rationale for phylogenetic definitions, with application to the higher-level taxonomy of Dinosauria. Neues Jahrbuch für Geologie und Paläontologie, Abhandlungen 210: 41–83.

Shepherd, J. D., P. M. Galton, and J. A. Jensen. 1977. Additional specimens of the hypsilophodontid dinosaur *Dryosaurus altus* from the Upper Jurassic of western North America. Brigham Young University Geological Studies 24: 11–15.

Stewart, D. J. 1978. The sedimentology and palaeoenvironment of the Wealden Group of the Isle of Wight, Southern England. Ph.D. thesis, Portsmouth Polytechnic, England, U.K.

Sues, H.-D., and D. Norman. 1990. Hypsilophodontidae, *Tenontosaurus*, Dryosauridae; pp. 498–509 in D. B. Weishampel, P. Dodson, and H. Osmólska (eds.), The Dinosauria. University of California Press, Berkeley.

Sullivan, R. M. 2006. A taxonomic review of the Pachycephalosauridae (Dinosauria:

Ornithischia). New Mexico Museum of Natural History Bulletin 35: 347–365.

Swinton, W. E. 1936a. The dinosaurs of the Isle of Wight. Proceedings of the Geological Society of London 47: 204–220.

———. 1936b. Notes on the osteology of *Hypsilophodon*, and on the family Hypsilophodontidae. Proceedings of the Zoological Society of London 1936: 555–578.

———. 1936c. A new exhibit of *Hypsilophodon*. Natural History Magazine 5: 331–336.

———. 1970. The dinosaurs. Wiley, New York, 331 pp.

Taquet, P. 1976. Géologie et Paléontologie du Gisement de Gadoufaoua (Aptien du Niger). Cahiers de Paléontologie. Editions du Centre National de la Recherche Scientifique, Paris, France, 191 pp.

Taquet, P., and D. A. Russell. 1999. A massively-constructed iguanodont from Gadoufaoua, Lower Cretaceous of Niger. Annales de Paléontologie 85: 85–96.

Thulborn, R. A. 1974. Thegosis in herbivorous dinosaurs. Nature 250: 729–731.

———. 1982. Speeds and gaits of dinosaurs. Palaeogeography, Palaeoclimatology, Palaeoecology 38: 227–256.

Vickaryous, M. K. 2006. New information on the cranial anatomy of *Edmontia rugosidens* Gilmore, a Late Cretaceous nodosaurid dinosaur from Dinosaur Provincial Park, Alberta. Journal of Vertebrate Paleontology 26: 1011–1013.

Vickaryous, M. K., T. Maryańska, and D. B. Weishampel. 2004. Ankylosauria; pp. 363–392 in D. B. Weishampel, P. Dodson, and H. Osmólska (eds.), The Dinosauria. 2nd ed. University of California Press, Berkeley.

Walls, G. L. 1942. The vertebrate eye and its adaptive radiation. Bulletin of the Cranbrook Institute of Science 19: 1–785.

Weishampel, D. B. 1984. Evolution of jaw mechanisms in ornithopod dinosaurs. Advances in Anatomy and Cell Biology 87: 1–110.

———. 1990. Dinosaurian distribution; pp. 63–139 in D. B. Weishampel, P. Dodson, and H. Osmólska (eds.), The Dinosauria. University of California Press, Berkeley.

Weishampel, D. B., and D. B. Norman. 1989. Vertebrate herbivory in the Mesozoic: Jaws, plants, and evolutionary metrics; pp. 87–100 in J. O. Farlow (ed.), Paleobiology of the dinosaurs. Geological Society of America Special Paper 238.

Weishampel, D. B., P. M. Barrett, R. A. Coria, J. Le Loeuff, X. Xu, X. Zhao, A. Sahni, E. M. P. Gomani, and C. R. Noto. 2004. Dinosaurian distribution; pp. 517–613 in D. B. Weishampel, P. Dodson, and H. Osmólska (eds.), The Dinosauria. 2nd ed. University of California Press, Berkeley.

Winkler, D. A. 2006. Ornithopod dinosaurs from the Early Cretaceous Trinity Group, Texas, USA; pp. 169–181 in J. C. Lü, Y. Kobayashi, D. Huang, and Y.-N. Lee (eds.), Papers from the 2005 Heyuan International Dinosaur Symposium. Geological Publishing House, Beijing.

Winkler, D. A., P. A. Murry, and L. J. Jacobs. 1997. A new species of *Tenontosaurus* (Dinosauria: Ornithopoda) from the Early Cretaceous of Texas. Journal of Vertebrate Paleontology 17: 330–348.

You, H., Z. Luo, N. H. Shubin, L. M. Witmer, Z. Tang, and F. Tang. 2003. The earliest-known duck-billed dinosaur from deposits of late Early Cretaceous age in northwest China and hadrosaur evolution. Cretaceous Research 24: 347–355.

Young, J. Z. 1962. The life of vertebrates. 2nd ed. Oxford University Press, Oxford, 820 pp.

Zheng, X.-T., H.-L. You, X. Xu, and Z.-M. Dong. 2009. An Early Cretaceous heterodontosaurid dinosaur with filamentous integumentary structures. Nature 458: 333–336.

18.1. Gadoufaoua (Aptian of Niger). Skeleton of *Ouranosaurus nigeriensis* prepared by eolian erosion.

The African Cousins of the European Iguanodontids

18

Philippe Taquet

In 1966, excavations in the Gadoufaoua locality (Niger) allowed us to describe two new genera of Early Cretaceous basal Iguanodontoidea from Africa: the first one, *Ouranosaurus nigeriensis* Taquet, 1976, was a gracile and facultative bipedal Iguanodontoidea, with bumped nasals on the skull and long neural spines on the dorsal vertebrae. The second one, *Lurdusaurus arenatus* Taquet and Russell, 1999, was a heavy, quadrupedal basal Styracosterna, with very short and robust limbs and a hippopotamus-like body. These two African basal iguanodontoids are closely related to the European iguanodontids, including *Iguanodon bernissartensis* and *Mantellisaurus atherfieldensis* from Bernissart locality; basal Iguanodontoidea probably migrated from Europe to Africa during the Early Cretaceous. Considering the Aptian dating of the Gadoufaoua locality and the morphologic evolution of the Nigerian iguanodontians, a Barremian age for the Bernissart iguanodons was suggested in 1975. This age was recently confirmed by palynological dating of the Wealden sediments of Bernissart.

Introduction

In the summer of 1822, Gideon Mantell, after the discovery of fossil teeth and bones of an unknown reptile in the sandstone of Tilgate forest (Sussex), sent a letter with drawings of the teeth to Georges Cuvier in Paris to get some information about this mysterious animal (Taquet, 1983). Cuvier wrote to Mantell and explained him that the teeth were unknown to him, but probably belonged to an herbivorous reptile. Mantell was then able to compare these teeth with those of the living lizard *Iguana* and published in 1825 a "Notice on the *Iguanodon*, a Newly Discovered Fossil Reptile from the Sandstone of Tilgate Forest, in Sussex." It was the first chapter of the saga of *Iguanodon*. Then, the discovery in 1834 of the Maidstone *Iguanodon* in Kent provided the early investigators of the anatomy and relationships of these reptiles with a partial associated skeleton in which it was possible to identify many of the key osteological characteristics of one of the three founder members of the Dinosauria by Richard Owen (Norman, 1995).

The second chapter in the history of *Iguanodon* began in 1878 with the spectacular discovery of complete skeletons in the Sainte-Barbe pit at Bernissart (Mons Basin, Belgium). In his "First Note on the Dinosaurs of Bernissart," Louis Dollo (1882) already distinguished the presence of two *Iguanodon* species at Bernissart. A robust species, named *Iguanodon bernissartensis* Boulenger *in* Van Beneden, 1881, is represented by at least 33 specimens. Two more gracile specimens from Bernissart were attributed to the same species as the Maidstone *Iguanodon*, named *Iguanodon mantelli* by Heinrich von Meyer in 1832.

But as I wrote in 1976, the systematic status of the different species of *Iguanodon* was confused: "The Wealdian swamps where Europeans Iguanodons were splashing were just water puddle compared to the taxonomic swamp in which they are paddle today" (Taquet, 1976, 159). Ostrom (1970) was the first to put some order in the classification of iguanodontids. From 1977 to 1986, David Norman redescribed the *Iguanodon* fauna of Bernissart (Norman, 1980, 1986), with a reappraisal of the anatomy and paleobiology of the two species of *Iguanodon*, and renamed the gracile form *Iguanodon atherfieldensis* Hooley, 1825.

More recently Paul (2007, 2008) discussed the systematics of "iguanodont" dinosaur genera and species, and named two new taxa for the gracile iguanodonts from the Wealden of western Europe: *Mantellisaurus atherfieldensis* (Hooley, 1825) for the specimens from the Isle of Wight, and *Dollodon bampingi* Paul, 2008, for the specimens from Bernissart. The validity of *Dollodon bampingi* is refuted by Norman (see Chapter 15 in this book).

Institutional abbreviation. MNHN, Musée national d'Histoire naturelle, Paris, France.

The African Cousins

The third chapter in the history of *Iguanodon* and related species was written in Africa. After the description of some fragments (a tooth and some postcranial elements) of an iguanodontid in South Africa by Haughton in 1915 (they were subsequently redescribed by Cooper in 1985; they belong to a small ornithopod related to *Dryosaurus* and *Valdosaurus*, but their exact position is still uncertain) and the notice of one tooth of an "*Iguanodon mantelli*" from the Lower Cretaceous of Remada in southern Tunisia (Lapparent, 1960, pl. V, fig. 23), new and important discoveries were made in Niger from 1965 to 1973.

In 1954, the geologist H. Faure discovered dinosaur remains in a region of the southern Sahara called Gadoufaoua (pronounced *gah-doo-fa-wa*), situated about 170 km east of the town of Agades (Niger). From 1963, geologists of the French Atomic Energy Agency actively prospected this area, and their work led to a delimitation of the fossil beds. These geologists then invited the Paleontology Institute of the Muséum National d'Histoire Naturelle in Paris to send a paleontologist at Gadoufaoua. Therefore, in 1965, I saw for the first time what became one the most important dinosaur localities in Africa, together with Tendaguru in Tanzania.

The Gadoufaoua locality is stratigraphically a member of the sandstone deposits of the *continental intercalaire*, a term defined by the geologist C. Kilian in 1931 and comprising all the continental formations in Sahara between the marine Late Carboniferous and the marine Late Cretaceous. In the Republic of Niger, this continental series is particularly rich in vertebrates remains: 26 vertebrate localities and three footprint or trackway localities have been discovered so far, distributed throughout 15 geologic horizons ranging in age from the Late Permian to the Late Cretaceous.

Gadoufaoua is the most famous of the dinosaur localities in the *continental intercalaire* of Niger. This site is made up of sandstones and silts, reaching a thickness of about 50 m. After tectonic movements during the Tertiary, the geologic beds in the area were slightly uplifted; this tectonic movement rejuvenated the topography, and erosion excavated part of the fossiliferous beds. Moreover, the present action of sandstorms, which blow for months each year, blasts away an additional part of the matrix and overburdens the vertebrate fossils: fishes, crocodiles, and dinosaurs (Fig. 18.1). The fossiliferous site is 93 miles long and about 1 mile wide. It was the bed of an immense swampy river area in the Aptian (Lower Cretaceous) period.

The enormous cemetery of Gadoufaoua is now famous thanks to its size, as well as the abundance and quality of its vertebrate fauna. I organized seven expeditions, in 1965, 1966, 1967, 1969, 1970, 1972, and 1973, and we collected abundant material in the field. Actinopterygians fishes are represented at Gadoufaoua by two semionotids: the first one, *Lepidotes*, is known by one nearly complete skeleton, close to 6 feet in length, and by numerous isolated scales. The other one is represented by numerous skeletons of a new form: *Pliodetes nigeriensis* (Wenz, 1999). Among the sarcopterygians, the coelacanth *Mawsonia* is present with a new species, *M. tegamensis* Wenz, 1975. Two ceratodontid dipnoans, *Ceratodus africanus* and *C. tiguidensis*, are known. Turtles are abundant, with complete carapaces of Pelomedusidae described by de Broin (1980, 1988): *Teneremys lapparenti*, *Platycheloides* cf. *nyasae*, *Taquetochelys decorata*, and *Araripemys* sp. Particularly impressive were giant, long-snouted crocodilians with skulls up to 1.7 m long and a total body length of about 11 m. Named *Sarcosuchus imperator* by de Broin and Taquet (1966), these pholidosaurid mesosuchians were among the largest crocodiles ever to exist and have been nicknamed "supercrocs" by American paleontologists. Huge theropod dinosaurs are represented by a fish-eating spinosaurid, named *Cristatusaurus lapparenti* by Taquet and Russell, 1998 (called also *Suchomimus tenerensis* by Sereno et al., 1998). Sereno and Brusatte (2008) recently described two new theropods, *Kryptos palaios* and *Eocarcharia dinops*, from this formation. A small dryosaurid, *Valdosaurus nigeriensis*, was described by Galton and Taquet (1982) and renamed *Elrhazasaurus nigeriensis* by Galton (2009). A strange rebbachisaurid sauropod with several functional tooth rows, *Nigersaurus taqueti*, was named by Sereno et al. (1999).

In January 1965, I was exploring for the first time the fossiliferous locality of Gadoufaoua. During this first expedition, it was possible to locate two finely preserved dinosaur skeletons in the same area. These skeletons were collected in 1966 during the first expedition (Fig. 18.2). The first one was very well preserved and nearly complete. It was described and named *Ouranosaurus nigeriensis* by Taquet (1976). It is a gracile iguanodontoid (sensu Norman, 2004) dinosaur. The second one is the rather complete skeleton of a robust iguanodontoid, with a hippopotamus-like body, named *Lurdusaurus arenatus* Taquet and Russell, 1999.

The diagnosis of both *Ouranosaurus nigeriensis* and *Lurdusaurus arenatus* was recently amended by Greg Paul (2008) in his paper devoted to a revised taxonomy of the iguanodont dinosaur genera and species (an asterisk precedes unambiguous autapomorphies, according to Paul, 2008):

Iguanodontia Dollo, 1888 [sensu Sereno, 1986]
Ankylopollexia Sereno, 1986
Styracosterna Sereno, 1986
Ouranosaurus nigeriensis Taquet, 1976
(Fig. 18.3)

18.2. Gadoufaoua (Aptian of Niger). Well-preserved skeleton of *Ouranosaurus nigeriensis* in the field.

Holotype. MNHN GDF300.

Locality and horizon. Gadoufaoua (Niger); Upper Elrhaz Formation (Late Aptian).

Amended diagnosis (Taquet, 1976, amended by Paul, 2008; *potential autapomorphies). Adults large at 8+ m and 2+ tonnes. Overall lightly constructed. Premaxillary tip to anterior orbital rim/latter to paroccipital process tip length ratio *~1.9; dentary precoronoid process length/minimum depth ration over 5. Rostrum and beak flattened, strongly subtriangular in lateral view. External nares* retracted posteriorly. Maxillary process of premaxilla moderately deep. Antorbital fossa and fenestra reduced. Dorsal apex of maxilla sited very posteriorly. Lacrimal long, contacts nasals. Accessory palpebral absent. Posterior border of occiput deeply indented. Lateral temporal fenestra small. Posterior portion of jugal long. Quadratojugal tall. Quadrate short, transversely broad, shaft curved, lateral foramen set high, dorsoposterior buttress absent. Diastema long. Tooth position *22 in maxilla and dentary. Dorsosacral/hindlimb length ratio ~1.2. Posterior dorsal centra not compressed anteroposteriorly. Neural spines of dorsal, sacrals and caudals *very tall, forming prominent sail. Six fused sacrals. Scapula blade narrow and constricted at middle of blade, base rather narrow, acromion process placed ventrally and directed distally. Forelimb ~55% of hind limb length. Deltopectoral crest of humerus distally placed, modest in size. Manus short and broad, phalanx 1 of digit I absent, pollex spike and other unguals small. Main body of ilium deep. Prepubic process of pubis *very deep. Femoral shaft straight. Metatarsal I absent, II long.

18.3. Gadoufaoua (Aptian of Niger). *Ouranosaurus nigeriensis* (type specimen, MNHN GDF300) in the field: dorsal vertebrae with long neural spines.

Discussion

The precise position of *Ouranosaurus* within iguanodontoids phylogeny is still controversial. Norman (1984, 1986) first regarded it as the sister taxon of *Iguanodon* in a monophyletic family Iguanodontidae, a hypothesis also retained by Norman and Weishampel (1990), Godefroit et al. (1998), and Kobayashi and Azuma (2003). Later Norman (1998) and You et al. (2003) considered *Ouranosaurus* as the sister taxon of *Altirhinus* within the iguanodontid clade. Conversely, Sereno (1986) considered that *Ouranosaurus* shares synapomorphies with Late Cretaceous Hadrosauridae (expanded narial fossa, widened anterior premaxillary bill margin, reflected rim on premaxilla, dorsoventrally expanded rostral tip of jugal, and presence of an antitrochanter on the ilium) and consequently considered it as the most basal member of his clade Hadrosauroidea (=all hadrosauriforms closer to *Parasaurolophus* than to *Iguanodon* [Sereno, 1998]). This opinion was subsequently followed by many authors, including Sereno (1997, 1998, 1999), Head (1998), Norman (2002, 2004), Dalla Vecchia (2009), and McDonald et al. (2010). Sues and Averianov (2009) estimated that *Ouranosaurus* occupied a more basal position within the phylogeny of Iguanodontoidea, as the sister taxon of *Iguanodon* + Hadrosauroidea (sensu Sereno, 1998). Wu and Godefroit (Chapter 19 in this book) and Godefroit et al. (Chapter 20 in this book) retain a monophyletic Iguanodontidae clade, including

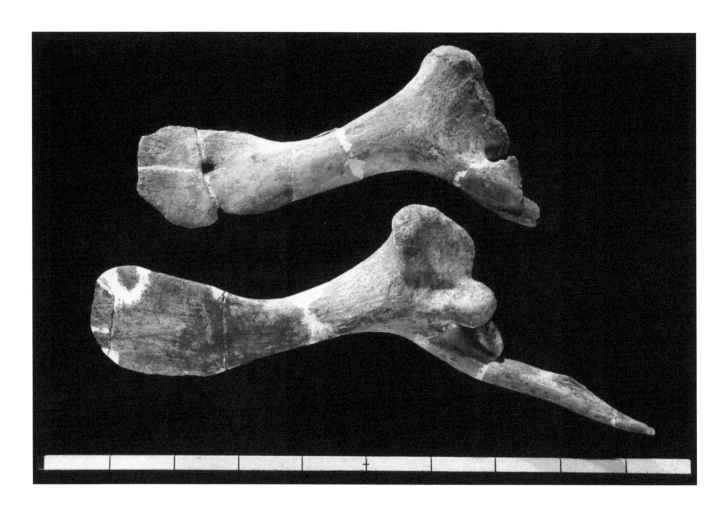

18.4. Gadoufaoua (Aptian of Niger). Pubes of *Lurdusaurus arenatus* (type specimen, MNHN GDF1700). The hole in the specimen (top) exhibits an injury due to an attack by a theropod claw.

Iguanodon, Mantellisaurus, and *Ouranosaurus,* and place *Ouranosaurus* and *Mantellisaurus* as sister taxa.

Lurdusaurus arenatus Taquet and Russell, 1999
(Figs. 18.4 and 18.5)

Holotype. MNHN GDF1700.
Locality and horizon. Gadoufaoua (Niger); Upper Elrhaz Formation (late Aptian).
Amended diagnosis (Taquet and Russell, 1999, amended by Paul, 2008; *potential autapomorphies). Adults large. Overall *extremely massively constructed. Occipital condyle *directed posteriorly. Quadrate *very short, shaft straight, transversely broad, dorsoposterior buttress large. Cervical series *elongated, *well-developed hypapophyses on posterior cervical centra. Dorsal neural spines short, ossified vertebral ligaments *probably absent. Rib cage *rotund. Forelimb ~60% length of hind limb. Coracoïd large. Olecranon process large. Manus short and broad. Pollex spike very large. Anterior ilia *diverge widely. Prepubic process of public shallow, postpubic process very short.

Discussion

Norman (2004) considered *Lurdusaurus* to be a basal member of the clade Styracosterna, defined as comprising *Lurdusaurus* and all more derived

18.5. Reconstruction of *Lurdusaurus arenatus* (copyright Michel Fontaine, 2009).

iguanodontians and characterized by the possession of hatchet-shaped sterna bones and flattened manus unguals, as well as by the structures of the pelvis. *Lurdusaurus* therefore occupies a more basal position in the phylogeny of Iguanodontia than *Iguanodon* and *Ouranosaurus* and is no longer considered as an Iguanodontidae (contra Taquet and Russell, 1999, and Paul, 2008). Because it is still incompletely described, *Lurdusaurus* is not included in the phylogeny of Iguanodontia presented in this book (Chapters 19 and 20).

The discovery of these two new genera led me to publish a paper on the differences between European and African iguanodonts: "It has allowed one to distinguish within the Iguanodontidae at least two groups of species: one comprising slender forms and the other more stocky forms. This confirms at that time the validity of the separation of the iguanodontids found at Bernissart (Belgium) into two separate species. It also reveals that the iguanodonts were more numerous and more varied than previously thought and that they attained their maximum of diversification at the end of the early Cretaceous. The replacement of the iguanodontids by the hadrosaurids must have taken place between the Aptian and the Cenomanian from the slender iguanodontids forms" (Taquet, 1975, 503). It was in this paper that a Barremian age was proposed for the Cretaceous beds of the Bernissart locality—an age that was recently confirmed by palynological data (Yans et al., 2005; Chapter 8 in this book).

Both in the Gadoufaoua and in Bernissart localities, the basal Iguanodontia fauna includes a robust and a gracile form (see also Chapter 15 in this book). The gracile *Mantellisaurus atherfieldensis* (including "*Dollodon bampingi*"), present in various Wealden localities from western Europe including Bernissart, and *Ouranosaurus nigeriensis* (from Gadoufaoua) are apparently closely related: Wu and Godefroit (Chapter 19 in this book) and Godefroit et al. (Chapter 20 this book) even regard them as sister taxa. Other phylogenies (e.g., Sereno, 1986, 1997, 1998, 1999; Head, 1998; Norman, 2002, 2004; Dalla Vecchia, 2009; McDonald et al., 2010) suggest that *Ouranosaurus nigeriensis* was the most basal representative of the clade Hadrosauroidea. The robust forms *Iguanodon bernissartensis* (western Europe, including Bernissart) and *Lurdusaurus arenatus* (Gadoufaoua)

Conclusions

probably do not form a monophyletic group, the latter being a basal, non-iguanodontoid, Styracosterna (Norman, 2004).

During the Early Cretaceous, descendants of the European basal Iguanodontia could cross the Tethys during some episodes of regression of the sea. Numerous dinosaur footprint localities discovered in Italy during the last 10 years (Dalla Vecchia, 1998, 2000) and recently in Algeria (Mahboubi et al., 2007; Regagba et al., 2007) on carbonate platform deposits indicate low levels of the marine waters and suggest that connections existed between Europe and Africa from the Hauterivian to the Albian.

Acknowledgment

I warmly thank Pascal Godefroit for his kind invitation to participate in the Brussels symposium in 2009 and for his useful remarks and suggestions.

References

Broin, F. de. 1980. Les tortues de Gadoufaoua (Aptien du Niger), aperçu sur la paléogéographie des Pelomedusidae (Pleurodira). Mémoires de la Société géologique de France (ser. 7) 13: 445–452.

———. 1988. Les tortues et le Gondwana. Examen des rapports entre le fractionnement du Gondwana et la dispersion géographique des tortues pleurodires à partir du Crétacé. Studia Palaeocheloniologica. Studi Geologica Salmanticena II 3: 103–142.

Broin, F. de, and P. Taquet. 1966. Découverte d'un Crocodilien nouveau dans le Crétacé inférieur du Sahara. Comptes rendus de l'Académie des Sciences (ser. 2) 262: 2326–2329.

Dalla Vecchia, F. M. 1998. Theropod footprints in the Cretaceous Adriatic–Dinaric carbonate platform (Italia and Croatia). Gaia 15: 355–367

———. 2000. Dinosauri della plattaforma carbonatica Adriatico–Dinarica: implicazioni paleoambientali; pp.186–187 in Riassunti delle Comunicazioni Orali e del Posters. 80 Riunione Estiva S.G.I. Trieste.

———. 2009. Tethyshadros insularis, a new hadrosauroid dinosaur (Ornithischia) from the Upper Cretaceous of Italy. Journal of Vertebrate Paleontology 29: 1100–1116.

Dollo, L. 1882. Première note sur les dinosauriens de Bernissart. Bulletin du Musée royal d'Histoire naturelle de Belgique 1: 55–80.

———. 1888. Iguanodontidae et Camptonotidae. Comptes Rendus hebdomadaires de l'Académie des Sciences, Paris 106: 775–777.

Cooper, M. R. 1985. A revision of the Ornithischian dinosaur Kangnasaurus coetzeei Haughton, with a classification of the Ornithischia. Annals of the South African Museum 95: 281–317.

Galton, P. M. 2009. Notes on Neocomian (Lower Cretaceous) ornithopod dinosaurs from England—Hypsilophodon, Valdosaurus, "Camptosaurus," "Iguanodon"—and referred specimens from Romania and elsewhere. Revue de Paléobiologie 28: 211–273.

Galton, P. M., and P. Taquet. 1982. A Hypsilophodontid dinosaur from the Lower Cretaceous of Europe and Africa. Geobios 15: 147–159.

Godefroit, P., Z.-M. Dong, P. Bultynck, H. Li, and L. Feng. 1998. New Bactrosaurus (Dinosauria: Hadrosauroidea) material from Iren Dabasu (Inner Mongolia, P.R. China). Bulletin de l'Institut royal des Sciences naturelles de Belgique, Sciences de la Terre 68 (supplement): 3–70.

Haughton, S. H. 1915. On some Dinosaur remains from Bushmanland. Transactions of the Royal Society of South Africa 5: 259–264.

Head, J. J. 1998. A new species of basal hadrosaurid (Dinosauria, Ornithischia) from the Cenomanian of Texas. Journal of Vertebrate Paleontology 28: 718–738.

Hooley, R. W. 1925. On the skeleton of Iguanodon atherfieldensis from the Wealden. Quarterly Journal of the Geological Society of London 81: 1–61.

Kilian, C. 1931. Des principaux complexes continentaux du Sahara. Comptes rendus sommaires de la Société Géologique de France 1931: 109–111.

Kobayashi, Y., and Y. Azuma. 2003. A new iguanodontian (Dinosauria: Ornithopoda) from the Lower Cretaceous Kitadani Formation of Fukui Prefecture, Japan. Journal of Vertebrate Paleontology 23: 166–175.

Lapparent, A. F. de. 1960. Les Dinosauriens du "Continental Intercalaire du Sahara central." Mémoires de la Société géologique de France (n.s.) 88A: 1–56.

Mahboubi, M., M. Bessedik, L. Belkebir, M. Adaci, H. Hebib, M. Bensalah, C. Mammeri, B. Mansour, and M. E. H. Mansouri. 2007. Premières découvertes d'empreintes de pas de dinosaures dans le Crétacé inférieur de la région d'El Bayadh (Algérie). Bulletin du Service géologique national 18: 127–139

Mantell, G. A. 1825. Notice on the *Iguanodon*, a newly discovered fossil reptile from the sandstone of Tilgate Forest, in Sussex. Philosophical Transactions of the Royal Society of London 115: 179–186.

McDonald, A. T., D. G. Wolfe, and J. I. Kirkland. 2010. A new basal hadrosauroid (Dinosauria: Ornithopoda) from the Turonian of New Mexico. Journal of Vertebrate Paleontology 30: 799–812.

Meyer, H. von. 1832. Palaeologica zur Geschichte der Erde. S. Scherrber, Frankfurt am Main, 560 pp.

Norman, D. B. 1980. On the Ornithischian dinosaur *Iguanodon bernissartensis* from Belgium. Mémoires de l'Institut Royal des Sciences Naturelles de Belgique 178: 1–105.

———. 1984. On the cranial morphology and evolution of ornithopod dinosaurs. Symposia of the Zoological Society of London 52: 521–547.

———. 1986. On the anatomy of *Iguanodon atherfieldensis* (Ornithischia : Ornithopoda). Bulletin de l'Institut royal des Sciences naturelles de Belgique, Sciences de la Terre 56: 281–372.

———. 1995. Gideon Mantell's "Mantel Piece": the earliest well-preserved ornithischian dinosaur; pp. 223–243 in W. A. S. Sargeant (ed.), Vertebrate fossils and the evolution of scientific concepts: a tribute to L. Beverly Halstead. Gordon and Breach, London.

———. 1998. On Asian ornithopods (Dinosauria: Ornithischia). 3. A new species of iguanodontid dinosaur. Zoological Journal of the Linnean Society 122: 291–348.

———. 2002. On Asian ornithopods (Dinosauria: Ornithischia). 4. *Probactrosaurus* Rozhdestvensky, 1966. Zoological Journal of the Linnean Society 136: 113–144.

———. 2004. Basal Iguanodontia; pp. 413–437 in D. B. Weishampel, P. Dodson, and H. Osmólska (eds.), The Dinosauria. 2nd ed. University of California Press, Berkeley.

Norman, D. B., and D. B. Weishampel. 1990. Iguanodontidae and related ornithopods; pp. 510–533 in D. B. Weishampel, P. Dodson, and H. Osmólska (eds.), The Dinosauria. University of California Press, Berkeley.

Ostrom, J. H. 1970. Stratigraphy and Paleontology of the Cloverly Formation (Lower Cretaceous) of the Bighorn Basin area, Wyoming and Montana. Bulletin of the Peabody Museum of Natural History 35: 1–234.

Paul, G. S. 2007. Turning the old into the new: a separate genus for the gracile iguanodont from the Wealden of England; pp. 69–77 in: K. Carpenter (ed.), Horns and beaks: ceratopsian and ornithopod dinosaur. Indiana University Press, Bloomington.

———. 2008. A revised taxonomy of the iguanodont dinosaur genera and species. Cretaceous Research 29: 192–216.

Regagba, A., L. Mekahli, M. Benhamou, N. Hammadi, and A. Zekri. 2007. Découverte d'empreintes de pas de dinosauriens (théropodes et sauropodes) dans les "Grès de Ksour" du Crétacé inférieur de la région d'El Bayadh (Atlas saharien central, Algérie). Bulletin du Service géologique national 18: 141–159.

Sereno, P. 1986. Phylogeny of the bird-hipped dinosaurs (order Ornithischia). National Geographic Research 2: 234–256.

———. 1997. The origin and evolution of dinosaurs. Annual Review of Earth and Planetary Sciences 25: 435–489.

———. 1998. A rationale for phylogenetic definitions, with application to the higher-level taxonomy of Dinosauria. Neues Jahrbuch für Geologie und Paläontologie, Abhandlungen 201: 41–83.

———. 1999. The evolution of dinosaurs. Science 284: 2137–2147.

Sereno, P. C., and S. L. Brusatte. 2008. Basal abelisaurid and carcharodontosaurid theropods from the Lower Cretaceous Elrhaz Formation of Niger. Acta Palaeontologica Polonica 53: 15–46.

Sereno P. C., A. L. Beck, D. B. Dutheil, B. Gado, H. C. E. Larsson, G. H. Lyon, J. D. Marcot, O. W. M. Rauhut, R. W. Sadleir, C. A. Sidor, D. D. Varrichio, G. P. Wilson, and J. A. Wilson. 1998. A long-snouted predatory dinosaur from Africa and the evolution of spinosaurids. Science 282: 1298–1302.

Sereno P. C., A. L. Beck, D. B. Dutheil, H. C. E. Larsson, G. H. Lyon, B. Moussa, R. W. Sadleir, C. A. Sidor, D. D. Varrichio, G. P. Wilson, and J. A. Wilson. 1999. Cretaceous sauropods from the Sahara and the uneven rate of skeletal evolution among dinosaurs. Science 286: 1342–1347.

Sues, H.-D., and A. Averianov. 2009. A new basal hadrosauroid dinosaur from the Late Cretaceous of Uzbekistan and the early radiation of duck-billed dinosaurs. Proceedings of the Royal Society B 276: 2549–2555.

Taquet, P. 1975. Remarques sur l'évolution des iguanodontidés et l'origine des hadrosauriens; pp. 503–511 in Colloque international CNRS 218, Problèmes actuels de Paléontologie—Evolution des vertébrés. CNRS, Paris.

———. 1976. Géologie et paléontologie du gisement de Gadoufaoua (Aptien du Niger). Cahiers de Paléontologie. CNRS, Paris.

Taquet P. 1983. Cuvier, Buckland, Mantell et les Dinosaures; pp. 475–494 in E. Buffetaut, J.-M. Mazin, and E. Salmon (eds.), Actes du Symposium paléontologique Georges Cuvier. Montbéliard, France.

Taquet, P., and D. E. Russell. 1998. New data on spinosaurid dinosaurs from the early Cretaceous of the Sahara. Comptes Rendus de l'Académie des Sciences (ser. 2) 327: 347–353.

———. 1999. A massively-constructed iguanodont from Gadoufaoua, Lower Cretaceous of Niger. Annales de paléontologie 85: 85–96.

Van Beneden, P. J. 1881. Sur l'arc pelvien chez les dinosauriens de Bernissart. Bulletin de l'Académie Royale des Sciences, des Lettres et des Beaux-Arts de Belgique (ser. 3) 1: 600–608.

Wenz, S. 1975. Un nouveau Coelacanthidé du Crétacé inférieur du Niger. Remarques sur la fusion des os dermiques; pp. 175–190 in Colloque international CNRS 218, Problèmes actuels de Paléontologie—evolution des vertébrés. CNRS, Paris.

———. 1999. *Pliodectes nigeriensis* gen. nov. et sp. nov., a new semionotid fish from the Lower Cretaceous of Gadoufaoua (Niger Republic): phylogenetic comments; pp. 107–120 in G. Arratia and H. P. Schultze (eds.), Mesozoic fishes 2—systematics and fossil record. Verlag Dr. Friedrich Pfeil, Munich.

Yans, J., J. Dejax, D. Pons, C. Dupuis, and P. Taquet. 2005. Implications paléontologiques et géodynamiques de la datation palynologique des sédiments à faciès wealdien de Bernissart (bassin de Mons, Belgique). Comptes rendus Palevol 4: 135–150.

You, H.-L., Z.-X. Luo, N. H. Shubin, L. M. Witmer, Z.-L. Tang, and F. Tang. 2003. The earliest-known duck-billed dinosaur from deposits of late Early Cretaceous age in northwest China and hadrosaur evolution. Cretaceous Research 24: 347–355.

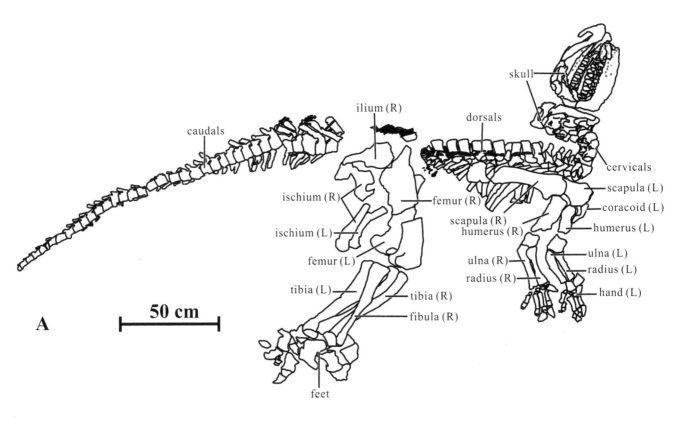

A

caudals

ilium (R)

dorsals

skull

cervicals

ischium (R)

femur (R)

scapula (L)

coracoid (L)

ischium (L)

scapula (R)

humerus (R)

humerus (L)

femur (L)

ulna (R)

ulna (L)

radius (R)

radius (L)

tibia (L)

tibia (R)

hand (L)

fibula (R)

feet

50 cm

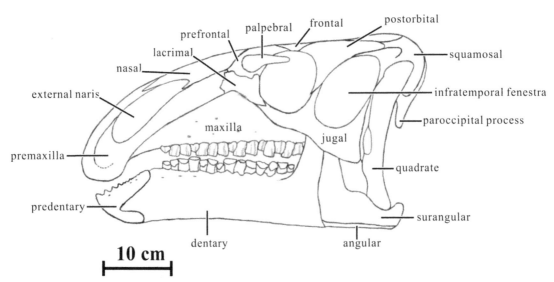

prefrontal

palpebral

frontal

postorbital

lacrimal

nasal

squamosal

external naris

infratemporal fenestra

maxilla

jugal

paroccipital process

premaxilla

quadrate

predentary

surangular

dentary

angular

B

10 cm

Anatomy and Relationships of *Bolong yixianensis,* an Early Cretaceous Iguanodontoid Dinosaur from Western Liaoning, China

19

Wu Wenhao* and Pascal Godefroit

The skeleton (YHZ-001) of a new basal iguanodontoid was discovered in the middle part of the Yixian Formation in western Liaoning, China. *Bolong yixianensis* Wu, Godefroit, and Hu, 2010, is characterized by cranial, dental, and postcranial autapomorphies, as well as a unique combination of characters. A phylogenetic analysis reveals that *Bolong* is the most primitive Hadrosauroidea described so far. During the Lower Cretaceous, Iguanodontoidea were subdivided into Iguanodontidae, which mainly occupied Neopangean territories, and Hadrosauroidea in Asia. The presence of Iguanodontoidea in the middle part of the Yixian Formation indicates that connections between Asia and western North America and/or Europe were already established during or before the Barremian.

19.1. A, *Bolong yixianensis,* YHZ-001 (holotype). Skull and partially articulated postcranial skeleton, as discovered, in right lateral view. *Abbreviations:* L, left; R, right. B, Skull reconstruction in left lateral view.

Introduction

Iguanodontoidea (=Hadrosauriformes sensu Sereno, 1997) is defined as *Iguanodon, Parasaurolophus,* their most recent common ancestor, and all descendants (Sereno, 1998, amended). During the Early Cretaceous, iguanodontoids had achieved a pan-Laurasian distribution and were also represented in Africa (Norman, 2004). During the Upper Cretaceous, advanced Iguanodontoidea, or Hadrosauridae (a node-based taxon defined as the most recent common ancestor of *Bactrosaurus* and *Parasaurolphus,* plus all the descendants of this common ancestor; see Norman, 2004, fig. 11.22), replaced basal iguanodontoids and became the most diverse and abundant large vertebrates of Laurasia during the Campanian and the Maastrichtian.

Many new basal iguanodontoids were described during the 2000s, most of which have been recovered from Early Cretaceous deposits in China and Mongolia, including *Altirhinus kurzanovi,* from Khuren Dukh (late Aptian–Early Albian) of Mongolia (Norman, 1998), *"Probactrosaurus" mazongshanensis* and *Equijubus normani,* from the Ximinbao Group (Aptian–Albian) of Gansu province (Lü, 1997; You et al., 2003b), *Jintasaurus meniscus,* from the Xinminpu Group (?Albian) of Gansu province (You and Li, 2009), *Nanyangosaurus zhugeii,* from the Sangping Formation (?Albian) of Henan province (Xu et al., 2000), and *Penelopognathus weishampeli,* from the Bayan Gobi Formation (Albian, Lower Cretaceous) of Inner Mongolia (Godefroit et al., 2005). *Lanzhousaurus magnidens,* represented by an incomplete skeleton from the Hekou Group (Early Cretaceous) of Gansu province, may represent a more basal iguanodontian (You et al., 2005).

The Early Cretaceous Jehol Biota of western Liaoning province in China is famous for its abundant, extraordinarily diversified, and exquisitely preserved fossils. The dinosaur fauna of the Jehol Biota is dominated by small-bodied taxa (<3 m in body length), including a variety of coelurosaurian theropods, basal ceratopsians, the basal ornithopods *Jeholosaurus*, and the ankylosaur *Liaoningosaurus* (Xu and Norell, 2006; Zhou, 2006). Larger herbivorous dinosaurs are rare in the Jehol fauna and include the titanosauriform *Dongbutitan dongi* (Wang et al., 2007) and the basal iguanodontoid *Jinzhousaurus yangi* (Wang and Xu, 2001; Barrett et al., 2009; Wang et al., 2011). *Shuangmiaosaurus gilmorei*, on the basis of an incomplete and deformed skull, was also collected in western Liaoning province, but in the younger Sunjiawan Formation (late Early or early Late Cretaceous; You et al., 2003a).

Here we describe the skeleton of a new basal iguanodontoid collected in 2000 in the middle part of the Yixian Formation to supplement the initial report by Wu et al. (2010). The fossil site is located at Bataigou, Toutai county. This incomplete skeleton was fossilized lying on its left flank, with the limbs roughly perpendicular to the vertebral column and parallel to each other, and with the skull and neck retracted over the back (Fig. 10.1). The estimated length for the complete skeleton (the tip of the tail is missing) is ~4 m (smaller than the holotype of *Jinzhousaurus*, which is 5–5.5 m). For the sake of convenience, all nonhadrosaurid Iguanodontoidea will be termed "basal iguanodontoids" hereafter.

Comparisons are made to other Iguanodontia on the basis of published descriptions of *Tenontosaurus* spp. (Ostrom, 1970; Forster, 1990; Winkler et al., 1997), *Dryosaurus* spp. (Janensch, 1955; Galton, 1983), *Zalmoxes* spp. (Weishampel et al., 2003; Godefroit et al., 2009), *Camptosaurus dispar* (Gilmore, 1909; Erickson, 1988), *Iguanodon bernissartensis* (Norman, 1980), *Mantellisaurus atherfieldensis* (Norman, 1986), *Ouranosaurus nigeriensis* (Taquet, 1976), *Lanzhousaurus magnidens* (You et al., 2005), *Lurdusaurus arenatus* (Taquet and Russell, 1999), *Nanyangosaurus zhugeii* (Xu et al. 2000), *Jinzhousaurus yangi* (Barrett et al., 2009), *Equijubus normani* (You et al., 2003b), *Altirhinus kurzanovi* (Norman, 1998), *Penelopognathus weishampeli* (Godefroit et al., 2005), *Fukuisaurus tetoriensis* (Kobayashi and Azuma, 2003), *Probactrosaurus gobiensis* (Norman, 2002), *"Probactrosaurus" mazongshanensis* (Lü, 1997), *Eolambia caroljonesa* (Kirkland, 1998), *Protohadros byrdi* (Head, 1998), *Bactrosaurus johnsoni* (Gilmore, 1933; Godefroit et al., 1998), *Levnesovia transoxiana* (Sues and Averianov, 2009), *Shuangmiaosaurus gilmorei* (You et al., 2003a), *Tethyshadros insularis* (Dalla Vecchia, 2009), *Telmatosaurus transsylvanicus* (Weishampel et al., 1993), and various Euhadrosauria. *Jintasaurus meniscus* You and Li, 2009, was published after the completion of this chapter and has not been included in the analysis.

Systematic Paleontology

Dinosauria Owen, 1842
Ornithischia Seeley, 1887
Ornithopoda Marsh, 1881
Iguanodontia Dollo, 1888 [Sereno, 1986]
Iguanodontoidea Cope, 1869
Hadrosauroidea Cope, 1869

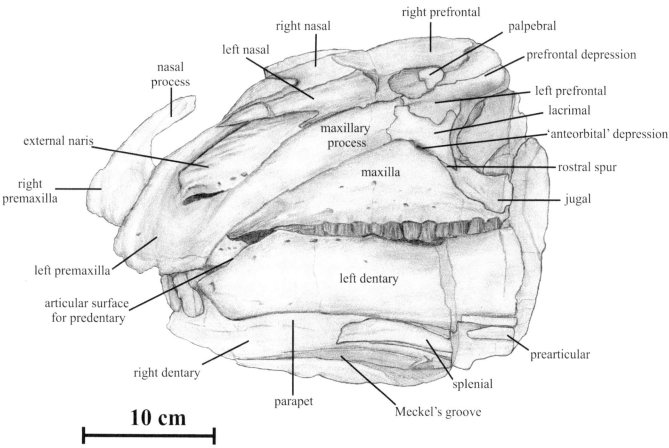

right nasal

right prefrontal

palpebral

prefrontal depression

left nasal

nasal
process

left prefrontal

lacrimal

maxillary
process

'anteorbital' depression

external naris

rostral spur

right
premaxilla

maxilla

jugal

left premaxilla

left dentary

articular surface
for predentary

right dentary

prearticular

splenial

parapet

Meckel's groove

10 cm

19.2. *Bolong yixianensis,* YHZ-001 (holotype).
Rostral portion of the skull in left lateral view.

Bolong yixianensis Wu, Godefroit, and Hu, 2010
(Figs. 19.1–19.11)

Etymology. The generic name is in honor of Bo Haichen and Bo Xue, who discovered and excavated the holotype; *long* means "dragon" in Chinese. The specific name refers both to the Yixian Formation, where the holotype was discovered, and to the city of Yixian, where the holotype is housed and displayed.

Holotype. YHZ-001, housed in Yizhou Fossil Museum, Yixian City, Liaoning province (P.R. China).

Locality and horizon. Bataigou, Toutai, Yixian County, western Liaoning province, P.R. China. GPS coordinates: N41°36'6.79", E121°7'43.1". Dakangpu Member (equivalent to the Dawangzhangzi Beds) of the middle part of Yixian Formation, Late Barremian–Early Aptian (Smith et al., 1995; Swisher et al., 1999, 2002).

Diagnosis. Differs from all other iguanodontoid taxa in possessing the following autapomorphies: depressed area at the junction between maxilla and lacrimal, corresponding to the position of the antorbital fenestra in *Iguanodon, Mantellisaurus,* and *Ouranosaurus;* caudal ramus of prefrontal forming a rostrocaudally depressed area above the orbital margin; ventral process of predentary extending caudally parallel to the ventral margin of the predentary main body; rostrodorsal articular surface for the predentary occupying less than two-thirds of the height of the rostral part of the dentary and rostral tip of dentary therefore situated above the ventral third of the bone; primary ridge deflected distally on maxillary crowns; ulna and radius proportionally short and robust (ratio craniocaudal height of proximal part/length = 0.28 for ulna and 0.33 for radius); proximal and distal ends of radius nearly symmetrically enlarged craniocaudally and triangular in medial and lateral views; postacetabular process of ilium dorsoventrally narrow (ratio length/maximal height = 2); metatarsals proportionally short: ratio metatarsal III/femur = 0.18.

Description:

All the measurements taken on YHZ-001 are compiled in Appendix 19.1.

The preorbital region of the skull is well preserved, and both sides were prepared. Like the rest of the skeleton, it was laterally crushed during fossilization. Bones are usually preserved in natural connection on the left side of the skull (Fig. 19.2), in direct contact with the sediment after the death of the animal, whereas they were slightly displaced on the right side of the skull (Fig. 19.3). A fault bisects the skull, and the fronto-orbital region of the skull has consequently been lost. Only the right side of the back of the skull has been prepared (Fig. 19.4), but this region is crushed and poorly preserved.

External naris. The external naris is drop shaped and proportionally larger (45% of preorbital length) than in *Jinzhousaurus* (38%), *Iguanodon, Mantellisaurus* (approximately 28–30%), and *Ouranosaurus* (approximately 18%). However, it is proportionally much shorter than in *Altirhinus* (approximately 56%) and Hadrosauridae. The maximum height of the external naris is located at the level of the rostral tip of the maxilla (Fig. 19.2).

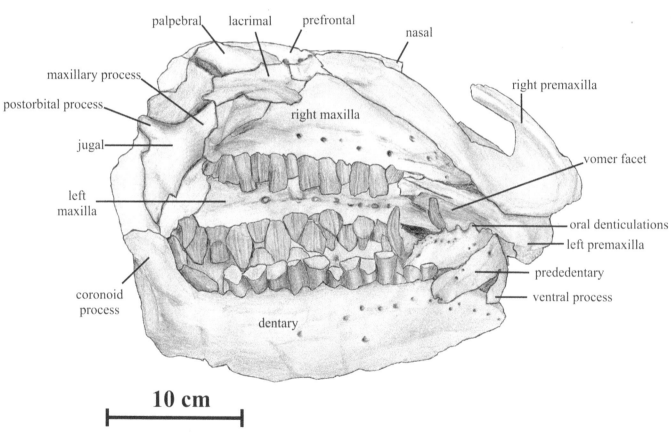

palpebral   lacrimal   prefrontal   nasal

maxillary process

postorbital process

jugal

right maxilla

right premaxilla

vomer facet

left
maxilla

oral denticulations

left premaxilla

prededentary

ventral process

coronoid
process

dentary

10 cm

19.3. *Bolong yixianensis,* YHZ-001 (holotype).
Rostral portion of the skull in right lateral view.

Nasal. Both nasals are preserved (Fig. 19.2). These elongated straplike elements are slightly arched in lateral view. The external surface of the nasal is gently convex, whereas its internal surface is concave. Rostrally, the nasal divides into two processes: an elongated rostromedial process that forms the longest part of the dorsal margin of the external naris and a small subtriangular rostrolateral process that forms the caudal fifth of the ventral margin of the external naris. Therefore, the caudal corner of the external naris is entirely circumscribed by the nasal in *Bolong*. The rostrolateral process is developed in *Jinzhousaurus* but absent in *Iguanodon*, *Mantellisaurus*, *Ouranosaurus*, *Equijibus*, and *Altirhinus* (contra Barrett et al., 2009). It is developed in some hadrosaurines, including *Gryposaurus* spp. and *Brachylophosaurus canadensis* (Gates and Sampson, 2007). Rostrally, the ventral margin of the nasal is grooved where it contacted the maxillary process of the premaxilla. There is apparently no contact between the nasal and the lacrimal, contrary to the situation observed in *Ouranosaurus* and *Altirhinus*. The articulation with the prefrontal is extensive on the caudolateral margin of the nasal, but the caudal part of the nasals is too crushed and deformed to be adequately described.

Premaxilla. The oral margin of the premaxilla extends well below the level of the maxillary tooth row (Fig. 19.2). Its external margin is slightly everted, as occurs in many other basal iguanodontoids (e.g., *Iguanodon*, *Mantellisaurus*, *Jinzhousaurus*, *Equijubus*, and *Probactrosaurus*). The rostral part of the oral margin is thickened and rugose, indicating that it was covered by a keratinous ramphotheca in life (Ostrom, 1961). Above this area, the external surface of the premaxilla is obliquely depressed and forms a narrow internarial septum that leads caudally into the narial chamber. Two small foramina are positioned close to the rostral margin of the internarial septum. The nasal process is gently curved and participates in the rostrodorsal margin of the external naris. It is mediolaterally compressed. Its external surface has an elongated articular surface covered by the rostromedial process of the nasal. Its medial surface is perfectly flat where it contacted the nasal process of the paired premaxilla. The lateral surface of the maxillary process is straight and planar. Rostrally, it is separated from the narial fossa by a blunt oblique ridge. Its ventromedial margin contacts the maxilla along its whole length. Its dorsal border floors the external naris. It reaches its maximum dorsoventral depth at the level of the caudal corner of the external naris and forms a thin sheet along the side of the snout between the nasal and the maxilla. This is similar to the condition seen in *Jinzhousaurus*, *Altirhinus*, *Eolambia*, *Ouranosaurus*, and *Probactrosaurus*, but contrasts with that seen in other basal iguanodontoids (e.g., *Iguanodon*, *Mantellisaurus*, and *Equijubus*), in which the maxillary process is much more slender and tapers distally (Barrett et al., 2009). The thin caudal part of the maxillary process covers the lacrimal and apparently contacts the prefrontal along a short distance. This contact, excluding the lacrimal from the nasal, appears more extensive in *Iguanodon*, *Mantellisaurus*, *Jinzhousaurus*, and, to a lesser extent, in *Equijubus*.

Maxilla. As usual in basal iguanodontoids, the maxilla is shaped like a low isosceles triangle in lateral view (Figs. 19.2–19.3). There are 15 tooth columns. The apex of the maxilla is caudal to the center of the bone, located at the level of the 10th tooth row. Rostrally, the maxilla is wedge shaped,

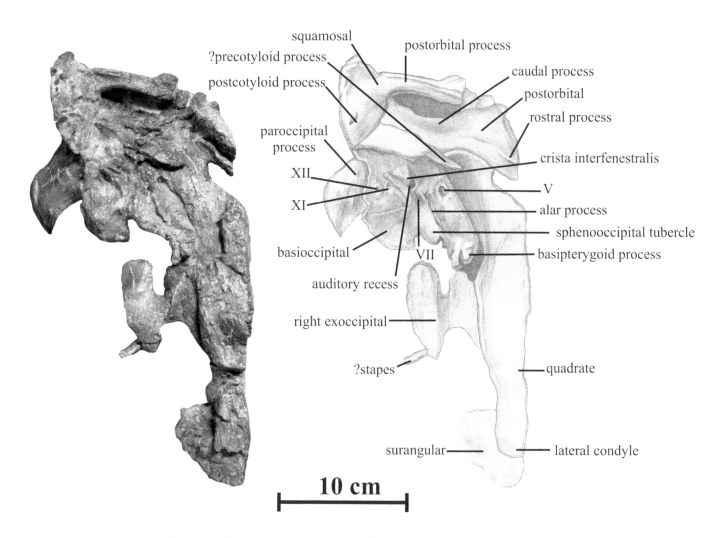

squamosal
?precotyloid process
postcotyloid process
paroccipital process
XII
XI
basioccipital
auditory recess
right exoccipital
?stapes
surangular

postorbital process
caudal process
postorbital
rostral process
crista interfenestralis
V
alar process
sphenooccipital tubercle
basipterygoid process
VII
quadrate
lateral condyle

10 cm

19.4. *Bolong yixianensis*, YHZ-001 (holotype). Caudal portion of the skull in right lateral view.

but its rostral tip is less distinctly downturned than in *Jinzhousaurus*, *Altirhinus*, *Equijubus*, *Shuangmiaosaurus*, and *Protohadros*. However, this character cannot easily be quantified phylogenetically. The dorsal surface of this rostral end forms an oblique dorsolaterally directed flange against which articulates the maxillary process of the premaxilla. The caudodorsal articulation with the lacrimal is much longer than in *Jinzhousaurus* and extends rostrally to the apex of the maxilla. There is no evidence for an antorbital opening, but a depressed area is present at the junction between the lacrimal and the maxilla, just caudal to the apex of the latter bone (Fig. 19.2). This depression, absent in *Jinzhousaurus*, *Equijubus*, and *Altirhinus*, corresponds to the emplacement of the antorbital opening in the basal iguanodontoids *Iguanodon*, *Mantellisaurus*, and *Ouranosaurus*. The lateral side of the maxilla is gently convex dorsoventrally, contrasting with the situation observed in *Jinzhousaurus* and *Equijubus*, in which a well-developed, acute, and rostrocaudal ridge defines the dorsal boundary of a deep buccal emargination (You et al., 2003b, fig. 1; Barrett et al., 2009). In *Bolong*, the slight buccal emargination is limited dorsally by six elliptical foramina. The dorsal part of the lateral surface is also pierced by six irregularly distributed foramina, mainly concentrated on the rostral half of the maxilla. The foramina are not as numerous and large as in *Jinzhousaurus* (Barrett et al., 2009) and particularly in *Equijubus* (You et al., 2003b). A hooklike jugal process, similarly developed in *Iguanodon*, *Mantellisaurus*, *Ouranosaurus*,

Jinzhousaurus, *Altirhinus*, *Probactrosaurus*, and *Bactrosaurus*, projects ventrolaterally and caudally. The jugal process is absent or poorly developed in *Equijubus* (You et al., 2003b) and *Fukuisaurus* (Kobayashi and Azuma, 2003). At its rostral end, the dorsomedial border of the maxilla forms a quite elongated and rugose sutural surface for the vomer (Fig. 19.3).

Lacrimal. This bone is rectangular and rostrocaudally elongate in *Bolong*, extending in front of the apex of the maxilla (Fig. 19.2). It is rather thick caudally, where it participates in the rostral margin of the orbit. It forms a robust caudoventral process that contacts the rostral process of the jugal. At this level, its lateral side is pierced by a small foramen at approximately midheight. It extends rostrally as a thin sheet of bone between the maxilla, the premaxilla, and the prefrontal. The right lacrimal was displaced during fossilization (Fig. 19.3). The caudal part of its ventral margin is strongly arched, corresponding to the dorsal margin of the antorbital fenestra in *Mantellisaurus* (Norman, 1986, fig. 12). More rostrally, the ventral margin has a bevelled and elongate contact for the maxilla. In *Jinzhousaurus*, the lacrimal is triangular in shape and more reduced than in *Bolong*.

Prefrontal. In lateral view, the prefrontal is crescentic in outline. Its dorsal margin has a straight contact with the nasal along its whole length. The caudal ramus of the prefrontal, which forms the rostrodorsal portion of the orbital margin, is particularly robust. Around the orbital margin, it forms a thickened ridge. Above this ridge, the lateral side of the caudal ramus forms a rostrocaudally elongate depression (Fig. 19.2). The rostral plate of the prefrontal is higher and much thinner mediolaterally. Ventrally, it overlaps the lacrimal. Its rostrodorsal corner apparently contacts the premaxilla. A roughened surface on the rostrolateral corner of the prefrontal forms the articular surface for the palpebral.

Palpebral (supraorbital). The palpebral of *Bolong* is robust. The rostral part of the left palpebral contacts the rostral ramus of the prefrontal (Fig. 19.2). It is slightly expanded and gently convex dorsoventrally.

Postorbital. The right postorbital is poorly preserved (Fig. 19.4). It is crushed inside the infratemporal fenestra and its rostral part is missing. The jugal process is robust, inclined rostrally and it tapers ventrally. The caudal process is thickened and straight along its whole length.

Squamosal. The right squamosal is poorly preserved (Fig. 19.4). The postorbital process is perfectly straight, particularly elongate, and dorsoventrally narrow. Its lateral side bears two prominent and parallel horizontal ridges that limit ventrally and dorsally an extended articular surface for the caudal process of the postorbital. The postcotyloid process appears triangular in lateral view and is relatively long, as in *Mantellisaurus* (Norman, 1986, fig. 14). The precotyloid process is tentatively identified as a thin straplike element displaced in the infratemporal fenestra.

Jugal. Both jugals are incompletely preserved. The maxillary process is proportionally shorter than in *Ouranosaurus* (Taquet, 1976, fig. 19) and *Equijibus* (You et al., 2003b). Rostrally, a small triangular projection from the maxillary process overlies the junction between the maxilla and the lacrimal (Fig. 19.2). This rostral spur is less developed than in *Altirhinus* (Norman, 1998, fig. 9). Above this projection, the dorsal side of the maxillary process forms an extended laterally facing articular facet for the lacrimal. The robust postorbital process is incompletely preserved on the right side of the skull (Fig. 19.3); it forms a 110-degree angle with the maxillary process. At the junction between the maxillary and the postorbital processes, the main body of the jugal appears dorsoventrally higher than in *Jinzhousaurus*. Below the infratemporal fenestra, the ventral margin of the jugal is gently concave.

Quadrate. The right quadrate is completely crushed and its rostral part is missing. The quadrate is proportionally high and rostrocaudally narrow in lateral view (Fig. 19.4). Its caudal margin is regularly concave, so that it looks deflected caudally, as observed in numerous iguanodontoids including *Jinzhousaurus* (Wang and Xu, 2001). The distal end of the quadrate forms a large hemispherical lateral condyle that articulates with the surangular component of the mandibular glenoid.

Exoccipital. The left exoccipital is partially preserved in anatomical position. The right one is ventrally displaced between the quadrate and the cervical series (Fig. 19.4). The medial side of the exoccipital condyloid is visible, but crushed. Caudal to the strong crista tuberalis that extends the entire height of the condyloid, two large foramina probably represent the passages for the hypoglossal nerve rostrally (XII) and the accessory nerve (XI) caudally. Rostral to the oblique pillar, the wall of the braincase forms a wide and depressed auditory recess. An oblique crista interfenestralis divides the auditory recess into a rostral fenestra ovalis (stapedial recess) and a caudal metotic foramen. The paroccipital process is massive and projects ventrolaterally, reaching about the level of the base of the occipital condyle.

Basioccipital. The occipital condyle is massive and its articular surface is not perfectly vertical as in hadrosauroids, but oriented caudoventrally, as observed in the basal iguanodontoids *Iguanodon*, *Mantellisaurus*, and *Ouranosaurus*. A prominent pyriform facet on the dorsolateral side of the basioccipital marks the contact with the right exoccipital. The basal tubera are separated from the occipital condyle by a distinct neck.

Basisphenoid. The caudolateral aspect of the basisphenoid is visible (Fig. 19.4). In front of the basal tubera, the basipterygoid processes project well below the occipital condyle as in Hadrosauridae (Godefroit et al., 1998). In *Iguanodon*, *Mantellisaurus*, and *Ouranosaurus*, on the other hand, the distal end of the basipterygoid processes is nearly on the same horizontal plane as the ventral border of the occipital condyle. On the lateral wall of the basisphenoid, the alar process is found directly above the base of the basipterygoid process, from which it is distinctly separated.

Predentary. The predentary is laterally crushed (Fig. 19.3). Its oral border bears four pairs of conical projections, better developed than in *Jinzhousaurus*, which are replaced further laterally by a rounded ridge. The lateral edge borders a moderately developed lingual shelf, broader caudally than rostrally. Beneath the denticulate border, the bone is punctured by foramina and grooves associated with the attachment of a keratinous beak. The ventral surface of the predentary forms a shelf that fitted on the rostral edge of the dentary. A well-developed ventral process helped to secure the dentaries in position. The ventral process is clearly bifurcate, contrary to that of *Jinzhousaurus* (Barrett et al., 2009), and is perfectly parallel to the ventral margin of the predentary, embracing the depressed area along the rostroventral border of the dentary. The ventral process of the predentary is usually more vertical, ventrocaudally inclined, in other basal iguanodontoids, as observed in *Iguanodon, Mantellisaurus, Ouranosaurus, Jinzhousaurus,* and *Altirhinus.* The rostral surface of the predentary has two deep diagonal vascular grooves that connect the dorsolateral corner of the ventral process to the base of the median crenulation. These grooves are also well developed in *Iguanodon, Mantellisaurus, Ouranosaurus, Jinzhousaurus, Altirhinus, Probactrosaurus,* and *Protohadros.*

Dentary. Both dentaries are preserved in the holotype, but their caudal part is damaged (Figs. 19.2–19.3). The dentary of *Bolong* is more robustly built than the elongate dentary of *Penelopognathus* (Godefroit et al., 2005), but more slender than the exceptionally deep dentary of *Fukuisaurus* (Kobayashi and Azuma, 2003). In lateral view, its ventral and dorsal borders are subparallel, and as in *Iguanodon, Mantellisaurus, Fukuisaurus,* and *Equijibus,* the rostral end of the dentary is not significantly downturned, differing from the strongly deflected dentary symphysis described in *Altirhinus* and *Protohadros.* The morphology of the rostral end of the dentary appears unique in *Bolong.* Usually in iguanodontoids, the articular surface for the predentary extends the entire height of the rostral margin of the dentary; in lateral view, the rostral end of the dentary is scoop shaped and the rostral tip of the dentary is situated at or close to the ventral margin. Conversely, in *Bolong,* the rostrodorsal articular surface for the predentary occupies less than two-thirds of the height of the dentary in lateral view, and the rostral tip is therefore situated above the ventral third of the dentary (Fig. 19.2). This peculiar morphology cannot be explained by diagenetic crushing or distortion of the rostral end of the left dentary; moreover, the left dentary is exposed in lateral view, without rotation that could influence the angle of view. The ventral and rostroventral borders of the dentary form a 140-degree angle, and below the articular surface for the predentary, the rostroventral and rostrodorsal borders form an 80-degree angle. The lateral side of the rostroventral border is depressed, marking contact with the wide and caudally directed ventral process of the predentary. The main body of the dentary is gently convex dorsoventrally along most of its length. The buccal emargination separating the tooth row from the lateral aspect of the bone increases caudally. Caudodorsally, the coronoid process is inclined slightly caudally to the long axis of the dentary. In *Jinzhousaurus,* it forms a 90-degree angle with the long axis of the bone (Barrett et al., 2009). Its apex is not rostrocaudally expanded as in more advanced hadrosauroids. The coronoid process is laterally offset with respect to the tooth row, and its apex is located slightly caudally to the last dentary tooth. As in *Iguanodon, Mantellisaurus, Ouranosaurus,* and *Jinzhousaurus,* there is no extended buccal shelf separating the coronoid process from the tooth row, thereby differing from the wide buccal shelf present in more derived hadrosauroids *Altirhinus, Probactrosaurus,* and *Protohadros.* The caudal part of the dentary is excavated by the large adductor fossa, which extends rostrally as a deep Meckel's groove (Fig. 19.2). Fourteen tooth positions are present. The diastema between the last tooth position and the articular surface for the predentary is not larger than two crown widths. A thin parapet, limited ventrally by a series of interconnected nutritional foramina, conceals the lingual aspect of the tooth rows (Fig. 19.2). Because the dental battery is less developed than in hadrosaurids, the parapet is also much lower.

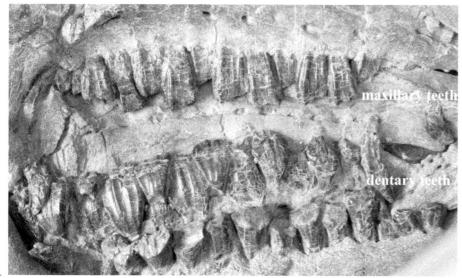

maxillary teeth

dentary teeth

A

2 cm

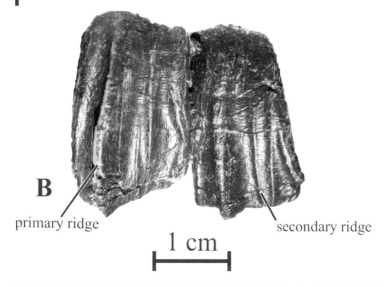

B

primary ridge

secondary ridge

1 cm

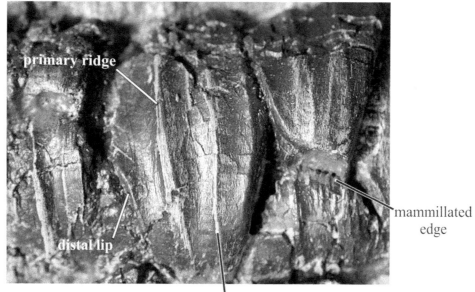

primary ridge

distal lip

secondary ridge

mammillated edge

C

Surangular. Only the retroarticular process of the right surangular is preserved (Fig. 19.4). As is usual in iguanodontoids, it is lobate and upturned above the glenoid. It is mediolaterally compressed, with a thin ventral edge.

Splenial. The splenial is tentatively identified as a thin elongate plate of bone that lies against the caudomedial aspect of the dentary (Fig. 19.2). It is displaced dorsally from its natural position, across the Meckel's groove.

Prearticular. A thin rodlike element that lies dorsal to the splenial and against the caudomedial surface of the dentary is tentatively identified as the rostral portion of the prearticular (Fig. 19.2).

Teeth. The teeth of *Bolong* resemble those of *Iguanodon* and *Mantellisaurus.* There is only one functional tooth for each tooth position. Both the maxillary and the dentary teeth have long, tapering roots with shallow grooves on their mesial and distal edges to accommodate the closely packed crowns of the adjacent successional teeth. The mesial teeth are the smallest in both the maxillary and dentary tooth rows (Fig. 19.5A).

The maxillary teeth are proportionally narrower than the dentary teeth (Fig. 19.5B). The maxillary crown is lozenge shaped. An extensive dorsolaterally inclined wear facet truncates its lingual side. The labial side of the crown is thickly enameled, and the margins of the unworn partially erupted teeth are strongly denticulate. A prominent primary ridge extends along the enameled labial side of the crown. It is placed more distally than in *Iguanodon, Mantellisaurus,* and *Jinzhousaurus.* The primary ridge is deflected distally on all the maxillary teeth. Up to four secondary ridges are positioned mesial to the primary ridge; all extend to the apex of the crown.

The labial side of the dentary crowns is always strongly truncated because of heavy abrasion. The enameled lingual side of the crowns is asymmetrical and leaf shaped. Although the proportions of the crown are variable within the dental battery, the height/width ratio of the crown is less than 2. A primary ridge, less prominent than on the maxillary teeth, extends the entire height of the crown, dividing the crown surface into two unequal halves (Fig. 19.5C). A less prominent secondary ridge, parallel to the primary ridge, bisects the larger mesial half of the crown and always reaches the upper part of the mesial margin. One or two less clearly defined ridges extend up the lingual side of the crown mesial to the primary ridge, but they tend to merge with the surface of the crown before reaching the apex. In addition, there may be a variable number of more or less parallel subsidiary ridges, which are extensions of the bases of the marginal denticles. On the upper part of the edge of the crown, the denticulations form simple tongue-shaped structures. The structure of the marginal denticulations becomes more elaborated further down the sides of the crown. The edge becomes thickened, and each of the denticulations forms a curved and crenulated ledge, with additional mammillations, which wrap around the edge of the crown. Marginal denticulations are not developed on the edges along the lower part of the crown. At this level, the mesial edge is only slightly thickened. The distal edge forms an everted, oblique lip, as if the edge of the crown had been pinched inward. However, this distal lip is less prominent than in *Altirhinus, Penelopognathus, Batyrosaurus,* and *Probactrosaurus,* and it does not bear mammillations.

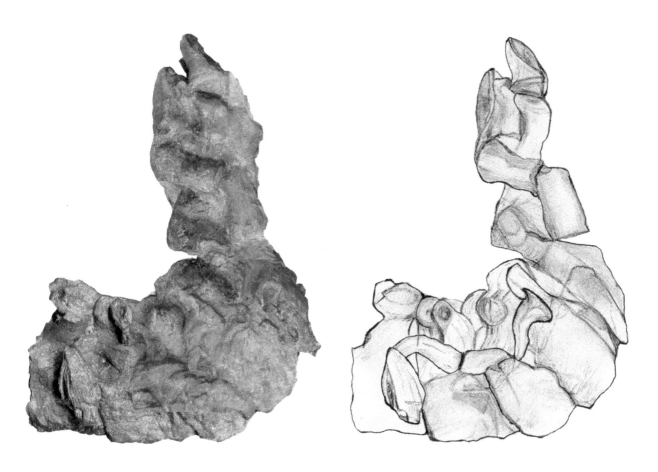

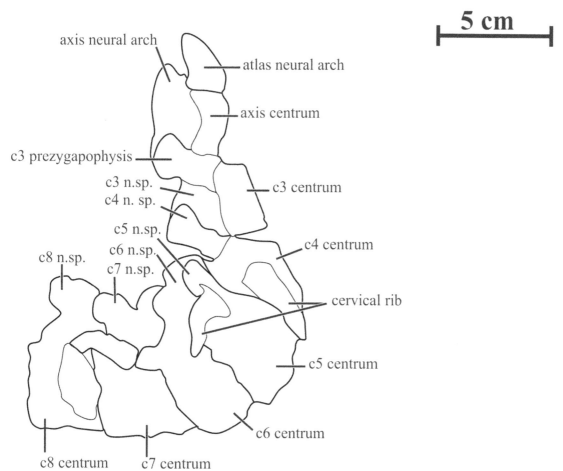

axis neural arch

atlas neural arch

axis centrum

c3 prezygapophysis

c3 n.sp.

c4 n. sp.

c3 centrum

c5 n.sp.

c6 n.sp.

c8 n.sp.

c7 n.sp.

c4 centrum

cervical rib

c5 centrum

c6 centrum

c8 centrum

c7 centrum

5 cm

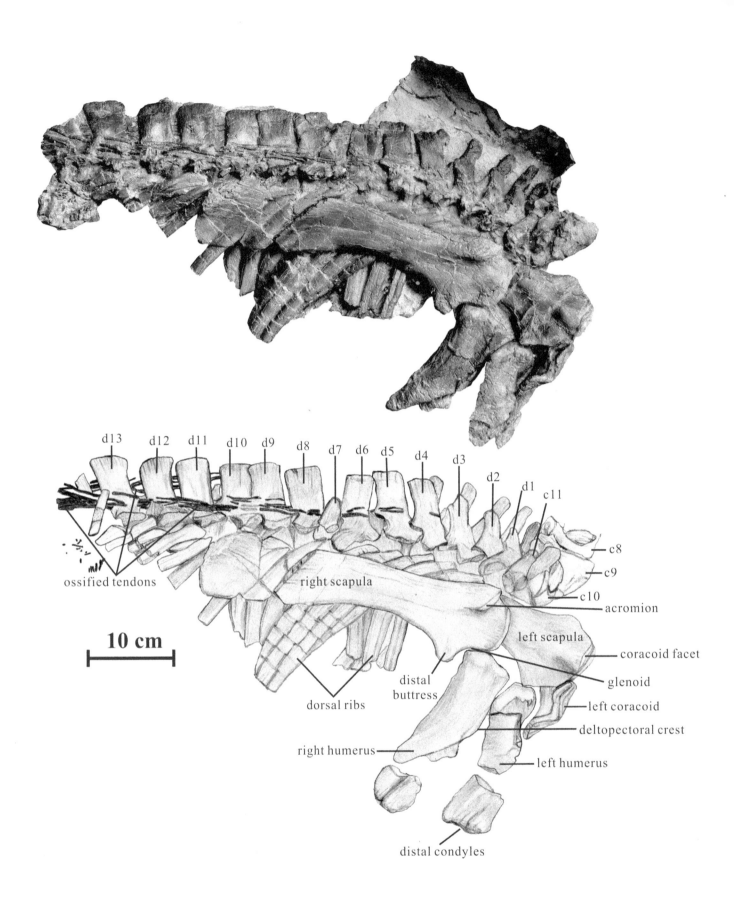

Most of the vertebral column is preserved in connection. However, the caudal part of the dorsal series and the sacrum are poorly preserved and cannot be adequately described. The distal part of the tail is missing.

Cervical vertebrae. There are 11 cervical vertebrae as in *Iguanodon*, *Mantellisaurus*, *Ouranosaurus*, and probably *Equijubus* (Figs. 19.6–19.7). The atlas neural arch is robust (Fig. 19.6). It expands cranially into robust prezygapophyseal processes. Caudally, the postzygapophyses are much smaller.

The axial centrum is poorly preserved, and its limits with the odontoid process and the axial intercentrum cannot be discerned. It appears elongated. The axial neural arch forms a large, craniocaudally expanded, neural spine. The cranioventral margin of the neural arch has an elongated prezygapophyseal process. Caudally, it forms a transverse process that projects caudolaterally and terminates in a diapophysis. Caudally, the spine bifurcates into two divergent buttresses that support the postzygapophyses.

The centra of the postaxial cervical vertebrae are opisthocoelous, with a hemispherical cranial articular surface and a deeply concave caudal surface. The centrum of the third cervical is craniocaudally elongate and longer than wide. But the centra of the succeeding cervicals appear proportionally shorter. Because of postmortem deformation and intimate articulation between adjacent opisthocoelous centra, the vertebrae cannot be adequately measured. Ventrally, the centra are transversely compressed and form a thick and rugose keel. A rounded foramen usually pierces the centrum above the keel. The parapophysis is located on the lateral side of the centrum, close to the cranial border. The neural arch is robust. Two paired processes are developed from the outer surface of the arch. The more cranially placed transverse process supports large prezygapophyses on its rostrodorsal side and also provides, at its caudal ends, the diapophyses for articulation with the cervical rib heads. The size of the transverse process regularly increases passing through the cervical series. The postzygapophyseal processes are particularly long and stout. They diverge caudally and laterally to cover the transverse process of the succeeding adjacent vertebra. The size of the postzygapophyseal processes also increases progressively along the cervical series. The neural spines of the cervical vertebrae are not preserved.

Cervical ribs. Fragments of cervical ribs are poorly preserved on the lateral aspect of the neck. All appear double-headed. A thin bony rod between the atlantal neural arch and the occipital condyle is tentatively interpreted an atlantal rib.

Dorsal vertebrae. The first 13 dorsal vertebrae are visible (Fig. 19.7). However, all the centra are completely hidden by the scapula or the dorsal ribs, and the transverse processes are eroded or destroyed. Consequently, only the neural spines can be adequately described. The neural spine of the first dorsal is slender, inclined caudally, and hook shaped. The size and the robustness of the spine increase regularly until the fifth dorsal. From this level, the spine is typically rectangular and slightly inclined caudally. The apex of the spine is slightly enlarged transversely and rugose, suggesting the presence of a cartilaginous cap in life. As in *Iguanodon*, the neural spines of the dorsals remain relatively low, contrasting with the situation observed in *Mantellisaurus*, where they are more than 2.5 times higher than

19.7. *Bolong yixianensis,* YHZ-001 (holotype). Dorsal part of the axial skeleton, pectoral girdle, and humeri in right lateral view. *Abbreviations:* c, cervical vertebra; d, dorsal vertebra;

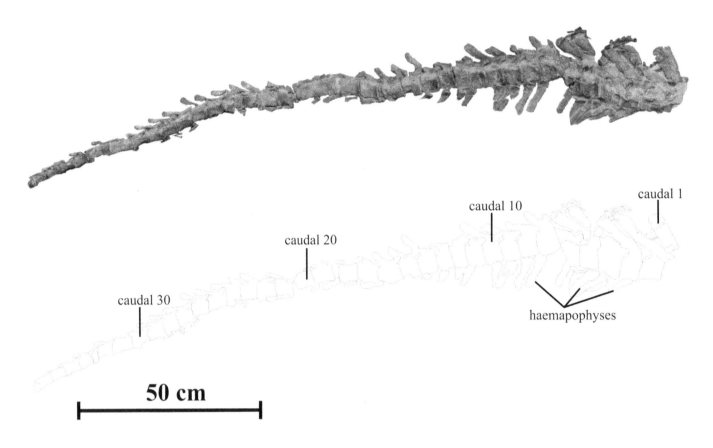

caudal 1

caudal 10

caudal 20

caudal 30

haemapophyses

50 cm

19.8. *Bolong yixianensis,* YHZ-001 (holotype). Tail in right lateral view.

the vertebral centrum and especially in *Ouranosaurus,* where the spine is as much as nine times the height of the centrum.

Dorsal ribs. Approximately 15 dorsal ribs are preserved in articulation (Fig. 19.7). However, their proximal head is hidden by the scapula and their distal portion is broken off. Therefore, nothing notable can be observed on them, except that they are curved and quite robust.

Caudal vertebrae. Thirty-seven caudal vertebrae are preserved in connection (Fig. 19.8). The centra of the proximal 15 caudals are proportionally short, higher than long, and subquadrangular in lateral view (length/height ratio = 0.82 at the level of the seventh caudal centrum). Distally, the centra become proportionally longer than high (length /height ratio = 1.86 at the level of the 36th preserved caudal centrum). The lateral sides of the centra are slightly depressed. Their ventral side is concave and forms four (two proximal and two distal) large hemapophyseal facets; the distal facets are usually better developed than the proximal ones. The dorsolateral sides of the proximal 14 centra have an elliptical and rugose articular facet for the caudal ribs. The fact that the ribs are not fused to the centrum suggests that YHZ-001 was not completely mature when it died. On the neural arch, the prezygapophyses are inclined medially, whereas the postzygapophyses are similarly inclined laterally. The neural spines of the proximal caudal vertebrae are about 1.5 as high as the centra. Although the neural spines are straight and roughly rectangular on the proximal half of the preserved portion of the tail, they become progressively curved and lobate on the distal half.

Hemapophyses. About 27 hemapophyses are preserved either articulated or not to the corresponding facets of the centra (Fig. 19.8). The first hemal arch articulates between the second and third preserved caudals.

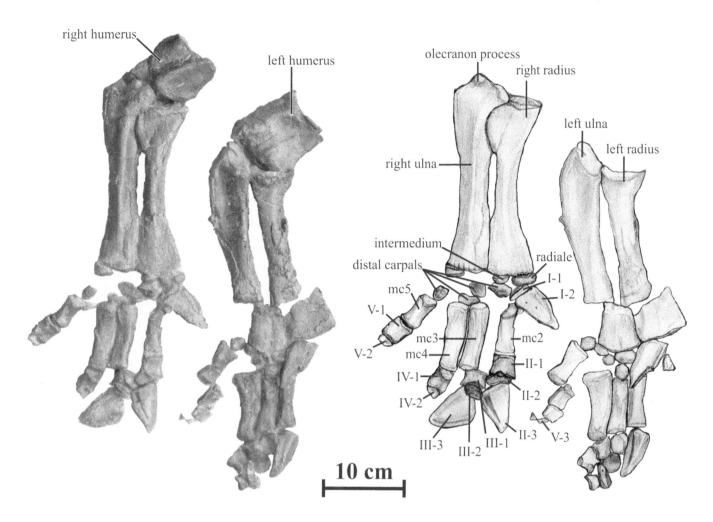

10 cm

19.9. *Bolong yixianensis,* YHZ-001 (holotype). Forearms and mani in right lateral view. *Abbreviations:* mc, metacarpal; I–V, digit numbers; 1–3, phalanx numbers.

It articulates between caudal vertebrae 2–3 in *Iguanodon, Mantellisaurus,* and lambeosaurines (Norman, 1980, 1986), between vertebrae 3–4 in *Ouranosaurus* (Taquet, 1976), distal to caudal vertebra 4 or 5 in hadrosaurines (Horner et al., 2004), and between vertebrae 7–8 in *Tethyshadros* (Dalla Vecchia, 2009). On the proximal part of the tail, the height of the chevron is roughly equivalent to the height of the corresponding neural spine; distally, it is always much lower and more slender. Their caudoventral orientation was also apparently equivalent to the caudodorsal angle of the corresponding neural spine.

Ossified epiaxial tendons. In *Bolong,* the ossified epiaxial tendons extend from dorsal 4 to about the level of the sixth preserved caudal (Fig. 19.7). Similarly, ossified tendons are to be found from dorsal 3 up to caudal 6 in the hadrosaurid *Brachylophosaurus* (Prieto-Marquez, 2007). The ossified tendons extend further distally on the tail in the basal hadrosaurid *Tethyshadros* (from dorsal 4 to caudal 20; Dalla Vecchia, 2009), in the hadrosaurine *Gryposaurus* (from dorsal 4 to caudal 19; Parks, 1920), and in the lambeosaurine *Corythosaurus* (Brown, 1916). Ossified tendons have never been reported along the cervical series in hadrosaurids (Brown, 1916). In the basal iguanodontoids *Iguanodon* and *Mantellisaurus* ossified tendons are present from cervical 10 up to about caudal 20. The arrangement of the ossified tendons appears quite primitive in *Bolong:* they are longitudinally arranged along the epaxial region of the vertebrae, between the transverse process and the base of the neural spine. On the dorsal vertebrae, they do not extend above the proximal third of the spine. This is the plesiomorphic condition also encountered in *Hypsilophodon*, *Gasparinisaura*, *Parksosaurus*, *Thescelosaurus*, *Dryosaurus*, and *Tenontosaurus*. In *Camptosaurus*, *Iguanodon*, and *Mantellisaurus*, the epaxial tendons are arranged in a complex double-layered rhomboidal structure, and the tendons extend to the distal portion of the neural spine on the dorsal vertebrae (Norman, 1980, 1986). This is also the case in hadrosaurids, in which the epaxial tendons even form a three-layered lattice (Adams and Organ, 2005).

Scapula. The scapula is described as horizontal, as presented in Figure 19.7. The right scapula is nearly completely preserved. The proximal plate of the left one is visible in medial view. The scapula of *Bolong* is typical for basal iguanodontoids. As in, for example, *Iguanodon*, *Mantellisaurus*, *Ouranosaurus*, *Altirhinus*, and *Probactrosaurus*, the proximal plate is expanded dorsoventrally (distinctly higher than distal scapula) to support the glenoid and to provide a sutural surface for the coracoid, the acromial process is directed dorsally, and the articular facet for the coracoid is extensive. In typical hadrosaurids, the proximal plate is dorsoventrally narrow (no higher than distal scapula), the acromial process projects horizontally, and the coracoid articulation is restricted. The coracoid facet is straight and occupies much of the proximal end of the scapula, with the glenoid forming a deep embayment on the caudal corner of this surface. The glenoid is supported distally by a prominent quadrangular buttress that faces ventrally. The acromial process is thickened laterally and gradually merges with the blade. Between the coracoid suture and the acromial process, the dorsal corner of the proximal plate forms a deep notch. Distal to the proximal plate, the scapular blade has its minimal height, but it remains comparatively thick transversely. The scapular blade is slightly curved both medially (to fit against the rib cage) and ventrally. It appears proportionally better developed than in *Iguanodon*, but less curved ventrally than in *Mantellisaurus* (although it cannot be excluded that this impression results from slight postmortem deformation). It progressively becomes mediolaterally thinner distally. The dorsal and ventral borders of the scapular blade remain subparallel along most of their length; only the distal third of the blade is slightly broadened dorsoventrally.

Coracoid. The left coracoid is deformed diagenetically (Fig. 19.7). It roughly resembles that of the basal iguanodontoids *Iguanodon*, *Mantellisaurus*, *Ouranosaurus*, and *Probactrosaurus*: it is large (coracoid/scapula lengths >0.2), and the articular surface for the scapula is longer than the glenoid. In hadrosaurids, the coracoid is proportionally shorter (coracoid/scapula lengths <0.2), and the glenoid is equal to or longer than the articulation (Evans and Reisz, 2007). A deep foramen pierces the coracoid in the angle between the glenoid and the scapular suture. This foramen is not closed, resembling the condition in *Iguanodon* and contrasting with the fully enclosed coracoid foramen in *Mantellisaurus* and *Bactrosaurus*. The ventral border of the coracoid is strongly concave, but the hook is not preserved in the holotype.

Humerus. Both humeri are crushed and diagenetically deformed, and their middle parts are missing (Fig. 19.7). In general proportions, the humerus of *Bolong* resembles that of *Probactrosaurus* in being rather slender. The proximal end is craniocaudally compressed and mediolaterally expanded. The humeral head is globular. The inner proximal tuberosity is poorly developed, whereas the outer tuberosity is more salient. The medial side of the humerus is regularly concave. The deltopectoral crest extends for approximately half the total length of the humerus. Distally, the crest progressively merges with the shaft and does not form a prominent angle on the humeral shaft as it does in hadrosaurids. Distally, the humerus slightly expands mediolaterally as it approaches the distal articular surfaces for the radius and the ulna. The radial condyle appears mediolaterally wider than the ulnar condyle, but this may be due to postmortem deformation. The intercondylar groove is well developed on the cranial side of the humerus.

Ulna. The ulna is notable for being proportionally short but particularly robust in *Bolong* (Fig. 19.9), with ratio of craniocaudal height of proximal part/length = 0.28. The olecranon process of the ulna is prominent, blunt, and rounded. The humerus articulated against its convex cranial surface. The medial and lateral proximal processes are also particularly high and robust, and the articular facet for the proximal part of the radius is correspondingly deep. The caudal border of the proximal ulna forms a prominent keel. The ulna progressively tapers but is craniocaudally expanded distally.

Radius. With ratio of craniocaudal height of proximal part/length = 0.28, the radius is also robust. The proximal and distal ends of the ulna are nearly symmetrically expanded craniocaudally and are triangular in mediolateral view (Fig. 19.9). Between the proximal and distal expansions, the shaft of the radius is circular in cross section. Because of the craniocaudal expansions of the extremities of these bones, a gap is present between the radius and the ulna, probably allowing for some ability for pronation/supination of the antebrachium.

Carpus. Six carpal elements are preserved in the right manus (Fig. 19.9). However, it is difficult to identify them precisely because they have been displaced. Some elements can also be observed in the left wrist, but they are too poorly preserved to be adequately described. The carpals remain separate elements in *Bolong*, and contrary to *Iguanodon*, *Mantellisaurus*, and *Ouranosaurus*, they do not co-ossify into two blocks. The ulnare is not preserved. The radiale is a mediolaterally elongated element that closely fits the cranial part of the distal surface of the radius. Its convex distal surface articulates with the flattened first phalanx of digit I. It is possible that metacarpal I is completely fused to the radiale, as observed in *Camptosaurus* (Erickson, 1988), although this cannot be conclusively demonstrated. The intermedium is tentatively interpreted as a small, flattened element that fits against the caudal portion of the articular surface of the radius. Distal carpals 2–5 form flattened and convexoconcave bones close to the proximal end of the corresponding metacarpals.

The hand of *Bolong* shows functional similarities with that of *Iguanodon*, *Mantellisaurus*, and *Ouranosaurus*. The first digit has an enlarged, probably divergent, spinelike ungual, the middle three digits form a rather compact unit that presumably allows for some degree of hyperextension, and the fifth digit was apparently long, flexible, and opposable (Norman, 1980, 1986).

Metacarpals. The metacarpals are proportionally much shorter and more robust than in *Iguanodon*, *Mantellisaurus*, and *Ouranosaurus* (Fig. 19.9).

Metacarpal II is moderately well preserved on both hands. It is more slender than metacarpals III and IV and somewhat transversely compressed. Its medial side is regularly convex, whereas its lateral side is slightly concave where it was closely bound to metacarpal III. The distal articular surface is slightly expanded mediolaterally and forms a shallow intercondylar groove.

Metacarpal III is longer and more robust than metacarpal II. Its shaft is quadrangular, with a flattened dorsal side. In craniocaudal view, its medial side is slightly sigmoidal where it closely articulates with metacarpal II. The proximal articular surface is not expanded and weakly convex. The distal articular surface is slightly expanded mediolaterally, with a shallow intercondylar groove.

Metacarpal IV is as long as metacarpal III but appears more robust because of the development, at midheight on its dorsal side, of a medial lip that partially covered metacarpal III, reinforcing the attachment between these metacarpals. The proximal articular surface is well expanded and mediolaterally convex. The distal articular surface is more reduced than the proximal one and convex; there is no trace of an intercondylar groove.

Metacarpal V is the shortest of the series. Its proximal articular surface is mediolaterally expanded and shallowly concave. The distal articular surface is only slightly expanded, broadly convex, and devoid of an intercondylar groove. Between the articular surfaces, the shaft of metacarpal V is strongly constricted.

Manus digits. If the first three digits of the right manus of YHZ-001 are complete, distal phalanges may be missing on digits IV and V (Fig. 19.9). The fifth digit of the left hand is apparently complete. From what can be observed in the holotype, the phalangeal formula of the manus of *Bolong* is 2–3–3–(2+)–4. It is 2–3–3–2–4 in *Iguanodon* and 2–3–3–3–3 in *Mantellisaurus*. In general, the proximal and intermediate phalanges in the manus of *Bolong* are proportionally much shorter than in *Iguanodon, Mantellisaurus, Ouranosaurus, Altirhinus,* and *Probactrosaurus.*

The first phalanx of digit I is a thin, mediolaterally elongated oval disc between the radiale and the large ungual phalanx. The ungual phalanx is quite characteristic for basal iguanodontoids: it is straight, triangular, and mediolaterally compressed, as in *Altirhinus.* Ungual grooves are well developed on either side of the dorsal surface and are supported by a prominent shelf. In *Iguanodon, Mantellisaurus,* and *Ouranosaurus,* the pollex is more conical and spikelike.

Digit II is formed by three phalanges. The first phalanx is short and trapezoidal in dorsal view, contrasting with the elongated and slender first phalanx in the second digit of *Mantellisaurus* and *Altirhinus.* Both the proximal and distal articular surfaces are dorsoventrally concave and extend onto the dorsal surface of the phalanx, suggesting that the second digit could be hyperextended. The second phalanx is a short, wider than long, blocklike element. It is roughly triangular in dorsal view. Its proximal articular surface is oblique and extends onto the ventral surface of the phalanx. Its distal articular surface is dorsoventrally convex. The ungual is nearly as large as that of the first digit, triangular in dorsal view, and dorsoventrally flattened. The ungual grooves are deep on either side of the dorsal surface, limiting a triangular central surface.

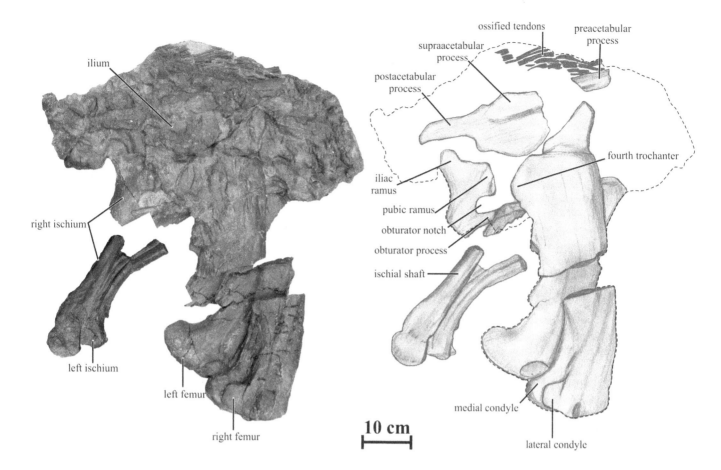

The third digit comprises three phalanges. The first and second phalanges have the same morphology as the corresponding phalanges of digit II but are proportionally shorter. The orientation of the articular surfaces shows that this digit could also be hyperextended. The ungual appears slightly more massive than in digit II and terminates in a bluntly rounded point. However, it is proximodistally longer than mediolaterally wide, contrasting with the condition seen in *Iguanodon* and *Mantellisaurus*, in which the ungual of digit III is hooflike and as wide as long.

Only the proximal phalanx of digit IV is preserved in both mani. A small fragment of the second phalanx can be observed on the right hand. The first phalanx is trapezoidal in dorsal view. Its dorsal side is regularly convex mediolaterally. It is asymmetrical in dorsal view. The proximal articular surface is shallowly concave and extends onto the dorsal side of the phalanx. The distal articular surface is dorsoventrally convex and also invades the dorsal side of the phalanx; it forms a well-developed intercondylar groove. This morphology suggests that digit IV could also be hyperextended. The proximoventral surface forms well-developed condyles marked by rugosities, which probably represent the scars of some ligaments that helped to bind the adjacent phalanges.

Digit V has four phalanges, as in *Iguanodon*. The first phalanx is robust and quadrangular in dorsal view. Its dorsal surface is shallowly concave. Its proximal articular surface is only slightly expanded mediolaterally and shallowly concave. Its distal articular surface is slightly convex dorsoventrally; it forms a shallow intercondylar groove. The second phalanx is much eroded and cannot be adequately described. It is about half the length of

19.10. *Bolong yixianensis*, YHZ-001 (holotype). Sacral part of the axial skeleton, pelvic girdle, and femora in right lateral view. The dashed lines indicate the limits of the block.

the first phalanx and trapezoidal in dorsal view. The third phalanx is poorly preserved and appears subtriangular in dorsal view. The fourth phalanx is a tiny triangular element that forms a shallowly convex proximal articular surface and bears a strong median ridge on its dorsal side.

Ilium. The right ilium is crushed (Fig. 19.10). The postacetabular process is remarkably narrow dorsoventrally (ratio length/maximal height [taken at the level of its cranial end] = 2). This ratio is >2 in *Tethyshadros* and in some hadrosaurines (e.g., *Gryposaurus*, *Edmontosaurus*, and *Anatotitan*) as a result of elongation of the postacetabular process of these taxa (Dalla Vecchia, 2009, fig. 5B). The dorsal edge of the postabular process is straight; however, the postacetabular notch is much deeper than in *Iguanodon*, *Mantellisaurus*, *Ouranosaurus*, *Altirhinus*, *Bactrosaurus*, and *Tethyshadros*. The main body of the ilium is about two times higher than the postacetabular process. Although it is crushed, its dorsal margin is convex in lateral view, without a distinct depression over the supracetabular process as is usually observed in hadrosaurids. The supracetabular process is poorly developed. The ventral part of the ilium (including the ischiac and pubic peduncles) is destroyed. The preacetabular process is also incompletely preserved. It is robust and weakly deflected ventrally, with a thick dorsal border.

Ischium. The proximal end of the ischium is dorsoventrally expanded, transversely flattened, and triradiate, formed by an iliac ramus, a pubic ramus, and an obturator process (Fig. 19.10). The iliac ramus is the largest. Its dorsal end is thickened laterally and rugose, and it abuts the ischial peduncle of the ilium. The pubic ramus is also transversely compressed and hatchet shaped. A large triangular oburator process is present on the ventral side of the proximal portion of the ischial shaft. The salient and thin caudoventral corner of the pubic ramus and the obturator process enclose a deep obturator notch. The semicircular shape of this notch closely resembles the condition that can be observed in *Altirhinus*, *Probactrosaurus*, *Bactrosaurus*, and Hadrosauridae. In *Iguanodon* and *Tethyshadros*, the obturator process is located more distally on the ventral side of the ischial shaft and the obturator notch is proportionally wider and less deep. Although it is incompletely preserved, the ischial shaft is robust, transversely compressed, and straight. It terminates distally as a moderately expanded knob.

Femur. The right femur is crushed against the left one and is consequently completely deformed. Moreover, its proximal half is largely destroyed, and a 7-cm portion of its shaft is missing (probably lost during excavation; Fig. 19.10). The femur of *Bolong* appears proportionally short and wide (although it may be an artifact of deformation) and is slightly recurved in lateral view (unlike the straight form seen in hadrosaurids). Above the midheight of the shaft, a prominent triangular fourth trochanter projects from the caudomedial edge. The distal portion of the shaft is bowed cranially and the distal condyles are prominently expanded, mainly caudally. The medial condyle appears larger and more robust than the lateral one. The lateral side of the lateral condyle bears a prominent rounded vertical ridge that separates the flattened caudal heel region from the more cranial lateral surface of the condyle. The flexor intercondylar groove extends proximally along the caudal side of the femoral shaft as a deep furrow. The extensor intercondylar groove is not visible.

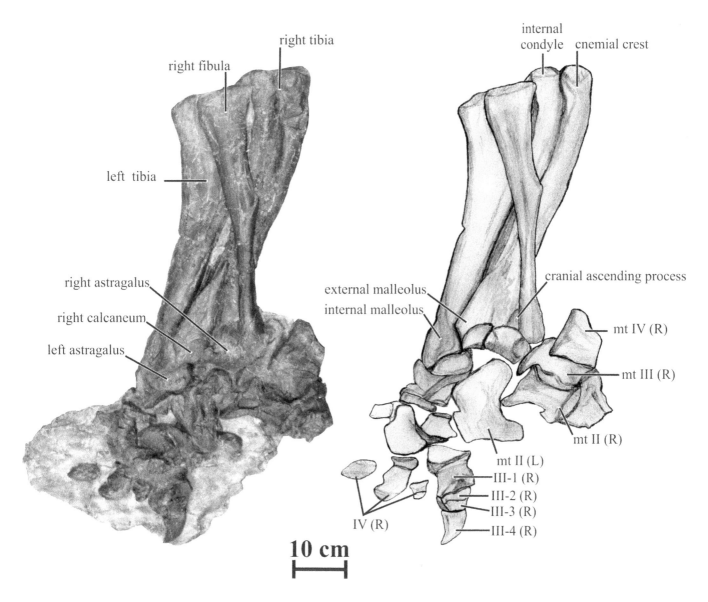

right fibula

right tibia

left tibia

right astragalus

right calcaneum

left astragalus

internal condyle

cnemial crest

external malleolus

internal malleolus

cranial ascending process

mt IV (R)

mt III (R)

mt II (R)

mt II (L)

III-1 (R)

III-2 (R)

III-3 (R)

III-4 (R)

IV (R)

10 cm

Tibia. The left tibia can be observed in medial view and the right one in cranial view (Fig. 19.11). Compared with other basal hadrosauroids, the tibia of *Bolong* is proportionally short and robust. Its proximal end is craniocaudally enlarged. The cnemial crest is weakly developed. It is oriented laterally and supported by a rounded ridge down the cranial side of the proximal third of the tibial shaft. The caudal corner of the medial side of the tibia forms a large internal proximal condyle. The deformed cnemial crest of the right tibia hides the lateral proximal condyles. In medial view, the caudal margin of the tibia is regularly concave. The distal half of the tibia is craniocaudally constricted, whereas its distal third expands transversely. The external malleolus is prominent distally. Its craniodistal surface articulates both with the calcaneum and the astragalus. Above the external malleolus, the lateral side of the distal tibia has a deep, triangular, and rugose area, marking the contact with the distal fibula. The internal malleolus is more prominent than the external malleolus and its distal surface articulates with the astragalus. Above the distal malleoli is a deep triangular depression on the distal third of the cranial side of the tibia.

19.11. *Bolong yixianensis,* YHZ-001 (holotype). Tibiae, right fibula, and pes. *Abbreviations:* L, left; R, right.

Fibula. The right fibula of YHZ-001 is visible in lateral view (Fig. 19.11). It is more robustly built than in *Mantellisaurus* and *Probactrosaurus*. Proximally, the fibula is craniolaterally expanded and transversely compressed. The fibular shaft progressively narrows distally, and its distal end forms a moderately developed and slightly everted bootlike structure, which articulated against the craniolateral edge of the distal end of the tibia.

Astragalus. The right astragalus, closely appressed to the distal articular surface of the tibia, is visible in cranial and distal views. The left one is visible only in medial view (Fig. 19.11). It contacts the calcaneum laterally, but these two bones remain unfused. Its distal articular surface is regularly convex in mediolateral view, concave in craniocaudal view, and rugose. The triangular cranial ascending process, which lies against the cranial side of internal distal malleolus of the tibia, is high compared to the low ascending process seen in *Mantellisaurus* (Norman, 1986, fig. 59B). In cranial view, it is subtriangular and distinctly skewed medially.

Calcaneum. The right calcaneum is visible in cranial view and partly in distal view (Fig. 19.11). It is closely appressed to the external distal malleolus of the tibia and to the astragalus. Its distal articular surface is rugose and convex in mediolateral view. The craniomedial corner of the calcaneum forms a high, triangular, and medially skewed ascending process, which lies against the cranial side of the external distal malleolus of the tibia. This process is unusual in iguanodontoids but has been described in *Bactrosaurus* (Godefroit et al., 1998).

Metatarsals. Few elements of the left foot (including metatarsal II) of the left foot can be observed in medial view. The right foot is more complete, but most bones are eroded and disarticulated, making identification difficult (Fig. 19.11). The metatarsals are proportionally shorter and more robust than in other basal iguanodontoids where it can be observed; ratio metatarsal III/femur = 0.18 in *Bolong*, 0.39 in *Iguanodon* (Norman, 1980), 0.37–0.39 in *Mantellisaurus* (Norman, 1986), 0.37 in *Ouranosaurus* (Taquet, 1976), 0.37 in *Probactrosaurus* (P.G., pers. obs.), 0.36 in *Tenontosaurus* (Winkler et al., 1997), 0.37 in *Nanyangosaurus* (Xu et al., 2000), 0.35 in *Tethyshadros* (Dalla Vecchia, 2009), 0.37–0.41 in *Edmontosaurus* (Lull and Wright, 1942), and 0.37–0.41 in *Corythosaurus* (Lull and Wright, 1942).

Metatarsal II is mediolaterally compressed. Its proximal articular end is expanded dorsoventrally. Proximomedially, a large concave surface closely abuts the shaft of metatarsal III. The shaft of metatarsal II is strongly constricted dorsoventrally. Its dorsal side is more concave than its ventral side. In distal view, its distal end expands to form a large saddle-shaped articular surface for the first phalanx.

Metatarsal III is poorly preserved. Proximally, its lateral side forms a wide, deep, and triangular depression for tight articulation with metacarpal IV.

Metatarsal IV is also incomplete and poorly preserved. It appears much more slender than metatarsal III, and its distal end is only slightly expanded dorsoventrally and smoothly convex.

Pedal phalanges. Only the distal end of phalanx II-1 is preserved in connection with metatarsal II (Fig. 19.11). No other phalanges were identified from digit II.

Digit III has four phalanges, as is usual in Iguanodontia. The first phalanx is robust and mediolaterally wider than proximodistally long, contrasting with the more elongated first phalanx in *Mantellisaurus* and *Probactrosaurus*. The short shaft is mediolaterally constricted between the proximal and distal articular surfaces. The distal articular surface is broad and convex, with a shallow intercondylar groove, and the surface extends on the dorsal (extensor) side of the phalanx. The second and third phalanges are short disklike elements. The ungual phalanx is robust and arrowhead shaped in dorsal view. It is not as proximodistally shortened and genuinely hoof shaped as in the majority of hadrosaurids. From its wide and thick proximal end, it becomes dorsoventrally flattened toward its tip, and its distal end is strongly arched ventrally. The distal end is eroded, but it was apparently less blunt than in *Iguanodon* and *Mantellisaurus*. The claw grooves are better developed than in *Probactrosaurus* and are supported by a shelf.

Several phalanges are visible in digit IV (the ungual is not preserved). However, their state of preservation is poor, and they cannot be formally identified. They are much smaller (about half the size) than the phalanges of digit III.

Size and Posture of *Bolong*

Although histological analyses have not been carried out, the complete closure of the sutures between the neural arches and centra throughout the vertebral column is suggestive of skeletal maturity in the holotype specimen of *Bolong*. However, the caudal ribs are not fused to the centra, indicating that sexual maturity was probably not completely achieved. As discussed above, we estimate a body length of 3 m for this individual, which is particularly small by the standard of known basal iguanodontoids. *Jinzhousaurus*, also from the middle part of the Yixian Formation, is distinctly larger, reaching a maximal length of about 5–5.5 m (R. Pan, pers. comm.). *Bolong* has about the same size as the dwarf basal Iguanodontia *Zalmoxes robustus*, from the Maastrichtian of Romania (Weishampel et al., 2003).

It is well known that the Jehol Biota dinosaur fauna is dominated by small-bodied taxa (<3m in body length; see above). Barrett and Wang (2007) proposed two hypotheses to account for this skewed body size distribution: (1) genuine scarcity of large taxa, perhaps due to resource limitations or local physical conditions that created habitats inappropriate for large animals; and (2) presence of a taphonomic bias that precluded the preservation of large taxa. Resource limitation and climatic constraints are unlikely because of the extraordinary diversity of the fauna. The great abundance of fossil tree trunks in different localities indicates that part of this area was covered by forest, limiting the movements of larger dinosaurs. However, many different ecosystems are probably represented in the Jehol Biota. Additional work is required to determine paleoenvironmental conditions during Jehol times in order to examine how these may have influenced the composition and evolution of this biota (Barrett et al., 2009).

Assessment of the relative proportions of fore- and hind limbs in *Bolong* is complicated by the poor preservation of the humeri and femora, preventing accurate measurements. The forelimb appears proportionally much shorter than in typical quadrupedal dinosaurs and in *Iguanodon*

19.12. Strict consensus tree of Iguanodontia, showing the phylogenetic relationships of *Bolong yixianensis*. Letters correspond to nodes defined in Appendix 19.4.

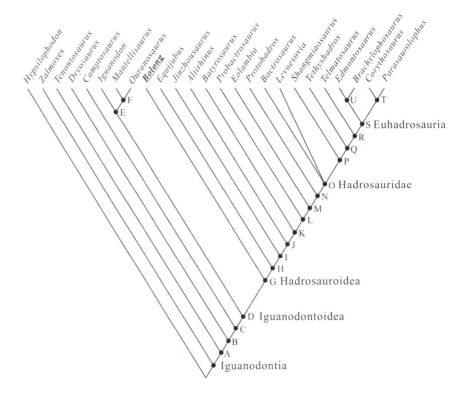

bernissartensis (humerus is 77–80% of femoral length; Norman, 1980) and more closely resembles the condition encountered in *Mantellisaurus atherfieldensis* (humerus is 58–56% of femoral length). However, the distal portion of the forelimb is particularly short in *Bolong*, similar to vertebrates that use their forelimbs to support weight (Norman, 1980). The absence of coossified carpals suggests that the hand was, in any case, less well adapted for supporting weight that that of *Iguanodon* and *Mantellisaurus*. It may be tentatively hypothesized that Bolong spent considerable periods of time walking or running bipedally, but it may have been a facultative quadruped, as in many other nonhadrosaurid iguanodontians (Norman, 2004).

However, the proportionally short metatarsals suggest that *Bolong* was a quadrupedal dinosaur. Indeed, Galton (1970) showed that graviportal mammals and typical quadrupedal dinosaurs are characterized by a low metatarsal III/femur ratio (respectively, 0.10–0.26 and 0.12–0.26). Although the femur cannot be adequately measured in *Bolong*, the metatarsal III/tibia ratio is 0.18, indicating that the metatarsal III/femur ratio probably fell within the range of ratios found in quadrupedal dinosaurs and graviportal mammals. Therefore, the posture and gait of *Bolong* remain problematic.

Phylogenetic Relationships of *Bolong*

We undertook a numerical cladistic analysis of Iguanodontia to resolve the phylogenetic position of *Bolong*. According to Norman (2004), Iguanodontia is a stem-based taxon defined as all euornithopods closer to *Edmontosaurus* than to *Thescelosaurus*. This analysis is based on 25 terminal species whose skeletal remains are well preserved and/or well documented in the literature. *Hypsilophodon* was chosen as outgroup because it is known by abundant material and particularly adequately described. Other taxa (among them *Rhabdodon, Planicoxa, Valdosaurus, Draconyx, Lurdusaurus,*

Fukuisaurus, Muttaburasaurus, "Probactrosaurus" mazongshanensis, Penelopognathus, and *Nanyangosaurus*) were initially included, but they were subsequently removed because of their paucity of character information, which considerably reduced the phylogenetic robustness of the analysis or obscured the general structure of the consensus tree. In this analysis, Iguanodontia are treated at the generic level, although several taxa genera are widely considered to contain more than one species. As a general rule, the type species was chosen as reference for multispecific genera. This analysis is based on 108 cranial, dental, and postcranial characters (character list in Appendix 19.2 and data matrix in Appendix 19.3), mainly compiled from the literature (Norman, 2002, 2004; Weishampel et al., 2003; You et al., 2003a, 2003b; Horner et al. 2004; Godefroit et al., 2004, 2008; Evans and Reisz, 2007; Butler et al., 2008; Sues and Averianov, 2009). Postcranial characters of *Jinzhousaurus* were not included in this analysis because the description of the postcranium of this iguanodontoid (Wang et al., 2011) was published after the present analysis was completed. The 108 characters were equally weighted and analyzed by PAUP*4.0b10 software (Swofford, 2000), both with accelerated (ACCTRAN) and delayed (DELTRAN) transformations. To assess the repeatability of tree topology, a bootstrap analysis was performed (1,000 replicates using the heuristic algorithm in PAUP). Three equally parsimonious trees of 185 steps resulted from a heuristic search, with a consistency index of 0.72, a retention index of 0.90, and a rescaled consistency index of 0.65. The strict consensus tree is presented in Fig. 19.12 and the tree description in Appendix 19.4.

According to the present phylogenetic analysis, Iguanodontoidea form a particularly robust robust (bs = 97%) clade, characterised by the following unambiguous synapomorphies (that diagnose a clade under both ACCTRAN and DELTRAN optimizations): the supraoccipital is excluded from the dorsal margin of the foramen magnum by the paired exoccipitals (character 24); the dorsal process of the maxilla is caudal to center (character 29); the maxilla and jugal form a "finger-in-recess" suture (character 32); 18–30 tooth positions are present in the dentary tooth row (character 55); the marginal denticles along the tooth crowns are developed as curved, mammillated ledges (character 62); the sternal is hatchet shaped (character 82); the unguals on manual digits II and III are flattened, twisted, and hooflike (character 90); digit V is elongated, with three or four phalanges (character 91); the caudal ramus of the pubis is shorter than the ischium and there is no pubic symphysis (character 99): the prepubic process is expanded distally (character 101); the fourth trochanter of the femur is tab shaped (character 103); pedal digit I is absent (with the exception of metatarsal I; character 106); and pedal unguals are hoof shaped, but not broader than long, with prominent claw grooves retained (character 108).

According to our phylogenetic analysis, Iguanodontoidea can be subdivided into two clades: Iguanodontidae and Hadrosauroidea. Contrary to other recent phylogenetic analyses (e.g., Head, 1998; Norman, 2002, 2004; Sues and Averianov, 2009), we retain a monophyletic Iguanodontidae, including *Iguanodon, Mantellisaurus,* and *Ouranosaurus.* This clade is characterized by the following autapomorphies: the ossified epaxial tendons extend cranially from cervical 10 (character 76; the polarity of this character cannot be observed in *Ouranosaurus*) and the pollex is spikelike and conical (character 89, 0 to 2). Within Iguanodontoidae, *Mantellisaurus* and *Ouranosaurus* are united by the presence of elongated neural spines on the posterior dorsals and sacrals, which are more than 2.5 times the centrum height (character 73). The low bootstrap value (bs = 53%) clearly indicates that the relationships within Iguanodontidae remain highly conjectural in the current state of our knowledge. Moreover, newly named iguanodontid genera (see Chapter 15 in this book) need to be tentatively included in this phylogenetic analysis.

Hadrosauroidea is defined as all Iguanontoidea closer to *Parasaurolophus* than to *Iguanodon* (Sereno, 1997, amended). In our analysis, this clade is characterized by the following unambiguous synapomorphies: the antorbital fenestra is not open on the lateral side of the skull (character 26); the basipterygoid process extend ventrally well below the level of the ventral border of the occipital condyles (character 42, 1 to 0); the carpals are well ossified but remain unfused (character 86, 1 to 0); the ischium is nearly straight in lateral view (character 96, 1 to 0). *Bolong* is here regarded as the most basal Hadrosauroidea. Indeed, the antorbital fenestra is still present as a small depression between the lacrimal and maxilla, although it is not exposed laterally in other hadrosauroids (character 26, 1 to 2), and the tooth roots are not cemented at all, contrary to more advanced hadrosauroids (character 61). *Jinzhousaurus*, also from the Yixian Formation of western Liaoning province, shares an additional dental synapomorphy with more advanced hadrosauroids: the maxillary teeth bear a single faint accessory ridge (character 65). Thus, *Bolong* apparently occupies a more basal position in the phylogeny of hadrosauroids than *Jinzhousaurus*. The phylogeny of more advanced Hadrosauroidea is discussed in Chapter 20 of this book.

Paleogeography of Basal Iguanodontia and Iguanodontoidea

The oldest Iguanodontia is *Callovosaurus leedsi*, which is based on an isolated femur from the Callovian of England (Ruiz-Omeñaca et al., 2007). During the Kimmeridgian, Iguanodontia had a Neopangean (Pangea without China and Siberia; Russell, 1993) distribution: diagnosable iguanodontians have been discovered in North America (*Dryosaurus altus, Camptosaurus dispar*), Europe (*Camptosaurus prestwichii*), and Africa (*Dryosaurus lettowvorbecki*). Isolated femora were also discovered in the Kimmeridgian–Tithonian of Portugal (Galton, 1980). Iguanodontia are apparently absent from Late Jurassic localities in Asia. During the Lower Cretaceous, Iguanodontia achieved a virtually cosmopolitan distribution (Norman, 2004). Basal Iguanodontia and Iguanodonidae apparently kept a Neopangean distribution, in Europe (*Iguanodon, Mantellisaurus*, and taxa recently named from England; see Norman, Chapter 15 in this book), Africa (*Ouranosaurus, Lurdusaurus*), North America (*Tenontosaurus*, "*Iguanodon*" *ottingeri, Dakotadon, Planicoxa, Cedrorestes*, and forms to be described from Cedar Mountain Formation in Utah; McDonald et al., 2009), Australia (*Muttaburasaurus*), and possibly South America (the phylogenetic position of *Talenkauen* and *Macrogryphosaurus*, regarded as Iguanodontia by Calvo et al., 2007, is still problematic according to Butler et al., 2008). Norman (1996) tentatively identified fragmentary remains ("*Iguanodon orientalis*") from the Lower Cretaceous of Mongolia as *I. bernissartensis*, but these fossils could just as well belong to some basal hadrosauroid; the maxillary teeth more closely resemble those of *Altirhinus*, for example, than those of *Iguanodon*.

Although basal Iguanodontia and Iguanodontidae are not confidently represented in Lower Cretaceous deposits from Asia, Hadrosauroidea have clearly an Asian origin. Indeed, Figure 19.12 shows that the most basal Hadrosauroidea (successively *Bolong, Equijubus, Jinzhousaurus, Altirhinus, Batyrosaurus,* and *Probactrosaurus*) are all from Asia. "*Probactrosaurus*" *mazongshanensis, Fukuisaurus, Nanyangosaurus, Penelopognathus,* and *Jintasaurus* have not been included in the phylogenetic analysis, but they are also potential basal Hadrosauroidea from the Lower Cretaceous of Asia. According to Russell (1993), the isolation of Central Asia came to an end during the Aptian–Albian, when a land route opened across the Bering Strait, although another intermittently emergent route appeared at about the same time toward Europe (Doré, 1991). The presence of the iguanodontoids *Bolong* and *Jinzhousaurus* in the middle part of the Yixian Formation indicates that connections between Asia and North America and/or Europe were already established during the lower Aptian or even during the upper Barremian. Information on the paleogeography of basal Ornithomimosauria (see Chapter 26 in this book) also suggests dinosaur immigrations from Europe into Asia during or before the Barremian. The oldest nonhadrosaurid Hadrosauroidea outside Asia is *Eolambia*, from the latest Albian (Garrison et al., 2007) Mussentuchit Member of the Cedar Mountain Formation in Utah that includes other taxa with Asian affinities, such as tyrannosauroids, ceratopsians, and pachycephalosaurs. These records are the earliest known representatives of these groups in western North America (Cifelli et al., 1997; McDonald et al., 2010). These occurrences can be attributed to an immigration of Asian taxa, including Hadrosauroidea, into western America following the establishment of the Bering land bridge in the Aptian–Albian (Russell, 1993). The paleogeography of more derived Hadrosauroidea is discussed in Chapter 20 of this book.

Although Hadrosauridae were the most abundant and diversified large vertebrates in Laurasia during the closing stages of the Late Cretaceous (Horner et al., 2004), basal Iguanodontia (*Rhabdodon, Zalmoxes*) subsisted in Europe during the Campanian and the Maastrichtian (see Chapters 31 and 31 in this book), implying a long ghost lineage duration for this small clade (Weishampel et al., 2003).

L, left; R, right.

Appendix 19.1. Measurements

Preorbital length: 284 mm
External naris (L): length: 129 mm
External naris (L): maximum height: 39 mm
Maxilla (L): length: 211 mm
Maxilla (L): maximum height: 66 mm
Quadrate (R): height: 200 mm
Dentay (L): length (from rostral tip to apex of coronoid process): 275 mm
Dentary (L): height (at midlength of dentary ramus): 63 mm
Dentary (L): height (at apex of coronoid process): 146 mm

Scapula (R): height (proximodistal): 495 mm
Scapula (R): maximum width (craniocaudal) of proximal plate: 127 mm
Scapula (R): minimum width (craniocaudal) of blade: 65 mm
Scapula (R): maximum width (craniocaudal) of blade: 86 mm
Humerus (R): mediolateral width of proximal end: 78 mm
Ulna (R): length: 237 mm
Ulna (R): craniocaudal height of proximal part: 66 mm
Ulna (R): craniocaudal height of distal part: 50 mm
Radius (R): length: 198 mm
Radius (R): craniocaudal height of proximal part: 65 mm
Radius (R): craniocaudal height of distal part: 58 mm
Metacarpal II (L): length: >66 mm
Metacarpal III (R): length: 81 mm
Metacarpal IV (R): length: 85 mm
Metacarpal V (L): length: 55 mm
Metacarpal V (L): width at proximal end: 29 mm
Digit I, ungual (R): length: 58 mm
Digit II, phalanx 1 (R), length: 25 mm
Digit II, phalanx 1 (R), width at distal end: 30.5 mm
Digit II, phalanx 2 (R), length: 14 mm
Digit II, phalanx 2 (R), width at distal end: 29 mm
Digit II, ungual (R): length: 61 mm
Digit III, phalanx 1 (R), length: 20 mm
Digit III, phalanx 1 (R), width at distal end: 32 mm
Digit III, phalanx 2 (R), length: 9 mm
Digit III, ungual (R): length: 56 mm
Digit III, ungual (R): maximal width: 41 mm
Digit IV, phalanx 1 (R), length: 24 mm
Digit IV, phalanx 1 (R), width at distal end: 24 mm
Digit V, phalanx 1 (R), length: 32 mm
Digit V, phalanx 1 (R), width at distal end: 25 mm
Digit V, phalanx 2 (R), length: 16 mm
Digit V, phalanx 2 (R), width at distal end: 19 mm
Ilium (R), length: >400 mm
Ilium (R), postacetabular process, length: 100 mm
Ilium (R), postacetabular process, maximal height (taken at its cranial point): 50 mm
Tibia (L): height: 510 mm
Tibia (L): width (mediolateral) of distal end: 140 mm
Fibula (R): height: 448 mm
Fibula(R): width (craniocaudal) of proximal end: 94 mm
Fibula(R): minimal width (craniocaudal) of shaft: 27 mm

Appendix 19.2. Character List

Characters are compiled from Norman (2002, 2004), Weishampel et al. (2003), You et al. (2003a, 2003b), Horner et al. (2004), Godefroit et al. (2004, 2008), Evans and Reisz (2007), Butler et al. (2008), Sues and Averianov (2009), and Dalla Vecchia (2009).

1. Preorbital skull length: approximately 50% of total skull length (0); more than 50% of total skull length (1).
2. Premaxilla, oral margin: smooth (0); denticulate (1).
3. Premaxilla, oral margin with a "double layer" morphology consisting of an external denticles-bearing layer seen externally and an internal palatal layer of thickened bone set back slightly from the oral margin and separated from the denticulate layer by a deep sulcus bearing vascular foramina: absent (0); present (1).
4. Lateral expansion of premaxilla: narial portion of premaxilla slopes steeply from the external naris to the oral margin (0); the ventral body of the premaxilla flares laterally, so as to form a partial floor of the narial fossa, but lateral expansion less than two times of the width at postoral constriction (1); widely expanded laterally, more than two times of the width at postoral constriction (2). Character treated as ordered.
5. Reflected rim of premaxilla: absent (0); present (1).
6. Premaxillary foramen ventral to anterior margin of external nares: absent (0); present (1).
7. Premaxillary accessory narial fossa: absent (0); present (1).
8. Premaxilla–lacrimal contact: absent (0); present (1).
9. Caudal premaxillary process: short, not meeting the lateral premaxillary process posterior to external naris, external naris surrounded by both nasal and premaxilla (0); long, meeting the lateral premaxillary process behind the external naris to exclude the nasal from the external naris, nasal passage enclosed ventrally by folded, divided premaxillae (1).
10. Position of nasal cavity: nasal restricted to area rostral to braincase, nasal cavity rostromedial to orbits (0); nasals retracted caudally to lie over braincase in adults resulting in convoluted, complex narial passage and hollow crest (1).
11. Circumnarial depression: absent (0); present (1).
12. External naris: confined to the area immediately above the oral margin of the premaxilla (0); extending caudally so as to lie above the maxilla (1).
13. External naris/basal skull length ratio: : <0.2 (0); >0.3 (1).
14. Caudal portion of prefrontal: oriented horizontally (0); participating in ventrolateral border of hollow crest (1).
15. Frontal, platform for nasal articulation: absent (0); present (1).
16. Frontal at orbital margin: forming part of the margin (0); excluded by prefrontal–postorbital contact (1).
17. Frontal, upward doming over braincase in adults: absent (0); present (1).
18. Supraorbital articulation: freely articulating on orbital rim (0); fused to orbital rim or absent (1).
19. Parietal, length/minimum width ratio: greater than 2 (0); less than 2 (1).
20. Parietal sagittal crest, shape: straight and level with skull roof or slightly downwarped along length (0); deepens caudally, strongly downwarped (1).

21. Parietal sagittal crest, length: short, less than ⅔ length of parietal (o); long, more than ⅔ length of parietal (1).
22. Squamosals, separation on skull roof: completely separated by the parietal (o); squamosals in broad contact with each other (1).
23. Median ramus of squamosal lower than paroccipital process (o); higher than paroccipital process (1).
24. Supraoccipital: participates in the dorsal margin of the foramen magnum (o); excluded from the dorsal margin of the foramen magnum by the paired exoccipitals (1).
25. Supraoccipital/exoccipital shelf above foramen magnum: limited (o); very extended (1).
26. Antorbital fenestra: exposed laterally (o); present as a small depression between lacrimal and maxilla (1); not exposed laterally (2). Character treated as ordered.
27. Maxilla, rostral end: single process lodged under or along the medial aspect of the premaxilla (o); second, rostrolateral process developed on the maxilla (1); development of a wide sloping maxillary shelf (2). Character treated as unordered.
28. Maxilla, dorsal process shape: low and gently rounded (o); tall and sharply peaked (1).
29. Maxilla, dorsal process position: apex of maxilla at or rostral to center (o); caudal to centre (1).
30. Maxillary foramen: absent (o); on rostrolateral maxilla (1); on dorsal maxilla along maxilla-premaxilla suture (2). Character treated as ordered.
31. Maxilla, ectopterygoid ridge: rostrocaudally short, ending at half or less length of caudal region of maxilla (o); well developed into a lateral and well demarked border continuous along caudal region of maxilla (1).
32. Maxilla–jugal suture: scarf joint (o); "finger-in-recess" (1); butt joint (2). Character treated as unordered.
33. Maxilla, number of maxillary teeth: less than 12 (o); 12–30 (1); more than 30 (2). Character treated as ordered.
34. Jugal–ectopterygoid articulation: large subtriangular ectopterygoid facet on medial side of jugal (o); small vertical ectopterygoid facet (1); absent, palatine–jugal contact enhanced (2). Character treated as ordered.
35. Jugal, expansion of rostral end below lacrimal: dorsoventrally narrow, forms little or the rostral orbital rim (o); expanded dorsoventrally in front of orbit, lacrimal pushed dorsally to lie completely above the level of maxilla, jugal forms lower portion of orbital rim (1).
36. Jugal, development of free ventral flange: absent, jugal expands gradually below infratemporal fenestra to meet quadratojugal–quadrate (o); present, jugal dorsoventrally constricted beneath infratemporal fenestra to set off flange rostral to constriction (1).
37. Jugal, development of caudal process: caudal process of jugal forms only the rostral part of the ventral margin of the infratemporal fenestra (o); extends caudally to form the ventral margin of the infratemporal fenestra (1).

38. Jugal, acute angle between postorbital bar and jugal bar: absent (0); present (1).
39. Paraquadratic foramen: present, two distinct articulation facets along margin of quadratojugal notch (0); absent, a single elongated articulation facet along margin of quadratojugal notch (1).
40. Quadrate, articular condyle: transversely expanded (0); dominated by a large hemispheric lateral condyle (1).
41. Occipital condyle, inclination: caudoventral (0); vertical (1).
42. Basipterygoid process, length: extending ventrally well below level of ventral border of occipital condyles (0); extending nearly horizontally (1).
43. Predentary, proportion: length of lateral process larger or subequal than mediolateral width of rostral margin (0); length of lateral process less than half mediolateral width of rostral margin (1).
44. Predentary, oral margin: relatively smooth (0); denticulate (1).
45. Predentary, ventral process: simple (0); bilobate (1).
46. Dentary, diastema: short, no more than width of 4 or 5 teeth (0); moderate, equal to approximately ⅕ to ¼ of tooth row (1); long, more than ⅓ of tooth row (2). Character treated as ordered.
47. Dentary ramus: parallel dorsal and ventral margins (0); deepens rostrally (1).
48. Dentary, position of coronoid process: laterally offset and dentition curves into its base (0); laterally offset and separated from dentition by a shelf (1).
49. Dentary, configuration of coronoid process: apex only slightly expanded rostrally, surangular large and forms much of caudal margin of coronoid process (0); dentary forms nearly all of rostrocaudally greatly expanded apex, surangular reduced to thin sliver along caudal margin and does not reach dorsal end of coronoid process (1).
50. Dentary, inclination of coronoid process: slightly inclined caudally or subvertical (0); inclined rostrally (1).
51. Dentary, caudal extension of tooth row: terminates even with or rostral to apex of coronoid process (0); terminates posterior to apex (1).
52. Dentary, shape of tooth row in occlusal view: bowed lingually (0); divergent caudally and convergent rostrally relative to lateral side of dentary (1); tooth row parallel with lateral side of dentary (2). Character treated as ordered.
53. Dentary, alveolar through grooves: shaped by dentary crowns (0); narrow, parallel-sided grooves (1).
54. Dentary, number of replacement teeth per tooth family: one (0); two–three (1); more than three (2). Character treated as ordered.
55. Number of tooth positions in dentary tooth row (adult specimens): less than 18 (0); 18–30 (1); more than 30 (2). Character treated as ordered.
56. Dentary, maximum number of functional teeth forming occlusal plane: one (0); two (1); three (2). Character treated as ordered.
57. Surangular foramen: present (0); absent (1).
58. Accessory surangular foramen or embayment close to the dentary suture: present (0); absent (1).

59. Angular, lateral exposure: present (0); absent (1).

60. Premaxillary teeth: 3 or more (0); absent (1).

61. Teeth roots: not cemented (0); partially cemented (1); rugose, angular-sided roots (2). Character treated as ordered.

62. Teeth, marginal denticles: single, tongue shaped (0); developed as curved, mammillated ledges (1); absent, or reduced to small papillae (2). Character treated as ordered.

63. Maxillary tooth crowns, height/width ratio at center of tooth row: ratio less than 2.4 (0); ratio at least 2.5 (1).

64. Maxillary tooth crowns, primary ridge: labial surface of maxillary crowns either smooth or exhibit an undifferentiated pattern of ridges (0); primary ridge developed (1).

65. Maxillary tooth crowns, number of accessory ridges: one primary ridge with up to three accessory ridges (0); one primary ridge with a single faint accessory ridge (1); one primary ridge, lacks accessory ridges (2). Character treated as ordered.

66. Maxillary tooth crowns, symmetry: asymmetrical, primary ridge offset from midline (0); symmetrical, primary ridge on midline (1).

67. Dentary tooth crowns, shape: broad and shieldlike, height/width ratio <3.0 (0); small, narrow, and lanceolate, height/width ratio >3.1 (1).

68. Dentary tooth crowns, symmetry: asymmetrical, apex and primary ridge offset from midline (0); symmetrical, apex and primary ridge on midline (1).

69. Dentary tooth crowns, development of secondary ridge mesial to primary ridge: well developed, reaches the apex of the crown (0); faintly developed, does not reach the apex of the crown, absent on some teeth (1); absent or occasional (2). Character treated as ordered.

70. Dentary tooth crowns, development of tertiary ridges mesial or distal to primary ridge: at least one tertiary ridge reaches the apex of the crown (0); faintly developed, do not reach the apex of the crown (1); absent (2). Character treated as ordered.

71. Dentary tooth crowns, distal shelf at the base of the crown: absent (0), well developed but smooth (1); well developed and mammillated (2). Character treated as unordered.

72. Number of cervical vertebrae: nine or fewer (0); 10 or more (1).

73. Posterior dorsal and sacral neural spines: relatively short, less than 2.5 times centrum height (0); elongate, more than 2.5 times centrum height (1).

74. Sacrum, number of vertebrae: a maximum of 7 (0); a minimum of 8 (1).

75. Ossified hypaxial tendons in the caudal skeleton: present (0); absent (1).

76. Ossified epaxial tendons, extension: do not extend on cervical vertebrae (0); extend from cervical 10 (1).

77. Ossified epaxial tendons, organisation: longitudinally arranged (0); arranged in a double- or triple-layered lattice (1).

78. Scapula, shape of proximal region: dorsoventrally deep, acromion process directed dorsally, prominent on rostral margin of scapula, articulation extensive (0); dorsoventrally narrow (no higherer than

distal scapula), acromion process projects horizontally, reflected laterally, and articulation restricted (1).

79. Coracoid, size: large, coracoid/scapula lengths more than 0.2, length of articular surface greater than length of glenoid (0); coracoid reduced in length relative to scapula, glenoid equal or longer than articulation (1).

80. Coracoid, cranioventral process: short and weakly developed (0); long, extends well below the glenoid (1).

81. Coracoid, biceps tubercle size: small (0); large and laterally projecting (1).

82. Sternal, shape: reniform (0); hatchet shaped (1).

83. Sternal, proximal plate: shorter than distal handle (0); longer than distal handle (1).

84. Proportions of humerus and scapula: length of humerus greater than or equal to length of scapula (0); scapula longer than humerus (1).

85. Humerus, deltopectoral crest shape: relatively low (0); angular and enlarged (1).

86. Carpus structure: fully ossified, with more than two small bones present, but carpals unfused (0); carpals fused into two blocks (1); reduced to no more than two small carpals (2). Character treated as unordered.

87. Manus, digit I: present, nearly parallel orientation relative to the long axis of antebrachium (0); present, diverges approximately 45 degrees from the antebrachial axis (1); absent (2). Character treated as unordered.

88. Metacarpal I, size: only slightly shorter than metacarpal II, not fused with radiale (0); less than 50% the length of metacarpal II, fused with radiale in adults (1).

89. Ungual of manual digit I: clawlike (0); spikelike, but flattened (1); spikelike and conical (2); absent (3). Character treated as unordered.

90. Unguals of manual digits II and III: clawlike (0); flattened, twisted, and hooflike (1).

91. Manual digit V: short, 2 phalanges (0); elongated, 3 or 4 phalanges (1); lost (2). Character treated as unordered.

92. Ilium, supracetabular process: small, projects only as a lateral swelling (0); large, broadly overhangs the lateral side of the ilium and usually extends at least half way down the side of ilium (1).

93. Ilium, shape of dorsal margin: regularly convex in lateral view (0); distinctly depressed over supracetabular process and dorsally bowed over base of preacetabular process (1).

94. Ilium, ischial peduncle: as a single large knob (0); formed by two small protrusions separated by a shallow depression (1).

95. Ilium, postacetabular process: tapers caudally to nearly a point, wide brevis shelf (0); postacetabular process subrectangular, no brevis shelf (1).

96. Ischium, shape of shaft in lateral view: nearly straight (0); strongly curved ventrally (1).

97. Ischium, position of obturator process: positioned beyond the proximal 25% of the length of the ischium (0); positioned within the proximal 25% (1); absent (2). Character treated as unordered.

98. Ischium, expansion of distal end: terminates without a significant increase in size (0); ischial terminus bears an expansion that is 50% larger than the adjacent ischial shaft (1).

99. Pubis, caudal ramus: terminates adjacent to distal end of ischium (0); shorter than ischium, no pubic symphysis (1).

100. Pubis, prepubic process proportion: rod shaped, wider transversely than dorsoventrally (0); transversely flattened (1).

101. Pubis, prepubic process shape in lateral view: bladelike, unexpanded distally (0); expanded distally (1).

102. Pubis, iliac peduncle: relatively small (0); has the shape of a large and dorsally directed process (1).

103. Femur, shape of fourth trochanter: pendent (0); tab shaped (1); curved, laterally compressed eminence (2). Character treated as unordered.

104. Femur, development of intercondylar extensor groove: moderately deep, groove fully open (0); deep, edges of groove meet or nearly meet cranially to enclose an extensor tunnel (1).

105. Tibia, cnemial crest: restricted to proximal head of tibia (0); extends on diaphysis (1).

106. Pedal digit 1, number of phalanges: 2 (0); 0 (1).

107. Metatarsal V: present (0); absent (1).

108. Pes, ungual phalanges: claw shaped (0); hoof shaped, but not broader than long, with prominent claw grooves retained (1); broad, short with rounded shield hooflike shape and reduced or absent claw grooves (2). Character treated as ordered.

Appendix 19.3. Data Matrix

Hypsilophodon	00000	00000	00000	00000	00?00	00000	00000
	00000	00000	00000	00000	00000	00000	00000
	00000	00000	00000	00000	?0000	00000	00000
	000						
Zalmoxes	?0000	00000	00000	00?00	?0?00	00000	00000
	01010	10001	00000	00000	00101	00000	00000
	0?000	0?000	000?0	0????	?0000	121?0	00000
	??0						
Tenontosaurus	10010	00100	01000	00000	00000	01000	000?0
	00000	0?010	00000	00000	00??1	00000	00000
	01000	00000	00010	00000	00000	00001	00000
	0?0						
Dryosaurus	00010	00100	01000	00000	00?00	01000	001?0
	01000	01011	00000	00000	0?101	00010	00000
	0?001	00000	0?010	000?0	?0000	11101	00000
	?00						
Camptosaurus	11010	00100	01000	00000	00000	01000	001?0
	01010	01011	00000	00000	00?01	00010	00000
	10001	01000	00010	11110	00000	11101	00000
	010						
Iguanodon	11010	00100	01000	00000	00010	01010	01100
	01000	01011	00000	00001	00001	01010	00000
	21001	11000	01010	11121	10000	11111	10100
	111						

Mantellisaurus	11010	00100	01000	00000	00010	01010	01100
	01000	01011	00000	00001	00001	01010	00000
	21101	11000	01010	11121	10000	11111	10100
	111						
Ouranosaurus	11010	00100	01000	00010	00010	01010	01100
	01000	01111	11000	00001	00001	01010	00000
	2110?	??000	01010	11121	?0000	01111	10100
	111						
Bolong	11010	??100	0110?	?00??	?????	1?01?	011?0
	??0??	00011	00000	0???1	0???1	01010	00000
	210?1	0000?	???10	01111	100?0	011??	????0
	1?1						
Jinzhousaurus	110?0	??100	01000	10000	000??	2?01?	????0
	010??	??710	?0000	?????	00001	?1?11	000??
	?????	?????	?????	?????	?????	?????	?????
	???						
Equijubus	11010	??100	01000	00?00	0?0??	2?01?	?0100
	01000	??711	10?00	??011	00101	11000	00000
	?1?01	?????	?????	?????	?00??	?????	?????
	???						
Altirhinus	110?0	00100	011?0	000??	?????	21010	11100
	11000	??011	00100	0?0?1	00001	11011	00000
	2????	??000	01010	01?11	?0000	01?11	10100
	??1						
Batyrosaurus	?????	?????	????0	00?00	00?10	?????	?1?00
	11?0?	1?011	00100	01111	?0?0?	11012	00001
	2????	?????	?10?0	???11	?????	?????	?????
	???						
Probactrosaurus	11010	00?00	010?0	00?00	00010	??010	01110
	010?0	1?011	00100	01111	10?00	11012	10001
	1??0?	??000	01010	?1?11	?0000	01111	10100
	??1						
Eolambia	?1010	10?00	010??	???00	0?0?0	2??10	01110
	0?0?0	10???	01100	0111?	?0000	11012	10012
	1????	?????	????0	?????	?00??	0?1??	??10?
	???						
Protohadros	?1010	10?00	01000	001??	?????	21010	01110
	01010	??011	11100	01111	?0100	11012	10012
	0????	?????	?????	?????	?????	?????	?????
	???						
Shangmiaosaurus	?????	?????	?????	?????	?????	??010	021??
	?????	?????	00100	011?1	?????	21?12	1????
	0????	?????	?????	?????	?????	?????	?????
	???						
Bactrosaurus	11010	10?00	01000	00000	00010	21010	01120
	01000	10011	00100	01111	11100	21012	10012
	0?001	??100	01010	????1	?0000	11111	11211
	??2						
Levnesovia	?????	?????	??700	00?01	00010	??010	01120
	010?0	10???	001??	?11?1	?110?	?1012	10012
	0????	???00	0???0	?????	?00??	?????	?1?1?
	??2						
Tethyshadros	11010	?0100	01000	00000	0?0??	1?000	?22?0
	01010	??211	00101	??11?	111?1	20111	10001
	01011	01101	01010	22?31	210?1	01011	11??0
	112						
Telmatosaurus	11010	10?00	010?0	00?00	00010	21011	1?2?0
	?1?10	10?11	00100	11122	21100	21012	10012
	0?0??	??1?1	??010	????1	?????	?1?11	1?21?
	???						
Brachylophosaurus	11121	11100	11100	00100	10011	21001	12221
	11111	10111	20111	12122	21110	22112	11122

	01011	01111	11010	22131	11111	01011	11211
	112						
Edmontosaurus	11121	11100	11100	00100	10011	21001	12221
	11111	10111	20111	12122	21110	22112	11122
	01011	01111	11010	22131	11111	01011	11211
	112						
Parasaurolophus	11110	00111	01011	11111	01110	22112	12221
	11111	10111	10111	12122	21110	22112	11122
	01111	01111	11111	22131	11111	01111	11211
	112						
Corythosaurus	11110	00111	01011	11111	01110	22112	12221
	11111	10111	10111	12122	21110	22112	11122
	01111	01111	11111	22131	11111	01111	11211
	112						

Appendix 19.4. Tree Description

The "describetrees" option of PAUP*4.0b10 was used to interpret character state transformations. All transformations are based on the derivative strict reduced consensus tree (see Fig. 19.12). Transformation was evaluated under accelerated transformation (ACCTRAN) and delayed transformation (DELTRAN) options; unambiguous synapomorphies are those that diagnose a node under both ACCTRAN and DELTRAN optimizations. Node numbers refer to Figure 19.12. For simple 0–1 state changes, only the character number is given; for other state changes, the type of change is specified in parentheses. bs, bootstrap values (1,000 replicates).

Node A (bs = 77): Unambiguous: 4, 8, 12, 27, 44, 84, 100; ACCTRAN: 1, 42, 72.

Node B (Dryomorpha; bs = 85): Unambiguous: 33, 37, 45, 64, 75, 96, 97, 98; ACCTRAN: 89; DELTRAN: 42.

Node C (Ankylopollexia ; bs = 94): Unambiguous: 2, 77, 86, 87, 88, 107; ACCTRAN: 71; DELTRAN: 1, 89.

Node D (Iguanodontoidea; bs = 97): Unambiguous: 24, 29, 32, 55, 62, 82, 90, 91, 99, 101, 103, 106, 108; ACCTRAN: 71 (1 to 2); DELTRAN: 71 (0 to 2), 72.

Node E (Iguanodontidae; bs = 76): Unambiguous: 76, 89 (0 to 2).

Node F: Unambiguous: 73.

Node G (Hadrosauroidea; bs = 53): Unambiguous: 26, 42 (1 to 0), 86 (1 to 0), 96 (1 to 0); ACCTRAN: 52, 54.

Node H (bs = 60): Unambiguous: 26 (1 to 2), 61; ACCTRAN: 41; DELTRAN: 54, 62.

Node I (bs = 57): Unambiguous: 65.

Node J (bs = 57): Unambiguous: 48; ACCTRAN: 36.

Node K (bs = 75): Unambiguous: 53, 65 (1 to 2), 70; ACCTRAN: 56, 60 (1 to 0), 86 (0 to 2); DELTRAN: 41, 52.

Node L (bs = 97): Unambiguous: 34, 66; ACCTRAN: 36 (1 to 0), 39, 71 (2 to 1); DELTRAN: 56, 60 (1 to 0), 71.

Node M (bs = 94): 6, 69, 70 (1 to 2); ACCTRAN: 47, 78, 87 (1 to 2), 89 (1 to 3), 102, 105, 108 (1 to 2).

Node N (bs = 62): 58, 71 (1 to 0); ACCTRAN: 103 (1 to 2), 104.

Node O (Hadrosauridae; bs = 75): Unambiguous: 34 (1 to 2), 57, 61 (1 to 2); ACCTRAN: 47 (1 to 0); DELTRAN: 78, 102, 103 (1 to 2), 104, 105, 108 (1 to 2).

Node P (bs = 63): Unambiguous: 32 (1 to 2); ACCTRAN: 63, 74, 80, 81, 92, 94, 95, 98 (1 to 0).

Node Q (bs = 79): Unambiguous: 33 (1 to 2); ACCTRAN: 31, 50, 51, 55 (1 to 2); DELTRAN: 39, 74, 86, 87, 89, 92, 95.

Node R (bs = 62): Unambiguous: 30, 54 (1 to 2), 56 (1 to 2); ACCTRAN: 18, 36, 38, 79, 93; DELTRAN: 31, 51, 55 (1 to 2), 80.

Node S (Euhadrosauria; bs = 100): Unambiguous: 3, 35, 40, 46, 49, 52 (1 to 2), 59, 62 (1 to 2), 67, 68, 69 (1 to 2); DELTRAN: 18, 36, 38, 43, 50, 63, 79, 81, 93, 94.

Node T (Lambeosaurinae; bs = 100): Unambiguous: 6 (1 to 0), 9, 10, 14, 15, 16, 17, 19, 20, 22, 23, 27 (1 to 2), 28, 30 (1 to 2), 73, 83, 85; ACCTRAN : 98.

Node U (Hadrosaurinae; bs = 100): Unambiguous: 4 (1 to 2), 5, 7, 11, 13, 21, 25, 29 (1 to 0), 46 (1 to 2); DELTRAN: 98 (1 to 0).

Acknowledgments

The authors are particularly grateful to Bo Haichen for access to the holotype specimen of *Bolong yixianensis*, and to Sun Ge for organizing this study. H. De Potter helped with the drawings. P. R. Bell and D. B. Norman read an earlier version of this chapter and made many helpful comments.

References

Adams, J. S., and C. L. Organ. 2005. Histologic determination of ontogenetic patterns and processes in hadrosaurian ossified tendons. Journal of Vertebrate Paleontology 25: 614–622.

Barrett, P. M., and X.-L. Wang. 2007. Basal titanosauriform (Dinosauria, Sauropoda) teeth from the Lower Cretaceous Yixian Formation of Liaoning province, China. Palaeoworld 16: 265–271.

Barrett, P. M., R. J. Butler, X.-L. Wang, and X. Xu. 2009. Cranial anatomy of the iguanodontoid ornithopod *Jinzhousaurus yangi* from the Lower Cretaceous Yixian Formation of China. Acta Palaeontologica Polonica 54: 35–48.

Brown, B. 1916. A new crested trachodont dinosaur, *Prosaurolophus maximus*. Bulletin of the American Museum of Natural History 35: 701–708.

Butler, R. J., P. Upchurch, and D. B. Norman. 2008. The phylogeny of the ornithischian dinosaurs. Journal of Systematic Palaeontology 6: 1–40.

Calvo, J. O., J. D. Porfiri, and F. E. Novas. 2007. Discovery of a new ornithopod dinosaur from the Portezuelo Formation (Upper Cretaceous), Neuquén, Patagonia, Argentina. Arquivos do Museu Nacional, Rio de Janeiro 65: 471–483.

Cifelli, R. L., J. I. Kirkland, A. Weil, A. L. Deino, and B. J. Kowallis. 1997. High-precision ^{40}Ar/^{39}Ar geochronology and the advent of North America's Late Cretaceous terrestrial fauna. Proceedings of the National Academy of Sciences of the United States of America 94: 11163–11167.

Cope, E. D. 1869. Synopsis of the extinct Batrachia, Reptilia and Aves of North-America. Transactions of the American Philosophical Society 14: 1–252.

Dalla Vecchia, F. M. 2009. *Tethyshadros insularis*, a new hadrosauroid dinosaur (Ornithischia) from the Upper Cretaceous of Italy. Journal of Vertebrate Paleontology 29: 1100–1116.

Dollo, L. 1888. Iguanodontidae et Camptonotidae. Comptes Rendus hebdomadaires de l'Académie des Sciences, Paris 106: 775–777.

Doré, A. G. 1991. The structural foundation and evolution of Mesozoic seaways between Europe and the Arctic. Palaeogeography, Palaeoclimatology, Palaeoecology 87: 441–492.

Evans, D. C., and R. R. Reisz. 2007. Anatomy and relationships of *Lambeosaurus magnicristatus*, a crested hadrosaurid dinosaur (Ornithischia) from the Dinosaur Park Formation. Alberta. Journal of Vertebrate Paleontology 27: 373–393.

Erickson, B. R. 1988. Notes on the postcranium of *Camptosaurus*. Scientific Publications of the Science Museum of Minnesota (n.s.) 6: 3–13.

Forster, C. A. 1990. The postcranial skeleton of the ornithopod dinosaur *Tenontosaurus tilletti*. Journal of Vertebrate Paleontology 10: 273–294.

Galton, P. M. 1970.The posture of hadrosaurian dinosaurs Journal of Paleontology 44: 464–473.

———. 1980. European Jurassic ornithopod dinosaurs of the families Hypsilophodontidae and Camptosauridae. Neues Jahrbuch für Geologie und Paläontologie, Abhandlungen 160: 73–95.

———. 1983. The cranial anatomy of *Dryosaurus*, a hypsilophodontid dinosaur from the Upper Jurassic of North America and East Africa, with a review of hypsilophodontids from the Upper Jurassic of North America. Geologica et Palaeontologica 17: 207–243.

Garrison, J. R., D. Brinkman, D. J. Nichols, P. Layer, D. Burge, and D. Thayn. 2007. A multidisciplinary study of the Lower Cretaceous Cedar Mountain Formation, Mussentuchit Wash, Utah: a determination of the paleoenvironment and paleoecology of the *Eolambia caroljonesa* dinosaur quarry. Cretaceous Research 28: 461–494.

Gates, T. A., and S. D. Sampson. 2007. A new species of *Gryposaurus* (Dinosauria: Hadrosauridae) from the late Campanian Kaiparowits Formation, southern Utah, USA. Zoological Journal of the Linnean Society 151: 351–376.

Gilmore, C. W. 1933. On the dinosaurian fauna of the Iren Dabasu Formation. Bulletin of the American Museum of natural History 67: 23–78.

Godefroit, P., Z.-M. Dong, P. Bultynck, H. Li, and L. Feng. 1998. New *Bactrosaurus* (Dinosauria: Hadrosauroidea) material from Iren Dabasu (Inner Mongolia, P.R. China). Bulletin de l'Institut royal des Sciences naturelles de Belgique, Sciences de la Terre 68 (supplement): 3–70.

Godefroit, P., V. Alifanov, and Y. L. Bolotsky. 2004. A re-appraisal of *Aralosaurus tuberiferus* (Dinosauria, Hadrosauridae) from the Late Cretaceous of Kazakhstan. Bulletin de l'Institut royal des Sciences naturelles de Belgique, Sciences de la Terre 74 (supplement): 139–154.

Godefroit, P., H. Li, and C.-Y. Shang. 2005. A new primitive hadrosauroid dinosaur from the Early Cretaceous of Inner Mongolia. Comptes rendus Palevol 4: 697–705.

Godefroit, P., S.-L Hai, T.-X. Yu, and P. Lauters. 2008. New hadrosaurid dinosaurs from the uppermost Cretaceous of northeastern China. Acta Palaeontologica Polonica 53: 47–74.

Godefroit, P., V. Codrea, and D. B. Weishampel. 2009. Osteology of *Zalmoxes shqiperorum* (Dinosauria: Ornithopoda), based on new specimens from the Upper Cretaceous of Nălaţ-Vad, Romania. Geodiversitas 30: 525–553.

Head, J. J. 1998. A new species of basal hadrosaurid (Dinosauria, Ornithischia) from the Cenomanian of Texas. Journal of Vertebrate Paleontology 28: 718–738.

Horner, J. R., D. B. Weishampel, and C. A. Forster. 2004. Hadrosauridae; pp. 438–463 in D. B. Weishampel, P. Dodson, and H. Osmólska (eds.), The Dinosauria. 2nd ed. University of California Press, Berkeley.

Janensch, W. 1955. Der Ornithopode *Dysalotosaurus* der Tendaguruschichten. Palaeontographica (supplement 7) 3: 105–176.

Kirkland, J. I. 1998. A new hadrosaurid from the Upper Cedar Mountain Formation (Albian–Cenomanian Cretaceous) of Eastern Utah—the oldest known hadrosaurid (lambeosaurine?); pp. 283–295 in S. G. Lucas, J. I. Kirkland, and J. W. Estep (eds.), Lower and Middle Cretaceous terrestrial ecosystems. New Mexico Museum of Natural History and Science, Albuquerque.

Kobayashi, Y., and Y. Azuma. 2003. A new iguanodontian (Dinosauria: Ornithopoda) from the Lower Cretaceous Kitadani Formation of Fukui Prefecture, Japan. Journal of Vertebrate Paleontology 23: 166–175.

Lü, J.-C. 1997. A new Iguanodontidae (*Probactrosaurus mazongshanensis* sp. nov.) from Mazongshan Area, Gansu province, China; pp. 27–47 in Z.-M. Dong, (ed.), Sino-Japanese Silk Road dinosaur expedition. China Ocean Press, Beijing.

Lull, R. S. an N. E. Wright. 1942. Hadrosaurian dinosaurs of North America. Geological Society of America Special Papers 40: 1–242.

Marsh, O. C. 1881. Classification of the Dinosauria. American Journal of Science (ser. 3) 23: 81–86.

McDonald, A. T., J. I. Kirkland, J. Bird, and D. Deblieux. 2009. Thumb-spiked dinosaurs large, small, and strange: new information on basal iguanodonts from the Cedar Mountain Formation of Utah. Journal of Vertebrate Paleontology 29 (supplement to 3): 145A.

McDonald, A. T., D. G. Wolfe, and J. I. Kirkland. 2010. A new basal hadrosauroid (Dinosauria: Ornithopoda) from the Turonian of New Mexico. Journal of Vertebrate Paleontology 30: 799–812.

Norman, D. B. 1980. On the ornithischians dinosaur *Iguanodon bernissartensis* of Bernissart (Belgium). Mémoires de l'Institut royal des Sciences naturelles de Belgique 178: 1–103.

———. 1986. On the anatomy of *Iguanodon atherfieldensis* (Ornithischia: Ornithopoda). Bulletin de l'Institut Royal des Sciences Naturelles de Belgique, Sciences de la Terre 56: 281–372.

———. 1996. On Mongolian ornithopods (Dinosauria: Ornithischia). 1. *Iguanodon orientalis* Rozhdestvensky 1952. Zoological Journal of the Linnean Society 116: 303–313.

———. 1998. On Asian ornithopods (Dinosauria: Ornithischia). 3. A new species of iguanodontid dinosaur. Zoological Journal of the Linnean Society 122: 291–348.

———. 2002. On Asian ornithopods (Dinosauria: Ornithischia). 4. *Probactrosaurus* Rozhdestvensky, 1966. Zoological Journal of the Linnean Society 136: 113–144.

———. 2004. Basal Iguanodontia; pp. 413–437 in D. B. Weishampel, P. Dodson, and H. Osmólska (eds.), The Dinosauria. 2nd ed. University of California Press, Berkeley.

Ostrom, J. H. 1961. Cranial morphology of the hadrosaurian dinosaurs of North America. Bulletin of the American Museum of Natural History 122: 33–186.

———. 1970. Stratigraphy and paleontology of the Cloverly Formation (Lower Cretaceous) of the Bighorn Basin area, Wyoming and Montana. Bulletin, Peabody Museum of Natural History 30: 1–234.

Owen, R. 1842. Report on British Fossil Reptiles. Part 2. Report of the British Association for the Advancement of Science (Plymouth) 11: 60–204.

Parks, W. A. 1920. The osteology of the trachodont dinosaur *Kritosaurus incurvimanus*. University of Toronto Studies, Geological Series 11: 1–74.

Prieto-Marquez, A. 2007. Postcranial osteology of the hadrosaurid dinosaur *Brachylophosaurus canadensis* from the Late Cretaceous of Montana; pp. 91–115 in K. Carpenter, (ed.), Horns and beaks: ceratopsian and ornithopod dinosaurs. Indiana University Press, Bloomington.

Ruiz-Omeñaca, J.-I, Pereda-Suberbiola, X., and Galton, P. M. 2007. *Callovosaurus leedsi*, the earliest dryosaurid dinosaur (Ornithischia: Euornithopoda) from the Middle Jurassic of England; pp. 3–16 in K. Carpenter (ed.), Horns and beaks: ceratopsian and ornithopod dinosaurs. Indiana University Press, Bloomington.

Russell, D. A. 1993. The role of Central Asia in dinosaurian biogeography. Canadian Journal of Earth Sciences 30: 2002–2012.

Seeley, H. G. 1887. On the classification of the fossil animals commonly called Dinosauria. Proceedings of the Royal Society of London 43: 165–171.

Sereno, P. C. 1986. Phylogeny of the bird-hipped dinosaurs (Order Ornithischia). National Geographic Society Research 2: 234–256.

———. 1997. The origin and evolution of dinosaurs. Annual Review of Earth and Planetary Sciences 25: 435–489.

———. 1998. A rationale for phylogenetic definitions, with application to the higher-level taxonomy of Dinosauria. Neues Jahrbuch für Geologie und Paläontologie. Abhandlungen 210: 41–83.

Smith, P. E., N. M. Evensen, D. York, M.-M. Chang, F. Jin, J.-L. Li, S. L. Cumbaa, and D. A. Russell. 1995. Dates and rates in ancient lakes: $^{40}Ar/^{39}Ar$ evidence for an Early Cretaceous age for the Jehol Group, northeast China. Canadian Journal of Earth Sciences 32: 1426–1431.

Sues, H.-D., and A. Averianov. 2009. A new basal hadrosauroid dinosaur from the Late Cretaceous of Uzbekistan and the early radiation of duck-billed dinosaurs. Proceedings of the Royal Society B 276: 2549–2555.

Swisher, C. C., III, Y.-Q. Wang, X.-L. Wang, X. Xu, and Y. Wang. 1999. Cretaceous age for the feathered dinosaurs of Liaoning, China. Nature 400: 58–61.

Swisher C. C., III, X. Wang, Z. Zhou, Y. Wang, F. Jin, J. Zhang, X. Xu, F. Zhang, and Y. Wang. 2002. Further support for a Cretaceous age for the feathered-dinosaur beds of Liaoning, China: new $^{40}Ar/^{39}Ar$ dating of the Yixian and Tuchengzi formations. Chinese Science Bulletin 47: 135–138

Swofford, D. L. 2000. Phylogenetic analysis using parsimony (and other methods). Version 4.0b10. Sinauer Associates, Sunderland, Mass., 40 pp.

Taquet, P. 1976. Géologie et paléontologie du gisement de Gadoufaoua (Aptien du Niger). Cahiers de Paléontologie, CNRS, Paris.

Taquet P., and D. A. Russell. 1999. A massively-constructed iguanodont from Gadoufaoua, Lower Cretaceous of Niger. Annales de Paléontologie 85: 85–96.

Wang, X.-L., and X. Xu. 2001. A new iguanodontid (*Jinzhousaurus yangi* gen. et sp. nov.) from the Yixian Formation of western Liaoning, China. Chinese Science Bulletin 46: 419–423.

Wang, X.-L., H.-L. You, Q. Meng, C.-L. Gao, X.-D. Cheng, and J.-Y. Liu. 2007. *Dongbeititan dongi*, the first sauropod dinosaur from the Lower Cretaceous Jehol Group of western Liaoning province, China. Acta Geologica Sinica 81: 911–916.

Wang, X.-L., R. Pan, R. J. Butler, and P. M. Barrett. 2011. The postcranial skeleton of the iguanodontian ornithopod *Jinzhousaurus yangi* from the Lower Cretaceous Yixian Formation of western Liaoning, China. Earth and Environmental Science Transactions of the Royal Society of Edinburgh 101: 135–159.

Weishampel, D. B., D. B. Norman, and D. Grigorescu. 1993. *Telmatosaurus transsylvanicus* from the Late Cretaceous of Romania: the most basal hadrosaurid dinosaur. Palaeontology 36: 361–385.

Weishampel, D. B., C.-M. Jianu, Z. Csiki, and D. B. Norman. 2003. Osteology and phylogeny of *Zalmoxes* (n.g.), an unusual euornithopod dinosaur from the latest Cretaceous of Romania. Journal of Systematic Palaeontology 1: 65–123.

Winkler, D. A., P. A. Murry, and L. L. Jacobs. 1997. A new species of *Tenontosaurus* (Dinosauria: Ornithopoda) from the Early Cretaceous of Texas. Journal of Vertebrate Paleontology 17: 330–348.

Wu, W.-H., P. Godefroit, and D.-Y. Hu. 2010. A new iguanodontoid dinosaur (*Bolong yixianensis* gen. et sp. nov.) from the Yixian Formation of western Liaoning, China. Geology and Resources 19: 127–133.

Xu, X., and M. A. Norell. 2006. Non-avian dinosaur fossils from the Lower Cretaceous Jehol Group of western Liaoning, China. Geological Journal 4: 419–437.

Xu, X., X.-Y. Zhao, J.-C. Lü, W.-B. Huang, and Z.-M. Dong. 2000. A new iguanodontian from the Sangping Formation of Neixiang, Henan and its stratigraphical implications. Vertebrata PalAsiatica 38: 176–191.

You, H.-L., and D.-Q. Li. 2009. A new basal hadrosauriform dinosaur (Ornithischia: Iguanodontia) from the Early Cretaceous of northwestern China. Canadian Journal of Earth Sciences 46: 949–957.

You, H.-L., Q. Ji, J.-L. Li, and Y.-X. Li. 2003a. A new hadrosauroid dinosaur from the Mid-Cretaceous of Liaoning, China. Acta Geologica Sinica 77: 148–154.

You, H.-L., Z.-X. Luo, N. H. Shubin, L. M. Witmer, Z.-L. Tang, and F. Tang. 2003b. The earliest-known duck-billed dinosaur from deposits of late Early Cretaceous age in northwest China and hadrosaur evolution. Cretaceous Research 24: 347–355.

You, H.-L., Q. Ji, and D.-Q. Li. 2005. *Lanzhousaurus magnidens* gen. et sp. nov. from Gansu province, China: the largest-toothed herbivorous dinosaur in the world. Geological Bulletin of China 24: 785–794.

Zhou, Z.-H. 2006. Evolutionary radiation of the Jehol Biota: chronological and ecological perspectives. Geological Journal 41: 377–393.

20.1. Geographical locations of sites where hadrosauroid dinosaurs have been recovered in Kazakhstan.

A New Basal Hadrosauroid Dinosaur from the Upper Cretaceous of Kazakhstan

20

**Pascal Godefroit*, François Escuillié,
Yuri L. Bolotsky, and Pascaline Lauters**

AEHM 4/1, a fragmentary specimen from the Bostobinskaya Svita (Santonian–Campanian) of Central Kazakhstan, is described as the holotype of a new genus and species of hadrosauroid dinosaur. *Batyrosaurus rozhdestvenskyi* gen. et sp. nov. is characterized by cranial autapomorphies, as well as a unique combination of characters. The curvature and size of the semicircular canals of the inner ear suggest that *Batyrosaurus rozhdestvenskyi* was an agile biped. A phylogenetic analysis reveals that *Batyrosaurus* is a basal hadrosauroid more derived than *Bolong, Equijubus, Jinzhousaurus,* and *Altirhinus,* but more primitive than *Probactrosaurus, Eolambia,* and *Protohadros.* Basal hadrosauroids were diversified in Asia during the Early Cretaceous and probably migrated into North America after the establishment of the Bering land bridge in the Aptian–Albian. *Batyrosaurus* is the youngest nonhadrosaurid hadrosauroid described so far according to the definition of Hadrosauridae here adopted. A phylogenetic analysis also confirms that hadrosaurids probably had an Asian origin.

Introduction

Hadrosauroidea is a stem-based taxon defined as all Iguanodontoidea closer to *Parasaurolophus* than to *Iguanodon* (Sereno, 1997, amended). Basal hadrosauroids (=nonhadrosaurid Hadrosauroidea) were particularly diversified in Asia during the Early Cretaceous. Some of them (*Jinzhousaurus, Bolong*) are represented by fairly complete skeletons, although most basal hadrosauroids are known from partial and disarticulated material. During the closing stages of the Late Cretaceous, Hadrosauroidea were represented by Hadrosauridae, a node-based taxon defined as the most recent common ancestor of *Bactrosaurus* and *Parasaurolphus,* plus all the descendants of this common ancestor (see Norman, 2004, fig. 11.22): they became the most diverse and abundant large vertebrates of Laurasia (Horner et al., 2004). In this chapter, we describe a new basal hadrosauroid from the Santonian–Campanian of Kazakhstan. This is the youngest basal hadrosauroid discovered so far, following the definition of Hadrosauridae here adopted.

The new taxon is based on a fragmentary skeleton, including both cranial and postcranial material, discovered in the Akkurgan locality, east of the Aral Sea in Central Kazakhstan, in the Kzyl-Orda Oblast', 135 km north of the Dzhusala railway station (Fig. 20.1). Shilin (1977) was the first to describe the Akkurgan locality. Shilin and Suslov (1982) allocated the fossiliferous beds in the Akkurgan locality to the Bostobinskaya Svita (=Bostobe Formation; Sues and Averianov, 2009) on the basis of the general similarity of the lithofacies with other outcrops of the formation in this area. Dinosaur fossils were discovered in greenish-gray shales with pink flecks about 2 m below purple shales (Shilin and Suslov, 1982; Norman and Kurzanov, 1997). Abundant plant remains were collected from Akkurgan (Shilin, 1977). This paleoflora is dominated by small and narrow leaves with denticulate margins attributed to the family Ulmaceae (Shilin, 1977). Comparisons with floras elsewhere in Kazakhstan support a Santonian–Campanian age because older Cenomanian to Turonian deposits are dominated by angiosperms belonging to the family Platanaceae, characterized by broader leaves (Shilin, 1978; Shilin and Suslov, 1982).

From this locality, Shilin and Suslov (1982) also described the caudal part of a left maxilla and the associated distal end of a femur that they attributed to a new hadrosaurid dinosaur, *Arstanosaurus akkurganensis*. Norman and Kurzanov (1997) restudied the holotype and one previously unreported maxillary tooth, and they regarded *Arstanasaurus akkurganensis* as a nomen dubium closely related to *Bactrosaurus johnsoni* from the Iren Dabasu Formation of Erenhot (P.R. China) and hadrosauroid material from Baunshin Tsav in Mongolia.

Institutional abbreviations. AEHM, Palaeontological Museum of the Institute of Geology and Nature Management, Far East Branch, Russian Academy of Sciences, Blagoveschensk, Russia; AMNH, American Museum of National History, New York, USA.

Systematic Paleontology

Dinosauria Owen, 1842
Ornithischia Seeley, 1887
Ornithopoda Marsh, 1881
Iguanodontia Dollo, 1888 [Sereno, 1986]
Iguanodontoidea Cope, 1869
Hadrosauroidea Cope, 1869

Batyrosaurus rozhdestvenskyi gen. et sp. nov.
(Figs. 20.2–20.12)

Etymology. The generic name refers to *batyrs*, Kazakh heroic knights; *sauros* (Greek) means "lizard." The specific name refers to the famous Russian paleontologist A. N. Rozhdestvensky for his pioneering works on Middle Asian Iguanodontia.

Holotype. AEHM 4/1, a partial skeleton, including partial skull and mandible, about 60 isolated teeth, paired sternals, a right humerus, a left radius, fragmentary metacarpals, and manual phalanges.

Locality and horizon. Akkurgan locality, east of the Aral Sea in Central Kazakhstan, in the Kzyl-Orda Oblast', 135 km north of the Dzhusala railway station (Fig. 20.1); Bostobinskaya Svita, Santonian–Campanian (Late Cretaceous).

Diagnosis. Nonhadrosaurid hadrosauroid characterized by the following autapomorphies: parietal forming long caudal lappets, which overlie the supraoccipital and are overlain by the squamosals, but remain well separated from the paroccipital wings (character observed in *Bactrosaurus johnsoni*); frontals rostrocaudally elongated, about 1.7 times longer than wide; lateral side of rostral process of jugal bearing a deep horizontal sulcus below the lacrimal facet; glenoid surface of surangular perforated by a foramen.

Osteology

Supraoccipital. The supraoccipital is a stout pyramidal bone that extends upward and forward in the occipital region of the skull (Fig. 20.2D). The inclination of the supraoccipital is sometimes regarded as an important character among Iguanodontoidea (e.g., Sues and Averianov, 2009). It is usually more vertical in basal Iguanodontoidea than in Hadrosauridae. However, this character has not been retained in the present phylogenetic analysis because inclination of the supraoccipital seems progressive in basal Hadrosauroidea. Surprisingly, its intraspecific variability appears high in the lambeosaurine *Amurosaurus riabinini* (P.G., pers. obs.): the apparent inclination might depend on the deformation of the skull as a whole. The supraoccipital–exoccipital contact is marked by a strong horizontal ridge, followed dorsally by a slitlike transverse groove. A tall median crest occupies the dorsal two-thirds of the posterior side of the bone and expands dorsally, forming a knoblike structure in contact with the parietals. This crest is bordered by two elongated and shallowly depressed areas for insertion of M. spinalis capitis. The lateral sides of the supraoccipital are covered by the caudal lappets of the parietals. Under this area, the lateral side of the supraoccipital is irregularly sculptured by grooves and ridges, marking the contact areas with the squamosals. Supraoccipital knobs are not developed on the caudal surface of the bone. Prominent knobs are developed in hadrosaurids (e.g., Ostrom, 1961, Godefroit et al., 2004), and *Levnesovia transoxiana* (see Sues and Averianov, 2009). However, the phylogenetic significance of this character remains doubtful among basal hadrosauroids and hadrosaurids because Godefroit et al. (1998) observed that the development of supraoccipital knobs is apparently an ontogenetic character in *Bactrosaurus johnsoni*. In rostral view, the rostrodorsal part of the supraoccipital is depressed to form the caudal portion of the roof of the myencephalon.

Fused exoccipital–opisthotic. In caudal view, the paired exoccipitals exclude the supraoccipital from the roof of the foramen magnum. The exoccipital condyloids are particularly massive and form the prominent dorsolateral portions of the occipital condyle. In lateral view, the exoccipital condyloids are pierced by three foramina for cranial nerves, caudal to the metotic strut (Fig. 20.3). The caudal foramen for the hypoglossal nerve (XII) is the largest, whereas the foramen for the accessory nerve (XI) is the smallest. Both open medially into the endocranial cavity. The vagus canal (for c.n X., "perilymphatic duct" of authors) is set more dorsally and opens into the metotic ("jugular") foramen. Cranial to the metotic strut, a large auditory recess is divided by an oblique crista interfenestralis into a rostral fenestra ovalis (stapedial recess) and a caudal metotic foramen (Fig. 20.3). The metotic foramen contains separate openings for c.n. IX (dorsal) and the internal jugular vein (ventral).

Prootic. The prootic takes an important part in the formation of the lateral wall of the braincase (Fig. 20.2C). Together with the opisthotic, the medial wall of the prootic is inflated to form the vestibular pyramid

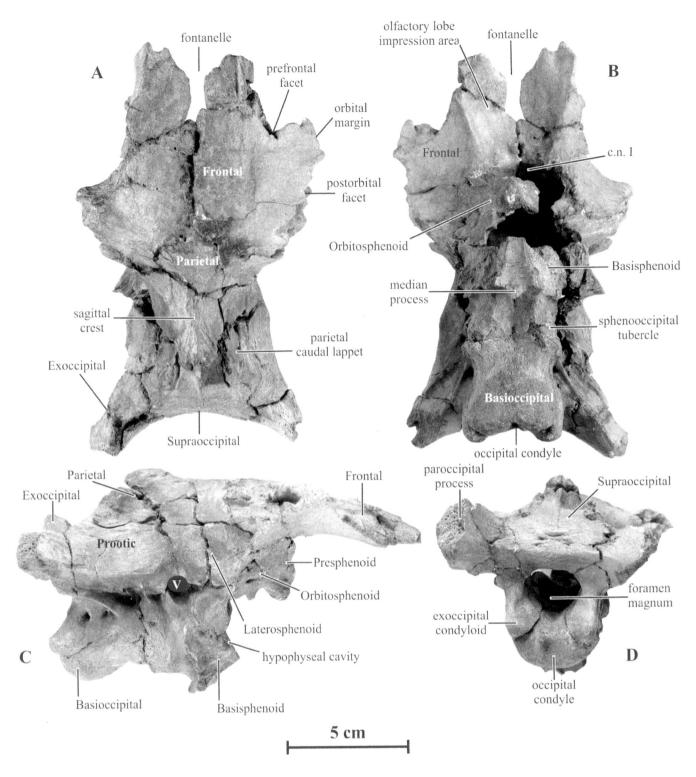

20.2. Braincase of *Batyrosaurus rozhdestvenskyi* gen. et sp. nov. (AEHM 4/1, holotype) in dorsal (A), ventral (B), right lateral (C), and caudal (D) views.

that contains the inner ear cavity. A broad and stout caudodorsal branch from the prootic covers the rostromedial part of the fused exoccipital–opisthotic. The dorsal border of this branch also contacts the supraoccipital. The lateral surface of the prootic has a prominent and horizontal crista otosphenoidale that extends into the rostrolateral side of the paroccipital process. The caudoventral portion of the prootic is notched by the rostral margin of the auditory recess, whereas its rostral border is deeply excavated by the caudal margin of the large and round foramen for the Gasserian ganglion of the trigeminal nerve (c.n. V). From this foramen, a vertical groove extends along the lateral surface of the prootic, just behind the basisphenoid process of the laterosphenoid (Fig. 20.3); this ventrally directed groove indicates the passage for ramus mandibularis of the trigeminal nerve (c.n. V_3). Between the notches for the auditory foramen and the trigeminal nerve, the lateral wall of the prootic is pierced by a small rounded foramen for transmission of the facial nerve (c.n. VII). From this foramen, a wide sulcus runs ventrally along the basisphenoid process of the prootic and connects the Vidian canal.

Laterosphenoid. The prootic process of the laterosphenoid forms a wide, triangular and caudally directed wing between the parietal and the prootic (Fig. 20.2C). The basisphenoid process forms a ventrally directed foot that covers the alar process of the basisphenoid and the rostrodorsal part of the ventral flange of the prootic. The angle between the prootic and the basisphenoid processes forms the rostral margin of the foramen for the trigeminal foramen. From this notch, a wide and deep groove extends rostrally along the lateral side of the laterosphenoid, indicating the forward passage of the deep ramus ophthalmicus of the trigeminal nerve (V_1). The rostral border of the basisphenoid process is notched by an elongated foramen for the occulomotor (c.n. III) and abducens (c.n. VI) nerves. The postorbital process is elongated and stout, but its lateral end that articulated with the postorbital is not preserved. The lateral side of the laterosphenoid bears a regularly round crest marking the separation between the orbit and the supratemporal fenestra.

Orbitosphenoid. This rounded bone participates in the rostral part of the lateral wall of the braincase and in the greatest part of the incomplete interorbital septum (Fig. 20.2B,C). Its surface is pierced by foramina for the trochlear (c.n. IV), dorsally, and the optic (c.n. II) nerves.

Presphenoid. The paired presphenoids form the rostral part of the interorbital septum and circumscribe the large median opening for the olfactory nerve (c.n. I; Fig. 20.2B).

Basioccipital. The basioccipital is longer than wide (Fig. 20.2B). The occipital condyle is proportionally wide and low. Its articular surface is perfectly vertical, perpendicular to the braincase, and it is incised by a vertical furrow (Fig. 20.2D). The median part of the dorsal surface of the basioccipital takes a large part in the formation of the foramen magnum. The sphenooccipital tubercles are separated from the occipital condyle by a distinct neck.

Basisphenoid. The basisphenoid is poorly preserved. Two large caudal processes, projecting slightly laterally from the basisphenoid and separated by a wide incision, form the rostral half of the sphenooccipital tubercles. The basipterygoid processes are broken off; between them, a median process projects caudally (Fig. 20.2B). Above the median process, the ventral surface of the basisphenoid is pierced by a small median foramen. Above the basipterygoid process and under the foramen for the trigeminal nerve, the lateral side of the basisphenoid forms the alar process (Fig. 20.3). The caudal part of this process conceals the Vidian canal, which carried the internal carotid artery through the basisphenoid into the hypophyseal cavity. Two other pairs of foramina pierce the caudodorsal wall of the hypophyseal cavity. The ventral openings correspond to the passages for the abducens

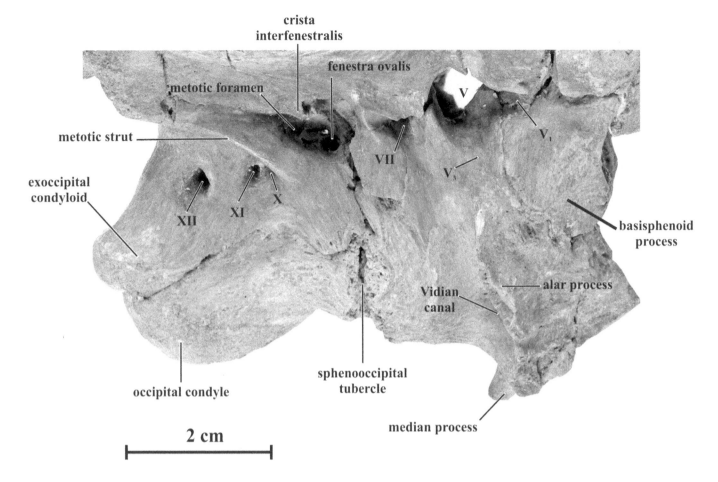

crista
interfenestralis

fenestra ovalis

metotic foramen

metotic strut

exoccipital
condyloid

V

V₁

VII

V₃

XII

XI

X

basisphenoid
process

alar process

Vidian
canal

sphenooccipital
tubercle

occipital condyle

median process

2 cm

20.3. Close-up of the right ventolateral portion of the braincase of *Batyrosaurus rozhdestvenskyi* gen. et sp. nov. (AEHM 4/1, holotype).

nerves (c.n. VI), whereas the dorsal openings are interpreted as passages for ramus caudalis of the internal carotid artery.

Parietal. The fused parietals closely resemble those of *Probactrosaurus gobiensis* and *Bactrosaurus johnsoni*. Rostral lappets of the parietal splay laterally to contact the frontals and the postorbitals (Fig. 20.2A). A wide rhomboid rostromedian process is wedged between the frontals. This process is limited caudolaterally by curved ridges that fuse together at the midlength of the parietal to form the sagittal crest. This crest is much lower than in *Levnesovia transoxiana* (see Sues and Averianov, 2009, fig. 1). Caudal lappets of the parietal cover the lateral sides of the supraoccipital. Irregular ridges on their dorsal surface indicate that they were covered by the squamosals. Similar lappets were also described in *Bactrosaurus johnsoni* (AMNH 6366) and *Probactrosaurus gobiensis* (see Norman, 2002). However, in the latter genus, they contact the paroccipital processes laterally, whereas they remain well separated from the paroccipital process in *Batyrosaurus* and *Bactrosaurus*. In ventral view, the impression for the myencephalon is a narrow and deep depression along the caudal third of the parietal. An ovoid and shallower depression at about the middle of the parietal corresponds to the impression of the cerebellum. The ventral side of the rostromedian process of the parietal has a deep depression corresponding to the location of the caudomedial portion of the cerebrum.

Frontal. The unfused frontals of *Batyrosaurus* appear more elongated rostrocaudally than in other hadrosauroids, being about 1.7 times longer than wide (Fig. 20.2A). Their elongated rostral processes have elongated

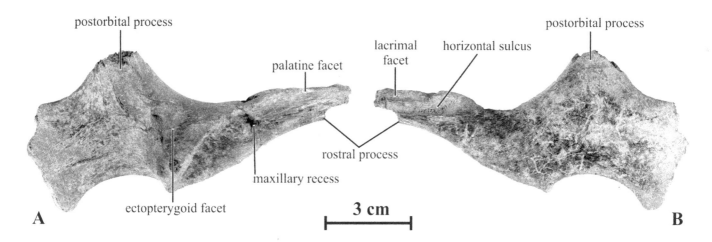

20.4. Left jugal of *Batyrosaurus rozhdestvenskyi* gen. et sp. nov. (AEHM 4/1, holotype) in medial (A) and lateral (B) views.

nasal facets and are separated from each other by a relatively large fontanelle. The rostrolateral margin of the frontal is notched by a long and thickened articular surface for the prefrontal. The caudomedial corner of the prefrontal notch is excavated by a deep recess for reception of the caudomedial process of the prefrontal, as also observed in other basal hadrosauroids and hadrosaurids, including *Probactrosaurus gobiensis* (P.G., pers. obs.), *Bactrosaurus johnsoni* (see Godefroit et al., 1998), and *Levnesovia transoxiana* (see Sues and Averianov, 2009). The orbital rim is short and thin. The articular surface for the postorbital, on the caudolateral border of the frontal, is particularly long, thick, and persillate. The caudal borders of the paired frontals form a wide midline notch where the rostral process of the parietal interfingers. The dorsal surface of the frontals is slightly domed above the cerebrum area, as also observed in juvenile hadrosauroid specimens, but doming is less pronounced than in juvenile lambeosaurines (Evans et al., 2005). In ventral view, the caudomedial part of the frontal is deeply excavated for the cerebrum. Rostrally, the ventral surface of the nasal process is also excavated by the long impression area of the olfactory lobe (Fig. 20.2B).

Jugal. The left jugal of *Batyrosaurus* is partially preserved and closely resembles that of *Altirhinus kurzanovi* (Fig. 20.4). The rostral process is elongate, asymmetrical, and tapers rostrally. In lateral view, its dorsal margin is nearly straight, whereas its ventral margin is sigmoidal. Its dorsal side has a prominent sulcus for insertion of the lacrimal. Below the lacrimal facet, the lateral side of rostral process has a deep horizontal sulcus that suddenly becomes dorsoventrally higher caudally (Fig. 20.4B). The medioventral surface of the rostral process is deeply excavated to form a long recess into which the jugal process of the maxilla slotted (Fig. 20.4A). Medially, a narrow palatine facet is located between the lacrimal facet and the maxillary recess. Immediately caudal to the maxillary recess, a large triangular depression on the medial side of the jugal represents the articular surface for the ectopterygoid. The base of the ectopterygoid facet is marked by a small knoblike protuberance. Basal Iguanodontoidea (sensu Norman, 2004) retain a large jugal–ectopterygoid joint, as observed in *Iguanodon bernissartensis*, *Mantellisaurus atherfieldensis* (see Norman, 2002), *Ouranosaurus nigeriensis* (see Taquet, 1976, fig. 19), and *Altrihinus kurzanovi* (see Norman, 1998, fig. 9). The ectopterygoid facet is distinctly less developed

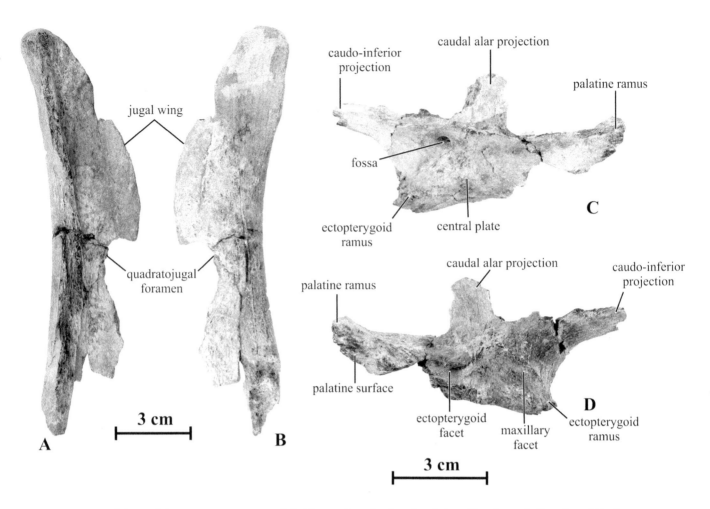

20.5. *Batyrosaurus rozhdestvenskyi* gen. et sp. nov. (AEHM 4/1, holotype). Left quadrate in medial (A) and lateral (B) views; left pterygoid in medioventral (C) and laterodorsal (D) views.

and slitlike in *Protohadros byrdi* (see Head, 1998, fig. 5) and *Probactrosaurus gobiensis* (see Norman, 2002, fig. 6). It is absent in *Bactrosaurus johnsoni* (see Godefroit et al., 1998, fig. 13), *Levnesovia transoxiana* (see Sues and Averianov, 2009), and Euhadrosauria. Both the postorbital and the caudal processes of the jugal are broken off. Under the postorbital process, the ventral margin of the jugal is moderately concave, as also observed in *Altirhinus kurzanovi* (see Norman, 1998, fig. 9), whereas it is distinctly straighter in *Protohadros byrdi* (see Head, 1998, fig. 5) and *Probactrosaurus gobiensis* (see Norman, 2002, fig. 6).

Quadrate. The dorsal portion of the left quadrate is preserved in the holotype. As is usual in basal hadrosauroids, it is slightly curved caudally (Fig. 20.5A,B). Its dorsal articular surface is triangular in cross section and only moderately developed. It is supported caudoventrally by a vertical buttress, also described in *Altirhinus kurzanovi* (see Norman, 1998) and *Probactrosaurus gobiensis* (see Norman, 2002). The pterygoid wing is completely destroyed. The jugal wing is smoothly convex, being interrupted by a fairly deep embayment for the quadratojugal foramen. Two distinct bevelled articular facets, respectively above and at the base of the quadratojugal notch, articulated with the quadratojugal. This situation can also be observed in *Altirhinus kurzanovi*, *Probactrosaurus gobiensis*, *Jeyawati rugoculus*, *Bactrosaurus johnsoni*, and *Gilmoreosaurus mongoliensis*, suggesting that the quadratojugal notch was not completely occluded by the quadratojugal in these taxa. In *Levnesovia transoxiana* (see Sues and

Averianov, 2009), probably *Telmatosaurus transsylvanicus* and Euhadrosauria, the quadratojugal facet is continuous along the whole height of the emargination.

Pterygoid. A partial left pterygoid is preserved in the holotype material (Fig. 20.5C,D). The palatine ramus forms an essentially horizontal, although slightly inclined dorsally, flattened plate. Its dorsolateral side forms an extensive contact surface for the palatine (Fig. 20.5D). At the base of the palatine ramus the pterygoid forms a thickened central plate. Caudal to the palatine suture, the lateral side of the central plate has a cup-shaped depression that accommodated a projection from the ectopterygoid. Immediately caudal to the ectopterygoid articulation, there is a shallow sulcus that articulated with the maxilla. The ectopterygoid ramus is short but robust and projects caudolaterally from the central plate. Its lateral side has a scarred attachment area for the ectopterygoid and the maxilla. The caudal alar projection is incompletely preserved; it extends dorsomedially from the central plate. The caudoinferior projection extends caudodorsally. Its medial side has a prominent buttressing flange, which widens as it merges with the central plate to form the area of the basal articulation. A deep elliptical fossa perforates the ventral side of the central plate just in front of the caudoinferior projection (Fig. 20.5C).

Predentary. This horseshoe-shaped median bone is similar in general structure to that of *Altirhinus kurzanovi* (see Norman, 1998, fig. 15). Its dorsal border has a graduated series of conical denticles, which are replaced further laterally by a rounded ridge (Fig. 20.6B). The edge of the predentary limits a moderately developed lingual shelf, broader caudally than rostrally. Under the denticles, the bone is punctured by foramina and grooves, probably associated with the attachment of a keratinous beak. The ventral surface of the predentary forms a shelf that fitted on the rostral edge of the dentary (Fig. 20.6A). A well-developed ventral process helped to secure the dentaries in position (Fig. 20.6C). The rostral surface of the predentary has two diagonal vascular grooves that connect the dorsolateral corner of the ventral process to the base of the median denticle. These grooves are also well developed in *Iguanodon bernissartensis*, *Mantellisaurus atherfieldensis* (P.G., pers. obs.), *Ouranosaurus nigeriensis* (see Taquet, 1976, fig. 28), *Jinzhousaurus yangi* (see Barrett et al., 2009, fig. 1), *Altirhinus kurzanovi* (see Norman, 1998, fig. 15), *Probactrosaurus gobiensis* (see Norman, 2002, fig. 10), *Protohadros byrdi* (see Head, 1998, fig. 10d), *Bactrosaurus johnsoni* (see Godefroit et al., 1998), *Levnesovia transoxiana* (see Sues and Averianov, 2009), and *Tethyshadros insularis* (see Dalla Vecchia, 2009). The development of these diagonal grooves is variable among Euhadrosauria, but they are usually less marked than in more basal Iguanodontoidea. As in *Altirhinus kurzanovi* (see Norman, 1998), there is no evidence of a median process immediately above the dentary symphysis. However, this part of the predentary is not perfectly preserved in *Batyrosaurus*, and the notion that it is an artefact of preservation cannot be excluded. This median process is, on the other hand, well developed in *Probactrosaurus gobiensis* (see Norman, 2002), *Bactrosaurus johnsoni* (see Godefroit et al., 1998), *Protohadros byrdi* (see Head, 1998), and *Levnesovia transoxiana* (see Sues and Averianov, 2009).

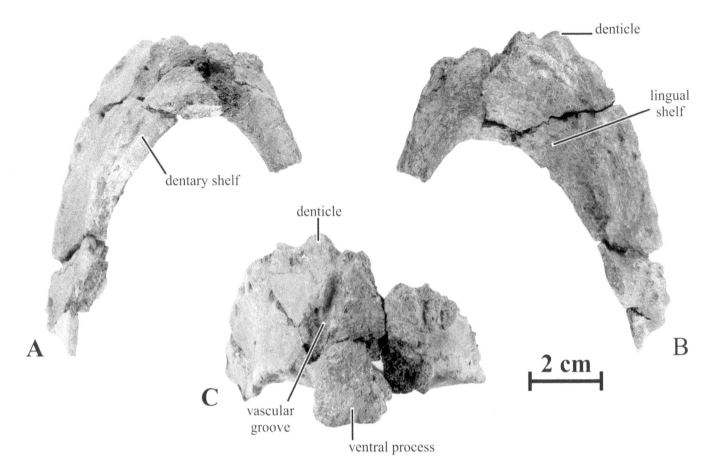

20.6. Predentary of *Batyrosaurus rozhdestvenskyi* gen. et sp. nov. (AEHM 4/1, holotype) in ventral (A), dorsal (B), and rostral (C) views.

Dentary. The left dentary is rather well preserved, although the right one is fragmentary. The dentary ramus is long, robust, and straight, with a ventral margin slightly arched at the level of the symphysis (Fig. 20.7B). The dentary teeth fitted into the array of narrow and parallel alveolar slots on the lingual side of the dentary ramus; 16 vertical slots are preserved in the left specimen, and the total number of slots was probably less than 20 (Fig. 20.7C). The alveolar slots are much deeper and straighter than in *Penelopognathus weishampeli* (see Godefroit et al., 2005, fig. 3) and *Altirhinus kurzanovi* (see Norman, 1998, fig. 16); in the latter taxa, the alveolar slots remain irregular and shaped by the dentary crowns. The dental battery did not extend beyond the level of the coronoid process. The lateral side of the dentary ramus is convex, and the labial shelf is well developed between the dental battery and the coronoid process (Fig. 20.7A). Two large foramina perforate the labial shelf. The coronoid process is high, and the right specimen is slightly rostrocaudally expanded near its apex; it is slightly inclined caudally. The rear edge of the coronoid is slotted for attachment of the surangular. In caudal view, the dentary is excavated by the large adductor fossa, which extends rostrally as a deep Meckel's groove (Fig. 20.7C).

Surangular. The retroarticular process of the surangular is lobate and upturned above the glenoid (Fig. 20.8B,C). It is mediolaterally compressed, with a thin ventral edge. In dorsal view, the glenoid is a long, shallow, and cup-shaped depression (Fig. 20.8A). A smooth transverse ridge bisects it. Rostral to this ridge, a small elliptical foramen perforates the glenoid. This foramen has not yet been described in other iguanodontoids. The lateral margin of the glenoid forms a prominent upturned lip, overhanging the

20.7. Left dentary of *Batyrosaurus rozhdest-venskyi* gen. et sp. nov. (AEHM 4/1, holotype) in dorsal (A), lateral (B), and medial (C) views.

lateral wall of the bone. Rostroventral of the lateral lip a rounded surangular foramen is present, as also seen in basal Iguanodontoidea, including *Bolong yixianensis* (see Chapter 19 in this book), *Jinzhousaurus yangi* (see Barrett et al., 2009), *Altirhinus kurzanovi* (see Norman, 1998, fig. 17), *Probactrosaurus gobiensis* (see Norman, 2002, fig. 13), and *Protohadros byrdi* (see Head, 1998, fig. 12). This foramen is absent in *Bactrosaurus johnsoni* (see Godefroit et al., 1998, fig. 19), *Levnesovia transoxiana* (see Sues and Averianov, 2009), *Telmatosaurus transsylvanicus* (see Weishampel et al., 1993), *Tethyshadros insularis* (see Dalla Vecchia, 2009), and Euhadrosauria. The medial lip of the glenoid is less elevated than the lateral lip, and its rostral end forms a prominent horizontal process. Beneath the medial lip, a deep recess marks the area for attachment of the articular. The attachment surface for the angular forms an elongated and deep slot on the ventral side of the surangular. This facet is slightly inclined laterally, indicating that the angular was well visible on the lateral aspect of the mandible. Rostrally, the surangular forms a curved and thin wall that formed the lateral wall of the adductor fossa beneath and behind the coronoid process.

Angular. The right angular is well preserved and closely resembles that of *Altirhinus kurzanovi*. The left one is more fragmentary. This bone is long and sinuous. Rostrally, its lateral side bears an extended, depressed, and scarred surface that was overlapped by the dentary (Fig. 20.9A). The medial side of its dorsal edge bears an elongated bevelled surface that fitted against the ventral slot of the surangular (Fig. 20.9B). A thin horizontal ridge bisects the rostral half of the medial side of the angular; according to

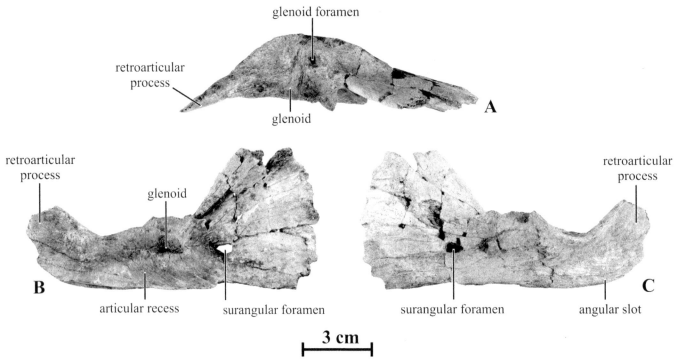

20.8. Left surangular of *Batyrosaurus rozhdestvenskyi* gen. et sp. nov. (AEHM 4/1, holotype) in dorsal (A), medial (B), and lateral (C) views.

Labels: glenoid foramen; retroarticular process; glenoid; A; retroarticular process; glenoid; B; articular recess; surangular foramen; surangular foramen; angular slot; retroarticular process; C; 3 cm

Norman (1998), this ridge would be associated with the muscle attachment to the floor of the adductor fossa. At the rear of the medial side, beneath the contact surface for the surangular, a roughened ledge probably represents the attachment for the articular. The ventral surface of the angular is essentially flat.

Dentary teeth. The enamel is restricted to the lingual side of the crown and to the marginal denticles. The root and the labial side of the crown are composed of dentine, smeared with rough areas of cementum. In mesial and distal views, the tooth is arched lingually (Fig. 20.10B). A vertical lingual facet and two pairs of mesial and distal facets run along the root and of the base of the crown. These facets represent the contact areas with adjacent teeth within the dental battery. The organization of these facets closely resembles the condition observed in *Probactrosaurus gobiensis* (see Norman, 2002, fig. 14) and in more advanced hadrosauroids. The facets are less clearly defined in *Penelopognathus weishampeli* (see Godefroit et al., 2005) and especially *Altirhinus kurzanovi* (see Norman, 2002), reflecting the fact that the dental battery as a whole was less compact in these taxa. The labial side is always strongly truncated because of intensive wearing. The enameled lingual side of the crown is asymmetrically diamond shaped (Fig. 20.10A). Although the proportions of the crown are variable within the dental battery, the height/width ratio of the crown is less than 2. A prominent primary ridge runs the entire height of the crown, dividing the crown surface into two unequal halves. A less prominent secondary ridge, parallel to the primary ridge, bisects the larger mesial half of the crown and always reaches the upper part of the mesial margin. This secondary ridge is situated halfway between the mesial edge of the crown and the primary ridge. A third ridge is faintly developed close to the mesial edge of the crown. A tiny tertiary ridge is developed on some teeth, close to the distal edge. The distal half of the crown also bears a vertical tertiary ridge. On the upper part

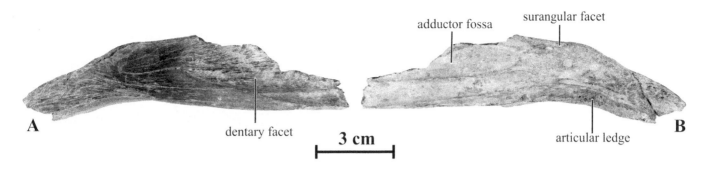

adductor fossa surangular facet

dentary facet

3 cm

articular ledge

A **B**

20.9. Right angular of *Batyrosaurus rozhdest-venskyi* gen. et sp. nov. (AEHM 4/1, holotype) in lateral (A) and medial (B) views.

of the edge of the crown, the denticulations form simple tongue-shaped structures. The structure of the marginal denticulations becomes more elaborated further down the sides of the crown. The edge becomes thickened, and each of the denticulations forms a curved and crenulated ledge, with additional denticulations, which wrap around the edge of the crown. On the mesial half of the crown, a variable number of subsidiary ridges are extensions of the bases of the marginal denticles. Marginal denticles are not developed on the edges along the lower part of the crown. At this level, the mesial edge is only slightly thickened. The distal edge forms an everted, oblique lip, as if the edge of the crown had been pinched inward. Small denticulations are developed on its surface, as also observed in *Iguanodon bernissartensis, Altirhinus kurzanovi,* and *Penelopognathus weishampeli,* but the surface of the ledge is apparently smooth in *Probactrosaurus gobiensis.* The distal everted edge is not developed in *Bactrosaurus johnsoni, Eolambia caroljonesa, Protohadros byrdi, Telmatosaurus transsylvanicus, Tethyshadros insularis,* and Euhadrosauria.

Maxillary teeth. About 30 maxillary teeth were discovered in association with the holotype material. Although no maxilla was discovered, it is highly probable that all belong to this specimen. The maxillary teeth are proportionally narrower and more lozenge shaped in labial view than the dentary teeth, and their structure is much simpler. The mesiolingual and distolingual side of the crows are channeled to accommodate the edges of adjacent replacement crowns. A prominent, keellike primary ridge (Fig. 20.10D) extends along the enameled labial side of the crown, limiting a larger mesial surface and a narrower distal surface (Fig. 20.10C). In *Probactrosaurus gobiensis, Bactrosaurus johnsoni, Levnesovia transoxiana, Protohadros byrdi, Telmatosaurus transsylvanicus, Tethyshadros insularis,* and Euhadrosauria, the maxillary crowns are symmetrical. The primary ridge is similarly keellike in *Altirhinus kurzanovi* (see Norman, 1998), *Levnesovia transoxiana* (see Sues and Averianov, 2009), and *Tethyshadros insularis* (see Dalla Vecchia, 2009). Contrary to *Altirhinus kurzanovi* (Norman, 1998), there is no subsidiary ridge. Denticulations are restricted to the apical half of the crown. Close to the apex, they have a simpler, tongue-shaped structure; they progressively form a curved and crenulated ledge.

Sternal. The sternals of *Batyrosaurus* are typically hatchet shaped (Fig. 20.11A). The proximal plate is about the same size as the distal "handle" and appears proportionally wider than in *Probactrosaurus gobiensis* (see Norman, 2002, fig. 21). It is thinner laterally than medially. Although it is incompletely preserved, the thin medial border of the proximal plate is distinctly convex. The distal handle of the sternal is massive and slightly

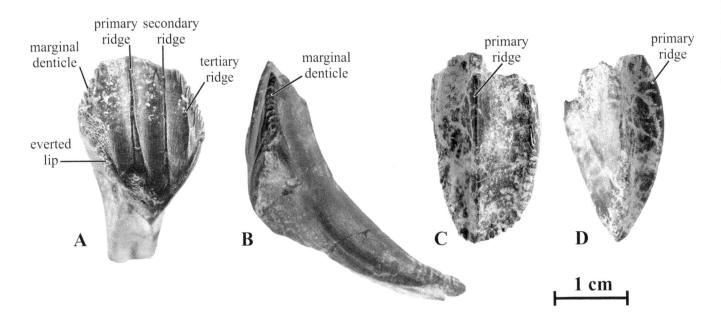

20.10. *Batyrosaurus rozhdestvenskyi* gen. et sp. nov. (AEHM 4/1, holotype). Left dentary tooth in lingual (A) and mesial (B) views; right maxillary tooth in labial (C) and distal (D) views.

curved dorsally; its distal end is slightly enlarged. The dorsal side of the handle has many longitudinal striations. Both the proximal and distal borders of the sternal are roughened, indicating the presence of cartilaginous caps in life.

Humerus. The right humerus is partially preserved in *Batyrosaurus*. The deltopectoral crest is proportionally longer than in *Probactrosaurus gobiensis* (see Norman, 2002, fig. 22), extending well below the middle of the bone, and its lateral edge is turned toward the cranial side of the bone (Fig. 20.11C). Nevertheless, the deltopectoral crest is not as wide as in lambeosaurines. The articular head is globular and separated from the outer tuberosity by a shallow sulcus. The medial side of the humerus is regularly concave. The cranial side of the humerus has a regularly concave and deep bicipital sulcus. Its regularly convex caudal side displays a well-marked scar generated by muscle attachments along its medial edge, as also described in *Probactrosaurus gobiensis* (see Norman, 2002, fig. 22). The distal portion of the humerus is heavily damaged and cannot be adequately described.

Radius. The left radius appears distinctly less robust than in *Altirhinus kurzanovi* (see Norman, 1998, fig. 28). In medial view, it is nearly perfectly straight (Fig. 20.11E). The proximal end of the radius is well expanded, but more compressed mediolaterally than in hadrosaurids; its caudolateral side forms an elongated and flattened surface where it articulated with the ulna (Fig. 20.11D). The distal end is expanded craniocaudally, and an elongated articulation facet on its lateral side faces cranially, suggesting that it fitted obliquely against the distal end of the ulna, as previously described in *Mantellisaurus atherfieldensis* (see Norman, 1986, fig. 49) and *Altirhinus kurzanovi* (see Norman, 1998, fig. 28).

Phalanges. Several fragmentary metacarpals and hand phalanges are preserved in the holotype specimen, but they are too poorly preserved to be adequately described. One eroded ungual phalanx (Fig. 20.11F) is tentatively identified as a pollex spike because it closely resembles that of *Probactrosaurus gobiensis* (see Norman, 1998, fig. 25d–g). Its base is rounded in cross section, but distally, it becomes compressed in the dorsopalmar

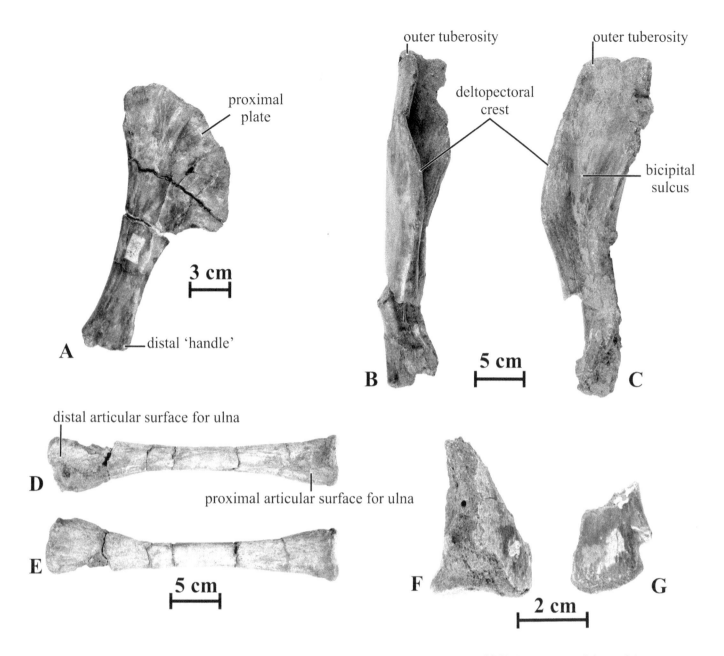

plane. Its proximal articular surface is concave. A well-developed ridge extends along the preserved portion of its lateral side. A second ungual is much more compressed in the dorsopalmar plane and has the tapering and asymmetric shape of an ungual phalanx of digit II (Fig. 20.11G). A well-developed mesial or lateral shelf supports a dorsal groove.

20.11. *Batyrosaurus rozhdestvenskyi* gen. et sp. nov. (AEHM 4/1, holotype). A, Right sternal in ventral view. Right humerus in lateral (B) and cranial (C) views. Left radius in lateral (D) and medial (E) views. F, Ungual phalanx of digit I in medial or lateral view. G, Partial ungual phalanx of digit II in dorsal view.

Paleoneurology

A silicone endocast of *Batyrosaurus rozhdestvenskyi* (AEHM 4/1) was made. The braincase was prepared by covering the smallest foramina and fractures with modeling clay. A thin layer of Vaseline was then sprayed onto the endocranial cavity. Silicone was then mixed with a catalyst and poured into the endocranial cavity to create a first endocranial silicone layer of the cast. This first layer was dried for at least 24 hours. Silicone was then again poured at intervals to create other layers. This technique allows the strengthening of the endocranial cast and prevents its tearing. When the last layer of silicone was completely dry, the endocranial cast was pulled out, and the braincase was then cleaned.

It has often been hypothesized that the brain of sauropsids does not fill the endocranial cavity, and the reasonable assumption is that the endocast is essentially a cast of dura mater (Osborn, 1912; Jerison, 1973; Hopson, 1979). Precisely how much of the endocranial cavity was occupied by the brain was a matter of great debate in paleoneurology (Jerison, 1973; Hopson, 1979; Hurlburt, 1996; Larsson, 2001). Recently, Evans (2005) and Evans et al. (2009) demonstrated for hadrosaurids and pachycephalosaurids that the portion of the endocast corresponding to the telencephalon faithfully represents the contours of the underlying brain. Indeed, valleculae can be observed on endocasts of these dinosaurs (Evans, 2005; P.L., pers. obs.). In *Batyrosaurus rozhdestvenskyi*, valleculae can also distinctly be observed on the anterior part of the telencephalon area (Fig. 20.12B), indicating that at least this part of the brain was in close contact with the bones and reflects the shape of the brain. It is the first time that valleculae are reported in non-hadrosaurid Iguanodontia. The presence of these valleculae indicates that the shape and dimensions of the endocast accurately reflects the general organization of the brain.

As is usual in ornithopods, the general organization of the brain in *Batyrosaurus rozhdestvenskyi* more closely resembles that observed in crocodilian than in birds or pterosaurs (Witmer et al., 2003). When the lateral semicircular canal is horizontal, the great axis of the brain has an inclination of approximately 30 degrees. The brain endocast measures 105 mm from the rostral margin of the cerebrum to the caudal hypoglossal branch, and measures 45.8 mm in maximal width, across the cerebrum. The endocast is relatively narrow, except the rostral part, which is round and enlarged. The telencephalon is round and pear shaped in dorsal view. The cerebral hemispheres are broad and slightly flattened dorsoventrally. The brain is marked by two constrictions: the first one is located just behind the telencephalon. The second and more marked constriction is located behind the large triangular peak above the midbrain region. This constriction housed the inner ear and semicircular canals. The width of the brain remains nearly constant through the medulla. The midbrain and hindbrain are undifferentiated.

A large triangular peak (Fig. 20.12A) projects dorsally above the midbrain region and extends rostrally and caudally over the cerebral and hindbrain regions, respectively. Similar peaks are observed on a wide variety of dinosaur brains, including *Allosaurus* and *Tyrannosaurus* (Hopson, 1979), *Majungasaurus* (Sampson and Witmer, 2007), and to a lesser extent in *Hypacrosaurus*, *Corythosaurus*, and *Lambeosaurus* (Evans et al., 2009). This feature can also be observed in *Iguanodon* and *Mantellisaurus* (see Chapter 16 in this book). It is possible that the dural space housed a well-developed pineal apparatus (epiphysis). Pineal glands are present in extant birds (Breazile and Hartwig, 1989), and evidence for pineal-like tissue in alligators was presented by Sedlmayr et al. (2004).

The ventral part of the telencephalon cannot be described because of the incompleteness of the braincase. As it is always the case in basal iguanodontoid (Andrews, 1897) and hadrosaurid endocasts (Lull and Wright, 1942; Evans et al., 2009), the optic lobes and the cerebellum cannot be

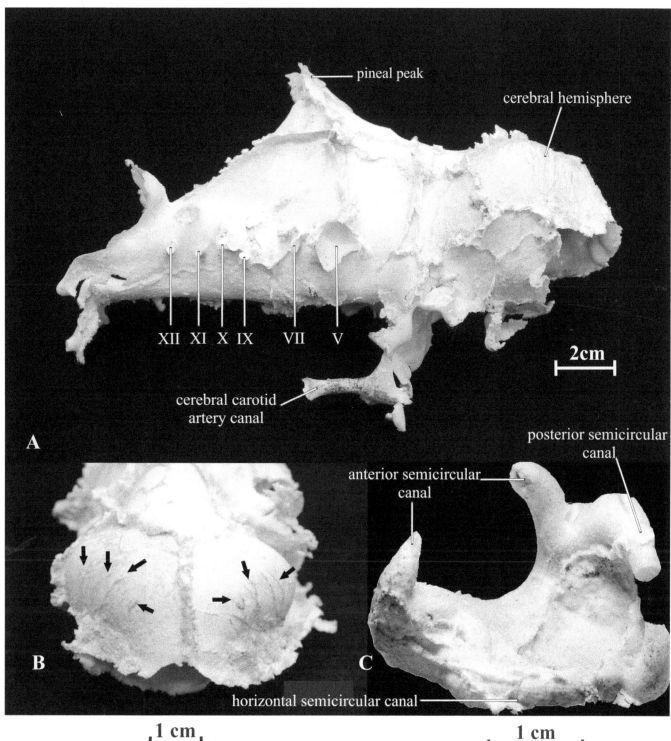

20.12. *Batyrosaurus rozhdestvenskyi* gen. et sp. nov. (AEHM 4/1, holotype). A, Endocranial cast in right lateral view. B, Telencephalon in rostral view; black arrows indicate the valleculae. C, Endosseous labyrinth of the left inner ear in lateral view.

adequately described. This can be partly explained by the impossibility of casting the rostroventral part of the endocranial cavity.

The caudal part of the pituitary fossa and the paired internal carotid arteries coming ventrodorsally through the fossa can be observed on the endocast.

The left inner ear is well preserved and was molded (Fig. 20.12C). The three semicircular canals form regular loops. The anterior semicircular canal is longer than the posterior semicircular canal, and the horizontal is the shortest of the three. As noted by Sipla et al. (2004), large, well-curved semicircular canals, as observed in *Batyrosaurus*, are consistent with agile locomotion. Moreover, an enlarged anterior semicircular canal reflects adaptation to bipedal locomotion, as is the case in theropods and in bipedal basal ceratopsians (*Psittacosaurus*; Zhou et al., 2007). Therefore, the curvature and relative size of the semicircular canals suggest that *Batyrosaurus rozhdestvenskyi* was an agile biped. However, this hypothesis needs to be supported by other inferences that can be drawn from the postcranial skeleton and from trackways—data that are not available so far.

Phylogenetic Relationships of *Batyrosaurus*

A phylogenetic analysis was conducted in order to assess the relationships of *Batyrosaurus rozhdestvenskyi* within Iguanodontia. This analysis was already described in detail in Chapter 19 in this book. The character list is presented in Appendix 19.2, the data matrix in Appendix 19.3, and the description of the strict consensus tree in Appendix 19.4 of Chapter 19. Three most parsimonious trees of 183 steps, with a consistency index (CI) of 0.73, a retention index (RI) of 0.90, and a rescaled consistency index (RCI) of 0.66 resulted from a heuristic search performed by PAUP 4.0 beta 10 (Swofford, 2000). The strict consensus tree (Fig. 20.13) shows that *Batyrosaurus rozhdestvenskyi* occupies a more advanced position within Hadrosauroidea than the Early Cretaceous taxa *Bolong yixianensis* and *Jinzhousaurus yangi* from the Yixian Formation of Liaoning province (China; Wang and Xu, 2001; Wu and Godefroit, in this book), *Equijubus normani* from the Xinminbao Group of Gansu province (China; You et al., 2003b), and *Altirhinus kurzanovi* from the Hühteeg Svita of Mongolia (Norman, 1998). With *Probactrosaurus gobiensis* and more advanced Hadrosauroidea, it forms a clade (k, according to Fig. 20.13), characterized by the following unambiguous (i.e., characters that do not change placement under both ACCTRAN and DELTRAN optimizations) synapomorphies: dentary teeth are set in narrow, parallel-sided grooves (character 53), maxillary teeth lack secondary ridges (character 65 [1 to 2]), and the dentary teeth have a faintly developed tertiary ridge that does not reach the apex of the crown (character 70). The bootstrap value for clade k is not particularly high (75%). Although it is not included in the present phylogenetic analysis, *Penelopognathus weishampeli*, from the Bayan Gobi Formation of Inner Mongolia (China; Godefroit et al., 2005), appears more primitive than *Batyrosaurus rozhdestvenskyi* because the alveolar grooves are less clearly developed on the medial side of the dentary and because the dentary teeth have a well-developed tertiary ridge that reaches the apex of the crown. Although *Batyrosaurus rozhdestvenskyi* closely resembles *Probactrosaurus gobiensis* from the Early Cretaceous Dashuiguo Formation of Inner Mongolia (China; Norman, 2002), the phylogenetic analysis suggests that the latter taxon is more derived because it shares the following unambiguous synapomorphies with more advanced Hadrosauroidea: the jugal–ectopterygoid articulation is more reduced, forming small vertical ectopterygoid facet on the medial side of the jugal (character 34); the primary ridge is closer to the midline on the maxillary crowns (character 66); and the distal shelf at the base of the dentary crowns is well developed but smooth (character 71). The bootstrap support for this node is high (97%).

20.13. Strict consensus tree of Iguanodontia, showing the phylogenetic relationships of *Batyrosaurus rozhdestvenskyi*. Letters correspond to nodes defined in Appendix 19.4.

In our analysis, the monophyly of Hadrosauridae is supported by the following unambiguous synapomorphies: the jugal–ectopterygoid articulation is absent (character 34 [1 to 2]), the surangular foramen is absent (character 57), and the marginal denticles are reduced on both the maxillary and dentary teeth (character 62 [1 to 2]). The relative position of *Bactrosaurus johnsoni* and *Levnesovia transoxiana* at the base of Hadrosauridae cannot be resolved in our analysis. Euhadrosauria (sensu Horner et al., 2004 = Hadrosaurinae + Lambeosaurinae) is a robust clade (bootstrap value = 100%) characterized by 11 unambiguous synapomorphies.

Paleogeography of Hadrosauroidea

Several authors (e.g., You et al., 2003b; Godefroit et al., 2005; Prieto-Márquez, 2008) have already pointed out that Asia occupied a central place in the radiation of Hadrosauroidea during the Lower Cretaceous. The phylogenetic analysis presented here confirms this hypothesis: the Asian taxa *Bolong yixianensis*, *Equijubus normani*, *Jinzhousaurus yangi*, *Altirhinus kurzanovi*, *Batyrosaurus rozhdestvenskyi*, and *Probactrosaurus gobiensis* represent successive steps in the evolution of nonhadrosaurid Hadrosauroidea. With the exception of *Batyrosaurus rozhdestvenskyi*, all come from Lower Cretaceous deposits. "*Probactrosaurus*" *mazongshanensis* Lü, 1997, *Nanyangosaurus zhugeii* Xu, Zhao, Lü, Huang, Li, and Dong, 2000, *Penelopognathus weishampeli* Godefroit et al., 2005, and *Jintasaurus meniscus* You and Li, 2009, have not been included in the phylogenetic analysis but are also potential basal Hadrosauroidea from the Lower Cretaceous of Asia. With its Santonian–Campanian age, *Batyrosaurus rozhdestvenskyi* is the youngest nonhadrosaurid Hadrosauroidea. In our analysis, the North American taxa *Eolambia caroljonesa* and *Protohadros byrdi* are the most derived nonhadrosaurid Hadrosauroidea and are successive sister taxa for hadrosaurids. *Eolambia caroljonesa* was discovered in the lower

Cenomanian (Garrison et al., 2007; Gradstein et al., 2008) Mussentuchit Member of the Cedar Mountain Formation in Utah, which includes other taxa with Asian affinities, such as tyrannosauroids, ceratopsians, and pachycephalosaurs. These records are the earliest known representatives of these groups in western North America (Cifelli et al., 1997; McDonald et al., 2010b). These occurrences can be attributed to an immigration of Asian taxa, including Hadrosauroidea, into western North America after the establishment of the Bering land bridge in the Aptian–Albian (Russell, 1993). Older members of the Cedar Mountain Formation have yielded remains of Iguanodontia incertae sedis, including *"Iguanodon" ottingeri* Galton and Jensen, 1979, *Planocoxa venenica* DiCroce and Carpenter, 2001, *Cedrorestes crichtoni* Gilpin, DiCroce, and Carpenter, 2007, *Iguanocolossus fortis* McDonald et al., 2010 (a), and *Hippodraco scutodens* McDonald et al., 2010 (a). *Protohadros byrdi*, from the Cenomanian Woodbine Formation of Texas, is a bit younger than *Eolambia caroljonesa*. McDonald et al. (2010b) recently described *Jeyawati rugoculus*, a new hadrosauroid dinosaur from the Moreno Hill Formation (middle Turonian) of New Mexico. This new hadrosauroid was described after the completion of the present phylogenetic analysis, and for that reason, it could not be included in the cladogram here (Fig. 12.13). However, the phylogenetic analysis presented by McDonald et al. (2010b) shows that *Jeyawati* is a basal hadrosauroid more derived that *Probactrosaurus*, *Eolambia*, and *Protohadros*, but more basal than *Shuangmiaosaurus*, *Bactrosaurus*, and *Telmatosaurus*. It confirms that basal hadrosauroids became diversified in North America at the beginning of the Late Cretaceous.

The results of the phylogenetic analysis here performed suggest an Asian origin for hadrosaurids and confirms an important evolutionary radiation for this clade around the Turonian (Sues and Averianov, 2009; McDonald et al., 2010b). Indeed, the most basal hadrosaurids are *Bactrosaurus johnsoni* Gilmore, 1933, from the Iren Dabasu Formation of Inner Mongolia, China (Turonian–Coniacian, according to Averianov, 2002, but early Maastrichtian, according to Van Itterbeeck et al., 2005), *Levnesovia transoxiana* Sues and Averianov, 2009, from the Bissekty Formation (middle–late Turonian) of Uzbekistan, and *Shuangmiaosaurus gilmorei* You, Ji, Li, and Li, 2003a, from the Sumjawan Formation (Cenomanian–Turonian) of western Liaoning province, China. According to Prieto-Márquez and Norell (2010), *Gilmoreosaurus mongoliensis* (Gilmore, 1933), from the Iren Dabasu Formation of Inner Mongolia, is also a hadrosaurid as defined here. According to McDonald et al. (2010b), the sister taxon of hadrosaurids, as defined here, would be *Jeyawati rugoculus*, also from the Turonian from New Mexico. *Telmatosaurus transsylvanicus* (Nopcsa, 1900), from the early Maastrichtian Sănpetru and Densus-Ciula Formations of the Haţeg Basin in Romania, and *Tethyshadros insularis* Dalla Vecchia, 2009, from the late Campanian–early Maastrichtian Liburnian Formation of northeastern Italy, are more advanced basal hadrosaurids. According to Dalla Vecchia (2009), hadrosaurids most probably reached the eastern part of the European Archipelago by dispersal through insular hopping along the southwestern margin of Asia during the late Campanian.

Prieto-Márquez (2010a) recently published a new global phylogeny of hadrosauroids (also after the analysis here presented was completed). He found that the nearest outgroups for Euhadrosauria (his Hadrosauridae) are *Claosaurus agilis* Marsh, 1872, from the Coniacian of Kansas, and *Lophorhothon atopus* Langston, 1960, from the Santonian of Alabama. He also considers *Hadrosaurus foulkii* Leidy, 1858, from the Campanian of New Jersey, as the most basal Euhadrosauria. Thus, Prieto-Márquez (2010b) hypothesized that Euhadrosauria originated in eastern North America at least as early as the Coniacian. However, eastern North American hadrosaurids are fragmentary and remain poorly understood, and we personally consider that this interesting hypothesis still requires support by the discovery and analysis of more complete material.

Godefroit et al. (2008) described *Wulagasaurus dongi* (also known from fragmentary material) from the Maastrichtian of northeastern China, as the most basal hadrosaurine described so far. This hypothesis implies a long ghost lineage of basal hadrosaurines in Asia. However, this hypothesis was not retained by Prieto-Márquez (2010b), who considered that *Wulagasaurus* is nested within "kritosaurs," and that hadrosaurines (his Saurolophinae) originated in North America.

If the origin of hadrosaurines remains conjectural, lambeosaurine hadrosaurids clearly have an Asian origin, according to recent phylogenies (e.g., Godefroit et al., 2003, 2004; Evans and Reisz, 2007; Evans, 2010). The oldest lambeosaurines were discovered in Kazakhstan, including *Jaxartosaurus aralensis* Riabinin, 1939, from the Syusyuk Formation (Santonian; Averianov and Nessov, 1995) of the Alimtau range (=Kyrk-Kuduk) in the Chuley Region of southeastern Kazakhstan, and the juvenile skeleton named *Procheneosaurus convincens* from the Dabrazinskaya Svita (Santonian; Rozhdestvensky, 1974) at the Syuk-Syuk Wells site, in the same area. According to Godefroit et al. (2004), Evans and Reisz (2007), Evans (2010), and Prieto-Márquez (2010a), *Aralosaurus tuberiferus* Rozhdestvensky, 1968, from the Beleutinskaya Svita (Turonian–Santonian, Rozhdestvensky, 1968), or the Bostobinskaya Svita (according to Nessov, 1995 and Sues and Averianov, 2009) of the Sakh-Sakh locality is the most basal lambeosaurine; conversely, Sues and Averianov (2009) regard this taxon as the sister taxon for Euhadrosauria (their Hadrosauridae).

Around the Santonian, the iguanodontoid fauna was thus diversified in Kazakhstan, including basal lambeosaurines (*Jaxartosaurus*, "*Procheneosaurus*" convincens, ?*Aralosaurus*) and relict nonhadrosaurid Hadrosauroidea (*Batyrosaurus*). The possibility exists that *Batyrosaurus rozhdestvenskyi* and *Arstanasaurus akkurganensis*, which are from the same locality, belong in fact to the same taxon because the holotypes do not possess diagnostic overlapping elements. In fact, the holotype maxilla of *Arstanosaurus akkurganensis* displays a series of plesiomorphic characters (dorsal process caudally placed, ectopterygoid ridge relatively weak and curved ventrally, jugal process moderately developed into a diagonal projection) that are not incompatible with the assessed phylogenetic position of *Batyrosaurus rozhdestvenskyi*. However, the isolated tooth associated with the holotype of *A. akkurganensis* (see Norman and Kurzanov, 1997, fig. 3) is different from the maxillary teeth associated with the holotype of *B. rozhdestvenskyi*: the crown is more lanceolate and the primary ridge is much less prominent and is closer to the midline. These observations support a separate taxonomic status.

Because Turonian–Santonian dinosaur faunas are still poorly known worldwide, further investigations in Upper Cretaceous dinosaur localities of Kazakhstan will be important for a better understanding of the biogeographical history of advanced Iguanodontoidea and of the early radiation of hadrosaurids. During the Campanian and Maastrichtian, Euhadrosauria reached their widest geographical distribution: they entered South America and Antarctica, both Lambeosaurinae and Hadrosaurinae were particularly diversified in Asia and North America, and at least Lambeosaurinae were represented in Europe.

Acknowledgments

We are grateful to T. Hubin for the photographs of the material. Thanks to F. Dalla Vecchia and A. T. McDonald for reading an earlier version of this chapter and offering useful comments. P. L.'s research is supported by the Fonds pour la Formation à la Recherche dans l'Industrie et dans l'Agriculture (FRIA).

References

Andrews, C. W. 1897. Note on a cast of the brain-cavity of *Iguanodon*. Annals and Magazine of Natural History (ser. 6) 19: 585–591.

Averianov, A. O. 2002. An ankylosaurid (Ornithischia: Ankylosauria) braincase from the Upper Cretaceous Bissekty Formation of Uzbekistan. Bulletin de l'Institut royal des Sciences naturelles de Belgique, Sciences de la Terre 72: 97–110.

Averianov, A., and L. Nessov. 1995. A new Cretaceous mammal from the Campanian of Kazakhstan. Neues Jahrbuch für Geologie und Paläontologie, Monatshefte 1995: 65–74.

Barrett, P. M., R. J. Butler, X.-L.Wang, and X. Xu. 2009. Cranial anatomy of the iguanodontoid ornithopod *Jinzhousaurus yangi* from the Lower Cretaceous Yixian Formation of China. Acta Palaeontologica Polonica 54: 35–48.

Breazile, J. E., and H.-G. Hartwig. 1989. Central nervous system; pp. 485–566 in A. S. King and J. McLelland (eds.), Form and function in birds, vol. 4. Academic Press, New York.

Cifelli, R. L., J. I. Kirkland, A. Weil, A. L. Deino, and B. J. Kowallis. 1997. High-precision ^{40}Ar/^{39}Ar geochronology and the advent of North America's Late Cretaceous terrestrial fauna. Proceedings of the National Academy of Sciences of the United States of America 94: 11163–11167.

Cope, E. D. 1869. Synopsis of the extinct Batrachia, Reptilia and Aves of North-America. Transactions of the American Philosophical Society 14: 1–252.

———. 2009. *Tethyshadros insularis*, a new hadrosauroid dinosaur (Ornithischia) from the Upper Cretaceous of Italy. Journal of Vertebrate Paleontology 29: 1100–1116.

DiCroce, T., and K. Carpenter. 2001. New ornithopod from the Cedar Mountain Formation (Lower Cretaceous) of eastern Utah; pp. 183–196 in D. H. Tanke and K. Carpenter (eds.), Mesozoic vertebrate life. Indiana University Press, Bloomington.

Dollo, L. 1888. Iguanodontidae et Camptonotidae. Comptes Rendus hebdomadaires de l'Académie des Sciences, Paris 106: 775–777.

Evans, D. C. 2005. New evidence on brain–endocranial cavity relationships in ornithischian dinosaurs. Acta Palaeontologica Polonica 50: 617–622.

———. 2010. Cranial anatomy and systematic of *Hypacrosaurus altispinus*, and a comparative analysis of skull growth in lambeosaurine hadrosaurids (Dinosauria: Ornithischia). Zoological Journal of the Linnean Society 159: 398–434.

Evans, D. C., and R. R. Reisz. 2007. Anatomy and relationships of *Lambeosaurus magnicristatus*, a crested hadrosaurid dinosaur (Ornithischia) from the Dinosaur Park Formation. Alberta. Journal of Vertebrate Paleontology 27: 373–393.

Evans, D. C., C. A. Forster, and R. R. Reisz. 2005. The type specimen of *Tetragonosaurus erectofrons* (Ornithischia: Hadrosauridae) and the identification of juvenile lambeosaurines; pp. 349–366 in P. J. Currie and E. B. Koppelhus (eds.), Dinosaur Provincial Park, a spectacular ancient ecosystem revealed. Indiana University Press,

Bloomington.

Evans, D. C., R. Ridgely, and L. M. Witmer. 2009. Endocranial anatomy of lambeosaurine hadrosaurids (Dinosauria: Ornithischia): a sensorineural perspective on cranial crest function. Anatomical Record 292: 1315–1337.

Galton, P. M., and J. A. Jensen. 1979. Remains of ornithopod dinosaurs from the Lower Cretaceous of North America. Brigham Young University, Geology Studies 25: 1–10.

Garrison, J. R., D. Brinkman, D. J. Nichols, P. Layer, D. Burge, and D. Thayn. 2007. A multidisciplinary study of the Lower Cretaceous Cedar Mountain Formation, Mussentuchit Wash, Utah: a determination of the paleoenvironment and paleoecology of the *Eolambia caroljonesa* dinosaur quarry. Cretaceous Research 28: 461–494.

Gilmore, C. W. 1933. On the dinosaurian fauna of the Iren Dabasu Formation. Bulletin of the American Museum of natural History 67: 23–78.

Gilpin, D., T. DiCroce, and K. Carpenter. 2007. A possible new basal hadrosaur from the Lower Cretaceous Cedar Mountain Formation of Eastern Utah; pp. 79–89 in K. Carpenter (ed.), Horns and beaks: ceratopsian and ornithopod dinosaurs. Indiana University Press, Bloomington.

Godefroit, P., Z.-M. Dong, P. Bultynck, H. Li, and L. Feng. 1998. New *Bactrosaurus* (Dinosauria: Hadrosauroidea) material from Iren Dabasu (Inner Mongolia, P.R. China). Bulletin de l'Institut royal des Sciences naturelles de Belgique, Sciences de la Terre 68 (supplement): 3–70.

Godefroit, P., Y. L. Bolotsky, and V. Alifanov. 2003. A remarkable hollow-crested hadrosaur from Russia: an Asian origin for lambeosaurines. Comptes Rendus Palevol 2: 143–151.

Godefroit, P., V. Alifanov, Y. L. Bolotsky. 2004. A re-appraisal of *Aralosaurus tuberiferus* (Dinosauria, Hadrosauridae) from the Late Cretaceous of Kazakhstan. Bulletin de l'Institut royal des Sciences naturelles de Belgique, Sciences de la Terre 74 (supplement): 139–154.

Godefroit, P., H. Li, and C.-Y. Shang. 2005. A new primitive hadrosauroid dinosaur from the Early Cretaceous of Inner Mongolia. Comptes Rendus Palevol 4: 697–705.

Godefroit, P., S. Hai, S., T. Yu, and P. Lauters. 2008. New hadrosaurid dinosaurs from the uppermost Cretaceous of northeastern China. Acta Palaeontologica Polonica 53: 47–74.

Gradstein, F. M., J. G. Ogg, and M. van Kranendonk. 2008. On the geologic time scale, 2008. Newsletters on Stratigraphy 43: 5–13.

Head, J. J. 1998. A new species of basal hadrosaurid (Dinosauria, Ornithischia) from the Cenomanian of Texas. Journal of Vertebrate Paleontology 28: 718–738.

Hopson, J. A. 1979. Paleoneurology; pp. 39–146 in C. Gans (ed.), Biology of the Reptilia. Vol. 9. Academic Press, London.

Horner, J. R., D. B.Weishampel, and C. A. Forster. 2004. Hadrosauridae; pp. 438–463 in D. B. Weishampel, P. Dodson, and H. Osmólska (eds.), The Dinosauria. 2nd ed. University of California Press, Berkeley.

Hurlburt, G. 1996. Relative brain size in recent and fossil amniotes: determination and interpretation. Ph.D. thesis, University of Toronto, 250 pp.

Jerison, H. J. 1973. Brain evolution and dinosaur brains. American Naturalist 103: 575–588.

Langston, W., Jr. 1960. The vertebrate fauna of the Selma Formation of Alabama, part VI: the dinosaurs. Fieldiana: Geology Memoirs 8: 319–360.

Larsson, H. C. E. 2001. Endocranial anatomy of *Carcharodontosaurus saharicus* (Theropoda: Allosauridea) and its implications for theropod brain evolution; pp. 19–33 in D. H. Tanke and K. Carpenter (eds.), Mesozoic vertebrate life: new research inspired by the paleontology of Philip J. Currie. Indiana University Press, Bloomington.

Leidy, J. 1858. *Hadrosaurus foulkii*, a new saurian from the Cretaceous of New Jersey, related to *Iguanodon*. Proceedings of the Academy of Natural Sciences of Philadelphia 10: 213–218.

Lü, J.-C. 1997. A new Iguanodontidae (*Probactrosaurus mazongshanensis* sp. nov.) from Mazongshan Area, Gansu province, China; pp. 27–47 in Z.-M. Dong, (ed.), Sino-Japanese Silk Road Dinosaur Expedition. China Ocean Press, Beijing.

Lull, R. S., and N. E. Wright. 1942. Hadrosaurian dinosaurs of North America. Special Paper 40. Geological Society of America.

Marsh, O. C. 1872. Notice on a new species of *Hadrosaurus*. American Journal of Science (ser. 3) 3: 301.

———. 1881. Classification of the Dinosauria. American Journal of Science (ser. 3) 23: 81–86.

Mc Donald, , A. T., J. I. Kirkland, D. Deblieux, S. K. Madsen, J. Cavin, A. R. C. Milner, and L. Panzarin. 2010a. New basal iguanodonts from the Cedar Mountain Formation of Utah and the evolution of thumb-spiked dinosaurs. Plos One 5 (11): 1–35.

McDonald, A. T., D. G. Wolfe, and J. I. Kirkland. 2010b. A new basal hadrosauroid (Dinosauria: Ornithopoda) from the Turonian of New Mexico. Journal of Vertebrate Paleontology 30: 799–812.

Nessov, L. A. 1995. Dinosaurs of northern Eurasia: new data about assemblages, ecology and palaeobiogeography. Izdatel'stvo Sankt-Peterburgskoi Universiteta, Sankt Peterburg, 156 pp.

Nopcsa, F. 1900. Dinosaurierreste aus Siebenbürgen (Schädel von *Limnosaurus transsylvanicus* nov. gen. et spec.). Denkschriften der königlichen Akademie der Wissenschaften, Wien 68: 555–591.

Norman, D. B. 1986. On the anatomy of *Iguanodon atherfieldensis* (Ornithischia: Ornithopoda). Bulletin de l'Institut royal des Sciences naturelles de Belgique, Sciences de la Terre 56: 281–372.

———. 1998. On Asian ornithopods (Dinosauria: Ornithischia). 3. A new species of iguanodontid dinosaur. Zoological Journal of the Linnean Society 122: 291–348.

———. 2002. On Asian ornithopods (Dinosauria: Ornithischia). 4. *Probactrosaurus* Rozhdestvensky, 1966. Zoological Journal of the Linnean Society 136: 113–144.

———. 2004. Basal Iguanodontia; pp. 413–437 in D. B. Weishampel, P. Dodson, and H. Osmólska (eds.), The Dinosauria. 2nd ed. University of California Press, Berkeley.

Norman, D. B., and S. M. Kurzanov. 1997. On Asian ornithopods (Dinosauria: Ornithischia). 2. *Arstanosaurus akkurganensis* Shilin and Suslov, 1982. Proceedings of the Geologists' Association 108: 191–199.

Osborn, H. F. 1912. Crania of *Tyrannosaurus* and *Allosaurus*. Memoirs of the American Museum of Natural History 1: 1–30.

Ostrom, J. H. 1961. Cranial morphology of the hadrosaurian dinosaurs of North America. Bulletin of the American Museum of Natural History 122: 33–186.

Owen, R. 1842. Report on British Fossil Reptiles. Part 2. Report of the British Association for the Advancement of Science (Plymouth) 11: 60–204.

Prieto-Márquez, A. 2008. Phylogeny and historical biogeography of hadrosaurid dinosaurs. Ph.D. dissertation, Florida State University, Tallahasse, 936 pp.

———. 2010a. Global phylogeny of hadrosauridae (Dinosauria: Ornithopoda) using parsimony and Bayesian methods. Zoological Journal of the Linnean Society 159: 135–502.

———. 2010b. Global historical biogeography of hadrosaurids dinosaurs. Zoological Journal of the Linnean Society 159: 503–525.

Prieto-Márquez, A., and M. A. Norell. 2010. Anatomy and relationships of *Gilmoreosaurus mongoliensis* (Dinosauria: Hadrosauroidea) from the Late Cretaceous of Central Asia. American Museum Novitates 3694: 1–49.

Riabinin, A. N. 1939. The Upper Cretaceous vertebrate fauna of south Kazakhstan. I. Pt. 1. Ornithischia. Tsentralnyy Nauchno-issledovatelnyy geologicheskii Institut Trudy 118: 1–40.

Rozhdestvensky, A. K. 1968. Hadrosaurs of Kazakhstan; pp. 97–141 in L. P. Tatarinov et al. (eds.), Upper Paleozoic and Mesozoic amphibians and reptiles. Akademia Nauk S.S.S.R., Moscow.

———. 1974. The history of dinosaur faunas in Asia and other continents and some problems of palaeogeography. Trudy Sovemestnoi Sovetsko-Mongol'skoi Paleonyologicheskoi Ekspeditsii 1: 107–131.

Russell, D. A. 1993. The role of Central Asia in dinosaurian biogeography. Canadian Journal of Earth Sciences 30: 2002–2012.

Sampson, S. D., and L. M. Witmer. 2007. Craniofacial anatomy of *Majungasaurus crenatissimus* (Theropoda: Abelisauridae) from the Late Cretaceous of Madagascar. Journal of Vertebrate Paleontology 27 (memoir 8, supplement to 2): 32–102

Sedlmayr, J. C., S. J Rehorek, E. J. Legenzoff., and J. Sanjur. 2004. Anatomy of the circadian clock in avesuchian archosaurs. Journal of Morphology 260: 327.

Seeley, H. G. 1887. On the classification of the fossil animals commonly called Dinosauria. Proceedings of the Royal Society of London 43: 165–171.

Sereno, P. C. 1986. Phylogeny of the bird-hipped dinosaurs (Order Ornithischia). National Geographic Society Research 2: 234–256.

———. 1997. The origin and evolution of dinosaurs. Annual Review of Earth and Planetary Sciences 25: 435–489.

Shilin, P. V. 1977. The development of the Late Cretaceous flora of Kazakhstan. Botanical Journal (NAUK) 62: 1404–1414.

———. 1978. The Senonian flora of southern and central Kazakhstan; pp. 78–93 in P. V. Shilin (ed.), Senonskiye Flory Kazakhstana. Nauka, Alma-Ata.

Shilin P. V., and Y. V. Suslov. 1982. A hadrosaur from the northeastern Aral region. Palaeontological Journal 1982: 131–135.

Sipla, J., J. Georgi, and C. Forster. 2004. The semicircular canals of dinosaurs: tracking major transitions in locomotion. Journal of Vertebrate Paleontology 24 (supplement to 3): 113A.

Sues, H.-D., and A. Averianov. 2009. A new basal hadrosauroid dinosaur from the Late Cretaceous of Uzbekistan and the early radiation of duck-billed dinosaurs. Proceedings of the Royal Society B 276: 2549–2555.

Swofford, D. L. 2000. Phylogenetic analysis using parsimony (and other methods). Version 4.0b10. Sinauer Associates, Sunderland, Mass., 40 pp.

Taquet, P. 1976. Géologie et paléontologie du gisement de Gadoufaoua (Aptien du Niger). Cahiers de Paléontologie, CNRS, Paris: 1–191.

Van Itterbeeck, J., D. J. Horne, P. Bultynck, and N. Vandenberghe. 2005. Stratigraphy and palaeoenvironment of the dinosaur-bearing Upper Cretaceous Iren Dabasu Formation, Inner Mongolia, People's Republic of China. Cretaceous Research 26: 699–725.

Wang, X., and X. Xu. 2001. A new iguanodontid (*Jinzhousaurus yangi* gen. et sp. nov.) from the Yixian Formation of western Liaoning, China. Chinese Science Bulletin 46: 419–423.

Weishampel, D. B., D. B. Norman, and D. Grigorescu. 1993. *Telmatosaurus transsylvanicus* from the Late Cretaceous of Romania: the most basal hadrosaurid dinosaur. Palaeontology 36: 361–385.

Witmer, L. M., S. Chatterjee, J. Franzosa, and T. M. Rowe. 2003. Neuroanatomy of flying reptiles and implications for flight, posture and behaviour. Nature 425: 950–953.

Xu, X., X.-Y. Zhao, J.-C. Lü, W.-B. Huang, and Z.-M. Dong. 2000. A new iguanodontian from the Sangping Formation of Neixiang, Henan and its stratigraphical implications. Vertebrata PalAsiatica 38: 176–191.

You, H.-L., and D.-Q. Li. 2009. A new basal hadrosauriform dinosaur (Ornithischia: Iguanodontia) from the Early Cretaceous of northwestern China. Canadian Journal of Earth Sciences 46: 949–957.

You, H.-L., Q. Ji, J. Li, and Y. Li. 2003a. A new hadrosauroid dinosaur from the Mid-Cretaceous of Liaoning, China. Acta Geologica Sinica 77: 148–154.

You, H.-L., Z.-X. Luo, N. H. Shubin, L. M. Witmer, Z.-L. Tang, and F. Tang. 2003b. The earliest-known duck-billed dinosaur from deposits of late Early Cretaceous age in northwest China and hadrosaur evolution. Cretaceous Research 24: 347–355.

Zhou, C.-F., K.-Q. Gao, R. C. Fox, and X.-K. Du. 2007. Endocranial morphology of psittacosaurs (Dinosauria: Ceratopsia) based on CT scans of new fossils from the Lower Cretaceous, China. Palaeoworld 16: 285–293.

Early Cretaceous Terrestrial Ecosystems In and Outside Europe

3

21.1. The exploitation of phosphate nodules from the Albian Sables verts in the Argonne in the late nineteenth century. Trenches are dug into the greensand to reach the phosphate-bearing layer. The nodules are then cleaned from their matrix in running water from a nearby stream (from Meunier, 1898).

Dinosaur Remains from the "Sables Verts" (Early Cretaceous, Albian) of the Eastern Paris Basin

21

Eric Buffetaut* and Laetitia Nori

Dinosaur remains have been known from the early Albian marine "Sables verts" (greensand) of the Argonne region of the eastern Paris Basin since the 1870s. The scanty material available was obtained in the course of commercial phosphate exploitation, an activity that ceased in the early twentieth century. This chapter describes dinosaur bones and teeth from the Sables verts collected in the late nineteenth century that have not hitherto been described. The distal end of a humerus is the first indisputable record of an ankylosaur from the Sables verts. Limb bones and teeth are referred to the enigmatic theropod *Erectopus*, previously described from the Sables verts. Sauropod caudal vertebrae are the first record of that group of dinosaurs from the Sables verts. Dinosaur diversity in the Sables verts is thus higher than previously recognized, but this is certainly only a fraction of the dinosaur fauna from which this assemblage is derived, which probably inhabited the Anglo-Brabant landmass to the north. In contrast to earlier Cretaceous assemblages from Europe, ornithopods are conspicuous by their absence. The mid-Cretaceous dinosaurs of Europe remain poorly known.

Introduction

Dinosaur remains have been known from the Albian "Sables verts" (greensand) of the Argonne region of the eastern Paris Basin since the 1870s, with most of the known material described in a monograph by Sauvage (1882). This material had been found in the course of the commercial exploitation of phosphate nodules contained in the greensand, which were used for the production of fertilizer (Buffetaut, 2006). This exploitation declined sharply at the beginning of the twentieth century because of competition from cheaper phosphate from other sources and completely stopped in the 1930s, so that the chances of finding new fossil specimens are now extremely small. Nevertheless, unpublished material can still be found in old collections. The specimens described here were found in the collections of the Ecole Nationale Supérieure de Géologie of Nancy, currently housed at the Zoological Museum in Nancy. They consist of a relatively small number of bones and teeth, which are interesting as an addition to our knowledge of the poorly known dinosaur assemblage from the Sables verts.

Institutional abbreviation. MNHN, Muséum national d'Histoire naturelle, Paris, France.

The specimens described below come from the hilly and wooded region of the eastern Paris Basin known as the Argonne, made famous by heavy fighting during World War I, in the southeastern part of *département* Ardennes and the northwestern part of *département* Meuse. There, a greensand formation (Sables verts) underlies the middle to late Albian Gault Clay. These glauconitic sands are 20 to 45 m thick and contain phosphate nodules that are concentrated in a 5- to 25-cm-thick layer (Nivoit, 1874). The nodules, which according to Sauvage and Buvignier (1842) were locally known as *coquins* or *crottes du diable* ("devil's shit"), were identified as a potential source of fertilizer in the 1850s, and their commercial exploitation soon began. In the Ardennes alone, 41,000 tonnes of phosphate were extracted in the year 1872. The extraction of the nodules was carried out in trenches or pits, or in underground shafts (Nivoit, 1874; Meunier, 1898; Bestel, 1908). The nodules were then screen-washed in running water to clean them of their clayey and sandy matrix (Fig. 21.1). During the heyday of phosphate exploitation in the Argonne in the 1870s, this activity employed hundreds of workers, and the value of phosphate-producing land had risen sharply. However, the phosphate boom in the Argonne did not last long, and by the first decade of the twentieth century, the industry had considerably declined (Collet, 1904), mainly because of the competition of other sources of phosphate—first the phosphate-bearing Chalk of the Somme in northern France, and then the large phosphate deposits of the French colonies and protectorates in North Africa (Buffetaut, 2006). Fighting in the Argonne during World War I resulted in the destruction of several factories and contributed to the decline of the industry. The last phosphate exploitation seems to have ceased its activity in 1936 (Lapparent, 1964).

The Sables verts of the eastern Paris Basin are 20 to 45 m thick and are overlain by the Gault Clay. They are glauconitic sands, with frequent clayey intercalations, and are interpreted as having been deposited in a quiet shallow marine environment. Their age is early Albian (Magnez-Jannin and Demonfaucon, 1980). Abundant fossils were found with the phosphate nodules during their commercial exploitation, including plant remains, mollusks, crustaceans, and fish and tetrapod remains. Both the plant material and the remains of terrestrial organisms (including dinosaurs) were probably derived from a landmass located a relatively short distance to the north, corresponding to the Paleozoic massif of the Ardennes, then part of the so-called Anglo-Brabant landmass (Hancock and Rawson, 1992). Barrois (1874, 1875) was the first to give brief descriptions of the vertebrate remains from the Sables verts, followed by Sauvage (1876). They were studied at greater length by Sauvage in a large memoir (Sauvage, 1882), in which ichthyosaurs, pliosaurs, crocodilians, and dinosaurs were described. After the decline and end of phosphate exploitation in the Argonne, few new vertebrate remains were found in the Sables verts during the twentieth century, and consequently, few new descriptions have been published. Despite occasional brief mentions of new finds (Collet, 1904) after Sauvage's original descriptions, most papers on the dinosaurs from the Sables verts have been revisions of specimens described by Sauvage (Huene, 1926, 1932; Allain, 2005) or reviews based on Sauvage's work (Corroy, 1922). The present chapter deals with specimens that were collected during the late nineteenth century but have never been described (Buffetaut, 2002). They were found by one of us (L.N.) in the collections of the Ecole Nationale Supérieure de Géologie de Nancy, currently kept at the Zoological Museum (Museum–Aquarium) in Nancy. Provisional collection numbers with the prefix SV ("Sables verts") have been attributed to them.

Few details are available about how the specimens we describe were collected and by whom, but their labels bear names of villages in the part of the Argonne in *département* Meuse. According to their labels, some of the specimens belonged to the collection brought together by Mr. Guillaumot, a tax collector in Meuse, who collected fossils from the phosphate pits in the 1880s (Fliche, 1896). His collection, which included pliosaur remains described by Sauvage (1903), was eventually donated to the University of Nancy.

The fossils described below are generally fragmentary and broken (presumably because they were found in the course of industrial exploitation), but the preservation is generally good. Patches of glauconite-rich greensand adhere to several of them. Pyrite is occasionally present.

Ankylosauria: Nodosauridae indet.

Systematic Description

Humerus. Ankylosaurs are represented in the Nancy collection by the distal end of a left humerus, from Auzéville-en-Argonne (SV1, Fig. 21.2). The specimen has suffered some abrasion, so the surface of the condyles is poorly preserved, exposing spongy bone. The cross section of the shaft at the level of the break shows a thick cortex of dense bone and a relatively small medullary cavity containing bony trabeculae. The cross section is subtriangular as a result of the presence of a weak, slightly oblique ridge on the cranial face of the shaft, directed toward the lateral condyle, which may be the inception of the deltopectoral crest. The distal articular region of the bone is much expanded relative to the shaft. The lateral condyle is much larger than the medial condyle. It projects strongly cranially and only slightly caudally. It has an ovoid outline in distal view. There are some indications of a well-marked ectepicondyle, but that area is poorly preserved, so no details are visible. The medial condyle is ball shaped and less prominent cranially and caudally than the lateral condyle. A short, weak ridge arises from the medial condyle on the cranial face, but there is no indication of an entepicondyle. The condyles are separated from each other by a distinct notch on the cranial side and by a shallow broad depression (olecranon fossa) on the caudal side.

The robust appearance of the bone, its marked distal expansion, and the shape and arrangement of the distal condyles allow referring it to an ankylosaur. A precise identification on the basis of such an incomplete specimen is not easy, although the relative slenderness of the shaft may suggest a nodosaurid rather than an ankylosaurid (Coombs, 1978a, 1978b).

Comparisons with ankylosaurs of similar geological age are difficult because there is little comparable material. Although several ankylosaur taxa have been described from the Early Cretaceous of Europe, including *Hylaeosaurus* and *Polacanthus*, from the Wealden of England, and *Anoplosaurus* from the Cambridge Greensand, their humeri are not easily comparable with the fragment from Auzéville. The humerus from the Barremian of the Isle of Wight tentatively referred to *Hylaeosaurus* by Hulke

(1874), then placed in *Polacanthoides* by Nopcsa (1928) and more recently considered as belonging to *Polacanthus* by Pereda-Suberbiola (1994), does not clearly show the distal end but seems to have condyles that project more strongly cranially than in the specimen from Auzéville. Ankylosaurs from the Cambridge Greensand are late Albian in age (see Unwin, 2001, for a discussion of the age of the Cambridge Greensand) and therefore close in age to the specimen from the Sables verts, but their humerus is poorly known. Seeley (1879) and Pereda-Suberbiola and Barrett (1999) described the fragmentary distal part of the humerus of *Anoplosaurus curtonotus*. Although the condyles are incompletely preserved, the general shape of the specimen is reminiscent of that of the bone from Auzéville, but little can be said beyond that.

Comparisons can also be made with Early Cretaceous ankylosaurs from North America, in particular *Sauropelta edwardsi*, from the late Aptian Cloverly Formation of Wyoming, for which well-preserved humeri are known (Ostrom, 1970). The distal part of the bone is fairly similar to the specimen from Auzéville, but the condyles appear to be more prominent than in the French specimen (see figures in Coombs, 1978a).

In conclusion, the humerus fragment from Auzéville cannot be identified with great precision, but it can probably be referred to a small nodosaurid, possibly related to *Anoplosaurus*, itself a poorly known taxon. Sauvage (1882) tentatively referred to the ankylosaur *Hylaeosaurus* a supposed dermal scute from the Sables verts of Grandpré (Ardennes), but this small, buttonlike specimen, housed in the Natural History Museum in Lille, is not really reminiscent of an ankylosaur scute. The humerus from Auzéville thus provides the first incontrovertible evidence of ankylosaurs in the Sables verts of the Argonne. A nodosaurid scute was described by Knoll et al. (1998) from the Sables verts of Mesnil Saint-Père (Aube), some 130 km to the southeast of the Argonne.

Measurements:

Maximum diameter of shaft at level of break: 52 mm.
Mediolateral width of distal end: 120 mm.

Theropoda: Carnosauria: cf. Erectopus

The theropods are represented in the Nancy collection by an incomplete right tibia (SV2), an incomplete left third metatarsal (SV3), and three teeth (SV4, SV5, and SV6).

Tibia. A right tibia (SV2, Fig. 21.3) lacking its proximal end bears a small paper sticker indicating that it comes from Varennes. Although the shaft is damaged, the distal end is well preserved. It is expanded mediolaterally, its rounded lateral corner jutting out both laterally and distally. The medial margin protrudes much less. In distal view, the bone has a more or less triangular outline, with the apex of the triangle placed medially, and is thicker medially than laterally. The cranial face of the bone shows strong reliefs that form the limits of the facet for the ascending process of the astragalus. The facet is bounded proximomedially by a prominent oblique ridge (suprastragalar buttress) running from the mediodistal corner of the

21.2. Distal end of a left ankylosaur humerus (SV1) from the Albian of Auzéville-en-Argonne (Meuse) in cranial (A), caudal (B), and distal (C) views. Scale bar = 50 mm.

bone in a proximolateral direction; however, its mediodistal part is not well preserved. That ridge increases in height proximally. Proximolaterally, the facet is bounded by a much less prominent but distinct oblique ridge. The facet is therefore triangular in outline, not very tall relative to the length of the bone (which can only be estimated because the proximal part is missing), and does not cover the whole width of the tibia. More proximally, the cranial face of the shaft is flat, whereas the caudal face is markedly convex, which gives the bone a D-shaped cross section. Distally, a small depression on the caudal face suggests that the astragalus extended slightly onto the caudal surface of the tibia. The shaft is straight, without any significant bowing, at least in the preserved part. Close to the break, a distinct ridge can be seen near the lateral margin of the cranial face. It is the distal inception of the fibular crest. The shaft is hollow, with fairly thick walls (thickness up to 15 mm).

The tibia from Varennes differs from that of ceratosaurs in the triangular outline of its distal end; in ceratosaurs, the outline is quadrangular (Rauhut, 2003). It is different from that of coelurosaurs in that the facet for the ascending process of the astragalus is bounded by a strong ridge; in coelurosaurs, which show the derived condition, this area is flat (Rauhut, 2003). The characters of the distal end of this bone thus lead to refer it to a noncoelurosaurian tetanuran. Among this group, it differs from sinraptorids, which have a more convex suprastragalar buttress at a lower angle to the distal margin (Buffetaut and Suteethorn, 2007). From this point of view, the tibia from Varennes is closer to *Allosaurus* and other forms having a suprastragalar buttress that forms a high angle with the distal margin. However, in *Allosaurus fragilis,* the buttress is less prominent (Madsen,

1976). Among Early Cretaceous large theropods from Europe, comparisons can be made mainly with *Neovenator salerii*, from the Wealden (Barremian) of England (Brusatte et al., 2008), and *Erectopus superbus*, from the Albian of France (Sauvage, 1882; Allain, 2005). The tibia of *Neovenator salerii* (Brusatte et al., 2008) differs from the specimen from Varennes in having a much weaker suprastragalar buttress and a facet for the ascending process of the astragalus that extends farther laterally. *Erectopus superbus* is of special interest because it was described on the basis of material from the Sables verts of the Argonne, which leads to suspect that the theropod remains in the Nancy collection may belong to it (for a review of the rather complex taxonomical history of *Erectopus* Huene, 1923, see Allain, 2005, who convincingly argues that the binomen *Erectopus superbus* (Sauvage, 1882) is valid). The distal end of a tibia, originally misidentified as a radius by Sauvage (1882), was redescribed by Allain (2005) on the basis of a cast. It is part of what was apparently a partly articulated skeleton found in an underground phosphate mine at Louppy-le-Château (Sauvage, 1882); most of the original material, which was in a private collection, is considered lost. The tibia of *Erectopus* is similar to the specimen from Varennes in its general shape, in particular in the great development of the laterodistal part, which forms a distally protruding lobe. The distal articular area for the astragalus (figured by Sauvage, 1882) is also quite similar. The probable extension of the astragalus onto the caudal face of the tibia is shared by the specimen from Varennes and *Erectopus superbus*. The main difference may be in the shape of the suprastragalar buttress, which in *Erectopus superbus* is first at an angle of about 30 degrees to the horizontal in its mediodistal part then becomes nearly vertical, whereas it seems to have been more regularly oblique in the specimen from Varennes. However, differences in preservation, and possibly individual variation, may account for part of this difference.

In conclusion, the incomplete tibia from Varennes apparently shows closer similarities with *Erectopus superbus* than with any other theropod, and is therefore tentatively referred to that taxon. Because of the fragmentary character of the available material, *Erectopus* remains a poorly known theropod. According to Allain (2005), it can be referred to Carnosauria.

Measurements:

Length as preserved: 372 mm.
Mediolateral width of distal end: 127 mm.
Height of facet for ascending process of astragalus: 70.5 mm.

Metatarsal III. An incomplete left third metatarsal (Fig. 21.4), lacking its proximal end, is the only specimen in this collection that has no clear indication of origin. However, a large cardboard plate bearing the mention "*Megalosaurus superbus*. Sauvage. Varennes. Froidos" was associated with the theropod limb bones and probably once served as a label in a display cabinet. Varennes and Froidos are two villages in the Argonne. The mention of Varennes probably refers to the incomplete tibia from Varennes described above, in which case it is likely that the metatarsal comes from Froidos. Unlike the other bones described here, it shows pyrite encrustations.

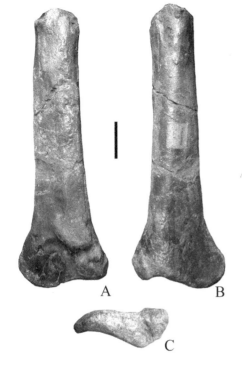

21.3. Distal part of a right theropod tibia (SV2; cf. *Erectopus*) from the Albian of Varennes-en-Argonne (Meuse) in cranial (A), caudal (B), and distal (C) views. Scale bar = 50 mm.

The distal end of the bone is well preserved, forming a mediolaterally and craniocaudally expanded articular area. The articular surface extends for a considerable distance onto the cranial surface of the bone, forming a well-defined lip, separated from the shaft by a distinct convex step followed proximally by a crescent-shaped depression. The articular surface does not extend so far on the caudal surface, where it is limited by a distinct concave line. In distal view, the articular area is thicker medially than laterally. Its proximal margin is more or less straight and slopes laterally, whereas its caudal margin is markedly concave. Both the lateral and medial margins are concave. The whole articular surface is slightly pulley shaped, with a shallow median groove. Both sides of the articular region are concave and excavated by deep pits for the insertion of the collateral ligaments. The shaft is distinctly narrower than the articular end both mediolaterally and craniocaudally. In its distal part, its cranial surface is flat and its caudal surface convex, which gives it a D-shaped cross section. More proximally, the section changes and becomes more or less diamond shaped at the level of the break. This is because of the development of an extensive facet for metatarsal II, which begins on the medial side of the bone and gradually extends onto its cranial surface. The caudal face of the shaft has two more or less parallel ridges, the medial one bounding the above-mentioned facet caudally. Next to the break, a small oval depression on the ventrolateral side of the shaft probably is a facet for metatarsal IV. Also next to the break, there is a distinct bulge on the craniolateral margin of the shaft, which may be a muscle insertion.

This bone shows the usual features of the third metatarsal of large theropods such as *Allosaurus*, but is remarkable for its elongation and consequent slenderness: the third metatarsal of *Allosaurus fragilis* appears much stouter (Madsen, 1976). Comparison with the Early Cretaceous allosauroid *Neovenator salerii* is difficult because only the proximal part of metatarsal III is known in that form (Brusatte et al., 2008). Metatarsal III is not known in *Erectopus superbus*, but metatarsal II, from the Louppy-le-Château skeleton, was described by Sauvage (1882) and Allain (2005). Interestingly, it is elongate and slender. Allain (2005, 81) noted that "it is longer relative to the femur than in most other large theropods." This cannot be directly estimated in the case of the metatarsal III from Froidos, as it is not associated with a femur. However, the elongation and slenderness of that third metatarsal suggest that it belonged to a theropod with peculiar hind limb proportions reminiscent of those of *Erectopus sauvagei*. As noted by Allain (2005), the exact significance of these unusual proportions is unclear: they may be used to diagnose *Erectopus* or may reflect immaturity. The discovery in a different locality in the Sables verts of a second theropod metatarsal that is also remarkable for its elongation and slenderness suggests that the specimen from Froidos probably belongs to *Erectopus*, and that that taxon is indeed characterized by a long metatarsus because it is relatively unlikely that both metatarsals, coming as they do from distinct localities, should belong to immature individuals.

Measurements:

Length as preserved: 300 mm.
Mediolateral width of distal end: 64 mm.
Maximum craniocaudal thickness of distal end: 52.4 mm.

21.4. Distal part of a left theropod metatarsal III (SV3; cf. *Erectopus*) from the Albian of Froidos (Meuse) in medial (A), cranial (B), lateral (C), caudal (D), and distal (E) views. Scale bar = 50 mm.

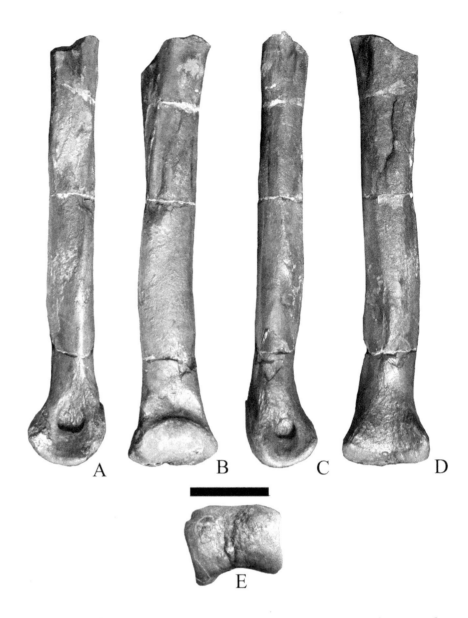

A   B   C   D

E

Teeth. Three isolated teeth (SV4, SV5, SV6, Fig. 21.5) from two localities in the Argonne are present in the collection. A fairly complete tooth (SV4), with a well-preserved crown and part of the root, comes from Louppy-le-Château (Meuse). The Les Islettes locality (Meuse) has yielded a fairly large specimen (SV5) consisting of the crown and part of the root, and a smaller one, which is an incomplete crown (SV6). It may be mentioned that Collet (1904) reported teeth of *Megalosaurus Bucklandi* [*sic*] from the Sables verts of Les Islettes.

All three teeth show similar characters: they are blade shaped, strongly compressed linguolabially with distinct mesial and distal carinae that extend along the whole length of the crown, and are serrated from the base of the crown to the apex. The serrations are regular and small. Serration density is quite constant along the carinae, with three denticles per millimeter. The denticles have rounded tips. The surface of the enamel is smooth on both the labial and the lingual face. In all three teeth, the mesial margin of the crown is markedly convex, whereas the distal margin is moderately concave. Slight shape differences between these teeth

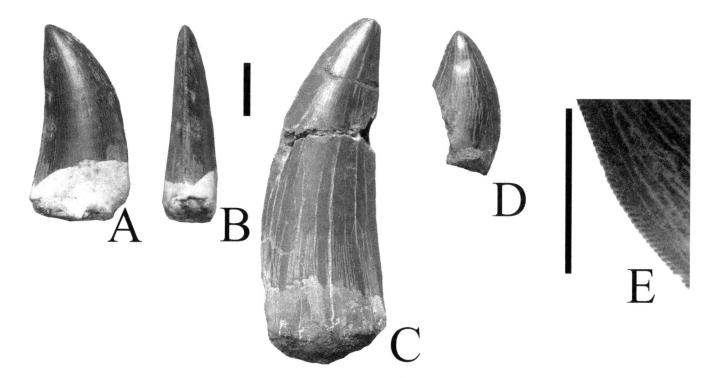

21.5. Theropod teeth (cf. *Erectopus*) from the Albian of the Argonne. A– B, Tooth from Louppy-le-Château (Meuse), SV4, in lingual (A) and distal (B) views; C, tooth from Les Islettes (Meuse), SV5, in lingual view; D, tooth from Les Islettes, SV6, in labial or lingual view; E, detail of serration of the mesial carina of SV6. Scale bars = 10 mm.

are related to the different heights of the crowns, which give the teeth a more or less slender appearance, but they do not exceed differences seen between teeth at various degrees of development or at different locations in a jaw, so there is good reason to suppose that these three teeth belong to the same taxon. They show great similarities with the teeth described and illustrated by Sauvage (1882) and by Allain (2005), including those still inserted in the maxilla from Louppy-le-Château (MNHN 2001–4) designated as the lectotype of *Erectopus superbus* by Allain (2005). Although isolated theropod teeth are not always easy to use for precise identifications, it seems legitimate to refer these teeth to *Erectopus*.

Measurements:

SV4: Total height of specimen: 37.7 mm.
    Labiolingual thickness at base of crown: 10.3 mm
    Mesiodistal width at base of crown: 18.6 mm.
SV5: Total height of specimen: 62.8 mm.
    Labiolingual thickness at base of crown: 12.4 mm.
    Mesiodistal width at base of crown: 21.8 mm.
SV6: Total height of specimen: 25 mm.
    Labiolingual thickness at base of crown: 8.5 mm.

In conclusion, in spite of uncertainties caused by the incompleteness of type material of *Erectopus superbus* and by the fragmentary nature of the theropod material described above, all these specimens are referred to as cf. *Erectopus* because of significant similarities with *Erectopus superbus*, the only theropod taxon hitherto reported from the Sables verts.

Sauropoda indet.

Caudal vertebrae. The sauropods are represented by two incomplete caudal vertebrae. Both are from Varennes-en-Argonne and belong to the Guillaumot collection. The larger specimen consists of a fairly well-preserved centrum, with the broken pedicels of the neural arch (SV7, Fig. 21.6). The centrum is amphicoelous, with cranial and caudal articular sides that are significantly wider than high and therefore oval in outline. The width of the centrum at its ends is greater than its length. In ventral or dorsal view, the centrum is hourglass shaped. The ventral face shows two blunt longitudinal ridges separated by a longitudinal groove. The ridges rise posteriorly to form well-defined chevron facets. The left one is broken, but the right one is triangular (with the apex pointing cranially) and has a concave articular area. The lateral sides of the centrum are concave anteroposteriorly, with a longitudinal ridge at midheight. Dorsally on the lateral side, the inception of the broken transverse process is visible; it is better preserved on the left side, oval in outline, and 40 mm in length. Comparison with sauropods in which the caudal series has been well described, such as *Camarasaurus* (Osborn and Mook, 1921), suggests a position in the anterior (but not anteriormost) part of the tail, because the transverse process disappears in more posterior vertebrae. Little can be said about the incomplete pedicels of the neural arch, except that they are inserted mainly on the anterior half of the centrum.

Measurements:

Length of centrum: 86.8 mm.
Width of anterior face: 103.1 mm.
Height of anterior face: 66.3 mm.
Width of posterior face: 96.6 mm.
Height of posterior face: 70 mm.

Although the smaller vertebra (SV8, Fig. 21.7) is associated with a label identifying it as a caudal vertebra of *Megalosaurus superbus*, it clearly belongs to a sauropod. Its preservation is not very good, as it shows signs of being waterworn. It consists of the centrum and a small part of the neural pedicels. The centrum is amphicoelous, with rounded cranial and caudal sides that are only slightly wider than high. The ventral face bears two faint longitudinal ridges separated by a nearly flat area. Posteriorly, the chevron facets are weak. The lateral sides are only slightly concave craniocaudally, so that the centrum appears almost cylindrical rather than clearly hourglass shaped. There is no indication of transverse processes. The pedicels are inserted mainly in the anterior half of the centrum.

Measurements:

Length of centrum: 55.3 mm.
Width of anterior face: 45.4 mm.
Height of anterior face: 44 mm.
Width of posterior face: 44.3 mm.
Height of posterior face: 34.9 mm.

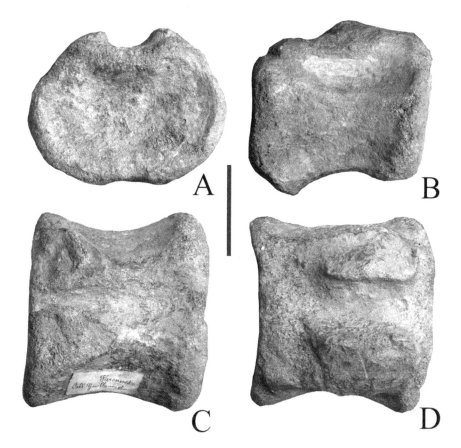

21.6. Sauropod caudal vertebra (SV7) from Varennes-en-Argonne (Meuse) in anterior (A), right lateral (B), ventral (C), and dorsal (D) views. Scale bar = 50 mm.

These two caudal vertebrae are referred to Sauropoda mainly because of the combination of rather low and broad centra and anterior position of the neural pedicels. Whether they may belong to the same taxon is a moot point. The smaller specimen is clearly from a more posterior part of the tail than the larger one, as shown by the absence of transverse processes. A precise identification is difficult, even at a high taxonomic level. No sauropod remains had hitherto been recorded from the Sables verts of the eastern Paris Basin—the vertebra from the Sables verts of Aube referred to a brachiosaur by Knoll et al. (2000) in fact belongs to a pliosaur (Buffetaut et al., 2005). The overlying Gault Clay has yielded a series of sauropod caudal vertebrae (Martin et al., 1993) at Pargny-sur-Saulx (Meuse), some 30 km south of the Argonne. Some of the more anterior ones resemble the larger vertebra from Varennes to some extent, with centra that are wider than high, but similarities are limited. A few sauropod caudal vertebrae are known from the Albian of northwestern France. They include a series of 10 caudals from the Gault Clay of the Pays de Bray at Moru, near Villers-Saint-Barthélémy (Oise), described by Lapparent (1946). Some of these vertebrae appear to be quite similar to the large specimen (SV7) from Varennes, with dorsoventrally flattened, hourglass-shaped centra that are wider than long and bear prominent chevron facets. The Albian of the Normandy coast near Le Havre has yielded a few sauropod remains, including an isolated caudal from Bléville (Seine-Maritime) described by Buffetaut (1984), which differs from the larger specimen from Varennes in having a higher, more cylindrical centrum. In the same area, associated sauropod bones found in 1990 by P. Gencey include two caudal vertebrae

21.7. Sauropod caudal vertebra (SV8) from Varennes-en-Argonne (Meuse) in anterior (A), left lateral (B), ventral (C), and dorsal (D) views. Scale bar = 20 mm.

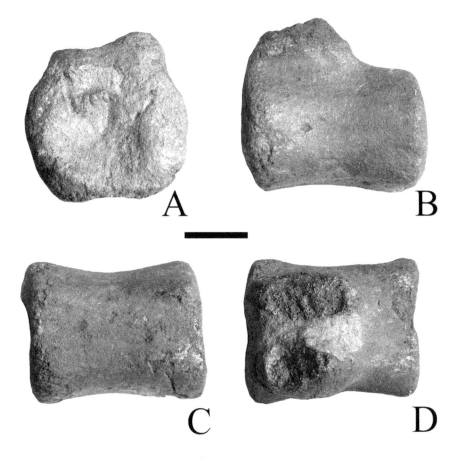

that have only been briefly mentioned by Buffetaut (1995). One is markedly procoelous, while the other, more posterior, one is amphicoelous, but with a higher centrum than the larger vertebra from Varennes. In England, sauropod caudal vertebrae from the Cambridge Greensand were described by Seeley (1876) as *Macrurosaurus semnus*. They include both procoelous and amphicoelous vertebrae. Le Loeuff (1993) concluded that *Macrurosaurus semnus* should be considered as a nomen dubium and that the material includes vertebrae of both a titanosaur and a nontitanosaurid indeterminate sauropod. Naish and Martill (2007, 499) considered that *Macrurosaurus semnus* "is a nomen dubium that cannot be identified beyond Titanosauria indet." The procoelous vertebrae bear little resemblance with the amphicoelous centra from Varennes (although the possibility that the Varennes sauropod had procoelous vertebrae in more anterior parts of its tail cannot be excluded, as shown by the above-mentioned instance from Normandy). Martin et al. (1993) concluded that at least four distinct sauropods occurred in the Albian of Europe, mainly on the basis of differences in the morphology of the caudal vertebrae, but admitted that most of them are of obscure affinities. Our knowledge of the Albian sauropods of Europe has not much improved since then, and in this context, it is difficult to identify the two isolated caudal vertebrae from Varennes beyond Sauropoda indet.

Conclusions

Although the fragmentary nature of the available material mostly precludes precise systematic identifications, the specimens described in the present chapter provide new data about the dinosaur assemblage from the

Albian Sables verts of the Argonne. The occurrence of ankylosaurs, which had been suggested by Sauvage (1882) on the basis of a highly doubtful specimen, is confirmed. "New" theropod material tentatively referable to *Erectopus* apparently confirms Allain's (2005) suggestion that an elongate metatarsus is one of the distinguishing features of this theropod. Finally, the occurrence of sauropods in the Sables verts is reported for the first time.

Our image of the dinosaurs that lived on a landmass close to the area of deposition of the Sables verts is thus getting slightly better. Theropods, sauropods, and ankylosaurs were present. However, it is highly unlikely that what is currently known reflects the real diversity of this fauna. Few specimens have been described, and they all correspond to middle-sized dinosaurs, both very large and small dinosaurs being conspicuous by their absence. Especially in the case of small forms, this may well reflect a collecting bias at the time of the phosphate exploitations. The lack of ornithopods is worth noting. Although the available dinosaur sample is still admittedly small, its composition contrasts with that of equally small assemblages from earlier stages of the Early Cretaceous of the eastern Paris Basin, in which iguanodontians are prominently present (Martin and Buffetaut, 1992), a difference already noted by Buffetaut (2004). The scarcity of ornithopod remains in the Cambridge Greensand (see Naish and Martill, 2008, for a review) may support the idea that the lack of iguanodontians in the Sables verts reflects a real faunal difference between Albian dinosaur assemblages and earlier faunas in western Europe.

Despite these additions to our knowledge, the dinosaurs from the Sables verts remain poorly known and to a large extent enigmatic. This in fact applies more generally to all the dinosaurs from the Albian of Europe. Even the assemblage from the Cambridge Greensand, from which a relatively large number of specimens is known, is still poorly understood (see Naish and Martill, 2007, 2008, for reviews). Because so little is known about these late Early Cretaceous forms from Europe, it is difficult to assess their phylogenetic or biogeographical relationships with better-known coeval dinosaurs in other parts of the world.

Vullo (2007) referred to *Carcharodontosaurus* isolated teeth from the Early Cenomanian of the Charentes, in western France. This raises the question of possible faunal links between Western Europe and North Africa during the mid-Cretaceous. In North Africa, Albian dinosaurs are mainly known from southern Tunisia (Bouaziz et al., 1988; Srarfi et al., 2004); the fossiliferous Kem Kem beds of southern Morocco are now considered as most probably Cenomanian (Cavin et al., 2010), rather than Albian as sometimes suggested. The Albian assemblage from Tunisia does not show significant similarities with that from the Sables verts. Spinosaurids, which are abundant in the Tunisian assemblage, are not known from the Sables verts, and *Erectopus* does not seem to be especially reminiscent of *Carcharodontosaurus*, another frequent theropod in North Africa. The sauropod material from the Sables verts is too scanty to warrant comparison with the Albian sauropods from Tunisia, which also remain poorly known. Ankylosaurs are not known from the Albian of Tunisia or from other mid-Cretaceous deposits in Africa. There is therefore no evidence of close links between the dinosaur assemblage from the Sables verts and those from

North Africa at the same period. However, the French assemblage is scanty and new finds might alter this conclusion.

Our insufficient knowledge of the Albian dinosaurs of Europe is part of the larger problem posed by the scarcity of the "mid-Cretaceous" record in that part of the world, between an Early Cretaceous pre-Albian record that is already good and steadily improving (with especially notable finds from the Wealden of England and Belgium and the nonmarine Early Cretaceous of Spain and Portugal) and a similarly improving record from the Santonian to Maastrichtian (particularly from Spain, France, Hungary, and Romania). Unfortunately, a large part of the available dinosaur material from the Albian of Europe came from phosphate-bearing levels that ceased to be exploited a century ago, whether in the Cambridge Greensand or in the Sables verts. In view of the present lack of outcrops in the Argonne and other parts of eastern France, additional finds from the Sables verts will probably have to come from old and hitherto overlooked collections.

Acknowledgment

Dale A. Russell reviewed an earlier version of this chapter and made useful comments.

References

Allain, R. 2005. The enigmatic theropod dinosaur *Erectopus superbus* (Sauvage 1882) from the Lower Albian of Louppy-le-Château (Meuse, France); pp. 72–86 in K. Carpenter (ed.), The carnivorous dinosaurs. Indiana University Press, Bloomington.

Barrois, C. 1874. Catalogue des poissons fossiles du terrain crétacé du Nord de la France. Bulletin biologique de la France et de la Belgique 6: 101–110, 130–136.

———. 1875. Les reptiles du terrain crétacé du N.-E. du Bassin de Paris. Bulletin scientifique, historique et littéraire du Nord 6: 1–11.

Bestel, F. 1908. Excursion de Juin 1908, à Novion-Porcien. Sables verts et phosphate dechaux. Bulletin de la Société d'Histoire naturelle des Ardennes 15: 123–134.

Bouaziz, S., E. Buffetaut, M. Ghanmi, J.-J. Jaeger, M. Martin, J.-M. Mazin, and H. Tong. 1988. Nouvelles découvertes de vertébrés fossiles dans l'Albien du Sud tunisien. Bulletin de la Société géologique de France 4: 335–339.

Brusatte, S. L., R. B. Benson, and S. Hutt. 2008. The osteology of *Neovenator salerii* (Dinosauria: Theropoda) from the Wealden Group (Barremian) of the Isle of Wight. Monographs of the Palaeontographical Society 162: 1–75.

Buffetaut, E. 1984. Une vertèbre de dinosaurien sauropode dans le Crétacé du Cap de la Hève (Normandie). Actes du Muséum de Rouen 7: 213–221.

———. 1995. Dinosaures de France. Editions du BRGM, Orléans.

———. 2002. New data from old finds: the dinosaurs from the Early Cretaceous Greensand ("Sables verts") of the eastern Paris Basin; p. 9 in 7th European Workshop on Vertebrate Palaeontology (Sibiu, Romania), Abstracts. Ars Docendi, Bucharest.

———. 2004. An *Iguanodon* jaw (Dinosauria, Ornithopoda) from the Lower Cretaceous of Aube (eastern Paris Basin, France). Oryctos 5: 63–68.

———. 2006. La "ruée vers les phosphates" du dix-neuvième siècle: une aubaine pour la paléontologie des vertébrés crétacés. Strata (ser. 1) 13: 11–23.

Buffetaut, E., and V. Suteethorn. 2007. A sinraptorid theropod (Dinosauria: Saurischia) from the Phu Kradung Formation of northeastern Thailand. Bulletin de la Société géologique de France 178: 497–502.

Buffetaut, E., C. Colleté, B. Dubus, and J.-L. Petit. 2005. The "sauropod" from the Albian of Mesnil-Saint-Père (Aube, France): a pliosaur, not a dinosaur. Carnets de Géologie 2005/01: 1–5.

Cavin, L., H. Tong, L. Boudad, C., Meister, A. Piuz, J. Tabouelle, M. Aarab, R. Amiot, E. Buffetaut, G. Dyke, S. Hua, and J. Le Loeuff. 2010. Vertebrate assemblages from the early Late Cretaceous of southeastern Morocco: an overview. Journal of African Earth Sciences 57: 391–412.

Coombs, W. P. 1978a. Forelimb muscles of the Ankylosauria. Journal of Paleontology 52: 642–657.

———. 1978b. The families of the ornithischian dinosaur order Ankylosauria. Palaeontology 21: 143–170.

Collet, P. 1904. Notices géologiques et paléontologiques pour servir à la géologie de l'arrondissement de Sainte-Ménehould. Bulletin de la Société d'Étude des Sciences naturelles de Reims 12: 17–87.

Corroy, G. 1922. Les reptiles néocomiens et albiens du Bassin de Paris. Comptes Rendus de l'Académie des Sciences de Paris 172: 1192–1194.

Fliche, P. 1896. Études sur la flore fossile de l'Argonne (Albien–Cénomanien). Bulletin de la Société des Sciences de Nancy 14: 114–306.

Hancock, J. M., and P. F. Rawson. 1992. Cretaceous; pp. 131–138 in J. C. W. Cope, J. K. Ingham, and P. F Rawson (eds.), Atlas of palaeogeography and lithofacies. Geological Society Memoirs 13.

Huene, F. von. 1923. Carnivorous Saurischia in Europe since the Triassic. Bulletin of the Geological Society of America 34: 449–458.

———. 1926. The carnivorous Saurischia in the Jura and Cretaceous formations principally in Europe. Revista del Museo de La Plata 29: 35–167.

———. 1932. Die fossile Reptil-Ordnung Saurischia, ihre Entwcklung und Geschichte. Monographien zur Geologie und Palaeontologie 4: 1–361.

Hulke, J. W. 1874. Note on a reptilian tibia and humerus (probably of Hylæosaurus) from the Wealden Formation in the Isle of Wight. Quarterly Journal of the Geological Society of London 30: 516–520.

Knoll, F., E. Buffetaut, and B. Dubus. 1998. Un ostéoderme d'ankylosaure (Ornithischia) dans l'Albien de l'Aube (France). Bulletin de l'Association Géologique Auboise 19: 61–65.

Knoll, F., C. Colleté, B. Dubus, and J.-L. Petit. 2000. On the presence of a sauropod dinosaur (Saurischia) in the Albian of Aube (France). Geodiversitas 22: 389–394.

Lapparent, A. F. de 1946. Présence d'un dinosaurien sauropode dans l'Albien du Pays de Bray. Annales de la Société géologique du Nord 66: 236–243.

———. 1964. Région de Paris. Excursions géologiques et voyages pédagogiques. Hermann, Paris.

Le Loeuff, J. 1993. European titanosaurids. Revue de Paléobiologie, Volume Spécial 7: 105–117.

Madsen, J. H. 1976. Allosaurus fragilis: a revised osteology. Bulletin of the Utah Geological Survey 109: 1–163.

Magnez-Jannin, F., and A. Demonfaucon. 1980. Sables verts; p. 277 in F. Mégnien (ed.), Synthèse géologique du Bassin de Paris. Vol. 3. Lexique des noms de formations. Mémoires du BRGM 103.

Martin, V., and E. Buffetaut. 1992. Les Iguanodons (Ornithischia–Ornithopoda) du Crétacé inférieur de la région de Saint-Dizier (Haute-Marne). Revue de Paléobiologie 11: 67–96.

Martin, V., E. Buffetaut, and J. Le Loeuff. 1993. A sauropod dinosaur in the Middle Albian of Pargny-sur-Saulx (Meuse, eastern France). Revue de Paléontologie, Volume Spécial 7: 119–124.

Meunier, S. 1898. Nos Terrains. Armand Colin, Paris.

Naish, D., and D. M. Martill. 2007. Dinosaurs of Great Britain and the role of the Geological Society of London in their discovery: basal Dinosauria and Saurischia. Journal of the Geological Society, London 164: 493–510.

———. 2008. Dinosaurs of Great Britain and the role of the Geological Society of London in their discovery: Ornithischia. Journal of the Geological Society, London 165: 613–623.

Nivoit, E. 1874. Notice sur le gisement et l'exploitation des phosphates de chaux fossiles dans le département de la Meuse. Rolin, Chuquet & Cie, Bar-le-Duc.

Nopcsa, F. von. 1928. Palaeontological notes on reptiles. Geologica Hungarica, series Paleontologica 1: 1–84.

Osborn, H. F., and C. C. Mook. 1921. Camarasaurus, Amphicoelias and other sauropods of Cope. Memoirs of the American Museum of Natural History 3: 249–387.

Ostrom, J. H. 1970. Stratigraphy and paleontology of the Cloverly Formation (Lower Cretaceous) of the Bighorn Basin area, Wyoming and Montana. Peabody Museum of Natural History Bulletin 35: 1–234.

Pereda-Suberbiola, J. 1994. Polacanthus (Ornithischia, Ankylosauria), a transatlantic armoured dinosaur from the Early Cretaceous of Europe and North America. Palaeontographica A 232: 133–159.

Pereda-Suberbiola, X., and P. M. Barrett. 1999. A systematic review of ankylosaurian dinosaur remains from the Albian–Cenomanian of England. Special Papers in Palaeontology 60: 177–208.

Rauhut, O. W. M. 2003. The interrelationships and evolution of basal theropod dinosaurs. Special Papers in Palaeontology 69: 1–213.

Sauvage, C., and A. Buvignier. 1842. Statistique minéralogique et géologique du département des Ardennes. Trécourt, Mézières.

Sauvage, H. E. 1876. Notes sur les reptiles fossiles. Bulletin de la Société géologique de France 4: 435–442.

———. 1882. Recherches sur les reptiles trouvés dans le Gault de l'Est du Bassin de Paris. Mémoires de la Société géologique de France 2: 1–43.

———. 1903. Sur la présence du genre Polyptychodon dans les Sables verts de la Meuse. Bulletin de la Société d'Histoire naturelle d'Autun 16: 321–323.

Seeley, H. G. 1876. On Macrurosaurus semnus (Seeley), a long-tailed animal with procoelous vertebrae from the Cambridge Upper Greensand, preserved in the Woodwardian Museum of the University of Cambridge. Quarterly Journal of the Geological Society of London 32: 440–444.

———. 1879. On the Dinosauria of the Cambridge Greensand. Quarterly Journal of the Geological Society of London 35: 591–636.

Srarfi, D., M. Ouaja, E. Buffetaut, G. Cuny, G. Barale, S. Ferry, and E. Fara. 2004. Position stratigraphique des niveaux à vertébrés du Mésozoïque du Sud-est de la Tunisie. Notes du Service Géologique de Tunisie 72: 5–16.

Unwin, D. M. 2001. An overview of the pterosaur assemblage from the Cambridge Greensand (Cretaceous) of Eastern England. Mitteilungen des Museums für Naturkunde Berlin, Geowissenschaftliche Reihe 4: 189–221.

Vullo, R. 2007. Les vertébrés du Crétacé supérieur des Charentes (Sud-Ouest de la France): biodiversité, taphonomie, paléoécologie et paléobiogéographie. Mémoires Géosciences Rennes 125: 1–302.

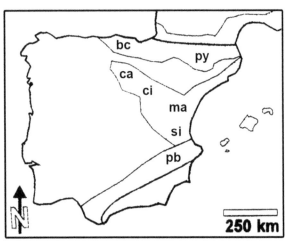

A

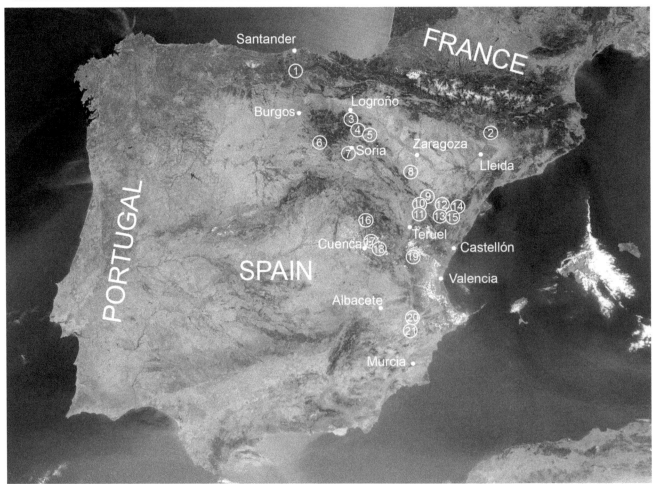

bc: Basque-Cantabrian Basin (1)
py: Pyrenees (2)
ca: Cameros Basin (3-7)
ci: Central Iberian Range (8)
ma: Maestrazgo Basin (9-15)
si: South Iberian Basin (16-19)
pb: Pre-Betic Zone (20, 21)

B

Dinosaur Faunas from the Early Cretaceous (Valanginian–Albian) of Spain

22

Xabier Pereda-Suberbiola*, José Ignacio Ruiz-Omeñaca,
José Ignacio Canudo, Fidel Torcida, and José Luis Sanz

The dinosaur skeletal record from the Early Cretaceous of Spain indicates a diverse fauna, including the richest assemblage known from continental Europe for the Hauterivian–Aptian interval. The Spanish record consists of theropods (all tetanurans: spinosaurids, carcharodontosaurian allosauroids, ornithomimosaurs, and several kinds of maniraptorans, including dromaeosaurids and enantiornithine birds), sauropods (basal macronarians and titanosauriforms, such as brachiosaurids, "euhelopodids" and possible titanosaurians, as well as rebbachisaurid diplodocoids), thyreophorans ("polacanthid" ankylosaurs and stegosaurs), and ornithopods (basal iguanodontoids, dryosaurids, and "hypsilophodontid"-like basal euornithopods). *Iguanodon* and other closely related basal iguanodontians are the most abundant dinosaurs in number of specimens, whereas the maniraptoriform theropods are the most diversified in number of taxa. Ten genera and species have been erected to date from the Spanish material: among theropods, the carcharodontosaur *Concavenator corcovatus*, the ornithomimosaur *Pelecanimimus polyodon* (the only record of this clade known in Europe), and the enantiornithine birds *Iberomesornis romerali*, *Concornis lacustris*, and *Eoalulavis hoyasi* from Cuenca, plus *Noguerornis gonzalezi* from Lleida (the most diverse Early Cretaceous avian assemblage reported out of Asia); among sauropods, the macronarian *Aragosaurus ischiaticus* and *Tastavinsaurus sanzi* from Teruel, and the rebbachisaurid diplodocoid *Demandasaurus darwini* from Burgos; and finally, the iguanodontian ornithopod *Delapparentia turolensis* from Teruel. All these taxa occurred exclusively in the Iberian Peninsula. Additional new taxa, including a basal titanosauriform sauropod and a basal ornithopod, are yet to be named. The most significant dinosaur discoveries have been made in the last 25 years, indicating that the Spanish outcrops have a great fossiliferous potential.

In 1872, Vilanova Piera published the discovery of bones of *Iguanodon* from the Early Cretaceous of Utrillas (Teruel), and one year later from the Early Cretaceous of Morella (Castellón). Vilanova Piera (1872, 1873) was the first to specifically mention the finding of dinosaur skeletal remains from Spain (but this is not the earliest discovery; see Pereda Suberbiola et al., 2010). A few years before, fossil bones were found near Morella; this material, now lost, may also belong to dinosaurs (Gasulla, 2005; Pérez-García et al., 2009b). The Vilanova collection, currently kept in the Museo Nacional de Ciencias Naturales of Madrid, contains a few dinosaur remains, but none is referable to *Iguanodon* (Pereda Suberbiola and Ruiz-Omeñaca, 2005).

22.1. A, Simplified geological map with the location of the sedimentary basins that have provided dinosaur remains (number in brackets corresponds to localities); modified from Martín-Chivelet (1996, fig. 2A). B, Main dinosaur outcrops from the Early Cretaceous (Hauterivian–Albian) of Spain: 1, Vega de Pas (Cantabria); 2, El Montsec (Lleida); 3, Soto de Cameros (La Rioja); 4, Enciso (La Rioja); 5, Igea (La Rioja); 6, Salas de los Infantes (Burgos); 7, Golmayo (Soria); 8, Villanueva de Huerva (Zaragoza); 9, Josa/Oliete (Teruel); 10, Utrillas (Teruel); 11, Galve (Teruel); 12, Castellote (Teruel); 13, Mirambel (Teruel); 14, Peñarroya de Tastavins (Teruel); 15, Cinctorres/Morella (Castellón); 16, Vadillos/Masegosa (Cuenca); 17, Buenache de la Sierra/Uña (Cuenca); 18, Las Hoyas (Cuenca); 19, Alpuente/Titaguas (Valencia); 20, Almansa (Albacete); 21, Yecla (Murcia).

Introduction

The study of dinosaurs in Spain was not highlighted during the nineteenth century, and the first significant discoveries were not made until the 1900s. Vidal (1902) reported the first Mesozoic avian remains (and their accidental destruction!) from a quarry of lithographic limestones in El Montsec, Lleida. Between 1918 and 1928, Royo Gómez published a series of papers on the dinosaur fauna from the Early Cretaceous (and probably also Late Jurassic) of eastern Spain, including material from Castellón, Teruel, and Valencia (Royo Gómez, 1926a, 1926b; see Sanz, 1996; Diéguez et al., 2004; Pérez-García et al., 2009a). In Burgos and Soria, Royo Gómez (1926c) reported the first dinosaur bones, but systematic excavations were not undertaken until recently (Fuentes-Vidarte et al., 2005; Torcida Fernández-Baldor, 2006). In Teruel, the first finds in the Galve area were made during the 1950s (Fernández-Galiano, 1958). Lapparent (1960) described the dinosaurs of Galve and noted the discovery of new Wealden remains in several localities of Teruel, Albacete, Cuenca, and Valencia (Lapparent, 1966; Lapparent et al., 1969). However, the majority of this material has never been described, and the stratigraphical context of the fossiliferous sites remains unknown. Some of them might in fact correspond to the Purbeck facies of the Jurassic–Cretaceous transition.

In an overview of Spanish dinosaurs, Sanz (1984) listed about 20 outcrops of Early Cretaceous age. During the last 25 years, the number of dinosaur sites has quintupled. A number of interesting discoveries from the Early Cretaceous of Spain has been made, providing valuable information in several respects (Ortega et al., 2006). The most noteworthy site is the *Konservat-Lagerstätten* of Las Hoyas (Cuenca) in the South Iberian Basin, where both avian and nonavian dinosaurs are known (Pérez-Moreno and Sanz, 1997; Sanz et al., 2001a; Sanz and Ortega, 2002). Another *Konservat-Lagerstätten* that has yielded avian remains is El Montsec in the south-central Pyrenees of Lleida (Chiappe and Lacasa Ruiz, 2002). Fossil birds from these Barremian localities provide new information about the early evolutionary history of birds (Chamero et al., 2009).

Other important Hauterivian to Aptian dinosaur localities are known in the vicinity of Salas de los Infantes in Burgos (Torcida Fernández-Baldor, 2006) and Golmayo in Soria (Fuentes-Vidarte et al., 2005), both in the Cameros Basin. All the sites of the Galve region in Teruel (Sanz et al., 1987; Ruiz-Omeñaca et al., 2004; Ruiz-Omeñaca, 2006) and the Els Ports area in Castellón (Suñer et al., 2008) are in the Maestrazgo Basin. These localities have produced abundant dinosaur fossils and are among the richest in Europe (see Weishampel et al., 2004).

The aim of this chapter is to review the dinosaur fauna from the Early Cretaceous of Spain on the basis of the skeletal fossil record. A brief summary of the Valanginian–Albian dinosaur-bearing localities and formations is presented below. Only the Wealden Beds (Valanginian–Barremian) and the deposits correlated with the Urgonian facies (Barremian–Aptian) and the Albian rocks are here considered. The Purbeck facies (Tithonian–Berriasian) are not taken into account in this work.

More than 100 dinosaur localities (only skeletal remains are here considered) of Valanginian–Albian age are known in Spain. The provinces of Burgos and Teruel have produced the highest concentration of sites. Additional records include those of Albacete, Cantabria, Castellón, Cuenca, La Rioja, Lleida, Murcia, Soria, Valencia, and Zaragoza. A list of the dinosaur-bearing locations by regions (autonomic communities) follows.

Aragón. In the province of Teruel, at least 30 localities are known. They are located in the townships of Castellote, Galve, Josa, Mirambel, Oliete, Peñarroya de Tastavins, and Utrillas (Lapparent, 1966; Sanz et al., 1987; Ruiz-Omeñaca et al., 1998a; Ruiz-Omeñaca and Canudo, 2003b; Ruiz-Omeñaca et al., 2004; Ruiz-Omeñaca, 2006, and references therein) (9–14, Fig. 22.1B). The material found in other localities (Alacón, Aliaga, Cantavieja, Ejulve, El Castellar, Mora de Rubielos, Muniesa, Las Parras de Castellote, Rubielos de Mora) has not been described in detail or is currently mislaid (Lapparent, 1966; Lapparent et al., 1969; Ruiz-Omeñaca and Canudo, 2003b; Ruiz-Omeñaca, 2006). In the province of Zaragoza, only one site (township of Villanueva de Huerva) has been reported to date (Infante et al., 2005a; Moreno-Azanza et al., 2009) (8, Fig. 22.1B).

Cantabria. Dinosaur remains have only been mentioned near Vega de Pas (Moratalla, 2004) (1, Fig. 22.1B).

Castilla y León. In the province of Burgos, at least 30 sites are known in the vicinity of Salas de los Infantes (6, Fig. 22.1B). They are situated in the townships of Aldea del Pinar, Barbadillo del Mercado, Cabezón de la Sierra, Hacinas, La Gallega, La Revilla-Ahedo, Pinilla de los Moros, Salas de los Infantes, and Villanueva de Carazo (Torcida Fernández-Baldor et al., 2003b; Ruiz-Omeñaca and Canudo, 2003b; Torcida Fernández-Baldor, 2006). In the province of Soria, at least four outcrops are known in Golmayo, near the city of Soria (Fuentes-Vidarte et al., 2005) (7, Fig. 22.1B).

Castilla–La Mancha. In the province of Cuenca, Las Hoyas is the most important dinosaur locality (Sanz et al., 2001a) (18, Fig. 22.1B). Other sites are Buenache de la Sierra, Masegosa, Uña, and Vadillos (Lapparent et al., 1969; Francés and Sanz, 1989; Rauhut, 2002; Buscalioni et al., 2008) (16–17, Fig. 22.1B). Several outcrops (Carrascosa de la Sierra, Beteta) are still undescribed. In the provincial of Albacete, a single locality (Almansa) has been reported (Lapparent, 1966) (20, Fig. 22.1B). This material is currently mislaid.

Catalonia. In the province of Lleida, two quarries (towns of Rubies and Santa Maria de Meià) are known from El Montsec (Chiappe and Lacasa Ruiz, 2002) (2, Fig. 22.1B).

La Rioja. Two outcrops are known near Igea (Torres and Viera, 1994; Viera and Torres, 1995). In addition, undescribed dinosaur material comes from Enciso and Soto de Cameros (Pérez-Lorente et al., 2001) (3–5, Fig. 22.1B).

Murcia. A single locality in the township of Yecla has yielded dinosaur material (Canudo et al., 2004c) (21, Fig. 22.1B).

Valencia. In the province of Castellón, about 20 dinosaur sites have been mentioned in the area of the Els Ports (townships of Cinctorres and Morella; 15, Fig. 22.1B); most of these sites are still unpublished

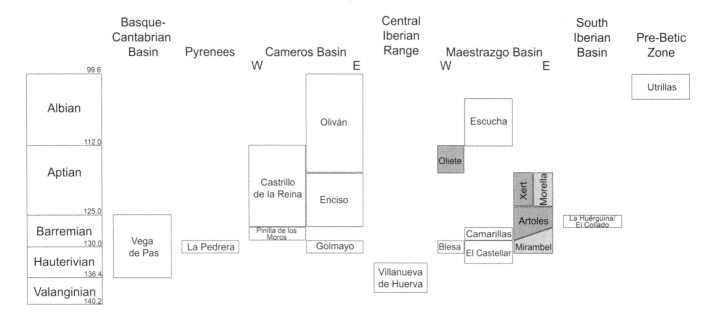

22.2. Schematic drawing showing the correlation of the stratigraphic units from the Early Cretaceous (Valanginian–Albian) of Spain that have yielded dinosaur skeletal remains. Continental formations are in white, marine formations are in dark gray, and the formations that contain both continental and marine beds are in light gray. Age of formations is mainly from Martín-Chivelet et al. (2002) and Vera (2004). Geological time scale (in Ma) from Gradstein et al. (2004).

(Santos-Cubedo et al., 2003). In the province of Valencia, Suñer et al. (2005) described three outcrops from the region of Los Serranos (township of Alpuente) (19, Fig. 22.1B). Previously, Lapparent (1966) listed other localities in Titaguas. This material has never been described. The age of these and other sites of Los Serranos is controversial: it may be Early Cretaceous or Late Jurassic (Casanovas-Cladellas et al., 1999; see below).

Geological Units

The fossiliferous formations range from the Valanginian to the Albian. They outcrop in different basins of the Iberian Chain (Cameros Basin, Central Iberian Range, Maestrazgo Basin, South Iberian Basin), the Prebetic Zone of the Betic Cordillera, the Pyrenean Basin, and the Basque–Cantabrian Region (Martín-Chivelet et al., 2002) (Fig. 22.1A).

Only a few marine units have yielded dinosaur remains (Artoles, Oliete, and Xert formations, besides some strata of the Mirambel Formation and the Morella Clays, all from the Maestrazgo Basin) (Fig. 22.2). The dating within the continental depositional sequences is based on charophyte biostratigraphy and low-resolution correlations (Soria et al., 1995; Martín-Closas and Alonso-Millán, 1998; Martín-Closas, 2000; Gómez et al., 2001; see also Schudack and Schudack, 2009). Consequently, the age of a number of dinosaur localities remains imprecise.

In the Cameros Basin of the Iberian Chain, the Hauterivian–Aptian geological units are the following: Castrillo de la Reina and Pinilla de los Moros formations in Burgos, Enciso and Oliván groups in La Rioja, and Golmayo Formation in Soria (Martín-Closas and Alonso-Millán, 1998).

In the Central Iberian Range, the Villanueva de Huerva Formation of Zaragoza is regarded as Valanginian–Hauterivian (Infante et al., 2005a).

In the Maestrazgo (or Maestrat) Basin, the Hauterivian–Aptian formations of Teruel are Artoles, Blesa, Camarillas, El Castellar, Mirambel, Oliete, and Xert (Soria et al., 1995). The Aptian Morella Clays crop out in Castellón (Salas et al., 2003; Gàmez et al., 2003).

22.3. The ornithomimosaur theropod *Pelecanimimus* and a couple of ornithopods closely related to *Iguanodon*. Landscape based on fossil evidence from the Barremian of Cuenca.

Courtesy of Raúl Martín.

In the South Iberian Basin, both El Collado and La Huérguina formations of Cuenca are Barremian in age (Fregenal-Martínez and Meléndez, 2000; Gómez et al., 2001).

Other units that have yielded dinosaur skeletal remains are the upper Hauterivian–lower Barremian Pedrera de Rubies Lithographic Limestones in the South Pyrenean of Lleida (Martínez-Delclòs et al., 1991), the Hauterivian–Barremian Vega de Pas Formation in the Basque–Cantabrian Region of Cantabria (Moratalla, 2004), and the Albian Utrillas Sandstone of the Prebetic Range (Canudo et al., 2004c). The latter unit and the Escucha Formation of Teruel are the only Albian formations that have yielded dinosaurs remains in Spain (Canudo et al., 2004b, 2005) (Fig. 22.2).

The stratigraphic position of the dinosaur sites of the Los Serranos area (Alpuente, Aras de los Olmos) in Valencia is controversial. Most of the material probably comes from the Villar del Arzobispo Formation (Purbeck facies, Tithonian–Berriasian in age) (see Ruiz-Omeñaca and Canudo, 2003b, and references therein), but it is possible that the outcrops of El Collado Formation (Barremian) in the north of Valencia have produced dinosaur remains as well (Suñer et al., 2005).

Ornithopods

Ornithopods are the most abundant dinosaurs from the Early Cretaceous of Spain. Large ornithopods are represented by several species of basal ankylopollexians (Fig. 22.3). More than 30 localities from Burgos, Castellón, Cuenca, and Teruel have yielded material that has been referred to

Dinosaur Diversity

Iguanodon (Royo Gómez, 1926b; Lapparent, 1960, 1966; Santafé et al., 1982; Sanz et al., 1984; Ruiz-Omeñaca et al., 1998a; Ruiz-Omeñaca and Canudo, 2003b, 2004; Fuentes-Vidarte et al., 2005; Torcida Fernández-Baldor, 2006). Ruiz-Omeñaca and Canudo (2004) recognized three species of *Iguanodon*: *I. bernissartensis*, *I.* cf. *atherfieldensis*, and *I.* cf. *fittoni*. Other material is referred to as *Iguanodon* sp. or Iguanodontoidea indet. (Ruiz-Omeñaca, 2006). Size and robustness have been commonly used as criteria to discriminate between the species of *Iguanodon*. Consequently, the taxonomy of basal ankylopollexians is confused, and there is no consensus about how many species belong to the genus *Iguanodon* (Sanz, 2005; Paul, 2008). Norman (2004) listed five valid species of *Iguanodon* in the Early Cretaceous of Europe (see Norman, 1980, 1986). Paul (2006, 2008) recognized at least three distinct genera in the Barremian–Aptian, distinguishing the gracile *Mantellisaurus* and *Dollodon* from the large, heavily built *Iguanodon*. Wealden ankylopollexian taxonomy of Britain of Belgium has recently been revised by Norman (2010; see also Chapter 15 in this book). The Lower Wealden (Valanginian) fauna is represented by two species: *Barilium dawsoni* and *Hypselospinus fittoni*, whereas the Upper Wealden (Barremian–Early Aptian) fauna comprises *Iguanodon bernissartensis* and *Mantellisaurus atherfieldensis* (Norman, 2010, and Chapter 15 in this book). In Spain, the presence of *I. bernissartensis* is well documented in the Early Aptian of Morella (Castellón), where a large collection of cranial, axial, and appendicular bones has been recovered from several outcrops (Santafé et al., 1982; Gasulla et al., 2007). This material exhibits the diagnostic characters of *I. bernissartensis* (sensu Norman, 1980). A partially articulated postcranial skeleton from the Early Barremian of Galve (Teruel), which was originally assigned to *I. bernissartensis* by Lapparent (1960), is referable to a new genus and species: *Delapparentia turolensis* Ruiz-Omeñaca, 2011, mainly on the basis of pelvic characters (Ruiz-Omeñaca, 2011). Additional material, including a basicranium, lower jaw remains, teeth, and one atlas, from the Early Barremian of Galve (Sanz et al., 1984, as *Iguanodon mantelli*; Ruiz-Omeñaca, 2006, as *Iguanodon* cf. *atherfieldensis*) needs to be revised. *Mantellisaurus atherfieldensis* has also been cited in the Early Hauterivian of Mirambel (Teruel, unnamed formation), but the material is nondiagnostic further away from the Iguanodontoidea or Iguanodontia level (see Ruiz-Omeñaca et al., 1998a). Thus, the occurrence of *Mantellisaurus* in Spain is not currently established. Partially articulated skeletons (still undescribed) of *Iguanodon*-like iguanodontoids are also known from the Hauterivian–Barremian of Soria and Burgos (Fuentes-Vidarte et al., 2005; Torcida Fernández-Baldor et al., 2006). Maisch (1997) assigned dorsal vertebrae with tall neural spines from the Berriasian–Valanginian of Salas de los Infantes (Burgos) to *Iguanodon* cf. *fittoni*. However, this material comes from the Late Hauterivian–Early Barremian and probably represents a medium-sized ornithopod distinct from *Iguanodon* and other known iguanodontians in having long dorsal neural spines that are vertically oriented (Torcida Fernández-Baldor et al., 2008; Pereda Suberbiola et al., 2011). Other Iberian remains have been referred to Iguanodontoidea indet. (Ruiz-Omeñaca and Canudo, 2004; Ruiz-Omeñaca et al., 2008).

A few isolated teeth from the Hauterivian–Barremian transition of Josa (Teruel) have been compared to those of basal hadrosauroids

(Ruiz-Omeñaca et al., 1997). Unlike *Iguanodon* and closely related iguanodontoids, the enameled surface of the crown bears a single prominent carina without subsidiary ridges; no marginal denticles are found. If correctly interpreted, the Josa teeth might represent one of the earliest occurrences of hadrosauroids worldwide (Ruiz-Omeñaca and Canudo, 2003b).

A collection of teeth from the Late Barremian–Aptian of Villanueva de Carazo (Burgos) shows similarities with those of Rhabdodontidae (Weishampel et al., 2003) and *Tenontosaurus* (Torcida Fernández-Baldor et al., 2005). Premaxillary teeth are present, and maxillary crowns have up to 10 subequal vertical ridges. These teeth, which were found associated with undescribed postcranial remains, have been provisionally regarded as Euornithopoda indet. (Torcida Fernández-Baldor et al., 2005).

Small ornithopods consist of "hypsilophodontids" and dryosaurids. "Hypsilophodontidae" appears to represent a paraphyletic assemblage of basal neornithischian and basal ornithopod taxa (Norman et al., 2004; Butler et al., 2008). This is the result of both the problematic definition of Iguanodontia and the instability of basal ornithopod phylogeny (Butler et al., 2008). For practical reasons, basal euornithopods is used here for noniguanodontian euornithopods (Galton, 2009). The Spanish record of basal euornithopods is one of the best for Europe; material is known from Burgos, Castellón, Cuenca, La Rioja, Soria, and Teruel (Sanz et al., 1983; Sanz et al., 1987; Torres and Viera, 1994; Fuentes-Vidarte and Meijide Calvo, 2001; Ruiz-Omeñaca, 2001, 2006; Rauhut, 2002; Torcida Fernández-Baldor et al., 2003b; Fuentes-Vidarte et al., 2005; Torcida Fernández-Baldor, 2006). At least three different taxa have been recognized (Ruiz-Omeñaca, 2001; Ruiz-Omeñaca and Canudo, 2003b, 2004): cf. *Hypsilophodon*, a new genus and species, and an indeterminate "hypsilophodontid." Material referred to cf. *Hypsilophodon* sp. was mainly assigned on the basis of isolated teeth and postcranial remains from the Late Hauterivian–Barremian of Teruel (Buscalioni and Sanz, 1984; Ruiz-Omeñaca, 2006), the Barremian–Aptian of Burgos (Ruiz-Omeñaca, 2001; Torcida Fernández-Baldor et al., 2003b; Torcida Fernández-Baldor, 2006), the Early Aptian of Castellón (Ruiz-Omeñaca, 2001), and the Late Aptian–Middle Albian of La Rioja (Torres and Viera, 1994). Although material of *Hypsilophodon foxii* has been reported from Spain, the presence of this taxon cannot be confirmed with the material at hand (Galton, 2009). A partially articulated skeleton from the Early Barremian of Galve (Teruel), which consists of vertebral elements as well as pelvic girdle and hind limb bones, represents a new genus and species; it may be diagnosable on the basis of characters observed in the pubis, femur, and fibula (Ruiz-Omeñaca, 2001, 2006 (Fig. 22.4). Skull and postcranial material from several individuals discovered in the Barremian of Salas de los Infantes (Burgos) could belong to the same taxon (Ruiz-Omeñaca, 2006; Torcida Fernández-Baldor, 2006). Isolated teeth from the Late Hauterivian–Early Barremian of Josa (Teruel) characterized by the absence of secondary ridges on the dentary crowns indicates the presence of a third basal euornithopod taxon; these teeth have been provisionally regarded as Hypsilophodontidae indet. (Ruiz-Omeñaca et al., 1997; Ruiz-Omeñaca, 2006). In addition, isolated teeth that possess subdivided denticles similar to those of *Drinker* from the Late Jurassic of North America have been reported from the Late Barremian of Uña in Cuenca

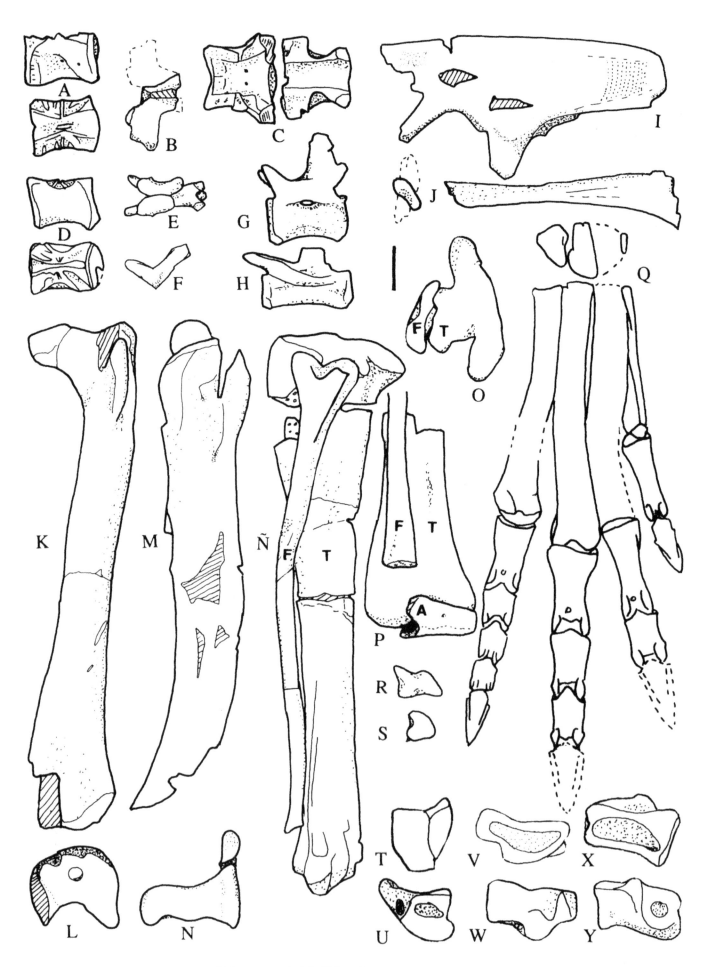

(Rauhut, 2002). It is possible that these teeth represent a basal ornithopod closely related to the new taxon from Teruel and Burgos (Ruiz-Omeñaca, 2006). Additional material has been referred to Hypsilophodontidae indet. (Ruiz-Omeñaca, 2001). Pending the evaluation of the taxonomic status of all this material, it is provisionally identified as basal Euornithopoda indet. (Galton, 2009).

Dryosaurids consist of at least two taxa from both Burgos and Teruel. A partially articulated skeleton from the Late Hauterivian–Early Barremian of Salas de los Infantes has been referred to "*Camptosaurus*" *valdensis* on the basis of femoral characters (Ruiz-Omeñaca, 2001). This material is still undescribed (Torcida Fernández-Baldor, 2006). "*C.*" *valdensis* is based on an isolated femur from the Wealden Group of the Isle of Wight (Galton, 1974, 2009). It has been regarded either as a valid taxon that requires a new generic name (Ruiz-Omeñaca, 2001, 2006) or as a nomen dubium (Galton, 2009). *Valdosaurus* sp. is represented by maxillary and mandibular remains with teeth and partially articulated postcranial bones, including femora, from the Barremian–Aptian near Salas de los Infantes (Ruiz-Omeñaca, 2001; Torcida Fernández-Baldor, 2006). As in the preceding case, this material has never been described in detail. Incomplete femora from the Barremian–Aptian of Salas de los Infantes and Galve may also belong to this taxon (Sanz et al., 1987; Ruiz-Omeñaca, 2001). According to Galton (2009), the record of *Valdosaurus* is restricted to England, and its occurrence in Spain is not yet well established.

Finally, an isolated tooth from the Hauterivian–Barremian transition of Galve has been tentatively regarded as ?Heterodontosauridae indet. (Ruiz-Omeñaca and Canudo, 2003b; Ruiz-Omeñaca, 2006). This tooth was previously assigned to as aff. *Echinodon* sp. (Estes and Sanchiz, 1982). Three other teeth from the Early Barremian of the same area have been referred to Heterodontosauridae indet. (Sánchez-Hernández et al., 2007); because of the fragmentary nature of the material, these teeth are best regarded as Ornithischia indet. (Galton, 2009).

Thyreophorans

Thyreophoran remains from the Early Cretaceous of Spain are rather scarce (Pereda Suberbiola and Galton, 2001). Ankylosaurs are mainly represented by dermal elements of *Polacanthus* sp. (Pereda Suberbiola, 2006). This taxon is known from the Hauterivian–Barremian transition of Golmayo in Soria (Pereda Suberbiola et al., 2007), the Late Barremian–Aptian of Salas de los Infantes in Burgos (Sanz, 1983; Pereda Suberbiola et al., 1999), and the basal Aptian of Morella in Castellón (Gasulla et al., 2003). Golmayo is currently the most productive site of *Polacanthus* outside England. Minor differences between the Soria material and the two species of *Polacanthus* found in England suggest the presence of a distinct species in the Iberian Peninsula (Pereda Suberbiola et al., 2007). However, additional material is needed to confirm this interpretation. The dermal remains of Morella are fragmentary but seem to be indistinguishable from those of the type species *Polacanthus foxii* from the Wealden Group of the Isle of Wight (Gasulla et al., 2003). Isolated teeth from the Hauterivian–Barremian transition of Josa in Teruel may also belong to *Polacanthus* or to

22.4. Ornithopoda nov. gen. et sp. from the Early Barremian of Poyales Barranco Canales, near Galve (Teruel). A, Last dorsal centrum; B, dorsal neural arch; C, first (left) and second (right) sacral centra; D; first caudal centrum; E, anterior caudal neural arch; F, first chevron; G, middle caudal vertebra; H, posterior caudal vertebra; I, left ilium; J, left prepubic process; K, L, left femur; M, N, right femur; Ñ, O, left tibia (T) and fíbula (F); P, right tibia (T), fibula (F), and astragalum (A); Q, right pes; R, left metatarsal IV; S, right metatarsal I; T–Y, right distal tarsal IV, in dorsal (A below, B, C, D below, Q below), ventral (E), left lateral (A above, D above, G, H, I, J right, Ñ), right lateral (F, M, Y), medial (X), anterior (J left, K, P, T), posterior (U), proximal (N, O, Q above, V) and distal (L, R, S W) views. Scale bar = 10 mm (A–S) or 5 mm (T–Y). From Ruiz-Omeñaca (2001, fig. 8).

22.5. Skeleton of the basal Carcharodontosaurian theropod *Concavenator corcovatus* from the Barremian of Las Hoyas in Cuenca.

Photograph courtesy of Museo de las Ciencias de Castilla-La Mancha, Cuenca.

a closely related "polacanthid" ankylosaur (Canudo et al., 2004a). In the same site, postcranial ankylosaurian remains (caudal vertebrae, ribs, scutes) have also been found (Ruiz-Omeñaca and Canudo, 2001). In addition, indeterminate ankylosaurian material is known from the Late Hauterivian–Early Barremian of Burgos (Izquierdo Montero et al., 2004; Torcida Fernández-Baldor, 2006) and the basal Aptian of Castellón (Santafé et al., 1982, as ?Nodosauridae indet.).

With regard to stegosaurs, all the material is indeterminate at the generic and species level. It consists of a few vertebrae and dermal plates from the Late Hauterivian–Early Barremian of Salas de los Infantes, Burgos (Pereda Suberbiola et al., 2003a), and a pair of dermal plates from the Galve area in Teruel, one from the latest Hauterivian and the other from the basal Barremian (Pereda Suberbiola et al., 2005). The Burgos remains have been referred to *Dacentrurus* sp. by Maidment et al. (2008). Finally, vertebral remains from the Barremian–basal Aptian of Castellote in Teruel have been tentatively referred to Stegosauria indet. (Ruiz-Omeñaca, 2000, 2006), but the material seems fragmentary for an accurate identification (Maidment et al., 2008).

22.6. Reconstruction of the basal Carcharodonto-saurian theropod *Concavenator corcovatus* from the Barremian of Las Hoyas in Cuenca.

Courtesy of Raúl Martín.

Theropods

Theropods from the Early Cretaceous of Spain are exclusively represented by tetanurans (Canudo and Ruiz-Omeñaca, 2003; Ruiz-Omeñaca, 2006). Though no ceratosaurian fossils have been found to date, their absence is probably due to taphonomical biases because abelisauroids are present in the Albian of southern Europe (Accarie et al., 1995; Carrano and Sampson, 2008). Among tetanurans, spinosaurids, allosauroids, ornithomimosaurs, and several kinds of maniraptorans, including dromaeosaurids and enantiornithine birds, are known.

Spinosaurid material from Spain consists of skull remains, postcranial elements, and, most frequently, isolated teeth. In La Rioja, a maxilla fragment from the Late Barremian–Aptian of Igea has been referred to *Baryonyx walkeri* (Viera and Torres, 1995). In Burgos, skull bones (postorbital, squamosal) and a tooth associated with vertebral remains, articulated metacarpals I–III, and a phalanx from an immature individual from the Late Barremian–Aptian of the vicinity of Salas de los Infantes have been ascribed to *Baryonyx* (Fuentes-Vidarte et al., 2001), even if the above-mentioned skull bones and metacarpus are unknown in the holotypic specimen of *B. walkeri* from the Barremian of England (Charig and Milner, 1986, 1997). Baryonychine teeth are relatively frequent in the Late Hauterivian–Early Aptian of Teruel, Burgos, and Castellón (Ruiz-Omeñaca et al., 1998b, 2005; Torcida Fernández-Baldor et al., 2003a; Gasulla et al., 2006; Ruiz-Omeñaca, 2006; Canudo et al., 2008a). Some of these teeth are similar to those of *Baryonyx* and have been regarded as cf. *Baryonyx* sp. (Canudo and Ruiz-Omeñaca, 2003). Other teeth, which differ from those of *B. walkeri* in having ornamented lingual and labial sides of the crown, have been attributed to Baryonychinae indet. These teeth resemble those known from the Hauterivian–Barremian of England that have been referred to cf. *Baryonyx* or *Baryonyx* sp. (Martill and Hutt, 1996; Charig and Milner,

1997; Martill and Naish, 2001). Moreover, a few baryonychine teeth from the Barremian–Early Aptian of Castellote (Teruel) are only serrated on the distal edge (Ruiz-Omeñaca et al., 1998b; Ruiz-Omeñaca, 2006). Finally, pelvic remains of an indeterminate spinosauroid have been described from the Early Aptian of Vallibona, Castellón (Gómez-Fernández et al., 2007). In conclusion, *Baryonyx* appears to be present in the Early Cretaceous of Spain (Milner, 2003); isolated teeth may represent additional baryonychine taxa (Ruiz-Omeñaca et al., 2005; Ortega et al., 2006).

Isolated material from a number of Spanish outcrops has been referred to Megalosauridae, but because it lacks diagnostic characters, it should be regarded as Theropoda indet. until it is restudied (Canudo and Ruiz-Omeñaca, 2003).

Basal carnosaurians (sensu Holtz et al., 2004b) have been reported from the Early Cretaceous of Spain. Isolated teeth from the Late Hauterivian to Early Aptian of Teruel and a tibia fragment from the Early Aptian of Castellón have been tentatively referred to Allosauroidea indet. (Infante et al., 2005b; Gasulla et al., 2006; Ruiz-Omeñaca, 2006). The teeth are smooth and show serrations of similar size on both mesial and distal carinae; the serrated mesial carina is shorter than the distal one, as the serrations do not reach the basal region of the crown (Ruiz-Omeñaca, 2006). The basal Carcharodontosaurian *Concavenator corcovatus* is known from the Barremian of Las Hoyas in Cuenca (Ortega et al., 2010). *Concavenator* is a medium-sized (approximately 6 m long) theropod that exhibits elongated neural spines in two presacral vertebrae forming a humplike structure and a series of small bumps on the ulna that could correspond to feather kill knobs (Figs. 22.5 and 22.6). A large isolated tooth from the Early Aptian of Castellón, which was originally referred to a megalosaurid, might represent a carcharodontosaur on the basis of minute denticles and wrinkles in the enamel next to the serrations (Santafé et al., 1982; Canudo and Ruiz-Omeñaca, 2003).

The Spanish record of maniraptoriforms is abundant. Smooth (or finely ornamented) and unserrated teeth from the Late Hauterivian–Early Aptian of Teruel have been referred to Maniraptoriformes indet. (Canudo and Ruiz-Omeñaca, 2003, as Coelurosauria indet.; Ruiz-Omeñaca, 2006).

Rauhut (2002) assigned several isolated teeth from the Late Barremian of Uña (Cuenca) to cf. *Paronychodon* sp. These teeth are small, with unserrated carinae, and they bear longitudinal ridges and grooves on both the lingual and labial sides of the crown (Zinke and Rauhut, 1994; Rauhut, 2002). They are similar to the teeth of *Paronychodon lacustris* from the Late Cretaceous of North America (Currie et al., 1990). The systematic position of this taxon is problematic; it has been provisionally regarded as Maniraptoriformes incertae sedis (Ruiz-Omeñaca, 2006). A few *Paronychodon*-like teeth from the Late Hauterivian–Early Barremian of Galve differ from those of Uña in the presence of denticles (Canudo and Ruiz-Omeñaca, 2003; Ruiz-Omeñaca and Canudo, 2003b; Ruiz-Omeñaca, 2006).

The ornithomimosaurs are represented by *Pelecanimimus polyodon* Pérez-Moreno, Sanz, Buscalioni, Moratalla, Ortega, and D. Rasskin-Gutman, 1994, from the Late Barremian of Las Hoyas, Cuenca (Fig. 22.3). The holotype consists of the articulated anterior half of the skeleton, including the skull, complete cervical, and an almost complete dorsal vertebral series, ribs, pectoral girdle, sternum, and both forelimbs (Pérez-Moreno et al., 1994; Pérez Pérez, 2004). Only a preliminary description of the material has been published. *Pelecanimimus* is unique among theropods in the large number of small teeth (approximately 220 teeth: seven premaxillary, about 30 maxillary, and about 75 in the dentary) (Pérez-Moreno et al., 1994). Soft tissue impressions reveal that it had an occipital crest and a gular pouch (Briggs et al., 1997). *Pelecanimimus* is regarded as a basal member of Ornithomimosauria and is the only representative of the group known from the Early Cretaceous of Europe (Pérez Pérez, 2004; Makovicky et al., 2004).

About 30 isolated teeth from the Late Hauterivian–Early Aptian of Teruel were previously referred to Dromaeosauridae indet. (Ruiz-Omeñaca et al., 1996). The crowns are smooth and exhibit distal denticles on the crown but lack mesial denticles; minor differences in the orientation and size of the denticles indicate the occurrence of several species. These teeth have been regarded as Maniraptora indet. because maniraptorans other than dromaeosaurids may have teeth bearing denticles on the distal carina and no denticles on the anterior one (Ruiz-Omeñaca, 2006).

Isolated teeth from the Late Hauterivian–Early Aptian of Teruel have been referred to "*Prodeinodon*" by Ruiz-Omeñaca (2006). These teeth are of large size and show denticles only in the distal edge. "*Prodeinodon*" is provisionally regarded as a valid taxon and classified as Maniraptora incertae sedis by Ruiz-Omeñaca and Canudo (2003a).

Dromaeosaurids are represented by numerous isolated teeth from the Late Hauterivian–Early Aptian of Teruel, Cuenca, Burgos, and Castellón (Ruiz-Omeñaca et al., 1996; Rauhut, 2002; Canudo and Ruiz-Omeñaca, 2003; Torcida Fernández-Baldor et al., 2003b; Ruiz-Omeñaca, 2006). The teeth with mesial denticles smaller than the distal ones are tentatively referred to as Velociraptorinae indet. (Ruiz-Omeñaca, 2006), resembling isolated teeth from England (Sweetman, 2004).

A collection of isolated teeth from the Barremian of Uña (Cuenca) was referred to cf. *Richardoestesia* by Rauhut and Zinke (1995) and Rauhut (2002). The current status of *Richardoestesia*, a taxon originally described from the Late Cretaceous of North America (Currie et al., 1990), is uncertain. The Uña teeth have elongate, slightly recurved crowns and bear small denticles in relation to the tooth size. According to Ruiz-Omeñaca (2006), these teeth may represent indeterminate velociraptorines.

The avian record consists of enantiornithines: *Iberomesornis romerali*, *Concornis lacustris*, *Eoalulavis hoyasi*, and Enantiornithes indet. from Las Hoyas, Cuenca (Sanz et al., 2002, and references therein); *Noguerornis gonzalezi* and Enantiornithes indet. are known from El Montsec, Lleida (Lacasa Ruiz, 1989; Chiappe and Lacasa Ruiz, 2002).

Iberomesornis romerali Sanz and Bonaparte, 1992, is based on an articulated specimen lacking the skull and the anterior cervical vertebrae (Sanz et al., 1988; Sanz and Bonaparte, 1992) (Fig. 22.7). It is characterized by a unique character combination, including strong coracoids, a furcula with a marked hypocleideum, a fused pelvis, sharply curved foot claws, and no evidence of metatarsal fusion (Sanz and Ortega, 2002; Padian, 2004). *Iberomesornis* was first placed in an intermediate position between *Archaeopteryx* and modern birds, but new phylogenetic analyses shows that it is a basal enantiornithine (Sereno, 2000; Chiappe, 2001; Sanz et al., 2002).

It was a sparrow-sized bird, with an estimated mass of about 15–20 g (Sanz and Buscalioni, 1992).

Concornis lacustris Sanz and Buscalioni, 1992, is known from a complete skeleton except the skull and neck, with evidence of some feather impressions (Sanz and Buscalioni, 1992; Sanz et al., 1995). It is diagnosed by a ribbonlike ischium, a transverse ginglymoid articulation of trochlea of metatarsal I, and a strongly excavated distal end of metatarsal IV (Sanz et al., 1995, 2002). *Concornis* is probably more derived than *Iberomesornis* and is regarded as a member of Euenantiornithes. It is about twice the size of *Iberomesornis*, and its mass is estimated to be 75–80 g (Sanz and Buscalioni, 1992).

Eoalulavis hoyasi Sanz, Chiappe, Pérez-Moreno, Buscalioni, Moratalla, Ortega, and Poyato-Ariza, 1996, is based on the anterior part of an articulated skeleton lacking the skull and anterior cervical vertebrae (Sanz et al., 1996). The specimen includes evidence of wing feathers in position (both primary and secondary ones) and an alula (feather attached to the first digit of the hand). Some body feathers are also visible in the vicinity of the humeri and pectoral girdle. Autapomorphic characters include the presence of laminar, keellike cervical and dorsal centra; an undulating ventral surface of the furcula; a distal humerus with a thick, posteriorly projected ventral margin; the presence of several small tubercles on the distal, posterior surface of the minor metacarpal; a depressed, spear-shaped sternum, with a footlike posterior expansion and a faint carina; and a deep, rostral cleft on the sternum (Sanz et al., 1996, 2002). As with *Concornis*, *Eoalulavis* is placed within Euenantiornithes (Sanz et al., 2002). The thoracic box of *Eoalulavis* contains elements of unidentified crustaceans, which is regarded as the earliest direct manifestation of trophic habits in Mesozoic birds (Sanz et al., 1996, 2002).

The holotype of *Noguerornis gonzalezi* Lacasa Ruiz, 1989, includes three trunk vertebrae, furcula, humeri, radii, ulna, three metacarpals of the hand, carpal, phalanx, ischium, tibia, and wing feather impressions; most bones are incomplete (Lacasa Ruiz, 1989; Chiappe and Lacasa Ruiz, 2002). *N. gonzalezi* is characterized by the presence of a strongly curved humerus and an ischiadic symphysis (Chiappe and Lacasa Ruiz, 2002). It is regarded as a basal enantiornithine bird closely related to *Iberomesornis* (Chiappe, 2001). It is intermediate in size between *Iberomesornis* and *Concornis*.

In addition, the lithographic limestones of the El Montsec (Lleida) have yielded a feathered hatchling of an enantiornithine bird (Sanz et al., 1997) and approximately 25 isolated feathers (Lacasa Ruiz, 1985). "*Ilerdopteryx viai*" Lacasa Ruiz, 1985, is based on an isolated feather from the same quarry and is regarded as a nomen dubium (Sanz et al., 1997).

Finally, Sanz et al. (2001b) described a regurgitated pellet from Las Hoyas containing four tiny juvenile bird individuals. On the basis of morphological and size data, at least three different species are present in the fossil bone assemblage. The most possible producer of the pellet was a small- to medium-sized nonavian theropod or a large pterosaur (Sanz et al., 2001b; Sanz and Ortega, 2002).

22.7. The basal enanthiornithine bird *Iberomesornis romerali* from the Barremian lithographic limestones of Las Hoyas (Cuenca). Holotype specimen and fluorescence-induced ultraviolet photograph. Scale bar in centimeters.

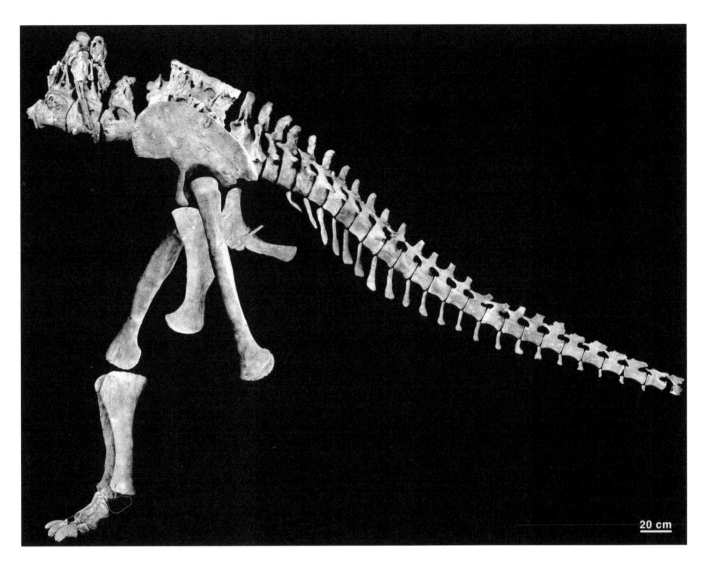

20 cm

22.8. Partial articulated skeleton of *Tastavinsaurus sanzi* from the Aptian of Peñarroya de Tastavins (Teruel).

Photograph courtesy of Grupo Aragosaurus–IUCA (Universidad de Zaragoza).

Sauropods

As is typical in Laurasian continents, macronarians were the most diversified sauropods from the Early Cretaceous of Spain (Royo-Torres and Canudo, 2003; Royo-Torres, 2009b). Two distinct genera have been described to date, both from Teruel: *Aragosaurus ischiaticus* Sanz, Buscalioni, Casanovas, and Santafé, 1987, from the Late Hauterivian of Galve, and *Tastavinsaurus sanzi* Canudo, Royo-Torres, and Cuenca-Bescós, 2008b, from the Early Aptian of Peñarroya de Tastavins. *Aragosaurus* is the earliest named Mesozoic dinosaur in Spain (Sanz et al., 1987). It is based on an isolated tooth and postcranial remains from a single individual, including cervical and dorsal ribs, caudal vertebrae and hemal arches, scapula, coracoid, forelimb, pelvis, femur, and several phalanges of the pes (Lapparent, 1960; Sanz et al., 1987; Royo-Torres and Canudo, 2003). The whole of the material is still to be adequately described and figured, and so this taxon is in need of further documentation and analysis. *Aragosaurus* was originally referred to the Camarasauridae (Sanz et al., 1987) and has been provisionally regarded as Eusauropoda incertae sedis by Upchurch at al. (2004). The presence of a lateral prominent bulge on the femur and other features support its attribution to Macronaria (Canudo et al., 2001; Barco-Rodríguez, 2009).

22.9. Dorsal vertebra and articulated caudal vertebrae from a partial skeleton of an unnamed titanosauriform sauropod from the Barremian–Aptian of Salas de los Infantes (Burgos) of El Oterillo, near Salas de los Infantes (Burgos).

Photograph courtesy of Colectivo Arqueológico-Paleontológico de Salas (CAS).

Tastavinsaurus is one of the most complete and best-preserved sauropods from the Early Cretaceous of Europe (Fig. 22.8): its partially articulated skeleton consists of vertebrae from the posterior thoracic region to the tail, hemal arches, complete pelvic girdle, femora, tibia, fibula, metatarsals, and phalanges (Canudo et al., 2008b; Royo-Torres, 2009a). *Tastavinsaurus* is characterized by a large number of autapomorphic traits on the vertebrae, pubis, tibia, fibula, and metatarsals. Among them are the presence of a honeycomb pattern in cross section of the thoracic vertebrae, a short axial lamina between the prezygapophyses of the anterior caudal vertebrae, rounded bumps on both articular surfaces of the midposterior caudal vertebrae, an acute anteroventral corner of the distal pubis, and a proximally expanded metatarsal V (Canudo et al., 2008b). *Tastavinsaurus* and *Aragosaurus* have been regarded as members of Laurasiformes—a clade of basal macronarians more derived than *Camarasaurus* whose representatives lived in Europe and in North America (Royo-Torres, 2009a, 2009b). Nevertheless, Laurasiformes probably includes a paraphyletic assemblage of basal macronarians (Barco-Rodríguez, 2009). Other phylogenetic analyses place *Tastavinsaurus* either as a basal, nonbrachiosaurid titanosauriform (Canudo et al., 2008a) or as a nontitanosauriform macronarian (Barco-Rodríguez, 2009).

Bones of a third macronarian taxon different from *Aragosaurus* and *Tastavinsaurus* have recently been unearthed from the Late Barremian–Early Aptian of Salas de los Infantes (Burgos). It consists of a partially articulated skeleton of a large-sized sauropod that includes a tooth, dorsal, sacral, and caudal vertebrae, dorsal ribs, hemal arches, both scapulae, ischia, and pubis (Fig. 22.9) (Torcida Fernández-Baldor, 2006). This material has not yet been formally described but probably represents a new genus and species of Titanosauriformes (Torcida Fernández-Baldor et al., 2009).

A few spoonlike teeth tentatively identified as camarasaurids have been reported from the Late Hauterivian–Early Barremian of Teruel (Sanz et al., 1987; Ruiz-Omeñaca, 2006; Royo-Torres and Cobos, 2007; Sánchez-Hernández et al., 2007). These teeth are similar to those of *Oplosaurus armatus* from the Wessex Formation of the Isle of Wight (regarded as a brachiosaurid by Martill and Naish, 2001; see Canudo et al., 2002), a taxon of unresolved affinity within Sauropoda (Naish and Martill, 2007, supplement). More diagnostic material is needed to confirm the presence of camarasaurids in the Wealden formations of Spain. Royo-Torres (2009b) regarded *O. armatus* as a basal macronarian.

Brachiosaurid remains (i.e., tooth, dorsal vertebrae and ribs, partial sacrum, caudal vertebrae and hemal arches, humerus, ilia, ischia, pubis, and femur) have been described from the Hauterivian–Barremian transition of Golmayo, Soria (Fuentes-Vidarte et al., 2005). The material has been provisionally referred to a nontitanosaurian Titanosauriformes (Royo-Torres et al., 2009), but a complete study is needed to determine its accurate relationships.

In addition, brachiosaurid-like titanosauriforms seem to be represented by old and new finds from the Early Aptian of Morella, Castellón (Santafé et al., 1982; Yagüe et al., 2001). Most of the material, which mainly consists of postcranial remains, is still to be published (Gasulla et al., 2008).

Isolated teeth from the Barremian of Teruel have been referred to "*Pleurocoelus*" *valdensis* (Sanz et al., 1987, as cf. *Astrodon* sp.; Ruiz-Omeñaca and Canudo, 2005). This taxon, originally described on the basis of teeth and vertebrae from the Wealden Group of Sussex and the Isle of Wight in England, is usually regarded as a nomen nudum (Martill and Naish, 2001). The name "*P.*" *valdensis* is reserved here to compressed cone–chisel–like teeth with enamel ornamented with irregular longitudinal ridges, probably pertaining to a diagnosable brachiosaurid (Ruiz-Omeñaca and Canudo, 2005).

Isolated spoon-shaped teeth with a lingual cingulum from the Late Hauterivan–Early Barremian of Teruel show a close resemblance with those of *Euhelopus* from eastern Asia (Euhelopodidae indet. in Canudo et al., 2002; Ruiz-Omeñaca, 2006). The phylogenetic relationships of *Euhelopus* have been controversial for quite a while; however, there is now a consensus that this taxon represents a basal titanosauriform (Wilson and Upchurch, 2009). The presence of *Euhelopus*-like teeth is the Early Cretaceous of Spain is noticeable from a biogeographical point of view (Canudo et al., 2002; see below).

Sánchez-Hernández et al. (2007) suggested the occurrence of derived titanosaurians in the Early Cretaceous of Teruel on the basis of a single caudal vertebra from the Barremian of Galve. This interpretation is worthy of consideration but needs to be confirmed on the basis of more complete material.

The only nonmacronarian sauropod found in the Early Cretaceous of Spain is a medium-sized rebbachisaurid diplodocoid from the Late Barremian–Early Aptian of Salas de los Infantes, Burgos (Pereda Suberbiola et al., 2003b). It is known by an incomplete but associated skeleton that includes premaxillary and dentary fragments, isolated teeh, cervical, dorsal, and caudal vertebrae, cervical and dorsal ribs, hemal arches, ischia, and femur. This material exhibits several autapomorphies in the teeth and vertebrae and represents a new genus and species of rebbachisaurid: *Demandasaurus darwini* (Torcida Fernández-Baldor et al., 2011). It is the first diplodocoid described from Spain. The new taxon shares its closest relationship with *Nigersaurus* from Africa (Sereno et al., 2007; Torcida Fernández-Baldor et al., 2011).

Additional sauropod records from the Early Cretaceous of Spain are listed by Royo-Torres and Canudo (2003) and Royo-Torres (2009a).

Discussion

The study of the dinosaur skeletal fossil record from the Early Cretaceous (Valanginian–Albian) of Spain provides significant information about the diversity and evolution of the dinosaur faunas in southern Europe. Spanish localities, which include the spectacular *Konservat-Lagerstätten* of Las Hoyas (Cuenca) and other important dinosaur-bearing sites that are mainly concentrated in the provinces of Burgos and Teruel, have provided a rich and diverse assemblage. All of the major dinosaur clades except the marginocephalians are represented (Table 22.1). The Hauterivian–Albian formations of Spain have yielded fossils from at least 30 dinosaur species; ornithopods and especially theropods are the most diversified clades, followed by the sauropods and to a lesser extent by thyreophorans. Ten new taxa (six theropods, including *Concavenator*, *Pelecanimimus* and the birds *Iberomesornis*, *Concornis*, *Eoalulavis*, and *Noguerornis*, the sauropods *Aragosaurus*, *Demandasaurus* and *Tastavinsaurus*, and the ornithopod *Delapparentia*) have been erected on the basis of skeletal remains found in Burgos, Cuenca, Lleida, and Teruel. The Late Hauterivian–Aptian dinosaur record of Spain is the richest one from mainland Europe and one of the richest of the world. Only the equivalent units of England have produced a greater number of dinosaur taxa (Weishampel et al., 2004). The fact that the most important discoveries have been realized in the last 25 years reveals the great potential of the Spanish record.

The Spanish assemblage shares a number of taxa with the dinosaur fauna of the Wealden Group and Wealden Clay Group of the Isle of Wight, Surrey, and Sussex (see Naish and Martill, 2007, 2008) (Table 22.1): the theropod *Baryonyx walkeri* (Milner 2003), the ornithopod *Iguanodon bernissartensis* (Santafé et al., 1982; Gasulla et al., 2007), and the ankylosaur *Polacanthus* (Pereda Suberbiola et al., 2007). The occurrence in both faunas of the sauropods *Oplosaurus armatus* and *"Pleurocoelus" valdensis* is also worthy of consideration (Ruiz-Omeñaca and Canudo, 2005; Royo-Torres, 2009b). Moreover, the occurrence of carcharodontosaurs in both Britain and Iberia is now confidently established (*Neovenator* in England, see Benson et al., 2010; and *Concavenator* in Spain, Ortega et al., 2010).

Table 22.1. List of dinosaur taxa from the Valanginian–Aptian of Spain and equivalent horizons
of the Wealden of southern England

Lower Wealden faunas (Valanginian–Early Hauterivian)

Spain		England
	Ornithopoda	
		Euornithopoda indet.
		Valdosaurus canaliculatus (Dr)
		Barilium dawsoni (Ig)
		Hypselospinus fittoni (Ig)
Iguanodontoidea indet.		Iguanodontoidea indet.
	Thyreophora	
		Hylaeosaurus armatus (An)
		Stegosauria indet. ("*Regnosaurus northamptoni*")
	Theropoda	
		cf. *Baryonyx sp.* (Sp)
Allosauroidea indet.		*Becklespinax altispinax* (Al)
		Valdoraptor oweni (Al)
		Maniraptora indet. (= "*Wyleyia valdensis*")
		?Dromaeosauridae indet.
	Sauropoda	
Sauropoda indet.		
		Diplodocidae indet.
		Titanosauriformes indet. (= "*Cetiosaurus*" *brevis*)
		"*Pleurocoelus*" *valdensis** (Br)
		Titanosauria indet. (= "*Pelorosaurus*" *becklesii*)

Upper Wealden faunas (Late Hauterivian/Early Barremian to Aptian)

Spain		England
	Ornithopoda	
Ornithopoda indet.		Euornithopoda indet.
Ornithopoda nov. gen. et sp. [Galve] (Hy)		
		cf. *Hypsilophodon* sp. (Hy)
		Hypsilophodon foxii (Hy)
Dryosauridae indet. (= *Valdosaurus* sp.)		*Valdosaurus canaliculatus* (Dr)
Dryosauridae indet. (different from *Valdosaurus*)		
Iguanodon bernissartensis (Ig)		*Iguanodon bernissartensis* (Ig)
?*Mantellisaurus atherfieldensis* (Ig)		*Mantellisaurus atherfieldensis* (Ig)
Delapparentia turolensis (Ig)		
Iguanodontia indet. [Salas]		Iguanodontoidea indet.
?Hadrosauroidea indet.		
	Thyreophora	
Polacanthus sp. (An)		*Polacanthus foxii* (An)
		Polacanthus rudgwickensis (An)
Stegosauria indet.		?Stegosauria indet.
	Theropoda	
Baryonyx walkeri (Sp)		*Baryonyx walkeri* (Sp)
Baryonychinae indet. (Sp)		cf. *Baryonyx* sp. (Sp)
Concavenator corcovatus (Ca)		*Neovenator salerii* (Ca)
Allosauroidea indet.		Allosauroidea indet.
		*Aristosuchus pusillus** (Co)
		Eotyrannus lengi (Ty)
		*Calamosaurus foxi** (Ty)
Maniraptoriformes indet.		
Maniraptoriformes incertae sedis (*Paronychodon*-like)		
Maniraptora indet. (including "*Prodeinodon*" sp.)		

Maniraptora incertae sedis "Prodeinodon"

Pelecanimimus polyodon (Or)

?Oviraptorosauria indet.
("*Thecocoelurus*")

Dromaeosauridae indet.

?Velociraptorinae indet.

Iberomesornis romerali (En)

Concornis lacustris (En)

Eoalulavis hoyasi (En)

Noguerornis gonzalezi (En)

Enantiornithes indet.

?Dromaeosauridae indet.

?Velociraptorinae indet.

Sauropoda

Demandasaurus darwini (Re)

Oplosaurus armatus* (Ma)

Aragosaurus ischiaticus (Ma)

Tastavinsaurus sanzi (Ma)

Titanosauriformes nov. gen. et sp. [Salas]

Titanosauriformes indet.

"*Pleurocoelus*" *valdensis** (Br)

Brachiosauridae indet.

Euhelopodidae indet.

?Titanosauria indet.

Rebbachisauridae indet.

Oplosaurus armatus* (Ma)

Ornithopsis hulkei*

"*Pleurocoelus*" *valdensis** (Br)

Brachiosauridae indet.
(= "*Eucamerotus foxi*")

Titanosauria indet.
(= "*Iuticosaurus valdensis*")

Abbreviations: Al, Allosauroidea; An, Ankylosauria; Br, Brachiosauridae; Ca, Carcharodontosauria ; Co, Compsognathidae; Dr, Dryosauridae; En, Enantiornithes; Hy, basal Euornithopoda (formerly "Hypsilophodon-tidae"); Ig, basal Iguanodontoidea; Ma, basal Macronaria; Or, Ornithomimosauria; Re, Rebbachisauridae; Sp, Spinosauroidea; St, Stegosauria; Ty, basal Tyrannosauroidea.

Note: See Martill and Naish (2001), Weishampel et al. (2004), Naish and Martill (2007, 2008), Galton (2009), Norman (2010) and Chapter 15 in this book. Note the poor fossil record of the Lower Wealden of Spain and the quality of the Upper Wealden dinosaur record, both in Britain and in Spain. An asterisk indicates provisionally valid taxa. Spanish localities that have yielded material of new taxa yet to be named are in square brackets.

Titanosaurians and rebbachisaurid diplodocoids are also present in both assemblages (Naish and Martill, 2007; Mannion, 2009), but they probably correspond to different genera. The identity of the dryosaurid material from Spain that is referred to *"Camptosaurus" valdensis* (Ruiz-Omeñaca, 2001, 2006) needs to be investigated in the future.

The presence in Spain of several typical taxa from the Barremian–Aptian formations of the Isle of Wight (i.e., the ornithopods *Hypsilophodon foxii* and *Valdosaurus canaliculatus*) has been suggested, but it cannot be confirmed on the basis of available descriptions (Galton, 2009). Further research in also needed to attest the presence in Spain of *Mantellisaurus atherfieldensis*. Though these taxa are currently not recognized in the Spanish material on the basis of diagnostic characters, their presence cannot be definitively excluded.

Ornithomimosaurs (*Pelecanimimus*), enantiornithine birds (*Iberomesornis, Concornis, Eoalulavis,* and *Noguerornis*) and *Euhelopus*-like sauropods found in Spain are taxa apparently absent in the English record (but isolated teeth from the Isle of Wight may represent an enantiornithine; see Sweetman and Martill, 2010). On the contrary, several theropod taxa that are documented in the Wealden Group of the Isle of Wight, such as a probable compsognathid (*Aristosuchus*), basal tyrannosauroids (*Eotyrannus* and *Calamosaurus*), and a caenagnathid-like oviraptorosaur (*"Thecocoelurus"*) (see Naish and Martill, 2007), are not represented in the Spanish outcrops.

The exceptional enantiornithine bird records from the Barremian of Las Hoyas (Cuenca) and El Montsec (Lleida) provide significant insights about the early steps of avian evolution (Sanz and Ortega, 2002; Sanz et al., 1997, 2002; Chamero et al., 2009). Las Hoyas has also provided the first evidence of an alular feather (Sanz et al., 1996), the earliest evidence that Mesozoic birds were usual prey animals (Sanz et al., 2001b), and one of the oldest direct evidences of trophic habits for birds (Sanz et al., 1996). The basal ornithomimosaur *Pelecanimimus* from Las Hoyas represents one of the earliest occurrences of this clade worldwide and the only record known so far in Europe (Pérez-Moreno et al., 1994; Makovicky et al., 2004). Other Iberian taxa that represent early occurrences of their respective clades are the new rebbachisaurid sauropod *Demandasaurus* from the Late Barremian–Early Aptian of Salas de los Infantes in Burgos (Pereda Suberbiola et al., 2003b; Torcida Fernández-Baldor et al., 2011), and the baryonychine theropods and, tentativaly, basal hadrosauroids from the Hauterivian–Barremian transition of Josa in Teruel (Ruiz-Omeñaca and Canudo, 2003b; Ruiz-Omeñaca et al., 2005).

From a biogeographical point of view, most of the dinosaur taxa from the Early Cretaceous of the Iberian Peninsula show Euramerican (covering Europe and North America) affinities (Russell, 1993; Pereda Suberbiola, 1994; Le Loeuff, 1997). These faunas are mainly dominated by basal iguanodontians, "polacanthid" ankylosaurs, and basal macronarians, including brachiosaurids (Holtz et al., 2004a; Ruiz-Omeñaca, 2006). The opening of the North Atlantic Ocean led to the interruption of the connections between these continents during the Aptian–Albian (Kirkland et al., 1997). The occurrence of *Euhelopus*-like teeth in the Barremian of Iberia provides additional support for faunal exchange between Europe and East Asia (Canudo et al., 2002; Barrett and Wang, 2007). Moreover, the occurrence in the Early Cretaceous of Iberia of dinosaur taxa in common with the Gondwanan landmasses (i.e., spinosaurids, rebbachisaurids, and dryosaurids) has been explained in terms of an intercontinental connection between the north of Africa and the south of Europe via the Apulian Route during the Barremian–Aptian (Canudo et al., 2009). This trans-Tethyan passageway was probably used by the dinosaurs of Laurasia to disperse into Gondwana and vice versa (Gheerbrant and Rage, 2006; Canudo et al., 2009).

Acknowledgments

We are grateful to Pascal Godefroit and the organizing committee of the Darwin–Bernissart meeting for the invitation to participate in these proceedings. We thank Peter M. Galton, whose helpful comments and suggestions have improved this chapter. Research work of the senior author was supported by the Spanish Ministerio de Ciencia e Innovación (MICINN, projects CGL2007-64061/BTE and CGL2010-18851/BTE) and the Gobierno Vasco/EJ (GIC07/14-361). The Aragosaurus–IUCA Group of the Universidad de Zaragoza is supported by the projects CGL2007-62469/BTE and CGL2010-16647/BTE (MICINN and European Regional Development Fund) and by "Grupos Consolidados" (Gobierno de Aragón). Research work of J. I. R.-O. is supported by Protocol CN-04-226 (Principado de Asturias). Thanks to the Colectivo Arqueológico-Paleontológico de Salas de los Infantes (CAS) and the Museo de Dinosaurios of this locality (Burgos, Spain) for their assistance.

References

Accarie, H., B. Beaudoin, J. Dejax, G. Friès, J. G. Michard, and P. Taquet. 1995. Découverte d'un dinosaure théropode nouveau (*Genusaurus sisteronis* n. g., n. sp.) dans l'Albien marin de Sisteron (Alpes-de-Haute-Provence, France) et extension au Crétacé inférieur de la lignée cératosaurienne. Comptes rendus de l'Académie des Sciences, Paris 320 (IIa): 327–334.

Barco-Rodríguez, J. L. 2009. Sistemática e implicaciones filogenéticas y paleobiogeográficas del saurópodo *Galvesaurus herreroi* (Formación Villar del Arzobispo, Galve, España). Ph.D. dissertation, Universidad de Zaragoza, Zaragoza, Spain, 389 pp. http://www.aragosaurus.com/secciones/publicaciones/artic/Barco2009.pdf.

Barrett, P. M., and X.-L. Wang. 2007. Basal titanosauriform (Dinosauria, Sauropoda) teeth from the Lower Cretaceous Yixian Formation of Liaoning province, China. Palaeoworld 16: 265–271.

Benson, R. B. J., M. T. Carrano, and S. L. A. Brusatte. 2010. A new clade of archaic large-bodied predatory dinosaurs (Theropoda: Allosauroidea) that survived to the latest Mesozoic. Naturwissenschaften 97: 71–78.

Briggs, D. E. G., P. R. Wilby, B. P. Pérez-Moreno, J. L. Sanz, and M. Fregenal-Martínez. 1997. The mineralization of dinosaur soft tissue in the Lower Cretaceous of Las Hoyas, Spain. Journal of the Geological Society, London 154: 587–588.

Buscalioni, A. D., and J. L. Sanz. 1984. Los arcosaurios (Reptilia) del Jurásico Superior–Cretácico Inferior de Galve (Teruel, España). Teruel 71: 9–30.

Buscalioni, A. D., M. A. Fregenal, A. M. Bravo, F. J. Poyato-Ariza, B. Sanchíz, A. M. Báez, O. Cambra Moo, C. Martín Closas, S. E. Evans, and J. Marugán Lobón. 2008. The vertebrate assemblage of Buenache de la Sierra (Upper Barremian of Serrania de Cuenca, Spain) with insights into its taphonomy and palaeoecology. Cretaceous Research 29: 687–710.

Butler, R. J., P. Upchurch, and D. B. Norman. 2008. The phylogeny of the ornithischian dinosaurs. Journal of Systematic Palaeontology 6: 1–40.

Canudo, J. I., and J. I. Ruiz-Omeñaca. 2003. Los restos directos de dinosaurios terópodos (excluyendo Aves) en España; pp. 347–374 in F. Pérez-Lorente (coord.), Dinosaurios y otros reptiles mesozoicos de España. Ciencias de la

Tierra 26. Instituto de Estudios Riojanos (IER), Logroño, Spain.

Canudo, J. I., J. L. Barco, R. Royo Torres, and J. I. Ruiz Omeñaca. 2001. Precisiones sobre la posición taxonómica de *Aragosaurus ischiaticus* (Dinosauria, Sauropoda); pp. 263–270 in G. Meléndez, Z. Herrera, G. Delvene, and B. Azanza (eds.), XVII Jornadas de la Sociedad Española de Paleontología: Los fósiles y la paleogeografía, Zaragoza, Spain. Publicaciones del Seminario de Paleontología de Zaragoza 5 (1).

Canudo, J. I., J. I. Ruiz Omeñaca, J. L., Barco, and R. Royo Torres. 2002. Saurópodos asiáticos en el Barremiense (Cretácico Inferior) de España? Ameghiniana 39: 443–452.

Canudo, J. I., J. I. Ruiz-Omeñaca, and G. Cuenca-Bescós. 2004a. Los primeros dientes de anquilosaurio (Ornithischia: Thyreophora) descritos en el Cretácico Inferior de España. Revista Española de Paleontología 19: 33–46.

Canudo, J. I., J. I. Ruiz-Omeñaca, and L. M. Sender. 2004b. Primera evidencia de un dinosaurio saurópodo en la Formación Escucha (Utrillas, Teruel), Albiense medio (Cretácico inferior); pp. 27–30 in C. Liesa Carrera, A. Pocovi Juan, C. Sancho Marcén, F. Colombo Piñol, A. González Rodríguez, and A. R. Soria de Miguel (eds.), VI Congreso Geológico de España, Zaragoza, Spain, July 12–15, 2004. Geo-Temas 6 (5).

Canudo, J. I., J. I. Ruiz-Omeñaca, A. Del Ramo, and F. Guillén Mondejar. 2004c. Primera evidencia de restos de dinosaurio en Murcia (Cretácico Inferior, Albiense). Geogaceta 35: 119–122.

Canudo, J. I., A. Cobos, C. Martín-Closas, X. Murelaga, X. Pereda-Suberbiola, R. Royo-Torres, J. I. Ruiz-Omeñaca, and L. M. Sender. 2005. Sobre la presencia de dinosaurios ornitópodos en la Formación Escucha (Cretácico Inferior, Albiense): redescubierto el "*Iguanodon*" de Utrillas (Teruel); pp. 51–56 in XVI Reunión Bienal de la Real Sociedad Española de Historia Natural, Fundación Conjunto Paleontológico de Teruel, Teruel, Spain. Fundamental 6.

Canudo, J. I., J. M. Gasulla, D. Gómez Fernández, F. Ortega, J. L. Sanz, and P. Yagüe. 2008a. Primera evidencia de dientes aislados atribuidos a Spinosauridae (Theropoda) en el Aptiano inferior (Cretácico Inferior) de Europa: Formación Arcillas de Morella (España). Ameghiniana 45: 649–662.

Canudo, J. I., R. Royo-Torres, and G. Cuenca-Bescós. 2008b. *Tastavinsaurus sanzi* gen. et sp. nov. (Saurischia, Sauropoda) from the Lower Aptian of Spain: new data on the radiation of Titanosauriformes in the Lower Cretaceous. Journal of Vertebrate Paleontology 28: 712–731.

Canudo, J. I., J. L. Barco, X. Pereda Suberbiola, J. I. Ruiz Omeñaca, L. Salgado, F. Torcida Fernández-Baldor, and J. M. Gasulla. 2009. What Iberian dinosaurs reveal about the bridge said to exist between Gondwana and Laurasia in the Early Cretaceous. Bulletin de la Société Géologique de France 180: 5–11.

Carrano, M. T., and S. D. Sampson. 2008. The phylogeny of Ceratosauria (Dinosauria: Theropoda). Journal of Systematic Palaeontology 6: 183–236

Casanovas-Cladellas, M. L., J. V. Santafé-Llopis, C. Santisteban-Bové, and X. Pereda-Suberbiola. 1999. Estegosaurios (Dinosauria) del Jurásico Superior–Cretácico Inferior de Los Serranos (Valencia, España). Revista Española de Paleontología, número extraordinario Homenaje al Prof. J. Truyols: 57–63.

Chamero, B., J. Marugán-Lobón, A. Buscalioni, and J. L. Sanz. 2009. The Mesozoic avian fossils of the Iberian Peninsula. Journal of Vertebrate Paleontology 29 (supplement to 3): 76A.

Charig, A. J., and A. C. Milner. 1986. *Baryonyx*, a remarkable new theropod dinosaur. Nature 324: 359–361.

———. 1997. *Baryonyx walkeri*, a fish-eating dinosaur from the Wealden of Surrey. Bulletin of the Natural History Museum of London, Geology Series 53: 11–70.

Chiappe, L. M. 2001. Phylogenetic relationships among basal birds; pp. 125–139 in J. Gauthier and L. F. Gall (eds.), New perspectives on the origin and early evolution of birds. Proceedings of the International Symposium in Honor of John H. Ostrom. Peabody Museum of Natural History Yale University, New Haven.

Chiappe, L. M., and A. Lacasa Ruiz. 2002. *Noguerornis gonzalezi* (Aves: Ornithothoraces) from the Early Cretaceous of Spain; pp. 230–239 in L. M. Chiappe and L. M. Witmer (eds.), Mesozoic birds: above the heads of dinosaurs. University of California Press, Berkeley.

Currie, P. J., J. K. Rigby Jr., and R. E. Sloan. 1990. Theropod teeth from the Judith River Formation of southern Alberta, Canada; pp. 107–125 in K. Carpenter and P. J. Currie (eds.), Dinosaur systematics: approaches and perspectives. Cambridge University Press, New York.

Diéguez, C., A. Perejón, and J. Truyols (coords.). 2004. Homenaje a José Royo Gómez, 1895–1961. Monografies, Consell Valencià de Cultura, Valencia, Spain, 322 pp.

Estes, R., and B. Sanchíz. 1982. Early Cretaceous lower vertebrates from Galve (Teruel), Spain. Journal of Vertebrate Paleontology 2: 21–39.

Fernández-Galiano, D. 1958. Descubrimientos de restos de Dinosaurios en Galve. Teruel 20: 1–3.

Francés, V., and J. L. Sanz. 1989. Restos de dinosaurios del Cretácico Inferior de Buenache de la Sierra (Cuenca); pp. 125–144 in J. L. Sanz (coord.), La fauna del pasado en Cuenca. Instituto Juan de Valdés, Cuenca, Spain.

Fregenal-Martínez, M. A., and N. Meléndez. 2000. The lacustrine fossiliferous deposits of the Las Hoyas subbasin (Lower Cretaceous, Serranía de Cuenca, Iberian Ranges, Spain); pp. 303–314 in E. H. Gierlowski-Kordesch and K. R. Kelts (eds.), Lake basins through space and time. American Association of Petroleum Geologists, Studies on Geology 46.

Fuentes-Vidarte, C., and M. Meijide Calvo. 2001. Presencia de un grupo de juveniles de *Hypsilophodon* cf. *foxii* (Dinosauria, Ornithopoda) en el Weald de Salas de los Infantes (Burgos, España); pp. 339–348 in Colectivo Arqueológico-Paleontológico de Salas (ed.), I Jornadas Internacionales sobre Paleontología de Dinosaurios y su Entorno, Actas, Salas de los Infantes, Burgos, Spain.

Fuentes-Vidarte, C, M. Meijide Calvo, L. A. Izquierdo, D. Montero, G. Pérez, F. Torcida, V. Urién, F. Meijide Fuentes, and M. Meijide Fuentes. 2001. Restos fósiles de *Baryonyx* (Dinosauria, Theropoda) en el Cretácico Inferior de Salas de los Infantes (Burgos, España); pp. 349–359 in Colectivo Arqueológico-Paleontológico de Salas (ed.), I Jornadas Internacionales sobre Paleontología de los Dinosaurios y su Entorno, Actas, Salas de los Infantes, Burgos, Spain.

Fuentes-Vidarte, C., M. Meijide Calvo, F. Meijide Fuentes, and M. Meijide Fuentes. 2005. Fauna de vertebrados del Cretácico Inferior del yacimiento de "Zorralbo" en Golmayo (Soria, España). Revista Española de Paleontología, número extraordinario 10: 83–92.

Galton, P. M. 1974. The ornithischian dinosaur *Hypsilophodon* from the Wealden of the Isle of Wight. British Museum (Natural History), Bulletin, Geology 25: 1–152.

———. 2009. Notes on Neocomian (Lower Cretaceous) ornithopod dinosaurs from England—*Hypsilophodon*, *Valdosaurus*, "*Camptosaurus*," "*Iguanodon*"—and referred specimens from Romania and elsewhere. Revue de Paléobiologie 28: 211–273

Gámez, D., P. Paciotti P., F. Colombo, and R. Salas. 2003. La Formación Arcillas de Morella (Aptiense inferior), Cadena Ibérica oriental (España): caracterización sedimentológica. Geogaceta 34: 191–194.

Gasulla, J. M. 2005. Los dinosaurios de Morella (Castellón, España): historia de su investigación. Revista Española de Paleontología, número extraordinario 10: 29–38.

Gasulla, J. M., F. Ortega, X. Pereda Suberbiola, and J. L. Sanz. 2003. Elementos de la armadura dérmica del dinosaurios anquilosaurio *Polacanthus* (Cretácico Inferior. Morella, Castellón, España); p. 83 in M. V. Pardo Alonso and R. Gozalo (eds.), XIX Jornadas de la Sociedad Española de Paleontología, Libro de Resúmenes, Morella, Castellón, Spain, October 16–18, 2003.

Gasulla, J. M., F. Ortega, F. Escaso, and J. L. Sanz. 2006. Diversidad de terópodos del Cretácico Inferior (Fm. Arcillas de Morella, Aptiense) en los yacimientos del Mas de la Parreta (Morella, Castellón); pp. 124–125 in E. Fernández-Martínez (ed.), XXII Jornadas de Paleontología de la Sociedad Española de Paleontología, Libro de resúmenes, León, Spain, September 27–30, 2006.

Gasulla, J. M., J. L. Sanz, F. Ortega, and F. Escaso. 2007. *Iguanodon bernissartensis* (Ornithopoda) del yacimiento CMP-5 (cantera Mas de la Parreta, Morella, Castellón) de la Formación Morella (Aptiense inferior, Cretácico Inferior); pp. 65–66 in Colectivo Arqueológico-Paleontológico de Salas (ed.), IV Jornadas Internacionales sobre Paleontología de Dinosaurios y su Entorno, Resúmenes, Salas de los Infantes, Burgos, Spain, September 13–15, 2007.

Gasulla, J. M., J. L. Sanz, F. Escaso, and P. Yagüe. 2008. Elementos de la cintura pélvica de dinosaurios saurópodos Titanosauriformes del Cretácico Inferior (Aptiense inferior) de Morella (Castellón); p. 129 in J. I. Ruiz-Omeñaca, L. Piñuela, and J. C. García-Ramos (eds.), XXIV Jornadas de la Sociedad Española de Paleontología, Libro de resúmenes, Colunga, Spain, 15–18 October 2008. Museo del Jurásico de Asturias.

Gheerbrant, E., and J.-C. Rage. 2006. Paleobiogeography of Africa: how distinct from Gondwana and Laurasia? Palaeogeography, Palaeoclimatology, Palaeoecology 241: 224–246.

Gómez, B., C. Martín-Closas, H. Méon, F. Thévenard, and G. Barale. 2001. Plant taphonomy and palaeoecology in the lacustrine Uña delta (Late Barremian, Iberian ranges, Spain). Palaeogeography, Palaeoclimatology, Palaeoecology 170: 133–148.

Gómez-Fernández, D., J. I. Canudo, and V. Cano-Llop. 2007. Descripción de la cintura pelviana de un nuevo dinosaurio terópodo de la Formación Morella (Aptiense inferior) en Vallibona (Castellón, España); pp. 71–72 in Colectivo Arqueológico Paleontológico de Salas (ed.), IV Jornadas Internacionales sobre Paleontología de Dinosaurios y su Entorno, Resúmenes, Salas de los Infantes, Burgos, Spain, September 13–15, 2007.

Gradstein, F. M., J. G. Ogg, and A. G. Smith (eds.). 2004. A geologic time scale, 2004. Cambridge University Press, Cambridge, 589 pp.

Holtz, T. R., Jr., R. E. Chapman, and M. C. Lamanna. 2004a. Mesozoic biogeography of Dinosauria; pp. 627–642 in D. B. Weishampel, P. Dodson, and H. Osmólska (eds.), The Dinosauria. 2nd ed. University of California Press, Berkeley.

Holtz, T. R., Jr., R. E. Molnar, and P. J. Currie. 2004b. Basal Tetanurae; pp. 71–110 in D. B. Weishampel, P. Dodson, and H. Osmólska (eds.), The Dinosauria. 2nd ed. University of California Press, Berkeley.

Infante, P., J. I. Canudo, M. Aurell, J. I. Ruiz-Omeñaca, L. M. Sender, and S. Zamora. 2005a. Primeros datos sobre los dinosaurios de Zaragoza (Theropoda, Valanginiense–Hauteriviense, Cretácico Inferior); pp. 119–120 in E. Bernárdez Sanchez, E. Mayoral Alfaro, and A. Guerrero dos Santos (eds.), XXI Jornadas de la Sociedad Española de Paleontología, October 4–8, 2005, Seville, Spain. Consejería de Cultura de la Junta de Andalucía.

Infante, P., J. I. Canudo, and J. I. Ruiz-Omeñaca. 2005b. Primera evidencia de dinosaurios terópodos en la Formación Mirambel (Barremiense inferior, Cretácico inferior) en Castellote, Teruel. Geogaceta 38: 31–34.

Izquierdo Montero, L. A., F. Torcida Fernández-Baldor, P. Huerta Hurtado, D. Montero Huerta, and G. Pérez Martínez. 2004. Nuevos restos de dinosaurios (Ankylosauria, Thyreophora) en el Cretácico inferior de Salas de los Infantes (Burgos, España); pp. 43–46 in C. Liesa Carrera, A. Pocovi Juan, C. Sancho Marcén, F. Colombo Piñol, A. González Rodríguez, and A. R. Soria de Miguel (eds.), VI Congreso Geológico de España, Zaragoza, Spain, July 12–15, 2004. Geo-Temas 6 (5).

Kirkland, J. I., B. Britt, D. L. Burge, K. Carpenter, R. Cifelli, F. DeCourten, J. Eaton, S. Hasiotis, and T. Lawton. 1997. Lower to Middle Cretaceous dinosaur faunas of the central Colorado Plateau; a key to understanding 35 million years of tectonics, sedimentology, evolution and biogeography. Brigham Young University Geology Studies 42: 69–103.

Lacasa Ruiz, A. 1985. Nota sobre las plumas fósiles del yacimiento eocretácico de "La Pedrera-La Cabrua" en la Sierra del Montsec (prov. Lleida, España). Ilerda 46: 217–278.

———. 1989. Nuevo género de ave fósil del yacimiento neocomiense del Montsec (provincia de Lérida, España). Estudios Geológicos 45: 417–425.

Lapparent, A. F. de. 1960. Los dos dinosaurios de Galve. Teruel 24: 177–197.

———. 1966. Nouveaux gisements de reptiles mésozoïques en Espagne. Notas y Comunicaciones del Instituto Geológico y Minero de España 84: 103–110.

Lapparent, A. F. de, R. Curnelle, B. Defaut, A. Miroschedji, and B. de Pallard. 1969. Nouveaux gisements de Dinosaures en Espagne centrale. Estudios Geológicos 25: 311–315.

Le Loeuff, J. 1997. Biogeography; pp. 51–56 in P. J. Currie and K. Padian (eds.), Encyclopedia of dinosaurs. Academic Press, San Diego.

Maidment, S. C. R., D. B. Norman, P. M. Barrett, and P. Upchurch. 2008. Systematics and phylogeny of Stegosauria (Dinosauria: Ornithischia). Journal of Systematic Palaeontology 6: 367–407.

Maisch, M. W. 1997. The Lower Cretaceous dinosaur *Iguanodon* cf. *fittoni* Lydekker 1998 (Ornithischia) from Salas de los Infantes (Province Burgos, Spain). Neues Jarhbuch für Geologie und Paläontologie, Monatshefte 1997: 213–222.

Makovicky, P. J., Y. Kobayashi, and P. J. Currie. 2004. Ornithomimosauria; pp. 137–150 in D. B. Weishampel, P. Dodson, and H. Osmólska (eds.), The Dinosauria. 2nd ed. University of California Press, Berkeley.

Mannion, P. D. 2009. A rebbachisaurid sauropod from the Lower Cretaceous of the Isle of Wight, England. Cretaceous Research 30: 521–526.

Martill, D. M., and S. Hutt. 1996. Possible baryonychid dinosaur teeth from the Wessex Formation (Lower Cretaceous, Barremian) of the Isle of Wight, England. Proceedings of the Geologists' Association 107: 81–84.

Martill, D. M., and D. Naish. 2001. Dinosaurs of the Isle of Wight.

Palaeontological Association Field Guide to Fossils 10. Palaeontological Association, London.

Martín-Chivelet, J. 1996. Late Cretaceous subsidence history of the Betic Continental Margin (Jumilla–Yecla region, SE Spain). Tectonophysics 265: 191–211.

Martín-Chivelet, J., X. Berasategui, I. Rosales, L. Vilas, J. A. Vera, E. Caus, K.-U. Gräfe, R. Mas, C. Puig, M. Segura, S. Robles, M. Floquet, S. Quesada, P. Ruiz-Ortiz, M. A. Fregenal-Martínez, R. Salas, C. Arias, A. García, A. Martín-Algarra, N. Meléndez, B. Chacón, J. A. Molina, J. L. Sanz, J. M. Castro, M. García-Hernández, B. Carenas, J. García-Hidalgo, J. Gil, and F. Ortega. 2002. Cretaceous; pp. 255–292 in W. Gibbons and T. Moreno (eds.), Geology of Spain. Geological Society, London.

Martín-Closas, C. 2000. Els caròfits del Juràssic superior i el Cretaci inferior de la Península Ibèrica. Arxius de les Seccions de Ciències 125. Institut d'Estudis Catalans, Barcelona, 304 pp.

Martín-Closas, C., and A. Alonso-Millán. 1998. Estratigrafía y bioestratigrafía (Charophyta) del Cretácico inferior en el sector occidental de la Cuenca de Cameros (Cordillera Ibérica). Revista de la Sociedad Geológica de España 11: 253–269.

Martínez-Delclòs, X., G. Barale, S. Wenz, R. Domènech, J. Martinell, L. Mercadé, and M. J. Ruiz de Loizaga. 1991. Els jaciments de calcàires litogràfiques del Montsec (Catalunya, Espanya). Estat actual dels coneixements; pp. 155–162 in X. Martínez-Delclòs (ed.), Les calcàires litogràfiques del cretaci inferior del Montsec. Deu anys de campanyes paleontològiques, Institut d'Estudis Ilerdencs, Lleida.

Milner, A. C. 2003. Fish-eating theropods: a short review of the systematics, biology and palaeobiogeography of spinosaurs; pp. 129–138 in Colectivo Arqueológico-Paleontológico de Salas (ed.), II Jornadas Internacionales sobre Paleontología de los Dinosaurios y su Entorno, Actas, Salas de los Infantes, Burgos, Spain.

Moratalla, J. J. 2004. El Cretácico inferior del área de Vega de Pas (Cantabria): primeras estimaciones de su biodiversidad; pp. 129–130 in A. Calonge, R. Gozalo, M. D. López Carrillo, and M. V. Pardo Alonso (eds.), XX Jornadas de la Sociedad Española de Paleontología, Libro de Resúmenes, Alcalá de Henares, Spain, October 20–23, 2004.

Moreno-Azanza, M., J. I. Canudo, and J. M. Gasca. 2009. Fragmentos de cáscara de huevo de Megaloolithidae en el Cretácico Inferior de la provincia de Zaragoza (Formación Villanueva de Hueva, España); pp. 253–262 in P. Huerta Hurtado and F. Torcida Fernández-Baldor (eds.), IV Jornadas Internacionales sobre Paleontología de los Dinosaurios y su Entorno, Actas, Salas de los Infantes, Burgos, Spain.

Naish, D., and D. M. Martill. 2007. Dinosaurs of Great Britain and the role of the Geological Society of London in their discovery: basal Dinosauria and Saurischia. Journal of the Geological Society, London 164: 493–510.

———. 2008. Dinosaurs of Great Britain and the role of the Geological Society of London in their discovery: Ornithischia. Journal of the Geological Society, London 165: 613–623.

Norman, D. B. 1980. On the ornithischian dinosaur *Iguanodon bernissartensis* from the Lower Cretaceous of Bernissart (Belgium). Mémoires de l'Institut royal des Sciences naturelles de Belgique 178: 1–105.

———. 1986. On the anatomy of *Iguanodon atherfieldensis* (Ornithischia: Ornithopoda). Bulletin de l'Institut royal des Sciences naturelles de Belgique, Sciences de la Terre 56: 281–372.

———. 2004. Basal Iguanodontia; pp. 413–437 in D. B. Weishampel, P. Dodson, and H. Osmólska (eds.), The Dinosauria. 2nd ed. University of California Press, Berkeley.

———. 2010. A taxonomy of iguanodontians (Dinosauria: Ornithopoda) from the lower Wealden Group (Cretaceous: Valanginian) of southern England. Zootaxa 2489: 47–66.

Norman, D. B., H.-D. Sues, L. M. Witmer, and R. A. Coria. 2004. Basal Ornithopoda; pp. 393–412 in D. B. Weishampel, P. Dodson, and H. Osmólska (eds.), The Dinosauria. 2nd ed. University of California Press, Berkeley.

Ortega, F., F. Escaso, J. M. Gasulla, P. Dantas, and J. L. Sanz. 2006. Dinosaurios de la Península Ibérica. Estudios Geológicos 62: 219–240.

Ortega, F., F. Escaso, and J. L. Sanz. 2010. A bizarre, humped Carcharodontosauria (Theropoda) from the Lower Cretaceous of Spain. Nature 467: 203–206.

Padian, K. 2004. Basal Avialae; pp. 210–231 in D. B. Weishampel, P. Dodson, and H. Osmólska (eds.), The Dinosauria. 2nd ed. University of California Press, Berkeley.

Paul, G. 2006. Turning the old into the new: a separate genus for the gracile iguanodont from the Wealden of England; pp. 69–77 in K. Carpenter (ed.), Horns and beaks: ceratopsian and ornithopod dinosaurs. Indiana University Press, Bloomington.

———. 2008. A revised taxonomy of the iguanodont dinosaur genera and species. Cretaceous Research 29: 192–216.

Pereda Suberbiola, J. 1994. *Polacanthus* (Ornithischia, Ankylosauria), a transatlantic armoured dinosaur from the Early Cretaceous of Europe and North America. Palaeontographica Abteilung A 232: 133–159.

Pereda Suberbiola, X. 2006. El dinosaurio acorazado *Polacanthus* del Cretácico Inferior de Europa y el estatus de los Polacanthidae (Ankylosauria); pp. 85–104 in Colectivo Arqueológico-Paleontológico de Salas (ed.), III Jornadas Internacionales sobre Paleontología de los Dinosaurios y su Entorno, Actas, Salas de los Infantes, Burgos, Spain.

Pereda Suberbiola, X., and P. M. Galton. 2001. Thyreophoran ornithischian dinosaurs from the Iberian Peninsula; pp. 147–161 in Colectivo Arqueológico-Paleontológico de Salas (ed.), I Jornadas Internacionales sobre Paleontología de los Dinosaurios y su Entorno, Actas, Salas de los Infantes, Burgos, Spain.

Pereda Suberbiola, X., and J. I. Ruiz-Omeñaca. 2005. Los primeros descubrimientos de dinosaurios en España. Revista Española de Paleontología, número extraordinario 10: 15–28.

Pereda Suberbiola, X., M. Meijide, F. Torcida, J. Welle, C. Fuentes, L. A. Izquierdo, D. Montero, G. Pérez, and V. Urién. 1999. Espinas dérmicas del dinosaurio anquilosaurio *Polacanthus* en las facies Weald de Salas de los Infantes (Burgos, España). Estudios geológicos 55: 267–272.

Pereda Suberbiola, X., P. M. Galton, F. Torcida, P. Huerta, L. A. Izquierdo, D. Montero, G. Pérez, and V. Urién. 2003a. First stegosaurian dinosaur remains from the Early Cretaceous of Burgos (Spain), with a review of Cretaceous stegosaurs. Revista Española de Paleontología 18: 143–150.

Pereda Suberbiola, X., F. Torcida, L. A. Izquierdo, P. Huerta, D. Montero, and G. Pérez. 2003b. First rebbachisaurid dinosaur (Sauropoda, Diplodocoidea) from the early Cretaceous of Spain: palaeobiogeographical implications. Bulletin de la Société Géologique de France 174: 471–479.

Pereda Suberbiola, X., P. M. Galton, J. I. Ruiz-Omeñaca, and J. I. Canudo. 2005. Dermal spines of stegosaurian dinosaurs from the Lower Cretaceous (Hauterivian–Barremian) of Galve (Teruel, Aragón, Spain). Geogaceta 38: 35–38.

Pereda Suberbiola, X., C. Fuentes, M. Meijide, F. Meijide-Fuentes, and M. Meijide-Fuentes Jr. 2007. New remains of the ankylosaurian dinosaur *Polacanthus* from the Lower Cretaceous of Soria, Spain. Cretaceous Research 28: 583–596.

Pereda Suberbiola, X., J. I. Ruiz-Omeñaca, N. Bardet, L. Piñuela, and J. C. García-Ramos. 2010. Wilhelm (Guillermo) Schulz and the earliest discoveries of dinosaurs and marine reptiles in Spain in R. Moody, E. Buffetaut, D. M. Martill, and D. Naish (eds.), Dinosaurs and other extinct saurians—a historical perspective. Geological Society, London, Special Publications 343: 155–160.

Pereda Suberbiola, X., J.I. Ruiz-Omeñaca, F. Torcida Fernández-Baldor, M. W. Maisch, P. Huerta, R. Contreras, L. A. Izquierdo, D. Montero Huerta, V. Urién Montero, and J. Welle. 2011. A tall-spined ornithopod dinosaur from the Early Cretaceous of Salas de los Infantes (Burgos, Spain). Comptes Rendus Palevol. 10: 551–558.

Pérez-García, A., B. Sánchez Chillón, and F. Ortega. 2009a. Aportaciones de José Royo y Gómez al conocimiento sobre los dinosaurios de España. Paleolusitana 1: 339–364.

Pérez-García, A., F. Ortega, and J. M. Gasulla. 2009b. Revisión histórica del primer registro de tetanuros basales (Theropoda) del Cretácico Inferior de Morella (Castellón). Geogaceta 47: 21–24.

Pérez-Lorente, F., M. M. Romero-Molina, E. Requeta Loza, M. I. Blanco Somovilla, and S. Caro Calatayud. 2001. Dinosaurios. Introducción y análisis de algunos yacimientos de sus huellas en La Rioja. Ciencias de la Tierra 24. Instituto de Estudios Riojanos (IER), Logroño, Spain, 102 pp.

Pérez-Moreno, B. P., and J. L. Sanz. 1997. Las Hoyas; pp. 398–402 in P. J. Currie and K. Padian (eds.), Encyclopedia of dinosaurs, Academic Press, San Diego.

Pérez-Moreno, B. P., J. L. Sanz, A. D. Buscalioni, J. J. Moratalla, F. Ortega, and D. Rasskin-Gutman. 1994. A unique multitoothed ornithomimosaur dinosaur from the Lower Cretaceous of Spain. Nature 370: 363–367.

Pérez Pérez, B. 2004. *Pelecanimimus polyodon*: anatomía, sistemática y paleobiología de un Ornithomimosauria (Dinosauria: Theropoda) de Las Hoyas (Cretácico Inferior; Cuenca, España).

Ph.D. dissertation, Universidad Autónoma de Madrid, Cantoblanco, 149 pp.

Rauhut, O. W. M. 2002. Dinosaur teeth from the Barremian of Uña, province of Cuenca, Spain. Cretaceous Research 23: 255–263.

Rauhut, O. W. M., and J. Zinke. 1995. A description of the Barremian dinosaur fauna from Uña with a comparison to that of Las Hoyas; pp. 123–126 in II International Symposium on Lithographic Limestones, Extended Abstracts, Lleida-Cuenca, July 9–16, 1995, Universidad Autónoma de Madrid, Cantoblanco Spain.

Royo Gómez, J. 1926a. Los vertebrados del Cretácico español de facies weáldica. Boletín del Instituto Geológico y Minero de España 47: 171–176.

———. 1926b. Los descubrimientos de reptiles gigantescos en Levante. Boletín de la Sociedad castellonense de Cultura 7: 147–162.

———. 1926c. Nuevos vertebrados de la facies weáldica de Los Caños (Soria) y Benagéber (Valencia). Boletín de la Real Sociedad Española de Historia Natural 26: 317–318.

Royo-Torres, R. 2009a. El saurópodo de Peñarroya de Tastavins. Monografías turolenses 6. Instituto de Estudios Turolenses-Fundación Conjunto Paleontológico de Teruel Dinópolis, Teruel, Spain.

———. 2009b. Los dinosaurios saurópodos en la Península Ibérica; pp. 139–166 in P. Huerta Hurtado and F. Torcida Fernández-Baldor (eds.), IV Jornadas Internacionales sobre Paleontología de Dinosaurios y su Entorno, Actas, Salas de los Infantes, Burgos, Spain.

Royo-Torres, R., and J. I. Canudo. 2003. Restos directos de dinosaurios saurópodos en España (Jurásico superior–Cretácico superior); pp. 313–334 in F. Pérez-Lorente (coord.), Dinosaurios y otros Reptiles Mesozoicos de España. Ciencias de la Tierra 26. Instituto de Estudios Riojanos (IER), Logroño, Spain.

Royo-Torres, R., and A. Cobos. 2007. Teeth of *Oplosaurus armatus* (Sauropoda) from El Castellar (Teruel, Spain); pp. 15–19 in J. Le Loeuff (ed.), 5th Meeting of the European Association of Vertebrate Paleontologists, Abstracts Volume, Carcassonne-Espéraza, May 15–19, 2007.

Royo-Torres, R., M. Meijide, C. Fuentes, F. Meijide-Fuentes, and M. Meijide-Fuentes. 2009. The Iberian Titanosauriformes of Zorralbo (Soria, Spain) from Lower Cretaceous; pp. 143–144 in A. D. Buscalioni and M. Fregenal Martínez (coords.), 10th International Symposium

on Mesozoic Terrestrial Ecosystems and Biota, Teruel, 2009. Ediciones UAM, Madrid, Spain, September 17–19, 2009.

Ruiz-Omeñaca, J. I. 2000. Restos de dinosaurios (Saurischia, Ornithischia) del Barremiense superior (Cretácico Inferior) de Castellote (Teruel) en el Muséum National d'Histoire Naturelle de París. Mas de las Matas 19: 39–119.

———. 2001. Dinosaurios hipsilofodóntidos (Ornithischia: Ornithopoda) en la Península Ibérica; pp. 175–266 in Colectivo Arqueológico-Paleontológico de Salas (ed.), I Jornadas Internacionales sobre Paleontología de Dinosaurios y su Entorno, Actas, Salas de los Infantes, Burgos, Spain.

———. 2006. Restos directos de dinosaurios (Saurischia, Ornithischia) en el Barremiense (Cretácico Inferior) de la Cordillera Ibérica en Aragón (Teruel, España). Ph.D. dissertation, Universidad de Zaragoza, Zaragoza, 432 pp. http://www.aragosaurus.com/secciones/publicaciones/panel/artic/ruizomenaca2006.pdf.

———. 2011. *Delapparentia turolensis* nov. gen et sp., un nuevo dinosaurio iguanodontoideo (Ornithischia: Ornithopoda) en el Cretácico Inferior de Galve. Estudios geológicos 67: 83–110.

Ruiz-Omeñaca, J. I., and J. I. Canudo. 2001. Dos yacimientos excepcionales con vertebrados continentales del Barremiense (Cretácico Inferior) de Teruel: Vallipón y La Cantalera. Naturaleza Aragonesa 8: 8–17.

———. 2003a. Un nuevo dinosaurio terópodo ("*Prodeinodon*" sp.) en el Cretácico Inferior de La Cantalera (Teruel). Geogaceta 34: 111–114.

———. 2003b. Dinosaurios (Saurischia, Ornithischia) en el Barremiense (Cretácico Inferior) de la Península Ibérica; pp. 269–312 in F. Pérez-Lorente (coord.), Dinosaurios y otros reptiles mesozoicos de España. Ciencias de la Tierra 26. Instituto de Estudios Riojanos (IER), Logroño, Spain.

———. 2004. Dinosaurios ornitópodos del Cretácico inferior de la Península Ibérica; pp. 63–65 in C. Liesa Carrera, A. Pocovi Juan, C. Sancho Marcén, F. Colombo Piñol, A. González Rodríguez, and A. R. Soria de Miguel (eds.), VI Congreso Geológico de España, Zaragoza, Spain, July 12–15, 2004. Geo-Temas 6 (5).

———. 2005. "*Pleurocoelus*" *valdensis* Lydekker, 1889 (Saurischia, Sauropoda) en el Cretácico Inferior (Barremiense) de la Península Ibérica. Geogaceta 38: 43–46.

Ruiz-Omeñaca, J. I., J. I. Canudo, and G. Cuenca-Bescós. 1996. Dientes de dinosaurios (Ornitischia, Saurischia) del Barremiense superior (Cretácico inferior) de Vallipón (Castellote, Teruel). Mas de las Matas 15: 59–103.

Ruiz-Omeñaca, J. I., J. I. Canudo, and G. Cuenca-Bescós. 1997. Primera evidencia de un área de alimentación de dinosaurios herbívoros en el Cretácico Inferior de España (Teruel). Monografías de la Academia de Ciencias Exactas, Físicas, Químicas y Naturales de Zaragoza 10: 1–48.

Ruiz-Omeñaca, J. I., J. I. Canudo, and G. Cuenca-Bescós. 1998a. Sobre las especies de *Iguanodon* (Dinosauria, Ornithischia) encontradas en el Cretácico Inferior de España. Geogaceta 24: 275–277.

Ruiz-Omeñaca, J. I., J. I. Canudo, and G. Cuenca-Bescós. 1998b. Primera cita de dinosaurios barionícidos (Saurischia: Theropoda) en el Barremiense superior (Cretácico Inferior) de Vallipón (Castellote, Teruel). Mas de las Matas 17: 201–223.

Ruiz-Omeñaca, J. I., J. I. Canudo, M. Aurell, B. Bádenas, J. L. Barco, G. Cuenca-Bescós, and J. Ipas. 2004. Estado de las investigaciones sobre los vertebrados del Jurásico Superior y Cretácico Inferior de Galve (Teruel). Estudios Geológicos 60: 179–202

Ruiz-Omeñaca, J. I., J. I. Canudo, P. Cruzado-Caballero, P. Infante, and M. Moreno-Azanza. 2005. Baryonychine teeth (Theropoda: Spinosauridae) from the Lower Cretaceous of La Cantalera (Josa, NE Spain). Kaupia, Darmstädter Beiträge zur Naturgeschichte 14: 59–63.

Ruiz-Omeñaca J. I., X. Pereda Suberbiola, F. Torcida Fernández-Baldor, M. Maisch, L. A. Izquierdo, P. Huerta, R. Contreras, D. Montero Huerta, G. Pérez Martínez, V. Urién Montero, and J. Welle. 2008. Resto mandibular de ornitópodo iguanodontoideo (Dinosauria) del Cretácico Inferior de Salas de los Infantes (Burgos) en las colecciones del Institut für Geowissenschaften de Tubinga (Alemania). Geogaceta 45: 63–66.

Russell, D. A. 1993. The role of Central Asia in dinosaurian biogeography. Canadian Journal of Earth Sciences 30: 2002–2012.

Salas, R., F. Colombo, D. Gámez, J. M. Gasulla, C. Martín Closas, J. Moratalla, P. Paciotti, and X. Querol. 2003; pp. 1–31 in M. V. Pardo Alonso and R. Gozalo (eds.), XIX Jornadas de la Sociedad Española de Paleontología, Guía de la Excursión, Morella, Castellón, Spain.

Sánchez-Hernández, B., M. J. Benton, and D. Naish. 2007. Dinosaurs and other fossil vertebrates from the Late Jurassic and Early Cretaceous of the Galve area, NE Spain. Palaeogeography, Palaeoclimatology, Palaeoecology 249: 180–215.

Santafé, J. V., M. L. Casanovas, J. L. Sanz, and S. Calzada. 1982. Geología y paleontología (Dinosaurios) de las Capas Rojas de Morella (Castellón, España). Diputación Provincial de Castellón and Diputación de Barcelona, 169 pp.

Santos-Cubedo, A., A. Galobart, R., Gaete, and M. Suñer. 2003. Nuevos yacimientos de vertebrados del Cretácico Inferior de la Comarca de Els Ports (Castellón); p. 156 in M. V. Pardo Alonso and R. Gozalo (eds.), XIX Jornadas de la Sociedad Española de Paleontología, Libro de Resúmenes, Morella, Castellón, Spain, October 16–18, 2003.

Sanz, J. L. 1983. A nodosaurid ankylosaur from the Lower Cretaceous of Salas de los Infantes (province of Burgos, Spain). Geobios 16: 615–621.

———. 1984. Las faunas españolas de dinosaurios; pp. 497–506 in I Congreso Español de Geología 1, vol. 1, April 9–14, 1984, Segovia, Spain. Ilustre Colegio Oficial de Geólogos.

———. 1996. José Royo y Gómez y los dinosaurios españoles. Geogaceta 19: 167–168.

———. 2005. Aproximación historica al género *Iguanodon*. Revista Española de Paleontología, número extraordinario 10: 5–14.

Sanz, J. L., and J. F. Bonaparte. 1992. A new order of birds (Class Aves) from the Lower Cretaceous of Spain; pp. 39–49 in K. E. Campbell Jr. (ed.), Papers in avian paleontology honoring Pierce Brodkorb. Science Series 36. Natural History Museum of Los Angeles County.

Sanz, J. L., and A. D. Buscalioni. 1992. A new bird from the Early Cretaceous of Las Hoyas, Spain, and the early radiation of birds. Palaeontology, 35: 829–845.

Sanz, J. L., and F. Ortega. 2002. The birds from Las Hoyas. Science Progress 85: 113–130.

Sanz, J. L., J. V. Santafé, and M. L. Casanovas. 1983. Wealden ornithopod dinosaur *Hypsilophodon* from the Capas Rojas formation (Lower Aptian, Lower Cretaceous) of Morella, Castellón, Spain. Journal of Vertebrate Paleontology 3: 39–42.

Sanz, J. L., M. L. Casanovas, and J. V. Santafé. 1984. Iguanodóntidos (Reptilia, Ornithopoda) del yacimiento del Cretácico inferior de San Cristóbal (Galve, Teruel). Acta Geológica Hispánica 19: 171–176.

Sanz, J. L., A. Buscalioni, M. L. Casanovas, and J. V. Santafé. 1987. Dinosaurios del Cretácico inferior de Galve (Teruel, España). Estudios Geológicos, volumen extraordinario Galve-Tremp: 45–64.

Sanz, J. L., J. F. Bonaparte, and A. Lacasa. 1988. Unusual Early Cretaceous birds from Spain. Nature 331: 433–435.

Sanz, J. L., L. M. Chiappe, and A. D. Buscalioni. 1995. The osteology of *Concornis lacustris* (Aves: Enantionithes) from the Lower Cretaceous of Spain and a reexamination of its phylogenetic relationships. American Museum Novitates 3133: 1–23.

Sanz, J. L., L. M. Chiappe, B. P. Pérez-Moreno, A. D. Buscalioni, J. J. Moratalla, F. Ortega, and F. J. Poyato-Ariza. 1996. A new Lower Cretaceous bird from Spain: implications for the evolution of flight. Nature 382: 442–445.

Sanz, J. L., L. M. Chiappe, B. P. Pérez-Moreno, J. J. Moratalla, F. Hernández-Carrasquilla, A. D. Buscalioni, F. Ortega, F. J. Poyato-Ariza, D. Rasskin-Gutman, and X. Martínez-Delclòs. 1997. A nestling bird from the Lower Cretaceous of Spain: implications for avian skull and neck evolution. Science 276: 1543–1546.

Sanz, J. L., M. A. Fregenal-Martínez, N. Meléndez, and F. Ortega. 2001a. Las Hoyas; pp. 356–359 in D. E. G. Briggs and P. R. Crowther (eds.), Palaeobiology II. Blackwell Science, Oxford.

Sanz, J. L., L. M. Chiappe, Y. Fernández-Jalvo, F. Ortega, B. Sánchez-Chillón, F. J. Poyato-Ariza, and B. P. Pérez-Moreno. 2001b. An Early Cretaceous pellet. Nature 409: 998–999.

Sanz, J. L., B. P. Pérez-Moreno, L. M. Chiappe, and A. D. Buscalioni. 2002. The birds from the Lower Cretaceous of Las Hoyas (province of Cuenca, Spain); pp. 209–229 in L. M. Chiappe and L. M. Witmer (eds.), Mesozoic birds: above the heads of dinosaurs. University of California Press, Berkeley.

Schudack, U., and M. Schudack. 2009. Ostracod biostratigraphy in the Lower Cretaceous of the Iberian chain (eastern Spain). Journal of Iberian Geology 35: 141–168

Sereno, P. C. 2000. *Iberomesornis romerali* (Aves, Ornithothoraces) reevaluated as an Early Cretaceous enantiornithine. Neues Jahrbuch fur Geologie und Palaontologie Abhandlungen 215: 365–395.

Sereno, P. C., J. A. Wilson, L. M. Witmer, J. A. Whitlock, A. Maga, O. Ide, and T. A. Rowe. 2007. Structural extremes in a Cretaceous dinosaur. PLoS One 2(11): e1230.

Soria, A. R., C. Martín-Closas, A. Meléndez, M. N. Meléndez and M. Aurell. 1995.

Estratigrafía del Cretácico Inferior continental de la Cordillera Ibérica Central. Estudios Geológicos 51: 141–152.

Suñer, M., C. de Santisteban, and A. Galobart. 2005. Nuevos restos de Theropoda del Jurásico Superior–Cretácico Inferior de la Comarca de los Serranos (Valencia). Revista Española de Paleontología, número extraordinario 10: 93–99.

Suñer, M., B. Poza, B. Vila, and A. Santos-Cubedo. 2008. Síntesis del registro fósil de dinosaurios en el Este de la Península Ibérica; pp. 397–420 in J. Esteve and G. Meléndez (eds.), Palaeontologica Nova 2008, vol. 8. Publicaciones del Seminario de Paleontología de Zaragoza, Zaragoza, Spain.

Sweetman, S. C. 2004. The first record of velociraptorine dinosaurs (Saurischia, Theropoda) from the Wealden (Early Cretaceous, Barremian) of southern England. Cretaceous Research 25: 353–364.

Sweetman, S. C., and D. M. Martill. 2010. Pterosaurs of the Wessex Formation (Early Cretaceous, Barremian) of the Isle of Wight, southern England: a review with new data. Journal of Iberian Geology 36: 225–242.

Torcida Fernández-Baldor, F. 2006. Restos directos de dinosaurios en Burgos (Sistema Ibérico): un balance provisional; pp. 105–128 in Colectivo Arqueológico-Paleontológico de Salas (ed.), III Jornadas Internacionales sobre Paleontología de los Dinosaurios y su Entorno, Actas, Salas de los Infantes, Burgos, Spain.

Torcida Fernández-Baldor, F., L. A. Izquierdo Montero, P. Huerta Hurtado, D. Montero Huerta, and G. Pérez Martínez. 2003a. Dientes de dinosaurios (Theropoda, Sauropoda), en el Cretácico inferior de Burgos (España); pp. 335–346 in F. Pérez-Lorente (coord.), Dinosaurios y otros reptiles mesozoicos de España. Ciencias de la Tierra 26. Instituto de Estudios Riojanos (IER), Logroño, Spain.

Torcida Fernández-Baldor, F., J. I. Ruiz-Omeñaca, L. A. Izquierdo Montero, P. Huerta Hurtado, D. Montero Huerta, and G. Pérez Martínez. 2003b. Nuevos restos de dinosaurios hipsilofodóntidos (Ornithischia: Ornithopoda) en el Cretácico inferior de Burgos (España); pp. 389–398 in F. Pérez-Lorente (coord.), Dinosaurios y otros reptiles mesozoicos de España. Ciencias de la Tierra 26. Instituto de Estudios Riojanos (IER), Logroño, Spain.

Torcida Fernández-Baldor, F., J. I. Ruiz-Omeñaca, L. A. Izquierdo Montero, D. Montero Huerta, G. Pérez Martínez, P. Huerta Hurtado, and V. Urién Montero. 2005. Dientes de un enigmático dinosaurio ornitópodo en el Cretácico Inferior de Burgos (España). Revista Española de Paleontología, número extraordinario 10: 73–81.

Torcida Fernández-Baldor, F., L. A. Izquierdo Montero, R. Contreras Izquierdo, P. Huerta, D. Montero Huerta, G. Pérez Martínez, and V. Urién Montero. 2006. Un dinosaurio "iguanodóntido" del Cretácico Inferior de Burgos (España); pp. 349–363 in Colectivo Arqueológico-Paleontológico de Salas (ed.), III Jornadas Internacionales sobre Paleontología de los Dinosaurios y su Entorno, Actas, Salas de los Infantes, Burgos, Spain.

Torcida Fernández-Baldor, F., J. I. Ruiz-Omeñaca, X. Pereda Suberbiola, M. W. Maisch, L. A. Izquierdo, P. Huerta, R. Contreras, D. Montero Huerta, G. Pérez Martínez, V. Urién Montero, and J. Welle. 2008. La colección de restos de dinosaurios del Cretácico Inferior de Salas de los Infantes (Burgos, España) depositada en el Institut und Museum für Geologie und Paläontologie de Tübingen (Alemania); pp. 205–206 in J. I. Ruiz-Omeñaca, L. Piñuela, and J. C. García-Ramos (eds.), XXIV Jornadas de la Sociedad Española de Paleontología, Libro de resúmenes, Colunga, Spain, October 15–18, 2008. Museo del Jurásico de Asturias.

Torcida Fernández-Baldor, F., J. I. Canudo, P. Huerta, D. Montero Huerta, R. Contreras, G. Pérez Martínez, and V. Urién Montero. 2009. Primeros datos sobre las vértebras caudales del saurópodo de El Oterillo II (Formación Castrillo de la Reina, Barremiense superior–Aptiense, Cretácico Inferior, Salas de los Infantes, España); pp. 311–319 in P. Huerta Hurtado and F. Torcida Fernández-Baldor (eds.), IV Jornadas Internacionales sobre Paleontología de los Dinosaurios y su Entorno, Actas, Salas de los Infantes, Burgos, Spain.

Torcida Fernández-Baldor, F., J. I. Canudo, P. Huerta, D. Montero, X. Pereda Suberbiola, and L. Salgado. 2011. *Demandasaurus darwini*, a new rebbachisaurid sauropod from the Early Cretaceous of the Iberian Peninsula. Acta Palaeontologica Polonica 56: 535–552.

Torres, J. A., and L. I. Viera. 1994. *Hypsilophodon foxii* (Reptilia, Ornitischia) en el Cretácico inferior de Igea (La Rioja, España). Munibe, Ciencias Naturales 46: 3–41.

Upchurch, P., P. M. Barret, and P. Dodson. 2004. Sauropoda; pp. 259–322 in D. B. Weishampel, P. Dodson, and H. Osmólska (eds.), The Dinosauria. 2nd ed. University of California Press, Berkeley.

Vera, J. A. (ed.) 2004. Geología de España. Sociedad Geológica de España and Instituto Geológico y Minero de España, Madrid, 884 pp.

Vidal, L. M. 1902. Sobre la presencia del tramo Kimeridgiense del Montsech y hallazgo de un batracio en sus hiladas. Memorias de la Real Academia de Ciencias y Artes de Barcelona 4: 263–267.

Viera, L. I., and J. A. Torres. 1995. Presencia de *Baryonyx walkeri* (Saurischia, Theropoda) en el Weald de La Rioja (España). Nota previa. Munibe, Ciencias Naturales 47: 57–61.

Vilanova Piera, J. 1872. Compendio de Geología. Imprenta de Alejandro Gómez Fuentenebro, Madrid, 588 pp.

———. 1873. Restos de *Iguanodon* en Utrillas (Sesión del 5 de Febrero de 1873). Anales de la Sociedad Española de Historia Natural, Actas 2: 8.

Weishampel, D. B., C. M. Jianu, Z. Csiki, and D. B. Norman. 2003. Osteology and phylogeny of *Zalmoxes* (n. g.), an unusual euornithopod dinosaur from the latest Cretaceous of Romania. Journal of Systematic Palaeontology 1: 65–123.

Weishampel, D. B., P. M. Barrett, R. Coria, J. Le Loeuff, X. Xu, X. Zhao, A. Shani, E. M. P. Gomani, and C. R. Noto. 2004. Dinosaur distribution; pp. 517–606 in D. B. Weishampel, P. Dodson, and H. Osmólska (eds.), The Dinosauria. 2nd ed. University of California Press, Berkeley.

Wilson, J. A., and P. Upchurch. 2009. Redescription and reassessment of the phylogenetic affinities of *Euhelopus zdanskyi* (Dinosauria: Sauropoda) from the Early Cretaceous of China. Journal of Systematic Palaeontology 7: 199–239.

Yagüe P., P. Upchurch, J. L. Sanz, and J. M. Gasulla. 2001. New sauropod material from Early Cretaceous of Spain; in 49th Annual Symposium of Vertebrate Palaeontology and Comparative Anatomy, Abstracts. Yorkshire Museum, York, September 3–7, 2001.

Zinke, J., and O. W. M. Rauhut. 1994. Small theropods (Dinosauria, Saurischia) from the Upper Jurasic and Lower Cretaceous of Iberian Peninsula. Berliner geowissenschaftliche Abhandlungen Reihe E, Paläobiologie 13: 163–177.

23.1. Geographic and geologic framework of the studied area. A, Simplified geologic map of the Iberian Peninsula with the location of the Iberian Ranges. B, Structural map of the Maestrazgo Basin (Teruel, Spain) and Aguilón Subbasin (Zaragoza, Spain) and the paleogeographic—and geologic—relationships with the Uña area (Cuenca, Spain) during the Late Jurassic–Early Cretaceous rifting, with the main depositional area in gray color. This rifting structured the Maestrazgo Basin into seven subbasins. Thirteen different paleontological sites, which have yielded mammal fossils, are indicated by stars. C, Litho- and chronostratigraphy of the Early Cretaceous multituberculate fossil record from the Iberian ranges. A and B modified from Liesa et al. (2006, fig. 1).

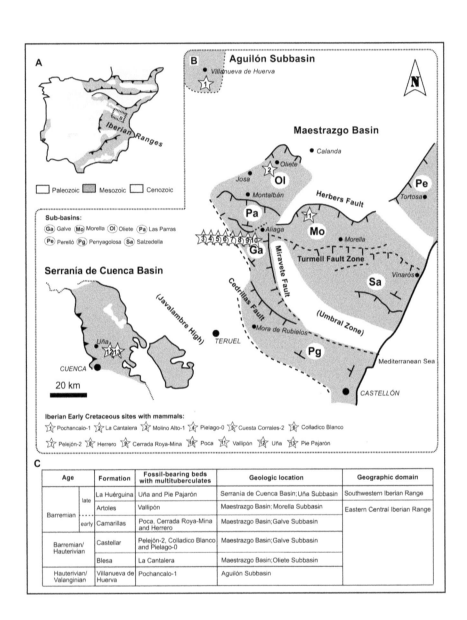

Age		Formation	Fossil-bearing beds with multituberculates	Geologic location	Geographic domain
Barremian	late	La Huérguina	Uña and Pie Pajarón	Serranía de Cuenca Basin; Uña Subbasin	Southwestern Iberian Range
		Artoles	Vallipón	Maestrazgo Basin; Morella Subbasin	Eastern Central Iberian Range
	early	Camarillas	Poca, Cerrada Roya-Mina and Herrero	Maestrazgo Basin; Galve Subbasin	
Barremian/ Hauterivian		Castellar	Pelejón-2, Colladico Blanco and Pielago-0	Maestrazgo Basin; Galve Subbasin	
		Blesa	La Cantalera	Maestrazgo Basin; Oliete Subbasin	
Hauterivian/ Valanginian		Villanueva de Huerva	Pochancalo-1	Aguilón Subbasin	

New Early Cretaceous Multituberculate Mammals from the Iberian Peninsula

23

Ainara Badiola*, José Ignacio Canudo,
and Gloria Cuenca-Bescós

The most abundant and continuous fossil record of the Early Cretaceous "Plagiaulacida" suborder of multituberculates comes from Western Europe and that of the Iberian Peninsula is becoming one of the richest and most continuous anywhere in the world. Evidence for that are the new finds from both the Valanginian–Hauterivian and Hauterivian–Barremian transitions, as well as from the Early and Late Barremian sites of the Central Iberian Range, in the Maestrazgo Basin and joined subbasin of Aguilón (province of Teruel and Zaragoza, northeastern Spain). The new discoveries include new species of Pinheirodontidae and Eobaataridae or Plagiaulacidae, new specimens of the Iberian eobaatarids *Eobaatar hispanicus* and *Eobaatar? pajaronensis* and of the Iberian pinheirodontid *Lavocatia*, and high numbers of upper and lower teeth of *Eobaatar*, including those poorly known within the genus, such as the upper incisors. The fossils we report have considerably increased the biodiversity of Iberian plagiaulacidan multituberculate and the biostratigraphic and paleobiogeographic resolution of these faunas during the Early Cretaceous.

Multituberculates were the most abundant and diversified mammals during the Mesozoic, mainly in the Northern Hemisphere (see Kielan-Jaworowska et al., 2004; and Hahn and Hahn, 2006, for the bibliography). Adapted for an omnivorous to herbivorous diet, multituberculates were the longest-lived order of the class Mammalia; they appeared in the Middle Jurassic (Bathonian), survived the mass extinction at the end of the Cretaceous, and persisted until the Early Cenozoic (Late Eocene). The order Multituberculata consists of two suborders (Kielan-Jaworowska and Hurum, 2001): the primitive and paraphyletic "Plagiaulacida," which range from the Middle Jurassic (Bathonian) until the Early Cretaceous (Aptian–Albian), and the apparently monophyletic suborder Cimolodonta, ranging mainly from the Late Cretaceous up to the Late Eocene (Kielan-Jaworowska et al., 2004; Butler and Hooker, 2005; Hahn and Hahn, 2006).

The Early Cretaceous is a key period in investigating the biodiversity, distribution, and evolution of multituberculates and of many continental vertebrate faunas in general. When Pangaea broke up at the end of the Jurassic, a significant separation between Laurasia and Gondwana took place, which led to an increasing differentiation between their respective faunas. Plagiaulacidans are mainly represented by isolated teeth and incomplete lower and upper jaws; consequently, these forms remain poorly understood from a biological point of view. Teeth and jaws nonetheless

Introduction

provide useful information about their biodiversity. The plagiaulacidan systematics, paleobiogeography, and evolutionary history, however, are still poorly known. Cladistic analysis has not provided reliable results regarding the interrelationships among the different representatives (e.g., Kielan-Jaworowska and Hurum, 2001; Hahn and Hahn, 2006). Families remain the real evolutionary and systematic units used by most of the authors, as in this work. The plagiaulacidan family Eobaataridae is considered to be closely related to the cimolodontans (Kielan-Jaworowska and Hurum, 2001; Kielan-Jaworowska et al., 2004). New eobaatarid finds from the Early Cretaceous of Asia support this hypothesis (Kusuhashi, 2008; Kusuhashi et al., 2009). This makes eobaatarids the key fossils for understanding the evolutionary history of the Plagiaulacida–Cimolodonta transition.

The most abundant and continuous fossil record for Plagiaulacida comes from Western Europe. It ranges from the Berriasian of Portugal, France, and England up to the Barremian of Spain and England (Cuenca-Bescós and Canudo, 2004; Kielan-Jaworowska et al., 2004; Kusuhashi, 2008; Pouech and Mazin, 2009; Sweetman, 2009). Fossils of Valanginian age are scarce, and no Hauterivian representatives are reported in Europe. The most abundant and diversified Barremian multituberculate faunas have been discovered in the Iberian Peninsula (see Cuenca-Bescós and Canudo, 2004, for a bibliography). Five species are described so far: the paulchoffatiids *Parendotherium herreroi* Crusafont-Pairó and Adrover, 1966, and *Galveodon nannothus* Hahn and Hahn, 1992; the eobaatarids *Eobaatar hispanicus* Hahn and Hahn, 1992, and *Eobaatar? pajaronensis* Hahn and Hahn, 2001; and the pinheirodontid *Lavocatia alfambrensis* Canudo and Cuenca-Bescós, 1996.

Because of the singular paleogeographical position of the Iberian Peninsula during the Early Cretaceous—when it was one of the islands of the European archipelago—situated between the southern part of Laurasia (Europe) and close to Gondwana (Africa), the study of Iberian Early Cretaceous multituberculate faunas is essential in investigating the distribution of these mammals between Laurasia and Gondwana (e.g., Europe and Africa) and among different areas of Laurasia (e.g., Europe and Asia). Fortunately for such investigations, the multituberculate fossil record of the Iberian Peninsula is increasing significantly. New finds are recorded in the Central Iberian Range in northeastern Iberia (Fig. 23.1A), which is becoming one of the most important areas in the study of Early Cretaceous vertebrate faunas. The singularity is based upon the numerous and well-correlated fossiliferous bonebeds, in long stratigraphic sequences (see Ruiz-Omeñaca et al., 2004, for a bibliography). Today, 13 sites with mammal fossils are recorded in this area (Fig. 23.1B). The new finds involve especially plagiaulacidan multituberculates both from the Valanginian–Hauterivian and Hauterivian–Barremian transitions as well as the Early and Late Barremian (Fig. 23.1C). This chapter summarizes these new discoveries and updates the systematics of the revised Early Cretaceous multituberculates from the Iberian Peninsula. Moreover, we analyze the new biostratigraphic and paleobiogeographic data at hand.

Institutional abbreviations. MPZ, Museo de Paleontología de la Universidad de Zaragoza, Zaragoza, Spain; FCTP, Fundación Conjunto Paleontológico de Teruel-Dinópolis, Teruel, Spain.

Localities. CAN, La Cantalera site.

Dentition. We use capital letters (I, C, P, M) for upper dentition and lowercase letters (i, c, p, m) for lower dentition. Cusps are labeled following Kielan-Jaworowska et al. (2004, fig. 8.28, modified from Hahn and Hahn, 1998, fig. 1a): cusps of the labial (b, B) and lingual (l, L) rows are numbered mesiodistally, indicating their corresponding number and letter; a capital letter (L, B) in the upper cheek tooth row and a lowercase letter (l, b) in the lower cheek tooth row (e.g., B3, third labial cusp of an upper tooth; l2, second lingual cusp of a lower tooth). The cusp formula is that proposed by Kielan-Jaworowska et al. (2004, 279), which is indicated by the number of cusps in consecutive rows given from labial to lingual, separated by a colon. We have here added the corresponding letter for each side for an easier reading (e.g., 3B:4L).

Historical Background

The discoveries of Mesozoic mammals in the Iberian Peninsula are relatively recent. Descriptions of the first fossils were published between the 1960s and 1980s as a result of the collaboration between Spanish and German teams led by Miquel Crusafont-Pairó from the Universidad de Barcelona and Walter Kühne from the Freie Universität Berlin, respectively (see Cuenca-Bescós and Canudo, 2004; and Hahn and Hahn, 2006, for the bibliography). The first multituberculate taxon from the Iberian Peninsula was described in the Late Jurassic (Kimmeridgian) site of the Guimarota coal mine (Leiria) in Portugal as *Paulchoffatia delgadoi* Kühne, 1961. Kühne started paleontological investigations in the Guimarota coal mine in the 1960s; the researches in this locality we pursued by Siegfried Henkel, Bernard Krebs, Thomas Martin, and others. More than 700 jaws, some partial and complete skulls, and a great quantity of isolated teeth from mammals have been recovered from the coal quarry. The Guimarota coal mine is the major Late Jurassic mammal site in Western Europe: abundant multituberculates were described, together with other mammals such as Docodonta (one genus), "Eupantotheria" including dryolestoids (three genera) and paurodontids (two genera), and Zatheria (one genus) (see Martin and Krebs, 2000; and Kielan-Jaworowska et al., 2004, for a bibliography). The multituberculate fauna of Guimarota is remarkably diverse and endemic. The fossil assemblage mainly comprises paulchoffatiids, such as *Paulchoffatia* Kühne, 1961, *Kuehneodon* Hahn, 1969, *Guimarotodon* Hahn, 1969, *Henkelodon* Hahn, 1977, *Pseudobolodon* Hahn, 1977, *Kielanodon* Hahn, 1987, *Meketibolodon* Hahn, 1993, *Meketichoffatia* Hahn, 1993, *Bathmochoffatia* Hahn and Hahn, 1998, *Plesiochoffatia* Hahn and Hahn, 1998, *Xenachoffatia* Hahn and Hahn, 1998, *Renatodon* Hahn, 2001, and many other forms with an open nomenclature, as well as the albionbaatarid *Proalbionbaatar plagiocyrtus* Hahn and Hahn, 1998. *Kuehneodon* is represented in two other Late Jurassic (Kimmeridgian–Tithonian) localities in Portugal: Pai Mogo and Porto Das Barcas. *Kuehneodon* is likely represented only in the Iberian Peninsula. The Early Cretaceous multituberculate fossil assemblages of Portugal come from the Berriasian site of Porto Pinheiro, Lourinhã, and consist exclusively of pinheirodontids: *Bernardodon* Hahn and Hahn, 1999, *Ecprepaulax* Hahn and Hahn, 1999, *Iberodon* Hahn and Hahn, 1999, *Pinheirodon* Hahn and Hahn, 1999, and many other specimens classified in an open nomenclature (see Hahn and Hahn, 1999, for a bibliography).

The first multituberculate taxon (*Parendotherium herreroi* Crusafont-Pairó and Adrover, 1966) from Spain was described from the Barremian of the Galve area (Teruel) in the Central Iberian Range of northeastern Spain (Fig. 23.1A). Crusafont-Pairó and Gibert (1976) described more Early Cretaceous multituberculate fossils from the Galve area, but subsequently, no Spanish paleontologists investigated Mesozoic mammals. In contrast, German paleontologists described further Barremian mammals in the Central and Southwestern Iberian Ranges (Fig. 23.1A,B). A "eupantotherian" dryolestoid mammal, *Crusafontia cuencana* Henkel and Krebs, 1969, and a "symmetrodontan" mammal, *Spalacotherium henkeli* Krebs, 1985, were reported from the Late Hauterivian–basal Barremian site of Colladico Blanco, in Galve, province of Teruel (Central Iberian Range, Galve Subbasin; Fig. 23.1B). *C. cuencana* was described for the first time in the Late Barremian site of Uña, province of Cuenca (Southwestern Iberian Range, Serranía de Cuenca Basin; Fig. 23.1B). In the 1990s and 2000s, Gerhard and Renate Hahn published a series of papers about Barremian plagiaulacidan multituberculates from the Iberian ranges (see Hahn and Hahn, 2006, for a bibliography). They described three taxa: the paulchoffatiid *Galveodon nannothus*, and the eobaatarids *Eobaatar hispanicus* and *Eobaatar? pajaronensis*. The first two were discovered in the upper Hauterivian and lower Barremian beds of the Galve Subbasin and in the above-mentioned Late Barremian bed at Uña. The latter site is dated as Early Barremian for Hahn and Hahn (2006 for a bibliography), but it is Late Barremian in age. Conversely, *E.? pajaronensis* is only known from the type locality, the Late Barremian site of Pie Pajarón, situated near Uña of the Serranía de Cuenca Basin (Hahn and Hahn, 2001). Most of the fossils described by the German paleontologists are kept in the Institut für Geowissenschaften Abteilung Paläontologie, Freie Universität Berlin, whereas others belong to a private fossil collection owned by José María Herrero in the village of Galve (Teruel, Spain).

At the beginning of the 1990s, the Aragosaurus research team from the Universidad de Zaragoza (Spain) took up the study of Iberian Early Cretaceous mammals, together with other Mesozoic vertebrate faunas. Exhaustive paleontological survey, screen washing, and excavation campaigns were carried out in many Mesozoic continental and transitional deposits of the Maestrazgo Basin and the adjacent Aguilón Subbasin, in the Central Iberian Range (Fig. 23.1B). As a result, more than 100 vertebrate fossil-bearing beds have been located in this area (e.g., Canudo et al., 1997, updated by Ruiz-Omeñaca et al., 2004; see also the bibliography at http://www.aragosaurus.com/). The pinheirodontid multituberculate *Lavocatia alfambrensis* and the "eupantotherian" peramurid *Pocamus pepelui* Canudo and Cuenca-Bescós, 1996, were added to the Iberian Early Cretaceous mammalian faunal list. More multituberculate specimens from the Late Barremian site of Vallipón, province of Teruel (Artoles Formation, Morella Subbasin; Fig. 23.1B,C), were reported in a preliminary work (Cuenca-Bescós et al., 1996), and a late eutriconodontan gobiconodontid mammal was described later from the same locality (Cuenca-Bescós and Canudo, 2003). Since then, new multituberculate finds have been reported in the abstracts of several congresses and meetings, including an overview of these discoveries presented at the Darwin–Bernissart meeting held in Brussels in February 2009. Some of the new finds have been already published (Badiola et al., 2008, 2011). In this work, a summary of the new discoveries is presented. The recent Early Cretaceous mammal finds also include a new taxon of "Eupantotheria." A new species of the dryolestoid *Crusafontia* from the El Castellar Formation in Galve has been described, and a revision of the specimens assigned to this genus has been presented (Cuenca-Bescós et al., 2011). The new finds and the revised collection from Galve are kept at the Museo Paleontológico de la Universidad de Zaragoza (MPZ), Zaragoza, Spain.

Geological and Paleontological Framework

Because of the intracontinental extensional tectonic deformation of the Iberian rifting process, related to the spreading of the Tethys and Central Atlantic during the Late Jurassic to Early Cretaceous, the wide Mesozoic Iberian Basin was structured into smaller sedimentary basins. These basins are situated in the paleogeographical domains of Cameros, Maestrazgo, and the Central and Southwestern Iberian ranges (e.g., Salas et al., 2001; Liesa et al., 2006; Buscalioni et al., 2008). Multituberculate fossils have been found in the continental and transitional deposits of the Maestrazgo Basin and the adjacent Aguilón Subbasin (Central Iberian Range) and the Serranía de Cuenca Basin (Southwestern Iberian Range) (Fig. 23.1B).

Central Iberian Range

AGUILÓN SUBBASIN

The only Late Valanginian–Early Hauterivian vertebrate fossil sites in the Iberian ranges are situated in the Villanueva de Huerva Formation, in the north of the Aguilón Subbasin (Fig. 23.1B,C). The latter is one of the subbasins adjacent to the Maestrazgo Basin (Central Iberian Range). Five main facies grouped in three different lithological units (A, B, and C) have been identified in the Villanueva de Huerva Formation (see Soria de Miguel, 1997, 103–110, for a bibliography). The multituberculate specimen that we report here comes from the Pochancalo-1 (or Poch-1) site. The fossiliferous level consists of gray lutites deposited in a shallow lacustrine system during anoxic conditions (at the top of the B section), which favored the processes of fossilization.

The Pochancalo-1 site consists of a microvertebrate bonebed, including hybodont and semionotiform fishes, postcranial bones from amphibians, turtle plates, and isolated teeth from crocodiles (Atoposauridae, Bernissartidae, and Goniopholidae), pterosaurs, mammals, and sauropod and theropod dinosaurs (Infante et al., 2005). Eggshell and plant fragments are also abundant (Moreno-Azanza et al., 2007). More vertebrate fossil sites exist in the same formation, with bone, tooth, and ichnite fossils of theropod dinosaurs.

MAESTRAZGO BASIN

The Late Jurassic to Early Cretaceous Iberian rifting process also shaped the Maestrazgo Basin into seven subbasins (Fig. 23.1B): Galve, Las Parras,

Morella, Oliete, Perelló, Penyagolosa, and Salzedella (e.g., Soria de Miguel, 1997; Salas et al., 2001; Liesa et al., 2006). Mammal fossils come from the Oliete, Galve, and Morella subbasins.

Oliete Subbasin

Multituberculate fossils come from the La Cantalera site (Teruel). The latter is an outcrop of Early Cretaceous clays (Wealden facies), specifically clays from the Blesa Formation, which is located in the Oliete Subbasin. The Blesa Formation includes a lower part with alluvial to lacustrine sedimentation, followed by an upper part with two episodes of coastal lagoonal influence. The lower and the upper parts of the formation are separated by a ferruginous and encrusted surface developed over lacustrine carbonates. In lithological terms, the La Cantalera site comprises gray clays deposited in the lowest part of the lower Blesa Formation, and the fossiliferous level outcrops over a broad area because its dip coincides with the slope. Laterally, these gray clays present a great abundance of rounded edges of Jurassic limestone with invertebrate marine fossils (ammonites, brachiopods, crinoids, etc.), which is proof of the contribution of rock fragments from the nearby Jurassic relieves when the site was being formed (Aurell et al., 2004). Among the microfossils, ostracods and charophytes are abundant, forming the greater part of the residue of the screen washing over 50 μm. Also frequent are plant fragments, microvertebrate remains (mainly teeth), and continental gastropods. The charophyte association, which includes oogonia attributed to *Atopochara trivolvis triquetra* (Grambast, 1968), dates the La Cantalera site as Late Hauterivian–Early Barremian in age (Martín-Closas, 1989; Schudack, 1989; Riveline et al., 1996) and by approximation the rest of the lower part of the Blesa Formation (Soria et al., 1995; Canudo et al., 2002; Aurell et al., 2004). The La Cantalera site has been interpreted as deposited in a marshy environment with periodic droughts resulting in a nonpermanent body of water and a marshy vegetated area (Aurell et al., 2004). The vertebrate fossils consist mainly of isolated teeth; disarticulated, complete, or fragmented bones; eggshell fragments; and coprolites. The vertebrate fossil assemblages comprise fishes (scarce fossils), frogs, lizards, crocodiles, turtles, pterosaurs, dinosaurs, and mammals (Ruiz-Omeñaca et al., 1997, 2005; Canudo et al., 2002, 2010; Gasca et al., 2009; Moreno-Azanza et al., 2009).

Galve Subbasin

The Galve Subbasin is a north–northwest to south–southeast elongated basin 40 km long and 20 km wide (Fig. 23.1B). There is a wide sedimentary succession of Late Jurassic and Early Cretaceous marine, transitional, and continental deposits, which range from the Kimmeridgian–Tithonian transition up to the Early Aptian (e.g., Soria et al., 1995; Canudo et al., 1997; Liesa et al., 2006). The vertebrate fossil sites are included in four formations: Higueruelas (Kimmeridgian–Tithonian), Villar de Arzobispo (Late Tithonian–middle? Berriasian), El Castellar (Late Hauterivian–Early Barremian), and Camarillas (Early Barremian). The vertebrate fossils consist mostly of isolated bones and teeth, although paleoicnologic and paleoologic remains are also present. Rich fossil assemblages are identified in the

Galve Subbasin, which consist of sharks, bony fishes, amphibians (frogs), squamates, crocodiles, turtles, dinosaurs, and mammals (Sanz et al., 1987; Estes and Sanchiz, 1982; Díaz Molina and Yébenes, 1987; Ruiz-Omeñaca et al., 2004; Canudo et al., 2006; Badiola et al., 2009, 2011; Gasca et al., 2007, 2009). The multituberculate fossils come from the El Castellar Formation (the Pielago-0, Corrales de Pelejón-2, and Colladico Blanco sites) and the Camarillas Formation (the Herrero, Cerrada Roya-Mina, and Poca sites) (Fig. 23.1C).

The El Castellar Formation discordantly overlies the Villar de Arzobispo Formation with a stratigraphic gap from the middle? Berriasian up to the Late Hauterivian (Díaz Molina and Yébenes, 1987; Soria de Miguel, 1997; Liesa et al., 2006; Meléndez et al., 2009). The El Castellar Formation represents the first synrift unit included in the Wealden facies of the Galve Subbasin and has a depth of roughly 100 m, consisting of a lower part with lutites, sandstones, conglomerates, and limestones (80 m thick), which represent alluvial, palustrine, and lacustrine subenvironments, and an upper part comprising alternating marls and limestones (20 m thick) typical of a lacustrine system in phases of expansion and retraction (see Meléndez et al., 2009, for a bibliography). In this carbonated upper part of the formation are located the above-mentioned multituberculate-bearing beds. In biostratigraphic terms, the same charophyte association as in the fossiliferous level of the La Cantalera site has been found in the upper part of the El Castellar Formation, and consequently it has been dated as Late Hauterivian–Early Barremian.

The bottom part of the Camarillas Formation is immediately above the uppermost part of the El Castellar Formation in the Galve Subbasin, whereas in other areas, the Camarillas Formation unconformably overlies the El Castellar Formation (see Soria de Miguel, 1997, for a bibliography). The Camarillas Formation comprises lutitic, silty, and sandy facies deposited in a low-sinuosity, multichannel fluvial system. The paleochannels, with mainly sandy facies, mostly exhibit vertical aggradation. The lutites, massive silts, and sandstones correspond to the floodplains and bank deposits (crevasses) of these fluvial systems. The latter pass upward into a nearshore deltaic plain characterized by lutitic facies with continental and marine fossils (charophytes, pollen, gastropods, vertebrates, ostreids, orbitolines, shark teeth, etc.) and sandy paleochannels that show sedimentary structures related to the tidal action in the fluvial paleocurrents (Díaz Molina and Yébenes, 1987). The fossil-bearing beds with multituberculate fossils are located at the bottom (the Herrero site), in the middle (Cerrada Roya-Mina), and in the uppermost part (the Poca site) of this formation. Revision of the charophyte and pollen spore fossil assemblages confirms the Lower Barremian stratigraphic position of the Camarillas Formation (Schudack, 1989; Martín-Closas, 1989).

Morella Subbasin

One of the richest and most diverse vertebrate fossil sites of the Maestrazgo Basin is the Vallipón site in Castellote (Teruel), which contains rich marine and continental vertebrate fossil assemblages, including mammals. The Vallipón site is included in the Vallipón section, which comprises three formations: from the bottom to top, these are designated the Mirambel, Artoles, and Utrillas formations, and they are dated as Early Barremian, Late Barremian–Early Aptian, and Albian, respectively (e.g., Querol et al., 1992; Salas et al., 1995; Canudo et al., 1996). The vertebrate fossils of the Vallipón site come from the base of the Artoles Formation, which contains red and yellowish sandstones and conglomerates. The fossiliferous level is easy to distinguish in the field because it consists mainly of vertebrate fossil remains firmly cemented by iron carbonate. Overlying the Vallipón site are gray marls and limestones, with abundant marine invertebrates, especially large foraminifers, and mollusks. The macroforaminifer species *Paleorbitolina lenticularis* gives a Late Barremian–Early Aptian age for the upper part of the Artoles Formation, so the Vallipón site might be at least Late Barremian (Canudo et al., 1996; Cuenca-Bescós and Canudo, 2003). The vertebrate fossil assemblages of the Vallipón site include mostly coprolites and isolated bones and teeth. The fossils do not bear the marks of obvious long-distance transport, although many mammal teeth show a high degree of etching, probably produced by the digestive corrosion of predators. A coastal environment has been proposed, with a hard substrate where the materials were carried by predators and accumulated by shallow streams and probably also tidal action (Ruiz-Omeñaca and Canudo, 2001). Forty different marine and continental vertebrate taxa have been identified so far, including sharks, fishes, amphibians (frogs), lizards, turtles, crocodiles, plesiosaurs, pterosaurs, dinosaurs, and mammals (see Ruiz-Omeñaca and Canudo, 2001, and Kriwet et al., 2008, for a bibliography). Multituberculates dominate the mammal fossil assemblage (Cuenca-Bescós et al., 1996), although representatives of Gobiconodontidae are also described (Cuenca-Bescós and Canudo, 2003).

Southwestern Iberian Range

SERRANÍA DE CUENCA BASIN

The Southwestern Iberian Range is oriented in accordance with the general northwest-to-southeast trend of the Iberian ranges. It opened toward the southeast (Valencia) to the Tethys Sea, where the deposition of marine shallow-water limestones dominated (Gómez et al., 2001). In this paleogeographic domain is located another main Mesozoic basin, designated as the Serranía de Cuenca Basin (Fig. 23.1B). The latter was in turn divided by the nothwest–southeast-extending Hesperic fault into two subsident troughs (Uña-Las Hoyas and la Huérguina). The Uña-Las Hoyas Trough is composed of at least five small subbasins: Uña, Buenache de la Sierra, Los Aliagares, Las Hoyas, and La Cierva (see Buscalioni et al., 2008, fig. 1C). During the Late Barremian, the Serranía de Cuenca Basin was filled by continental sediments without any direct marine influence (Poyato-Ariza et al., 1998). Upper Barremian sediments overlie unconformably Middle Jurassic (Bathonian) marine limestone and are distributed in two lithostratigraphic units related by a lateral change of facies: the El Collado Sandstone Formation and the La Huérguina Limestone Formation (see Buscalioni et al., 2008, for a bibliography). Abundant Early Cretaceous fossils, including charophytes, ostracods, gastropods, bivalves, plant remains, and bone fragments, have been recovered from the alluvial, palustrine, and lacustrine deposits of the La Huérguina Formation (e.g., Henkel and Krebs, 1969; Hahn and Hahn, 1992; Gómez et al., 2001; Buscalioni et al., 2008). Rich vertebrate faunas are described in the *Konservat-Lagerstätte* site of Las Hoyas (e.g., Sanz et al., 2001; see Chapter 22 in this book), the lacustrine coal-bearing bed at Uña (see Gómez et al., 2001, and Hahn and Hahn, 1992, for a bibliography), and the siliciclastic bonebeds of Buenache de la Sierra (Buscalioni et al., 2008). Mammal fossils have only been found at the Uña site and at another nearby site of the same age, the Pie Pajarón site (Fig. 23.1B,C). The age of both sites is Late Barremian according to paleontological data of Uña and the lateral equivalents of the same stratigraphic unit at the well-known paleontological site of Las Hoyas (see Gómez et al., 2001; Buscalioni et al., 2008; and Schudack and Schudack, 2009, for a bibliography). The Late Barremian mammal fossil assemblages from Uña comprise multituberculates and eupantotheres (Henkel and Krebs, 1969; Krebs, 1985, 1993; Hahn and Hahn, 1992), whereas in the Pie Pajarón site, only multituberculates are described (Hahn and Hahn, 2001).

Here we report the findings of plagiaulacidan multituberculates in several Early Cretaceous localities in the Central Iberian Range (the Maestrazgo Basin and Aguilón Subbasin), northeastern Spain, such as Pochancalo-1, La Cantalera, Galve collection (from the Colladico blanco and/or Herrero sites) and Vallipón (Fig. 23.1C).

New Early Cretaceous Multituberculate Fossils from Spain

Aguilón Subbasin (Villanueva de Huerva Formation)

For the first time we describe a multituberculate tooth from the Late Valanginian–Early Hauterivian of the Iberian Peninsula. The material consists of an upper anterior tooth found in the Pochancalo-1 site (Zaragoza, Spain). We provisionally identify it as an upper canine, and it is tentatively assigned to the Albionbaataridae (Fig. 23.2, 1, 2).

In the plagiaulacidan multituberculates the upper canine, if present, is premolariform with several cusps and differs from the anterior upper premolars by the presence of one instead of two roots. In the Pochancalo-1 specimen (MPZ 2010/858), only the crown is preserved, so we are unable to confirm it as an upper canine (with only one robust root) or an upper anterior premolar (with two roots). In the only albionbaatarid representative whose anterior teeth are known (only premolars), *Albionbaatar denisae* Kielan-Jaworowska and Ensom, 1994, a tiny species from the Berriasian of Dorset (South England), the prominent and radiating enamel ridges of the cusps continue onto the lingual slope as far as the crown margin and exhibit a subparallel disposition (Kielan-Jaworowska and Ensom, 1994, pls. 1, 2). Moreover, the enamel ridges are present on the entire occlusal surface. The tooth from Pochancalo-1 also belongs to a tiny form and shows the same disposition of radiating ridges (Fig. 23.2, 1, 2) but differs from the anterior upper premolars of *A. denisae* by having fewer cusps, which are not arranged in three rows as in *A. denisae*. Therefore, it may be hypothesized that MPZ 2010/858 is an albionbaatarid upper canine; however, upper canines are currently unknown in this family. The upper canines known from the

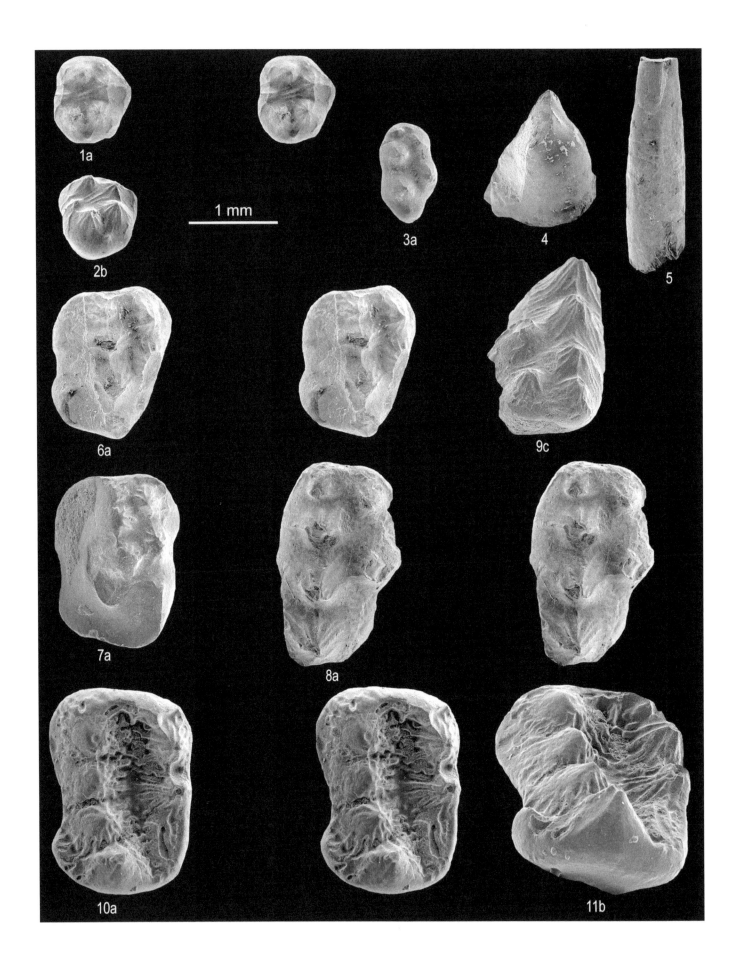

rest of the plagiaulacidan multituberculates are clearly different from the tooth from Pochancalo-1. In the current state of knowledge, MPZ 2010/858 is provisionally identified as an upper ?canine of ?Albionbaataridae indet.

Maestrazgo Basin

OLIETE SUBBASIN (BLESA FORMATION)

Recently, new multituberculate mammals from the Hauterivian–Barremian transition have been described in the La Cantalera site (Teruel). The fossils have been assigned to at least three different plagiaulacidan representatives on the basis of nine isolated teeth (Badiola et al., 2008): the pinheirodontid *Cantalera abadi* Badiola, Canudo, and Cuenca-Bescós, 2008 (P4/5 and two M1); the eobaatarid *Eobaatar* sp. (two P5); and two taxa classified in an open nomenclature, one as "Plagiaulacida" indet. (I2 and ?ix) and the other as Plagiaulacidae or Eobaataridae gen. et sp. indet. (P1/3 and a p4 fragment). The latter has recently been assigned to a new genus and species, *Iberica hahni* Badiola, Canudo, and Cuenca-Bescós, 2011 (see next section for this taxon), and is also present in the Galve area. We include a summary of the systematic section of Badiola et al. (2008) below.

The pinheirodontid *Cantalera abadi* is characterized by having abundant, long, and strongly marked radiating enamel ridges from the tip to the base of the cusps on P4/5 and M1 (Fig. 23.2, 8–11), which intermesh at their distal ends with the neighboring ridges; these form a labyrinth-like structure on the floor of the central valley on M1. The P4/5 is rectangular in outline in the occlusal view and considerably lengthened mesiodistally, with two rows of cusps, arranged parallel to the mesiodistal axis. The best-preserved M1 exhibits a 2B:4L cusp formula, L4 enlarged and B2 reduced, the central valley open distolabially, and small, prominent cuspules on the distolingual and mesiolabial corners of the crown; the first is located above a cingulum-like structure, which is strongly laterally protruding, and the second is placed above the thick cingulum that encircles the mesial margin. The M1 is morphologically close to the pinheirodontid *Iberodon quadrituberculatus* Hahn and Hahn, 1999, from the Berriasian of Porto Pinheiro (Portugal), but it differs from this species in having the L4 enlarged and the B2 reduced rather than both cusps reduced, and in lacking the additional row of small cuspules on the lingual wall of the crown. In addition to the crown outline and the arrangement of the cusp rows, the P4/5 of La Cantalera has the same cusp formula as ?P5 of *Sunnyodon notleyi* Kielan-Jaworowska and Ensom, 1992, from the Berriasian of the Purbeck Group of Dorset (South England), but it differs from this species in lacking the two labial cusps arranged symmetrically in the middle of the length of the crown with two small cuspules, one on the mesial margin and the other on the distal one (Kielan-Jaworowska and Ensom, 1992, pl. 5, figs. 3, 10). The P5 of *Lavocatia* from the Early Barremian site of Poca (Fig. 23.1C) resembles *Cantalera* in possessing cusps covered by prominent radiating enamel ridges (Canudo and Cuenca-Bescós, 1996, fig. 3), but *Cantalera* is considerably larger in size than *Lavocatia*. *Cantalera abadi* is also a larger pinheirodontid than *S. notleyi* and *I. quadrituberculatus*.

23.2. Scanning electron photomicrographs of Early Cretaceous plagiaulacidan multituberculate teeth from the Pochancalo-1 (1–2) and La Cantalera (3–11) sites, province of Zaragoza and Teruel, respectively, Spain. 1–2, Upper anterior tooth (upper ?canine; MPZ2010/858) tentatively assigned to ?Albionbaataridae indet. 3, Left P1/3 [FCPT (CAN 1/936), holotype] of *Iberica hahni* Badiola, Canudo, Cuenca-Bescós, 2011, classified as Plagiaulacidae or Eobaataridae indet. in Badiola et al. (2008, pl. 3). 4–5, Incisors classified in an open nomenclature as "Plagiaulacida" indet. 4, I2 [FCPT (CAN 1/939)]; and 5, a ?ix [FCPT (CAN 1/938)]. 6–7) two upper posterior premolars of *Eobaatar* sp. 6, left P5 [FCPT (CAN 1/935)] and 7, left P5 with severe wear on the lingual and distal margins of the crown [FCPT (CAN 1/1611)]. 8–11, Upper dentition of *Cantalera abadi* Badiola, Canudo, and Cuenca-Bescós, 2008. 8–9, Left P4/5 [FCPT (CAN 1/934), paratype]; and 10–11, left M1 [FCPT (CAN 1/1609), holotype]. (a) Occlusal, (b) distal, and (c) mesial views. 1, 6, 8, and 10, Stereopair photographs. Specimens from La Cantalera taken from Badiola et al. (2008, 2011).

The morphologic features of both P5s (Fig. 23.2, 6, 7) are typical for the eobaatarid *Eobaatar*: the cusp formula is 2B:4L; the lingual row of cusps is obliquely oriented to mesiodistal axis; the labial row of cusps is shortened, and the distal portion of the crown is considerably narrowed, comprising only the lingual row of cusps; and lateral to the cusps of the labial row, there is a small wing. The size of the better-preserved P5 from La Cantalera (Fig. 23.2, 6) is similar to those of *Eobaatar magnus* Kielan-Jaworowska, Dashzeveg, and Trofimov, 1987 from the Aptian or Albian beds at Khovboor (Gobi Desert, Mongolia) but differs from them by the absence of lingual cusps distinctly increasing in height distally and by having, lateral to the cusps of the lingual row, a smooth, wide wing that longitudinally covers the whole lingual margin and is strongly laterally protruding. By contrast, the lingual wing in *E. magnus* is not present or it is incipiently developed distolingually, although it is clearly less laterally protruding than in the teeth from La Cantalera (Kielan-Jaworowska et al., 1987, pl. 4, figs. 2 and 3, and pl. 5). The P5s from La Cantalera seem smaller than the unworn P5 (holotype) of *Eobaatar hispanicus* from Galve (the Herrero site; Fig. 23.1C) and differs from it in the absence of lingual cusps distinctly increasing in size distally and by having a narrower occlusal surface, which exhibits a clearly developed triangular outline, absent in *E. hispanicus*, and strongly laterally protruding lingual and labial wings, not observed in *E. hispanicus* (Hahn and Hahn, 1992, fig. 5). Nevertheless, the differences between the P5s under study and those of *E. magnus* and *E. hispanicus* could be variations within *Eobaatar*, and more material and more complete cheek teeth of this genus are required to investigate whether these differences are sufficiently great to regard the P5s from La Cantalera as belonging to a different species. The P5s of La Cantalera, therefore, are provisionally classified as *Eobaatar* sp.

The I2 and ?ix (Fig. 23.2, 4, 5) seem to correspond in size to the same multituberculate taxa described as *Eobaatar* sp. or Plagiaulacidae or Eobaataridae gen. et sp. indet. (Badiola et al., 2008), recently assigned to *Iberica hahni* (Badiola et al., 2011). Because of their fragmentary stage, they are classified as "Plagiaulacida," indet.

GALVE SUBBASIN
(EL CASTELLAR AND CAMARILLAS FORMATIONS)

Here we present a systematic reassessment of isolated teeth previously studied by Crusafont-Pairó and Adrover (1966) and Crusafont-Pairó and Gibert (1976), and a study of other unpublished teeth found in the revised collection. The fossils were collected during the 1960s and 1970s by German and Spanish teams under Kühne and Crusafont-Pairó from the Late Hauterivian–Early Barremian site of Colladico Blanco and the Early Barremian site of Herrero, in the Galve Subbasin (Fig. 23.1B,C). There is no mention in Crusafont-Pairó and Gibert (1976) whether each specimen comes from the Colladico Blanco or Herrero site. Several synonymies have been proposed for the teeth that appear figured in the above-mentioned papers (see Hahn and Hahn, 2006, for a bibliography), but their descriptions have not been updated, nor has the figuration of all the specimens been published. Badiola et al. (2011) presented a more systematic study of this material with the emended descriptions and comparisons as well as all the specimen illustrations. Here we summarize the main systematic results of this work.

Five teeth, referred to as *Forma 1* (p4), *Forma 5* (P4/5), and *Forma 6* (M2 figured in pl. 3, fig. 1), as well as *Forma 2* (P5) and *Forma 3* (I2 figured in pl. 4, fig. 2) in Crusafont-Pairó and Gibert (1976), which were assigned to *Eobaatar hispanicus* and *Parendotherium herreroi*, respectively, by Hahn and Hahn (2006), are missing in the revised collection. The rest of the specimens (17 teeth) have been assigned to at least four different plagiaulacidans: *Eobaatar hispanicus*; *Iberica hahni*, also present in the La Cantalera site as mentioned above; a paulchoffatiid classified provisionally in an open nomenclature as Paulchoffatiidae indet.; Plagiaulacidae or Eobaataridae indet.; and "Plagiaulacida" indet. The latter could belong to any of the above-mentioned taxa.

Iberica hahni. Five P1/3s (Fig. 23.3, 1–4), previously described as *conjunto II* of *Forma 4* by Crusafont-Pairó and Gibert (1976, 62, pl. 2, fig. 1), together with a further unpublished P1/3 fragment found in the collection, and another P1/3 from La Cantalera (Fig. 23.2, 3), previously described as Plagiaulacidae or Eobaataridae gen. et sp. indet. in Badiola et al. (2008, pl. 3, A–E) have been assigned to a new genus and species *Iberica hahni* in Badiola et al. (2011). This new taxon could be either an eobaatarid or an plagiaulacid. If the tentatively referred teeth, like the p4 fragment from La Cantalera, classified together with the aforementioned P1/3 as Plagiaulacidae or Eobaataridae gen. et sp. indet. in Badiola et al. (2008, pl. 3, F–J), and two M2s from the revised Galve collection are definitively assigned to *I. hahni*, the latter will be included in Plagiaulacidae.

A mesiodistally elongated crown and cusps not forming clear longitudinal labial and lingual rows are only described in the P1/3s under study (Figs. 23.2, 3 and 23.3, 1–4) and in *Eobaatar? pajaronensis* from the Late Barremian site of Pie Pajarón (Cuenca, Spain; Fig. 23.1B,C), and in the plagiaulacid *Parabolodon elongatus* (Simpson, 1928) from the Berriasian of Dorset (South England). The specimens under study differ from the P1/3s of *E.? pajaronensis* and *P. elongatus* (Hahn and Hahn, 2001, figs. 1, 2, and Hahn and Hahn, 2004, fig. 5b, respectively) in having four cusps rather than three, with cuspules on the mesial and distal margins, which are absent in the other two taxa, and in having both a cusp arrangement and morphology of the distal margin that are clearly different. *Iberica hahni* resembles *P. elongatus* in having P1/3s with a talonidlike heel on the distal margin of the crown, but this structure is distally much more protruding in *I. hahni*. The P1/3s of the latter taxon also differs from those of *P. elongatus* by having L1 confined to the mesiolingual corner and B2 between its lateral lingual counterparts L1–L2. By contrast, in *P. elongatus*, L1 is situated almost at (P1) or at (P2–P3) the longitudinal midpoint of the mesial margin, and the B2 is transversally oriented with respect to its lingual counterpart L2 in P2–P3. Moreover, *P. elongatus* corresponds to a much larger species. The size of *E.? pajaronensis* is similar to that of *I. hahni.*, but the arrangement of the main cusps on P1/3s is clearly different from that of the Galve and La Cantalera specimens. The P1/3s of *I. hahni* differ from those of *E.? pajaronensis* by having B2 situated more or less longitudinally at the midpoint of the labial margin, between its lingual counterparts L1–L2 in most of the specimens, rather than near the mesiolabial corner, and by having a talonidlike heel structure on the distal margin, which is absent in *E.? pajaronensis*. In the latter species a thick distal cingulum is present, but it does not protrude notably as in *I. hahni*.

Eobaatar hispanicus. Two P1/3s described as *conjunto Ib* of *Forma 4* by Crusafont-Pairó and Gibert (1976, 61–63) correspond roughly in size to a P1/3 described by the same authors as *conjunto Ia* of *Forma 4* (op. cit., 61, pl. 1, fig. 2). However, like Crusafont-Pairó and Gibert (1976), we also differentiate the two groups because of their different crown outline and cusp arrangement. In Badiola et al. (2011), the P1/3s of *conjunto Ib* have been assigned to *Eobaatar hispanicus* and two other unpublished upper posterior premolars (probably P4) have been also tentatively assigned to *E. hispanicus* (Fig. 23.3, 5–7); the P1/3 of *conjunto Ia* (Fig. 23.3, 8) has been classified in an open nomenclature as Plagiaulacidae or Eobaataridae indet. (see below).

The best-preserved ?P4 (Fig. 23.3, 7) fragment shows similar morphology to that of *E. hispanicus* (Hahn and Hahn, 1992, fig. 5) and *E. magnus* (Kielan-Jaworowska et al., 1987, pl. 4, figs. 2 and 3, pl. 5), but its fragmentary state prevents detailed comparisons. The P1/3s also resembles the P1/3 fragment from Galve described by Hahn and Hahn (2002, fig. 1) as *E. hispanicus* in having a similar cusp arrangement, with B2 confined to near the mesiolabial corner and L2 situated almost transversally midway along the distal margin of the crown, with a wide and quite protruding distal cingulum that continues along the labial margin, quite enlarged laterally. Because of the enlargement of the distal and labial cingula, the P1/3s under study show a subquadrangular crown outline in the occlusal view, which also seems to be present in the other P1/3 fragment from Galve described by Hahn and Hahn (2002).

Plagiaulacidae or Eobaataridae indet. The above-mentioned P1/3 described as *conjunto Ia* of *Forma 4* by Crusafont-Pairó and Gibert (1976, pl. 1, fig. 2) shows a somewhat mesiodistally elongated crown and three cusps triangularly arranged, in which B2 is confined to the longitudinal midpoint of the labial margin, between its lingual counterparts L1–L2 (Fig. 23.3, 8); in the P1/3s of *conjunto Ib*, by contrast, B2 is situated near the mesiolabial corner (Fig. 23.3, 5, 6). The P1/3 of *conjunto Ia* also differs from other three-cusped teeth with a mesiodistally elongated crown, such as the above-mentioned *Eobaatar? pajaronensis* and *Parabolodon elongatus*, in having a different cusp arrangement. In the specimen under study, B2 is confined to the longitudinal midpoint of the labial margin, between its lingual counterparts L1–L2, whereas in *E.? pajaronensis* it is situated near the mesiolabial corner (Hahn and Hahn, 2001, figs. 1 and 2) and in *P. elongatus* it is situated almost at (P1) or at the distolabial corner, transversally oriented with respect to its lingual counterpart L2. Moreover, in *P. elongatus* L1 is situated almost at (P1) or at (P2–P3) the longitudinal midpoint of the mesial margin (Hahn and Hahn, 2004, fig. 5b); by contrast, in the P1/3 under study, it is confined to the mesiolingual corner (Fig. 23.3, 8). The P1/3 from Galve exhibits thicker and more prominent enamel ridge on the cusps and thinner and less mesially and distally protruding cingula than the P1/3 of *E.? pajaronensis*, and there is no talonidlike heel structure on the distal margin like in *P. elongatus*. Because of the scantiness of the sample, this anterior upper premolar is classified as Plagiaulacidae or Eobaataridae indet.

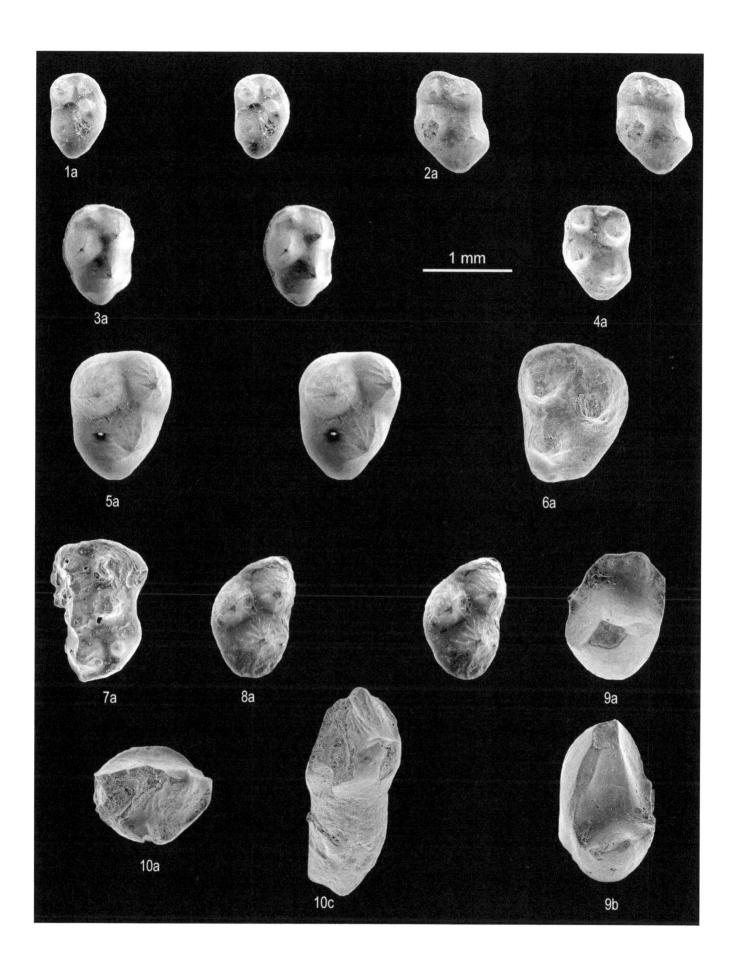

23.4. Scanning electron micrographs of Early Cretaceous plagiaulacidan multituberculate from the Vallipón site, province of Teruel, Spain. 1–4, Upper incisors tentatively assigned to *Eobaatar* sp. 1, I1 (MPZ96/156), 2–3, I2 (MPZ96/158 and MPZ96/157, respectively), and 4, I3 (MPZ96/159). 5–8, Upper premolars of aforementioned taxon *Eobaatar* sp. 5, right P1/3 (MPZ2010/859), 6–7, left P1/3s (MPZ2010/860 and MPZ2010/861, respectively), and 8, left P5 (MPZ2010/862). 9, 10, Upper teeth of *Lavocatia* sp. 9, a right M1 fragment (MPZ96/165); 10, right P5 highly abraded (MPZ96/175). 11–14, Upper anterior premolars of *Eobaatar? pajaronensis* Hahn and Hahn, 2001. 11 and 14, left P1/3s (MPZ96/160 and MPZ2010/863, respectively); 12 and 13, right P1/3s (MPZ96/164 and MPZ2010/865); (a) occlusal, (b) distolingual, (c) occlusal–mesial, (d) occlusal–distal, and (e) distal views. 6–10, Stereopair photographs.

Paulchoffatiidae indet. Three I2s were described as *conjunto I, II,* and *III* of *Forma* 3 in Crusafont-Pairó and Gibert (1976, 60–61), but only the *conjunto III* has been identified (Fig. 23.3, 9). The presence of more than one distal cusp on the I2 under study allows us to assign it either to Paulchoffatiidae or Pinheirodontidae (Hahn and Hahn, 2004, 124). The crown outline, the size of the distal cusps, and their configuration on the distal margin in the I2 under study are closer to those of Paulchoffatiidae than Pinheirodontidae; in the I2 in question, the distal cusps show similar size and are arranged in a transverse row that encircles the distal margin like in many Paulchoffatiidae (Hahn and Hahn, 2004, fig. 2). An unpublished I3 fragment (Fig. 23.3:10) has also been found in the collection, which corresponds in size to the aforementioned I2 and resembles the paulchofaatiid I3 in possessing a roughly subquadrangular crown outline in the occlusal view, with the main ridge subdivided into several cusps, and zero to three cusps in the labial and lingual margins. In the I3 fragment in question, the main ridge is subdivided into two cusps, and two lingual cusps are present at the midway point of the lingual margin, slightly separated from each other, and no cusp is observed on the labial margin; the latter, however, is surrounded by a prominent crescentic ridge. The size of I2 under study is closer to that of *Parendotherium herreroi* than to the I2s of the tiny species *Galveodon nannothus* described in an Early Barremian sites from Galve as well as in the Late Barremian site of Uña (Hahn and Hahn, 1992, figs. 2, 3). *P. herreroi*, however, is only known by the lost holotype (a I2 described by Crusafont-Pairó and Adrover, 1966, 30) and should perhaps be considered a *nomen dubium*. Because of the scantiness of I2 and I3 in the Paulchoffatiidae and the similarities between the incisors of the different representatives of this family, both teeth are classified in an open nomenclature as Paulchoffatiidae indet. (Badiola et al., 2011).

"Plagiaulacida" indet. Two highly abraded teeth, I2 and M2, cannot be described in detail and are classified as "Plagiaulacida" indet. (Badiola et al., 2011). The I2 shows similar morphological features to those described by Crusafont-Pairó and Gibert (1976, 60) for *conjunto II* of *Forma* 3, but these authors did not mention its state of preservation, and we do not know whether it belongs to *conjunto II* or to an unpublished specimen. The same question applies to M2. Three M2s were described in Crusafont-Pairó and Gibert (1976, 63–64) as *Forma* 6, and these authors mentioned that one of them was highly abraded. The M2 in question could belong to this specimen.

MORELLA SUBBASIN (ARTOLES FORMATION)

Here we report at least three Late Barremian plagiaulacidan multituberculates on the basis of a study of 29 isolated teeth. Some of them were reported in a preliminary work of Cuenca-Bescós et al. (1996). In this work, brief comments of the specimens and the figuration of some of them are included. A paper that includes the detailed descriptions and figurations of all the specimens is in preparation.

Eobaatar? pajaronensis. Seven P1/3s resemble those of *Eobaatar? pajaronensis* from the Late Barremian site of Pie Pajarón (Fig. 23.1B,C) in having a similar, somewhat mesiodistally elongated crown outline in the

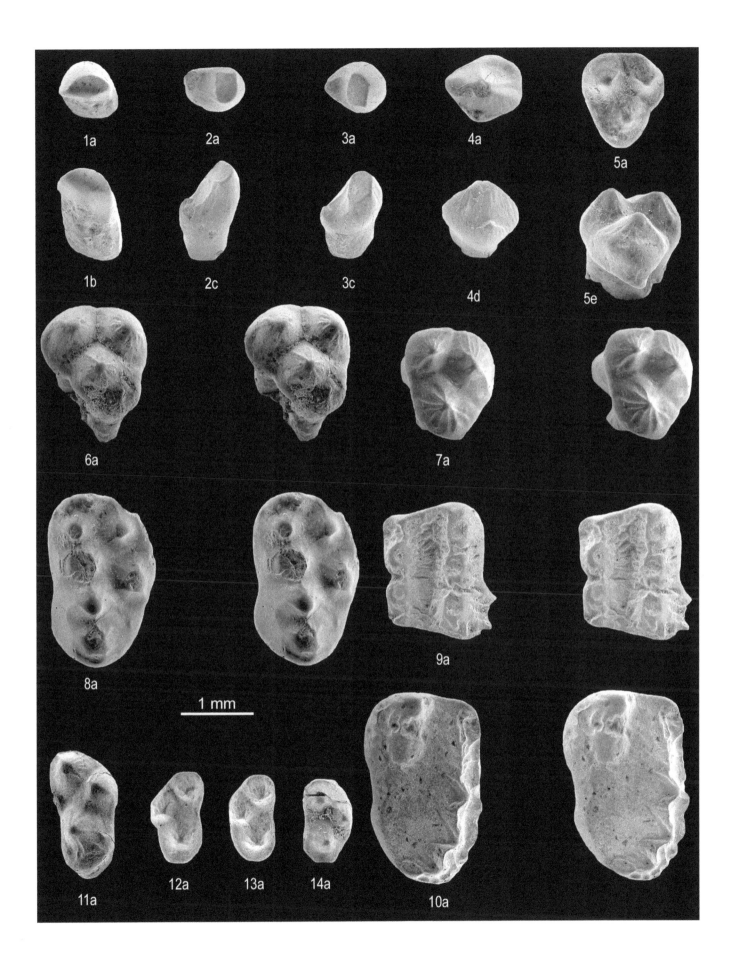

occlusal view, thick mesial, and distal cingula, the latter being somewhat enlarged, and both the same cusp formula 1B:2L and cusp arrangement; L2 situated transversally midway along the distal margin, and B2 confined near the mesiolabial corner (Fig. 23. 4, 11–14). The size variability observed among the different P1/3s from Vallipón is probably related to the tooth position in the P1–P3 series.

Eobaatar sp. The size and morphology of the three P1/3s under study (Fig. 23.4, 5–7) resemble the two subquadrangular P1/3s found in the revised Galve collection and assigned to *E. hispanicus* and the other P1/3 fragment from Galve (Hahn and Hahn, 2002, fig. 1) in possessing a subquadrangular crown outline in occlusal view rather than subtriangular as in the P1/2s from Uña (Hahn and Hahn, 1992, figs. 7, 8) and enlarged cingula surrounding the labial and distal margin. The Vallipón teeth, however, differ from the Galve specimens in having L2 situated transversally at or near the midway along the distal margin, and enlarged cingula lateral to L2 on the labial margin as well as on the lingual margin. By contrast, there is no cingulum surrounding the lingual margin in the Galve specimens, and L2 is situated in a more distolingual position. We do not know whether these differences in both the cingulum development and cusp arrangement are sufficient to regard them as being a different species from *E. hispanicus* because a ?P5 of *Eobaatar* from Vallipón (Fig. 23.4, 8), which may belong to this *Eobaatar* species, is similar in size and morphology to that of *E. hispanicus* (Hahn and Hahn, 1992, fig. 5). Other teeth, such as the three upper incisors (Fig. 23.4, 1–4), and upper and lower first and second molars may also belong to this species of *Eobaatar*.

Lavocatia sp. The P5 and M1 (Fig. 23.4, 9, 10) resemble the Iberian pinheirodontids *Cantalera abadi* and *Lavocatia alfambrensis* (Badiola et al., 2008, pl. 1, and Canudo and Cuenca-Bescós, 1996, fig. 3, respectively) in having long and strongly marked radiating enamel ridges from the tip to the base of the cusps, which intermesh at their distal ends with the neighboring ridges. However, this feature is more pronounced in *C. abadi* than in *L. alfambrensis* and in the Vallipón specimens. Moreover, the M1 of the Vallipón site differs from the M1 of *C. abadi* in having three labial cusps instead of two, and the morphology of P5 is closer to that of *L. alfambrensis* than *C. abadi*. The P5 under study is too abraded to observe the autapomorphic characters of *L. alfambrensis* represented only by a P5. Nevertheless, the main occlusal tooth pattern of the Vallipón P5 (Fig. 23.4, 10) seems similar to that from Galve, although some differences between them can be addressed. The cusp formula of the P5 under study seems slightly different (2–4?:4–5:6) from that of Galve (4:6:5), and the cusps of the medial row seem to increase in height distally rather than mesially as in *L. alfambrensis*. These differences need to be confirmed in an unworn P5 and are probably not sufficiently great for it to be included in another genus, because different cusp formulae are described in many teeth from different species of the same genus (e.g., *Sinobaatar* Hu and Wang, 2002, see Kusuhashi et al., 2009, for a bibliography). The Vallipón specimens, however, correspond to a larger species. Because of the fragmentary state of the specimens and the scantiness of the material, we do not assign the P5 and M1 to a new species, and they are provisionally classified as *Lavocatia* sp.

The new plagiaulacidan multituberculate fossils from the Iberian Peninsula have considerably increased the biodiversity and biostratigraphic resolution of the Iberian Early Cretaceous multituberculates (Fig. 23.5).

The oldest plagiaulacidans from the Iberian Peninsula come from the Berriasian deposits of the Porto Pinheiro site in Portugal and correspond to four genera of Pinheirodontidae. New finds of pinheirodontids in the Late Hauterivian–Early Barremian of La Cantalera and in the Late Barremian of Vallipón demonstrate a greater biodiversity for these faunas in the Iberian Peninsula than was previously known. Two other, younger taxa are known; one from the Hauterivian–Barremian transition, *Cantalera abadi*, and the other from the Early Barremian of Galve (Camarillas Formation: the Poca site), *Lavocatia alfambrensis*. A larger species than *L. alfambrensis*, classified as *Lavocatia* sp. because of the scantiness of the material at hand, may be present in the Late Barremian site of Vallipón.

The next in the stratigraphic section after the Portuguese site of Porto Pinheiro is the Spanish site Pochancalo-1. If the assignment of the putative albionbataarid upper canine from this site is correct, it will be the first representative of this family in the Iberian Peninsula and the first from the Valanginian–Hauterivian transition from Europe. In the Barremian sites of the Galve area (Teruel, Spain), the paulchoffatiid *Galveodon nannothus* and the eobaatarid *Eobaatar hispanicus* are described in both the El Castellar (Late Hauterivian–Early Barremian) and Camarillas (Early Barremian) formations. The presence of these species in the Late Barremian of Uña (Hahn and Hahn, 1992, 2002, 2006) is questionable. Biostratigraphically, there is an important difference between the Early Barremian sites of Herrero and Cerrada Roya-Mina, characterized by the last appearance of *Eobaatar hispanicus* and *Galveodon nannothus* in the Galve area (Fig. 23.5) and the Late Barremian site of Uña. Moreover, the scantiness of the specimens found in Uña and the morphological differences observed between the fossils from Uña and Galve allows us to question whether the specimens from these localities belong to the same species or whether they should be regarded as belonging to different species (Badiola et al., 2011). A larger species than *Galveodon nannothus* is also present in Galve, which is represented by I2 and I3 and is classified provisionally as Paulchoffatiidae indet. *Parendotherium herreroi* has been removed from the multituberculate faunal list of Galve, whereas another taxon has been added: *Iberica hahni*. The latter is also registered in the Late Hauterivian–Early Barremian site of La Cantalera. If the tentatively referred material (a p4 fragment and two M2s) can really be assigned to this new taxon, it will be the first representative of Plagiaulacidae in the Iberian Peninsula.

The biodiversity of Iberian eobaatarids has also increased. Currently, two species of *Eobaatar* are described: the Early Barremian species *E. hispanicus*, and the Late Barremian species *E.? pajaronensis*, tentatively assigned to the genus *Eobaatar*, which is present in the Spanish Late Barremian sites of Vallipón and Pie Pajarón. Another representative of *Eobaatar* is also known in the Late Hauterivian–Early Barremian site of La Cantalera, represented only by two P5s. The latter teeth are classified as *Eobaatar* sp. Currently it is difficult to ascertain whether the differences observed between the P5s of La Cantalera and that of *E. hispanicus* are intraspecific or interspecific—that is, whether the P5s from La Cantalera

23.5. Iberian Early Cretaceous plagiaulacidan multituberculate taxa, and chronology. See litho- and chronostratigraphy of the paleontological sites in Fig. 23.1C. A, The presence of *Eobaatar hispanicus* and *Galveodon nannothus* in the Uña site is questionable. B, *Parendotherium herreroi* is only known by the lost holotype (a I2 described by Crusafont-Pairó and Adrover, 1966, 30) and should perhaps be considered a *nomen dubium*.

AGE			SITES	TAXA	DATA SOURCE
BARREMIAN	LATE		Uña	*Eobaatar hispanicus*[a] *Galveodon nannothus*[a]	Hahn & Hahn, 1992 Hahn & Hahn, 2006
			Pie Pajarón	*Eobaatar? pajaronensis*	Hahn & Hahn, 2001 Hahn & Hahn, 2006
			Vallipón	*Eobaatar? pajaronensis* *Eobaatar* sp. *Lavocatia* sp.	This work
	EARLY	Galve	Poca	*Lavocatia alfambrensis*	Canudo & Cuenca-Bescós, 1996
			Cerrada Roya-Mina	*Galveodon nannothus*	Hahn & Hahn, 1992 Hahn & Hahn, 2006
			Herrero	*Eobaatar hispanicus* *Parendotherium herreroi*[b]	Crusafont-Pairó & Adrover, 1966
HAUTERIVIAN/BARREMIAN			Colladico Blanco	Plagiaulacidae or Eobaataridae gen. et sp. nov. Plagiaulacidae or Eobaataridae indet. Paulchoffatiidae indet.	Hahn & Hahn, 1992 Hahn & Hahn, 2006 This work Badiola et al. submitted
			Pelejón-2	*Eobaatar hispanicus* *Galveodon nannothus*	Hahn & Hahn, 1992 Hahn & Hahn, 2006
			Piélago 0	*Eobaatar hispanicus*	Hahn & Hahn, 2002 Hahn & Hahn, 2006
			La Cantalera	*Cantalera abadi* *Eobaatar* sp. Plagiaulacidae or Eobaataridae gen. et sp. nov.	Badiola, Canudo & Cuenca-Bescós, 2008 This work Badiola et al. submitted
VALANGINIAN/HAUTERIVIAN			Pochancalo-1	?Albionbaataridae indet.	This work
BERRIASIAN			Porto Pinheiro	*Bernardodon atlanticus* *Ecprepaulax anomala* *Iberodon quadrituberculatus* *Pinheirodon pygmaeus* *Pinheirodon vastus*	Hahn & Hahn, 1999 Hahn & Hahn, 2006

are to be regarded as belonging to a different species. The same problem exists with respect to the differences between the P1/3s from *E. hispanicus* from Galve and the P1/3s from the Late Barremian of Vallipón; because of the scantiness of the material, the specimens from Vallipón are provisionally classified in an open nomenclature as *Eobaatar* sp. *Iberica hahni* may also belong either to another eobaatarid representative or to the first of the Plagiaulacidae from the Iberian Peninsula.

On the basis of paleogeographical criteria and phylogenetic analyses (e.g., Luo et al., 2001; Martin and Rauhut, 2005), a dichotomous paleobiogeographical model of Gondwana versus Laurasia has been proposed for the repartition of Early Cretaceous vertebrates. Some groups, including the multituberculates, were considered to be restricted to Laurasia. However, the existence of multituberculates in North Africa (Morocco) is known since the beginning of the 1990s (Sigogneau-Russell, 1991, updated by Hahn and Hahn, 2003). Multituberculate fossils have also been recorded in the Early Cretaceous of Australia (Rich et al., 2009) and in the Late Cretaceous of Africa (Krause et al., 2006) and South America (Kielan-Jaworowska et al., 2007). These discoveries thus contradict a pan-Laurasian distribution for multituberculates and indicate dispersal pathways of the group into Gondwana from the Early Cretaceous on (Rich et al., 2009). Although the Laurasian Cretaceous multituberculate fossil record is more complete than that from Gondwana, knowledge of their dispersal pathways among the different areas of Laurasia during this period is poorly understood.

However, new plagiaulacidan multituberculate finds from the Iberian Peninsula provide more information about their paleobiogeographical distribution within Laurasia: the oldest representative of *Eobaatar* is described in the Late Hauterivian–Early Barremian of La Cantalera; if its assignment to Albionbaataridae is correct, the specimen found in the Late Valanginian–Early Hauterivian is the first representative of this family in the Iberian Peninsula and the first from the Valanginian–Hauterivian transition in Europe. If *Iberica hahni* can be assigned to Plagiaulacidae, it will be the first representative of this family in the Iberian Peninsula. And the presence of the eobaatarid *Eobaatar* from the Hauterivian–Barremian transition to the Late Barremian in the Iberian Peninsula, during the Barremian in Britain, and in the Aptian or Albian of Asia, suggests that geographical connections between Europe and Asia could have existed either sporadically or constantly for most of the Early Cretaceous (Badiola et al., 2008). This hypothesis is supported by other paleontological discoveries in the Early Cretaceous from Spain, such as the gobiconodontid mammals (Cuenca-Bescós and Canudo, 2003) and some sauropod and ornithopod dinosaurs (Canudo et al., 2002). The Early Cretaceous plagiaulacidan multituberculates from the Iberian Peninsula include endemic Jurassic survivors (Paulchoffatiidae), families with a Western European distribution (Pinheirodontidae, Albionbaataridae, and possibly Plagiaulacidae), and forms that are also registered in Asia (Eobaataridae). Recently, further eobaatarid taxa have been described in Aptian or Albian sites from Japan (Kuwajima Formation) and China (Shahai and Fuxin Formations): *Hakusanobaatar* Kusuhashi, 2008, *Tedoribaatar* Kusuhashi, 2008, and

Paleobiogeography

Liaobaatar Kusuhashi, Hu, Wang, Setoguchi, and Matsuoka, 2009, and to different species of *Sinobaatar* (Kusuhashi, 2008; Kusuhashi et al., 2009). These latter taxa have not been described in the Iberian Peninsula, but sites younger than Late Barremian have not yet been found.

The paucity of the fossil record for *Eobaatar* in the Iberian Peninsula and Asia, represented by isolated teeth and lacking the complete upper and lower cheek teeth, currently prevents the use of cladistic analysis to investigate the area of origin or/and the dispersals of different species of this genus. Neither phylogenetical nor paleobiogeographical relationships between the other plagiaulacidan multituberculates in Laurasia can be investigated with the material at hand. According to Kusuhashi et al. (2009), multituberculates are considered to have migrated from Europe to Asia probably in the Early Cretaceous and were already flourishing in Asia during the Aptian–Albian.

Acknowledgments

This chapter forms part of the project CGL2007-62469, subsidized by the Ministry of Science and Innovation of Spain, the European Regional Development Fund, the Government of Aragon ("Grupos Consolidados" and "Dirección General de Patrimonio Cultural"). Zofia Kielan-Jaworowska's comments have improved the quality of this chapter. A. B. acknowledges support from the Programa Juan de la Cierva of the Ministry of Education and Science. Rupert Glasgow revised the chapter's English language.

References

Aurell, M., B. Bádenas, J. I. Canudo, and J. I. Ruiz-Omeñaca. 2004. Evolución tectosedimentaria de la Formación Blesa (Cretácico Inferior) en el entorno del yacimiento de vertebrados de La Cantalera (Josa, Teruel). Geogaceta 35: 11–14.

Badiola, A., J. I. Canudo, and G. Cuenca-Bescós. 2008. New multituberculate mammals from the Hauterivian/Barremian transition of Europe (Iberian Peninsula). Palaeontology 51: 1455–1469 [Corrigendum: Palaeontology 52, p. 271].

———. 2009. Systematic reassessment of Early Cretaceous multituberculate from Galve (Teruel, Spain). Journal of Vertebrate Paleontology 29 (supplement to 3): 57A.

———. 2011. A systematic reassessment of Early Cretaceous multituberculates from Galve (Teruel, Spain). Cretaceous Research 32: 45–57.

Buscalioni, A. D., M. A. Fregenal, A. Bravo, F. J. Poyato-Ariza, B. Sanchíz, A. M. Báez, O. Cambra Moo, C. Martín-Closas, S. E. Evans, and J. Marugán Lobón. 2008. The vertebrate assemblage of Buenache de la Sierra (Upper Barremian of Serranía de Cuenca, Spain) with insights into its taphonomy and palaeoecology. Cretaceous Research 29: 687–710.

Butler, P. M., and J. J. Hooker. 2005. New teeth of allotherium mammals from the English Bathonian, including the earliest multituberculates. Acta Palaeontologica Polonica 50: 185–207.

Canudo, J. I., and G. Cuenca-Bescós. 1996. Two new mammalian teeth (Multituberculata and Peramura) from the Lower Cretaceous (Barremian) of Spain. Cretaceous Research 17: 215–228.

Canudo, J. I., G. Cuenca-Bescós, J. I. Ruiz-Omeñaca, and A. R. Soria. 1996. Estratigrafía y Paleoecología de los vertebrados del Barremiense superior (Cretácico inferior) de Vallipón (Castellote, Teruel). Mas de las Matas 15: 9–34.

Canudo, J. I., O. Amo, G. Cuenca-Bescós, A. Meléndez, J. I. Ruiz-Omeñaca, and A. R. Soria. 1997. Los vertebrados del Tithónico–Barremiense de Galve (Teruel, España). Cuadernos de Geología Ibérica 23: 209–241.

Canudo, J. I., J. I. Ruiz-Omeñaca, J. L. Barco, R. Royo-Torres. 2002. ¿Saurópodos asiáticos en el Barremiense inferior (Cretácico Inferior) de España? Ameghiniana 34: 443–452.

Canudo, J. I., J. I. Ruiz-Omeñaca, M. Aurell, J. L. Barco, and G. Cuenca-

Bescós. 2006. A megatheropod tooth from the late Tithonian–lower Berriasian (Jurassic–Cretaceous transition) of Galve (Aragon, NE Spain). Neues Jahrbuch für Geologie und Paläontologie, Monatshefte 239: 77–99.

Canudo J. I., J. M. Gasca, M. Aurell, A. Badiola, H.-A. Blain, D. Gómez-Fernández, M. Moreno-Azanza, J. Parrilla, R. Rabal, and J. I. Ruiz-Omeñaca. 2010. La Cantalera: an exceptional window onto the vertebrate biodiversity of the Hauterivian–Barremian transition in the Iberian Peninsula. Journal of Iberian Geology 32: 205–224.

Crusafont-Pairó, M., and R. Adrover. 1966. El primer mamífero del Mesozoico español. Fossilia 5–6: 28–33.

Crusafont-Pairó, M., and J. Gibert. 1976. Los primeros multituberculados de España. Nota preliminar. Acta Geologica Hispánica 11: 57–64.

Cuenca-Bescós, G., and J. I. Canudo. 2003. A new gobiconodontid mammal from the Early Cretaceous of Spain and its palaeobiogeographical implications. Acta Palaeontologica Polonica 48: 575–582.

———. 2004. Los mamíferos del Cretácico inferior de España. Geo-temas 6: 35–38.

Cuenca-Bescós, G., J. I. Canudo, and J. I. Ruiz-Omeñaca. 1996. Los mamíferos del Barremiense superior (Cretácico Inferior) de Vallipón (Castellote, Teruel). Mas de las Matas 15: 105–137.

Cuenca-Bescós, G., A. Badiola, J. I. Canudo, J. M. Gasca, and M. Moreno-Azanza. 2011. New dryolestidan mammal from the Hauterivian–Barremian transition of the Iberian Peninsula. Acta Palaeontologica Polonica 56: 257–267.

Díaz Molina, M., and A. Yébenes. 1987. La sedimentación litoral y continental durante el Cretácico Inferior. Sinclinal de Galve, Teruel; pp. 3–21 in J. L. Sanz (coord.), Geología y Paleontología (arcosaurios) de los yacimientos cretácicos de Galve (Teruel) y Tremp (Lérida). Estudios Geológicos, volumen extraordinario Galve-Tremp.

Estes, R., and B. Sanchiz. 1982. Early Cretaceous lower vertebrate from Galve (Teruel), Spain. Journal of Vertebrate Paleontology 2: 21–39.

Gasca, J. M., J. I. Canudo, and M. Moreno-Azanza. 2007. Restos de dinosaurios de la Formación El Castellar en Miravete de la Sierra (Cretácico Inferior, Teruel, España); pp. 63–64 in IV Jornadas Internacionales sobre Dinosaurios y su Entorno, September 13–15, Salas de los Infantes, Burgos, Spain.

Gasca, J. M., D. Gómez-Fernández, M. Moreno-Azanza, and J. I. Canudo. 2009. Un paseo por los yacimientos del tránsito Hauteriviense–Barremiense (Cretácico Inferior) de Aragón. Paleolusitana 1: 211–219.

Gómez, B., C. Martín-Closas, H. Méon, F. Thévenard, and G. Barale. 2001. Plant taphonomy and palaeoecology in the lacustrine Uña delta (Late Barremian, Iberian Ranges, Spain). Palaeogeography, Palaeoclimatology, Palaeoecology 170: 133–148.

Grambast, L. 1968. Evolution of the utricule in the Charophyte genera Perimneste Harris and Atopochara Peck. Journal of the Linnean Society of London (Botany) 61: 5–11.

Hahn, G. 1969. Beiträge zur Fauna der Grube Guimarota Nr. 3. Die Multituberculata. Palaeontographica 133 (⅓): 1–100.

———. 1977. Neue Schädel-Reste von Multituberculaten (Mammalia) aus dem Malm Portugals. Geologica et Palaeontologica 11: 161–186.

———. 1987. Neue Beobachtungen zum Schädel- und Gebiss-Bau der Paulchoffatiidae (Multituberculata, Ober-Jura). Palaeovertebrata 17: 155–196.

———. 1993. The systematic arrangement of the Paulchoffatiidae (Multituberculata) revisited. Geologica et Palaeontologica 27: 201–214.

———. 2001. Neue Beobachtungen an Schädel-Resten von Paulchoffatiidae (Multituberculata; Ober-Jura). Geologica et Palaeontologica 35: 121–143.

Hahn, G., and R. Hahn. 1992. Neue Multituberculaten-Zähne aus der Unter-Kreide (Barremium) von Spanien (Galve und Uña). Geologica et Paleontologica 26: 143–162.

———. 1998. Neue Beobachtungen an Plagiaulacoidea (Multituberculata) des Ober-Juras. 3. Der Bau des Molaren bei den Paulchoffatiidae. Berliner geowissenschaftliche Abhandlungen E 28: 39–84.

———. 1999. Pinheirodontidae n. fam. (Multituberculata, Mammalia) aus der tiefen Unter-Kreide Portugals. Palaeontographica 253: 77–222.

———. 2001. Multituberculaten-Zähne aus der Unter-Kreide (Barremium) von Pié Pajarón (Prov. Cuenca, Spanien). Paläontologische Zeitschrift 74: 587–589.

———. 2002. Neue Multituberculaten-Zähne aus dem Barremium (Unter-Kreide) von Galve (Spanien). Paläontologische Zeitschrift 76: 257–259.

———. 2003. New multituberculate teeth from the Early Cretaceous of Morocco.

Acta Palaeontologica Polonica 48: 349–356.

———. 2004. The dentition of the Plagiaulacida (Multituberculata, Late Jurassic to Early Cretaceous). Geologica et Palaeontologica 38: 119–159.

———. 2006. Catalogus Plagiaulacidorum cum figuris (Multituberculata suprajurassica et subcretacea)., Fossilium Catalogus I: Animalia Pars 140. Backhuys Publishers, Leiden, 344 pp.

Henkel, S., and B. Krebs. 1969. Zwei Säugetier-Unterkiefer aus der Unteren Kreide von Uña (Prov. Cuenca, Spanien). Neues Jahrbuch fur Geologie und Paläontologie, Monatshefte 8: 449–463.

Hu, Y., and Y. Wang. 2002. Sinobaatar gen. nov.: first multituberculate from the Jehol Biota of Liaoning, northeast China. Chinese Science Bulletin 47: 382–386.

Infante, P., J. I. Canudo, M. Aurell, J. I. Ruiz-Omeñaca, L. M. Sender, and S. A. Zamora. 2005. Primeros datos sobre los dinosaurios de Zaragoza (Theropoda, Valanginiense-Hauteriviense, Cretácico Inferior); pp. 119–120 in XXX Jornadas de la Sociedad Española de Paleontología. Gestión e Investigación de la Paleontología en el Siglo XXI, Sevilla, Spain, 4–8 October, 2005.

Kielan-Jaworowska, Z., and P. C. Ensom. 1992. Multituberculate mammals from the Upper Jurassic Purbeck Limestone Formation of southern England. Palaeontology 35: 95–126.

———. 1994. Tiny plagiaulacoid multituberculate mammals from the Purbeck Limestone Formation of Dorset, England. Palaeontology 37: 17–31.

Kielan-Jaworowska, Z., and J. H. Hurum. 2001. Phylogeny and systematics of multituberculate mammals. Palaeontology 44: 389–429.

Kielan-Jaworowska, Z., D. Dashzeveg, and B. A. Trofimov. 1987. Early Cretaceous multituberculates from Mongolia and a comparison with Late Jurassic forms. Acta Palaeontologica Polonica 32: 3–47.

Kielan-Jaworowska, Z., R. L. Cifelli, and Z.-X. Luo. 2004. Mammals from the Age of Dinosaurs: origins, evolution, and structure. Columbia University Press, New York, 630 pp.

Kielan-Jaworowska, Z., E. Ortiz-Jaureguizar, C. Vieytes, R. Pascual, and F. Goin. 2007. First ?cimolodontan multituberculate mammal from South America. Acta Palaeontologica Polonica 52: 257–262.

Krause, D. W., P. M. O'Connor, K. C. Rogers, S. D. Sampson, G. A. Buckley, and R. R. Rogers. 2006. Terrestrial vertebrates from Madagascar: implications for Latin

American biogeography. Annals of the Missouri Botanical Gardens 93: 178–208.

Krebs, B. 1985. Theria (Mammalia) aus der Unterkreide von Galve (Provinz Teruel Spanien). Berliner geowissenschaftliche Abhandlungen A 60: 29–48.

———. 1993. Das Gebiss von *Crusafontia* (Eupantotheria, Mammalia)—Funde aus der Unter-Kreide von Galve und Uña (Spanien). Berliner geowissenschaftliche Abhandlungen E 9: 233–252.

Kriwet, J., S. Klug, J. I. Canudo, and G. Cuenca-Bescós. 2008. A new Early Cretaceous lamniform shark (Chondrichthyes, Neoselachii). Zoological Journal of the Linnean Society 154: 278–290.

Kühne, W. G. 1961. Eine Mammaliafauna aus dem Kimeridge Portugals. Neues Jahrbuch für Geologie und Paläontologie, Monatshefte 7: 374–381.

Kusuhashi, N. 2008. Early Cretaceous multituberculate mammals from the Kuwajima Formation (Tetori Group), central Japan. Acta Palaeontologica Polonica 53: 379–390.

Kusuhashi, N., Y. Hu, Y. Wang, T. Setoguchi, and H. Matsuoka. 2009. Two eobaatarid (Multituberculata, Mammalia) genera from the Lower Cretaceous Shahai and Fuxin formations, northeastern China. Journal of Vertebrate Paleontology 29: 1264–1288.

Liesa, C. L., A. R. Soria, N. Meléndez, and A. Meléndez. 2006. Extensional fault control on the sedimentation patterns in a continental rift basin: El Castellar Formation, Galve sub-basin, Spain. Journal of the Geological Society, London 163: 487–498.

Luo, Z. X, R. L. Cifelli, and Z. Kielan-Jaworowska. 2001. Dual origin of tribosphenic mammals. Nature 409: 53–57.

Martín-Closas, C. 1989. Els caròfits del Cretaci inferior de les conques perifèriques del Bloc de l'Ebre. Ph.D. dissertation, Universitat de Barcelona, Barcelona, Spain, 581 pp.

Martin, T., and B. Krebs (eds.). 2000. Guimarota: a Jurassic ecosystem. Verlag Dr. Friedrich Pfeil, Munich, 155 pp.

Martin, T., and O. M. Rauhut. 2005. Mandible and dentition of *Asfaltomylos patagonicus* (Australophenida, Mammalia) and the evolution of tribosphenic teeth. Journal of Vertebrate Paleontology 25: 414–425.

Meléndez, N., C. L. Liesa, A. R. Soria, and A. Meléndez. 2009. Lacustrine system evolution during early rifting: El Castellar Formation (Galve sub-basin, Central Iberian Chain). Sedimentary Geology 222: 64–77.

Moreno-Azanza, M., J. I. Canudo, and J. M. Gasca. 2007. Primera evidencias de fragmentos de cáscara de huevo en el Valanginiense-Hauteriviense (Cretácico Inferior) de la Cordillera Ibérica (Zaragoza, Spain). XXIII Jornadas de la Sociedad Española de Paleontología, October 3–6, Caravaca de la Cruz, Spain; pp. 149–150 in J. C. Braga, A. Checa, and M. Company (eds.), Instituto Geológico y Minero de España y Universidad de Granada.

Moreno-Azanza, M., J. M. Gasca, and J. I. Canudo. 2009. A high-diversity eggshell locality from the Hauterivian–Barremian transition of the Iberia Peninsula. Journal of Vertebrate Paleontology 29 (supplement to 3): 151A.

Pouech, J., and J.-M. Mazin. 2009. Description and West European affinities of the Mammalian fauna of Cherves-de-Cognac; p. 81 in P. Godefroit and O. Lambert (eds.), Tribute to Charles Darwin and Bernissart Iguanodons: New Perspectives on Vertebrate Evolution and Early Cretaceous Ecosystems, Brussels, February 9–13, 2009.

Poyato-Ariza, F. J., M. R. Talbot, M. A. Fregenal Martínez, N. Meléndez, and S. Wenz. 1998. First isotopic and multidisciplinary evidence for nonmarine coelacanths and pycnodontiform fishes: paleoenvironmental implications. Palaeogeography, Palaeoclimatology, Palaeoecology 144: 65–84.

Querol, X., R. Salas, G. Pardo, and L. L. Ardévol. 1992. Albian coal-bearing deposits of the Iberian Range in northeastern Spain. Geological Society of America, Special Paper 267: 193–208.

Rich, T. H., P. Vickers-Rich, T. F. Flannery, B. P. Kear, D. J. Cantrill, P. Komarower, L. Kool, D. Pickering, P. Trusler, S. Morton, N. Van Klaveren, and E. M. G. Fitzgerald. 2009. An Australian multituberculate and its palaeobiogeographic implications. Acta Palaeontologica Polonica 54: 1–6.

Riveline, J., J. P. Berger, M. Feist, C. Martín-Closas, M. Schudack, and I. Soulié-Märsche. 1996. European mesozoic-cenozoic charophyte biozonation. Bulletin de la Société Geologique de France 167: 453–468.

Ruiz-Omeñaca, J. I., and J. I. Canudo. 2001. Vallipón y La Cantalera: dos yacimientos paleontológicos excepcionales. Naturaleza Aragonesa 8: 8–17.

Ruiz-Omeñaca, J. I., J. I. Canudo, and G. Cuenca-Bescós. 1997. Primera evidencia de un área de alimentación de dinosaurios herbívoros en el Cretácico

Inferior de España (Teruel). Monografías de la Academia de Ciencias Exactas, Físicas, Químicas y Naturales de Zaragoza 10: 1–48.

Ruiz-Omeñaca, J. I., J. I. Canudo, M. Aurell, B. Bádenas, J. L. Barco, G. Cuenca-Bescós, and J. Ipas. 2004. Estado de las investigaciones sobre los vertebrados del Jurásico Superior y Cretácico Inferior de Galve (Teruel). Estudios Geológicos 60: 179–202.

Ruiz-Omeñaca, J. I., J. I. Canudo, P. Cruzado-Caballero, P. Infante, M. Moreno-Azanza. 2005. Baryonychine teeth (Theropoda: Spinosauridae) from the Lower Cretaceous of La Cantalera (Josa, NE Spain). Kaupia. Darmstädter Beiträge zur Naturgeschichte 14: 59–63.

Salas, R., C. Martín-Closas, X. Querol, J. Guimera, and E. Roca. 1995. Evolución tectonosedimentaria de las cuencas del Maestrazgo y Aliaga-Penyagolosa durante el Cretácico Inferior; pp. 13–94 in R. Salas and C. Martín-Closas (eds.), El Cretácio inferior de Nordeste de Iberia. Universitat de Barcelona, Spain.

Salas, R., J. Guimera, R. Mas, C. Martín-Closas, A. Meléndez, and A. Alonso. 2001. Evolution of the Mesozoic Central Iberian Rift System and its Cainozoic inversion (Iberian chain). Mémoires du Museum d'Histoire Naturelle 186: 145–185.

Sanz, J. L., A. D. Buscalioni, M. L. Casanovas, and J. V. Santafé. 1987. Dinosaurios del Cretácico Inferior de Galve (Teruel, España). Estudios geológicos, volumen extraordinario Galve-Tremp: 45–64.

Sanz, J. L., M. A. Fregenal-Martinez, N. Meléndez, and F. Ortega. 2001. Las Hoyas; pp. 356–359 in D. E. G. Briggs, and P. R. Crowther (eds.), Paleobiology II. Blackwell Science, Oxford.

Sigogneau-Russell, D. 1991. First evidence of Multituberculata (Mammalia) in the Mesozoic of Africa. Neues Jahrbuch für Geologie und Paläontologie, Monstshefte 2: 119–125.

Simpson, G. G. 1928. A catalogue of the Mesozoic Mammalia in the Geological Department of the British Museum. British Museum (Natural History), London, 215 pp.

Soria, A. R., C. Martín-Closas, A. Meléndez, N. Meléndez, and M. Aurell. 1995. Estratigrafía del Cretácico inferior del sector central de la Cordillera Ibérica. Estudios Geológicos 51: 141–152.

Soria de Miguel, A. R. 1997. La sedimentación en las cuencas marginales del Surco Ibérico durante el Cretácico Inferior y su contorno estructural. Ph.D.

dissertation, Servicio de Publicaciones de la Universidad de Zaragoza, Spain, 363 pp.

Schudack, M. 1989. Charophytenfloren aus den unterkretazischen Vertebraten-Fundschichten bei Galve und Uña

(Otspanien). Berliner geowissenschaftliche Abhandlungen A 106: 409–443.

Schudack, U., and M. Schudack. 2009. Ostracod biostratigraphy in the Lower Cretaceous of the Iberian Chain (eastern Spain). Journal of Iberian Geology 35: 141–168.

Sweetman, S. C. 2009. A new species of the plagiaulacoid multituberculate mammal *Eobaatar* from the Early Cretaceous of southern Britain. Acta Palaeontologica Polonica 54: 373–384.

Addendum

Badiola et al. (2011) recently published a systematic reassessment of Late Hauterivian–Early Barremian multituberculates from Galve and a description of new specimens, including scanning electron photomicrographs of all the specimens previously studied by Crusafont-Pairó and Adrover (1966) and Crusafont-Pairó and Gibert (1976).

The multituberculate fossil assemblage from Galve comprises at least four taxa: the paulchoffatiid *Galveodon nannothus*, the eobaatarid *Eobaatar hispanicus*, a new taxon of Plagiaulacidae or Eobaataridae named *Iberica hahni*, and the pinheirodontid *Lavocatia alfambrensis*. Moreover, the species *Parendotherium herreroi* has been removed from the multituberculate faunal list of Galve.

In a recent revision of the dryolestoid mammals from the Early Barremian of Galve, Cuenca-Bescós et al. (2011) updated the faunal list of other Early Cretaceous vertebrate fossil assemblages from the Central Iberian Range. They allocated the material from Galve, previously described as *Crusafontia cuencana* by Henkel and Krebs (1969), to a new *Crusafontia* species and relegated *C. cuencana* to the Late Barremian site of Uña.

New data on the vertebrate paleodiversity and paleoenvironment of La Cantalera (Hauterivian–Barremian) have been published by Canudo et al. (2010).

References

Badiola, A., J. I. Canudo, and G. Cuenca-Bescós. 2011. A systematic reassessment of Early Cretaceous multituberculates from Galve (Teruel, Spain). Cretaceous Research 32: 45–57.

Canudo, J. I., J. M. Gasca, A. Aurell, A.Badiola, H-A. Blain, P. Cruzado-Caballero, D. Gómez-Fernández, M. Moreno-Azanza, J. Parrilla, R. Rabal-Garcés, and J. I. Ruiz-Omeñaca 2010. La Cantalera: an exceptional Windows onto the vertebrate biodiversity of the Hauterivian–Barremian transition in the Iberian Peninsula. Journal of Iberian Geology 32: 205–224.

Crusafont-Pairó, M. and R. Adrover. 1966. El primer mamífero del

Mesozoico español. Fossilia 5–6: 28–33 (corresponding to 1965 year).

Crusafont-Pairó, M. and J. Gibert. 1976. Los primeros multituberculados de España. Nota preliminar. Acta Geologica Hispánica 11: 57–64.

Cuenca-Bescós, G., A. Badiola, J. I. Canudo, J. M. Gasca, and M. Moreno-Azanza. 2011. New dryolestidan mammal from the Hauterivian–Barremian transition of the Iberian Peninsula. Acta Palaeontologica Polonica 56: 257–267.

Henkel, S., and B. Krebs. 1969. Zwei Säugetier-Unterkiefer aus der Unteren Kreide von Uña (Prov. Cuenca, Spanien). Neues Jahrbuch fur Geologie und Paläontologie, Monatshefte 8: 449–463.

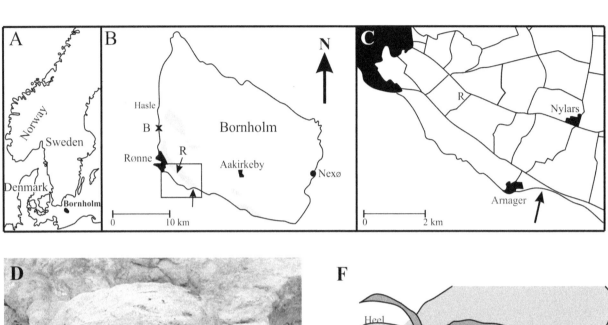

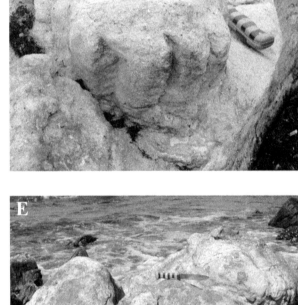

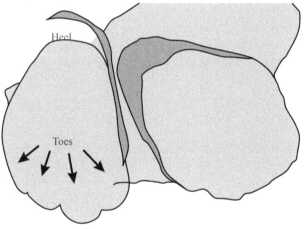

Danish Dinosaurs: A Review

24

Niels Bonde

In Denmark, continental deposits with dinosaur remains are only exposed on the island of Bornholm in the Baltic Sea. These fossiliferous formations are Aalenian (Middle Jurassic) and Berriasian (Early Cretaceous) in age. The Middle Jurassic Bagå Formation has produced large dinosaur footprints, tentatively attributed to stegosaurian-like thyreophorans and large sauropods in a dark clay from a swamp environment. Dinosaur footprints are also preserved as black mud infillings in light sand in the Rabekke Formation (Lower Berriasian) near Arnager. Associated burrows supposed to be dipnoan estivation cocoons are unlikely to have been made by such fishes. A microvertebrate assemblage in a sandy lens in the Rabekke Formation is dominated by small crocodile remains and has also yielded fish, amphibian, and small reptile fossils, a single multituberculate tooth, and tiny theropod teeth. The latter have been referred to at least three different taxa of dromaeosaurids, and one possibly to birds. The base of the Jydegaard Formation (Late Berriasian) at Robbedale has produced three dinosaurian tooth crowns. Two of these teeth have been described as a new moderately large dromaeosaurine, *Dromaeosauroides bornholmensis* Christiansen and Bonde, 2003. The third tooth belongs to a juvenile sauropod, maybe a titanosauriform. Critique of the naming and determination of the last two taxa is countered, a stratigraphic misunderstanding concerning the Robbedale fauna is cleared up, and the possible biogeographic affinities of the continental Berriasian faunas from Bornholm near the west coast of the large Scandinavian–Russian Continent are discussed, stressing possible Asian connections.

Dinosaurs in Denmark are unlikely to be found in mainland (western) Denmark, where the only exposed pre-Tertiary sediments are Maastrichtian chalks deposited in the Danish–Polish–North German Basin at some considerable distance from the coastline. The only tetrapods found are mosasaurian teeth (Bonde, 1997; Lindgren and Jagt, 2005; Bonde et al., 2008) and a turtle fragment (Karl and Lindow, 2009), and even fish fragments, including shark teeth (see Adolfssen, 2010), are rather uncommon. Slightly older deposits in Scania, south Sweden (mainly Campanian), are much richer in mosasaurs, plesiosaurs, and turtles (Persson, 1959; Lindgren, 2004a, 2004b, 2004d). Even dinosaur teeth are found in the Campanian of northeast Scania, with the first European neoceratopsian (Lindgren et al., 2007) published at about the same time as it was realized that the iguanodont *Craspedodon* from Belgium was also a first neoceratopsian (Godefroit and Lambert, 2007). However, Persson's supposed large theropod teeth from the Early Campanian on the Isle of Ivö belong to protosphyraenid fishes.

24.1. Dinosaur sites on Bornholm and Jurassic footprints. A, Bornholm in the Baltic Sea. B, Dinosaur localities; gray shading indicates the Nyker Group with Rabekke, Robbedale, and Jydegaard formations. B, Bagaa pit S. of Hasle; R, Robbedale east of Rønne; lower arrows, Arnager Bay east of Arnager. C, Closer at the Berriasian sites, road map. D, DK 426 mold of thyreophoran/stegosaurian (?) footprint upside down. E, F, DK 427 large sauropod footprints (molds) showing four toes, and their interpretation. From Bonde et al. (2008). A–C, Modified from Rees et al. (2005); D, E, photos taken west of Bagaa pit by J. Milan; his knife has a 10-cm-long handle.

Introduction: The Mesozoic of Bornholm

The geology of the Isle of Bornholm, the easternmost part of Denmark in the Baltic Sea, is completely different from the west because only pre-Maastrichtian sediments are exposed (Fig. 24.1A–C), including Upper Cretaceous marine deposits and mainly continental and lagoonal sediments from Early Cretaceous, Jurassic, and Late Triassic (Gravesen et al., 1982; Gravesen, 1996). There are few indications of marine horizons in this series. Shark teeth (Rees, 1998) and plesiosaurian (Bonde, 1993a; Rees and Bonde, 1999; Milan and Bonde, 2001; Bonde et al., 2008) remains are common finds in the marine Liassic Hasle Sandstone; a possible ichthyosaurian vertebral fragment) was also discovered in this formation. The Berriasian Robbedale Formation is a shallow marine deposit that contains many crustacean burrows, whereas the overlying Jydegaard Formation is brackish to fresh (Surlyk, 2006, figs. 9.32–35; Noe-Nygaard and Surlyk, 1988; Gravesen et al., 1982; Bonde, 2004; Bonde et al., 2008).

The neighboring Scania is, like Bornholm, situated in the northwest–southeast-striking Tornquist Zone (or Fenno-Scandian Border Zone) of tectonic instability between the Scandinavian–Russian Continent and the wider North Sea Basin, which in general sinks in (Larsen, 2006). Thus Bornholm and Scania are made up of similar geology (Jensen, 2002), stratigraphy, and faunas (e.g., Rees, 2001). When fossils of one sort are found in one of these areas, we expect them to be found in the other area too. Footprints and a few bones of dinosaurs found in the Late Triassic–Rhaetian of northwest Scania (Bölau, 1952, 1954; Ahlberg and Siverson, 1991; Gierlinski and Ahlberg, 1994) give rise to hopes of finding something similar in the small exposures of Late Triassic–earliest Jurassic in Bornholm (see Gravesen, 1996). The dinosaurs and other terrestrial and aquatic vertebrates found in the Berriasian of Bornholm makes it likely that similar faunas can be found in the near-contemporaneous deposits in south Scania (e.g., in Fyledalen; see Rees, 2001).

Institutional abbreviations. MGUH, Geological Museum/Statens Naturhistoriske Museum of Copenhagen University, Copenhagen, Denmark; DK, *Danekræ*, Geological Museum, Copenhagen, Denmark.

Middle Jurassic and Early Cretaceous Dinosaur Footprints from the Isle of Bornholm

Milan (2004) described the first dinosaur footprints from Denmark in the vicinity of the abandoned and water-filled clay pit at Bagå, south of Hasle, west Bornholm. The molds of the dinosaur footprints can be observed on the under surface of approximately 1-m-thick whitish sandstones (Milan and Bromley, 2005) that had been thrown out from the pit onto the beach as useless for tile production. These sandstones are included in the Middle Jurassic Bagå Formation (Aalenian–Bajocian; see Gravesen et al., 1982; Koppelhus and Nielsen, 1994; Surlyk, 2006). Those thick sandstones are well represented in this area and are interbedded with dark clay and thin coal beds; they are usually full of large plant debris but devoid of bones. They must be the result of catastrophic swamp flooding, and footprints were never observed during the tile production period because the beds are inclined about 20–30 degrees and were never properly exposed on the nearly horizontal floor of the pit, which is now filled with water.

Milan (2004) and Milan and Bromley (2005) identified these footprints as belonging to two different kinds of dinosaurs. A moderate-sized (25 cm in diameter and over 10 cm deep) footprint is characterized by the presence of five digits (Fig. 24.1D) and was tentatively identified as the manus of a fairly big stegosaurian-like thyreophoran. The two sauropod prints are large, about 70 cm in length, and quite deep, and one of them shows rounded imprints of four toes (Fig. 24.1E–F). They were identified as the prints of the posterior feet of a large sauropod, estimated to be about 20 m in length (Milan and Bromley, 2005). Both blocks are on exhibit in the event and exhibition center, NaturBornholm, at Åkirkeby, as part of the exhibit of the geological history of Bornholm.

Such finds at exactly this locality were predicted a decade before (Bonde, 1993b). Both blocks have been valuated as *Danekræ* (Bonde et al., 2008), and therefore they have official status as unique Danish fossils,[1] so it is a shame that they are not figured or even mentioned in a review of Danish Mesozoic geology (Surlyk, 2006).

Dinosaur footprints from the Rabekke Formation (Early Berriasian; Petersen et al., 1996) were discovered in a nearly vertical section along a small coastal cliff on the south coast, approximately 1 km east of Arnager. They form dark clay infilling imprints in the underlying light-colored sand (Fig. 24.2A; Surlyk et al., 2008). Surlyk first recognized these imprints as possible footprints, and his group subsequently cleaned a few of them to make it evident (Fig. 24.2C; Surlyk et al., 2008, fig. 6). Some of these footprints apparently belong to sauropods (Fig. 24.2B–C). The width of the footprints is usually between 20 and 45 cm, but the largest measure over 1 m in width. The depth of the footprints measures between 15 and 30 cm.

In association with the dinosaur footprints, Surlyk et al. (2008) also described some smaller subcylindrical burrows filled by both the dark clay and lighter sand, which they interpreted as estivation burrows, cocoons made by lungfishes (Fig. 24.2B). These burrows are quite narrow, measuring between 1 and 9 cm in width and up to 45 cm in length. However, no lungfish remains have been discovered in the Rabekke Formation so far. Moreover, it appears extremely unlikely that dipnoans should have survived into the Cretaceous in northern Europe. Indeed, the last occurrence of lungfishes in Europe date from the Middle Jurassic (Schultze, 1996), and dipnoan tooth plates, which preserve well, were never reported from the rich European Purbeck–Wealden faunas. With the exception of a few discoveries in North America, Cretaceous dipnoans were restricted to the southern continents (Schultze, 1992, 1996). Therefore, it is much more likely that these burrows were dug by other organisms such as crayfishes or other crustaceans, a possibility not favored by Surlyk et al. (2008).

In 2002, J. Rees from Lund University, assisted by a local amateur group called the Fossil Group,[2] dug out a sandy lens on the beach close to the section mentioned above approximately 1 km east of the Arnager village. This is within the type locality of the (upper) Skyttegaard Member of the Early Berriasian Rabekke Formation (Gravesen et al., 1982; Lindgren et al., 2008). In 2003, J. Lindgren screen washed sediments from this lens in Lund and collected vertebrate microremains. The vertebrate assemblage is dominated by the small crocodile *Bernissartia*, but it also includes numerous teeth, scales, and bone fragments of *Lepidotes*, pycnodonts, and other

Theropod Teeth from the Rabekke Formation (Early Berriasian)

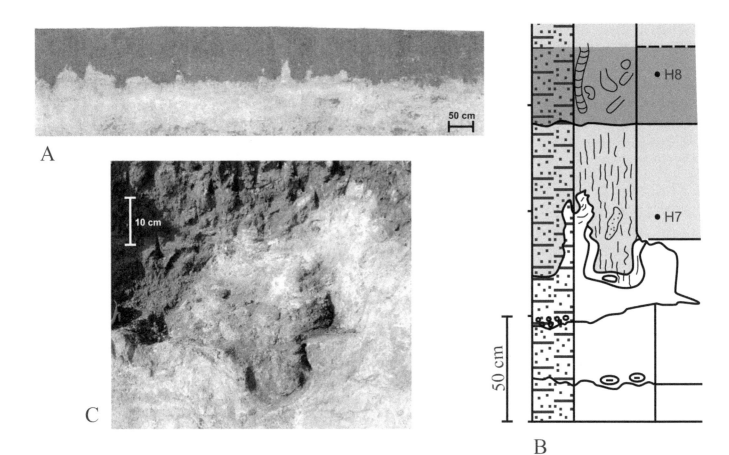

24.2. Early Berriasian footprints. A, Coastal cliff east of Arnager showing sections of dark clay penetrating into footprints in light silt and sand. Photograph by N. Bonde (color photograph in Surlyk et al., 2008). B, Schematic drawing of footprint section and a supposed dipnoan cocoon based on Surlyk et al. (2008) with outlines emphasized. C, One footprint infill in 3D with sand cleared away.

Photograph by J. Milan.

fishes (Lindgren et al., 2004), together with bones from amphibians (Rees et al., 2005) and turtles. The only Mesozoic mammal from Scandinavia, one multituberculate tooth, cf. *Sunnyodon* (Fig. 24.3Y), was also described from this fossiliferous lens (Lindgren et al., 2004). Lindgren et al. (2008) compared this assemblage broadly with the Late Jurassic fauna of the Guimarota coal mine in Portugal (Rauhut, 2000) and the English Purbeck and Wealden faunas as reviewed by Milner and Batten (2002) and Martill and Naish (2001). However, faunules dominated by crocodilians seem to be quite rare in the English Purbeck and Wealden (Salisbury, 2002; Freeman, 1975). These differences in the relative abundance of the represented taxa reflect either paleoecological differences or taphonomic processes that remain to be explained; in Guimarota, the abundance of crocodilian remains is stunning (Krebs and Schwarz, 2000).

From this microvertebrate assemblage in the Rabekke Formation, Lindgren et al. (2008) described 13 small theropod tooth crowns. They are between 1.5 and 5 mm high. The roots were never preserved, so they are probably shed teeth. Some of them show natural wear facets. Twelve of them are identified as Dromaeosauridae incertae sedis. Among them, three different morphotypes can be recognized. Two teeth (Fig. 24.3A–C and D–F) are tentatively attributed to velociraptorines (sensu Currie et al., 1990), and one tooth (Fig. 24.3O–R) perhaps belongs to a dromaeosaurine (sensu Currie et al., 1990). Lindgren et al. (2008) suggested that the best preserved of the remaining teeth (Fig. 24.3I–N, S) belonged to more basal dromaeosaurids and compared them specifically with *Nuthetes* (see

Milner, 2002), from the the English Purbeck, accepting her suggestion that *Nuthetes* is a basal velociraptorine (not accepted here; see Christiansen and Bonde, 2003, 294, and below). The last tooth (Fig. 24.3T–X) is characterized by a somewhat different type of serrations along the distal carina, with apicodistally directed oblique denticles of different shape and width. Lindgren et al. (2008, 253) referred the last tooth to as Maniraptora incertae sedis and finally suggested the possibility that it "may not be a non-avian theropod"—a crooked, or perhaps sophisticated, way of saying that it could belong to a bird. Descriptions follow the methods developed by P. J. Currie in many articles (Currie et al., 1990; Farlow et al., 1991; Sankey et al., 2002), supplemented by Rauhut and Werner's (1995) way of comparing the serrations on the carinae, and Smith et al. (2005).

The oldest potential dromaeosaurid fossils described so far are isolated teeth discovered in the Late Jurassic of Portugal (cf. "*Dromaeosaurus*"—a genus this long-lived is clearly unlikely—and the many velociraptorine teeth from Guimarota; Zinke, 1998). The diversity of dromaeosaurid-like teeth in the Rabekke Formation of the Isle of Bornholm suggests that these theropods were well diversified in Europe from the Early Berriasian. Other potential dromaeosaurid-like remains from the Berriasian of Europe include numerous isolated teeth collected at Cherves-de-Cognac, southwest France (Pouech et al., 2006), with small teeth of dromaeosaurs (one figured as *Nutethes* sp.) and birds (one figured as Archaeopterygidae indet.; both crowns under 3 mm), while teeth referred to as *Nuthetes destructor* from the contemporary Purbeck Limestone Formation of southern England (Milner, 2002) are probably not all dromaeosaurs. In the slightly younger Wealden deposits of the U.K., there are dromaeosaur/velociraptorine teeth (Sweetman, 2004), but larger.

Not all small dromaosaurs could fly (Zheng et al., 2009), and there is a certain variation among the teeth of these small microraptorines that is neither well characterized nor described in detail (apart from in *Bambiraptor* [Burnham, 2004], approximately 1 m long with slightly larger teeth), making precise determinations of such small teeth quite difficult. This counts for the many Chinese finds (Xu et al., 1999, 2000, 2003; P. Godefroit, pers. comm., 2010) of feathered and several possibly flying microraptorines (see Chatterjee and Templin, 2007). Nonetheless, the Bornholmian diversity, with perhaps three taxa of tiny dromaeosaurs, makes it quite likely that among them were also flying or gliding animals (Hansen and Bonde, 2006a, 2006b). This may also be true for other European localities of the Late Jurassic to Early Cretaceous, but so far, the teeth described are mostly somewhat larger (Canudo et al., 1997; Zinke, 1998; Rauhut, 2000, 2002; Naish et al., 2001; Milner, 2002; Sweetman, 2004), and as such less obviously from airborne animals; but compare the above-mentioned small teeth from Cherves-de-Cognac (Pouech et al., 2006), and there were also small dromaeosaurs, with 3–5-mm teeth, in that area as late as the Cenomanian (Vullo et al., 2007). The last find of these small microraptorines is much younger and recorded as a new genus and species from the Late Cretaceous of Alberta (Longrich and Currie, 2009, postcranial material). They show that this form and at least some of these small dromaeosaurids, of which some were flying, do in fact belong to a proper clade, Microraptorinae.

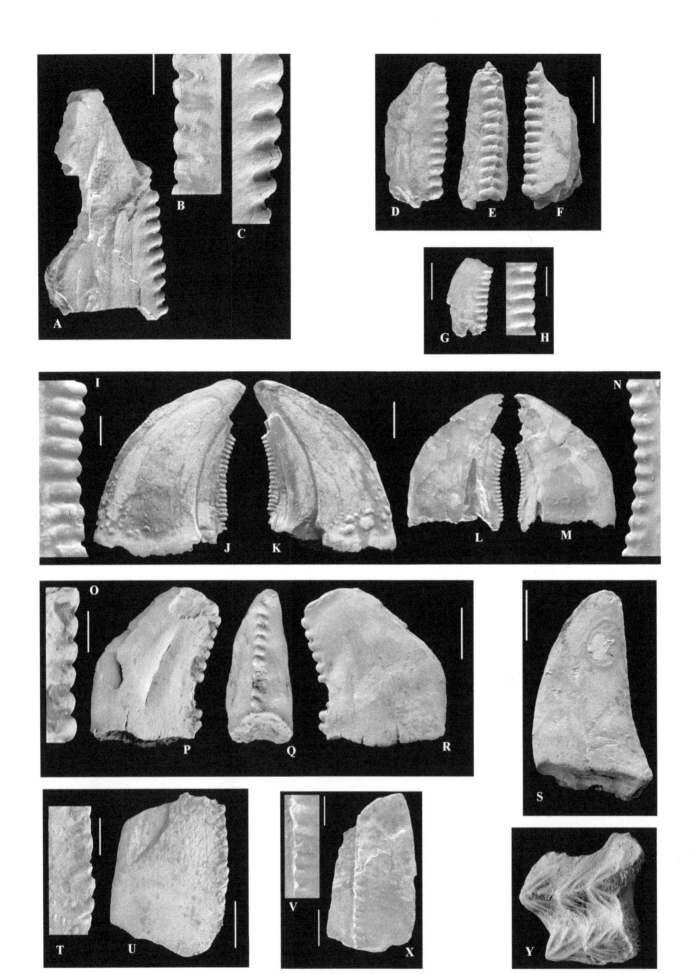

This diversity pattern should be carefully mapped in the future because of the principally different biogeographic situations of the Bornholmian(–Scanian) dinosaurs at the southwest-facing coast of the huge Scandinavian–Russian Continent and the other European localities on much smaller, mostly isolated islands of a large archipelago. Furthermore the Scandinavian–Russian Continent has sometimes been directly connected with East Asia, when the north–south-running Turgai Strait was dry, in the Middle Jurassic and earlier, but the Scandinavian–Russian Continent was probably nearly always separated from the European islands by sometimes narrow waterways (see maps in Smith and Briden, 1977; Smith et al., 1994; Blakey, 2001—or alternatively Howarth, 1981, partly based on Owen's 1981 expanding earth theory; see below).

24.3. Early Berriasian dromaeosaurid and bird (?) teeth (see Lindgren et al., 2008). A–C, MGUH 28406, "velociraptorine." D–F, MGUH 28407, "velociraptorine" (?). G– H, MGUH 28408. I–K and L–M, MGUH 28403 and 28404, primitive dromaeosaurids ("dromaeosaurines"?). O–R, MGUH 28405, dromaeosaurine (juvenile?). S, MGUH 28409, dromaeosaurid incertae sedis. T–X, MGUH 28410a, b, fragments of bird/avialan (?) tooth crown. Scales: A, D–G, P–S, U, X, 0.5 mm; B, C, H, T, V, 0.1 mm; I, N, 0.5 mm; J–M, 1 mm. Y, *Sunnyodon* sp., occlusal and mesial view (mesial at right) of an upper premolar, 1.2 by 0.9 mm; from Bonde et al. (2008; see Lindgren et al., 2004).

Dinosaur Teeth from the Jydegaard Formation (Early Berriasian)

The first dinosaur ever found in Denmark is a dromaeosaurid-like tooth crown (MGUH 27218/DK 315), named *Dromaeosauroides bornholmensis* by Christiansen and Bonde (2003), discovered in the year 2000 in the conservation site at the Robbedale locality 5 km east of Rønne (Fig. 24.1B,C; Fig. 24.4A). In 2002, a sauropod tooth was discovered in the same locality, but this fossil is still in the private collection of R. Benthien. In 2008, another isolated tooth crown (DK 559), which can be referred to *Dromaeosauroides bornholmensis*, was discovered by a nature guide, J. Kofoed, in sand from the same locality, and was brought to NaturBornholm. The fossiliferous sediments at the conservation site are exposed in a 2–3-m-high section (Fig. 24.4B) in a small corner of a large abandoned sand pit at Robbedale, now almost entirely overgrown and with a large lake at the bottom. The vertebrate fossils were found at the base of the Jydegaard Formation (see Gravesen et al., 1982; Gravesen, 1996; Bonde, 2004) in rusty, sideritic claystones, locally named the *Neomiodon* bed, and in the overlying 2–3 m of sand (Fig. 24.4B). The age of these sediments is Late Berriasian, probably corresponding to the middle part of the German Wealden (Middle Bückeberg Formation, W 3–W 4, according to Piasecki, 1984), to the Durlston Beds, Upper Purbeck (Milner and Batten, 2002), and to the lower Ashdown Formation at the base of the Hastings Beds of the type Wealden in U.K. (Allen and Wimbledon, 1991; Bonde, 2004). The remaining higher part of Jydegaard Formation might be of Valanginian age (Gravesen et al., 1982). The vertebrates could not have been collected from the few decimeters of sand below the *Neomiodon* bed, as claimed by Rees (2000) when describing a tiny lacertilian dentary, because these layers have not been exposed since late in the 1980s. Lindgren et al. (2008, fig. 2) indicated that the Robbedale vertebrate fauna was correlated to the Robbedale Formation that immediately underlies the Jydegaard Formation. However, the Robbedale Formation has not produced any vertebrate remain so far.

The fauna of the Jydegaard Formation includes bivalves, sharks (Rees, 2001), and fishes in clay sediments that probably were deposited in a lagoon environment (Noe-Nygaard et al., 1987; Bonde, 2004), whereas lacertilians (Rees, 2000) and dinosaurs (Bonde and Christiansen, 2003) were found in sands that probably represent land and beach deposits. Crocodile and turtle fossils are found in both environment, while clays with packed freshwater snails are from shallow, drying-out lakes on the beach on the back side of a sandy barrier protecting the lagoon from the southern sea (Noe-Nygaard and Surlyk, 1988), perhaps similar to the Florida keys or the southwest coast of Jutland today. There is some confusion and omissions about the fishes in Surlyk et al. (2008) because the relevant reference is not cited.

MGUH 27218/DK 315 (Fig. 24.4C) is the holotype of *Dromaeosauroides bornholmensis* Christiansen and Bonde, 2003. The overall morphology of the crown, its fairly wide and less bladelike mediolateral cross sections, and crown curvature bear substantial resemblance to maxillary and dentary teeth of the Late Cretaceous *Dromaeosaurus* from North America (Currie, 1995). Significantly, both the mesial and distal carinae are turned slightly medially, characteristic of *Dromaeosaurus* and many dromaeosaurids, although the distal carina is less medially directed in the specimen from Bornholm. The preserved crown height is 21.7 mm, the fore and aft basal length is 9.7 mm, and the basal mediolateral width is 6.6 mm. This is more than 25% larger than *Dromaeosaurus*, indicating an overall length of the animal of around 3 m. Both carinae are finely serrated. The apical part of the mesial carina bears 30.7 denticles per 5 mm (6.1/1 mm) and the distal carina bears 30.1 denticles per 5 mm (6.05/1 mm). Thus, the denticle index is 1.01, nearly identical to *Dromaeosaurus*, but the individual denticles are distinctly smaller, despite the larger size of the tooth: *Dromaeosaurus* only has 13–20 denticles per 5 mm (Rauhut and Werner, 1995). For that reason, it has been decided that the dromaeosaur from Bornholm most likely represents a new genus and species. DK 559 (Fig. 24.4D) is smaller (15 mm) than MGUH 27218 but has the same morphological characteristics. It is therefore attributed to the same dromaeosaurine-like theropod.

Both teeth have been valuated as *Danekræ* (Bonde et al., 2008), and the type figures prominently in Danish popular books (Christiansen, 2003; Benthien, 2003) and professional papers (Bonde, 2004), and also in a recent book that reviews Denmark's geology (Larsen, 2006). Surlyk (2006) figures it, misspells its name, and provides incorrect information on its relation to the smaller teeth from Rabekke Formation described above.

It is of some biogeographic interest that among the nearly 30 isolated teeth of *Nuthetes* described by Milner (2002), there are three tooth crowns much larger (15–18 mm) than the rest, which are all less than 1 cm, as in the type jaw. These three tooth crowns do not belong to *Nuthetes* but are in fact similar to the two Bornholmian teeth, probably conspecific, and are true dromaeosaurs, contrary to *Nuthetes* proper (see below; Bonde, in prep.).

Lindgren et al. (2004, 2008) and Weishampel et al. (2004) consider that *Dromaeosauroides bornholmensis* is indeterminate. Nevertheless, the dromaeosaurine-like teeth from Robbedale clearly differ from those described in other Early Cretaceous dromaeosaurids. Teeth of *Nuthetes* from the Berriasian of southern England do not possess the dental synapomorphies of dromaeosaurids (Christiansen and Bonde, 2003; Bonde and Christiansen, 2003), and it cannot be excluded that this taxon is a more generalized tetanuran/neotheropod ("Megalosaurian" as in Swinton, 1934). Sweetman (2004) described six velociraptorine-like teeth from the Wealden of the Isle of Wight. These teeth differ from the Robbedale teeth in the inclination of the denticles and in the rather large difference in size between the mesial and distal denticles (denticle index about 1.4–1.6). The Robbedale teeth are much larger than those in the microraptorine dromaeosaurids

(*Sinornithosaurus, Microraptor, Tianyuraptor*) from the Yixian and Jiufotang formations of western Liaoning province, China (Xu et al., 1999, 2000, 2003; Xu and Wu, 2001). Maxillary and dentary teeth are also clearly different in *Deinonychus antirrhopus*, from the late Aptian–middle Albian of the western United States; the crowns are more compressed linguolabially, both the mesial and distal carinae are perfectly straight, and the denticles are better developed on the distal carina than on the mesial one (Ostrom, 1969).

It can therefore be concluded that the theropod teeth from Robbedale possess the dental synapomorphies of dromaeosaurids and closely resemble those of *Dromaeosaurus*. However, significant differences exist with the teeth of *Dromaeosaurus* and those of other Early Cretaceous dromaeosaurids. For those reasons, *Dromaeosauroides bornholmensis* must be retained as a valid taxon, and it cannot be regarded as indeterminate, as by Lindgren et al. (2004, 2008) and Weishampel et al. (2004), who did not bring the name but provided misinformation about the locality, strata, and age. The naming of this tooth is no different from other descriptions of taxa based on isolated teeth (e.g., Lindgren, 2004c; Lindgren et al., 2004; Rees, 1998) or, say, otoliths.

The sauropod tooth from Robbedale is an approximately 15-mm-high crown with a slightly odd shape (Fig. 24.4E). The crown is semicylindrical with a large pulp cavity and with rounded edges that show remnants of weak serrations. It is much higher than mesodistally long (over 2.3 times) and labiolingually wide (3.6 times). The mesial or distal side is worn and slightly concave, and there is a long and deep natural wear facet on the opposite apical corner. Therefore, two teeth from the opposite jaw produced those wear facets, and one of them was long enough to reach the base of the crown. Among dinosaurs, such kinds of wear facets can be observed only in sauropods with well-spaced and rather slim teeth (often called pencillike teeth), such as diplodocoids and titanosaurs. However, serrated carinae are absent in diplodocoids (Upchurch, 1994; Wilson and Sereno, 1998). Therefore, a titanosaurian (sensu lato) affinity for the Robbedale sauropod is more likely because basal titanosauriforms have serrated carinae, and terminal wear facets on titanosaur teeth are usually produced by tooth-to tooth contact (Calvo, 1994; Wilson and Sereno, 1998). This tooth also resembles the smallest so-called brachiosaurid tooth figured from the Late Jurassic Guimarota fauna in Portugal by Rauhut (2000, fig. 11.4), and it shows some similarity to one of the teeth, 18 mm high, figured by Vullo et al. (2007) as a brachiosaurid or basal titanosaur from the Cenomanian of southwest France. Although taxon identification based on dental morphology is highly speculative in sauropod dinosaurs, we tentatively refer this tooth to a Titanosauriformes. This is contra the unsubstantiated claims (by Lindgren et al., 2004; Rees et al., 2005; and hinted at by Lindgren et al., 2008, 254) that the tooth is too worn to be determined even as a sauropod. This is even more peculiar because one of these coauthors, P. J. Currie, was the first to identify this tooth as that of a sauropod (see Bonde and Christiansen, 2003, 24).

One should not, however, expect to find larger dinosaur bones in the Robbedale quarry; if they were there, they would have been found long ago, during the quarry's industrial exploitation.

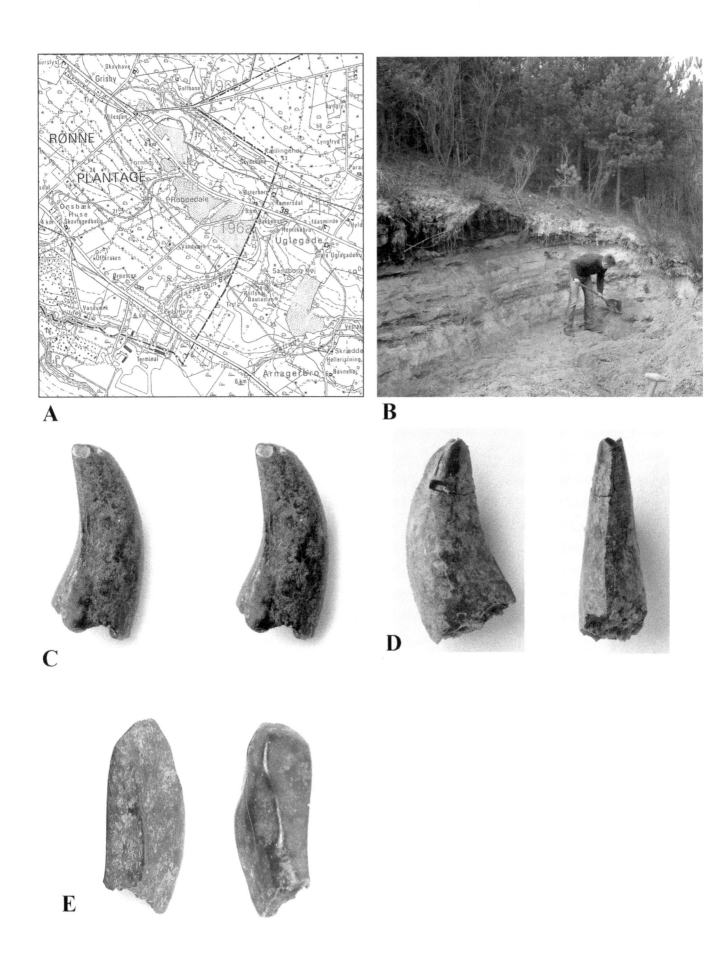

It is well known that in the European archipelago are classical examples of insular dwarfism among herbivorous dinosaurs (Weishampel et al., 1991; Benton et al., 2010)—for example in hadrosaurs—and several of the Romanian dinosaurs are rather small (Le Loeuff, 2005; Grigorescu, 2010). This also concerns the titanosaurian sauropods (Stein et al., 2010), but there is no reason to believe that such should be the case for the small sauropod described above from Robbedale, Bornholm. This is probably a juvenile (perhaps about 5 m long) living on an enormous continental island without insular limitations, and therefore presumably growing to a more typical size. Neither does dwarfism count for all insular sauropods (cf. Naish and Martill, 2001).

Dwarfism apparently did not occur in the theropods on the European islands, either in the Franco-Iberian region (Canudo et al., 1997; Allain and Taquet, 2000; Chantasit and Buffetaut, 2009), in northwest Europe (Naish et al., 2001; Sweetman, 2004, and *Dromaeosauroides* from the U.K., mentioned above), or in Central Europe (Grigorescu, 2010; Ősi et al., 2010a; no teeth are unfortunately preserved in the most complete European dromaeosaurid recently found in Romania; Csiki et al., 2010).

It may seem odd that the single sauropod tooth from Bornholm does not really match any of the many sauropod teeth described from the European Early Cretaceous (e.g., Naish and Martill, 2001), especially as the large dromaeosaurid teeth seem to have conspecifics in the Purbeckian of the U.K. It is, however, not necessarily with the remaining European archipelago that the biogeographical affinities should be found for Early Cretaceous fossils from the large Scandinavian–Russian Continent. During most of the Jurassic–Cretaceous (and Paleogene), the wide Turgai Strait (and Obik Sea) in the Uralian region and further east separated Scandinavian–Russian Continent from Asia. But for some periods, including the entire Early Jurassic (and earlier), this strait did not exist, so direct terrestrial connection was possible between the Scandinavian–Russian Continent and eastern Asia. This condition remained during the Middle Jurassic (Howarth, 1981; Smith et al., 1994; Blakey, 2001). On the contrary, published palaeogeography of the Jurassic–Cretaceous nearly always have the Scandinavian–Russian Continent separated from the small European islands by water ways. Sometimes these were quite narrow, as in the Berriasian, so there were possibilities for dispersal between the Scandinavian–Russian Continent and the Middle European Island (and further to the U.K., as witnessed by map 19 of Smith et al., 1994, and by *Dromaeosauroides*; see Bonde, in prep.), but only later and ending in the Aptian was there probably a narrow land bridge between the Scandinavian–Russian Continent and the central European island.

It is probably significant that Rees et al. (2005), in their search for relatives to the fragmentary amphibian fossils from the Rabekke Formation, could only find one vertebra that looks similar to one described by Averianov and Voronkevich (2002) from west Siberia (on the other side of the Turgai Strait). This late Early Cretaceous, west Siberian locality, Shestakovo, is regarded as Early Albian in age, and it has obvious affinities to the East Asiatic faunas. *Psittacosaurus* has its most western occurrence

24.4. Late Berriasian dinosaurs from Robbedale. A, C. Nielsen's sandpit. B, Section in basal Jydegaard Formation; Claus Bonde is 185 cm, and the *Neomiodon* bed is 30–40 cm below his feet; thick, dark layer at the top with *Viviparus* mass mortality. C, *Dromaeosauroides bornholmensis* holotype, 22-mm crown, MGUH 27218, DK 315, stereo. D, *D. bornholmensis*, second specimen, 15-mm crown, DK 559; labial and distal. E, Juvenile sauropod tooth, 15-mm crown; ? labial and ? labial views (R. Benthien collection).

here (Averianov et al., 2006), and there are other dinosaurs from this site. If any Early Cretaceous locality in Asia (see Averianov and Skutschas, 2000, for a review) should show specific European affinities, it would probably be this one. Faunal similarities might well be more obvious with the Scandinavian–Russian Continent than with the rest of Europe.

The Early Cretaceous localities between west Siberia and East Asia are located in the Transbaikal region (Averianov et al., 2003; Averianov and Shutschas, 2009) and are Barremian–Albian in age. They comprise both titanosaurs and dromaeosaurs and other theropods, with the top predator being "*Prodeinodon*" with figured 5-cm tooth crowns (there is no reason to believe that these large teeth belong to *Nutethes*, as claimed by Averianov and Shutschas, 2009, 368). Dromaeosaurids are small (Averianov and Shutschas, 2009, fig. 2), with crown height of between 7 and 9 mm. Other crowns, called "Theropoda indet.," might well belong to small dromaeosaurids; they are only 2–3 mm high, not unlike some of the Bornholmian teeth. (There is a real mess concerning scales in fig. 2 and the measurements in table 3 in Averianov and Shutschas, 2009, and some numbers must have been mixed up too.) The second tooth identified as "*Prodeinodon*" (Averianov and Shutschas, 2009, fig. 2E–G, with wrong scale) is 13 mm high and in my opinion could also belong to a dromaeosaurid. None of these dromaeosaurs are typical velociraptorines despite serration indices from 1.2–1.5, because the denticles more closely resemble those in dromaeosaurines. None of the titansaurian teeth resemble the Bornholmian specimen, but they are also likely to be from more mature individuals.

Moving forward to the Late Cretaceous, it has lately been realized that there are clear faunal affinities between Asia and both the Scandinavian–Russian Continent and the European archipelago. Neoceratopsians were mentioned in the introduction (Lindgren et al., 2007; Godefroit and Lambert, 2007), and lately (Ősi et al., 2010b) a leptoceratopsian from Hungary with Asian affinities. The biogeographical affinities of all Late Cretaceous tetrapods from Europe were analyzed by Pereda-Superbiola (2009), who found mixed affinities for the European islands (but did not separate the Scandinavian–Russian Continent). Some tetrapods had Asian affinities. The last direct connection with Asia via the Scandinavian–Russian Continent occurred during the Aptian–Albian (Smith et al., 1994, maps 16, 17). A few taxa have affinities with North America and some with Gondwana (North Africa). The latter affinities were especially recognized by Vullo et al. (2005) for the Cenomanian localities and by Buffetaut (1989) for the latest Cretaceous faunas in southwest France.

However, if the maps by Smith et al. (1994, maps 10–15) provide a correct overall picture of the paleogeography, the European islands were separated from the Scandinavian–Russian Continent and isolated from each other during the Late Cretaceous, with a rather wide Tethys Sea between Europe and Africa. It implied some serious island hopping to traverse that barrier (see suggestions and maps in Ősi et al., 2010a), and typically the early Gondwanan affinities seem limited to aquatic forms (hybodonts, ziphodont crocodiles; Vullo et al., 2005). However, the latest Cretaceous Gondwanan element in Europe is an abelisaurian theropod (Le Loeuff and Buffetaut, 1991), which would have an easier route on a smaller Earth (as in Owen, 1981, and Howarth, 1981).

Concluding this review of biogeographic terrestrial relations of Europe during the Early and Late Cretaceous (see also Milner et al., 2000), it appears that Asian affinities in the earliest Cretaceous would have their roots in connections between the Scandinavian–Russian Continent and Central Europe and the Asian continent in Bajocian and earlier, when the Turgai Strait did not exist. North American affinities could be based on a connection via an early Thulean land bridge (unlikely as it may seem) during the earliest Cretaceous (Smith et al., 1994, map 19), when the strait between the Central European Island and the Scandinavian–Russian Continent was also quite narrow. The Late Cretaceous Asian affinities could be based on Aptian land connections between the Scandinavian–Russian Continent and both Asia and Central European islands (Smith et al., 1994, map 17).

Conclusions and Future Research

This points to another surprising aspect: apparently the Bornholmian–Scanian area is the only window for the terrestrial faunas of the Scandinavian–Russian Continent during the Early Cretaceous (e.g., no dinosaurs of that age are mentioned from this huge area by Weishampel et al., 2004, and Benton et al., 2000). This may confer a certain significance to the Bornholmian–Scanian area, especially in future biogeographic investigations, as we can expect to find more dinosaur footprints in the water outside the Bagå pit (see Bonde, 1993b), and with a rigorous search for microvertebrates, it is likely that more mammalian teeth, microdinosaurs including birds, perhaps pterosaurs, and certainly more reptiles, amphibians, and fishes will turn up. It will then be possible to attempt more serious comparisons between these Bornholmian–Scanian area faunas and those from Asia and the European archipelago. My guess would be more Asian affinities.

Acknowledgments

The Fossil Group, with Regitze Benthien and Tom Ipsen, among others, is thanked for many years of collegial work on Bornholmian localities, especially the Robbedale pit and Hasle Cliff, and their fossils. I also thank my son, Claus; my wife, Jane Richter; and Jane's family for many years of joint collection of such fossils. Thanks to Eliza Jarl Estrup for finding the first tooth and her continuous interest, and to NaturBornholm for its never-failing interest in promoting and exhibiting Bornholmian fossils. I owe my institute thanks for the use of field station Gravgærde (now abandoned) and for some financial support for trips to London in connection with studies of *Nuthetes* (and plesiosaurs and *Archaeopteryx*). Dr. Angela Milner and S. Chapman, London, are thanked for their hospitality and permission to study material in their care. Thanks to J. Lindgren, Lund, Jesper Milan, Geological Museum, Faxe, and Sten L. Jakobsen, Geological Museum, Copenhagen, for several photographs. Extra thanks to go Sten for combining the figures for the plates and for his interest in and collaboration concerning the *Danekræ*. Thanks also to Fur Museum for its financial support of some of my journeys. I extend my warmest thanks to Dr. Per Christiansen, now the zoological director of Aalborg Zoo, North Jutland, for years of collaboration on the problems of mosasaurs, dinosaurs, and *Archaeopteryx*, and for his friendship over the years. We both thank Dr. Phil Currie, Alberta, for friendship, many good discussions, and advice. Finally, thanks to Drs. P. Godefroit, O. Lambert, and E. Steurbaut and colleagues for their hospitality and for a splendid meeting in Brussels and Bernisart in 2009, and to P. Godefroit for a most informative review, several discussions, and a lot of help with references, from which I learned a lot.

Notes

1. According to the museum law of 1990, Danish fossils or other natural non-living objects of scientific or museographic importance must be presented to the state (*in casu* Geological Museum, Copenhagen—in practice often through local museums) for evaluation. If found important enough, they are declared *Danekræ* ("Danish creatures") and a reimbursement is received (Bonde et al., 2008).

2. The Fossil Group was a group of self-activating unemployed people (active from the 1990s to about 2005) led by Regitze Benthien, who successfully applied for funds in Denmark and EEC, so that a few persons could be employed for a period to maintain and clean up geological localities on Bornholm, especially the Robbedale conservation site, situated partly on R. B.'s ground. They even paid for part of J. Lindgren's postdoctoral study at the Geological Museum. They found several *Danekræ*, and R. B. published a booklet on finding the first Danish dinosaurs (Benthien, 2003). The group insisted that we call a five-day course for amateur collectors in 2000 the Hunt for the Danish Dinosaur, during which the first tooth was found by Eliza Jarl Elstrup.

References

Adolfssen, J. 2010. The end of the Mesozoic sharks from the White Chalk and the KT-boundary at Stevns Klint, Denmark; p. 17 in K. Gonzales-Rodrigues and G. Arratia (comp.), 5th International Meeting on Mesozoic Fishes, Mexico, August 1–7, 2010, abstract book and field guides. Universidad Autonoma del Estado de Hidalgo, Mexico.

Ahlberg, A., and M. Siverson. 1991. Early Jurassic dinosaur footprints in Helsingborg, southern Sweden. Geologiska Föreningen för Stockholms Förhandlingar 113: 339–340.

Allain, R., and P. Taquet. 2000. A new genus of Dromaeosauridae (Dinosauria, Theropoda) from the Upper Cretaceous of France. Journal of Vertebrate Paleontology 20: 404–407.

Allen, P., and W. A. Wimbledon. 1991. Correlation of NW European Purbeck-Wealden (nonmarine, Lower Cretaceous) as seen from the English type area. Cretaceous Research 12: 511–526.

Averianov, A., and P. P. Skutschas. 2000. A eutherian mammal from the Early Cretaceous of Russia and biostratigraphy of the Asian Early Cretaceous vertebrate assemblages. Lethaia 33: 330–340.

———. 2009. Additions to the Early Cretaceous dinosaur Fauna of Transbaikalia, eastern Russia. Proceedings of the Zoological Institute RAS 313: 663–678.

Averianov, A. O., and A. V. Voronkevich. 2002. A new crown-group salamander from the Early Cretaceous of western Siberia. Russian Journal of Herpetology 9: 209–214.

Averianov, A. O., A. Starkov, and P. P. Skutschas. 2003. Dinosaurs from the Early Cretaceous Murtoi Formation in Buryatia, eastern Russia. Journal of Vertebrate Paleontology 23: 586–594.

Averianov, A. O., A. V. Voronkevich, S. V. Leshchinskiy, and A. V. Fayngertz. 2006. A ceratopsian dinosaur, *Psittacosaurus sibiricus*, from the Early Cretaceous of West Siberia, Russia, and its phylogenetic relationships. Journal of Systematic Palaeontology 4: 359–395.

Benthien, R. 2003. Jagten på den danske dinosaur [The hunt for the Danish dinosaur]. Privately printed, Rønne, Denmark.

Benton, M. J., M. A. Shishkin, D. A. Unwin, and E. N. Kurochkin. 2000. The age of dinosaurs in Russia and Mongolia. Cambridge University Press, Cambridge, 696 pp.

Benton, M. J., Z. Csiki, D. Grigorescu, R. Redelstorff, K. Sander, K. Stein, and D. B. Weishampel. 2010. Dinosaurs and the island rule: the dwarfed dinosaurs from Hateg Island. Palaeogeography, Palaeoclimatology, Palaeoecology 293: 438–454.

Blakey, R. C. 2001. Regional paleogeographic views of earth history. University of Northern Arizona, Geological Department. http://jan.ucc.nau.edu/~rcb7/globaltext.html.

Bölau, E. 1952. Neue Fossilfunde aus dem Rhät Schones und ihre paläogeographisch-ökologische Auswartung. Geologiska Föreningen för Stockholms Förhandlingar 74: 44–50.

————. 1954. The first find of dinosaur skeletal remains in the Rhaetic–Liassic of N.W. Scania. Geologiska Föreningen för Stockholms Förhandlingar 76: 501–503.

Bonde, N. 1993a. Bornholmske svaneøgler [Plesiosaurs of Bornholm]. Varv 1993: 35–41.

————. 1993b. Bornholms fortidsøgler—om svaneøgler, krokodiller m.m. [Prehistoric reptiles of Bornholm—plesiosaurs, crocodiles etc.]. Bornholms Natur 17: 55–69.

————. 1997. En kæmpemæssig mosasaur fra Israel [A gigantic mosasaur from Israel]. Varv 1997: 26–30.

————. 2004. An Early Cretaceous (Ryazanian) fauna of "Purbeck–Wealden" type at Robbedale, Bornholm, with references to other Danish fossil vertebrates; pp. 507–528 in G. Arratia and A. Tintori (eds.), Mesozoic fishes 3. Verlag Pfeil, Munich.

Bonde, N., and P. Christiansen. 2003. New dinosaurs from Denmark. Comptes Rendus Palevol 2: 13–26.

Bonde, N., S. Andersen, N. Hald, and S. L. Jakobsen. 2008. Danekræ—Danmarks bedste fossiler [Danekrae—Denmark's best fossils]. Gyldendal, København, Denmark, 226 pp.

Buffetaut, E. 1989. Archosaurian reptiles with Gondwana affinities in the Upper Crertaceous of Europe. Terra Nova 1: 69–74.

Burnham, D. A. 2004. New information on Bambiraptor feinbergi (Theropoda; Dromaeosauridae) from the Late Cretaceous of Montana; pp. 67–111 in P. J. Currie, E. B. Koppelhus, M. A. Shugar, and J. L. Wright (eds.), Feathered dragons. Indiana University Press, Bloomington.

Calvo, J. A. 1994. Jaw mechanics in sauropod dinosaurs. Gaia 10: 183–193.

Canudo, J. I., G. Cuenca-Bescos, and J. I. Ruiz-Omenaca. 1997. Dinosaurios dromeosauridos (Saurischia: Theropoda) en el Barremiense Superior (Cretacico Inferior) de Castellote, Teruel. Geogaceta 22: 39–42.

Chantasit, P., and E. Buffetaut. 2009. New data on the Dromaeosauridae (Dinosauria: Theropoda) from the Late Cretaceous of southern France. Bulletin de la Société Géologique de France 180: 145–154.

Chatterjee, S., and R. J. Templin. 2007. Biplane wing planform and flight performance of the feathered dinosaur Microraptor gui. Proceedings of the National Academy of Sciences of the United States of America 104: 1576–1580.

Christiansen, P. 2003. Rovdinosauren fra Bornholm [The theropod from Bornholm]. Carlsen, København, Denmark, 48 pp.

Christiansen, P., and N. Bonde. 2003. The first dinosaur from Denmark. Neues Jahrbuch für Geologie und Paläontologie, Abhandlungen 227: 287–299.

Csiki, Z., M. Vermir, S. L. Brusatte, and M. A. Norell. 2010. An aberrant island-dwelling theropod dinosaur from the Late Cretaceous of Romania. Proceedings of the National Academy of Sciences of the United States of America 107: 15357–15361.

Currie, P. J. 1995. New information on the anatomy and relationships of Dromaeosaurus albertensis (Dinosauria: Theropoda). Journal of Vertebrate Paleontology 15: 576–591.

Currie, P. J., J. K. Rigby, and R. E. Sloan. 1990. Theropod teeth from the Judith River Formation of southern Alberta, Canada; pp. 107–125 in K. Carpenter and P. J. Currie (eds.), Dinosaur systematics: perspectives and approaches. Cambridge University Press, Cambridge.

Farlow, J. O., D. L. Brinkman, W. L. Abler, P. J. Currie. 1991. Size, shape and serration density of theropod dinosaur lateral teeth. Modern Geology 16: 161–198.

Freeman, E. F. 1975. The isolation and ecological implications of the microvertebrate fauna of a Lower Cretaceous lignite bed. Proceedings of the Geological Association, London 86: 307–312.

Gierlinski, G., and A. Ahlberg. 1994. Late Triassic and Early Jurassic dinosaur footprints in the Höganäs Formation of southern Sweden. Ichnos 3: 99–105.

Godefroit, P., and O. Lambert. 2007. A re-appraisal of Craspedodon lonzeensis Dollo, 1883 from the Upper Cretaceous of Belgium: the first record of a neoceratopsian in Europe? Bulletin de l'Institut royal des Sciences naturelles de Belgique, Sciences de la Terre 77: 83–93.

Gravesen, P. 1996. Geologisk set: Bornholm. Geografforlaget, København, Denmark, 168 pp.

Gravesen, P., F. Rolle, and F. Surlyk. 1982. Lithostratigraphy and sedimentary evolution of the Triassic, Jurassic and Lower Cretaceous of Bornholm, Denmark. Geological Survey of Denmark, Series B (7): 1–51.

Grigorescu, D. 2010. The latest Cretaceous fauna with dinosaurs and mammals from the Hațeg Basin—a historical overview. Palaeogeography, Palaeoclimatology, Palaeoecology 293: 271–282.

Hansen, K., and N. Bonde. 2006a. Tetrapteryx og flyvningens oprindelse [Tetrapteryx and the origin of flight]. Varv 2005 (4): 24–32.

————. 2006b. Tetrapteryx og flyvning [Tetrapteryx and flight]. Varv 2006 (3): 1–10.

Howarth, M. K. 1981. Palaeogeography of the Mesozoic; pp. 197–220 in L. R. M. Cocks (ed.), The evolving Earth. British Museum (Natural History), London.

Jensen, E. S. 2002. Skåne gennem 1800 millioner år [Scania during 1800 million years]. Varv 2002 (1): 3–31.

Karl, H.-V., and B. E. K. Lindow. 2009. First evidence of a late Cretaceous marine turtle (Testudines: Chelonioidea) from Denmark. Studia Geologica Salmanticensa 45: 175–180.

Koppelhus, E. B., and L. H. Nielsen. 1994. Palynostratigraphy and palaeoenvironmenrt of the Lower to Middle Jurassic Bagå Formation of Bornholm, Denmark. Palynology 18: 139–194.

Krebs, B., and D. Schwarz. 2000. The crocodiles from the Guimarota mine; pp. 69–74 in T. Martin and B. Krebs (eds.), Guimarota: a Jurassic ecosystem. Pfeil, Munich.

Larsen, G. (ed.). 2006. Naturen i Danmark: Geologien. Gyldendal, København, Denmark.

Le Loeuff, J. 2005. Romanian Late Cretaceous dinosaurs: big dwarfs or small giants? Historical Biology 17: 15–17.

Le Loeuff, J., and E. Buffetaut. 1991. Tarascosaurus salluvicus nov. gen., nov. sp., dinosaure theropode du Crétacé superieur du Sud de la France. Geobios 25: 585–594.

Lindgren, J. 2004a. Skåne under dinosauriernes era [Scania during the era of dinosaurs]. Geologisk Forum 43: 48–53.

————. 2004b. Stratigraphical distribution of Campaniam and Maastrichtian mosasaurs in Sweden—evidence of an intercontinental marine extinction event? Geologiska Föreningen för Stockholms Förhandlingar 126: 221–229.

————. 2004c. The utility of isolated teeth in mosasaur taxonomy; p. 56 in A. S. Schulp and J. W. M. Jagt (eds.), First Mosasaur Meeting, Maastricht, May 2004, abstracts and Field Guide. Naturhistorische Museum, Maastricht.

————. 2004d. Early Campanian Mosasaurs (Reptilia, Mosasauridae) from the Kristianstad Basin, southern Sweden. Thesis, Lund University, 89 pp.

Lindgren, J., and J. W. M. Jagt. 2005. Danish mosasaurs. Netherlands Journal of Geoscience 84: 315–320.

Lindgren, J., J. Rees, M. Siverson, and G. Cuny. 2004. The first Mesozoic mammal from Scandinavia. Geologiska

Föreningen för Stockholms Förhandlingar 126: 325–330.

Lindgren, J., P. J. Currie, M. Siverson, J. Rees, P. Cederström, and F. Lindgren. 2007. The first neoceratopsian dinosaur remains from Europe. Palaeontology 50: 929–938.

Lindgren, J., P. J. Currie, J. Rees, M. Siverson, S. Lindström, and C. Alwmark. 2008. Theropod dinosaur teeth from the lowermost Cretaceous Rabekke Formation on Bornholm, Denmark. Geobios 41: 253–263.

Longrich, N. R., and P. J. Currie. 2009. A microraptine (Dinosauria, Dromaeosauridae) from the Late Cretaceous of North America. Proceedings of the National Academy of Sciences of the United States of America 106: 5002–5007.

Martill, D. M., and D. Naish (eds.). 2001. Dinosaurs of the Isle of Wight. Palaeontological Association, London, 433 pp.

Milan, J. 2004. På sporet af de bornholmske dinosaurer [Tracking Bornholmian dinosaurs]. Naturens Verden 2004 (11/12): 54–61.

Milan, J., and N. Bonde. 2001. Svaneøgler, nye fund på Bornholm [Plesiosaurs: new finds on Bornholm]. Varv 2001 (4): 3–8.

Milan, J., and R. Bromley. 2005. Dinosaur footprints from the Middle Jurassic Bagå Formation, Bornholm, Denmark. Bulletin of the Geological Society of Denmark 52: 7–16.

Milner, A. C. 2002. Theropod dinosaurs of the Purbeck Limestone Group, southern England. Special Papers in Palaeontology 68: 191–201

Milner, A. C., and D. Batten (eds.). 2002. Life and environment in Purbeck times. Special Papers in Palaeontology (London) 68. Palaeontological Association, London, 208 pp.

Milner, A. C., A. R. Milner, and S. E. Evans. 2000. Amphibians, reptiles and birds: a biogeographical review; pp. 316–332 in S. J. Culver and P. F. Rawson (eds.), Biotic response to global change—the last 145 million years. Cambridge University Press, Cambridge.

Naish, D., and D. M. Martill. 2001. Saurischian dinosaurs 1: Sauropods; pp. 185–241 in Martill and Naish (eds.), Dinosaurs of the Isle of Wight. Palaeontological Association, London.

Naish, D., S. Hutt, and D. M. Martill. 2001. Saurischian dinosaurs 2: Theropods; pp. 242–309 in Martill and Naish (eds.), Dinosaurs of the Isle of Wight. Palaeontological Association, London.

Noe-Nygaard, N., and F. Surlyk. 1988. Wash over fan and brackish bay sedimentation in the Berriasian–Valanginian of Bornholm, Denmark. Sedimentology 35: 197–217.

Noe-Nygaard, N., F. Surlyk, and S. Piasecki. 1987. Bivalve mass mortality caused by dinoflagellate blooms in a Berriasian–Valanginian lagoon, Bornholm, Denmark. Palaios 2: 263–273.

Ősi, A., S. M. Apesteguía, and M. Kowalewski. 2010a. Non-avian theropod dinosaurs from the early Late Cretaceous of Central Europe. Cretaceous Research 31: 304–320.

Ősi, A., R. J. Butler, and D. B. Weishampel. 2010b. A Late Cretaceous ceratopsian dinosaur from Europe with Asian affinities. Nature 465: 466–468.

Ostrom, J. H. 1969. Osteology of *Deinonychus antirrhopus*, an unusual theropod from the Lower Cretaceous of Montana. Peabody Museum of Natural History Bulletin 30: 1–165.

Owen, H. G. 1981. Constant dimensions or an expanding Earth?; pp. 179–196 in L. R. M. Cocks (ed.), The evolving Earth. British Museum of Natural History, London.

Pereda-Superbiola, X. 2009. Biogeographical affinities of Late Cretaceous continental tetrapods of Europe: a review. Bulletin de la Société Géologique de la France 180: 57–71.

Persson, P. O. 1959. Reptiles from the Senonian (U. Cret.) of Scania (S. Sweden). Arkiv för Mineralogie och Geologie 2: 431–478.

Petersen, H. I., J. A. Bojesen-Koefoed, and H. P. Nytoft. 1996. Depositional environment and burial history of a Lower Cretaceous carbonaceous claystone, Bornholm, Denmark. Bulletin of the Geological Society of Denmark 43: 133–142.

Piasecki, S. 1984. Dinoflagellate cyst stratigraphy of the Lower Cretaceous Jydegaard Formation, Bornholm, Denmark Bulletin of the Geological Society of Denmark 32: 145–161.

Pouech, J., J.-M. Mazin, and J.-P. Billon-Bruyat. 2006. Microvertebrate biodiversity from Cherves-de-Cognac (Lower Cretaceous, Berriasian: Charente, France); pp. 96–100 in the 9th International Symposium on Mesozoic Terrestrial Ecosystems and Biota, abstracts and proceedings.

Rauhut, O. W. M. 2000. The dinosaur fauna from the Guimarota mine; pp. 75–82 in in T. Martin and B. Krebs (eds.), Guimarota: a Jurassic ecosystem. Pfeil, Munich.

———. 2002. Dinosaur teeth from the Barremian of Una, province of Cuenca, Spain. Cretaceous Research 23: 255–263.

Rauhut, O. W. M., and C. Werner. 1995. First record of the family Dromaeosauridae (Dinosauria, Theropoda) in the Cretaceous of Gondwana (Wadi Milk Formation, northern Sudan. Paläontolologische Zeitschrift 69: 475–489.

Rees, J. 1998. Early Jurassic selachians from the Hasle Formation on Bornholm, Denmark. Acta Palaeontologica Polonica 43: 439–462.

———. 2000. An Early Cretaceous scincomorph lizard dentary from Bornholm, Denmark. Bulletin of the Geological Society of Denmark 48: 105–109.

———. 2001. Jurassic and Early Cretaceous selachians—focus on southern Scandinavia. Lund Publikationer i Geologie 153: 1–19.

Rees, J., and N. Bonde. 1999. Plesiosaur remains from the Early Jurassic Hasle Formation, Bornholm, Denmark [abstract]; p. 70 in E. Hoch and A. M. Brantsen (eds.), Secondary adaptations to life in water. Geological Museum, Copenhagen, Denmark.

Rees, J., J. Lindgren, and S. E. Evans. 2005. Amphibians and small reptiles from the Berriasian Rabekke Formation on Bornholm, Denmark. Geologiska Föreningens för Stockholms Förhandlinger 127: 233–238.

Salisbury, S. W. 2002. A taxonomic review of crocodilians from the Lower Cretaceous (Berriasian) Purbeck Limestone Group of Dorset, Southern England. Special Papers in Palaeontology 68: 180–190.

Sankey, J. T., D. B. Brinkman, M. Guenther, and P. J. Currie. 2002. Small theropod and bird teeth from the Late Cretaceous (Late Campanian) Judith River Group, Alberta. Journal of Paleontology 76: 751–763.

Schultze, H.-P. 1992. Dipnoi; in F. Westphal (ed.), Fossilium Catalogus I: Animalia, Pars 131. Kugler Publications, Amsterdam, New York, 464 pp.

———. 1996. Mesozoic sarcopterygians; pp. 463–492 in G. Arratia and A. Tintori (eds.), Mesozoic fishes 3. Pfeil, Munich.

Smith, A. G., and J. C. Briden. 1977. Mesozoic and Cenozoic palaeocontinental maps. Cambridge University Press, Cambridge .

Smith, A. G., D. G. Smith, and B. M. Funnell. 1994. Atlas of Mesozoic and Cenozoic coastlines. Cambridge University Press, Cambridge, 99 pp..

Smith, J. B., D. R. Vann, and P. Dodson. 2005. Dental morphology and variation in theropod dinosaurs: implications for the taxonomic identification of isolated teeth. Anatomical Record A 285: 699–736.

Stein, K., Z. Csiki, K. Curry Rogers, D. B. Weishampel, R. Redelstorff, J. L. Carballido, and P. M. Sanders. 2010. Small body size and extreme cortical bone remodelling indicate phyletic dwarfism in *Magyarosaurus dacus* (Sauropoda: Titanosauria). Proceedings of the National Academy of Sciences of the United States of America 107: 9258–9263.

Surlyk, F. 2006. Fra ørkener til varme have: Trias, Jura og Kridt; pp 139–180 in G. Larsen (ed.), Naturen i Danmark: Geologien [From deserts to warm seas]. Gyldendal, København, Denmark.

Surlyk, F., J. Milan, and N. Noe-Nygaard. 2008. Dinosaur tracks and possible lungfish aestivation burrows in a shallow water coastal lake; lowermost Cretaceous, Bornholm, Denmark. Palaeogeography, Palaeoclimatology, Palaeoecology 267: 292–304.

Sweetman, S. C. 2004. The first record of velociraptorine dinosaurs (Saurischia, Theropoda) from the Wealden (Early Cretaceous, Barremian) of southern England. Cretaceous Research 25: 353–564.

Swinton, W. E. 1934. The dinosaurs. Allen and Unwin, London, 233 pp.

Upchurch, P. 1994. Sauropod phylogeny and palaeoecology. Gaia 10: 249–260.

Vullo, R., D. Néraudeau, R. Allain, and H. Capetta. 2005. Un nouveau gisement à microrestes de vertébres continentaux dans le Cénomanien inferieur de Fouras (Charente-Maritime, Sud-Ouest de la France. Comptes Rendus Palevol 4: 95–107.

Vullo, R., D. Néraudeau, and T. Lenglet. 2007. Dinosaur teeth from the Cenomanian of Charente, western France: evidence for a mixed Laurasian–Gondwanan assemblage. Journal of Vertebrate Paleontology 27: 931–943.

Weishampel, D. B., D. Grigorescu, and D. B. Norman. 1991. The dinosaurs of Transylvania: island biogeography in the Late Cretaceous. National Geographic Research and Exploration 7: 68–87.

Weishampel, D. B., P. M. Barrett, R. A. Coria, et al. 2004. Dinosaur distribution; pp. 517–606 in D. B. Weishampel, P. Dodson, and H. Osmólska, The Dinosauria,

2nd ed. University of California Press, Berkeley.

Wilson, J. A., and P. C. Sereno. 1998. Early evolution and higher level phylogeny of sauropod dinosaurs. Journal of Vertebrate Paleontology 18 (supplement to 2): 1–72.

Xu, X., X.-L. Wang, and X.-C. Wu. 1999. A dromaeosaurid dinosaur with a filamentous integument from Yixiang Formation of China. Nature 401: 262–266.

Xu, X., Z. H. Zhou, and X.-L. Wang. 2000. The smallest known non-avian theropod dinosaur. Nature 408: 705–708.

Xu, X, Z. H. Zhou, X.-L. Wang, X. Kuang, F. Zhang, and X. Du. 2003. Four-winged dinosaurs from China. Nature 421: 335–340.

Zheng, X., X. Xu, H. You, Q. Zhao, and Z. Dong. 2009. A short-armed dromaeosaurid from the Jehol Group of China with implications for early dromaeosaurid evolution. Proceedings of the Royal Society B, 277 (1679): 211–217.

Zinke, J. 1998. Small theropod teeth from the Upper Jurassic coalmine of Guimarota (Portugal). Paläontologische Zeitschrift 72: 179–189.

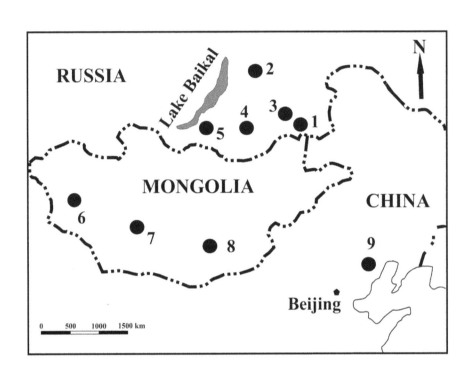

25.1. Map showing the distribution of localities of *Lycoptera* beds. 1, Middendorf's outcrop; 2, Baisa; 3, Shiviya; 4, Semion; 5, Gusinoe Ozero; 6, Gurvan-Eren; 7, Bon-Tsagan; 8, Manlaj; 9, main localities of Yixian Formation.

The Age of *Lycoptera* Beds (Jehol Biota) in Transbaikalia (Russia) and Correlation with Mongolia and China

25

Evgenia V. Bugdaeva* and Valentina S. Markevich

Volcanogenic sedimentary deposits, which contain abundant fossils of lacustrine fauna and flora, are widespread in the territory of Transbaikalia (Russia), Mongolia, and northeastern China. These fossil ecosystems are known as the Jehol Biota. The Jehol Biota of the Yixian Formation in western Liaoning province (China) are particularly famous because of the discovery, from 1996, of feathered dinosaurs and of the earliest angiosperms. However, the age of the Yixian Formation is still the object of discussions. The possibilities of correlation of the fossil flora of *Lycoptera* beds of Transbaikalia with potential coeval floras of Yakutia, Mongolia, and northeastern China are considered. Comparison with the Transbaikalia flora allows us dating the beds with Jehol Biota, including the Yixian Formation, as Barremian to Aptian.

Introduction

The volcanogenic sedimentary deposits, which contain abundant remains of lacustrine ecosystems, are widespread in the territory of Transbaikalia (Russia), Mongolia, and northeastern China. In the nineteenth century, A. P. Gerassimov was the first to discover fossils from the Jehol Biota in so-called fish shales, outcropping in the basin of the Turga and Byrka rivers. Gerassimov transmitted his findings to A. Th. Middendorf, who accomplished a journey through Siberia between 1842 and 1845. It is possible to say that the history of paleontology of the region began from that time. The outcrop in the bank of the Turga River (Figs. 25.1 and 25.2) is often called Middendorf's outcrop, although it was the less famous local, Gerassimov, who discovered and collected fossils. Also in Alexander Theodore Middendorf's honor are the name of the fish from these layers, *Lycoptera middendorfii* Müller (Fig. 25.3), and the representative of the conchostracans, *Bairdestheria middendorfii* Jones.

The fossiliferous layers in this locality are dominated by diverse insects, the most abundant of which is *Ephemeropsis trisetalis* (Fig. 25.4). Subsequently, Reis (1909) published a description of new animal and plant taxa from the fish shales in the basins of the Turga and Vitim rivers from materials collected during Middendorf's expedition. The author noted the scarcity and fragmentary nature of the plant remains. He came to the conclusion that these sediments were Late Jurassic to Early Cretaceous in age. This dating of fossil-bearing strata was the first in a chain of opinions of researchers. Stratigraphic correlation of the Upper Mesozoic deposits was hampered by complicated tectonics of Transbaikalia, facies diversity, and

25.2. The Middendorf's outcrop locality (Transbaikalia, Russia).

Photo by R. Korostovsky.

high endemism of the biota. Curiously, the high diversity of animals and plants at numerous sites did not facilitate the age determination of the layers containing organic remains, but resulted in a debate among specialists about the dating and position of beds in the sequence. Paradoxically, some paleontologists who studied the flora or the same groups of faunas came to diametrically opposite conclusions.

A more complete study of fossil plants of Transbaikalia was undertaken by Prynada (1950, 1962). He regarded the Turga Formation as the horizon of the same name and considered it as facies of coal-bearing formation completing the sequence of the continental Mesozoic in the isolated basins of Transbaikalia. The localities cropping out on the banks of the Turga and Vitim rivers, and near Shivia town (Fig. 25.1), were included by Prynada in the Turga facies. In his conclusion, "there are no plants in the Turga flora, which can be considered as guide fossils for a certain time," so he considered the age of the coal-bearing formation to be the Late Jurassic or Early Cretaceous (Prynada, 1962, 15). In the assemblage of the Turga Formation, Prynada included fossil plants from Bukachacha, Holbon, Chernovskoe, Duroi, Kharanor, Tarbagatai, Tugnui, Gusino-Uda, and Bayangol coal mines, as well as from deposits outcropping in the basins of Alyangui, Bukukun, and Chikoi rivers. Martinson (1961) substantiated a different age for the above-mentioned localities. He thought that the Turga–Vitim Formation, set up by him, was Early Cretaceous in age. However, Kolesnikov

25.3. *Lycoptera middendorfii* Müller, actinopterygian fish, Transbaikalia, Middendorf's outcrop, Turga Formation, Barremian–Aptian, ZBT531/3-5522, IBSS FEB RAS.

(1964) recognized four correlative biostratigraphic horizons in Transbaikalia and considered that the Turga horizon was Late Jurassic in age. Perhaps his opinion had a major impact on the development during the 1950s and 1960s of the Mesozoic stratigraphy in northeastern China. Many Chinese researchers still consider the fish shales to be Late Jurassic in age.

The studies of Vachrameev and Kotova (1977) played an important role in the stratigraphy of Transbaikalia. These authors described the earliest angiosperms (leaf *Dicotylophyllum pussilum* Vachr., fruit, pollen *Asteropollis asteroides* Hedl. et Norris) from Baisa locality (Zaza Formation; Fig. 25.1) and substantiated the late Neocomian–Aptian age for the plant-bearing deposits (Vachrameev and Kotova, 1977). On the basis of their results, it became possible to date the flora from the Turga Formation, which closely resembles the flora from the Zaza Formation. In addition, Kotova (1964, 1968) substantiated the Early Cretaceous age of coal-bearing deposits from Gusinoe Ozero and Bukachacha basins (Fig. 25.1), shattering the positions of Martinson (1961) and Kolesnikov (1964), who considered a Late Jurassic age.

Institutional abbreviation. IBSS FEB RAS, Institute of Biology and Soil Science, Far East Branch of Russian Academy of Sciences, Vladivostok, Russia.

During the Early Cretaceous, the Transbaikalia and Mongolia area was a vast ecotone between the temperate and subtropical zones (Krassilov, 1981, 1982). Transbaikalia then was part of the Amur province of the Siberian–Canadian realm and was enriched by elements from the Euro-Sinian realm (Vachrameev, 1964; Vachrameev et al., 1970; Bugdaeva, 1989). On the one hand, the Turga flora included representatives of the flora from the Siberian–Canadian realm such as *Gleichenia lobata* Vachr. (Fig. 25.5),

Correlation of *Lycoptera* Beds in Transbaikalia and Mongolia

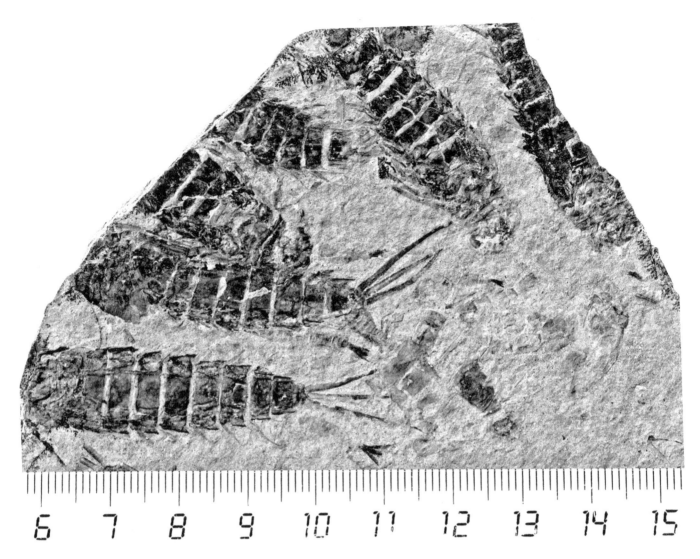

25.4. The burial of *Ephemeropsis trisetalis,* mayflies, Transbaikalia, Baisa locality, Zaza Formation, Barremian–Aptian, ZBB15–01, IBSS FEB RAS.

Czekanowskia, Phoenicopsis, Coniopteris setacea (Pryn.) Vachr., *Neozamites verchojanensis* Vachr. (Fig. 25.6), and *Pityolepis oblonga* Samyl. (Fig. 25.7). On the other hand, representatives of the Euro-Sinian realm (*Otozamites, Cladophlebidium, Onychiopsis*) occur in this flora. All the material figured in this chapter is deposited at the Institute of Biology and Soil Science, Far East Branch of Russian Academy of Sciences, Vladivostok, Russia.

The transitional location of the Transbaikalian flora during the Early Cretaceous provides an excellent opportunity to carry out its correlation with the floras from both the temperate and subtropical zones. When compared with synchronous paleofloras, the flora from the Turga Formation shows distinctive features (Bugdaeva, 1989).

The paleoflora from the Lena Basin is one of the best-studied Early Cretaceous floras from temperate regions. Kiritchkova (1985) singled out four stages in the evolution of the Early Cretaceous flora: Batylykh, Eksenyakh, Khatyrykh, and Agrafenovo. These stages correspond to horizons established on a paleobotanic basis traceable from south to north over almost 200 km from the lower reaches of the Aldan River to the delta of the Lena River (Vachrameev, 1991). The Eksenyakh horizon is of Aptian age; in the south, it encompasses the Eksenyakh Formation from Vilyui Syneclise, whereas it encompasses the upper parts of the Chonkogor, Bulun, and

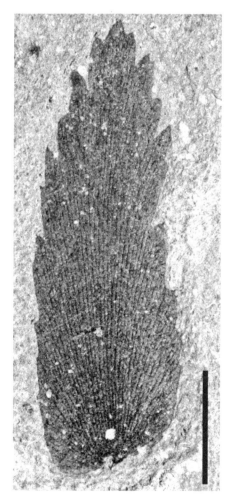

25.5. *Gleichenia lobata* Vachr., Gleicheniaceae, Transbaikalia, Semion locality, Semion unit, Barremian–Aptian, ZB325/I-01, IBSS FEB RAS; scale bar = 1 cm.

Bakh formations in the middle and northern parts of the region. The upper part of this horizon is characterized by the occurrence of the bennettite *Neozamites verchojanensis*, whose remains were found in the Transbaikalian localities of Baisa and Semion. The fern *Gleichenia lobata*, found in the Semion locality in the Chita-Ingoda Basin, is also typical for the Eksenyakh horizon.

Common or similar species are found in floristic assemblages from the Barremian–Aptian Chegdomyn and Chemchukin formations in the Bureya Basin (Vachrameev and Doludenko, 1961; Krassilov, 1973), from the Ussuri (Barremian) and Lipovtsy (Aptian) formations in the Razdolnoe Basin, and from the Starosuchan (Barremian) and Severosuchan (Aptian) formations in the Partizan Basin of South Primorye (Krassilov, 1967).

25.6. *Neozamites verchojanensis* Vachr., Bennettitales, Transbaikalia, Semion locality, Semion unit, Barremian–Aptian, ZB325/III-3033, IBSS FEB RAS; scale bar = 1 cm.

25.7. *Pityolepis oblonga* Samyl., Pinaceae, Transbaikalia, Baisa locality, Zaza Formation, Barremian–Aptian, ZB31/87, IBSS FEB RAS; scale bar = 1 cm.

Krassilov (1982) defined four paleofloristic assemblages in the Early Cretaceous of Mongolia: (1) a zone with *Baiera manchurica* (Berriasian?), (2) a zone with *Otozamites lacustris–Pseudolarix erensis* (Barremian–Aptian), (3) a zone with *Baierella hastata–Araucaria mongolica* (Aptian), and (4) a zone with *Limnothetis–Limnoniobe* (Aptian, and perhaps Early Albian).

Study of the Early Cretaceous flora of Mongolia has revealed its undoubted affinities with the Transbaikalian flora (Krassilov, 1982; Bugdaeva, 1989). In the localities of Middendorf's outcrop and Semion, besides *Pseudolarix*, remains of the bennettite *Otozamites lacustris* Krassil. have been found (Fig. 25.8). Both morphological and epidermal studies showed absolute identity of the Mongolian and Transbaikalian *O. lacustris*. These plants allowed correlation of the Turga flora with an assemblage from the second zone defined in the Early Cretaceous flora of Mongolia, which includes plants from the Bon-Tsagan, Manlaj, and Gurvan-Eren localities (Fig. 25.1).

25.8. *Otozamites lacustris* Krassil., Bennettitales, Transbaikalia, Baisa locality, Zaza Formation, Barremian–Aptian, ZB31/173, IBSS FEB RAS; scale bar = 5 mm.

In the Baisa locality, the remains of *Samaropsis aurita* Krassil. (Fig. 9) are abundant. This plant is peculiar to the Early Cretaceous Mongolian localities Bon-Tsagan, Holbotu-Gol, Shin-Khuduk, and Manlaj.

Thus, the correlation of the Turga flora with the Barremian–Aptian flora of Mongolia confirms its age.

In the second half of the twentieth century, intensive geological research began to be carried out in China. In the Mesozoic sediments of northeastern China, abundant fossil remains were revealed. In the mass burials, diverse fishes, insects, mollusks, vertebrates, and plants were discovered. This biota, which undoubtedly had a high level of endemism, was named Jehol Biota.

Correlation of Jehol Biota in Northeastern China

25.9. *Samaropsis aurita* Krassil., Coniferales, Transbaikalia, Baisa locality, Zaza Formation, Barremian–Aptian, ZB31/476, IBSS FEB RAS; scale bar = 5 mm.

The Jehol Biota has attracted the attention of the world science community since 1996, when the discoveries of feathered dinosaurs from sediments of the Yixian Formation (Liaoning province) were published. Later, exquisitely preserved fossils of pterosaurs, birds, mammals, and plants (including earliest angiosperms) were also described (e.g., Cao et al., 1998; Gibbons, 1998; Sun et al., 1998, 2001; Swisher et al., 1999; Unwin, 1998). The amazing findings and their unique preservation contributed to detailed studies of these extraordinary animals and plants, which answered many questions about the evolution of the Cretaceous biota and allowed reconstructions of the ecosystem. However, the fundamental question—what is the age of the fossil-bearing beds?—remained unsolved.

According to Chinese stratigraphers, the age of the Yixian Formation is Late Jurassic (Cao et al., 1998; Chen and Chang, 1994; Lo et al., 1999; Sun et al., 1998, 2001), although some paleontologists advocate for an Early Cretaceous age (Li and Liu, 1994, 1999; Mao et al., 1990; Pu and Wu, 1992; Swisher et al., 1999). The most detailed biostratigraphic analysis of the Yixian Formation was undertaken by Smith et al. (2001, 559), who presented an overview of all faunistic and floristic groups found in this stratigraphic unit and concluded that "biostratigraphy is a poor means by which to determine the age of the Yixian Formation." The most reliable are radiometric dates consistently given late Early Cretaceous ages. $^{40}Ar/^{39}Ar$ analyses of volcanic rocks from the Yixian Formation give ages ranging from 121.1 ± 0.2 to 122.9 ± 0.3 Ma (Smith et al., 1995) and 124.6 ± 0.3 Ma (Swisher et

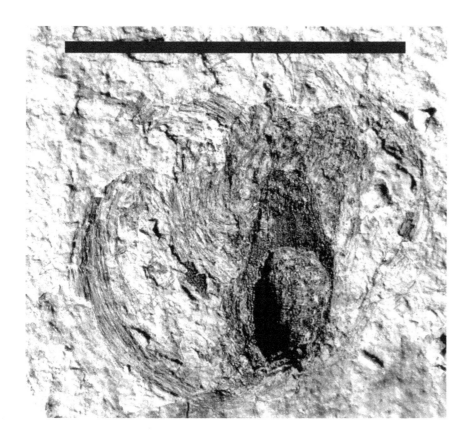

25.10. *Baisia hirsuta* Krassil., Proangiospermae, Transbaikalia, Baisa locality, Zaza Formation, Barremian–Aptian, ZB31/209, IBSS FEB RAS; scale bar = 5 mm.

al., 1999). According to these authors, the age estimations correspond to the Barremian, although according to the Gradstein et al. (2004), it rather corresponds to the Aptian. Recently Chang et al. (2009) obtained an age of 129.7 ± 0.5 Ma for a basaltic lava from the bottom of the Yixian Formation and an age of 122.1 ± 0.3 Ma for a tuff from the lowermost part of the overlying Jiufotang Formation. Their age results provide an age calibration of the whole Yixian Formation and show that the whole formation was deposited during the Early Cretaceous, from the Barremian to the Early Aptian, within an interval of ~7 Ma.

Fossils of flowering plants and plants having affinities with angiosperms were found in Early Cretaceous deposits from Transbaikalia, Mongolia, and northeastern China (Vachrameev and Kotova, 1977; Krassilov, 1982, 1986; Krassilov and Bugdaeva, 1982, 2000). Krassilov called the latter plants proangiosperms. In proangiosperms, the features typical for flowering plants are not yet fully developed but appear discretely, not forming stable combinations. According to his opinion, angiospermization was expressed in whole plant communities. It can only be assumed that it was the response of biota to some external influence.

In June 2006, E. V. Bugdaeva visited the main localities in the Yixian Formation in Liaoning and Hebei provinces and collected plant remains (Fig. 25.1). She was particularly attentive to plants regarded as guide fossils for the Turga assemblage of Transbaikalia.

The most important plant is *Baisia hirsuta* Krassil., which belongs to proangiospermous plants with bennettitalean affinities (Krassilov and Bugdaeva, 1982). This plant consists of dispersed cupulate ovules, which occupy an apical position on a short rounded to triangular receptacle that is persistent in disseminules and bears minute bracts and tufts of long bristles

25.11. *Baikalophyllum lobatum* Bugd., Cycadales, Transbaikalia, Semion locality, Semion unit, Barremian–Aptian, ZB325/III-2880, IBSS FEB RAS; scale bar = 1 cm.

(Fig. 25.10). In Transbaikalia, this species has been found in abundance in the Middendorf's outcrop, Baisa, Semion (Elizavetino Basin) and Shiviya (Unda-Daya Basin) localities.

We visited the Museum of Natural History in Beijing and the Museum of National Geopark around Sihetun locality in Chaoyang city (Liaoning province). In the exhibition halls of these museums, cycad leaves are displayed. The label on the specimens from the Sihetun locality provides the name *Pityolepis larixiformis*, which is clearly erroneous. This species was described from the Semion locality in Transbaikalia as *Baikalophyllum lobatum* Bugd. (Bugdaeva, 1983). The Yixian flora has also yielded the conifer *Nageiopsis transbaikalica* Srebr., originally described on the basis of material from the Semion locality (Srebrodolskaya, 1983) and also present in the Baisa locality in Transbaikalia, but Chinese paleobotanists referred it to as *Podocarpidites reheensis* (Wu) Sun et Zheng (Sun et al., 2001) (Figs. 25.11 and 25.12). *Botrychites reheensis* Wu, *Neozamites verchojanensis* Vachr., *Pityolepis pseudotsugaoides* Sun et Zheng, *Brachyphyllum longispicum* Sun, Zheng, et Mei, *Scarburgia hilii* Harris, *Ephedrites chenii* (Cao et Wu) Guo et Wu X.W., *Carpolithus multiseminalis* Sun et Zheng, and *C. pachythelis* Sun et Zheng are present both in the Turga and in the Yixian floras.

In addition, the endemic plant *Gurvanella* occurs in the Yixian flora (Sun et al., 2001). This genus was found in the Early Cretaceous Gurvan-Eren locality in Mongolia (Krassilov, 1982). *Gurvanella* has not been discovered in Transbaikalia yet, but floras of the Gurvan-Eren, Manlaj, and

25.12. *Nageiopsis transbaikalica* Srebr., Coniferales, Transbaikalia, Semion locality, Semion unit, Barremian–Aptian, ZB31/341, IBSS FEB RAS; scale bar = 1 cm.

Bon-Tsagan localities are similar with the Turga flora of Transbaikalia (Bugdaeva, 1989).

Conclusions

Paleofloras from the Yixian Formation in northeastern China and from the Turga Formation in Transbaikalia have the same age. During the Barremian–Aptian, a high level of endemism characterized the biota over a vast area that included the modern territories of Transbaikalia, Mongolia, and northeastern China. Floras from these regions were similar and included elements from temperate and subtropical floras, which allow for broad cross-regional correlations. Therefore, paleobotanical data support a Barremian–Aptian age for the Yixian Formation and the Jehol Biota, as also indicated by $^{40}Ar/^{39}Ar$ analyses.

Acknowledgments

Our research was supported by the Presidium of the Far Eastern Branch of the Russian Academy of Sciences and the Presidium of the Russian Academy of Sciences (grant 09-I-P15-02). We are grateful to Sun Ge, D. Dilcher, P. Grote, M. Nelson, R. Korostovsky, L. Starukhina, V. Scoblo, N. Lyamina, A. Averianov, and N. Yadrishchenskaya. Thanks to V. Krassilov and H. Pfefferkorn for useful comments and to P. Godefroit for his support.

References

Bugdaeva, E. V. 1983. A new genus based on leaves from the Cretaceous deposits of Eastern Transbaikalia; pp. 44–47 in V. A. Krassilov (ed.), Paleobotany and phytostratigraphy of eastern USSR. DVNC AN SSSR, Vladivostok. [In Russian]

———. 1989. Correlation of the Lower Cretaceous deposits of isolated basins of Transbaikalia based on fossil flora; pp. 162–168 in V. A. Solovyov (ed.), Stage and zonal scales of the boreal Mesozoic of the USSR. Nauka, Moscow. [In Russian]

Cao, Z. Y., S. I. Wu, P. A. Zhang, and J. R. Li. 1998. Discovery of fossil monocotyledons from Yixian Formation, Western Liaoning. Chinese Science Bulletin 43: 230–233.

Chang, S. C., H. C. Zhang, P. R. Renne, and Y. Fang. 2009. High-precision

^{40}Ar-^{39}Ar age for the Jehol Biota. Palaeogeography, Palaeoclimatology, Palaeoecology 280: 94–104.

Chen, P. J., and Z. L. Chang. 1994. Non-marine Cretaceous stratigraphy of eastern China. Cretaceous Research 15: 245–257.

Gibbons, A. 1998. Dinosaur fossils, in fine feathers, show link to birds. Science 280: 2051.

Gradstein, F. M., J. G. Ogg, A. G. Smith, W. Bleeker, and L. J. Lourens. 2004. A new geologic time scale, with special reference to Precambrian and Neogene. Episodes 27 (2): 83–100.

Kiritchkova, A. I. 1985. Phytostratigraphy and flora of Jurassic and Lower Cretaceous deposits of Lena Basin. Nedra, Leningrad, 223 pp. [In Russian]

Kolesnikov, C. M. 1964. Stratigraphy of the continental Mesozoic of Transbaikalia; pp. 5–138 in G. G. Martinson (ed.), Stratigraphy and paleontology of the Mesozoic and Cenozoic deposits of East Siberia and the Far East. Nauka, Moscow. [In Russian]

Kotova, I. Z. 1964. Age of continental deposits of Gusinoe Ozero depression and features of the Early Cretaceous floras of the Transbaikalia. Izvestiya AN USSR, Seriya geologicheskaya 8: 84–93. [In Russian]

———. 1968. About the age of coal-bearing deposits in Eastern Transbaikalia (Bukachacha depression). Izvestiya AN USSR, Seriya geologicheskaya 11: 95–103. [In Russian]

Krassilov, V. A. 1967. The Early Cretaceous flora of South Primorye and its significance for stratigraphy. Nauka, Moscow, 364 pp. [In Russian]

———. 1973. Materials on stratigraphy and paleofloristics of coal-bearing strata of Bureya Basin; pp. 28–51 in V. A. Krassilov (ed.), The fossil flora and phytostratigraphy of the Far East. DVNC AN SSSR, Vladivostok. [In Russian]

———. 1981. Changes of Mesozoic vegetation and the extinction of dinosaurs. Palaeogeography, Palaeoclimatology, Palaeoecology 34: 207–224.

———. 1982. Early Cretaceous flora of Mongolia. Palaeontographica, B 181: 1–43.

———. 1986. New floral structure from the Lower Cretaceous of Lake Baikal area. Review of Palaeobotany and Palynology 47: 9–16.

Krassilov, V. A., and E. V. Bugdaeva. 1982. Achene-like fossils from the Lower Cretaceous of the Lake Baikal area. Review of Palaeobotany and Palynology 36: 279–295.

———. 2000. Gnetophyte assemblage from the Early Cretaceous of Transbaikalia. Palaeontographica, B 253: 139–151.

Li, W. B., and Z. S. Liu. 1994. The Cretaceous palynofloras and their bearing on stratigraphic correlation in China. Cretaceous Research 15: 333–365.

———. 1999. Sporomorph assemblage from the basal Yixian Formation in Western Liaoning and its geological age. Palaeoworld 11: 68–79.

Lo, C. H., P. J. Chen, T. Y. Tsou, S. S. Sun, and C. Y. Lee. 1999. ^{40}Ar/^{39}Ar laser single-grain and K-Ar dating of the Yixian Formation, NE China. Palaeoworld 11: 341–338.

Mao, Z. Z., J. X. Yu, and J. K. Lentin. 1990. Palynological interpretation of Early Cretaceous non-marine strata of northeast China; pp. 115–118 in E. M. Truswell and J. A. K. Owen (eds.), Proceedings of the 7th International Palynological Congress, Amsterdam, Part II.

Martinson, G. G. 1961. The Mesozoic and Cenozoic mollusks of the continental deposits of the Siberian platform, Transbaikalia and Mongolia. AN SSSR, Moscow, 322 pp. [In Russian]

Prynada, V. D. 1950. The Mesozoic flora of Transbaikalia and its stratigraphic distribution. Materials on geology and commercial minerals of Eastern Siberia, Irkutsk, 31 pp. [In Russian]

———. 1962. The Mesozoic flora of Eastern Siberia and Transbaikalia. Gosgeoltekhizdat, Moscow, 368 pp. [in Russian]

Reis, O. M. 1909. Die Binnenfauna der Fischschieffer in Transbaikalien; pp. 1–68 in Geological researches and explorations along the Siberian railway. Buchdruckerei M.M. Stassjulewitsch, St. Petersburg.

Pu, R. G., and H. Z. Wu. 1992. Mesozoic sporo-pollen assemblages in Western Liaoning and their stratigraphic significance. Acta Palaeontologica Sinica 10: 121–212.

Smith, J. B., Harris, J. D., Omar, G. I., Dodson, P., and Y. Hailu. 2001. Biostratigraphy and avian origins in northeast China; pp. 549–574 in J. Gauthier and L. F. Gauthier (eds.), New perspectives on the origin and early evolution of birds. Yale University Press, New Haven.

Smith, P. E., N. M. Evensen, D. York, M. M. Chang, F. Jin, J. L. Li, S. Cumbaa, and D. Russel. 1995. Dates and rates in ancient lakes: ^{40}Ar-^{39}Ar evidence for an Early Cretaceous age for the Jehol Group, northeast China. Canadian Journal of Earth Sciences 32: 1426–1431.

Srebrodolskaya, I. N. 1983. A new Early Cretaceous plants from Transbaikalia. Paleontologicheskii zhurnal 4: 117–120. [in Russian]

Sun, G., D. L. Dilcher, S. L. Zheng, and Z. K. Zhou. 1998. In search of the first flower: a Jurassic angiosperm, Archaeofructus, from Northeast China. Science 282: 1692–1695.

Sun, G., S. L. Zheng, D. L. Dilcher, Y. D. Wang, and S. W. Mei. 2001. Early angiosperms and their associated plants from Western Liaoning, China. Shanghai Scientific and Technological Education Publishing House, Shanghai, 227 pp.

Swisher, C. C., Y. Q. Wang, X. L. Wang, X. Xu, and Y. Wang. 1999. Cretaceous age for the feathered dinosaurs of Liaoning, China. Nature 400: 58–61.

Unwin, D. M. 1998. Feathers, filaments and theropod dinosaurs. Nature 391: 119–120.

———. 1964. The Jurassic and Early Cretaceous floras of Eurasia and paleofloristic provinces of this time. Nauka, Moscow, 263 pp. [in Russian]

———. 1991. Jurassic and Cretaceous floras and climates of the Earth. Cambridge University Press, Cambridge, 318 pp.

Vachrameev, V. A., and M. P. Doludenko. 1961. The Upper Jurassic and Lower Cretaceous flora of Bureya Basin and its significance for stratigraphy. AN SSSR, Moscow, 136 pp. [In Russian]

Vachrameev, V. A., and I. Z. Kotova. 1977. The early angiosperms and the associated plants from the Lower Cretaceous deposits of Transbaikalia. Paleontologicheskii zhurnal 4: 101–109. [In Russian]

Vachrameev, V. A., I .A. Dobruskina, E. D. Zaklinskaya, and S. V. Meyen. 1970. The Paleozoic and Mesozoic floras of Eurasia and phytogeography of this time. Nauka, Moscow, 426 pp. [in Russian]

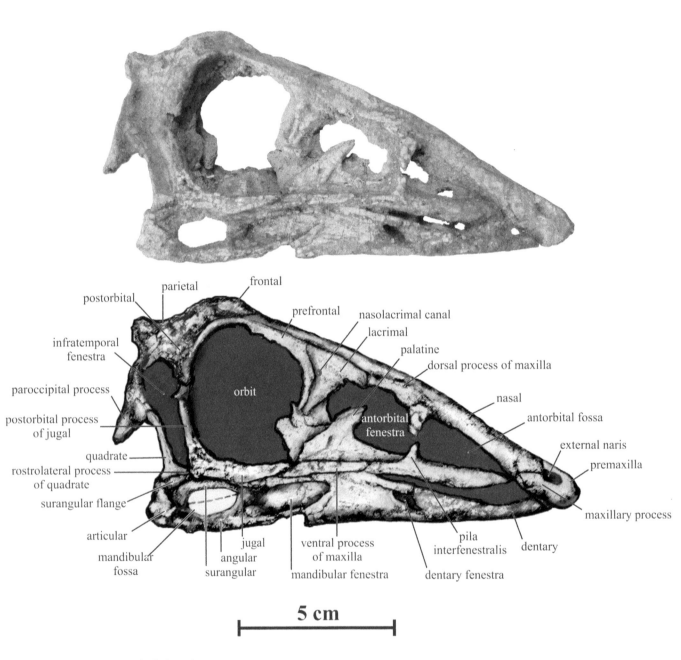

26.1. *Hexing qingyi*, JLUM-JZ07b1 (holotype). Skull in right lateral view. L, left; R, right.

A New Basal Ornithomimosaur (Dinosauria: Theropoda) from the Early Cretaceous Yixian Formation, Northeast China

26

Jin Liyong, Chen Jun, and Pascal Godefroit*

We describe a new basal Ornithomimosauria from a specimen (JLUM-JZ07b1) discovered in the lowest beds (Lower Valanginian–Lower Barremian, Lower Cretaceous) of the Yixian Formation in western Liaoning province, P.R. China. *Hexing qingyi* gen. et sp. nov. is characterized by a series of cranial and postcranial autapomorphies: the rostral portion of the premaxilla deflected ventrally in front of the lower jaw, so that its oral surface is level with the ventral border of the dentary, a deep antorbital fossa that invades the whole lateral surface of the maxilla, a sagittal crest on the parietal, pendant paroccipital processes that extend ventrally below the level of the foramen magnum, a dentary fenestra, a phalangeal formula for manus of 0-(1 or 2)-3-3-0, and elongated (>85% length of corresponding metacarpal) proximal phalanges of digits III and IV. A phylogenetic analysis shows that *Hexing* is a basal ornithomimosaur more derived than *Pelecanimimus* but basal to *Harpymimus, Beishanglong,* and the edentulous clade formed by *Garudimimus* and the Ornithomimidae. Incomplete overlap between preserved body parts across specimens prevents full resolution of the relationships between *Hexing, Shenzhousaurus* (also from the Yixian Formation of western Liaoning province), and more advanced ornithomimosaurs. Ornithomimosauria had an extensive evolutionary history in eastern Asia from the Valanginian–Barremian, until the early Maastrichtian, and a single dispersal across the Bering Strait is required to account for the distribution of advanced Ornithomimidae in Asia and North America.

Introduction

Ornithomimosauria (ostrich-mimic dinosaurs) is a group of lightly built and cursorial theropods that are mainly known from Cretaceous localities of Asia and North America (Makovicky et al., 2004). Although phylogenetically nested among carnivorous theropods, advanced Ornithomimidae lack teeth but had beaks and gastoliths (Kobayashi et al., 1999), suggesting that their diet was completely different from that of typical theropods. Ornithomimosauria have been viewed as carnivores of small prey, insectivores, and even herbivores (Nicholls and Russell, 1981; Kobayashi et al., 1999; Makovicky et al., 2004; Barrett, 2005). They may have exhibited gregarious habits (Kobayashi and Lü, 2003; Varricchio et al., 2008).

Thousands of exceptionally preserved dinosaurs, birds, early mammals, amphibians, aquatic reptiles, pterosaurs, flowering plants, and insects have been unearthed during the last 15 years in western Liaoning province and constitute the famous Early Cretaceous Jehol Biota. Among dinosaurs, feathered theropods from the Jehol Biota have received particular attention, allowing for the reconstruction of the early evolution of feathers and flight as well as reconstruction of the phylogenetic relationships between coelurosaurian dinosaurs and birds.

Ji et al. (2003) described *Shenzhousaurus orientalis*, the oldest known Ornithomimosauria, from the lowest more fluvial, volcanoclastic beds of Yixian Formation (Lower Valanginian–Lower Barremian; Swisher et al., 2002). Here, we describe a second ornithomimosaur from the same beds. As it is usually the case for fossils discovered in the Yixian Formation, this specimen was discovered by a local farmer, who began its preparation. The fossil was subsequently carefully prepared in the laboratory of the Geological Museum of Jilin University. All the dubious parts of the skeleton, which may have been affected by the preparation of the local farmer, have been removed and are not included in the following description.

Institutional abbreviations. CMN, Canadian Museum of Nature, Ottawa, Canada; JLUM, Geological Museum of the Jilin University, Changchun, P.R. China.

Systematic Paleontology

Dinosauria Owen, 1842
Theropoda Marsh, 1881
Tetanurae Gauthier, 1986
Coelurosauria von Huene, 1914
Ornithomimosauria Barsbold, 1976

Hexing qingyi gen. et sp. nov.
(Figs. 26.1–26.7)

Etymology. From Mandarin, *Hexing*, "like a crane," and *quingyi*, "with thin wings."

Holotype. JLUM-JZ07b1, housed in the Geological Museum of the Jilin University, Changchun (Jilin province, P.R. China).

Locality and horizon. Xiaobeigou locality, Lujiatun, Shangyuan, Beipiao City, western Liaoning province, P.R. China; lowest more fluvial, volcanoclastic beds of Yixian Formation, older than 128 million years and younger than 139 million years (Lower Valanginian–Lower Barremian; Swisher et al., 2002).

Diagnosis. Ornithomimosauria with the following autapomorphies: rostral portion of premaxilla deflected ventrally in front of lower jaw, so that its oral surface is level with the ventral border of dentary; deep antorbital fossa invades the whole lateral surface of maxilla; sagittal crest on parietal; pendant paroccipital processes that extend ventrally below the level of the foramen magnum; dentary fenestra; phalangeal formula for manus: 0-(1 or 2)-3-3-0; proximal phalanges of manus digits III and IV elongated (> 75% length of corresponding metacarpal); high (137%) hind limb ratio for tibiotarsus/femur.

Description

JLUM-JZ07b1 is incompletely preserved, including the skull, a portion of the cervical series, and most of the appendicular skeleton. All the measurements taken on JLUM-JZ07b1 are compiled in Appendix 26.1. The small size of JLUM-JZ07b1 roughly corresponds to that of juvenile *Sinornithomimus dongi* specimens, from the Upper Cretaceous of Inner Mongolia (Kobayashi and Lü, 2003). The dimensions of JLUM-JZ07b1 are also smaller than those of the holotype of the basal ornithomimid *Shenzhousaurus orientalis*, also from the Yixian formation of western Liaoning (Ji et al., 2003). However, JLUM-JZ07b1 is not a juvenile: most cranial sutures are completely fused; and complete fusion of the scapula and coracoid is present, as is fusion of the astragalus and calcaneum.

Skull. The right side of the skull is well preserved (Fig. 26.1). However,, the left side was crushed during fossilization. Palatal elements and the lateral wall of the braincase are poorly exposed. The muzzle is long and triangular, gradually tapering rostrally, and particularly slender. The skull is proportionally large when compared with other Ornithomimosauria (Table 26.1). The skull is about as long as the femur, as in *Shenzousaurus orientalis*, also from the Yixian Formation of western Liaoning province (Ji et al., 2003), although it is much shorter in other ornithomimids. However, the larger proportion of the skull in both basal ornithomimosaurs from the Yixian Formation probably mainly reflects the small size of these specimens. With a femur length of 135 mm (*Hexing*) and 181 mm (*Shenzhousaurus*), they are less than half the size of the other ornithomimosaurs taken into consideration. Further allometric studies of the ornithomimosaur skeleton should clarify this problem.

Skull openings. The whole skull is highly pneumatized (Fig. 26.1). The orbit is distinctly higher than long. The antorbital fossa is extremely developed, invading the whole lateral surface of the maxilla. The antorbital fenestra forms more than half the length of the antorbital fossa. The external narial opening is particularly small. As is usual in Ornithomimosauria (Makovicky et al., 2004), it is separated from the maxilla by the premaxillary–nasal contact. An internarial septum, as in *Harpymimus* (Kobayashi and Barsbold, 2005a), cannot be observed. The dimensions of both the supratemporal and infratemporal fenestrae cannot be measured because of lateral crushing against the lateral wall of the braincase of the postorbital and squamosal and because of the absence of the quadratojugal.

Premaxilla. The edentulous premaxilla is formed by thin nasal and maxillary processes that form the dorsal, rostral, and ventral margins of the small external opening (Fig. 26.1). The maxillary process terminates rostral to the rostral end of the antorbital fossa, as in *Shenzousaurus* (Ji et al., 2003) and *Harpymimus* (Kobayashi and Barsbold, 2005a). In more advanced Ornithomimosauria, the maxillary process extends further caudally (Kobayashi and Lü, 2003). The dorsal border of the maxillary process contacts the nasal, excluding the maxilla from the external narial opening. The rostral portion of the premaxilla, which forms the tip of the snout, is deflected ventrally in front of the rostral end of the lower jaw, so that its oral surface is level with the ventral margin of the dentary. The external surface of the premaxilla does not appear pitted by neurovascular exits that supplied the horny beak, as it is usually observed in advanced ornithomimosaurs (Norell et al., 2001; Makovicky et al., 2004).

Maxilla. The edentulous maxilla is elongate (Fig. 26.1). Its dorsal process contact the rostral process of the lacrimal at the midpoint of the antorbital fenestra, whereas its longer ventral process contacts the jugal at the level of the caudal border of the antorbital fenestra. The exact limits between the maxilla and the nasal cannot be discerned because these bones are intimately fused together. A stout pila interfenestralis, perpendicular to the ventral process of the maxilla, forms the vertical rostral margin of the antorbital fenestra. In front of pila interfenestralis, the whole lateral surface of the maxilla is deeply excavated by the antorbital fossa. The medial wall of the anterior part of the maxillary fossa is destroyed, and it is therefore not possible to observe the presence of the maxillary and promaxillary fenestrae. The oral margin of the maxilla is sinuous. Its rostral portion is distinctly convex, corresponding to a concavity on the dorsal margin of the dentary. However, the ventral expansion of oral margin of the maxilla is not as prominent as in *Sinornithomimus*, *Garudimimus*, and *Gallimimus* (Kobayashi and Lü, 2003), and the oral margins of the maxilla and dentary do not form a cutting edge as this level.

Jugal. The jugal is a straplike bone that forms the ventral margin of the orbit (Fig. 26.1). Its rostral end covers the caudal end of the maxilla at the level of the antorbital bar. It resembles the condition observed in *Sinornithomimus*, but it contrasts with the bifurcated rostral end for contacts with both the maxilla and lacrimal observed in *Struthiomimus* and *Ornithomimus* (Kobayashi and Lü, 2003). Under the orbit, it becomes slightly deeper. Caudally, it bifurcates to form a caudodorsally projecting process that contacts the caudal border of the ventral process of the postorbital, as well as a short process that contacts the distal head of the quadrate. The postorbital process participates in the rostroventral border of the infratemporal fenestra and terminates below the midheight of the orbit, as usually observed in Onithomimosauria (Makovicky et al., 2004).

Nasal. The nasal is transversely narrow but rostrocaudally long: it extends from the caudal margin of the external narial opening up to the level of the midpoint of the orbit (Fig. 26.1). The nasal is slightly transversely vaulted. As is usual in ornithomimosaurs (Makovicky et al., 2004), the nasal is excluded from the dorsal margin of the antorbital fenestra by the maxilla and the lacrimal. Its caudolateral border contacts the prefrontal, and its caudal end, the frontal. Foramina on the dorsal surface of the nasal are not observed in JLUM-JZ07b1.

Lacrimal. The lacrimal is roughly T shaped (Fig. 26.1). Its long ventral process, which forms the caudal margin of the antorbital fenestra, is broken off ventrally. The caudal border of the preserved portion of the ventral process is bounded by the prefrontal along its whole height. The rostral process of the lacrimal extends forward to contact the dorsal process of the maxilla above the midpoint of the antorbital fenestra. A shorter caudal process inserts between the nasal and the prefrontal on the dorsal aspect of the skull. Close to the junction with the prefrontal, the lateral side of the lacrimal is pierced by a round foramen, also observed in *Garudimimus* and *Gallimimus* (Kobayashi and Barsbold, 2005b), and which corresponds to a caudal opening for the nasolacrimal canal.

Table 26.1. Skeletal Proportions in Selected Ornithomimosauria Genera

Taxon	Femur (mm)	Sk/F (% femur length)	H/F (% femur length)	M/F (% femur length)	T/F (% femur length)	MTIII/F (% femur length)	P/F (% femur length)
Hexing qingyi	135	101	67	76	>137	62	49
Shenzousaurus orientalis[a]	191	97	—	—	—	—	—
Sinornithomimus dongi[b]	323	57	66	—	102	66	—
Garudimimus brevipes[c]	371	68	—	—	105	62	—
Struthiomimus altus[d]	480–513	50	65–72	70–77	109–111	75–79	46–50
Ornithomimus sp.[d]	435–500	54	64	64	108–109	71–76	49
Dromiceiomimus sp.[d]	378–468	51	75	—	119–124	78–86	50–56
Gallimimus bullatus[d]	192–665	51–62	80	47	108–113	78–81	38–47

Sk/F, ratio skull length/femur length; H/F, ratio humerus length/femur length; M/F, ratio manus length (including metacarpals)/femur length; T/F, ratio tibiotarsus length/femur length; MTIII/F, ratio metatarsal III length/femur length; P/F, ratio pes length (excluding metatarsals)/femur length.

[a] Ji et al. (2003).
[b] Kobayashi and Lü (2003).
[c] Kobayashi and Barsbold (2005b).
[d] Nicholls and Russell (1981).

Prefrontal. The prefrontal is a narrow but long curved bone that forms the rostrodorsal margin of the orbit. It starts ventrally below the midpoint of the rostral orbital margin and extends dorsally behind the midpoint of the dorsal orbital margin, well beyond the nasal (Fig. 26.1). The long ventral process of the prefrontal closely adheres to the ventral process of the lacrimal on the antorbital bar. The caudal process of the prefrontal excludes the caudal part of the nasal and he rostral part of the frontal from the orbital margin. In dorsal view, the prefrontal appears slightly larger than the lacrimal.

Frontal. The frontal of *Hexing* is relatively short, less than half the length of the nasal (Fig. 26.2). It contrasts with the situation observed in *Sinornithomimus* (Kobayashi and Lü, 2003, fig. 5), in which the frontal is only slightly shorter than the nasal. Of course, this character is likely correlated with the relative development of the snout in the different taxa. Because of the lateral crushing of the postorbital and squamosal against the lateral wall of the braincase, the lateral side of the frontal is partly obscured. However, the frontal apparently formed only a short portion of the orbital rim because of the caudal extension of the prefrontal. The planar rostral part of the frontals is sloped rostrally and meets the nasal at the midpoint of the dorsal orbital margin. As is usual in Ornithomimosauria, the frontals are domed near the caudal part of the orbit, forming a flexure between the flat rostral parts of the frontals and parietals. The paired frontals form a single dome, as in *Garudimimus* (Kobayashi and Barsbold, 2005b), but unlike *Gallimimus* (Osmólska et al., 1972) and *Sinornithomimus* (Kobayashi and Lü, 2003), which have a dome on each frontal separated by a midline depression. The caudal border of the frontal that contacts the parietal is concave. In *Shenzhousaurus* (Ji et al., 2003) and *Galliminus* (Osmólska et al., 1972), the frontoparietal suture is sinuous in dorsal view.

Parietal. As is usual in Ornithomimosauria, the parietal of *Hexing* is much shorter than the frontal (Fig. 26.2). Its rostral portion forms a wide tonguelike process that inserts between the caudal parts of the paired frontals. Contrary to other Ornithomimosauria (Ji et al., 2003; Makovicky et al., 2004), the dorsal surface of the parietal forms a well-developed sagittal crest. The parietal forms a pair of thin caudal processes that extend ventrally against the paroccipital processes. In *Garudimimus* (Kobayashi and

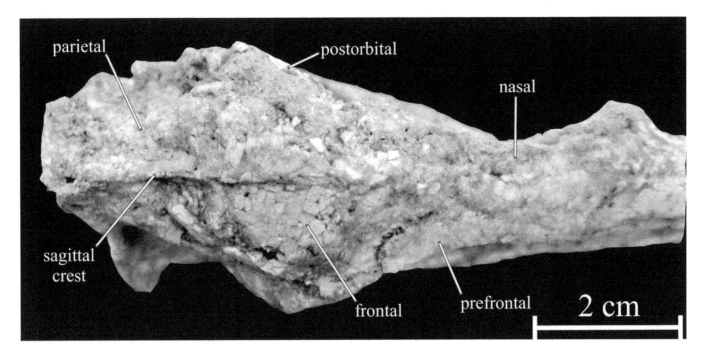

26.2. *Hexing qingyi*, JLUM-JZ07b1 (holotype). Skull roof in dorsal view.

Barsbold, 2005b) and *Sinornithomimus* (Kobayashi and Lü, 2003), these processes are straight and extend caudolaterally beyond the caudal end of the skull table.

Postorbital. The postorbital of JLUM-JZ07b1 is poorly preserved on both sides of the skull (Fig. 26.1). The rostral process is relatively short and the postorbital–frontal suture originates at the caudodorsal part of the orbit. The postorbital forms a long hooklike caudal process that participates in the intertemporal bar. The jugal process of the postorbital is long, narrow, and regularly curved, nearly forming the whole caudal margin of the orbit.

Squamosal. The complex structure of the squamosal cannot be adequately described in JLUM-JZ07b1 because this bone is completely crushed.

Caudal aspect of the skull (Fig. 26.3). The supraoccipital, exoccipitals, and opisthotics are completely fused together, and their limits cannot be discerned. The foramen magnum appears much taller than wide, but this could be due to lateral crushing of the dorsal part of the skull. Above the foramen magnum, the supraoccipital forms a prominent, dorsoventrally elongated knob. On either side of this knob, depressed areas border the dorsolateral margins of the foramen magnum. The fused exoccipitals and opisthotics form the large, pendant paroccipital processes, which extend ventrally well below the foramen magnum. They appear much more expanded ventrally than in *Gallimimus* (Osmólska et al., 1972), *Sinornithomimus* (Kobayashi an Lü, 2003, fig. 7), and *Garudimimus* (Kobayashi and Barsbold, 2005b). In any case, this unusual orientation of the processes cannot be explained by the eventual mediolateral postmortem crushing of the skull. Medially the exoccipitals border the foramen magnum. A large slitlike foramen perforates the caudal surface of the paroccipital process at midheight, close to its lateral margin. A caudal foramen on the paroccipital process can also be observed in juvenile specimens of *Gallimimus* (Makovicky et al., 2004), in *Sinornithomimus* (Kobayashi and Lü, 2003), and in *Garudimimus* (Kobayashi and Barsbold, 2005b). The medioventral portion of the caudal aspect of the paroccipital process is depressed. The

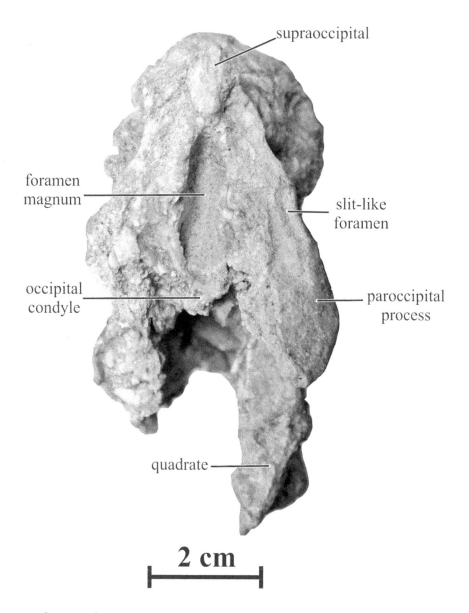

supraoccipital

foramen
magnum

slit-like
foramen

occipital
condyle

paroccipital
process

quadrate

2 cm

26.3. *Hexing qingyi,* JLUM-JZ07b1 (holotype). Skull in caudal view.

medioventral corner of the paroccipital process participates in the dorsal part of the occipital condyle

Quadrate. The lateral (Fig. 26.1) and caudal (Fig. 26.3) sides of the right quadrate can be observed. This bone is rather high, narrow in lateral view, and slightly curved caudally. The proximal articular head is quite simple and rounded, articulating with the squamosal. The distal head of the quadrate forms a small triangular rostrolateral process that articulates with the dorsolateral flange of the surangular just rostral to the mandibular glenoid and contacts the caudal end of the jugal. In caudal view, the fossa that encloses the quadratic foramen forms a slitlike aperture, at about mid-height of the main quadrate body, close to its medial margin.

Palatine. A large, hooklike, and rostrally inclined plate inside the antorbital fenestra is interpreted as the dorsal process of the right palatine (Fig. 26.1). It appears much better developed than in other Ornithomimosauria described so far. Its lateral side bears a well-developed ridge, surrounded by triangular depressed areas.

Basioccipital. The basioccipital forms the major part of the occipital condyle. Although it is much deformed, the latter appears particularly small, as is common in ornithomimosaurs (Kobayashi and Barsbold, 2005a). The occipital condyle has a long and slightly constricted neck, like in *Struthiomimus* (Kobayashi and Barsbold, 2005a) and the Ukhaa Tolgod ornithomimid described by Makovicky and Norell (1998). The ventral side of the basioccipital bears a strong median carina along its whole length. Between the basal tubera, the median crest forms a small tubercle, also known in the Ukhaa Tolgod ornithomimid (Makovicky and Norell, 1998). The basal tubera are moderately developed. The basisphenoid is too poorly preserved to be adequately described.

Mandible. The mandible of *Hexing* is lightly built and distinctly shorter than the skull. Its tip terminates caudal to the ventrally deflected oral surface of the premaxilla (Fig. 26.1). The large opening at the back of the mandible of JLUM-JZ07b1 is not natural but was accidentally made during the early stage of the preparation of the specimen. Indeed, the whole lateral surface of the angular and surangular is deeply excavated by a long caudal mandibular fossa and the lateral surface of both bones is therefore extremely thin and easily breakable. The "true" external mandibular fenestra begins in front of this mandibular fossa. As is usual in ornithomimosaurs, it is relatively small, extending from the level of the midheight of the orbit until the caudal third of the antorbital fenestra.

Articular. Only the lateral side of the right articular can be observed. It is trapezoidal in outline and exposed between the caudoventral border of the surangular and the caudodorsal border of the angular. It does not form a short retroarticular process that curves dorsocaudally, as observed in *Garudimimus* and Ornithomimidae (Makovicky et al., 2004).

Surangular. The lateral surface of the surangular is deeply excavated by the wide caudal mandibular fossa. Its caudodorsal border forms a small flange that extends rostrolaterally from the glenoid and articulates with the rostrolateral process on the distal head of the quadrate. This surangular flange is regarded as a synapomorphy for Ornithomimosauria (e.g., Makovicky et al., 2004). The thin rostral process of the surangular forms the dorsal margin of the external mandibular fenestra. It is overlapped by the caudodorsal process of the dentary along a short distance. Because an important part of the lateral surface of the surangular was destroyed during the early phase of the preparation of JLUM-JZ07b1, it is not possible to know whether a minute caudal foramen is present, as in *Gallimimus* and *Struthiomimus* (Makovicky et al., 2004).

Angular. The angular is slightly shorter than the surangular. If its dorsal is particularly thin because of the important development of the caudal mandibular fossa, its ventral part, which forms the caudal third of the ventral margin of the mandible, is robust. The angular and the surangular form a straight suture. They are separated caudally by the articular. Rostrally, the angular forms the caudal corner and the ventral margin of the external mandibular fenestra. The rostral process of the angular overlaps the caudoventral process of the dentary along a short distance.

Dentary. The dentary is elongated, forming the rostral two-thirds of the lateral side of the lower jaw, and triangular in lateral view (Fig. 26.1). Its bifid caudal end forms the rostral margin of the external mandibular fenestra and contacts the surangular dorsally and the angular ventrally. The ventral edge of the dentary is perfectly straight, like in *Pelecanimimus* (Pérez-Moreno et

26.4. *Hexing qingyi,* JLUM-JZ07b1 (holotype). Detail of left dentary teeth in lingual view.

al., 1994) and *Shenzhousaurus* (Ji et al., 2003). In *Harpymimus*, the ventral edge of the dentary is gently concave (Kobayashi and Barsbold, 2005a). In other Ornithomimosauria, the ventral edge of the dentary is deflected ventrally at the level of the symphysis. Rostrally, the dorsal edge of the dentary is deflected ventrally, as in other ornithomimosaurs except *Pelecanimimus* (Ji et al., 2003). This deflection corresponds to the convex margin of the rostral part of the maxilla. However, the rostrodorsal border of the dentary does not form a cutting edge. A rostroventrally sloping elliptical fenestra pierces the lateral surface of the dentary at midlength. Because the edges of this fenestra are regular, with a finished aspect, and slightly thickened, it is unlikely that this is an artifact of preservation or a pathology. Above this fenestra, the dorsal edge of the dentary is thicker. Such aperture has not been observed in other ornithomimosaurs so far. Teeth are not preserved on the right dentary. However, three or four small conical teeth can be observed in lingual view on the rostral end of the left dentary (Fig. 26.4). They are unfortunately too poorly preserved to be adequately described.

Cervical vertebrae. Five cervical vertebrae are poorly preserved in JLUM-JZ07b1 (Fig. 26.5). The centrum is cylindrical, elongated, and dorsoventrally compressed. The diapophyses form large triangular winglike processes, attached on the lateral side of the cranial half of the centrum (Fig. 26.5D). The infradiapophyseal fossae are particularly wide below the diapophyses. Caudal to the infradiapophyseal fossa, the lateral side of the centrum has a smaller elliptical central pneumatic fossa (Fig. 26.5A). On the lateroventral sides of the centrum, the parapophyses are less expanded both craniocaudally and laterally. Cervical ribs are fused on the distal articular surfaces of the diapophyses and parapophyses (Fig. 26.5B,C). Small infraparapophyseal fossae are developed under the proximal part of the parapophyses (Fig. 26.5A). The postzygapophyses form particularly well-developed triangular winglike process, nearly symmetrical to the diapophyses on the distal half of the cervical vertebra (Fig. 26.5D). They widely cover the prezygapophyses, which consequently cannot be described. The

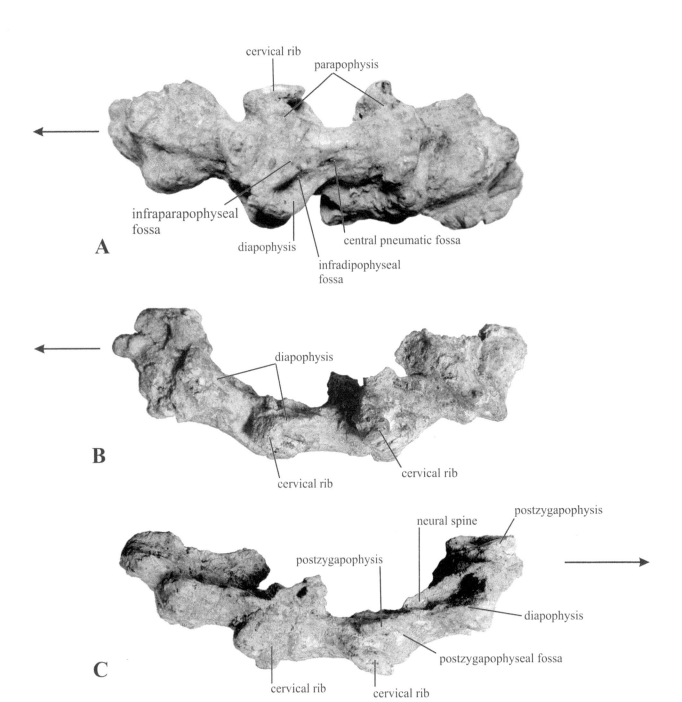

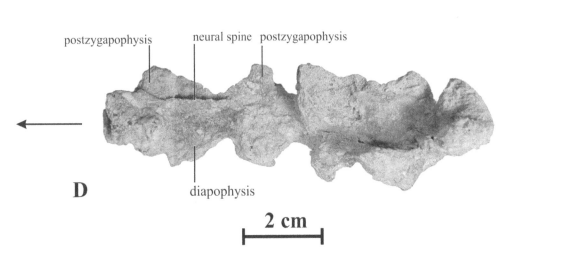

2 cm

infrapostzygapophyseal fossa is wide and deep (Fig. 26.5A). As is usual in ornithomimosaurs, the neural spine is low (Fig. 26.5C). However, contrary to the situation in ornithomimids except *Archaeornithomimus* (Kobayashi and Lü, 2003), it remains relatively elongated and occupies more than the cranial half of the neural arch.

Scapula. The right scapula and coracoid are better preserved than their left counterparts. The scapula and coracoid appear fused together, and their respective limits are not easily discernable (Fig. 26.6A). This suggests that JLUM-JZ07b1, despite of its small size, was an adult individual. The scapula is particularly elongated and thin. It is slightly longer than the humerus, like in *Harpymimus* (Kobayashi and Barsbold, 2005a), and its blade does not expand distally. The scapular blade is gently curved cranially. The acromion process is weakly developed and truncated. The supraglenoid buttress is particularly strongly developed, and the flange of the buttress extends on the cranial surface of the scapula. Dorsal to the buttress, the depression for attachment of *M. scapulotriceps* is weakly developed. As in *Harpymimus* (Kobayashi and Barsbold, 2005a), there is a low ridge dorsal to the depression along the caudal edge of the blade. The glenoid faces posterolaterally.

Coracoid. Both coracoids are poorly preserved, and only few details can be distinguished. The postglenoid process ("posterior process" of Pérez-Moreno et al., 1994; Kobayashi and Lü, 2003) is less elongated than in *Beishanlong* (Makovicky et al., 2010), and contrary to *Harpymimus* (Kobayashi and Barsbold, 2005a), the infraglenoid buttress is much less developed than the supraglenoid buttress on the scapula (Fig. 26.6A). The infraglenoid buttress is aligned with the postglenoid process and is not offset laterally, as in some ornithomimids (Kobayashi and Lü, 2003). Although the lateral surface of the coracoid is eroded, a subglenoid shelf along the dorsal aspect of the postglenoid process, as observed in *Beishanlong* and *Archaeornithomimus* (Makovicky et al., 2010), is apparently absent.

Humerus. The humerus is slightly shorter than the scapula and slender, especially when compared with *Harpymimus* (Kobayashi and Barsbold, 2005a, fig. 6.6). Its proximal portion is partly destroyed (Fig. 26.6B). The deltopectoral crest is particularly short and weakly developed, extending distally up to one-quarter of the total length. The humeral shaft is subcircular and straight. As described in *Harpymimus*, the proximal and distal ends of the humerus are roughly aligned in the same plane, and the twist between the ends known in Late Cretaceous taxa is absent (Kobayashi and Barsbold, 2005a). Both distal condyles are about the same size. The entepicondyle (contra *Anserimimus* and *Gallimimus*; Kobayashi and Lü, 2003) and ectepicondyle (contra *Beishanlong*; Makovicky et al., 2010) are not developed at all.

Ulna and radius. The forearm bones are tightly appressed along their whole length (Fig. 26.6B), indicating a limited range of pronation and supination, as is usual in Ornithomimosauria (Nicholls and Russell, 1985; Makovicky et al., 2004). The ulna and radius are only slightly shorter than the humerus. Both are particularly gracile and perfectly straight. The radius is only slightly narrower than the ulna. The olecranon appears less prominent than in other ornithomimosaurs, but it might be an artefact of preservation. The distal end of the ulna is subquadrangular in cross section,

26.5. *Hexing qingyi*, JLUM-JZ07b1 (holotype). Cervical vertebrae in ventral (A), left lateral (B), right lateral (C), and dorsal (D) views. The arrows indicate the cranial direction.

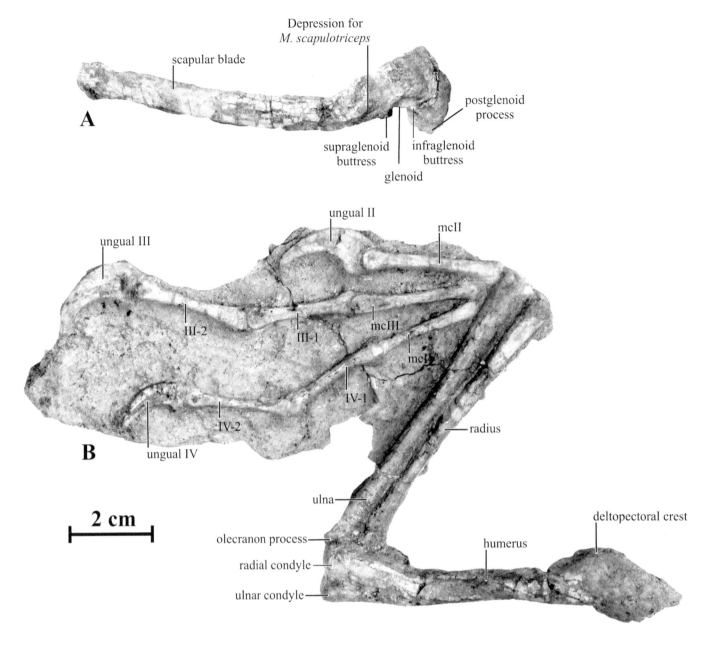

26.6. *Hexing qingyi*, JLUM-JZ07b1 (holotype). A, Right scapulocoracoid in lateral view. B, Left humerus (caudal view), forearm (lateral view), and hand (lateral view); mc, metacarpal.

whereas the distal end of the radius is craniocaudally flattened. The distal end of the ulna extends slightly more distally than that of the radius.

Manus. We follow the nomenclature of Xu et al. (2009) for manual digits (digits II, III, and IV are preserved in Tetanurae). The manus of *Hexing* appears highly modified when compared with that in other Ornithomimosauria. Our careful preparation reveals that the left hand has not been restored at all by the local farmer who discovered the fossil and is thus trustable for description. The hand of *Hexing* is much elongated, being distinctly longer than the humerus (Fig. 26.6B), like in *Struthiomimus* (Makovicky et al., 2004). Metacarpals III and IV are similar in size and morphology: they are elongated, perfectly straight, and particularly slender. Nothing indicates that they were tightly appressed in life, contrasting with the condition observed in other ornithomimosaurs. The proximal end of metacarpals III and IV is hidden under the ulna. The distal end of the metacarpals forms a ginglymoidal surface, and there is a collateral ligament fossa on the medial side. The proximal element of digit II is elongated and slender, closely resembling metacarpals III and IV. Moreover, its distal end is slightly rotated laterally, as is usually observed on metacarpal II of ornithomimosaurs except *Harpymimus* (Kobayashi and Lü, 2003; Kobayashi and Barsbold, 2005a). For these reasons, this bone is tentatively identified as metacarpal II. However, it must be noted that in basal ornithomimosaurs, metacarpal II is a small element, approximately half or less than metacarpal III (Kobayashi and Lü, 2003; Makovicky et al., 2004); on the contrary, the first phalanx of digit II is usually the longest in the ornithomimosaur manus (Makovicky et al., 2004). Therefore, it cannot be excluded that the first element of digit II in JLUM-JZ07b1 is the first phalanx and that metacarpal II is lost.

The phalangeal formula depends on the identification of this proximal element in digit II: it is therefore 0-(1 or 2)-3-3-0. In any case, it is unique is unique among Ornithomimosauria and other basal Tetanurae (basal phalangeal formula: 0-2-3-4-0; Xu et al., 2009). It implies the disappearance probably of phalanx II-1 and of one of the phalanges in digit IV. The proximal and intermediate phalanges of digits III and IV closely resemble each other in size and morphology and are also similar with the metacarpals. All are particularly elongated and slender (length of proximal phalanx in digit III and IV >75% length of corresponding metacarpal), although the proximal phalanges of digits III and IV are usually short and robust in other ornithomimosaurs. Their distal ends have well-developed ginglymoid articulations that fit into the concave proximal surface of the adjacent phalanx. The collateral ligament fossa is also well developed on the lateral side of the distal end and faces quite laterally. The ungual phalanges are well curved in lateral view. As is usual in ornithomimosaurs (e.g., Makovicky et al., 2004), the plantar surface of the unguals is not trenchant but flattened, and it bears a small flexor tubercle distal to the proximal articulation. The groove for the claw sheath extends parallel to the plantar side of the unguals.

Femur. The left femur is poorly preserved in JLUM-JZ07b1, its medial side being heavily damaged (Fig. 26.7A,B). The right one was apparently restored and is therefore not described here. The femur is much shorter than the tibia. In medial view, it appears more distinctly bowed than in other Ornithomimosauria (Fig. 26.7B). The partially preserved lesser trochanter is alariform and robust; it is placed more laterally than the greater trochanter (Fig. 26.7A). A rounded ridge divides the lesser trochanter into cranially and caudally oriented surfaces. The apex of the greater trochanter is rounded. It lateral side is extensively depressed, delimited cranially by the lesser trochanter, and caudally by a low poorly preserved proximodistally extending ridge. The distal end of the femur is slightly enlarged mediolaterally and the medial condyle appears better developed than the lateral condyle (Fig. 26.7A).

Tibiotarsus and fibula. The right tibia and fibula are also poorly preserved and cannot be adequately described. The tibia, fibula, and astragalus are completely fused together (Fig. 26.7C), suggesting that this specimen was an adult individual. The tibia and fibula appear elongated and slender. The hind limb ratio for tibiotarsus/femur is more than 137%—much higher than in other Ornithomimosauria (Table 26.1). However, it cannot be excluded that this character is size related among ornithomimosaurs. Indeed,

Currie (1998) shows that juvenile tyrannosaurids have proportionally longer tibiae than older individuals: bivariate comparisons of tibial versus femoral lengths indicate an important negative allometry of the tibia length during growth. Although Nicholls and Russell (1981, table 2) indicate that the hind limb ratio for tibiotarsus/femur varies between 108–109 in larger *Ornithomimus* specimens (femur lengths 435–500 mm), this ratio is much higher (147) in a smaller *Ornithomimus* individual (CMN 8636, femur length 310 mm; Currie, 1998, table 2). It is therefore possible that smaller ornithomimosaurs taxa have proportionally longer tibiae than larger individuals and that the elongated tibia in the small JLUM-JZ07b1 mainly reflects this negative allometry. Of course, this statement uses intraspecific scaling relations to extrapolate interspecific scaling, which is a weak justification; further allometric investigations in ornithomimosaurs would help in clarifying this situation. The distal end of the fibula articulates with the lateral side of the tibia, whereas it is placed on the cranial side of the tibia in some advanced ornithomimids. The astragalus is firmly attached to the distal part of the tibia, and the calcaneum is imperceptibly fused with the astragalus.

Metatarsals. Both metatarsal I and digit I are preserved in JLUM-JZ07b1, like in *Garudimimus* (Kobayashi and Barsbold, 2005b). There is no trace of a metatarsal V, as it is also preserved in the latter taxon, but it might be an artifact of preservation. The reduction of metatarsal I is not as drastic as in *Garudimimus*, in which it only one fifth of the length of metatarsal III (Kobayashi and Barsbold, 2005b). Although it is incompletely preserved in JLUM-JZ07b1, it is a bit more than half of the length of metatarsal III (Fig. 26.7C). Metatarsals II and III have nearly the same size, while metatarsal IV is a bit shorter. In contrast to the tibia, the metatarsals appear relatively short when compared with the femur. According to Kobayashi and Barsbold (2005b), a relatively low (<0.7) hind limb ratio for metatarsal III/femur can be observed in other Asian ornithomimids, including *Sinornithomimus*, *Anserimimus*, *Garudimius*, and also some *Gallimimus* specimens (although this ratio seems highly variable in the latter) and appears to be the retention of ancestral nonornithomimosaur theropod metatarsal proportions for taxa of that body size (Holtz, 1994). Kobayashi and Barsbold (2005b) also observed that metatarsal proportions are apparently not size related among ornithomimosaurs. The proximal end of metatarsal III is well exposed in cranial view, as in basal ornithomimosaurs. It is slightly mediolaterally compressed when compared with the proximal end of metatarsal II. Although it is incompletely preserved, the lateral side of metatarsal IV looks shallowly concave. A deep collateral ligament fossa is developed on the medial side of metatarsal I. The distal surfaces of the metatarsals look only slightly convex.

Pedal phalanges. The pes phalangeal formula is 2-3-4-5-0 (Fig. 26.7C). The proportions of the foot are similar to those in other ornithomimosaurs in which they can be measured (Table 26.1). Digit III is the longest, while the second and the fourth toes are subequal in length. All proximal and intermediate phalanges closely resemble each other. Their size decreases distally and the phalanges from the fourth toe are smaller than those from digits I to III (Appendix 26.1). Like in *Harpymimus*, the intermediate phalanges remain relatively long in comparison with the proximal ones, whereas they are proportionally much shorter in *Garudimimus* and ornithomimids.

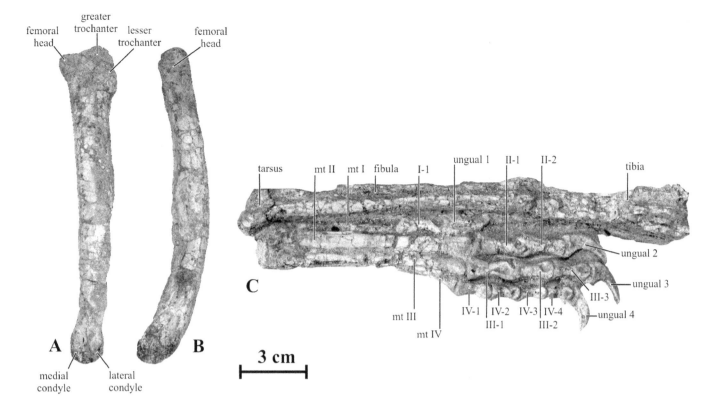

26.7. *Hexing qingyi,* JLUM-JZ07b1 (holotype). Left femur in cranial (A) and medial (B) views. C, Left tibiotarsus, fibula and pes (elements exposed in various views); mt, metatarsal.

For example, phalanx II-2 is 81% of the length of II-1, which is much longer than phalanx II-2 in ornithomimids (<60%; Kobayashi and Lü, 2003). The distal surfaces of the phalanges (except the unguals) apparently form ginglymoidal articulations. The collateral ligament fossa is deep on the lateral side of the distal articular surface. The ungual of digit II is the longest, but it appears slightly less recurved than on the third toe. As in other ornithomimosaurs, the plantar surface of the unguals is flat. Like in *Harpymimus* (Kobayashi and Barsbold, 2005a), the flexor tubercle is well developed on the caudal part of the ventral surface of the pedal unguals.

Phylogenetic Analysis

We undertook a numerical cladistic analysis of Ornithomimosauria in order to resolve the phylogenetic position of *Hexing*. This phylogenetic analysis is based on the matrix of Kobayashi and Lü (2003). However, a few characters were slightly modified, and nine supplementary characters, culled from Nicholls and Russell (1981), Ji et al. (2003), Makovicky et al. (2004, 2010), and Kobayashi and Barsbold (2005a) have been added to the original matrix (see Appendix 26.2). The data matrix (Appendix 26.3) was analyzed using PAUP*4.0b10 (Swofford, 2000), with Branch and Bound search, and both with accelerated (ACCTRAN) and delayed (DELTRAN) transformations. The analysis produced nine most parsimonious trees of 71 steps, with CI = 0.69, RI = 0.82, and RC = 0.57. The strict consensus tree is presented in Fig. 26.8, and the tree description appears in Appendix 26.4. It must be noted that character 28 (length of metacarpal II: approximately half of less than metacarpal III [0], slightly shorter [1] or longer [2]; Kobayashi and Lü, 2003) was coded as 1 for *Hexing*, assuming that the proximal element in digit II is metacarpal II. However, the general topology of the consensus tree does not change when character 28 is coded as 0 in *Hexing*.

This cladogram of course closely resembles that obtained by Kobayashi and Lü (2003), but it is also perfectly compatible with those recovered by Ji et al. (2003) and Makovicky et al. (2010), and there is thus a general consensus concerning the phylogeny of Ornithomimosauria. *Pelecanimimus*, from the late Barremian of Calizas de la Huergina Formation (Cuenca, Spain: Pérez-Moreno et al., 1994; see Chapter 22 in this book), retains plesiomorphic states such as an upper dentition (characters 1 and 3), a prominence on the lateral surface of the lacrimal (character 6), and a straight dorsal edge of the dentary (character 39). This analysis also posits *Hexing* as a basal ornithomimosaur more derived than *Pelecanimimus* but basal to *Harpymimus*, from the Shinekhudag Svita (Late Albian) of Mongolia (Kobayashi and Barsbold, 2005a), *Beishanlong*, from the Xinminpu Group (Aptian–Albian) of Gansu province in China (Makovicky et al., 2010), and the edentulous clade formed by *Garudimimus* and Ornithomimidae. Incomplete overlap between preserved body parts across specimens prevents full resolution of relationships between *Hexing*, *Shenzhousaurus*, also from the Yixian Formation of western Liaoning province, and more derived ornithomimosaurs. *Hexing* and *Shenzhousaurus* are particularly small ornithomimosaurs. As in *Hexing* and *Harpymimus*, the manus digit II of *Shenzhousaurus* is apparently much smaller than digits III and IV. However, because only the ungual phalanx is preserved in digit II, it cannot be decided whether this shortening results from shortening of metacarpal II, like in *Harpymimus*, or loss of the proximal phalanx, as in *Hexing*.

Harpymimus, *Beishanlong*, and the edentulous clade share the following unambiguous (that diagnose a node under both ACCTRAN and DELTRAN optimization) synapomorphies: rostral portion of the ventral edge of the dentary deflected ventrally (character 9), presence of two antorbital fenestrae (character 16), sigmoid ischial shaft (character 44), proximal end of metatarsal III partly or completely covered by metatarsals II and IV (character 37), and truncated postacetabular process of ilium (character 45). *Garudimimus*, from the Bayanshiree Svita (Cenomanian–Campanian: Makovicky et al., 2004), and Ornithomimidae are characterized, for example, by the complete loss of their dentition (character 10). The phylogeny of Ornithomimosauridae has already been discussed in detail by Kobayashi and Lü (2003); North American Ornithomimidae (*Strutiomimus*, *Dromiceiomimus*, and *Ornithomimus*) are apparently monophyletic and are best characterized by the strongly convex ventral border of their pubic boot, with ventral expansion (character 35).

Paleogeographic Implications

With a presumed Early Valanginian to Early Barremian age, *Hexing* and *Shenzhousaurus*, from the lower beds of the Yixian Formation of western Liaoning province, are the oldest known ornithomimosaurs. *Pelecamimimus*, from the Calizas de la Huergina Formation of Cuenca (Spain), is the most primitive ornithomimosaur known so far but is Late Barremian in age (Pérez-Moreno et al., 1994; see Chapter 22 in this book). This temporal paradox indicates that the place for origin of Ornithomimosauria (Europe or Asia?) remains conjectural. Whatever it may be, the presence of basal ornithomimosaurs in western Europe and eastern Asia during the

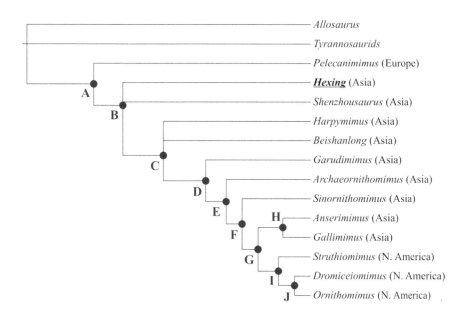

26.8. Cladogram of Ornithomimosauria, showing the phylogenetic relationships of *Hexing qingyi*. Letters correspond to nodes defined in Appendix 26.4.

Barremian indicates that connections between Europe and Asia through the Turgai Straight were already established, well before the Aptian (contra Doré, 1991; Russell, 1993; Smith et al., 1994).

Examining distributions within the context of the phylogeny (Fig. 26.8) reveals that ornithomimids have an extensive evolutionary history, from the Valanginian–Barremian until the Early Maastrichtian, in eastern Asia, as also described for Iguanodontoidea (see also Chapters 19 and 20 in this book). With the exception of *Pelecanimimus* and of the doubtful *Timimus* from the Eumeralla Formation (Early Albian) of Dinosaur Cove (Victoria, Australia: Rich and Vickers-Rich, 1994), recently reidentified as a Dromaeosauridae? indet. cf. Unenlaiinae (Agnolin et al., 2010), all Early Cretaceous ornithomimosaurs have been discovered in Eastern Asia. The oldest ornithomimid is *Sinornithomimus*, from the Ulansuhai Formation (early Late Cretaceous according to Kobayashi and Lü, 2003, but Aptian–?Albian according to Makovicky et al., 2004) of Inner Mongolia (China) and the most primitive representative of this clade, is *Archaeonithomimus*, from the Iren Dabasu Formation (Turonian–Coniacian according to Averianov, 2002, but Early Maastrichtian according to Van Itterbeeck et al., 2005) of Inner Mongolia. The three North American ornithomimosaurid genera are Late Campanian and Maastrichtian in age. Because they are regarded as monophyletic, a single dispersal across the Bering Strait is required to account for their distribution. Because ornithomimids have not been reported yet from older Late Cretaceous strata in North America, Makovicky et al. (2004) hypothesized that ornithomimids may have dispersed less readily across the Bering Strait than did clades such as hadrosaurids and pachycephalosaurs.

Ornithomimosaurs have recently been reported in Europe on the basis of a fragmentary humerus discovered in Late Maastrichtian marine deposits from Bulgaria (Mateus et al., 2010). However, this identification is uncertain, and the presence of ornithomimosaurs in Upper Cretaceous deposits from Europe, following the separation of Asia and Europe by the Turgai Strait after the Albian (Smith et al., 1994), therefore remains speculative.

Appendix 26.1. Measurements taken on JLUM-JZ07b1

L, left; R, right.

Skull: length: 136 mm
maximum height: 53 mm
Orbit (R): length: 31.5 mm
height: 37.8 mm
Snout: length (rostral point of orbit–tip of premaxilla): 80 mm
Antorbital fossa (R): length: 55 mm
Antorbital fenestra (R): length: 29.5 mm
Height: 24 mm
External naris (R): length: 7 mm
Mandible (R): length: 115 mm
Scapula (R): length: ~104 mm
height at distal end: 10 mm
height of proximal plate: 14 mm
Humerus (L): length: ~90 mm
width at deltopectoral crest: 19 mm
width at distal end: 16 mm
Ulna (L): length: 81 mm
width at distal end: 9 mm
height (craniocaudal diameter) at distal end: 9 mm
Radius (L): length: ~76 mm
height (craniocaudal diameter) at distal end: 6 mm
width at distal end: 10 mm
Manus (L): length: ~103 mm
"Metacarpal II" (L): length: 33 mm
proximal height: 4.5 mm
Metacarpal III (L): length: > 31 mm
Metacarpal IV (L): length: > 30 mm
Manual ungual II (L): length: 22 mm
proximal height: 8.5 mm
Manual phalanx III-1 (L): length: 25 mm
proximal height: 5 mm
Manual phalanx III-2 (L): length: 26 mm
proximal height: 5 mm

Manual ungual III (L): length: 23 mm
proximal height: 9 mm
Manual phalanx IV-1 (L): length: 23 mm
proximal height: 4.5 mm
Manual phalanx IV-2 (L): length: 24 mm
proximal height: 5 mm
Manual ungual IV (L): length: 18 mm
proximal height: 7 mm
Femur (L): length: 135 mm
Tibia (R): length: >185 mm
Metatarsal II (R): length: 86 mm
Metatarsal III (R): length: 84 mm
Metatarsal IV (R): length: ~79 mm
Pes phalanx I-1 (R): length: 23 mm
Pes ungual I (R): length: 21 mm
Pes phalanx II-1 (R): length: 22 mm
proximal height: 9 mm
Pes phalanx II-2 (R): length: 18 mm
proximal height: 8 mm
Pes ungual II (R): length: 20 mm
proximal height: 9 mm
Pes phalanx III-1 (R): length: 18 mm
proximal height: 9 mm
Pes phalanx III-2 (R): length: 16 mm
proximal height: 7.5 mm
Pes phalanx III-3 (R): length: 15 mm
proximal height: 7.5 mm
Pes ungual III (R): length: 16 mm
proximal height: 9 mm
Pes phalanx IV-1 (R): length: 13.5 mm
proximal height: 7 mm
Pes phalanx IV-2 (R): length: 13 mm
proximal height: 6 mm
Pes phalanx IV-3 (R): length: 12 mm
proximal height: 6 mm
Pes phalanx IV-4 (R): length: 12 mm
proximal height: 6 mm
Pes ungual IV (R): length: 16 mm
proximal height: 7.5 mm

The following list only includes characters that have been modified from Kobayashi and Lü (2003) or that are not included in Kobayashi and Lü's (2003) original matrix.

9. Ventral edge of dentary: straight (0), or rostral portion reflected ventrally (1).
37. Proximal end of metatarsal III: completely exposed in cranial view (0), subarctometatarsal metatarsus (1), or arctometatarsal metatarsus (2). Character treated as ordered.
39. Dorsal edge of dentary: straight (0), or rostral portion concave resulting in a gap between upper and lower jaws when jaws are closed (1).
40. Retroarticular process of mandible: deflected ventrally (0), or curved caudodorsally (1) (Makovicky et al., 2004).
41. Surangular flange: absent (0) or present (1) (Makovicky et al., 2004).
42. Proximal and distal ends of humerus: in the same plane (0), or twisted (1). (Kobayashi and Barsbold, 2005a).
43. Radius and ulna: well separated distally (0), or closely appressed distally (1) (Nicholls and Russell, 1985; Makovicky et al., 2004).
44. Ischial shaft: straight (0), or sigmoid (1) (Ji et al., 2003).
45. Postacetabular process of ilium: gently curved (0), or truncated (1) (Ji et al., 2003).
46. Medial expansion of metatarsal III diaphysis: absent (0), or present (1) (Makovicky et al., 2010).
47. Pedal unguals: curved (0), or straight (1) (Makovicky et al., 2010).

Characters 1–38: see Kobayashi and Lü (2003, appendix 1), characters 9 and 37 modified—see Appendix 26.1; characters 39–47: see Appendix 26.1.

Allosaurus	00000	00000	00000	00000	00000	00000	00000
	00000	00000	00				
Tyrannosaurids	00010	00000	00000	00000	00000	10000	?0000
	02000	00000	00				
Pelecanimimus	0?0??	0?100	10???	0?1?1	????1	??110	111??
	???00	101??	??				
Hexing	1011?	10?00	10101	?1?01	?1001	10110	?01??
	00010	101??	00				
Shenzhousaurus	10111	1??00	10???	0????	?????	?????	??110
	???10	1??00	??				
Harpymimus	10111	10?10	10101	1110?	01001	10000	0110?
	?1010	10111	00				
Beishanlong	?????	?????	?????	????1	01001	10??0	??1??
	010??	?011?	00				
Garudimimus	11111	11111	11111	11?0?	?????	?????	???00
	01111	1???1	10				
Archaeornithomimus	?????	?????	?????	???01	01001	20110	1?110
	?2???	??111	11				
Sinornithomimus	11111	11111	111?0	1?111	00001	20110	11100
	12111	11111	11				
Anserimimus	?????	?????	?????	???11	10100	21211	01100
	121??	?1?1?	1?				
Gallimimus	11111	11111	11100	00111	11101	21111	11100
	12111	11111	11				

Struthiomimus	11110	01?11	11110	11111	01001	?0111	11101
	12111	11111	11				
Dromiceiomimus	1111?	11?11	1??10	1???1	00011	10???	??101
	12111	11?11	11				
Ornithomimus	11110	11?11	11111	11111	00011	20211	11101
	12111	11111	11				

Appendix 26.4. Tree Description

The "describetrees" option of PAUP*4.0b10 was used to interpret character state transformations. All transformations are based upon the derivative strict reduced consensus tree (see Fig. 26.8). Transformation was evaluated under accelerated transformation (ACCTRAN) and delayed transformation (DELTRAN) options; unambiguous synapomorphies are those that diagnose a node under both ACCTRAN and DELTRAN optimization. Node numbers refer to Fig. 26.8. For simple 0–1 state changes, only the character number is given; for other state changes, the type of change is specified in parentheses.

Node A (Ornithomimosauria): Unambiguous: 8, 11, 18, 20, 25, 28, 29, 33, 41, 43; ACCTRAN: 5, 13, 15, 17, 22, 32.
Node B: Unambiguous: 1, 3, 6, 39; DELTRAN: 5.
Node C: Unambiguous: 9, 16, 37, 44, 45; DELTRAN: 32.
Node D: Unambiguous: 2, 7, 10, 12, 38, 40, 46; ACCTRAN: 14, 26, 42.
Node E (Ornithomimidae): Unambiguous: 37 (1 to 2), 47; ACCTRAN: 15, 36; DELTRAN: 26, 31.
Node F: Unambiguous: 19; ACCTRAN: 22 (1 to 0); DELTRAN: 15 (1 to 0), 36, 42.
Node G: Unambiguous: 30.
Node H: Unambiguous: 21, 23, 27; ACCTRAN: 14 (1 to 0), 16 (1 to 0), 17 (1 to 0).
Node I: Unambiguous: 5 (1 to 0), 35; DELTRAN: 14.
Node J: Unambiguous: 24; ACCTRAN: 28 (0 to 2); DELTAN: 22 (1 to 0).

Acknowledgments

The authors are grateful to P. J. Makovicky for reviewing an earlier version of this paper and making useful recommendations.

References

Agnolin, F. L., M. D. Ezcurrah, D. F. Pais, and S. W. Salisbury. 2010. A reappraisal of the Cretaceous non-avian dinosaur faunas from Australia and New Zealand: evidence for their Gondwanan affinities. Journal of Systematic Palaeontology 8: 257–300.

Averianov, A. O. 2002. An ankylosaurid (Ornithischia: Ankylosauria) braincase from the Upper Cretaceous Bissekty Formation of Uzbekistan. Bulletin de l'Institut royal des Sciences naturelles de Belgique, Sciences de la Terre 72: 97–110.

Barrett, P. M. 2005. The diet of ostrich dinosaurs (Theropoda: Ornithomimosauria). Palaeontology 48: 347–358.

Barsbold, R. 1976. On the evolution and systematics of the late Mesozoic dinosaurs. Paleontologi i biostratigrafi Mongolii, Sovmestna Svetsko–Mongolska Paleontologiceska Ekspedici Trudy 3: 68–75.

Currie, P. J. 1998. Possible evidence of gregarious behavior in tyrannosaurids. Gaia 15: 271–277.

Doré, A. G. 1991. The structural foundation and evolution of Mesozoic seaways between Europe and the Arctic. Palaeogeography, Palaeoclimatology, Palaeoecology 87: 441–492.

Gauthier, J. A. 1986. Saurischian monophyly and the origin of birds; pp. 1–55 in K. Padian (ed.), The origin of birds and the evolution of flight. Memoirs of the California Academy of Sciences 8.

Holtz, T. R., Jr. 1994. The phylogenetic position of Tyrannosauridae: implication for theropod systematic. Journal of Paleontology 68: 1100–1117.

Huene, F. von. 1914. Das natürliche System der Saurischia. Zentralblatt für Mineralogie, Geologie und Paläontologie B 1914: 69–82.

Ji, Q., M. A. Norell, P. J. Makovicky, K.-Q. Gao, S. Ji, and C. X. Yuan. 2003. An early ostrich dinosaur and implications for ornithomimosaur phylogeny. American Museum Novitates 3420: 1–19.

Kobayashi, Y., and R. Barsbold. 2005a. Reexamination of a primitive ornithomimosaur, *Garudimimus brevipes* Barsbold, 1981 (Dinosauria: Theropoda), from the Late Cretaceous of Mongolia. Canadian Journal of Earth Sciences 42: 1501–1521.

———. 2005b. Anatomy of *Harpymimus okladnikovi* Barsbold and Perle 1984 (Dinosauria: Theropoda) of Mongolia; pp. 97–126 in K. Carpenter (ed.), The carnivorous dinosaurs. Indiana University Press, Bloomington.

Kobayashi, Y., and J.-C. Lü. 2003. A new ornithomimids dinosaur with gregarious habits from the Late Cretaceous of China. Acta Palaeontologica Polonica 48: 235–259.

Kobayashi, Y., J.-C. Lü, Z.-M. Dong, R. Barsbold, Y. Azuma, and Y. Tomida. 1999. Palaeobiology: herbivorous diet in an ornithomimid dinosaur. Nature 402: 480–481.

Makovicky, P. J., and M. A. Norell. 1998. A partial ornithomimids braincase from Ukhaa Tolgod (Upper Cretaceous, Mongolia). American Museum Novitates 3247: 1–16.

Makovicky, P. J., Y. Kobayashi, and P. J. Currie. 2004. Ornithomimosauria; pp. 137–150 in D. B. Weishampel, P. Dodson, and H. Osmólska (eds.), The Dinosauria. 2nd ed. University of California Press, Berkeley.

Makovicky, P. J., D. Li, K.-Q. Gao, M. Lewin, G. M. Erickson, and M. A. Norell. 2010. A giant ornithomimosaur from the Early Cretaceous of China. Proceedings of the Royal Society B 277: 191–198.

Marsh, O. C. 1881. Classification of the Dinosauria. *American Journal of Science* (ser. 3) 23: 81–86.

Mateus, O., G. J. Dyke, N. Motchurova-Dekova, J. D. Kamenov, and P. Ivanov. 2010. The first record of a dinosaur from Bulgaria. Lethaia 43: 88–94.

Nicholls, E. L., and A. P. Russell. 1981. A new specimen of *Struthiomimus altus* from Alberta with comments on the classificatory characters of Upper Cretaceous ornithomimids. Canadian Journal of Earth Sciences 18: 518–526.

Norell, M. A., P. J. Makovicky, and P. J. Currie. 2001. The beaks of ostrich dinosaurs. Nature 412: 873–874.

Osmólska, H., E. Roniewicz, and R. Barsbold. 1972. A new dinosaur, *Gallimimus bullatus* n. gen., n. sp. (Ornithomimidae), from the Upper Cretaceous of Mongolia. Palaeontologica Polonica 27: 103–143.

Owen, R. 1842. Report on British fossil reptiles. Part 2. Report of the British Association for the Advancement of Science (Plymouth) 11: 60–204.

Pérez-Moreno, B. P., J. L. Sanz, A. D. Buscalioni, J. J. Moratalla, F. Ortega, and D. Rasskin-Gutman. 1994. A unique multitoothed ornithomimosaur dinosaur from the Lower Cretaceous of Spain. Nature 370: 363–367.

Rich, T. H., and P. Vickers-Rich. 1994. Neoceratopsians and ornithomimosaurs: dinosaurs of Gondwanan origin? National Geographic Research and Exploration 10: 129–131.

Russell, D. A. 1993. The role of Central Asia in dinosaurian biogeography. Canadian Journal of Earth Sciences 30: 2002–2012.

Smith, A. G., D. G. Smith, and B. M. Funnell. 1994. Atlas of Mesozoic and Cenozoic coastlines. Cambridge University Press, Cambridge, 99 pp.

Swisher, C. C., III, X. Wang, Z. Zhou, Y. Wang, F. Jin, J. Zhang, X. Xu, F. Zhang, and Y. Wang. 2002. Further support for a Cretaceous age for the feathered-dinosaur beds of Liaoning, China: new ^{40}Ar/^{39}Ar dating of the Yixian and Tuchengzi Formations. Chinese Science Bulletin 47: 135–138.

Swofford, D. L. 2000. Phylogenetic Analysis Using Parsimony (and other methods). Version 4.0b10. Sinauer Associates, Sunderand, Mass., 40 pp.

Van Itterbeeck, J., D. J. Horne, P. Bultynck, and N. Vandenberghe. 2005. Stratigraphy and palaeoenvironment of the dinosaur-bearing Upper Cretaceous Iren Dabasu Formation, Inner Mongolia, People's Republic of China. Cretaceous Research 26: 699–725.

Varricchio, D. J., P. C. Sereno, X.-J. Zhao, L. Tan, J. A. Wilson, and G. H. Lyon. 2008. Mud-trapped herd captures evidence of distinctive dinosaur sociality. Acta Palaeontologica Polonica 53: 567–578.

Xu, X., J. M. Clark, J. Mo, J. Choiniere, C. A. Forster, G. M. Erickson, D. W. E. Hone, C. Sullivan, D. A. Eberth, S. Nesbitt, Q. Zhao, R. Hernandez, C.-K. Jia, F.-L. Han, and Y. Guo. 2009. A Jurassic ceratosaur from China helps clarify avian digital homologies. Nature 459: 940–944.

27.1. Outline map showing the location of the fluviatile Early Cretaceous coastal outcrops in southeastern Australia together with the lacustrine Early Cretaceous Koonwarra site.

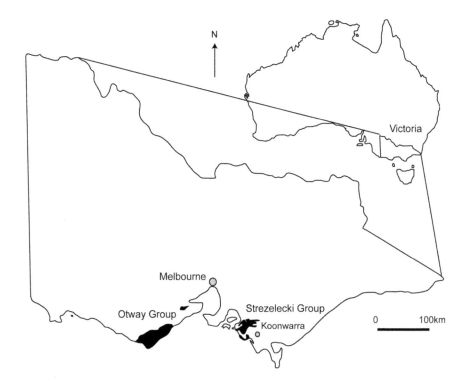

Australia's Polar Early Cretaceous Dinosaurs

27

Thomas H. Rich* and Patricia Vickers-Rich

Although meager in terms of the number and completeness of the specimens known of it, the Early Cretaceous dinosaurs of southeastern Australia are notable for being a significant part of a polar tetrapod assemblage that during the Aptian occupied the coldest terrestrial region known on earth for the entire Mesozoic Era. As such, it provides a unique perspective regarding the physiology of dinosaurs.

Facing the Southern Ocean in the southeastern corner of Australia are coastal exposures locally termed shore platforms. Fossiliferous Early Cretaceous fluviatile deposits extending for just under 150 km form a significant part of these shore platforms (Fig. 27.1). Because the typical width of these shore platforms is about 20 m, there are roughly 3 km² of outcrop that has yielded a variety of polar Cretaceous fossil vertebrates. Although the same rock units occur inland, they are typically vegetated, and where they do occasionally crop out, they are deeply weathered, unlike the fresh coastal outcrops, which are typically unweathered and well exposed as a result of active erosion by the ocean waves pounding against them (Figs. 27.2, 27.3).

Institutional abbreviation. NMV P, Paleontology Collection of the Museum Victoria, Melbourne, Australia.

Background

There are two main areas of continental dinosaur-bearing Cretaceous rocks in southeastern Australia. They flank the city of Melbourne, with one to the east and a more extensive one to the west (Fig. 27.1). Those yielding fossil vertebrates to the east make up the Wonthaggi Formation of the Strzelecki Group (Fig. 27.4). Where fossil tetrapods occur in this unit, palynomorphs date it as Early Aptian (Rich et al., 1999). More extensive in extent than the coastal outcrops of the Wonthaggi Formation are those of the Eumerella Formation, Otway Group, to the west of Melbourne (Fig. 27.5). Where fossil tetrapods occur in that unit, palynomorphs date it as Early Albian (Wagstaff and McEwen-Mason, 1989).

Although fossil tetrapods have been recovered from a number of sites on these coastal outcrops, the bulk of specimens have come from two localities, Dinosaur Cove in the Eumerella Formation of the Otway Group (Figs. 27.6–27.7) and Flat Rocks in the Wonthaggi Formation of the Strzelecki Group (Figs. 27.8–27.9). Except for 1988, the former was worked annually from 1984 to 1994. The latter has been worked annually from 1994 to the present.

Geology and Biota

Taphonomy

27.2. Unweathered outcrop of the Eumerella Formation of the Otway Group in Dinosaur Cove. The device pictured is the basket of a flying fox, 305 m long, utilized to lower equipment 90 m vertically into Dinosaur Cove and raise those same items as well as the excavated fossiliferous rocks.

Photograph by Sally Cohen.

The Cretaceous rocks of Victoria have yielded bones and teeth of fossil tetrapods in four different facies. Fossil feathers have been found in a fifth one. The general setting of these fossil sites was a floodplain that formed the floor of the rift valley between Australian and Antarctica as those two landmasses were in the initial stage of separation from one another during the Late Mesozoic Era.

Across this floodplain flowed broad rivers in a generally westward direction, much like the modern Lena River of northern Siberia. The sea lay about 2,000 km to the west of the fossil sites during the Aptian but had encroached eastward by the Albian to only a few hundred kilometers west of them (Fig. 27.10). The physical evidence for this environmental interpretation was described by Bryan et al. (1997, 91) as a "series of multi-storied sheet flood to braided-river-like channel complexes, up to 200 m thick, separated by overbank sequences ranging between 5 and 100 m in thickness. Internally, the channel complexes are characterized by thick packages of planar cross-stratified, massive, and low-angle cross-stratified sandstone. The range of lithofacies coupled with the remarkable uniformity in sand grain size suggests the sediments were deposited by a fluvial system subject to high energy discharge events." Isolated fossil tetrapod bones have been found in the sheet flood to braided river–like channel facies, but only rarely.

The bulk of the fossil tetrapod bones and teeth found in Victoria have been recovered from lag concentrations in the overbank deposits mentioned by Bryan et al. (1997). These lag deposits are thought to have formed as the result of evulsion events occurring when levees of large rivers were

27.3. Weathered outcrop of the Eumerella Formation, Otway Group, about 1 km inland from the Southern Ocean and 4 km from Dinosaur Cove.

breached during floods, and as a consequence, narrower side channels were cut into the surrounding floodplain (Rich et al., 2010). During the initial phase of the cutting of side channels, debris on the floodplain was swept up and came to rest at the base of the newly formed watercourses as the water velocity, and hence its carrying capacity, declined. This type of lag consist for the most part of clay gall clasts, pieces of carbon, and, rarely, isolated tetrapod bones and teeth. These lag elements in the sandstones are confined exclusively to the 10 cm immediately above the contact between the base of the paleochannels and the underlying finer-grained sediments into which the newly formed watercourses were cut. Above such basal lag deposits, the paleochannel deposits typically grade into a more uniform sandstone containing no larger clasts. This type of deposit was particularly well developed at Dinosaur Cove and Flat Rocks.

In this facies, the most frequently recovered elements of dinosaurs are the isolated femora, vertebrae, teeth, maxillae, and jaws of small ornithopods. By contrast, mammals are known almost exclusively from mandible fragments. Fossils of larger dinosaurs are rare; when present at all, they are represented by the smallest diagnostic elements of such animals—for example, the astragalus of cf. *Australovenator* (Fig. 27.11). Presumably, only such relatively small skeletal elements could be transported to the sites, where the bones were eventually concentrated by relatively small streams incapable of transporting larger elements. It is abundantly clear that in all the various fossil-yielding facies, the different depositional mechanisms all resulted in a highly biased sample.

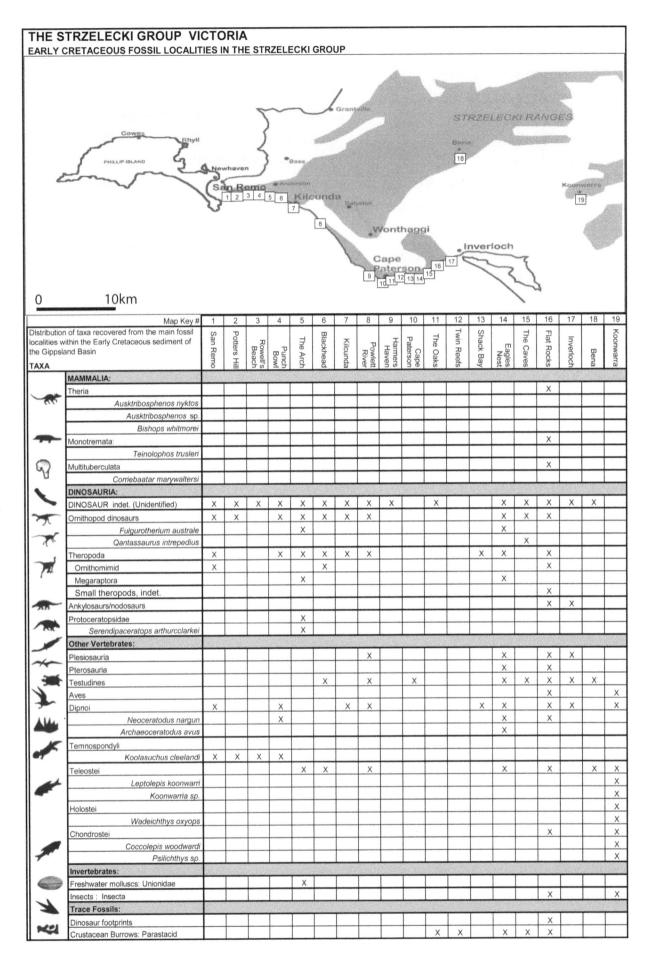

0   10km

Map Key #	1	2	3	4	5	6	7	8	9	10	11	12	13	14	15	16	17	18	19
Distribution of taxa recovered from the main fossil localities within the Early Cretaceous sediment of the Gippsland Basin — TAXA	San Remo	Potters Hill	Rowell's Beach	Punch Bowl	The Arch	Blackhead	Kilcunda	Powlett River	Harmers Haven	Cape Paterson	The Oaks	Twin Reefs	Shack Bay	Eagles Nest	The Caves	Flat Rocks	Inverloch	Bena	Koonwarra
MAMMALIA:																			
Theria																X			
Ausktribosphenos nyktos																			
Ausktribosphenos sp.																			
Bishops whitmorei																			
Monotremata:																X			
Teinolophos trusleri																			
Multituberculata																X			
Corriebaatar marywaltersi																			
DINOSAURIA:																			
DINOSAUR indet. (Unidentified)	X	X	X	X	X	X	X	X	X		X			X	X	X	X	X	
Ornithopod dinosaurs	X	X		X	X	X	X	X						X	X	X			
Fulgurotherium australe					X									X					
Qantassaurus intrepedius															X				
Theropoda	X			X	X	X	X						X	X		X			
Ornithomimid	X				X											X			
Megaraptora					X									X					
Small theropods, indet.																X			
Ankylosaurs/nodosaurs																X	X		
Protoceratopsidae					X														
Serendipaceratops arthurcclarkei					X														
Other Vertebrates:																			
Plesiosauria								X						X		X	X		
Pterosauria														X		X			
Testudines						X		X		X				X	X	X	X	X	
Aves																X			X
Dipnoi	X			X			X						X	X		X	X		
Neoceratodus nargun				X										X	X				
Archaeoceratodus avus														X					
Temnospondyli																			
Koolasuchus cleelandi	X	X	X	X															
Teleostei						X	X		X					X		X		X	X
Leptolepis koonwarri																			X
Koonwarria sp.																			X
Holostei																			X
Wadeichthys oxyops																			X
Chondrostei																X			X
Coccolepis woodwardi																			X
Psilichthys sp.																			X
Invertebrates:																			
Freshwater molluscs: Unionidae					X														
Insects : Insecta																X			X
Trace Fossils:																			
Dinosaur footprints																X			
Crustacean Burrows: Parastacid											X	X		X	X	X			

At the Slippery Rocks site within Dinosaur Cove, there is a variation on this general pattern. There, about 30 cm above the base of the paleochannel, sandstone grades upward into a massive claystone. The claystone is interpreted as an oxbow deposit, about 5 m wide, which developed after a significant quantity of water pooled in the paleochannel, which preceded it (Rich et al., 2010). In this oxbow facies, two partially articulated skeletons of the ornithopod *Leaellynasaura amicagraphica* were found (Fig. 27.12) (Gross et al., 1993; Rich and Rich, 1989; Rich et al., 2010). A third ornithopod partial articulated skeleton, tentatively regarded as a taxon other than *Leaellynasaura amicagraphica*, was discovered in the Eumerella Formation about 20 km southeast of Dinosaur Cove at the Eric the Red West locality (Fig. 27.5). In contrast to the two partial skeletons of *L. amicagraphica*, it was found in a high-energy channel deposit. This at least partially intact corpse evidently became entrapped in debris snagged around an upright tree or stump while floating in an actively flowing stream, rather than being part of the bed load of isolated bones (Rich et al., 2009a).

In yet another facies, half a dozen fossil feathers have been reported from lacustrine deposits at the Aptian Koonwarra locality (Waldman, 1971). Both the lithology and the associated biota are remarkably similar to that on display at the visitor facility of the Sihetun Landscape Fossil Bird National Geopark of Chaoyang City, Liaoning, China (Rich et al., 2009b; see Chapter 28 in this book). At that facility, articulated skeletons of 30 tetrapods are on display in situ. Because of the depositional similarities, a program to search for sites similar to Koonwarra in the adjacent area is being undertaken in a deliberate attempt to find similarly preserved Cretaceous fossil tetrapods in Victoria (Rich et al., 2009b; see Chapter 28 in this book).

27.4. Fossiliferous sites in the Wonthaggi Formation of the Strzelecki Group with an indication of the faunal elements recovered from each.

Paleoenvironment

The paleolatitude of southeastern Australia during the Cretaceous has been estimated to be 70°–80° (Embleton, 1984) (Fig. 27.10). Specific determinations of the paleolatitude of the Otway and Strezlecki groups, based on paleomagnetic analysis of samples taken from these units, similarly indicate paleolatitudes of 67° and 78°, respectively (Whitelaw, 1993). Whether or not the temperature was frigid, the animals living in the area would certainly have had to deal with a prolonged period of polar winter darkness annually, because the inclination of the earth's axis relative to the plane of the ecliptic has only varied within about 2 degrees over geological time (Laplace, 1798–1827).

On the basis of a study of the ratio of $^{18}O/^{16}O$ in carbonate concretions, Gregory et al. (1989) estimated that the mean annual temperature that prevailed during the deposition of the Otway and Strzelecki groups was no more than +5°C, and possibly below the freezing point of water.

Furthermore, Constantine et al. (1998) recognized sedimentary structures in the Strzelecki Group indicative of the former presence of permafrost. One permafrost horizon lies only 3 m stratigraphically below the fossiliferous unit at Flat Rocks. Whether or not the climatic conditions cold enough to have generated permafrost continued during the time of deposition of the intervening 3 m of rock, thus indicative of extremely cold conditions at the time the animals lived whose fossils at found at Flat Rocks, is yet to be resolved.

THE OTWAY GROUP, VICTORIA
EARLY CRETACEOUS FOSSIL LOCALITIES IN THE OTWAY GROUP

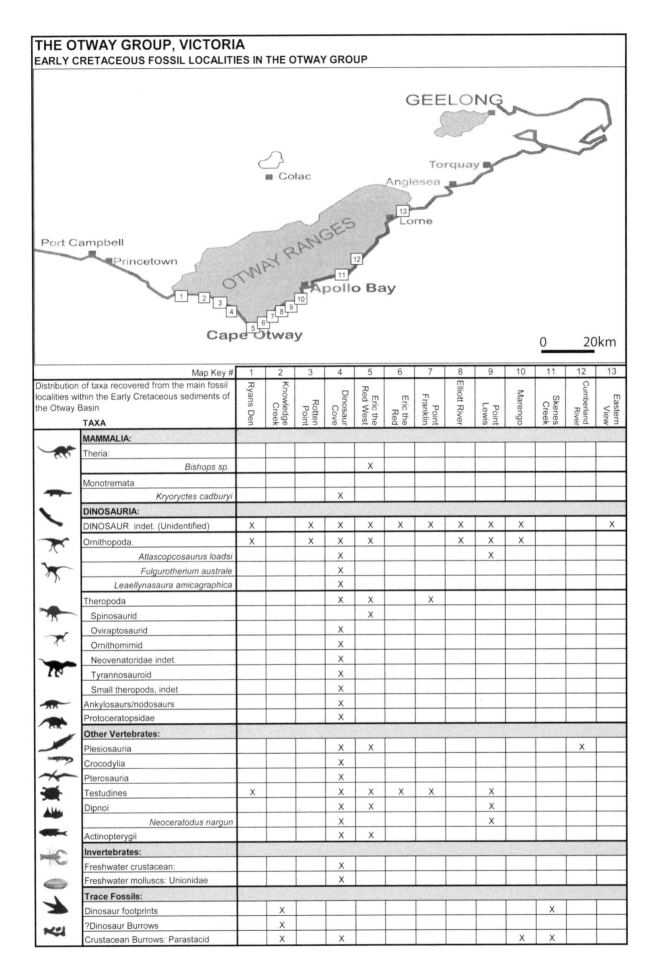

Map Key #	1	2	3	4	5	6	7	8	9	10	11	12	13
Distribution of taxa recovered from the main fossil localities within the Early Cretaceous sediments of the Otway Basin **TAXA**	Ryans Den	Knowledge Creek	Rotten Point	Dinosaur Cove	Eric the Red West	Eric the Red	Point Franklin	Elliott River	Point Lewis	Marengo	Skenes Creek	Cumberland River	Eastern View
MAMMALIA:													
Theria:													
Bishops sp.					X								
Monotremata													
Kryoryctes cadburyi				X									
DINOSAURIA:													
DINOSAUR indet. (Unidentified)	X		X	X	X	X	X	X	X	X			X
Ornithopoda.	X		X	X	X				X	X	X		
Atlascopcosaurus loadsi				X					X				
Fulgurotherium australe				X									
Leaellynasaura amicagraphica				X									
Theropoda				X	X		X						
Spinosaurid					X								
Oviraptosaurid				X									
Ornithomimid				X									
Neovenatoridae indet.				X									
Tyrannosauroid				X									
Small theropods, indet				X									
Ankylosaurs/nodosaurs				X									
Protoceratopsidae				X									
Other Vertebrates:													
Plesiosauria				X	X							X	
Crocodylia				X									
Pterosauria				X									
Testudines	X			X	X	X	X		X				
Dipnoi				X	X				X				
Neoceratodus nargun				X					X				
Actinopterygii				X	X								
Invertebrates:													
Freshwater crustacean:				X									
Freshwater molluscs: Unionidae				X									
Trace Fossils:													
Dinosaur footprints		X										X	
?Dinosaur Burrows		X											
Crustacean Burrows: Parastacid		X		X						X	X		

Forthcoming studies will address this problem in two ways. First is analysis by clumped-isotope geochemistry (Eagle et al., 2010; Eiler, 2007). Robert Eagle is analyzing fossil bone and enamel from the Flat Rocks site in an attempt to determine the mean annual temperature at the time the animals producing those fossils lived. Second, Barbara Wagstaff plans to investigate the variation in the ratio of the palyno-morphs in the 3-m stratigraphic interval between the permafrost and the overlying fossiliferous unit at Flat Rocks. Presumably a low degree of variation would indicate a continuation of those climatic conditions typical of the region at the time of permafrost formation. Conditions in Victoria during the Albian appear to have become warmer than during the Aptian. There are no indicators of extremely cold conditions: no ice wedges, no patterned ground, and no evidence of permafrost to the west in the Otway Group.

Thus far, the Early Cretaceous dinosaurs of southeastern Australia can be assigned to superfamilies, if not families, known elsewhere on the globe. There is no dinosaur as unique to Australia as the living koala. In agreement with Benson et al. (2010a), as more discoveries have been made of Cretaceous theropod groups in recent years, they are becoming known to be more globally widespread. Given the significant number of tetrapod taxa known from a single bone or tooth in Victoria, this generalization may certainly apply to other Cretaceous Australian dinosaurs, when more fossils are known (Fig. 27.11).

At present, two particularly controversial areas of research concerning the Early Cretaceous dinosaurs of Victoria are their systematic assignment

27.5. *(facing)* Fossiliferous sites in the Eumerella Formation of the Otway Group with an indication of the faunal elements recovered from each.

27.6. Dinosaur Cove looking west. The three fossil vertebrate sites within Dinosaur Cove were all located in the well-exposed rock of the intertidal area.

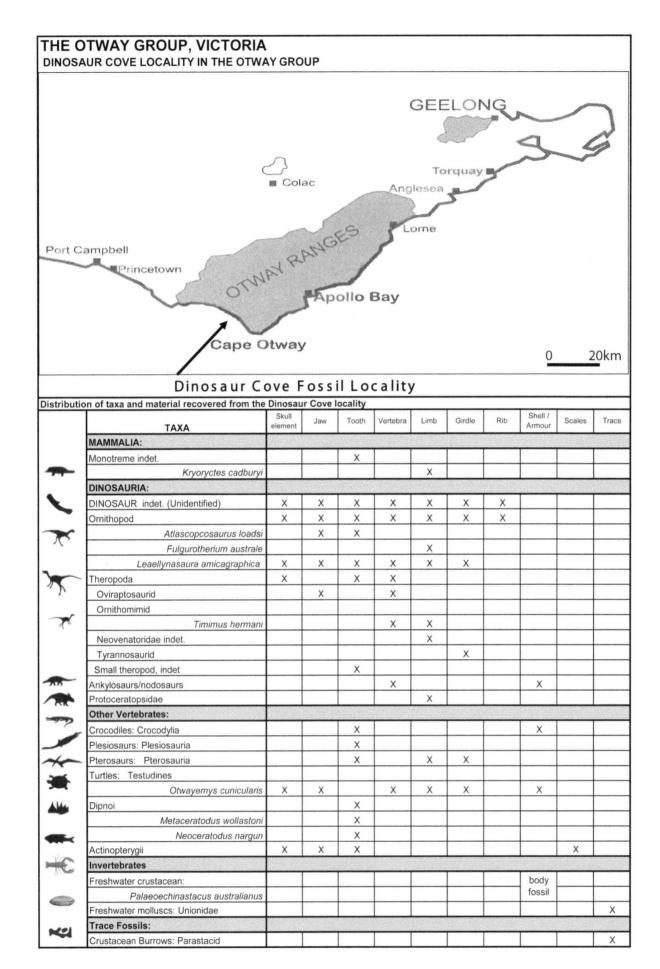

THE OTWAY GROUP, VICTORIA
DINOSAUR COVE LOCALITY IN THE OTWAY GROUP

GEELONG

■ Colac

Torquay ■

Anglesea ■

Lorne ■

Port Campbell

■ Princetown

OTWAY RANGES

■ Apollo Bay

Cape Otway

0   20km

Dinosaur Cove Fossil Locality

Distribution of taxa and material recovered from the Dinosaur Cove locality

TAXA	Skull element	Jaw	Tooth	Vertebra	Limb	Girdle	Rib	Shell / Armour	Scales	Trace
MAMMALIA:										
Monotreme indet.			X							
Kryoryctes cadburyi					X					
DINOSAURIA:										
DINOSAUR indet. (Unidentified)	X	X	X	X	X	X	X			
Ornithopod	X	X	X	X	X	X	X			
Atlascopcosaurus loadsi		X	X							
Fulgurotherium australe					X					
Leaellynasaura amicagraphica	X	X	X	X	X	X				
Theropoda	X		X	X						
Oviraptosaurid		X		X						
Ornithomimid										
Timimus hermani				X	X					
Neovenatoridae indet.					X					
Tyrannosaurid						X				
Small theropod, indet			X							
Ankylosaurs/nodosaurs				X				X		
Protoceratopsidae					X					
Other Vertebrates:										
Crocodiles: Crocodylia			X					X		
Plesiosaurs: Plesiosauria			X							
Pterosaurs: Pterosauria			X		X	X				
Turtles: Testudines										
Otwayemys cunicularis	X	X		X	X	X		X		
Dipnoi			X							
Metaceratodus wollastoni			X							
Neoceratodus nargun			X							
Actinopterygii	X	X	X						X	
Invertebrates										
Freshwater crustacean:								body fossil		
Palaeoechinastacus australianus										
Freshwater molluscs: Unionidae										X
Trace Fossils:										
Crustacean Burrows: Parastacid										X

and the interpretation of their biogeographic significance (Agnolin et al., 2010; Barrett et al., 2010; Benson et al., 2010a, 2010b, 2010c; Herne et al., 2010; Rich et al., 2010; Smith et al., 2008).

The temnospondyl *Koolasuchus cleelandi* Warren et al., 1997, from the Aptian Wonthaggi Formation of Victoria is much younger than any other known member of this group (Warren et al., 1997). For this reason, Warren et al. (1997) suggested that members of this amphibian group survived at high polar latitudes simply because they were not subject to competition by the ecologically and structurally similar eusuchian crocodilians. No record of temnospondyls is known in the Albian Eumeralla Formation of Victoria, where scutes and a quadratojugal assigned to eusuchian crocodilians have been recovered (R. E. Molnar, pers. comm.; Warren et al., 1997; Willis, 1997). Perhaps eusuchians were able to enter Victoria in the Albian when temperatures rose, as indicated by a study of oxygen isotopes (Gregory et al., 1989). Temnospondyls, if they had a physiology at all comparable to those modern amphibians, some of which are capable of being active in temperatures near 0°C (Brattstrom, 1970), may have been favored in polar Victoria after being completely displaced by eusuchians at lower and warmer paleolatitudes.

The ornithopod *Leaellynasaura amicagraphica* Rich and Rich, 1989, had enlarged optic lobes on the brain that may have enabled it to see under low-light conditions such as would be encountered during a polar winter (Rich and Rich, 1988). Although the specimen that forms the basis for this

Adaptations to Polar Conditions

27.7. (*facing*) Detailed faunal list by element of the tetrapod assemblages from Dinosaur Cove.

27.8. The Flat Rocks locality. Dinosaurs and other fossils occur in the excavation denoted by where the group of people standing furthest to the right and extending for another 20 m further to the right. Because sand is swept over the site with each tide, removing it from or keeping it off the fossiliferous rock is an ongoing technical difficulty of working this locality.

Photograph by Leley Kool.

THE STRZELECKI GROUP VICTORIA
FLAT ROCKS FOSSIL LOCALITY IN THE STRZELECKI GROUP

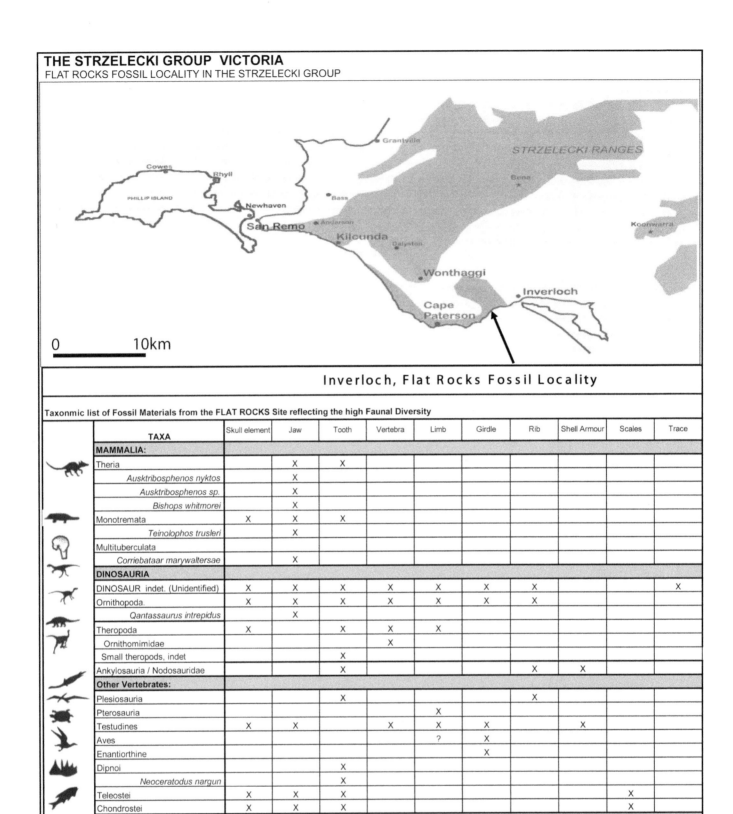

Inverloch, Flat Rocks Fossil Locality

Taxonmic list of Fossil Materials from the FLAT ROCKS Site reflecting the high Faunal Diversity

TAXA	Skull element	Jaw	Tooth	Vertebra	Limb	Girdle	Rib	Shell Armour	Scales	Trace
MAMMALIA:										
Theria		X	X							
Ausktribosphenos nyktos		X								
Ausktribosphenos sp.		X								
Bishops whitmorei		X								
Monotremata	X	X	X							
Teinolophos trusleri		X								
Multituberculata										
Corriebataar marywaltersae		X								
DINOSAURIA										
DINOSAUR indet. (Unidentified)	X	X	X	X	X	X	X			X
Ornithopoda.	X	X	X	X	X	X	X			
Qantassaurus intrepidus		X								
Theropoda	X		X	X	X					
Ornithomimidae				X						
Small theropods, indet			X							
Ankylosauria / Nodosauridae			X				X	X		
Other Vertebrates:										
Plesiosauria			X				X			
Pterosauria					X					
Testudines	X	X		X	X	X		X		
Aves					?	X				
Enantiorthine					X					
Dipnoi			X							
Neoceratodus nargun			X							
Teleostei	X	X	X						X	
Chondrostei	X	X	X						X	
Invertebrates:										
Unionidae										X
Trace Fossils:										
Dinosaur footprints										X
Crustacean Burrows: Parastacid										X

27.9. Detailed faunal list by element of the tetrapod assemblages from Flat Rocks.

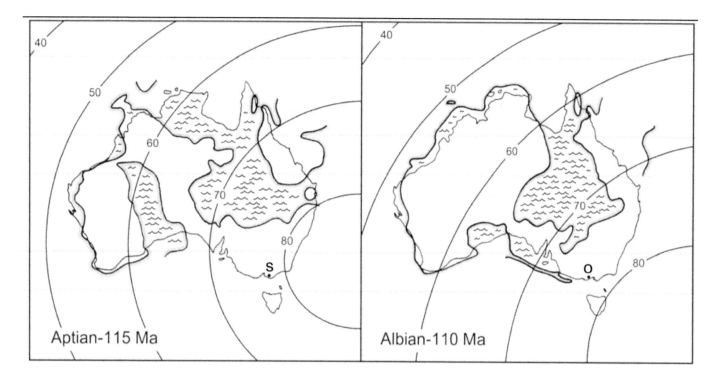

27.10. Late Early Cretaceous Australia showing the location of Strzelecki Group (S) on the Aptian paleogeographic map and Otway Group (O) on the Albian paleogeographic map. Paleoshorelines are in bold from Kear et al. (2006), and paleolatitudes are from Embleton (1984).

assertion is a juvenile, and thus the enlargement of the optic lobes might be interpreted as an ontogenetic effect, because vertebrate brains approach an adult size early in their growth (Gould, 1966), the large size of the optic lobes in *L. amicagraphica* cannot be simply dismissed as an exclusively juvenile feature.

As a consequence of the chemically intractable nature of the laumonite cement that tightly binds the clasts of the fluviatile deposits of the Wonthaggi and Eumerella formations, only one way has proven practical to extract fossils from these rocks. That technique is the tried and true (if admittedly primitive) method of physically breaking these rocks down (Fig. 27.13). Once the presence of fossil mammals was recognized at the Flat Rocks locality on March 8, 1997, the practice has been to break the fossiliferous rock to small fragments about 1 cm across. As this is done, with each break of a rock, the freshly exposed surfaces are examined for fossil bones and teeth. Although this method damages a significant number of fossils, it is frequently possible to glue the two pieces of rock containing a fossil so discovered back together with little loss, because a freshly broken surface seldom crumbles unless it is directly struck by a tool.

The rate of processing fossiliferous rock by this method is not high. At Flat Rocks, about 5 tonnes is examined during a field season of typically six weeks' duration, with usually 10–15 volunteers on site. Previously, at Dinosaur Cove, the rate of recovery was even slower than that, owing to the necessity of first cutting tunnels to access the fossiliferous rock at the most productive fossil site there, Slippery Rock. The ratio of overburden to fossiliferous rock at Slippery Rock was about 30:1. For about the same amount of annual effort as at Flat Rocks, approximately 3 tonnes of fossiliferous rock was processed at Dinosaur Cove (Rich and Vickers-Rich, 2000).

Collection Techniques

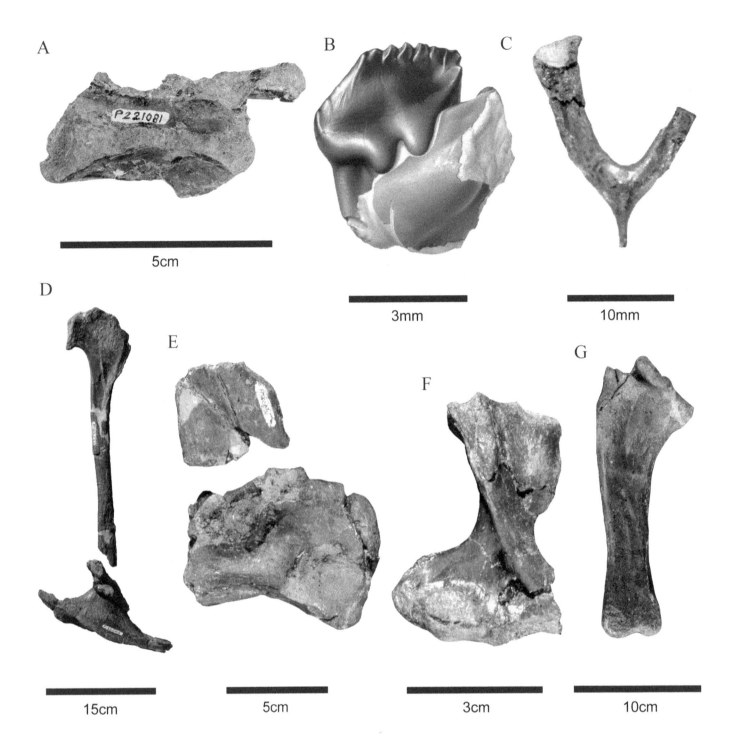

27.11. Cretaceous tetrapod taxa represented in Victoria by single specimens. That these seven taxa are each only known from the single fossil illustrated is indicative of how incomplete the record of Cretaceous tetrapods from Victoria is. If these few specimens were unknown, the diversity of Cretaceous tetrapods in southeastern Australia would be reduced by 32%! For that matter, with the exception of the astragalus referred to as cf. *Australovenator*, these few spemens are also the only records of their groups in the Cretaceous of Australia. A, Spinosaur cervical vertebra, NMV P221081. B, *Corriebaatar* premolar, NMV P216655. C, Enantiornithine furculum, NMV P208103. D, Tyrannosauroid pubis, NMV P186046. E, cf. *Australovenator* astragalus, NMV P150070. F, *Kryoryctes* humerus, NMV P208094. G, *Serendipaceratops,* NMV P186385.

27.12. The corpse of the holotype individual of *Leaellynasaura amicagraphica* as it might have appeared at an initial stage of becoming a fossil on the convex side of an oxbow about 5 m wide.

Artwork by Peter Trusler.

Although the Early Cretaceous dinosaur assemblage from southeastern Australia is modest in terms of the number and completeness of specimens, it has produced the most diverse polar dinosaur assemblage known to date (Rich et al., 2002). Because of the polar nature of this assemblage, it offers unique insights into dinosaur physiology that cannot be gleaned by analysis of specimens found at lower paleolatitudes.

Conclusions

We thank Lesley Kool, David Pickering, and John Wilkins for compiling the bulk of the information presented in Figs. 27.4, 27.5, 27.7, and 27.9, and Timothy Holland and Ben Kear for assistance with producing the necessary graphics. The fossils central to this report have been collected over three decades; this would not have happened without the assistance of literally hundreds of volunteers. Major financial support has been provided by the National Geographic Society (currently grant 8967-11) and the Australian Research Council (currently grant LP100100339), together with equally critical in-kind support by Atlas Copco. This work would not have been possible without their support.

Acknowledgments

Agnolin, F. L., M. D. Ezcurrah, D. F. Pais, and S. W. Salisbury. 2010. A reappraisal of the Cretaceous non-avian dinosaur faunas from Australia and New Zealand: evidence for their Gondwanan affinities. Journal of Systematic Palaeontology 8: 257–300.

Barrett, P., T. H. Rich, P. Vickers-Rich,

References

27.13. Physically breaking down the fossiliferous rock after it had been removed from the tunnels at the Slippery Rock site, Dinosaur Cove.

Photograph by Steve Morton.

T. A. Tumanova, M. Inglis, D. Pickering, L. Kool, and B. Kear. 2010. Ankylosaurian dinosaur remains from the Early Cretaceous of southeastern Australia. Alcheringa 34: 205–217.

Benson, R. B. J., M. T. Carrano, and S. L. Brusatte. 2010a. A new clade of archaic large-bodied predatory dinosaurs (Theropoda: Allosauroidea) that survived to the latest Mesozoic. Naturwissenschaften 97: 71–78.

Benson, R. B. J., P. M. Barrett, T. H. Rich, and P. Vickers-Rich. 2010b. First tyrant reptile from the southern continents. Science 327: 1613.

Benson, R. B. J., P. M. Barrett, T. H. Rich, and P. Vickers-Rich, D. Pickering, and T. Holland, T. 2010c. Response to comment on "A southern tyrant reptile." Science 329: 1013.

Brattstrom, B. H. 1970. Chapter 4. Amphibia; pp. 135–166 in G. C. Whittow (ed.), Comparative physiology of thermoregulation. Vol. 1, Invertebrates and nonmammalian vertebrates. Academic Press, New York.

Bryan, S. E., A. E. Constantine, C. J. Stephens, A. Ewart, R. W. Schoön, and J. Parianos. 1997. Early Cretaceous volcano-sedimentary successions along the eastern Australian continental margin: implications for the break-up of eastern Gondwana. Earth and Planetary Science Letters 153: 85–102.

Constantine, A., A. Chinsamy, P. Vickers-Rich, and T. H. Rich. 1998. Periglacial environments and polar dinosaurs. South African Journal of Science 94: 137–141.

Eagle, R. A., E. A. Schauble, A. K. Tripati, T. Tutken, R. C. Hulbert, and J. M. Eiler. 2010. Body temperatures of modern and extinct vertebrates from ^{13}C-^{18}O bond abundances in bioapatite. Proceedings of the National Academy of Sciences of the United States of America 107: 10377–10382.

Eiler, J. M. 2007. "Clumped-isotope" geochemistry—the study of naturally-occurring, multiply-substituted isotopologues. Earth and Sciences Planetary Letters 262: 309–327.

Embleton, B. J. 1984. Australia's global setting: past geological settings; pp. 11–17 in J. J Veevers (ed.), Phanerozoic earth history of Australia. Clarendon Press, Oxford.

Gould, S. J. 1966. Allometry and size in ontogeny and phylogeny. Biological Reviews 41: 587–640.

Gregory, R. T., C. B. Douthitt, I. R. Duddy, P. V. Vickers-Rich, and T. H. Rich. 1989. Oxygen isotopic composition of carbonate concretions from the lower Cretaceous of Victoria, Australia: implications for the evolution of meteoric waters on the Australian continent in a paleopolar environment. Earth and Planetary Science Letters 92: 27–42.

Gross, J. D., T. H. Rich, and P.Vickers-Rich. 1993. Dinosaur bone infection. National Geographic Society Research Reports 9: 286–293.

Herne, M. C., J. P. Nair, and S. W. Salisbury. 2010. Comment on "A southern tyrant reptile." Science 329: 1013.

Kear, B. P., N. I. Schroeder, P. Vickers-Rich, and T. H. Rich. 2006. Early

Cretaceous high latitude marine reptile assemblages from southern Australia. Paludicola 5: 200–205.

Laplace, P. S. 1798–1827. Traité de Mécanique Céleste. Vol. 2. Duprat, Paris.

Rich, T. H., and P. Vickers-Rich. 1988. A juvenile dinosaur brain from Australia. National Geographic Society Research Reports 4: 149.

———. 1989. Polar dinosaurs and biotas of the Early Cretaceous of southeastern Australia. National Geographic Society Research Reports 5: 15–53.

———. 2000. The dinosaurs of darkness. Indiana University Press, Bloomington, 222 pp.

Rich, T. H., P. Vickers-Rich, A. Constantine, T. F. Flannery, L. Kool, and N. van Klaveren. 1999. Early Cretaceous mammals from Flat Rocks, Victoria, Australia. Records of the Queen Victoria Museum 106: 1–34.

Rich, T. H., P. Vickers-Rich, and R. A. Gangloff. 2002. Polar dinosaurs. Science 295: 979–980.

Rich, T. H., P. Vickers-Rich, T. F. Flannery, D. Pickering, L. Kool, A. M. Tait, and E. M. G. Fitzgerald. 2009a. A fourth Australian Mesozoic mammal locality. Bulletin of the Museum of Northern Arizona 65: 677–681.

Rich, T. H., X.-B. Li, and P. Vickers-Rich. 2009b. A potential Gondwanan polar Jehol Biota lookalike in Victoria, Australia. Transactions of the Royal Society of Victoria 121: 5–13.

Rich, T. H., P. M. Galton, and P. Vickers-Rich. 2010. The holotype individual of the ornithopod dinosaur Leaellynasaura amicagraphica Rich & Rich, 1989 (late Early Cretaceous, Victoria, Australia). Alcheringa 34: 385–396.

Smith, N. D., P. J. Makovicky, F. L. Agnolin, M. D. Ezcurra, D. F. Pais, and S. W. Salisbury. 2008. A Megaraptor-like theropod (Dinosauria: Tetanurae) in Australia: support for faunal exchange across eastern and western Gondwana in the Mid-Cretaceous. Proceedings of the Royal Society of London B 275: 1085–1093.

Wagstaff, B. E., and J. McEwen-Mason. 1989. Palynological dating of lower Cretaceous coastal vertebrate localities, Victoria, Australia. National Geographic Research 5: 54–63.

Waldman, M. 1971. Fish from the freshwater Lower Cretaceous of Victoria, Australia, with comments on the palaeo-environment. Special Papers in Palaeontology: 1–129.

Warren, A., T. H. Rich, and P. Vickers-Rich. 1997. The last last labyrinthodont? Palaeontographica A247: 1–24.

Whitelaw, M. J. 1993. Paleomagnetic paleolatitude determinations for the Cretaceous vertebrate localities of southeastern Australia-high latitude dinosaur faunas. Journal of Vertebrate Paleontology 13 (supplement to 3): 62A.

Willis, P. M. A. 1997. Review of fossil crocodilians from Australia. Australian Zoologist 30: 287–298.

28.1. Comparison of preservation of different types of fossils from the Koonwarra site with comparable forms from various sites that have yielded the fossils in the Jehol Biota.

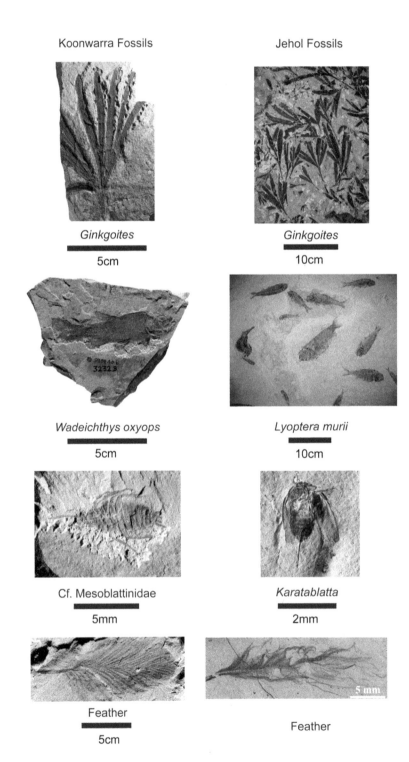

Koonwarra Fossils

Jehol Fossils

Ginkgoites

5cm

Ginkgoites

10cm

Wadeichthys oxyops

5cm

Lyoptera murii

10cm

Cf. Mesoblattinidae

5mm

Karatablatta

2mm

Feather

5cm

Feather

5 mm

Assessment of the Potential for a Jehol Biota–like Cretaceous Polar Fossil Assemblage in Victoria, Australia

28

Thomas H. Rich*, Li Xiao-Bo, and Patricia Vickers-Rich

An abundance of exquisitely preserved birds, mammals, and feathered dinosaurs, among other fossils, has been recovered from the Early Cretaceous Jehol Biota of northeastern China. The similarities, both in the nature of those deposits and the plant, arthropod, and fish fossils they contain, to those from the lacustrine facies of the Strzelecki Group of southwestern Gippsland, Victoria, Australia, suggest that a prolonged, systematic search of the latter could yield tetrapods of similar quality.

Although fossil remains of Early Cretaceous fish, arthropods, and plants have been known from the Jehol fossil province of northeastern China for more than a century, it has only been in the past two decades that the presence in those deposits of exquisitely preserved tetrapods have been recognized (Chang, 2003).

A visit in June 2009 by one of us (T.R.) to one site in particular where the Jehol Biota occurs—the visitor facility of the Sihetun Landscape Fossil Bird National Geopark of Chaoyang City—demonstrated that a striking similarity exists between it and the Aptian (Early Cretaceous) Koonwarra site in southwestern Gippsland, Victoria, Australia (Rich et al., 2009; Fig. 28.9, Fig. 27.1). In both are seen a suite of similar fossil arthropods, small fish, and plant matter (Fig. 28.1). The Koonwarra fossils have been described by Waldman (1971), Drinnan and Chambers (1986), and Jell and Duncan (1986), and the age determination was made by Dettmann (1986).

The fossils at the Sihetun Visitor Facility for the most part occur in fine-grained lacustrine deposits laid down in a quiet, anoxic environment with little or no sign of bioturbation. Specifically, the fossiliferous units are alternating dark and light layers of fine-grained shales, each individual layer between 0.1 and 1 mm thick. The laminar nature of the fossiliferous rock suggests that there was little or no bioturbation, and presumably the water column was stratified for at least part of the year, rather than mixed, and the bottom water was anoxic (Liu et al., 2002; Fürsich et al., 2007).

The lithology of the Koonwarra site is much the same with finely laminated shales indicative of a lacustrine environment, which likewise suggests an anoxic environment with little or no sign of bioturbation (Waldman 1971).

Half a dozen feathers from the Koonwarra site are the only evidence thus far for the presence of tetrapods. However, although the feathers are few in number, they are significant for demonstrating not only that

Background

28.2. Visitor facility of the Sihetun Landscape Fossil Bird National Geopark of Chaoyang City, Liaoning province, People's Republic of China. The area of approximately 400 m², where the 31 fossil vertebrates are exposed, is under the part of the building with the arched roof to the left of the main entrance. The area to the right has two floors of exhibits.

tetrapods were present in the area, but also that the preservation of such fossils did occur in the depositional environment at Koonwarra, just as it often did where the Jehol Biota is preserved. The similarities between the Jehol and Koonwarra sediments, together with the nature of the fossils that occur at both, suggest that tetrapod skeletons might be found at Koonwarra or elsewhere in the lacustrine facies of the Strzelecki Group if a concerted effort is made to do so.

Investigation of the Jehol Biota

At the visitor's center of the Sihetun Landscape Fossil Bird National Geopark of Chaoyang City, under cover in a quite substantial building, are exposed 31 fossil vertebrates found over an area of approximately 400 m² (Fig. 27.2). The fossils were seemingly distributed at random (Fig. 27.3). There were 26 plastic boxes enclosing the fossil tetrapods and one large fish (Figs. 27.4, 27.5). Most often there was one fossil vertebrate in each box, but three of these boxes each had two specimens, and there was another box containing three specimens. Because the purpose of this exhibit was to show the most prominent fossil vertebrates in place, none had been removed, and thus the distribution of the fossils and their abundance is accurately known.

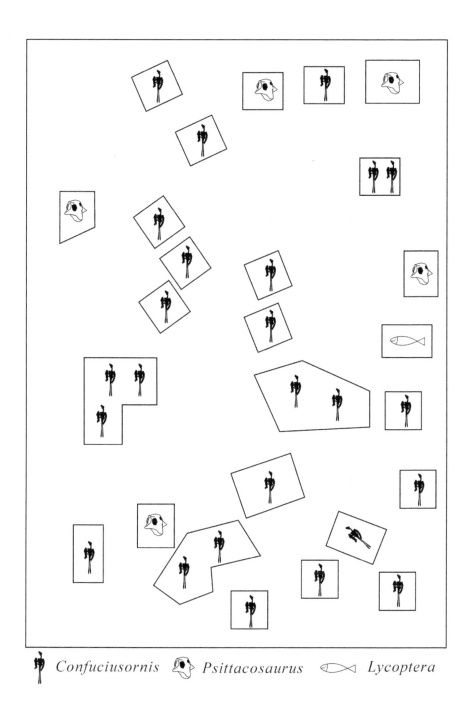

28.3. Map of the Sihetun fossil site. East is at the top, south is to the right. Each polygon represents one case containing one to three in situ fossils on the floor of the Sihetun Visitor Facility. In addition to the large specimens in cases, the fossil remains of many small fishes, arthropods, and plants are visible on the exposed rock surface. Not to scale as the vertical dimension represents about 40 m and the horizontal dimension, approximately 10 m.

Confuciusornis   *Psittacosaurus*   *Lycoptera*

All 31 of the encased vertebrate fossils on display in situ at the Sihetun Visitor Facility are complete or virtually complete skeletons. There are no isolated bones. Numerous fish, plants, and arthropod fossils were found, as well as the tetrapods, which, together with the fossiliferous rock (a fine-grained fissile shale), are all aspects reminiscent of the Koonwarra locality. The fossils on display at the Sihetun Visitor Facility occur within a dark shale rather than in one of the numerous tuffs so common in the area.

In 1998 and 1999, the Institute of Vertebrate Paleontology and Paleo-anthropology, Beijing, carried out excavations about 100 m north of the Sihetun Visitor Facility. Wang et al. (2000) indicated that the vertebrates collected during the course of these excavations occurred primarily through a thickness of 2 m of gray or grayish black shale.

28.4. Looking east at the gallery at the Sihetun Visitor Facility showing 26 illuminated plastic boxes, each containing one or more fossil vertebrate skeletons. Each skeleton is still in the place where it was uncovered. No fossil tetrapods have been removed from this area. The width of this space is about 10 m, and the depth from the point where the image was made is about 35 m.

Jiang and Sha (2007: 184; Figs. 28.6, 28.7) carried out a study of the depositional environments in the Sihetun area. One rock unit in their section, where "abundant non-marine . . . vertebrates" occur, is "3.2.8 Laminated mudstones (M2)" (Fig. 28.7). Their only other mention of non-marine vertebrates is in the unit "3.2.11 Channel-fill, cross-stratified pebbly sandstones (S3)," where turtle fragments occur locally. No fossil vertebrates were indicated in any of the other 10 lithologies identified in their three sections close to the Sihetun Visitor Facility or mentioned in their descriptions of them, including the tuffs.

Jiang and Sha (2007, 191) stated, "During volcanic eruptions, animals and plants living in or around the lake were killed, transported into the lake, and were rapidly buried by abundant volcaniclastic sediments." Rapid burial related to volcanic events, however, seems to be contrary to their characterization of the unit M2 where most of the fossil tetrapods occur: "greyish-green and dark grey, planar-laminated volcanic siltstones, silty claystones and claystones (shales), intercalated with gypsum and calcareous

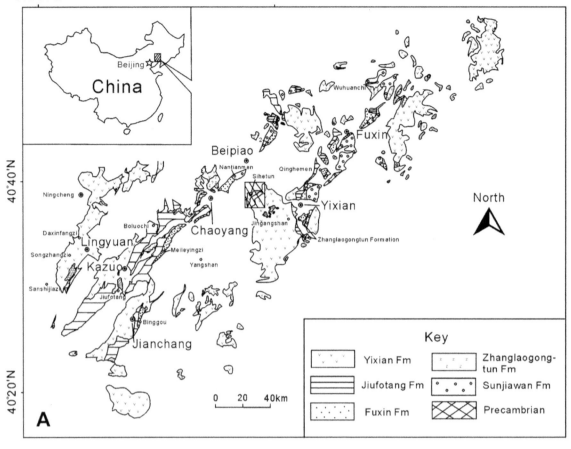

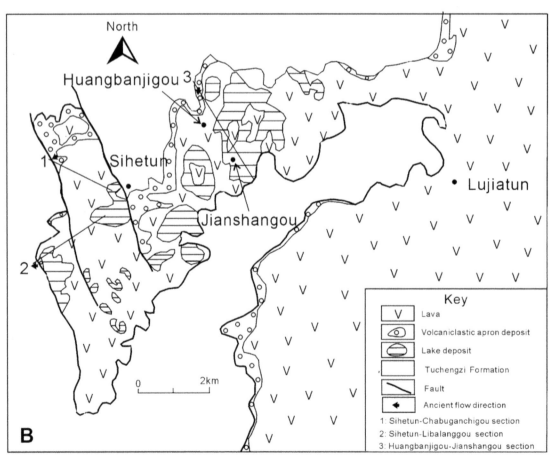

claystones, sometimes with convoluted bedding. . . . The laminae are 0.1–1 mm thick and normally graded."

Fürsich et al. (2007) carried out a detailed taphonomic analysis of a section in a test pit located less than 2 km east of the Sihetun Visitor Facility. The test pit was 3.5 m deep, about 9 m² at the top and 5.5 m² at the bottom. The focus of the study was a faunal assemblage dominated by two species, an insect nymph and a conchostrean. They occurred most abundantly in the mudstone component of the normally graded sandstones and mudstone facies of Jiang and Sha (2007). The conclusion of Fürsich et al. (2007) was that the mass mortality events was the result of anoxia that took place in a lake at the end of summer; these authors did not invoke any aspect of volcanism. Anoxia was hypothesized by Waldman (1971) to explain the mass death of fish at Koonwarra. However, he envisioned that the mass mortality events that took place there were the result of a winter-kill mechanism where a sheet of ice retarded the passage of oxygen from the atmosphere into the water of the lake.

In light of the evidence that anoxia may have played the same role to explain the accumulation of the arthropods and fishes, it seems reasonable to use the rock unit at the Sihetun Visitor Facility, where the 30 tetrapods are preserved, as a guide to developing a strategy for attempting to recover fossil tetrapods from the lacustrine deposits of the Strzelecki Group. This is because the two are so similar, both in the nature of the rocks in which the fossils occur and in the types of fossils represented.

If the frequency of tetrapod fossils preserved within the lacustrine facies of the Strzelecki Group is similar to that at the Sihetun Visitor Facility, a question arises: what is the probability of finding one or more tetrapod fossils if a given area is excavated of the lacustrine rock in the Strzelecki Group? Table 28.1 suggests an answer to that question.

These figures are generated by the simple equation $P = 1 - e^{-np}$, where P is the probability of one or more successes after n trials, each with a probability of success (p) for each independent trial. As used in this case, both n and p are functions of area: n is the number of square meters to be excavated in a lacustrine facies of the Stretzlecki Group that is presumed to be as rich in tetrapods as the fossil-bearing unit at the Sihetun Visitor Facility, and p in this instance is the number of specimens found at the Sihetun Visitor Facility divided by the area of the Sihetun Visitor Facility (30 specimens/400 m²).

The next question that comes to mind is, how much area of fossiliferous rock has been previously excavated at Koonwarra? T. H. Rich participated in the second major excavation at Koonwarra in 1982. On the basis of that excavation and the description of the work carried out there in 1965–1967 in Waldman (1971), less than 10 m² has been excavated at Koonwarra (Fig. 28.8). Assuming the circumstances of deposition at the Sihetun Visitor Facility and Koonwarra are comparable, the chances of having discovered a tetrapod skeleton to date in the small area excavated at the latter would not be even 50%. Rather, what is remarkable is that any trace at all of tetrapods was found in the small area uncovered to date at Koonwarra—namely, the half-dozen feathers. Their discovery bodes well for recovering more complete tetrapod specimens if a much larger area is excavated in the lacustrine facies of the Strzelecki Group, using the Sihetun Visitor Facility as a guide.

28.6. Figure 1 of Jiang and Sha (2007): "Distribution of the upper Mesozoic formations in western Liaoning (A) and the location of the study area (B)."

Table 28.1. Probability of Finding Tetrapod Fossils in Lacustrine Rock in the Strzelecki Group

Square Meters Excavated at Sihetun	Likelihood of Finding One or More Tetrapod Specimens in the Sihetun Visitor Facility
10	53%
20	78%
30	89%
40	95%
50	98%

The Way Ahead in the Strzelecki Group

On the basis of the information available concerning the occurrence of 30 tetrapods on display at the Sihetun Visitor Facility, we infer that the lacustrine deposits of the Strzelecki Group may have a similar accumulation of fossils. This is indicated by both the similarity of the most common fossils in the two (fishes, arthropods, and plants) and the similarity of the lithologies in which these fossils occur. In light of those considerations, the reason why no tetrapods have yet to be recovered to date from the lacustrine deposits of the Strzelecki Group would seem to be disarmingly straightforward: not enough rock has been excavated.

Generally when fossil tetrapods are excavated, there is an expectation that numerous isolated bones and teeth are found for every partial or complete skeleton recovered. In sharp contrast, the tetrapod fossils uncovered at the Sihetun Visitor Facility are all complete, or nearly so. Thus, while in the process of excavating in the lacustrine facies of the Strzelecki Group, a lack of isolated tetrapod bones and teeth during an exploratory excavation should not be interpreted as indicative that rarer partial or complete skeletons are unlikely to be found.

To excavate sufficient lacustrine rock in the Strzelecki Group in order to test whether tetrapods occur there in the frequency that they do at the Sihetun Visitor Facility requires that 50 m^2 be exposed to have a 98% chance of finding one or more tetrapods. This can be done in one of two ways. First, such an excavation can be done at Koonwarra itself. Second, such an excavation can be carried out elsewhere.

To uncover 50 m^2 at Koonwarra would entail excavating an area underground comparable to the Slippery Rock site at Dinosaur Cove (Rich and Vickers-Rich, 2000). Logistically, this would be in some ways easier and in some ways more difficult than cutting the tunnels at the latter site. The great difficulty in excavating underground extensively at Koonwarra is that the fossiliferous unit dips downward at about 37 degrees, while the hill above it is about as steep in the other direction. Because the surface of that hill is unstable, having been logged late in the nineteenth century, an adit is one way to excavate at Koonwarra. An open cut there would mean that the footwall would be more than 1 m higher for every 1 m further inward that such an excavation was extended. Unexpected collapse of the soil surface on the hill above could result in serious injury to those working the site unless a much larger area than the 50 m^2 desired is excavated by cutting steps in the hillside above the quarry so that the footwall immediately adjacent to the area where the fossils are to be excavated is kept to a minimum.

Finding another Koonwarra-like lacustrine occurrence elsewhere in the Strzelecki ranges in more favorable circumstances from the point of view of excavation logistics is a second approach. Lacustrine deposits similar to those at Koonwarra do occur elsewhere in the Strzelecki ranges (Fig.

28.7. Figure 2 of Jiang and Sha (2007): "Stratigraphic correlation and distribution of lithofacies of the Yixian Formation." *Facies abbreviations:* G1, matrix-supported conglomerates; G2, unstratified, clast-supported conglomerates; G3, stratified, clast-supported conglomerates; G4, stratified sandy conglomerates and pebbly sandstones; L1, lapillistones and lapilli tuffs; S1, cross-stratified sandstones; M1, horizontally stratified homogeneous mudstones; M2, laminated mudstones; S2, horizontally stratified, homogeneous sandstones and tuffs; S/M, normally graded sandstones and mudstones; S3, channel-fill, cross-stratified pebbly sandstones; S4, normally graded pebbly sandstones.

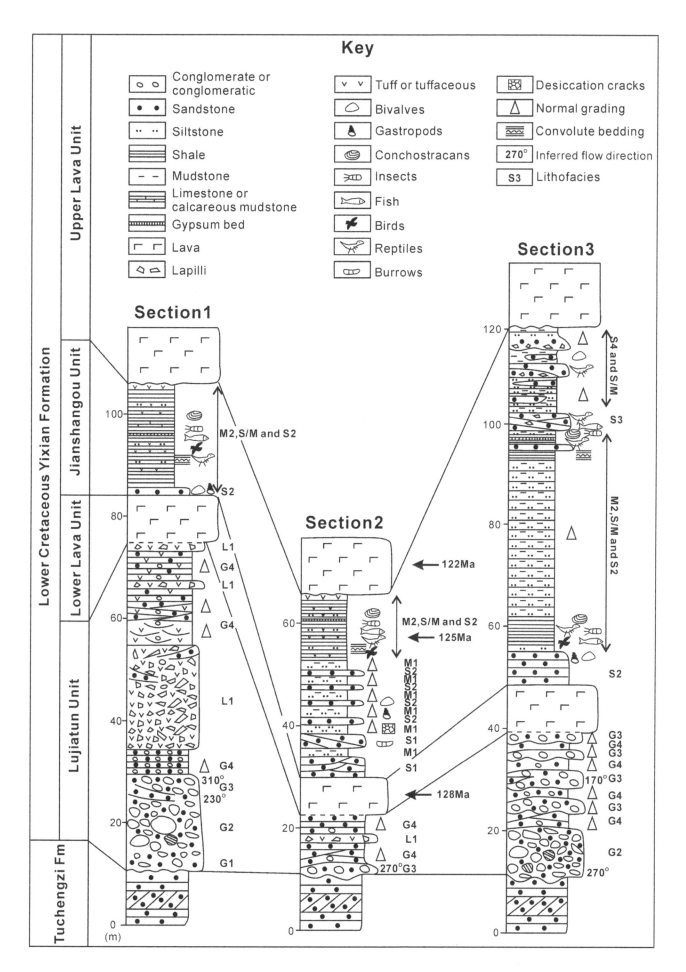

28.9). Fragments of fossil fish have been found in a lacustrine deposit in a road cut a few kilometers from the Koonwarra site. Because the Strzelecki ranges are heavily vegetated, locating other lacustrine deposits will not be easy, but artificial excavations made for other purposes such as road cuts and dams may help overcome this difficulty.

Conclusion

28.8. A, Koonwarra 1967, north of the South Gippsland Highway.

Courtesy of James Warren.

B, Koonwarra 1967, south of South Gippsland Highway.

Courtesy of James Warren.

C, Koonwarra 1982, east of 1967 sites.

Courtesy of Andrew Drinnan.

The most difficult fossil to find is the first one. The discovery in lacustrine deposits of the Strzelecki Group of a single tetrapod specimen similar in preservation to those on display at the Sihetun Visitor Facility would fundamentally transform the study of Mesozoic terrestrial vertebrates in Australia. The outcome could well be analogous to the flow on effect of the discovery of a single jaw of a Cretaceous mammal at a site not 20 km away from Koonwarra in 1997, which subsequently led to the collection of more than 85% of all Mesozoic mammal specimens now known from that continent. From that modest beginning, a number of quite unexpected insights as to the history of that group in Australia during that era have been forthcoming (Rich and Vickers-Rich, 2000).

Operationally, when carrying out exploratory excavations in the lacustrine facies of the Strzelecki Group, there are two particularly important aspects of the occurrence of tetrapods at the Chinese site to be borne in

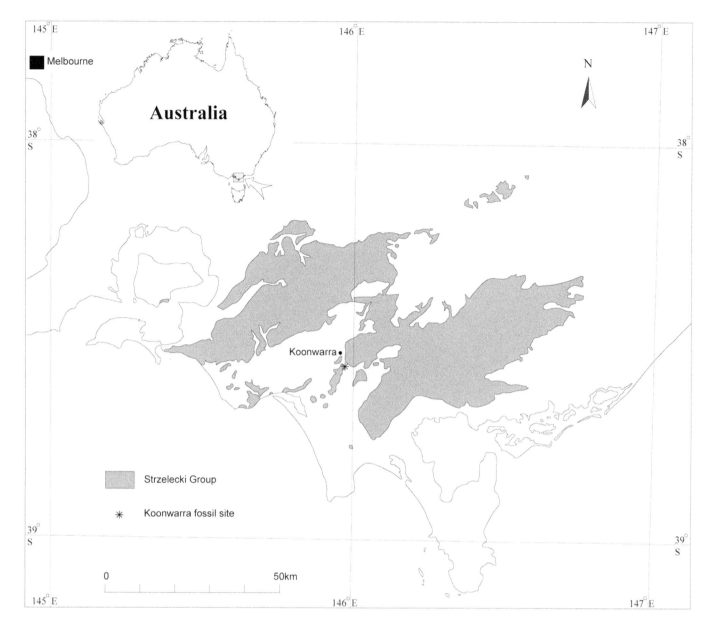

28.9. Surface outcrops of the Strzelecki Group, Early Cretaceous.

mind. First, an area of 50 m² must be uncovered in order to have a good chance of finding one or more fossil tetrapod skeletons if they occur at the frequency they do at the Sihetun locality. Second, in the process of carrying out such an excavation, just because no isolated tetrapod bones and teeth are found midway through an excavation, is no reason to abandon the effort before at least 50 m² are uncovered. This is because it is most likely that complete or nearly complete skeletons will be uncovered if any tetrapod fossil material is found at all.

We thank Yuanqing W., Qingguo L., and Huaquan S. for their making T.R.'s trip to the Jehol area possible. Xiaolin W., Institute of Vertebrate Paleontology and Paleoanthropology, provided useful information about fossil tetrapod sites in the Jehol area he did not visit. Ming L. H., who participated in the excavation at the Sihetun Visitor Facility, provided much useful information regarding the procedures used there to carry out

Acknowledgments

that work. Baoyu J., School of Earth Sciences and Engineering, Nanjing University, corresponded with us about aspects of the taphonomy of the deposits in the vicinity of Sihetun and kindly gave permission to reproduce two figures from Jiang and Sha (2007). F. T. Fürsich, S. Martin, E. Thompson, S. Turner, W. A. Clemens, and B. Kear made useful comments about the text.

References

Chang, M. (ed.). 2003. The Jehol Biota: the emergence of feathered dinosaurs, beaked birds and flowering plants. Shanghai Scientific & Technical Publishers, Shanghai, 209 pp.

Dettmann, M. E. 1986. Early Cretaceous palynoflora of subsurface strata correlative with the Koonwarra Fossil Bed, Victoria. Memoirs of the Association of Australian Palaeontologists 3: 79–110.

Drinnan, A. N., and T. C. Chambers. 1986. Flora of the Lower Cretaceous Koonwarra Fossil Beds (Korumburra Group), South Gippsland, Victoria. Memoirs of the Association of Australian Palaeontologists 3: 1–77.

Fürsich, F .T., J. Sha, B. Jiang, and Y. Pan. 2007. High resolution palaeoecological and taphonomic analysis of Early Cretaceous lake biota, western Liaoning (NE-China). Palaeogeography, Palaeoclimatology, Palaeoecology 253: 434–457.

Jell, P. A., and P. M. Duncan. 1986. Invertebrates, mainly insects, from the freshwater, Lower Cretaceous Koonwarra Fossil Beds (Korumburra Group), South Gippsland, Victoria. Memoirs of the Association of Australian Palaeontologists 3: 111–205.

Jiang, B., and J. Sha. 2007. Preliminary analysis of the depositional environments of the Lower Cretaceous Yixian Formation in the Sihetun area, western Liaoning, China. Cretaceous Research 28: 183–193.

Liu, T., L. Liu, and G. Chu. 2002. Early Cretaceous maars, depositional environments and their relationship to the fossil preservation in Sihetun, Liaoning, Northeast China; pp. 307–311 in Z. Zhou and F. Zhang (eds.), Proceedings of the 5th Symposium of the Society of Avian Paleontology and Evolution. Science Press, Beijing.

Rich, T. H., and P. Vickers-Rich. 2000. The dinosaurs of darkness. Indiana University Press, Bloomington, 222 pp.

Rich, T. H., Li Xiao-bo, and P. Vickers-Rich, P. 2009. A potential Gondwanan polar Jehol Biota lookalike in Victoria, Australia. Transactions of the Royal Society of Victoria 121: 5–13v–xiii.

Rich, T. H., P. Vickers-Rich, and R. A. Gangloff. 2002. Polar dinosaurs. Science 295: 979–980.

Waldman, M. 1971. Fish from the freshwater Lower Cretaceous of Victoria, Australia, with comments on the palaeo-environment. Special Paper 9. Palaeontology of the Palaeontological Association, London.

Wang, X., Y. Wang, Z. Zhou, F. Jin, J. Zhang, and F. Zhang. 2000. Vertebrate faunas and biostratigraphy of the Jehol Group in Western Liaoning, China. Vertebrata PalAsiatica 38 (supplement): 41–63.

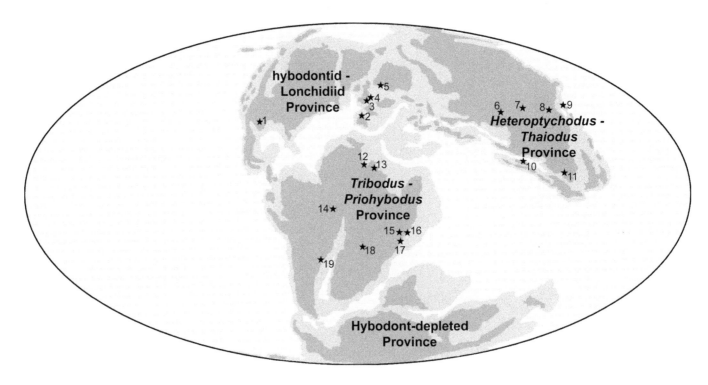

29.1. Paleogeographic map showing the distribution of freshwater hybodont sharks during the Early Cretaceous. The most common taxa in each province were used to characterize them: Hybodontid–Lonchidiid for North America and Europe, *Heteroptychodus–Thaiodus* for Asia, *Tribodus–Priohybodus* for West Gondwana, and hybodont depleted for East Gondwana. Note that *Tribodus* is younger than *Priohybodus,* and the two genera are actually never found together. Main fossiliferous areas used to define each province are indicated as follows: 1, Texas; 2, Spain; 3, England; 4, Belgium; 5, Sweden; 6, Kyrgyzstan; 7, Mongolia; 8, Liaoning province, China; 9, Japan; 10, Tibet; 11, Thailand; 12, Tunisia; 13, Libya; 14, Santana Formation, Brazil; 15, Ethiopia; 16, Yemen; 17, Somalia; 18, Democratic Republic of the Congo; 19, Uruguay.

Freshwater Hybodont Sharks in Early Cretaceous Ecosystems: A Review

29

Gilles Cuny

Hybodont sharks capable of entering rivers and lakes were quite common during the Early Cretaceous, and they showed diverse adaptations to a number of diets. However, the concept of freshwater sharks is difficult to define in paleontology. In some cases, the discovery of both young and adult remains in a fully freshwater environment demonstrates beyond a doubt that a shark was indeed able to reproduce in freshwater, but such a discovery has been documented so far only in Thailand. Most of the other records of freshwater sharks most likely correspond to euryhaline sharks, possibly living near the coast and occasionally entering large river systems. Moreover, it is quite difficult to precisely define the environment of deposition of many sites, as in Tunisia. The distribution of these euryhaline sharks in freshwater ecosystems can be documented, and I provide here a general overview of their diversity, demonstrating a distribution in three main provinces: North America and Europe, West Gondwana, and Asia. Absence of freshwater hybodont sharks in East Gondwana suggests a latitudinal control on their distribution. The European and North American records show a quite low diversity in term of adaptation to specialized diets, whereas specialized cutting and grinding dentition appeared independently in the two other provinces.

Introduction

Hybodont sharks represent one of the most successful chondrichthyan lineages of all time. On the basis of the record of isolated teeth, they appeared as early as the Late Devonian (Famennian; Ginter et al., 2002), and they disappeared at the end of the Cretaceous (Maastrichtian; Becker et al., 2004). The first fossils found in connection date from the Early Carboniferous (*Onychoselache traquairi* and *Tristychius arcuatus*; see Coates and Gess, 2007) and already show some affinities for a freshwater environment (Baird, 1978; Coates and Gess, 2007). Although often considered as conservative in term of body shape, hybodonts cover a wide array of size, from the small *Lissodus cassangensis* (20 cm long; Antunes et al., 1990) to the huge *Ptychodus rugosus* or *P. mortoni* (probably in excess of 10 m; Shimada et al., 2009, 2010). They have also developed diverse types of dentition, including clutching (many *Hybodus* species; Cappetta, 1987), tearing (*Egertonodus* species, *Planohybodus*; Cappetta, 1987; Rees and Underwood, 2008), cutting (*Pororhiza, Priohybodus, Thaiodus, Mukdahanodus*; Cuny et al., 2009), crushing (some *Ptychodus* species, like *Ptychodus whipplei, Lonchidion* species, *Acrorhizodus, Diplolonchidion*; Cappetta, 1987; Heckert, 2004; Cappetta et al., 2006), grinding (*Palaeobates, Acrodus*, some *Ptychodus* species, *Heteroptychodus*; Cappetta, 1987; Cappetta et al., 2006),

519

clutching–grinding (*Lissodus*; Duffin, 2001a; Rees and Underwood, 2002) and cutting–crushing (*Lonchidion humblei*; Heckert et al., 2007).

Hybodonts appear not to have been much diversified during the Paleozoic, but they survived the Permo-Triassic mass extinction and became the dominant sharks in the seas for much of the Triassic and Jurassic (Maisey et al., 2004). Although the diversification of the neoselachian sharks during the Early Jurassic was an opportunistic rather than a competitive event (Kriwet et al., 2009), by the end of the Jurassic, they were highly diversified and abundant in marine environments (Underwood, 2006). The impact of neoselachian diversification on hybodont evolution during the early Cretaceous is still unclear, but this is a time of increased diversification among nonmarine hybodont sharks (Cuny et al., 2008; Kriwet and Klug, 2008), whereas the oldest record of a freshwater neoselachian shark is currently from the Barremian of England (Sweetman and Underwood, 2006). I propose below a review of this diversification event of hybodont sharks in nonmarine environment during the Late Jurassic–Early Cretaceous time interval (Kimmeridgian–Albian).

Freshwater Elasmobranchs

The simplest and easiest definition of a freshwater elasmobranch is an animal that lives and reproduces in lakes and rivers. Nowadays, 32 species of elasmobranchs fit this definition. Most are batoids, belonging to the genera *Paratrygon*, *Plesiotrygon*, *Potamotrygon*, *Dasyatis*, and *Himantura*, whereas only one shark genus, *Glyphis*, possesses obligate freshwater species (Martin, 2005). However, many other elasmobranch species are euryhaline and can be encountered in freshwater environments without being restricted to them. The best example is *Carcharhinus leucas*, a species abundant in shallow marine environments under the tropics, which has been reported 4,200 km up the Amazon, 1,120 km up the Zambezi River, and 2,800 km up the Mississippi River (Martin, 2005). However, reproduction of *Carcharhinus leucas* (including copulation, development, and parturition) usually takes place in brackish water along the coast (Montoya and Thorson, 1982). The sawfish *Pristis perotteti*, on the other hand, gives birth to the young in freshwater, but they always keep a connection with the sea (Schultze and Soler-Gijón, 2004). To specify the difference between euryhaline and obligate freshwater species in the fossil record is therefore difficult, because the occurrence of a fossil in a freshwater environment is no proof that the animal was restricted to this environment. Moreover, such paleoenvironments are difficult to characterize (Anderson et al., 2007; Schultze, 2009). A good example of this situation is to be found with the xenacanth sharks. For a long time, these sharks were considered to be typical freshwater sharks, but many of them were in fact euryhaline (Schultze and Soler-Gijón, 2004). Xenacanth sharks appear to have been oviparous, a mode of reproduction that is today restricted to marine sharks. Modern obligate freshwater elasmobranchs are all viviparous, probably because osmoregulation of the embryo through the wall of the egg capsule is not possible (Schultze and Soler-Gijón, 2004). Modern sharks do lay egg capsules in estuaries, though, and fossil elasmobranchs might have developed the capacity of laying egg capsules in freshwater (Fischer et al., 2007; Schultze, 2009). There are currently few data that can be used to discuss the reproductive strategy of hybodont sharks, although an association between the egg capsule of *Palaeoxyris* and *Lissodus* has been proposed (Fischer and Kogan, 2008). Even if *Lissodus* was indeed oviparous, implying it could not reproduce in freshwater according to Schultze and Soler-Gijón (2004), it does not mean that all hybodont sharks were oviparous. Only the presence of very young individuals together with adults in freshwater can point to the possibility that hybodont sharks were able to reproduce in lakes and rivers. So far, such an association has been only tentatively reported from the Aptian of Thailand (Cuny et al., 2008).

In conclusion, it appears that we are currently unable to determine whether a fossil shark found in a freshwater environment was indeed restricted to such an environment. Therefore, the use of the term "freshwater hybodont sharks" in this chapter has to be understood as sharks spending at least part of their life in freshwater, without implying that they were actually restricted to this environment.

East Gondwana (Australia, Antarctica, and India)

Hybodont sharks have yet to be retrieved from the Late Jurassic–Early Cretaceous of East Gondwana, although both neoselachian and actinopterygian remains have been described from this area (López-Arbarello, 2004; Prasad et al., 2004). During the Late Cretaceous, only a possible *Ptychodus* has been mentioned in this area (López-Arbarello, 2004). The rarity of hybodont sharks in this province might be due to a latitudinal control of their distribution (Fig. 29.1), although such a hypothesis remains difficult to demonstrate.

West Gondwana (South America, Africa and Arabian Peninsula)

West Gondwana has yielded a number of hybodont genera that were endemic to the area during the Late Jurassic–Early Cretaceous time interval. These include the highly specialized genera *Tribodus*, *Priohybodus*, and *Pororhiza*.

Priohybodus arambourgi is the only hybodont shark to have developed a high-crowned, serrated dentition (Cuny et al., 2009). Its phylogenetic affinities lie with the European genera *Planohybodus* and *Secarodus* (Rees, 2008). It has been found in association with continental faunas in the Kimmeridgian–Tithonian of Yemen, the Tithonian of Ethiopia, the Late Jurassic of Somalia, the Late Jurassic–Early Cretaceous of Uruguay, and the Early Cretaceous of Tunisia and Libya (Goodwin et al., 1999; Duffin, 2001b; Perea et al., 2001; Cuny et al., 2004).

Pororhiza molimbaensis is another hybodont with a cutting dentition, but its teeth are different from those of *Priohybodus arambourgi*, and its phylogenetic affinities among hybodont sharks are currently unclear. It is known from isolated teeth and dorsal fin spines from the Albian of the Democratic Republic of the Congo (Casier, 1969).

Tribodus is a hybodont shark possessing a crushing dentition similar to what has been developed among some modern stingrays, and like the latter, it has developed a hyostylic jaw suspension (Maisey and de Carvalho,

1997). It is considered to belong to the subfamily Acrodontinae (Rees, 2008; Lane and Maisey, 2009). The type species *T. limae* is known from the Santana Formation (Albian) of Brazil, and the species *T. tunisiensis* has been described from the Aïn Guettar Formation (Albian) of Tunisia (Cuny et al., 2004). The genus might also be present in the Early Cretaceous of the Democratic Republic of the Congo, where it was described as *Hylaeobatis* (Casier, 1961; Maisey, 2000). It is also known during the Cenomanian in Egypt, Libya, Morocco, and France (Dutheil, 1999; Maisey, 2000; Rage and Cappetta, 2002; Vullo and Néraudeau, 2008).

Priohybodus and *Pororhiza* are mainly restricted to freshwater environments (Perea et al., 2001; Cuny et al., 2004; Cappetta et al., 2006; Anderson et al., 2007), indicating that the development of a serrated cutting dentition in hybodont sharks was somehow linked to their life in this environment (Cuny et al., 2009), whereas *Tribodus* is more euryhaline (Anderson et al., 2007). *Priohybodus* is often found together with teeth of *Hybodus* (Goodwin et al., 1999; Cuny et al., 2004), indicating that some species among this genus were also well adapted to life in freshwater. The *Hybodus* specimens associated with *Priohybodus* in the Tithonian of Ethiopia certainly represent a new species.

The hybodont assemblage from the Oum ed Diab Member of the Aïn Guettar Formation in Tunisia (*Lissodus* sp., *Diabodus tataouinensis*, and *Hybodus* sp.) appear more euryhaline than strictly restricted to freshwater (Anderson et al., 2007). It should be noted that the genus *Diabodus* is not a junior synonym of *Steinbachodus* (=*Bahariyodon*; see Rees and Underwood, 2002), contrary to the claim made by Cappetta (2006). *Steinbachodus* is easily identified by a concave and flared lingual face of the crown (Rees and Underwood, 2002), a character never present in the teeth of *Diabodus* (Cuny et al., 2004). *Diabodus*, like *Pororhiza* and *Priohybodus*, is so far endemic to West Gondwana. *Tribodus*, however, reached the European archipelago during the Cenomanian (Vullo and Néraudeau, 2008).

North America

In North America, hybodont sharks appear to retain mainly their original marine mode of life during the Early Cretaceous, and "*Polyacrodus*" has been described from fully marine environments (Everhart, 2004, 2009). They are also often associated with Lamniformes sharks: *Lonchidion anitae*, different *Hybodus* species, cf. *Hylaeobatis ornata* (the latter probably representing a genus of its own) were all found together with *Leptostyrax bicuspidatus* (Winkler et al., 1990), a species normally restricted to marine environments. The only occurrence of the genus *Leptostyrax* in freshwater appears to be in the Kem Kem Beds in Morocco (D. Ward, pers. comm.). Although they could have gone upriver from time to time, the North American hybodonts show no special affinity for freshwater environments. The only hybodont that appears to have been more restricted to a freshwater to brackish environment is *Meristodonoides butleri* from the Aptian– Albian of Texas (Thurmond, 1971; Welton and Farish, 1993; Underwood and Cumbaa, 2010).

29.2 Tooth of *Lonchidion striatum* (RBINS P8784, Collection of the Royal Belgian Institute of Natural Sciences, Brussels) from the Hautrage Clays Formation (Middle Barremian–earliest Apian) of Hautrage (Belgium) in labial (A), apical (B), and lingual (C) views. Scale bars = 1 mm.

Europe

Several localities in Europe have yielded hybodont remains in nonmarine environments. In Spain, Bermúdez-Rochas (2009) described from a Hauterivian–Barremian fluvio-lacustrine site in Cantabria province a hybodont assemblage containing *Egertonodus basanus, Planohybodus ensis,* "*Hybodus*" *parvidens, Lonchidion breve, Parvodus* sp., and *Lissodus* sp. The assemblage from Galve in Teruel province contains "*Hybodus*" *parvidens* and *Lonchidion microselachos* and appears to be no more than peripherally associated with marine environments (Estes and Sanchiz, 1982). However, "*Hybodus*" *parvidens* and *Lonchidion breve* were also found associated with marine neoselachians in the latter province (Canudo et al., 1995), and sites that appear to have been deposited under strict freshwater conditions (Las Hoyas, Uña, Buenache de la Sierra) never yielded hybodont sharks (Buscalioni et al., 2008). Faunas from the Berriasian of Sweden and the Purbeck of England (*Egertonodus basanus, Planohybodus ensis,* "*Hybodus*" *parvidens, Parvodus rugianus, Lonchidion crenulatum,* and *Lonchidion inflexum*; Rees, 2002; Underwood and Rees, 2002; Rees and Underwood, 2008) were all deposited in environments with varying salinity so that they cannot be considered as fully adapted to life in freshwater. One of these, *Egertonodus basanus,* was, however, found in a fully freshwater environment in the Wessex Formation (Wealden Group, Isle of Wight), together with *Lonchidion striatum* and possibly *Vectiselachos ornatus* (Sweetman and Underwood, 2006). The two latter species are also found in the lagoonal Vectis Formation (Rees and Underwood, 2002; Sweetman and Underwood, 2006), which overlies the Wessex Formation. Therefore, they cannot be considered as restricted to freshwater environments. *Lonchidon striatum* has, however, been found also in the freshwater Hautrage Clays Formation (Middle Barremian–earliest Aptian; Dejax et al., 2008) at Hautrage in Belgium (Fig. 29.2).

European hybodont sharks from the Early Cretaceous appeared to have been excluded from fully marine environments, as in the Valanginian of Poland or Hauterivian of England (Underwood et al., 1999; Rees, 2005), where only neoselachian sharks were found. They appear to show a maximum diversity in brackish waters and were able to enter full freshwater environments in, for example, Spain, England, and Belgium. However, none of them appears to have been restricted to such a freshwater environment. At the specific level, this hybodont fauna appears quite endemic to the European waters, with only *"Hybodus" parvidens*, which could have reached North America (Bermúdez-Rochas, 2009), and *Egertonodus basanus*, which has been mentioned in the Early Cretaceous of Japan and Morocco, but in marine settings (Yabe and Obata, 1930; Duffin and Sigogneau-Russell, 1993; Kozai et al., 2005).

Asia

A single fin spine attributed to *Hybodus* sp. has been found in a terrestrial environment in the Early Cretaceous Yixian Formation (Liaoning province), indicating the presence of hybodont sharks in Chinese freshwater (Zhang, 2007). It should be noted, however, that isolated fin spines are not diagnostic at the generic level (Bermúdez-Rochas, 2009), so its attribution to *Hybodus* is doubtful.

The richest assemblages of freshwater hybodont sharks from the Asian continent have been collected during the past 20 years in Thailand (Cuny et al., 2007). Three different assemblages have been identified so far. The first one comes from the Phu Kradung Formation, which is latest Jurassic–Early Cretaceous in age and was deposited in a lacustrine-dominated alluvial floodplain (Racey, 2009). The assemblage contains *Hybodus* spp., *Acrodus* sp., *Lonchidion* sp., *Lissodus* sp., and *Heteroptychodus* sp. (Cuny et al., 2007). Although this assemblage is quite diverse, it remains poorly known: the Phu Kradung Formation has so far mainly yielded poorly preserved and broken teeth. As a result, it is, for example, quite unclear whether a single species of *Hybodus* is present in this assemblage, and it has so far been impossible to identify any of the taxa down to the specific level. The presence of *Acrodus* is also questionable because this genus probably disappeared at the end of the Early Jurassic and was mainly a marine shark (Rees, 2000). More material is needed to settle these problems, and a promising new site, Phu Noi, has recently been discovered in Kalasin province. Hybodont shark teeth appear better preserved in this new site than in any other sites from the Phu Kradung Formation. It is therefore hoped that the Phu Kradung assemblage will soon become better understood.

The second assemblage comes from the Sao Khua Formation, which is of Early Cretaceous age and was deposited in an alluvial floodplain (Racey, 2009). Between the Phu Kradung and Sao Khua formations comes the Phra Wihan Formation, which corresponds to a braided river system. No shark remains have been found in the latter formation. The Sao Khua assemblage includes *Hybodus* sp., cf. *Egertonodus* sp., *Lonchidion khoratensis*, *Isanodus paladeji*, *Heteroptychodus steinmanni*, *Heteroptychodus kokutensis*, and *Mukdahanodus trisivakulii* (Cuny et al., 2007, 2009, 2010). *Heteroptychodus steinmanni* was originally described from the Ryoseki Formation in Japan (Yabe and Obata, 1930), which is Hauterivian in age and corresponds to a freshwater/brackish environment (Kozai et al., 2005).

The genus *Heteroptychodus* is also known in the Early Cretaceous Matsuo Group of Japan in a brackish environment (Tanimoto and Tanaka, 1998) and from the Aptian–Albian of Mongolia with the species *H. chuvalovi* (Cuny et al., 2008). This genus was recorded in a marine environment in Kyrgyzstan (Averianov and Skutschas, 2000).

The third assemblage is from the Khok Kruat Formation, which is separated from the Sao Khua Formation by the Phu Phan Formation, corresponding, like the Phra Wihan Formation, to a braided river system. The Khok Kruat Formation is Aptian in age and was deposited in fluvial to paralic conditions (Racey, 2009). It has yielded *Hybodus aequitridentatus*, *Heteroptychodus steinmanni*, *Thaiodus ruchae*, *Khoratodus foreyi*, and *Acrorhizodus khoratensis*. At least one additional genus is present but known so far by a single tooth (Cuny et al., 2008). Small teeth with weakly developed cusps have been attributed to juvenile *Heteroptychodus steinmanni* and could hint to the fact that *Heteroptychodus* was able to reproduce in freshwater (Cuny et al., 2008). *Thaiodus ruchae* is also known from the Aptian–Albian of Tibet, where it has been found in deltaic beds (Cappetta et al., 1990), indicating that it could live in brackish water like *Heteroptychodus*, unless the Tibetan tooth is allochthonous. Together with *Khoratodus foreyi*, it belongs to Thaiodontidae, a family endemic to Asia (Cuny et al., 2008).

Discussion

This review of Early Cretaceous freshwater hybodont sharks allows identification of three main assemblages: West Gondwana, Europe and North America, and Asia, whereas East Gondwana is hybodont depleted. The European–North American assemblage is massively dominated by Hybodontidae and Lonchidiidae and does not show major innovation regarding tooth morphology. It appears, therefore, as a somewhat conservative assemblage. Although found in full freshwater environments, none of these sharks was restricted to this environment. They were mostly living in brackish/coastal waters but commonly entered fluvial systems, where their abundance and diversity indicate that they played an important role in these ecosystems, both as top predators, as for example, *Egertonodus* and *Planohybodus*, but also as small, opportunistic predators like *Lonchidion*. *Egertonodus* is found in both full freshwater and marine environments and has a wide geographic distribution, being known in Western Europe, Morocco, Tunisia, Thailand, and Japan (Yabe and Obata, 1930; Duffin and Sigogneau-Russell, 1993; Cuny et al., 2007; Rees and Underwood, 2008). It seems, therefore, that it occupied an ecological niche similar to that of the modern bull shark, *Carcharhinus leucas*. It should be noted, however, that it has not been reported from North America so far, and that North America and Europe do not share species in common, except perhaps *"Hybodus" parvidens*. On the contrary, hybodont sharks are only known from the southern central part of North America (Texas, Oklahoma, and Kansas) during the Early Cretaceous, and geographical distance might suffice to explain these differences in faunal composition.

The West Gondwanan assemblage is more diversified than the European–North American one from the point of view of tooth morphology. Two taxa, *Priohybodus* and *Pororhiza*, developed fully serrated teeth for the first time in hybodont history. Interestingly, the appearance of this kind of dentition seems to be correlated with a mode of life more restricted to a freshwater environment. We also notice the appearance of a new type of crushing dentition with *Tribodus*, convergent with that of modern dasyatid rays. However, the vast majority of these hybodonts were euryhaline and not restricted to freshwater, and some, like *Tribodus*, were able to reach Europe in the Cenomanian.

Finally, the Asian assemblage has yielded the most unusual hybodont faunas so far with the convergent appearance of cutting dentition in *Mukdahanodus* and *Thaiodus*, as well as specialized grinding dentition with *Heteroptychodus*, *Isanodus*, and *Khoratodus*. The Asian hybodonts develop a unique pattern of tooth ornamentation, made of several transverse ridges showing small, perpendicular ridges in *Heteroptychodus* and *Isanodus*. A family, Thaiodontidae, is endemic to this part of the world, and *Acrorhizodus* tooth morphology is so unusual that its phylogenetic relationships among hybodont sharks remain currently impossible to decipher (Cuny et al., 2008). It is also in this Asian assemblage that full adaptation to life in freshwater might have been developed. Apart from *Thaiodus* and *Heteroptychodus*, all the species from the Sao Khua and Khok Kruat formations in Thailand are unknown outside freshwater settings. We also notice an increase of endemism through time in Thailand, with the appearance of at least one (Thaiodontidae) and probably two (*Acrorhizodus*) endemic families in the Aptian. It is possible that this increase of endemism is linked to the appearance of sharks, which were more restricted to freshwater environments. A major radiation of actinopterygian fishes in freshwater environment occurred also in China during the basal Cretaceous and is linked with the development of young lake systems (Cavin et al., 2007). Freshwater species spread subsequently in adjacent areas, including Thailand and Japan. The Chinese fossil record for hybodont sharks during the Early Cretaceous remains poorly documented, and it is therefore difficult to test whether the same scenario applies to these animals. Although there are records of *Heteroptychodus* teeth in marine environments, it is also the only genus for which an association between juvenile and adult teeth in freshwater settings is likely. This could hint to the fact that all known species of *Heteroptychodus* were anadromous, and comparison with modern sharks would therefore suggest that they were not oviparous (Schultze and Soler-Gijón, 2004).

Conclusions

Early Cretaceous hybodont faunas are distributed in three main provinces: North America and Europe, West Gondwana, and Asia. In these three provinces, they were common in freshwater environments, although it remains difficult to know whether some species were indeed restricted to this environment. None were in North America and Europe, and only two (*Priohybodus arambourgi* and *Pororhiza molimbaensis*) might have been in West Gondwana. It is in Thailand that there is the most chance that some species might have been restricted to freshwater, as *Isanodus paladeji*, *Mukdahanodus trisivakulii*, *Khoratodus foreyi*, and *Acrorhizodus khoratensis* are so far unknown outside freshwater settings, and it is also the only place where teeth of adults and juveniles of the same species, *Heteroptychodus steinmanni*, have been found together in a freshwater environment. The absence of hybodont sharks in East Gondwana suggests a latitudinal control for the distribution of these animals.

In North America and Europe, hybodonts appear somewhat unspecialized, whereas in West Gondwana and Asia, specialist lineages with highly derived cutting and grinding dentition evolved, which suggests that their ecological role was more diverse, and therefore more important, in the fluvial systems of the last two provinces.

Acknowledgments

I thank Pascal Godefroit for inviting me to write this essay. Comments from Sylvain Adnet and David Ward during the review process greatly improved this chapter. This work was funded by the Carlsberg Foundation and the Danish Agency for Science. Lionel Cavin provided me with the paleogeographic frame for Fig. 29.1. Fieldwork was made possible through the help of many (too many to be cited by name here) people and I would like to especially thank the wonderful crews I work with in Thailand and Tunisia.

References

Anderson, P. E., M. J. Benton, C. N. Trueman, B. A. Paterson, and G. Cuny. 2007. Palaeoenvironments of vertebrates on the southern shore of Tethys: the nonmarine Early Cretaceous of Tunisia. Palaeogeography, Palaeoclimatology, Palaeoecology 243: 118–131.

Antunes, M. T., J. G. Maisey, M. M. Marques, B. Schaeffer, and K. S. Thomson. 1990. Triassic fishes from the Cassange depression (R.P. de Angola). Ciências da Terra (UNL) Número Especial: 1–64.

Averianov, A., and P. Skutschas. 2000. A eutherian mammal from the Early Cretaceous of Russia and biostratigraphy of the Asian Early Cretaceous vertebrate assemblages. Lethaia 33: 330–340.

Baird, D. 1978. Studies on Carboniferous freshwater fishes. American Museum Novitates 1641: 1–22.

Becker, M. A., J. A. J. Chamberlain, and D. O. J. Terry. 2004. Chondrichthyans from the Fairpoint Member of the Fox Hills Formation (Maastrichtian), Meade County, South Dakota. Journal of Vertebrate Paleontology 24: 780–793.

Bermúdez-Rochas, D. D. 2009. New hybodont shark assemblage from the Early Cretaceous of the Basque–Cantabrian Basin. Geobios 42: 675–686.

Buscalioni, A. D., M. A. Fregenal, A. Bravo, F. J. Poyato-Ariza, B. Sanchíz, A. M. Báez, O. Cambra Moo, C. Martín Closas, S. E. Evans, J. Marugán Lobón. 2008. The vertebrate assemblage of Buenache de la Sierra (Upper Barremian of Serrania de Cuenca, Spain) with insights into its taphonomy and palaeoecology. Cretaceous Research 29: 687–710.

Canudo, J. I., G. Cuenca-Bescós, and J. I. Ruiz-Omeñaca. 1995. Tiburones y rayas (Chondrichthyes, Elasmobranchii) del Barremiense superior (Cretácico inferior) de Vallipón (castellote, Teruel). Beca del Museo de Mas de la Matas 1995: 35–57.

Cappetta, H. 1987. Chondrichthyes II. Mesozoic and Cenozoic Elasmobranchii; in H.-P. Schultze (ed.), Handbook of Paleoichthyology, vol. 3B. Gustav Fischer Verlag, Stuttgart, 193 pp.

———. 2006. Elasmobranchii Post-Triadici (index specierum et generum): Fossilium Catalogus I: Animalia, vol. 142. Backhuys Publishers, Leiden, 472 pp.

Cappetta, H., E. Buffetaut, and V. Suteethorn. 1990. A new hybodont from the Lower Cretaceous of Thailand. Neues Jahrbuch für Geologie und Paläontologie Monatshefte 1990: 659–666.

Cappetta, H., E. Buffetaut, G. Cuny, and V. Suteethorn. 2006. A new elasmobranch assemblage from the Lower Cretaceous of Thailand. Palaeontology 49: 547–555.

Casier, E. 1961. Matériaux pour la la faune ichthyologique Eocrétacique du Congo. Annales du Musée royal de l'Afrique centrale (ser. 8) 39: 1–96.

———. 1969. Addenda aux connaissances sur la faune ichthyologique de la serie de Bokungu (Congo). Annales du Musée royal de l'Afrique Centrale (ser. 8) 62: 1–20.

Cavin, L., P. L. Forey, and C. Lécuyer. 2007. Correlation between environment and Late Mesozoic ray-finned fish evolution. Palaeogeography, Palaeoclimatology, Palaeoecology 245: 353–367.

Coates, M. I., and R. W. Gess. 2007. A new reconstruction of *Onychoselache traquairi*: comments on early chondrichthyan pectoral girdles and hybodontiform phylogeny. Palaeontology 50: 1421–1446.

Cuny, G., M. Ouaja, D. Srarfi, L. Schmitz, E. Buffetaut, and M. J. Benton. 2004. Fossil sharks from the Early Cretaceous of Tunisia. Revue de Paléobiologie, volume special 9: 127–142.

Cuny, G., V. Suteethorn, S. Khama, K. Lauprasert, P. Srisuk, and E. Buffetaut. 2007. The Mesozoic fossil record of sharks in Thailand; pp. 349–354 in W. Tantiwanit (ed.), Proceedings of the International Conference on Geology of Thailand: Towards Sustainable Development and Sufficiency Economy. Department of Mineral Resources, Bangkok.

Cuny, G., V. Suteethorn, S. Khamha, and E. Buffetaut. 2008. Hybodont sharks from the Lower Cretaceous Khok Kruat Formation of Thailand, and hybodont diversity during the Early Cretaceous; pp. 93–107 in L. Cavin, A. Longbottom, and M. Richter (eds.), Fishes and the break-up of Pangaea. Special Publications 295. Geological Society, London.

Cuny, G., L. Cavin, and V. Suteethorn. 2009. A new hybodont with a cutting dentition from the Lower Cretaceous of Thailand. Cretaceous Research 30: 515–520.

Cuny, G., C. Laojumpon, O. Cheychiw, and K. Lauprasert. 2010. Fossil vertebrate remains from Kut Island (Gulf of Thailand, Early Cretaceous). Cretaceous Research 31: 415–423.

Dejax, J., D. Pons, and J. Yans. 2008. Palynology of the Wealden facies from Hautrage quarry (Mons Basin, Belgium). Memoirs of the Geological Survey 55: 45–52.

Duffin, C. J. 2001a. Synopsis of the selachian genus *Lissodus* Brough, 1935. Neues Jahrbuch für Geologie und Paläontologie Abhandlungen 221: 145–218.

———. 2001b. The hybodont shark, *Priohybodus* d'Erasmo, 1960 (Early Cretaceous, northern Africa). Zoological Journal of the Linnean Society 133: 303–308.

Duffin, C. J., and D. Sigogneau-Russell. 1993. Fossil shark teeth from the Early Cretaceous of Anoual, Morocco. Belgian Geological Survey, Professional Paper 264: 175–190.

Dutheil, D. B. 1999. An overview of the freshwater fish fauna from the Kem Kem Beds (Late Cretaceous: Cenomanian) of southeastern Morocco; pp. 553–563 in G. Arratia and H.-P. Schultze (eds.), Mesozoic fishes 2—systematics and fossil record. Verlag Dr. Friedrich Pfeil, Munich.

Estes, R., and B. Sanchíz. 1982. Early Cretaceous lower vertebrates from Galve (Teruel), Spain. Journal of Vertebrate Palaeontology 2: 21–39.

Everhart, M. J. 2004. First record of the hybodont shark genus, "*Polyacrodus*" sp. (Chondrichthyes; Polyacrodontidae) from the Kiowa Formation (Lower Cretaceous) of McPherson County, Kansas. Transactions of the Kansas Academy of Science 107: 83–87.

———. 2009. First occurrence of marine vertebrates in the Early Cretaceous of Kansas: Champion Shell Bed, basal Kiowa Formation. Transactions of the Kansas Academy of Science 112: 201–210.

Fischer, J., and I. Kogan. 2008. Elasmobranch egg capsules *Palaeoxyris*, *Fayolia* and *Vetacapsula* as subject of palaeontological research—an annotated bibliography. Freiberger Forschungsheft C 528: 75–91.

Fischer, J., S. Voigt, and M. Buchwitz. 2007. First elasmobranch egg capsules from freshwater lake deposits of the Madygen Formation (Middle to Late Triassic, Kyrgyzstan, Central Asia). Freiberger Forschungshefte C 524: 41–46.

Ginter, M., V. Hairapetian, and C. Klug. 2002. Fammennian chondrichthyans from the shelves of North Gondwana. Acta Geologica Polonica 52: 169–215.

Goodwin, M. B., W. A. Clemens, J. H. Hutchison, C. G. Wood, M. S. Zavada, A. Kemp, C. J. Duffin, and C. R. Schaff. 1999. Mesozoic continental vertebrates with associated palynostratigraphic dates from the northwestern Ethiopian plateau. Journal of Vertebrate Paleontology 19: 728–741.

Heckert, A. B. 2004. Late Triassic microvertebrates from the lower Chinle Group (Otischalkian–Adamanian: Carnian), southwestern USA. New Mexico Museum of Natural History and Science Bulletin 27: 1–170.

Heckert, A. B., A. Ivanov, and S. G. Lucas. 2007. Dental morphology of the hybodontoid shark *Lonchidion humblei* Murry from the Upper Triassic Chinle Group, USA. New Mexico Museum of Natural History and Science Bulletin 41: 45–48.

Kozai, T., K. Ishida, F. Hirsch, S.-O. Park, and K.-H. Chang. 2005. Early Cretaceous non-marine mollusc faunas of Japan and Korea. Cretaceous Research 26: 97–112.

Kriwet, J., and C. Klug. 2008. Diversity and biogeography patterns of Late Jurassic neoselachians (Chondrichthyes: Elasmobranchii); pp. 55–70 in L. Cavin, A. Longbottom, and M. Richter (eds.), Fishes and the break-up of Pangaea. Special Publications 295. Geological Society, London.

Kriwet, J., W. Kiessling, and S. Klug. 2009. Diversification trajectories and evolutionary life-history traits in early sharks and batoids. Proceedings of the Royal Society B 276: 945–951.

Lane, A. L., and J. G. Maisey. 2009. Pectoral anatomy of *Tribodus limae* (Elasmobranchii: Hybodontiformes) from the Lower Cretaceous of northeastern Brazil. Journal of Vertebrate Paleontology 29: 25–38.

López-Arbarello, A. 2004. The record of Mesozoic fishes from Gondwana (excluding India and Madagascar); pp. 597–624 in G. Arratia and A. Tintori (eds.), Mesozoic fishes 3—systematics, paleoenvironment and biodiversity. Verlag Dr. Friedrich Pfeil, Munich.

Maisey, J. G. 2000. Continental break-up and the distribution of fishes of Western Gondwana during the Early Cretaceous. Cretaceous Research 21: 281–314.

Maisey, J. G., and M. R. de Carvalho. 1997. A new look at old sharks. Nature 385: 779–780.

Maisey, J. G., G. J. P. Naylor, and D. J. Ward. 2004. Mesozoic elasmobranchs, neoselachian phylogeny and the rise of modern elasmobranch diversity; pp. 17–56 in G. Arratia and A. Tintori (eds.), Mesozoic fishes 3—systematics, paleoenvironment and biodiversity. Verlag Dr. Friedrich Pfeil, Munich.

Martin, R. A. 2005. Conservation of freshwater and euryhaline elasmobranchs: a review. Journal of the Marine Biology Association of the U.K. 85: 1049–1073.

Montoya, R. V., and T. B. Thorson. 1982. The bull shark (*Carcharhinus leucas*) and largetooth sawfish (*Pristis perotteti*) in Lake Bayano, a tropical man-made impoundment in Panama. Environmental Biology of Fishes 7: 341–347.

Perea, D., M. Ubilla, A. Rojas, and C. A. Goso. 2001. The West Gondwanan occurrence of the hybodontid shark *Priohybodus*, and the Late Jurassic–Early Cretaceous age of the Tacuarembo Formation, Uruguay. Palaeontology 44: 1227–1235.

Prasad, G. V. R., B. K. Manhas, and G. Arratia. 2004. Elasmobranch and actinopterygian remains from the Jurassic and Cretaceous of India; pp. 625–638 in G. Arratia and A. Tintori (eds.), Mesozoic fishes 3—systematics, paleoenvironment and biodiversity. Verlag Dr. Friedrich Pfeil, Munich.

Racey, A. 2009. Mesozoic red bed sequences from SE Asia and the significance of the Khorat Group of NE Thailand; pp. 41–67 in E. Buffetaut, G. Cuny, J. Le Loeuff, and V. Suteethorn (eds.), Late Palaeozoic and Mesozoic ecosystems in SE Asia. Special Publications 315. Geological Society, London.

Rage, J.-C., and H. Cappetta. 2002. Vertebrates from the Cenomanian, and the geological age of the Draa Ubari fauna (Libya). Annales de Paléontologie 88: 79–84.

Rees, J. 2000. A new Pliensbachian (Early Jurassic) neoselachian shark fauna from southern Sweden. Acta Palaeontologica Polonica 45: 407–424.

———. 2002. Shark fauna and depositional environment of the earliest Cretaceous Vitabäck Clays at Eriksdal, southern Sweden. Transactions of the Royal Society of Edinburg Earth Sciences 93: 59–71.

———. 2005. Neoselachian shark and ray teeth from the Valanginian, Lower Cretaceous, of Wawal, Central Poland. Palaeontology 48: 209–221.

———. 2008. Interrelationships of Mesozoic hybodont sharks as indicated by dental morphology—preliminary results. Acta Geologica Polonica 58: 217–221.

Rees, J., and C. J. Underwood. 2002. The status of the shark genus Lissodus Brough 1935, and the position of nominal Lissodus species within the Hybodontoidea (selachii). Journal of Vertebrate Paleontology 22: 471–479.

———. 2008. Hybodont sharks of the English Bathonian and Callovian (Middle Jurassic). Palaeontology 51: 117–147.

Schultze, H.-P. 2009. Interpretation of marine and freshwater paleoenvironments in Permo-Carboniferous deposits. Palaeogeography, Palaeoclimatology, Palaeoecology 281: 126–136.

Schultze, H.-P., and R. Soler-Gijón. 2004. A xenacanth clasper from the ?uppermost Carboniferous–Lower Permian of Buxières-les-Mines (Massif Central, France) and the palaeoecology of the European Permo-Carboniferous basins. Neues Jahrburg für Geologie und Paläontologie, Abhandlungen 232: 325–363.

Shimada, K., C. K. Rigsby, and S. H. Kim. 2009. Partial skull of Late Cretaceous durophagous shark, Ptychodus occidentalis (Elasmobranchii: Ptychodontidae), from Nebraska, USA. Journal of Vertebrate Paleontology 29: 336–349.

Shimada, K., M. J. Everhart, R. Decker, and P. D. Decker. 2010. A new skeletal remain of the durophagous shark, Ptychodus mortoni, from the Upper Cretaceous of North America: an indication of gigantic body size. Cretaceous Research 31: 249–254.

Sweetman, S. C., and C. J. Underwood. 2006. A neoselachian shark from the non-marine Wessex Formation (Wealden Group: Early Cretaceous, Barremian) of the Isle of Wight, southern England. Palaeontology 49: 457–465.

Tanimoto, M., and S. Tanaka. 1998. Heteroptychodus sp. (Chondrichthyes) from the Lower Cretaceous Matsuo Group of Arashima, Toba City, Mie Prefecture, Southwest Japan. Chigakukenkyu 47: 37–40.

Thurmond, J. T. 1971. Cartilaginous fishes of the Trinity Group and related rocks (Lower Cretaceous) of North Central Texas. Southeastern Geology 13: 207–227.

Underwood, C. J. 2006. Diversification of the Neoselachii (Chondrichthyes) during the Jurassic and Cretaceous. Paleobiology 32: 215–235.

Underwood, C. J., and S. L. Cumbaa. 2010. Chondrichthyans from a Cenomanian (Late Cretaceous) bonebed, Saskatchewan, Canada. Palaeontology 53(4): 903–944.

Underwood, C. J., and J. Rees. 2002. Selachian faunas from the lowermost Cretaceous Purbeck Group of Dorset, Southern England. Special Papers in Palaeontology 68: 83–101.

Underwood, C. J., S. F. Mitchell, and K. J. Veltkamp. 1999. Shark and ray teeth from the Hauterivian (Lower Cretaceous) of north-east England. Palaeontology 42: 287–302.

Vullo, R., and D. Néraudeau. 2008. When the "primitive" shark Tribodus (Hybodontiformes) meets the "modern" ray Pseudohypolophus (Rajiformes): the unique co-occurrence of these two durophagous Cretaceous selachians in Charentes (SW France). Acta Geologica Polonica 58: 249–255.

Welton, B. J., and R. F. Farish. 1993. The collector's guide to fossil sharks and rays from the Cretaceous of Texas. Before Time, Lewisville, Tex., 204 pp.

Winkler, D. A., P. A. Murry, and L. L. Jacobs. 1990. Early Cretaceous (Comanchean) vertebrates of central Texas. Journal of vertebrate Paleontology 10: 95–116.

Yabe, H., and T. Obata. 1930. On some fossil fishes from the Cretaceous of Japan. Japanese Journal of Geology and Geography 8: 1–8.

Zhang, J.-Y. 2007. Two shark finspines (Hybodontoidea) from the Mesozoic of North China. Cretaceous Research 28: 277–280.

Cretaceous Vertebrate Faunas after the Bernissart Iguanodons

4

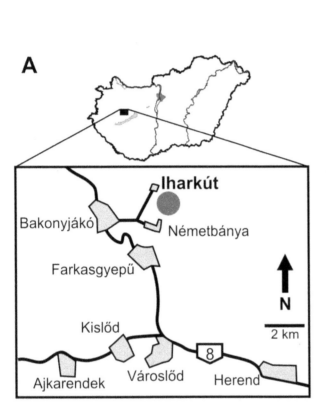

A

Iharkút

Bakonyjákó

Németbánya

Farkasgyepű

N

Kislőd

2 km

Ajkarendek

Városlőd

8

Herend

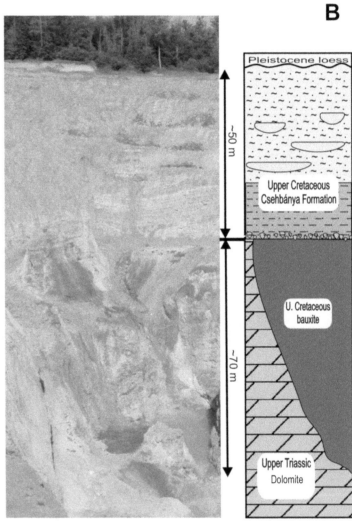

B

Pleistocene loess

~50 m

Upper Cretaceous
Csehbánya Formation

~70 m

U. Cretaceous
bauxite

Upper Triassic
Dolomite

30.1. A, Location map of the Iharkút vertebrate
locality, Upper Cretaceous (Santonian) Csehbánya
Formation, Bakony Mountains, western Hungary.
B, Schematic section of the open-pit Iharkút
(middle Eocene conglomerates and limestones and
late Eocene conglomerates occur northward; after
Ősi and Mindszenty, 2009).

The Late Cretaceous Continental Vertebrate Fauna from Iharkút (Western Hungary): A Review

30

**Attila Ősi*, Márton Rabi, László Makádi,
Zoltán Szentesi, Gábor Botfalvai, and Péter Gulyás**

The composition of the Late Cretaceous (Santonian) continental vertebrate fauna of Iharkút (Csehbánya Formation, Bakony, western Hungary, Central Europe) is reviewed here. In the last decade, fieldwork has produced almost 5,000 associated and isolated bones and teeth belonging to at least 24 different genera, represented by pycnodontiform and lepisosteid fishes, albanerpetontid and anuran amphibians, dortokid, bothremydid and cryptodiran turtles, scincomorphan and mosasauroid lizards, mesoeucrocodylian and eusuchian crocodilians, nodosaurid ankylosaurs, rhabdodontid ornithopods, basal tetanuran, abelisaurid, paravian, and enantionthine theropods, and azhdarchid pterosaurs. Remains of mammals are still unknown from the locality. Because of its Santonian age, the discovered fauna fills an important and still underrepresented temporal gap in the Cretaceous vertebrate record of Europe. The fauna is a mixture of Euramerican and Gondwanan forms. The first group consists of numerous taxa (e.g., hylaeochampsid crocodilians, nodosaurid ankylosaurs, rhabdodontid ornithopods, basal tetanurans), the closest relatives of which are stratigraphically much older (Late Jurassic–late Early Cretaceous) forms. These members of the fauna are suggested to be relict forms surviving in an insular habitat of the Iharkút area within the western Tethyan archipelago. At least bothremydid turtles further strengthen the immigration of Gondwanan forms into Europe during the Late Cretaceous. The supposed insular habitat of the Iharkút fauna is also supported by the presence of a peculiar small-bodied heterodont crocodilian with specialized feeding preference, and of pycondontiform fishes and mosasaurs that colonized freshwater environments.

Introduction

One of the greatest paleontological highlights over the last decade in Hungary has been the discovery of and the subsequent research on the Late Cretaceous (Santonian) vertebrate-bearing site near Iharkút in the Bakony Mountains of western Hungary. This is the only Mesozoic vertebrate locality in Hungary where continuous and productive excavations have been carried out, and it has provided the first evidence for several vertebrate groups (e.g., bothremydid turtles, mosasaurs, nonavian dinosaurs, pterosaurs) from the country (Ősi, 2004a; Ősi et al., 2005; Makádi, 2005; Rabi and Botfalvai, 2006).

Since its discovery in 2000, approximately 5,000 isolated and sometimes associated bones, teeth, and scales of at least 24 different taxa have been collected from Iharkút. The fossils represent a diverse fauna of lepisosteid and pycnodontiform fishes, albanerpetontids, anurans, turtles, lizards, crocodilians, nonavian theropod, ornithopod and nodosaurid dinosaurs, enantiornithine birds, and pterosaurs.

Thanks to its abundance of fossils and diversity of taxa, the Iharkút locality is critical for understanding the history of European Late Cretaceous continental vertebrate faunas. The Iharkút locality is of further significance because of its geographic position in the western Tethyan archipelago and its Santonian age (Knauer and Siegl-Farkas, 1992; Szalai, 2005). As the only Santonian-age locality, Iharkút fills an important gap in the Late Cretaceous record of nonmarine vertebrates in Europe.

In this chapter, we review our knowledge on the Iharkút vertebrate fauna. After discussing the geological and taphonomical aspects of the locality, we provide short descriptions and comparisons of the taxa identified so far at Iharkút. Finally, we discuss the paleoecological and paleobiogeographical significance of the locality and its fauna.

Materials and Methods

Materials

All the fossils (including plant, invertebrate, and vertebrate specimens) discovered at the Iharkút vertebrate site are housed in the Hungarian Natural History Museum (MTM) in Budapest. Except for five disarticulated but associated skeletons of the armored dinosaur *Hungarosaurus* (see Ősi and Makádi, 2009), all bones, teeth, and scales are isolated remains. Depending on which year specimens were collected, they would have been cataloged using any of four different types of catalog numbers (e.g., MTM V 01.49, MTM V 2003.12, MTM V 2007.30.1, and MTM Gyn/123) as a result of the different systems used over that interval in the museum. In this chapter, the abbreviation MTM has been omitted before these catalog numbers.

Preparation and Conservation

The fossils were prepared in the technical laboratories of the Department of Paleontology, Eötvös Loránd University, and of the Hungarian Natural History Museum. Most of the material was discovered while hand-quarrying the bone-yielding beds. However, some of the smaller bones, especially the scales and teeth, were recovered through extensive screen washing of excavated matrix. Specimens were prepared with vibro tools, needles, and localized use of 10% acetic acid. In the preparatory stage, if necessary, conservation was performed with polyvinyl acetate (PVA) or polyvinyl butyral (PVB). After removing matrix from the bones, cyanoacrylate (SuperGlue) was used to repair breaks. The bones are rich in pyrite and organic material (Tuba et al., 2006), and the oxidization of the pyrite within them is relatively fast. The specimens were thus soaked in PVA or PVB to stop or at least slow down this process.

Locality and Geological Setting

The vertebrate locality is situated in an open-pit bauxite mine near the villages of Iharkút and Németbánya, in the northern part of the Bakony Mountains (Transdanubian Central Range), western Hungary (Fig. 30.1A).

The exposed sequence overlies Paleozoic bedrocks and consists of Mesozoic and Paleogene sedimentary strata. The oldest formation that crops out in the Iharkút area is an Upper Triassic shallow marine dolomite (Main Dolomite Formation). During the Late Cretaceous (pre-Santonian), sediments forming the Nagytárkány Bauxite Formation were trapped in karstic cavities that descended to a depth of 60 to 80 m in the Main Dolomite Formation (Fig. 30.1B) (Jocha-Edelényi, 1988). The dolomite and the bauxite deposits are overlain by the Upper Cretaceous Csehbánya Formation, which contains the vertebrate fossils reported herein. The Csehbánya Formation contains fluvial and floodplain deposits that consist mainly of variegated clay, paleosol horizons, and silt with sand and sandstone layers, with the latter being interpreted as channel fills (Tuba et al., 2006; Ősi and Mindszenty, 2009). Most of the vertebrate fossils were found in the coarse-grained, pebbly sandy basal beds of the fluvial half-cycles, where bones, teeth, and scales were washed in and concentrated together (see the bonebed Sz-6 account below). The Csehbánya Formation has a maximum thickness of 200 m (in the Csehbánya Basin), but in the Iharkút area, it reaches only 50–60 m. The predominance of fine overbank sediments, the abundance of hydromorphic paleosols, the presence of extensive fine-sand sheets and only shallow channels, together with the absence of lateral accretion structures, indicates that the depositional environment must have been the floodplain of a low-gradient perennial river. Intercalated sand sheets (and the sz-6 site; see below) are interpreted as crevasse splays (Ősi and Mindszenty, 2009). The Santonian age for the Csehbánya Formation is supported by both palynological (*Occullopollis–Complexiopollis* zone) and paleomagnetic data (Knauer and Siegl-Farkas, 1992; Szalai, 2005).

At some places in the quarry, higher up in the stratigraphic sequence, Middle Eocene (Lutetian) conglomerates and limestones unconformably rest on the surface of the eroded Mesozoic formations. These Middle Eocene strata are separated by another unconformity from Upper Eocene conglomerates. In some parts of the Iharkút area, Oligocene clays, siltstones, sandstones, and conglomerates of the alluvial Csatka Formation cover the Eocene strata. A thin, discontinuous blanket of Pleistocene loess covers a large part of the area.

Depositional Environment of the Bonebeds (SZ-6 and SZ-7–8 Sites)

Vertebrate remains can be found in exposures of the Csehbánya Formation throughout the entire mine, but sz-6 and sz-7–8 (the latter is 300 m eastward from sz-6) sites are the most important levels because of the rich accumulation of vertebrate fossils.

The sz-6 site is the most important bone accumulation in the Iharkút mine because about 80% of the complete specimens were discovered in this layer (Fig. 30.2). Most of the vertebrate fossils were recovered from an approximately 3-m-thick sequence of beds made up of coarse, pebbly sand and organic-rich silt and clay; these beds are interpreted as crevasse splay deposits (Ősi and Mindszenty, 2009). The base of this sequence is clearly erosional where it cuts into the floodplain deposits (Fig. 30.2A). The bonebed in sz-6 is a 10- to 50-cm-thick, basal breccia composed of gray sand, siltstone, clay clasts, pebbles, and plant debris (also charcoal) that occasionally contains complete, but more frequently, fragmentary and highly abraded bones. The basal breccia, which would have been under high-energy conditions, is sometimes interrupted by finer sediments that settled out during calmer times. This means that energy conditions must have changed several times during the cyclic deposition of the sediments. As a result of the alternating energy conditions of the depositional environment, bones in different states of preservation can be found in the same beds. The depositional environment of these fossiliferous layers can be characterized by permanent hydromorphy and low energy; consequently, they indicate stagnant water conditions, presumably on the low-lying areas on the floodplain.

The sandstone bed above the basal breccia also contains vertebrate fossils, but the bones are fewer and more poorly preserved. However, two incomplete skeletons of the nodosaurid ankylosaur *Hungarosaurus tormai* have been found in this bed.

The overlying bed is a 30- to 50-cm-thick, laminated, grayish siltstone and contains plant debris and fewer bones. This sediment yielded two incomplete skeletons of *Hungarosaurus* (Fig. 30.2B), a large-sized partial turtle plastron with a possibly associated skull, as well as a few small bones of various taxa.

The second important site is a microvertebrate bonebed at sz-7 and sz-8. This layer is a 10- to 20-cm-thick dark gray silt and clay bed with high organic matter content. Another feature of the bed is that it contains 1- to 3-mm-sized fragments of amber. Microvertebrate fossils were recovered by screen washing and consist of fishes, albanerpetontids, anurans, turtles, and squamates, as well as teeth and scales of fishes, crocodilians, and dinosaurs.

Taphonomic Features of the Iharkút Vertebrate Material

The Late Cretaceous (Santonian) locality at Iharkút has already provided several thousand vertebrate fossils as a result of various excavations performed in the last 10 years (2000–2009). The large number of fossils collected from Iharkút and the differences in their states of preservation provided an opportunity to undertake a taphonomic evaluation (Botfalvai, in prep.). Most fossil bones recovered from the Csehbánya Formation are pyritized and fragmentary. The pyrite in the bones is interpreted as early diagenetic mineral and primary cement (Tuba et al., 2006). Currently the collection of vertebrate fossils from Iharkút contains 4,828 bones and 1,661 teeth (including incomplete skeletons of *Hungarosaurus*). Of these, 4,468 specimens were suitable for taphonomic investigation. These fossils are dominated by turtles and dinosaurs (about 67%), and 86% of the total number of bone material of dinosaurs belongs to *Hungarosaurus tormai* (Fig. 30.3). In this area, the most abundant dinosaur taxa are *Hungarosaurus* and Rhabdodontidae; the other dinosaur species are only subordinately present. The taphonomic studies pointed out that the quantitative and qualitative (bones of weathering and abrasion) differences in the fossil material of the two dinosaur species studied taphonomically are possibly in correlation with their different life habitats. The particular taphonomic study and description of the Iharkút remains are in progress; therefore, it is not possible to cover the data and results here.

The size distribution of the bones shows that larger-bodied species tend to be represented by larger bones, while the opposite is the case with the

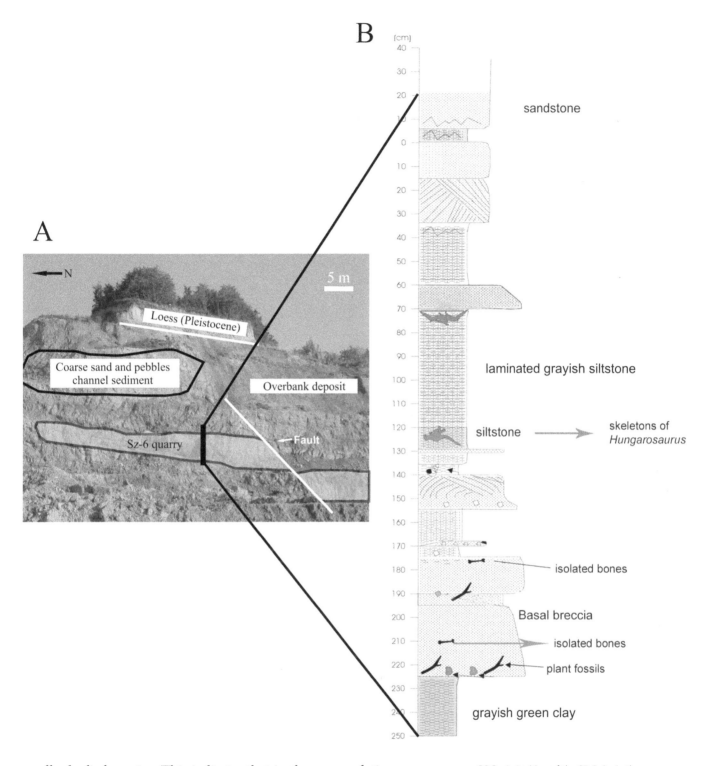

B (cm)

sandstone

laminated grayish siltstone

siltstone →→ skeletons of *Hungarosaurus*

isolated bones

Basal breccia

isolated bones

plant fossils

grayish green clay

A

← N

5 m

Loess (Pleistocene)

Coarse sand and pebbles channel sediment

Overbank deposit

Sz-6 quarry

Fault

smaller-bodied species. This indicates that in the accumulation process, there was no filter effect that caused sorting by size.

Abrasion on the bone surface indicates friction between the bone and sediment during transport (Behrensmeyer, 1982; Astibia et al., 1999; Pereda-Suberbiola et al., 2000). Consequently, on the basis of the degree of abrasion, the complete bone material from the sz-6 site can be divided into two groups. The first group contains small and highly abraded bone fragments (bone pebbles) transported with the sediment for a long distance. The second group consists of larger and better-preserved bones

30.2. A, Position of the SZ-6 site in the sequence at the Iharkút locality (Upper Cretaceous Cseh-bánya Formation). B, Shematic statigraphic section showing the main paleoenvironments and lithofacies associations at SZ-6 site (after Tuba et al., 2006).

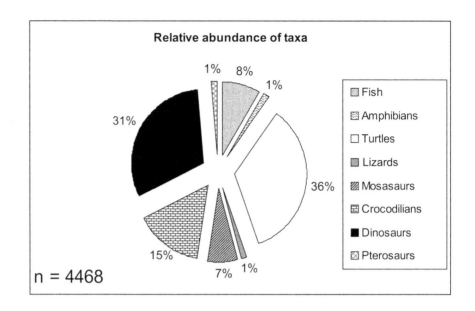

30.3. Relative abundance of taxa at Iharkút locality (Upper Cretaceous Csehbánya Formation) based on counts of number of identifiable remains; n indicates number of samples.

rapidly buried after short transportation with the sediment. The presence of both abrasional groups in the same layer is consistent with the crevasse splay accumulation model that was originally proposed on the basis of sedimentological evidence (Tuba et al., 2006; Ősi and Mindszenty, 2009). The small and highly abraded bone pebbles were transported in the channel bed from distant areas, whereas the better-preserved bones underwent little transportation or abrasion and thus were likely deposited during the formation of the crevasse splay.

The material can be regarded as fragmentary because 78% of the bones exhibit some kind of postmortem fracture. The data retrieved from the breakage angles suggest that most breaks occurred after the fossilization of the bones, which also suggests rapid burial and deposition shortly after the initiation of diagenetic processes.

Systematic Paleontology

Osteichthyes Huxley, 1880
Actinopterygii Klein, 1885
Pycnodontiformes Lehman, 1966
Pycnodontiformes indet.
(Fig. 30.4A,B)

Material. Right prearticulars (V 2010.131.1, V 2010.135.1); left prearticulars (V 2010.139.1, V 2010.140.1); right fragmentary prearticulars (V.01.83, V 2010.143.1, V 2010.145.1, V 2010.146.1, V 2010.149.1, V 2010.151.1, V 2010.153.1); left fragmentary prearticulars (V 2010.132.1–V2010.134.1, V 2010.136.1–V 2010.138.1, V 2010.141.1, V 2010.142.1, V 2010.144.1, V 2010.147.1, V 2010.148.1, V 2010.150.1, V 2010.152.1, V 2010.154.1–V 2010.158.1).

Description and comparisons. On the basis of morphological differences between the well-preserved prearticulars, two morphotypes can be distinguished. Prearticulars of the first type (Fig. 30.4A) contain three longitudinally arranged, regular tooth rows similar to *Stemmatodus rhombus* (Kriwet, 2004) and in contrast to the irregular tooth arrangement of some pycnodontiforms (e.g., *Macromesodon macropterus*; Nursall, 1996). Tooth

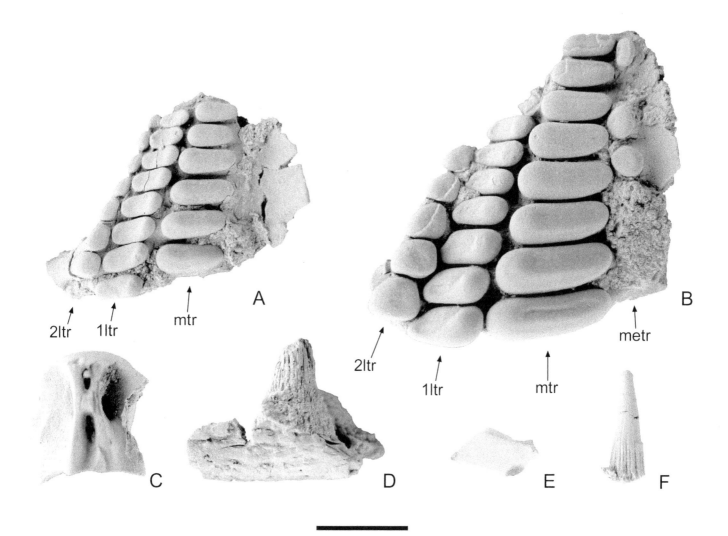

30.4. Fish remains from the Upper Cretaceous (Santonian) Csehbánya Formation, Iharkút locality, Bakony Mountains, western Hungary. A–B, Pycnodontiformes indet.: A, morphotype 1, left prearticular (V 2010.140.1) in lingual view; B, morphotype 2, left prearticular (V 2010.139.1) in lingual view. C–E, Lepisosteiformes indet.: C, vertebral centrum (V 2010.156.1) in dorsal view; D, jaw fragment (V 2010.155.1) in ?lingual view; E, scale (V 2010.158.1). F, *Atractosteus* sp. tooth (V 2010.157.1). *Abbreviations:* metr, medial tooth row; mtr, main tooth row; 1ltr, first lateral tooth row; 2ltr, second lateral tooth row. Scale bar = 10 mm.

size decreases anteriorly along each row, which is similar to the pattern in *Proscinetes hugii* (Kriwet, 2005). The teeth in the main row are the largest and are slightly bean-shaped in occlusal view, as in *Pycnodus pellei* (Kriwet, 2005), although they are transversely more elongate. The teeth in the first lateral row are transversely elongate and oval in occlusal view, and a transverse trench is visible on their occlusal surface. This trench is also present in *Anomoeodus pauciseriale*, although the teeth of the latter are irregular in their outline (Kriwet, 2002). Teeth in the second lateral row are the smallest, and their occlusal outlines vary from longitudinally oval in the anterior part of the row to subcircular in the posterior part of the row.

The second type (Fig. 30.4B) resembles the first, but with one important difference. It contains four tooth rows, similar to *Neoproscinetes penalvai*: besides the main and the first and second lateral rows, a medial row is also present (Maisey, 1991). The teeth in the medial row are rounded in occlusal view and are the smallest in the whole dentition, as in "*Coelodus*" *syriacus* (Hussakof, 1916).

Discussion. Makádi et al. (2006) noted similarities between the pycnodontiform fossils from Iharkút and those of *Anomoeodus* sp. On the basis of new finds, however, it is now evident that the Iharkút specimens clearly differ from that genus. Further investigation is needed to determine whether the Iharkút specimens belong to a single taxon or the two different morphotypes perhaps represent two different taxa (Gulyás, in prep.).

Iharkút is among the few localities where this dominantly shallow marine fish group occurred in nonmarine sediments (Kocsis et al., 2009). Freshwater pycnodontiform remains in Europe are known from Bernissart in Belgium and from the Lower Cretaceous Galve (Estes and Sanchíz, 1982) and Las Hoyas localities in Spain. The latter yielded *Macromesodon* aff. *bernissartensis* and *Eomesodon* sp. specimens, as well as the first known freshwater coelacanth (Poyato-Ariza et al., 1998). In addition, remains of pycnodontiforms from freshwater sediments were discovered in North America (Winkler et al., 1990; Eaton et al., 1999; Cifelli et al., 1999) and Thailand (Cavin et al., 2009).

Holostei Müller, 1845
Lepisosteiformes Hay, 1929
Lepisosteidae Matsubara, 1955
Atractosteus Rafinesque, 1820
Atractosteus sp.
(Fig. 30.4C–F)

Material. Jaw fragment (V 2010.155.1); isolated teeth (V 2010.157.1).

Description and comparisons. The well-preserved teeth are conical with longitudinal ridges basally (Fig. 30.4D) and a lanceolate apex occlusally. The morphology of the teeth (Fig. 30.4F) is typical of the genus *Atractosteus* (Grigorescu et al., 1999).

Discussion. Lepisosteiform remains have been found worldwide in various localities. From Africa, India, and South America (Gondwana), they are known since the Early Cretaceous (Gayet et al., 2002). From North America, they were described since the Late Cretaceous (Turonian–Coniacian) (Friedman et al., 2003), but in Europe, the oldest lepisosteiform fossils previously know were found in Ventabren (France) from the early Campanian (Grigorescu et al., 1999; Gayet et al., 2002). Thus, the new lepisosteiform remains from the Santonian Iharkút locality are the geologically oldest known occurrence of the group in Europe.

Remarks. Several fish vertebrae (V 2010.156.1) and scales (V 2010.158.1) were also discovered and can be assigned to Lepisosteiformes. The vertebral centra are opisthocoelous, and the scales are rhomboidal with a ganoin layer. The centra (Fig. 30.4C) and scales (Fig. 30.4E) are similar to the remains known from the Upper Cretaceous locality of Laño, Spain (Cavin, 1999). Their morphology is typical for lepisosteiforms (Wiley, 1976); however, similar to the Laño specimens, the Iharkút lepisosteiform vertebrae and scales cannot be determined at a specific level.

Amphibia Linneaus, 1758
Lissamphibia Haeckel, 1866
Allocaudata Fox and Naylor, 1982
Albanerpetontidae Fox and Naylor, 1982
Albanerpetontidae indet.
(Fig. 30.5A–D)

Material. Premaxilla (V 2008.22.1); dentaries (V 2008.25.1, V 2008.26.1–V 2008.29.1, V 2009.6.1–V 2009.9.1).

Description and comparison. Although the material from Iharkút cannot be identified beyond the family level, the remains exhibit a suite of features (Fox and Naylor, 1982; Gardner, 2000) that are diagnostic for albanerpetontids. No specimen preserves intact teeth, but the preserved tooth bases are typical for albanerpetontids in being pleurodont, closely packed, and slightly compressed mesiodistally. The best-preserved premaxilla (Fig. 30.5A,B) retains the base of a broad, dorsally projecting pars dorsalis and the lingual surface of that process accommodates a deep suprapalatal pit with an elliptical outline. Dentaries (Fig. 30.5C,D) also are typical for albanerpetontids in having a relatively tall dental parapet with a nearly straight or shallowly convex dorsal edge, a prominent subdental shelf that deepens posteriorly, and a Meckelian canal that is closed lingually along the anterior portion of the bone. Specimens from Iharkút primitively resemble the albanerpetontid specimens reported by Folie and Codrea (2005) from the Hațeg Basin of Romania but are too poorly preserved for further comparisons.

Discussion. Albanerpetontids are known from many Early Cretaceous and latest Cretaceous (Campanian–Maastrichtian) localities in Europe (e.g., Gardner and Böhme, 2008, table 2). The Iharkút occurrence extends the Late Cretaceous record for albanerpetontids in Europe back into the Santonian and demonstrates that albanerpetontids were present in the western Tethyan archipelago by at least that time. The only other records of albanerpetontids in the Central European region are from the Maastrichtian of central Romania (Folie and Codrea, 2005; Grigorescu et al., 1999), the Late Miocene of Vienna Basin, Austria (Harzhauser and Tempfer, 2004), and the Early Pliocene of south-central Hungary (Venczel and Gardner, 2005) and northeastern Italy (Delfino and Sala, 2007).

Remarks. Isolated amphibian bones are preserved in two layers at Iharkút: in the basal breccia bonebed deposit within a sand, pebble, and clay clast layer, and in an upper dark siltstone layer containing abundant amber grains and carbonized plant material. Although several tons of matrix have been screen washed from the two fossiliferous layers since 2005, relatively few amphibian bones have been recovered. Nevertheless, albanerpetontids (an extinct clade of salamander-like amphibians) and frogs are known from both layers.

Salientia Laurenti, 1768
Anura Fischer von Waldheim, 1813
Anura indet.
(Fig. 30.5E–G)

Material. Ilia (V 2009.22.1, V 2008.12.1–V 2008.18.1); tibiofibulae (V 2008.19.1, V 2008.32.1).

Description and comparison. The frog material from Iharkút consists of relatively well-preserved ilia and fragmentary tibiofibulae. The ilium (Fig. 30.5E,F) is heavily built and dorsally bears an extremely high iliac crest that is prominently sculptured laterally with longitudinal ridges and grooves; these features indicate that the pelvic and upper leg muscles were enhanced, presumably for jumping or swimming. The enlarged interiliac

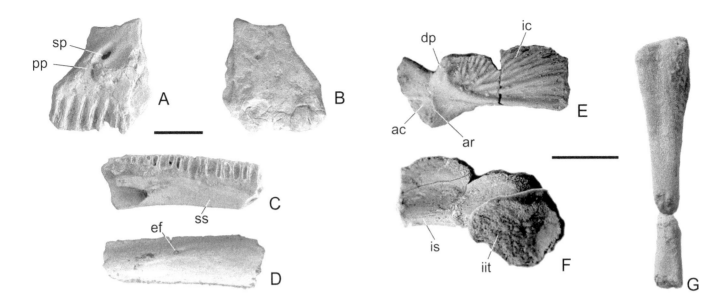

30.5. Amphibian remains from the Upper Cretaceous (Santonian) Csehbánya Formation, Iharkút locality, Bakony Mountains, western Hungary. A–D, Albanerpetontidae indet.: V 2008.22.1 left premaxilla, in (A) lingual and (B) labial view; V 2008.25.1, left dentary in (C) lingual and (D) labial view. E–G, Anura indet.: V 2009.22.1, ilium in (E) lateral and (F) medial view; G, V 2008.19.1, tibiofibula in ventral view. *Abbreviations:* ac, acetabulum; ar, acetabular rim; dp, dorsal protuberance; ef, external foramen; ic, iliac crest; iit, interiliac tubercle; is, iliac shaft; pp, pars palatinum; sp, suprapalatal pit; ss, subdental shelf. Scale bars = 1 mm for A–D and 5 mm for E–G.

tubercle covers almost the entire medial face of the acetabular region, it is wedge shaped in ventral or dorsal outline, and its medial surface is deeply pitted for strong sutural contact with the opposite ilium. This kind of enhanced interiliac contact is typical for frogs (e.g., pipids, palaeobatrachids) that are strong swimmers. One of the best-preserved tibiofibulae (Fig. 30.5G) is rather long and slightly curved laterally, and the epiphyses are completely ossified. The largest available ilia and tibiofibulae are probably from medium-sized frogs with snout–vent lengths in the 50-mm range. Only one kind of ilium and one kind of tibiofibula are represented; therefore, the specimens probably all pertain to the same species. Details of the crest and ilioischiadic junction of the ilia argue for the Iharkút frog being a member of the Neobatrachia (the so-called advanced frogs) and most likely closely related to ranoids (Szentesi and Venczel, 2010). Compared with unequivocal ranoids, ilia from Iharkút differ in having an iliac crest that is thick, tall, and sculptured laterally with longitudinal grooves and ridges and in having an enlarged interiliac tubercle.

Discussion. Previously the oldest record of neobatrachians in Europe was from the late Paleocene of France (Rage, 2003; Rage and Roček, 2003), which means that the Iharkút frog extends the European record for that clade several tens of millions of years back into the Santonian. Africa is widely accepted as the center of origin for ranoids (Savage, 1973; Báez and Werner, 1996; Feller and Hedges, 1998; Roelants and Bossuyt, 2005), and molecular phylogenetic work by Van der Meiden et al. (2005) estimated that the clade originated there by at least 92 Ma before the present. The presence of a ranoid or ranoidlike frog at Iharkút indicates that advanced frogs were present in the western Tethyan archipelago by at least Santonian time and possibly arrived from Africa across the Tethys (Szentesi and Venczel, 2010). This is an alternative to the recent suggestion by Bossuyt et al. (2006) that ranoids colonized Europe from India during the Maastrichtian.

Testudines Linneaus, 1758
Pleurodira Cope, 1864
Megapleurodira Gaffney, Tong, and Meylan, 2006

Dortokidae Lapparent de Broin and Murelaga, 1996
Dortokidae indet.
(Fig. 30.6A–G)

Material. Isolated costals (V 2010.196.1, V 2010.197.1), neural (V 2010.245.1), peripheral (V 2010.246.1), left hyoplastron (V 2010.222.1), and ilia (V 2010.160.1, V 2010.162.1).

Description and comparison. Remains of this small-sized turtle are not abundant at Iharkút. The plates have a characteristic microreticulated external ornamentation, making this form easily distinguishable from other taxa from the site. It consists of fine longitudinal ridges completed with round minute pits on neurals (Fig. 30.6A–G). The axillary process was sutured to the carapace on costal 1. These two characters allow an assignment to the family Dortokidae (Lapparent de Broin and Murelaga, 1999).

Discussion. The family Dortokidae is a primitive group of freshwater pleurodires endemic to Europe known only by shell material. Lapparent de Broin and Murelaga (1999) considered the Pelomedusoides to be the sister taxon of the Dortokidae, but later, both Lapparent de Broin et al. (2004) and Gaffney et al. (2006) suggested a basal position relative to Eupleurodira, but more derived than *Platychelys* and *Notoemys*. The family was first described from the Late Campanian–Early Maastrichtian of Spain and France, and originally contained a single genus and species, *Dortoka vasconica* (Lapparent de Broin and Murelaga, 1996, 1999). Later the family was also reported from the Late Barremian of Spain and the Late Paleocene of Romania; the latter occurrence provided a new taxon, *Ronella botanica* (Lapparent de Broin and Murelaga, 1999; Lapparent de Broin et al., 2004). Recently Vremir and Codrea (2009) identified new dortokid material from the Early to "middle" Maastrichtian of Romania and named a third taxon, *Muehlbachia nopcsai*, on the basis of isolated plates. Further investigation is needed to establish the identity of the Iharkút dortokid and its relationships with the three previously recognized species.

Eupleurodira Gaffney and Meylan, 1988 (sensu Gaffney, Tong, and Meylan, 2006)
Pelomedusoides Cope, 1868
Bothremydidae Baur, 1891
Bothremydidae indet.
(Fig. 30.7A–D)

Material. Fragmentary and almost complete skulls (V 2010.86.1, V 2010.87.1, V 2010.216.1); lower jaw (V 2010.219.1); left partial lower jaw (V 2010.89.1); anterior portion of carapace (V 2010.212.1); posterior portion of carapace (V 2010.211.1); posterior carapace fragment with left costal 7 and 8 (V 2010.214.1) anterior lobe of plastron (V 2010.207.1); scapula (V 2010.181.1); pubis (V 2010.169.1); humerus (V 2010.187.1); femur (V 2010.190.1).

Description and comparison. The most abundant turtle remains at the site belong to an as yet undescribed bothremydid. The wide triturating surface of the skull, the wide prefrontals, and the presence of a basisphenoid–quadrate contact collectively suggest that this turtle belonged to the Bothremydidae (sensu Gaffney et al., 2006). The skulls and lower jaws (Fig.

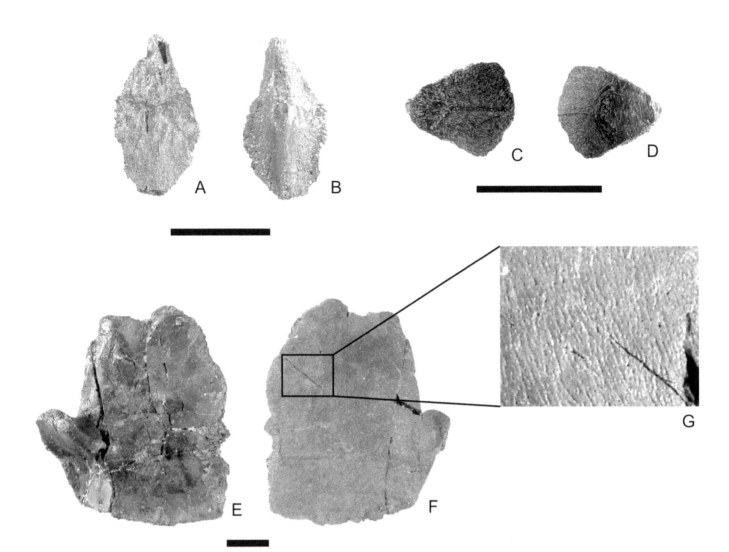

30.6. Isolated shell elements of Dortokidae indet. from the Upper Cretaceous (Santonian) Cseh-bánya Formation, Iharkút locality, Bakony Mountains, western Hungary. A, Neural, V 2010.245.1 in dorsal and B, ventral views; C, peripheral, V 2010.246.1 in dorsal, D, ventral views; E, left hyoplastron, V 2010.222.1 in dorsal, F, ventral view; G, close-up of decoration. Scale bar = 1 cm.

30.7A,B) are most similar to those of *Foxemys mechinorum* from the Late Campanian–Early Maastrichtian of southern France (Tong et al., 1998; Rabi et al., in prep.). An open incisura columellae auris further indicates close relationship with this taxon. However, the shell (Fig. 30.7C,D) differs from *F. mechinorum* in the presence of a nuchal notch.

Discussion. Bothremydidae is a diversified family of pleurodiran turtles that were widely distributed during the Early Cretaceous–Eocene in Gondwana (except Australia) and Euramerica. In Europe, they were previously known only from France, Spain, and Portugal, and they were represented by two closely related groups: the subtribe Foxemydina (including the genera *Foxemys*, *Polysternon*, and *Elochelys*) and the subtribe Bothremydina (including *Rosasia*) (Lapparent de Broin and Murelaga, 1996, 1999; Gaffney et al., 2006). As shown by the open incisura columellae auris of the quadrate, the bothremydid from Iharkút is closer to Foxemydina than to Bothremydina. Foxemydina inhabited freshwater environments and have been considered to be endemic to Western Europe. The new Iharkút specimens indicate that Foxemydina were also present outside Western Europe, in the Santonian of the central part of the western Tethyan archipelago.

This turtle indicates a Gondwanan influence on the Iharkút fauna because its outgroup taxa, *Cearachelys* and *Galianemys*, are known from

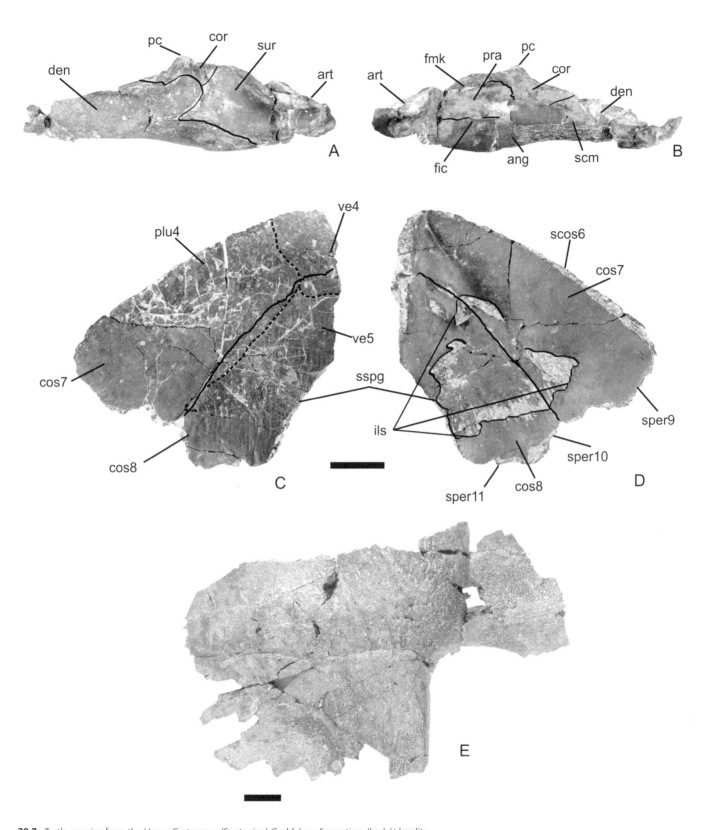

30.7. Turtle remains from the Upper Cretaceous (Santonian) Csehbánya Formation, Iharkút locality, Bakony Mountains, western Hungary. A–D, Bothremydidae indet.: A, left mandible, V 2010.219.1 in lateral, B, medial views; C, left costals 7–8, V 2010.214.1 showing iliac scars in dorsal, D, ventral view. E, Cryptodira indet. posteriorly incomplete left hypoplastron (V 2010.194.1) in ventral view. *Abbreviations:* ang, angular; art, articular; cor, coronoideum; cos, costal; den, dentary; fic, foramen intermandibularis caudalis; fmk, fossa Meckeli; ils, iliac scar; pc, processus coronoideus; plu, pleural; pra, prearticular; scm, sulcus cartilaginis Meckeli; scos, suture for costal; sper, suture for peripheral; sspg, suture for suprapygal; sur, surangular; ve, vertebral. Lines correspond to sutures; dotted lines correspond to sulci. Scale bar = 1 cm.

the Albian of Brazil and the Cenomanian of Morocco (Gaffney et al., 2006; Rabi et al., in prep.). The Santonian age of this taxon also means that it is one of the earliest reliable evidences for the presence of Gondwanan vertebrates in the Late Cretaceous of Europe (Cavin et al., 1996, 2005; Vullo et al., 2005; Pereda-Suberbiola, 2009).

Cryptodira Cope, 1868
Cryptodira indet.
(Fig. 30.7E)

Material. A single, posteriorly incomplete left hypoplastron (V 2010.194.1).

Description and comparison. This specimen is clearly distinguishable from the dortokids and bothremydids from Iharkút. The hypoplastron is relatively thick, being particularly swallowed at the beginning of the inguinal buttress. The external ornament consists of small, closely spaced, shallow pits and fine granulations or vermiculations. This ornament is reminiscent of *Kallokibotion bajazidi* from the Maastrichtian of Romania (M. Rabi, pers. obs., Gaffney and Meylan, 1992). Nonetheless, the available material from Iharkút is inadequate to refer it to *Kallokibotion*.

Discussion. The presence of a relatively large and probably primitive cryptodiran in the Iharkút fauna is noteworthy, especially because it does not seem to be a solemydid, which is a widespread Late Cretaceous cryptodiran family in continental faunas of Europe and has been reported from Spain, France, and possibly Romania (Lapparent de Broin and Murelaga, 1999; Company, 2004; Vremir and Codrea, 2009). When more material is available, it may be possible to compare the Iharkút cryptodiran with *Kallokibotion* from Romania (Nopcsa, 1923) and with the primitive cryptodiran recently reported by Garcia et al. (2009) from the Late Cretaceous Lo Hueco site in Spain.

It is noteworthy that the remains of these cryptodires are rare in Iharkút, followed by the dortokids in abundance, but the dominant turtles were far the bothremydids. This might indicate that the relatively high energetical conditions of the rivers principally favored the bothremydids, which are generally considered to be good swimmers (Gaffney et al., 2006; Lapparent de Broin and Murelaga, 1999). In other Late Cretaceous faunas of Europe (i.e., in Romania, Austria, southern France, and Spain), a significant difference can be observed in the relative abundance of certain taxa between the various localities (Seeley, 1881; Tong et al., 1998; Buffetaut et al., 1999; Lapparent de Broin and Murelaga, 1999; Tong and Gaffney, 2000; Gaffney et al., 2006; Rabi, 2009; Vremir and Codrea, 2009). This might reflect the facies dependence of turtles because the depositional environments vary from site to site, and the archipelago-like paleogeography probably less affected the distributional patterns.

Squamata Oppel, 1811
Scincomorpha Camp, 1923
Borioteiioidea Nydam, Eaton, and Sankey, 2007
Polyglyphanodontinae Estes, 1983
Bicuspidon Nydam and Cifelli, 2002

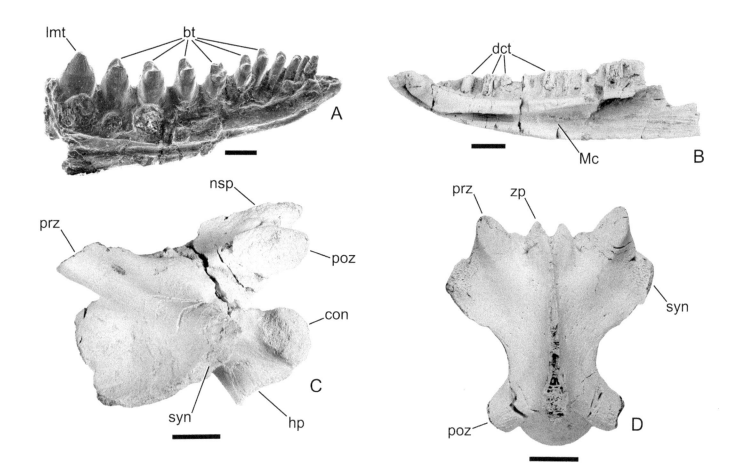

30.8. Squamate remains from the Upper Cretaceous (Santonian) Csehbánya Formation, Iharkút locality, Bakony Mountains, western Hungary: A, Left dentary of *Bicuspidon* aff. *hatzegiensis* (V 2006.112.1) in occlusal–lingual view. B, Fragmentary right dentary of Scincomorpha indet. A (V 01.80) in lingual view. C, D, Mosasauroidea indet.: V 01.149 cervical vertebra in left lateral view; D, V 01.222 dorsal vertebra of Mosasauroidea indet. in dorsal view. *Abbreviations:* lmt, last monocuspid tooth; bt, bicuspid teeth; dct, damaged cylindrical teeth; Mc, Meckelian canal; nsp, neural spine; prz, prezygapophysis; con, condyle; hp, hypapophyseal peduncle; syn, synapophysis; poz, postzygapophysis; zp, zygosphene. Scale bars = 1 mm for A, 5 mm for B, and 10 mm for C, D.

Bicuspidon aff. *hatzegiensis* Folie and Codrea, 2005
(Fig. 30.8A)

Material. Left dentary (V 2006.112.1); incomplete left lower jaw (V 2006.111.1); incomplete right lower jaw (V 01.27); fragmentary left dentary (V 2006.108.1); dentary fragments (V 01.24., V 2006.109.1 and V 2006.110.1); small jaw fragment with one tooth (V 2006.114.1).

Description and comparisons. These jaws belong to the most abundant lizard at the locality (Makádi, 2006). All of them bear teeth, and because the morphology of these teeth is so characteristic, the specimens can be easily assigned to the genus *Bicuspidon* (Nydam and Cifelli, 2002; Folie and Codrea, 2005).

Teeth in the middle part of the tooth row are the most typical (Fig. 30.8A). These teeth have a larger labial cusp and a slightly smaller lingual cusp that are linked together by a transverse ridge that divides the tip of the crown into a mesial and a distal facet; this pattern is typical for most polyglyphanodontines. The well-developed mesial and distal accessorial blades that border deep anterior and posterior facets in *Dicothodon* (Nydam, 1999, 2002) are missing in the Iharkút specimens. The teeth are not as broad as in *Polyglyphanodon* or *Paraglyphanodon*, and the ridge between the cusps is V shaped, contrasting with the horizontal or U-shaped ridge of the both latter genera (Estes, 1983; Nydam, 1999; Nydam and Cifelli, 2002). On some Iharkút *Bicuspidon* specimens (V 01.24, V 01.27, and V 2006.111.1), the distal part of the tooth row is preserved and shows different, monocuspid teeth. In all three specimens, the second to the last tooth is monocuspid and it is the largest, both in terms of its width and height, in the tooth row. Two specimens (V 01.27 and in V 2006.111.1) preserve an even more posterior tooth that is also monocuspid but significantly smaller, about one third of the size of the preceding tooth.

Discussion. Bicuspidon, along with other genera of Polyglyphanodontinae, belongs to Borioteiioidea, a clade established recently to unite several Cretaceous lizards previously included in Teiidae (Estes et al., 1988; Nydam et al., 2007). *Bicuspidon* has two species: *B. numerosus* from the Albian–Cenomanian of Utah, and *B. hatzegiensis* from the Maastrichtian of Transylvania (Nydam and Cifelli, 2002; Folie and Codrea, 2005). Compared to the Iharkút specimens, *B. numerosus* differs in having only bicuspid teeth along the part of the distal tooth row (Nydam and Cifelli, 2002). The presence of monocuspid teeth at the end of the tooth row is reminiscent of *B. hatzegiensis*, but because the end of the tooth row is not known for *B. hatzegiensis* (Folie and Codrea, 2005), it is unknown whether that species also had two monocuspid teeth distally. Although it is possible that the Iharkút *Bicuspidon* specimens belong to *B. hatzegiensis*, there is also a chance that they represent a new species. Accordingly, the *Bicuspidon* species from Iharkút is conservatively identified as *Bicuspidon* aff. *hatzegiensis*, until better-preserved specimens of *B. hatzegiensis* are found that clarify the structure of the distalmost tooth.

Borioteiioidea indet.

Material. Incomplete right lower jaw (V 2006.106.1).

Description and comparisons. The mandible preserves the posterior part of the dentary, the splenial, and parts of the surangular and coronoid. The large posterodorsal coronoid process hides the coronoid in labial view. It has an open and posteriorly wide Meckelian canal, a wide subdental shelf, and a deep sulcus dentalis. The robust teeth have cementum deposition and large subcircular resorption pits at their bases. The tooth crowns are monocuspid. Faint, blunt ridges extend from the mesial and distal labial sides of the apex lingually and wrap around the crown mesially and distally. The crowns bear distinct striation on both their lingual and labial sides.

Discussion. The combination of a deep sulcus dentalis, robust teeth with cementum deposition and characteristic resorption pits, and a coronoid process that overlaps the coronoid are features that Nydam et al. (2007) considered diagnostic for the Borioteiioidea. Moreover, the morphology of the teeth in the Iharkút mandible is similar to those of Chamopsiinae (Denton and O'Neill, 1995). Whereas the morphology of the teeth of the specimen differs from those of previously known species, it indicates that the Iharkút mandible possibly belongs to a previously undescribed species, but until the specimen is more thoroughly studied, it is identified as Borioteiioidea indet.

Scincomorpha indet. A
(Fig. 30.8B)

Material. Right dentary (V 2006.113.1); fragmentary right dentary (V V 01.80); right dentary fragment (V 2006.107.1).

Description and comparisons. The Meckelian canal is open and posteriorly wide and the dentition pleurodont with cylindrical teeth. The subdental shelf is wide, but the sulcus dentalis is shallow. The labial surface of each dentary is ornamented with vermiculate rugosities. On V 2006.107.1, the preserved tooth crowns are laterally compressed and bear a large distal cusp and a small mesial accessory cusp. Both the labial and lingual sides of the crowns are finely striated longitudinally. On V 2006.113.1, the crowns lack the small accessory cusp.

Discussion. The posteriorly wide and open Meckelian canal, the cylindrical teeth, and the shallow sulcus dentalis imply scincomorphan affinities for these specimens (Camp, 1923; Estes et al., 1988). Additional characters not discussed here indicate that these dentaries belong to a new species of scincoid. Pending further study of these specimens, we informally indentify them as Scincomorpha indet. A.

Scincomorpha indet. B

Material. Incomplete right lower jaw (V 2010.129.1).

Description and comparisons. The robust mandible consists of the dentary, splenial, and crushed postdentary bones (surangular?). The Meckelian canal is open and wide posteriorly, the subdental shelf is wide, and the sulcus dentalis is deep. The teeth have pleurodont attachment and are heterodont along the tooth row, with small and recurved teeth mesially, cylindrical and peglike teeth in the middle, and large, bulbous teeth distally.

Discussion. The shape of the Meckelian canal, subdental shelf, and sulcus dentalis are typical for Scincomorpha (Camp, 1923; Estes et al., 1988), and the robust construction of the dentary and its heterodont dentition are characteristic for Borioteiioidea (Nydam et al., 2007). However, other features, most notably the morphology of the teeth, resemble the condition in some scincoids (e.g., *Acontias*, *Hemisphaeriodon*; Kosma, 2004). Until this distinctive mandible is adequately studied, we informally identify it as Scincomorpha indet. B.

Anguimorpha Fürbringer, 1900
Mosasauroidea Camp, 1923
Mosasauroidea indet.
(Fig. 30.8C,D)

Material. Premaxillae (V 2007.24.1 and V 2007.25.1); maxillae (V 2007.26.1, V 2007.33.1 and V 2007.29.1); left postorbitofrontals (V 2007.22.1 and V 2007.28.1); quadrates (V 01.115 and V 2007.31.1); associated left and right dentaries (V 2007.37.1 and V 2007.37.2); left dentary fragment (V 2007.77.1); right splenials (V 01.37 and V 2007.38.1); right angulars (V 2007.34.1, V 2007.35.1, and V 2007.36.1); left coronoid (V 2007.23.1); left and right surangulars (V 01.49 and V 2007.30.1); right articular (V 2007.39.1); 91 isolated teeth (V 2007.78.1–V 2007.78.91); cervical vertebrae (Gyn/106, Gyn/108, Gyn/113, Gyn/120, Gyn/121, V 2000.19, V 01.149, V 2007.49.1, V 2007.52.1, V 2007.57.1, V 2007.58.1, V 2007.66.1, V 2007.68.1, V 2007.70.1, V 2007.71.1, V 2007.75.1, V 2007.76.1, V 2007.79.1, V 2007.93.1, and V 2007.94.1); dorsal vertebrae (Gyn/109, Gyn/111, Gyn/114,

Gyn/116, Gyn/123–Gyn/126, V 2000.21, V 01.212, V 01.222, V 2007.32.1, V 2007.44.1, V 2007.45.1, V 2007.47.1, V 2007.48.1, V 2007.50.1, V 2007.51.1, V 2007.53.1–V 2007.56.1, V 2007.59.1–V 2007.65.1, V 2007.67.1, V 2007.69.1, V 2007.72.1–V 2007.74.1, V 2007.80.1–V 2007.84.1, and V 2007.106.1); sacral vertebrae (V 2007.85.1, Gyn/122, and V 2007.86.1); caudal vertebrae (Gyn/104, Gyn/105, Gyn/118, Gyn/119, V 01.173, V 2007.46.1, V 2007.95.1–V 2007.105.1); fragmentary vertebrae (Gyn/112, Gyn/115, Gyn/117, V 2000.17, V 2007.90.1–V 2007.90.11, and V 2007.107.1); rib fragments (V 2007.87.1–V 2007.89.1); ilia (V 01.43, V 2007.40.1, V 2007.41.1, and V 2007.43.1); humerus proximal epiphysis (V 2007.42.1).

Description and comparisons. Isolated bones of different mosasaur individuals are abundant at the locality. These specimens, especially the vertebrae (Fig. 30.8C,D), are similar to those of *Tethysaurus nopcsai* from the Turonian of Morocco (Bardet et al., 2003), although they clearly differ from the Moroccan specimens. For example, in the quadrate, the stapedial pit is not as elongated as in *Tethysaurus*, and the quadrate condyle is more saddle shaped. As a strange and unique feature, the premaxilla of the Iharkút mosasaur is greatly flattened dorsoventrally, and it is wide and violin-shaped in dorsal and ventral outline. The mandible has an extremely high dorsal coronoid process and bicarinate, striated teeth, both differing from those of *Tethysaurus* (Bardet et al., 2003). Uniquely within Mosasauroidea, the precaudal vertebrae of the Iharkút mosasaur have well-developed precondylar constrictions, as in extant species of *Varanus* (Hoffstetter and Gasc, 1969). As in primitive members of Mosasauroidea (aigialosaurs), it has well-developed sacral vertebrae that are associated with backward-projecting, plesiopelvic ilia (Carroll and DeBraga, 1992; Caldwell and Palci, 2007).

Discussion. These characters indicate that the Iharkút mosasaur is closely related to *Tethysaurus nopcsai* but differs from it at the generic level and likely represents a new genus. This new primitive mosasauroid, together with *Tethysaurus* and other genera, can be regarded as an aigialosaur (previously known as Aigialosauridae, possibly a paraphyletic group consisting of primitive members of different mosasaur subfamilies, representing a lower degree of adaptation to aquatic life; Bell and Polcyn, 2005). Thus, these fossils provide valuable information about the evolution of mosasaurs for two reasons. First, because these primitive members of the group play a key role in our understanding of the evolution of the group, it may provide further insights into relationships among mosasauroids. Second, from an ecological point of view, it represents a previously unknown adaptation among mosasaurs to life in freshwater. Only exclusively marine members of the group were previously reported. By contrast, the Iharkút mosasaur is not known from marine sediments; instead, it is abundantly represented in the alluvial floodplain deposits of the Iharkút locality. The interpretation that the Iharkút mosasaur lived in freshwater is supported by geochemical analyses (Kocsis et al., 2009). Different individuals ranging in estimated body sizes from 70 cm to 6 m long can be identified, suggesting that these mosasaurs even reproduced in the freshwaters of the Iharkút area.

Crocodylomorpha Walker, 1970
Crocodyliformes Hay, 1930
Mesoeucrocodylia Whetstone and Whybrow, 1983

Doratodon Seeley, 1881
Doratodon sp.
(Fig. 30.9A–C)

Material. Isolated teeth (V 2010.226.1).

Description and comparisons. Currently, only ziphodont teeth can be confidently referred to this crocodile. An incomplete right dentary and a maxilla, both without teeth and not figured here, may also belong to this taxon, but more comparison is needed to confirm this assignment. Three morphotypes can be distinguished, all of which could be assigned to the same taxon (Fig. 30.9A–B). All isolated teeth are labiolingually compressed, with denticulated mesial and distal carinae (Fig. 30.9C). The first morphotype has a high, lingually curved crown (Fig. 30.9A); the second is the lowest, giving a triangular shape for the crown (Fig. 30.9B), whereas the third is lower, with symmetrically compressed labial and lingual surfaces (not figured). These morphotypes probably correspond to different positions along the tooth row.

Discussion. These serrated teeth are highly reminiscent of those of *Doratodon carcharidens* from the historical Muthmannsdorf locality (Campanian) in the Austrian Gosau Group. That species is known by an almost complete lower jaw, an incomplete maxilla, and some isolated teeth (Bunzel, 1871; Seeley, 1881; Buffetaut, 1979). Recently, a new species was described (Company et al., 2005) from the Late Cretaceous of Spain as *Doratodon ibericus*. Though Company et al. (2005) found a sister-taxon relationship between Gondwanan sebecosuchians and *Doratodon*, that result is rather poorly supported, and the generic identities of the Austrian and the Iberian species are uncertain. Clearly, the teeth from Iharkút are closer in morphology to *D. carcharidens* in their degree of labiolingual compression. On the basis of isolated teeth, Martin et al. (2006) reported *Doratodon* in the Late Cretaceous of Romania.

Mesoeucrocodylia Whetstone and Whybrow, 1983
Mesoeucrocodylia indet.
(Fig. 30.9D,E)

Material. Isolated teeth (V 2010.243.1).

Description and comparison. These teeth can be distinguished from those of other crocodilians at Iharkút in being labiolingually compressed and lacking true serrations. Two morphotypes can be distinguished. The first has a high crown that is lanceolate in outline with sharp but smooth mesial and distal carinae and smooth labial and lingual surfaces (Fig. 30.9D). The second morphotype is lower and leaf-shaped, and it has striae on both the labial and lingual surfaces that extend outward and terminate in the carinae (Fig. 30.9E). According to the classification of Prasad and Lapparent de Broin (2002), the latter morphotype is pseudoziphodont. The tooth morphology in general is reminiscent to that of *Theriosuchus pusillus*.

Discussion. These isolated teeth appear almost identical to in situ teeth in a maxilla that was briefly described and figured by Martin et al. (2006, fig. 9) from the Maastrichtian of Romania. This specimen, together with

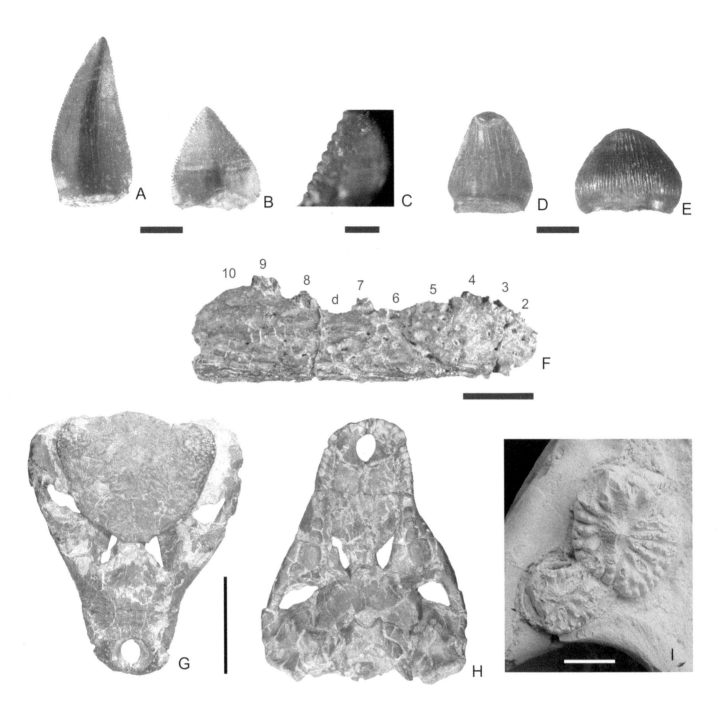

30.9. Crocodyliform remains from the Upper Cretaceous (Santonian) Csehbánya Formation, Iharkút locality, Bakony Mountains, western Hungary, Bakony. A–C, *Doratodon* sp., V 2010.226.1 isolated teeth in labial view, A, morphotype 1; B, morphotype 2; C, detail of carina showing minute denticules. D, E, Mesoeucrocodylia indet., V 2010.243.1 isolated teeth, in lingual view, D, morphotype 1 (with worn apex); E, morphotype 2. F, Neosuchia indet., V 01.128 partial right dentary. G–I, *Iharkutosuchus makadii* holotype skull in (G) dorsal and (H) ventral view; I, the larger, 17th and the smaller 18th multicusped maxillary teeth in detail. *Abbreviations:* d, diastema; numbers correspond to broken teeth or alveoli. Scale bar = 20 mm in A, B and D, E, 1 mm in C, 10 mm in F, 50 mm in G, H, and 5 mm in I.

others from the Haţeg Basin (Romania), is currently under study by Martin, Rabi, and Csiki, and the preliminary results suggest atoposaurid affinities.

Mesoeucrocodylia Whetstone and Whybrow, 1983
Neosuchia sensu Clark, 1994
Neosuchia indet.
(Fig. 30.9F)

Material. Fragmentary maxillae (V 2010.230.1, V 2010.229.1); dentaries (V 01.128., V 2010.234.1); surangular (V 2010.232.1); angular (V 2010.231.1); isolated teeth (V 2010.227.1).

Description and comparison. The fragmentary maxilla (V 2010.230.1, not figured) preserves the first six alveoli. This maxilla is horizontally expanded, with occlusal pits medial to the alveoli and with an incomplete conical, ca- niniform fourth tooth. Another fragmentary maxilla (V 2010.229.1) preserves the posterior part, and it seems to show that the ectopterygoid formed the medial wall of the posterior alveoli. The shape of the dentaries is sinusoid in lateral view along the tooth row, with the first peak at the level of the fourth tooth (which is caniniform) and the second at the level of the ninth tooth (Fig. 30.9F). A short symphysis reaches back to the third to fourth teeth, which sit in separate alveoli. A diastema is present between the seventh and eighth alveoli (Fig. 30.9F). The surangular has a posteriorly projecting pro- cessus articularis, and although incomplete, there is no sign of mandibular fenestra on either the surangular or on the angular. The foramen aërum on the articular seems to be shifted laterally, as in alligatoroids.

The teeth are conical, with sharp mesial and distal carinae, and wrin- kles are present on both the labial and lingual surfaces. Three morphotypes can be distinguished that may correspond to different positions along the tooth row. The first has a lingually curved high crown; the second and third are lower and have either a flat or a convex lingual surface, respectively.

Discussion. The shape of the maxillae and dentaries resembles that in aquatic neosuchian crocodiles. Originally, the Iharkút neosuchian was interpreted as a relatively advanced alligatoroid (Rabi, 2006), but the pos- terior part of a maxilla recently discovered at Iharkút casts doubt on that assignment. The newly discovered specimen shows that the medial wall of the posterior maxillary alveoli was formed by the ectopterygoid, instead of by the maxilla as in Alligatoroidea (Norell et al., 1994; Brochu, 1999, 2004). However, the laterally shifted foramen aërum on the Iharkút articular is diagnostic of Alligatoroidea (Brochu, 1999, 2004). Whether the presence of these two characters is evidence for mosaic evolution in stem crocodilians or whether these specimens represent different taxa cannot be resolved until more complete material is found.

Primitive eusuchians are known from a number of localities in the Up- per Cretaceous of Europe. *Allodaposuchus precedens* has been described from Romania, France, and Spain, but its presence in the Iberian Pennisula is doubtful (Nopcsa, 1928; Delfino et al., 2008; Buscalioni et al., 2001; Mar- tin, 2010). *Iharkutosuchus makadii* is a peculiar heterodont crocodile known only from Iharkút. Among the less well known, more fragmentary forms are *Musturzabalsuchus buffetauti* from Spain and *Massaliasuchus affuvelensis* from France (Buscalioni et al., 1997; Martin and Buffetaut, 2008). *Musturza- balsuchus* has been referred to Alligatoroidea by Buscalioni et al. (1997), but Brochu (1999) and Martin and Buffetaut (2008) cast doubts on that assign- ment. Similarly, material of *Massaliasuchus* has proved inadequate to clarify its relationship within Eusuchia (Martin and Buffetaut, 2008). The remains of the derived neosuchian taxon from Iharkút are fragmentary as well; there- fore, its phylogenetic position among early eusuchians cannot be solved. However, its unique morphology implies that it is taxonomically distinctive from all other Late Cretaceous European crocodiles described so far.

Eusuchia Huxley, 1875
Hylaeochampsidae Andrews, 1913
Iharkutosuchus makadii Ösi, Clark, and Weishampel, 2007
(Fig. 30.9G–I)

Material. Nearly complete skull (V 2006.52.1, holotype); paratypes: nearly complete skull (V 2006.53.1); three partial skulls (V 2006.54.1–V 2006.56.1); three skull fragments (V 2006.57.1, V 2006.65.1, V 2006.66.1); several different isolated skull elements (V 2006.67.1–V 2006.69.1, V 2006.79.1, V 2006.73.1, V 2006.77.1); fragmentary right mandible (V 2006.58.1); 12 isolated fragmentary mandibles (V 2006.58.1–V 2006.64.1, V 2006.71.1, V 2006.72.1, V 2006.74.1–V 2006.76.1); two mandible fragments (V 2006.78.1, V 2006.70.1); 148 isolated teeth (V 2006.80.1).

Description and comparisons. The sole autapomorphic feature of this small-bodied (total body length approximately 1 m) hylaeochampsid eusuchian is the presence of flat, molariform, multicuspid teeth (Fig. 30.9I) in a strongly heterodont dentition. The following combination of traits also characterizes *Iharkutosuchus:* the posterior maxillary process is extremely elongated caudally, the supratemporal fenestrae are closed (even in juveniles), and the lateral pterygoid flanges are relatively narrower lateromedially than in most crocodilians (Ősi et al., 2007; Ősi, 2008a). The rostrum is short and narrow transversely. The orbits face dorsally and slightly laterally. The undivided secondary choana is completely surrounded by the pterygoids. The posterior process of the pterygoid is long, extending well beyond the posterior margin of the skull. The ectopterygoids are wide and robust, and medially and posteriorly border the posteriormost three alveoli in the upper jaw. Ventrally, the extremely short quadrates bear well-developed protuberances for the origin of adductor muscles. The first 14 alveoli in the upper jaw and the first 12 in the dentary are of the same size. More posteriorly, the alveoli become larger, with the 17th in the maxilla and the 14th in the dentary being the largest. Enlarged caniniform teeth are absent from the upper and lower jaws.

Discussion. This small-bodied hylaeochampsid eusuchian is characterized by a complex dentition and a modified cranial adductor musculature. Wear facets oriented close to the transverse plane are frequent and extensive on all types of teeth (Ősi, 2008a). Adductor muscle reconstruction, dental wear pattern analysis, and study of the enamel microstructure clearly revealed the presence of unique oral food processing that was accompanied by precise dental occlusion (Ősi and Weishampel, 2009). The relatively narrow lateral pterygoid flanges and the presence of long transverse scratches on the worn dentine surface of the teeth are the main indicators of transverse mandibular movement during feeding. This feature is unique among eusuchian crocodilians and might also have been present in other hylaeochampsids, such as in *Hylaeochampsa vectiana* from the Wealden of the Isle of Wight (Owen, 1874; Clark and Norell, 1992).

Numerous eusuchian cranial and postcranial elements have been unearthed in Iharkút, but only cranial material can be referred with certainty to *Iharkutosuchus.* Postcranial elements are usually not diagnostic among crocodilians, so it is ambigous whether any of these bones from Iharkút are from the hylaeochampsid or the Neosuchia indet. crocodilians.

From a biogeographical point of view, the crocodilian fauna of Iharkút is partly composed of survivors from the Early Cretaceous of Europe (*Iharkutosuchus* and the atoposaurid-like taxon here referred to Mesoeucrocodylia indet). The Neosuchia indet. taxon from Iharkút is poorly known, although it shows affinities with European eusuchians such as

Allodaposuchus, Musturzabalsuchus, and *Massaliasuchus.* Because relationships of these taxa to North American and Early Cretaceous European eusuchians are controversial or ambiguous (cf. Delfino et al., 2008, versus Martin and Buffetaut, 2008; Martin, 2010; Martin and Delfino, 2010), their geographic origins also remain unclear.

Doratodon has been considered a sister taxon to Sebecosuchia by Company et al. (2005), but that relationship is not well supported as a result of the fragmentary nature of *Doratodon* material. Consequently, it is ambiguous whether *Doratodon* is a Gondwanan component of the Iharkút crocodilian fauna.

From a paleoecological point of view, the crocodilian fauna of Iharkút is composed of two semiaquatic and two rather terrestrial forms. Among the semiaquatic taxa, *Iharkutosuchus* likely had a specialized feeding preference (Ősi and Weishampel, 2009), while the Neosuchia indet. could have occupied the role of a generalist, similar to *Alligator. Doratodon* may have been slightly larger than the atoposaurid-like species, although both of them likely had a more terrestrial lifestyle; differences in their tooth morphologies (i.e., presence or absence of serrations) suggest distinct feeding strategies. A similar ecological distribution can also be postulated for the Romanian, French, and Spanish continental crocodilian faunas, which consist of a specialist (*Acynodon*), a generalist (*Allodaposuchus* or *Musturzabalsuchus*), and terrestrial carnivores (*Doratodon* and the atoposaurid-like taxon in Romania and France) (Buscalioni et al., 1997, 1999, 2001; Company, 2004; Martin et al., 2006, 2010; Delfino et al., 2008). It is worth noting that ziphodont crocodiles remain unknown from the Senonian of France, whereas they are present in other European Late Cretaceous crocodilian faunas.

Ornithischia Seeley, 1888
Ankylosauria Osborn, 1923
Nodosauridae Marsh, 1890
Hungarosaurus tormai Ősi, 2005
(Fig. 30.10)

Material. Holotype skeleton (V 2007.26.1–V 2007.26.34., V 2007.89.1, V 2007.89.2, see Ősi, 2005; Ősi and Makádi, 2009). Paratypes: three partial skeletons (V 2007.22.1, V 2007.23.1–V 2007.23.5, V 2007.90.1, V 2007.90.2, V 2007.24.1–V 2007.26.10). Referred specimens: fifth partial skeleton (V 2007.25.1–V 2007.25.30). Isolated specimens: premaxilla-maxilla fragment (V 2003.12); left fragmentary postorbital (V 2007.28.1).

Description and comparisons. Hungarosaurus tormai is a medium-sized (total body length approximately 4–4.5 m) nodosaurid ankylosaur that can be differentiated from other nodosaurids by having a postorbital bearing a high and anterodorsal–posteroventrally elongated crest, an inverted U-shaped premaxillary notch that is as high dorsoventrally as it is wide mediolaterally (Ősi, 2005) and a forelimb–hindlimb length ratio of 1.0 (humerus/femur = 0.92). The premaxilla is longer than wide, bears teeth, and is curved ventrally at an angle of about 40 degrees so it overlaps the complementary ventrally sloping anterior part of the dentary (Ősi and Makádi, 2009). In contrast to *Struthiosaurus* (Nopcsa, 1929), *Pawpawsaurus* (Lee, 1996), and *Gargoyleosaurus* (A. Ősi, pers. obs), the ventral part of the orbital rim is quite thin dorsoventrally and wider mediolaterally. The massive quadrate condyles are relatively much wider anteroposteriorly than those of *Struthiosaurus* (see Ősi, 2005, fig. 5). Vertebrae and ribs are quite similar to other nodosaurid ankylosaurs and they strongly resemble those of *Struthiosaurus* spp. (Pereda-Suberbiola and Galton, 2001; Garcia and Pereda-Suberbiola, 2003). Except for *Minmi paravertebra* (Molnar and Frey, 1987), *Hungarosaurus* is the only ankylosaur known to have paravertebral elements. The scapulocoracoid is massive, and the curvature of the scapular blade is highly similar to that of *Struthiosaurus*. Several limb elements of *Hungarosaurus* are more slender and elongate than in any other ankylosaur (Ősi and Makádi, 2009).

Discussion. The dorsal, posterior, and ventral margins of the orbital rim are well known from holotype material (V 2007.26.4) and a recently referred left postorbital (V 2007.28.1). This high and anterodorsally–posteroventrally elongated crest on the postorbital changed during ontogeny (Ősi and Makádi, 2009). The pattern is further supported by a newly discovered left orbital rim (V 2010.1.1) that, on the basis of its size, represents the same ontogenetic stage as V 2007.28.1. Similar to V 2007.28.1, but in contrast to the holotype specimen, the postorbital crest on this fragmentary specimen is separated from the ventral, rimlike part of the postorbital by deep grooves and pits both medially and laterally. On the contrary, the new orbital rim revealed that similar separation of the quadratojugal protuberance cannot be observed (Fig. 30.10B,E). The medial surface of this quadratojugal protuberance is different from that of the holotype specimen in having a deeper concave surface.

Cladistic analyses of various ankylosaurs (Ősi, 2005; Ősi and Makádi, 2009) indicate that the two European Late Cretaceous ankylosaurs (*Struthiosaurus* and *Hungarosaurus*) are basal nodosaurids, with *Hungarosaurus* being more derived than *Struthiosaurus* but more primitive than the North American genera. Placement of *Struthiosaurus* and *Hungarosaurus* at the base of the nodosaurid clade is interesting because they are younger (20–30 Myr) than some of the more derived North American genera.

Remarks. Hundreds of cataloged and uncataloged isolated cranial and postcranial elements (?nasal, teeth, vertebrae, ribs, appendicular and limb elements, and dermal osteoderms) have been unearthed in Iharkút, and they are most probably from *Hungarosaurus*. However, none of these remains bears any of the diagnostic features of the genus, so they are tentatively referred here to Nodosauridae indet.

Rhabdodontidae Weishampel, Jianu, Csiki, and Norman, 2003
Rhabdodontidae indet.

Material. Two right (V 2010.103.1, V2010.104.1) and one left dentaries (V 2010.105.1); two right (V 2010.107.1, V 2010.109.1) and two left fragmentary dentaries (V 2010.106.1, V 2010.108.1); one dentary fragment (V 2010.112.1); two right quadrates (V 2010.110.1, V 2010.111.1); teeth (V 2000.01., V 2000.32., V 2000.33., V 2003.10., V 01.161., V 2003.14,–V.2003.16, V 01.64); vertebrae (V 2010.115.1–V 2010.120.1); sacrum (V 2010.121.1); three fragmentary coracoids (V 01.53., V 2010.122.1, V 2010.123.1); right humerus (V 2010.128.1); ilium fragment (V 2010.124.1); left femur (V 01.225); left compressed femur

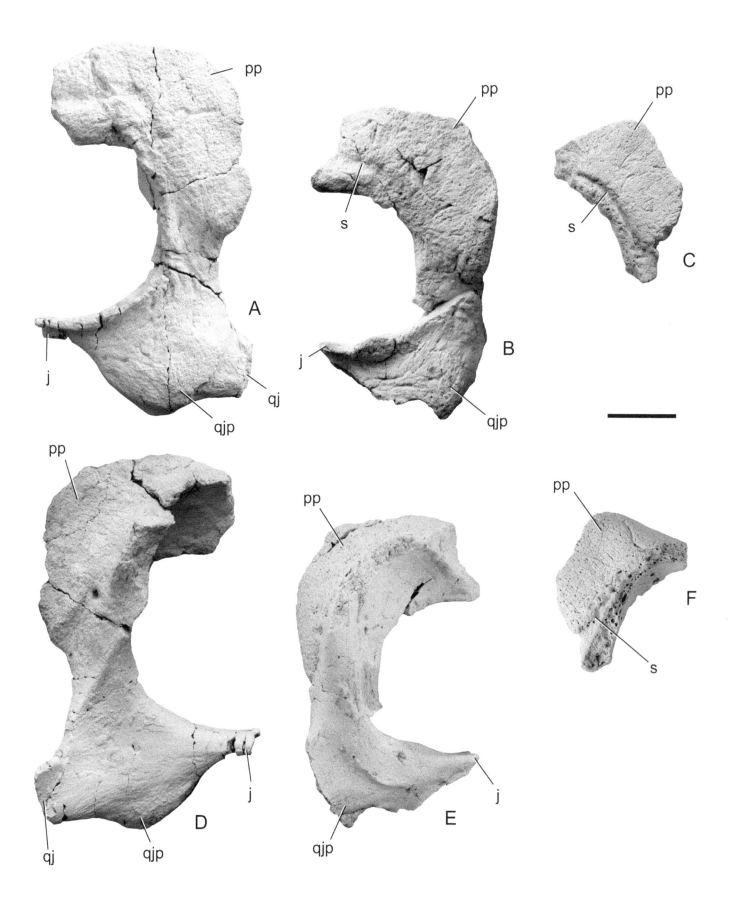

(V 2010.126.1); right ?femur fragment (V 2010.125.1); two fragmentary tibiae (V 01.101., V 2010.127.1).

Description and comparisons. A detailed description of the material is now in progress (Ősi et al, in prep.). The most abundant elements of this rhabdodontid ornithopod are teeth; these are highly similar to those of other rhabdodontid taxa (e.g., *Rhabdodon, Zalmoxes*; Buffetaut et al., 1999; Garcia et al., 1999; Weishampel et al., 2003; Sachs and Hornung, 2006). Dentary teeth possess a strongly developed primary ridge lingually, which divides two U-shaped surfaces. These surfaces bear sharp and thin secondary ridges nearly parallel with the primary ridge. The labial surface of the maxillary teeth does not bear a primary ridge as on dentary teeth, but it is ornamented by nearly parallel thin ridges (Ősi et al., 2003; Ősi, 2004a). A cingulum is developed only on the labial side of the base of the maxillary tooth crown. For the postcranial material, the strongly bent femur differs in several aspects from those of other rhabdodontid femora, thus suggesting that the Iharkút rhabdodontid is a new taxon.

Discussion. In contrast with the other main Late Cretaceous continental vertebrate localities in Europe (e.g., Buffetaut et al., 1999; Garcia et al., 1999; Weishampel et al., 2003), rhabdodontid remains are rare in the Iharkút assemblage, dominated by ankylosaur remains. Ősi (2004b) observed that fine-grained sediments are dominant and coarse-grained channel deposits are much rarer in Iharkút and suggested that the low abundance of the rhabdodontid remains might be related to the different environmental circumstances of the Iharkút area. On the contrary, embedding rocks of rhabdodontid remains in Spain, southern France, eastern Austria, and the Haţeg Basin deposited in variable depositional environments. Remains of rhabdodontids have been discovered, for example, from a gray marl level (Garcia et al., 1999), from channel, overbank, and floodplain deposits (Weishampel et al., 2003), but also from multispecies layers of variegated clays (paleosoils; Buffetaut el al., 1999). Although this question is difficult to answer at present, because of its taxonomic position (the Hungarian rhabdodontid appears to be a new taxon, Ősi et al., 2003), different ecological preferences (e.g., in the feeding habit) for the Hungarian rhabdodontid can be suggested.

The Iharkút specimens are of great importance because they are the earliest unequivocal (Santonian) occurrence for the exclusively European Rhabdodontidae and therefore have the potential for clarifying relationships within this endemic ornithopod clade.

Saurischia Seeley, 1888
Theropoda Marsh, 1881
Tetanurae Gauthier, 1986
Tetanurae indet.
(Fig. 30.11A,B)

Material. Fifty-eight isolated teeth and tooth fragments, V 01.54, V 01.30, V 01.20, V 2003.04–08, V 2008.36.1–V 2008.36.51.

Description and comparisons. Teeth are labiolingually compressed and gently curved distally, and have serrated distal and mesial carinae. The distal carina always extends along the entire height of the crown. The mesial carina generally extends only about one-half or one-third of the way down the crown, and in this regard, these teeth are similar to those of the "megalosaurid" *Megalosaurus dunkeri* from the Lower Cretaceous (Wealden) of England and *Poekilopleuron? valesdunensis* from the Middle Jurassic of France (Allain, 2002; Benson et al., 2008). In addition, the Hungarian teeth are practically identical to two fragmentary teeth from the lower Campanian of Muthmannsdorf (Austria). Bands of growth along the labial and lingual surfaces are frequent and well developed (Ősi et al., 2010).

Discussion. On the basis of statistical analyses of numerous dental characters listed by Smith et al. (2005), the Hungarian tetanuran teeth are clearly separated on the different plots from all other theropod taxa used in the matrix, and as was suggested, they were always found together with "megalosaurid" teeth from England and Austria. The tetanuran teeth from Iharkút (Santonian) and from Austria (early Campanian) are also most similar to much older (i.e., Early Cretaceous) forms.

Abelisauridae Bonaparte and Novas, 1985
Abelisauridae indet.

Material. Ungual phalanx (V 2008.43.1).

Description and comparisons. To date, the only evidence for abelisauroids in the Iharkút assemblage is a fragmentary right femur (Ősi and Buffetaut, in press) and a single, well-preserved ungual (Ősi and Apesteguía, 2008). The latter bears the two diagnostic characters of Abelisauridae (Novas and Bandyopadhyay, 2001): a shallow groove on its ventral and bifurcated grooves on its lateral surfaces delimit a convex, triangular area.

Discussion. The oldest definitive European abelisauroid remains were described as *Genusaurus sisteronensis* and are from the Albian of France (Accarie et al., 1995). Although those authors determined *Genusaurus* to be a neoceratosaurian, a recent review of the material (Carrano and Sampson, 2008) revealed that it is an abelisauroid likely related to the noasaurids. Besides *Genusaurus*, scanty remains of Late Cretaceous abelisaurid theropods are known from Campano-Maastrichtian sediments in Spain, France, and probably Romania (Buffetaut, 1989; Astibia et al., 1990; Le Loeuff and Buffetaut, 1991). The Hungarian material is of great importance because its Santonian age helps to reduce a temporal gap in the Cretaceous abelisauroid record of Europe.

Paraves Sereno, 1997, sensu Holtz and Osmólska (2004)
Pneumatoraptor fodori Ősi, Apesteguía, and Kowalewski, 2010

Material. Left scapulacoracoid (V 2008.38.1, holotype), and see remarks below.

Description and comparisons. As is characteristic for all paravians (Turner et al., 2007), the scapulocoracoid is L shaped. The scapula is narrow and distally flattened. The acromion process is not complete, but it is well developed and anteroposteriorly more elongate than that in *Velociraptor*. The glenoid fossa is oriented laterally and slightly ventrally, similar to *Velociraptor* (Norell and Makovicky, 1999). An autapomorphic feature of the genus is a large and deep pneumatic foramen (2 mm in diameter) that

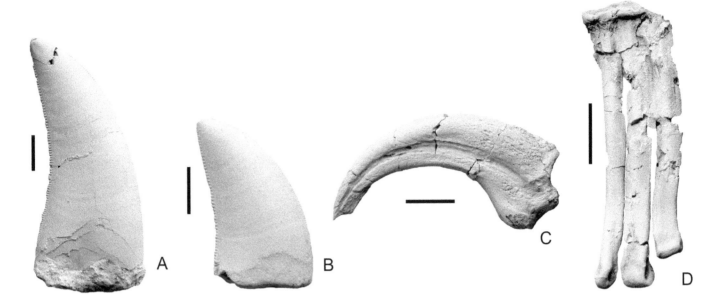

30.11. Theropod dinosaur remains from the Upper Cretaceous (Santonian) Csehbánya Formation, Iharkút locality, Bakony Mountains, western Hungary. A, B, Basal tetanuran: A, tooth (V.01.54) in lateral view; B, posteriorly positioned basal tetanuran tooth (V 2008.36.30) in lateral view. C, Paravian manual ungual phalanx in lateral view. D, *Bauxitornis mindszentyae* tarsometatarsus (V 2009.38.1) in dorsal view. Scale bar = 5 mm in A–C and 10 mm in D.

is located ventral to the coracoid foramen and opens toward the coracoid tubercle and probably enters the scapular blade.

Discussion. The complete fusion of the scapula and coracoid indicates an ontogenetically mature animal. On the basis of the size of this fossil, however, *Pneumatoraptor* would have been a small-bodied animal, approximately three times smaller than the Campanian *Velociraptor* and 1.2–1.5 times smaller than the Barremian *Sinornithosaurus* (Ősi et al., 2010). Although no diagnostic character supports the more precise affinity of *Pneumatoraptor* and the other paravian remains, here we suggest that similarly to other dromaeosaurid-like theropod remains from the Cretaceous of Europe (see, e.g., Sánchez-Hernández et al., 2007; Vullo et al., 2007; Csiki and Grigorescu, 1998; Ősi et al., 2010) represents dromaeosaurid theropods.

Remarks. Numerous remains (isolated teeth: V 01.215, V 01.231, V 2000.35, V 2008.37.1–V 2008.37.10, V.2000.02–V.2000.06; three isolated distal caudal vertebrae V 2009.46.1–3; one metacarpal V 2009.45.1; three ungual phalanges V 2008.40.1–3 [Fig. 30.11C]; phalanges V 2008.42.1–2; proximal half of a left tibia V 2008.31.1) have been described and referred to Paraves indet. by Ősi et al. (2010). This material may also belong to *Pneumatoraptor fodori*.

In addition, a fragmentary sacrum (V 2009.48.1), six isolated caudals (V 2009.47.1–6), and a left distal femur (V 2009.49.1) from Iharkút have been described as belonging to indeterminate theropods (Ősi et al., 2010).

Enantiornithes Walker, 1981
Avisauridae Brett-Surman and Paul, 1985
Bauxitornis mindszentyae Dyke and Ősi, 2010
(Fig. 30.11D)

Material. Nearly complete right tarsometatarsus (holotype, V 2009.38.1).

Description and comparisons. Ősi (2008b) and Dyke and Ősi (2010) provided detailed descriptions of this well-preserved avisaurid tarsometatarsus. It lacks the prominent insertion surface for M. tibialis cranialis that is normally present on the dorsal surface of metatarsal II. Metatarsals are fused only at their proximal ends, as is characteristic for enantiornithine birds (Chiappe and Walker, 2002). Metatarsal II is markedly short, and its distal end has a J-shaped profile in dorsal and plantar views (Dyke and Ősi, 2010).

Discussion. The Hungarian enantiornithine remains are of great importance because they encompass a wide range of sizes. The tarsometatarsus of *Bauxitornis* represents a relatively large animal (i.e., the size of a buzzard, *Buteo* sp.). A similarly sized individual is known by the distal part of an enantiornithine humerus (V 2009.39.1, Dyke and Ősi, 2010). The other end of the size spectrum is represented by a fragmentary femur (V 2002.05), also referred to Enantiornithes indet. by postcranial avian elements that may belong to the same group; collectively, they indicate the occurrence of thrush-sized birds in the Iharkút ecosystem.

Remarks. The proximal end of a right humerus (V 2009.40.1); distal half of a left femur (V 2009.41.1); distal end of a femur of a subadult (V 2002.06); distal portion of an isolated metatarsal III (V 2003.19); abraded distal end of a left femur (V 2009.42.1); and medioproximal portion of a right femur (V 2009.43.1) are among the other avian remains from Iharkút. Unfortunately, these fossils cannot be identified more precisely than Aves indet.

Pterosauria Kaup, 1834
Azhdarchidae Nessov, 1984 (emend. Padian, 1986)
Bakonydraco galaczi Ősi, Weishampel, and Jianu, 2005
(Fig. 30.12)

Material. Fragmentary premaxilla (V 2010.80.1, Ősi et al., 2011); complete left and right mandible (V 2007.110.1 = Gyn/3); 21 symphyseal fragments (V 2007.111.1 = Gyn/4, paratypes); 22 uncataloged symphyseal fragments (V 2010.74.1–22).

Description and comparisons. The fragmentary premaxilla is noncrested and shows a slightly lower profile than the beaklike part of the mandible (Ősi et al., 2011). Whereas the mandibular ramus is extremely thin walled and fragile, the fused, triangular, beaklike symphysis is massive. The dentary does not bear teeth, and the occlusal surface of the symphyseal region is ornamented by several pairs of small foramina that most probably served as openings for nutritive vessels that supplied the keratinous beak. The symphyseal region is similar to that of *Azhdarcho lancicollis* (Nessov, 1984; Buffetaut, 1999); however, their lateral profiles are different. *Bakonydraco* has a deeper symphysis, and the dorsal surface of the anterior half of the symphysis curves posterodorsally (quite strongly in older individuals), ending in a transverse ridge posteriorly. The glenoid on the posterior end of the mandible is not divided into medial and lateral cotyles as in *Quetzalcoatlus* (Ősi et al., 2005).

Discussion. One of the most interesting aspects of the complete mandible depicted in Figure 30.12 is the circumstances of its preservation. The sediment that contained the bone was a coarse, clastic basal brecchia in which even large bones are sometimes eroded or crushed. Most of the walls of the mandibular rami are not thicker than 0.5 mm. Whatever conditions

30.12. Mandible of *Bakonydraco galaczi* (V 2007.110.1 = Gyn/3) from the Upper Cretaceous (Santonian) Csehbánya Formation, Iharkút locality, Bakony Mountains, western Hungary. Dorsal (A) and lateral (B) views.

ruled the depositional environment of this delicate bone, it could not have been transported any distance, and it must have been buried quickly.

Remarks. Several postcranial elements such as cervical vertebrae (V 2010.100.1 = Gyn/448 in Ősi et al., 2005, V 01.51. = Gyn/449, 2010.101.1 = Gyn/450, V 2003.21 = Gyn/451, V 2010.81.1, V 2010.82.1); scapulocoracoid fragment (V 2010.79.1); distal humerus fragment (V 2010.77.1); a right radius (V 2010.102.1 = Gyn/452); two distal ulna fragments (V 2010.78.1 and V 2010.75.1 = Gyn/453); a proximal half of a first phalanx of the wing finger (V 2002.04); a ?third or ?fourth wing phalanx (V 2010.83.1); and a proximal tibiotarsus fragment (V 2010.76.1) from Iharkút appear to be from azhdarchid pterosaurs (Ősi et al., 2005; Ősi et al., in prep.). At least some of these might belong to *Bakonydraco*. However, none possesses any features that are diagnostic at the generic level.

Discussion

As a result of ongoing fieldwork over the last decade, the east Central European Iharkút locality has became one of the most important Late Cretaceous continental vertebrate-bearing sites in Europe. Besides the great abundance of vertebrate remains and its taxonomically diverse faunal composition, the Iharkút fauna is of great significance for the following four reasons (Ősi, 2004b):

1. It is Santonian in age. Supported by both palynological (Knauer and Siegl-Farkas, 1992) and paleomagnetic studies (Szalai, 2005), the Iharkút fauna is older than the other main Late Cretaceous nonmarine vertebrate sites in Europe (in Spain, France, Austria, and Romania; see Pereda-Suberbiola, 2009); thus, it helps to narrow a significant temporal gap between the late Early Cretaceous and younger Campano-Maastrichtian faunas.
2. Many taxa are closely related to stratigraphically much older forms, and this is related to a peculiar paleobiogeographical scenario within the Late Cretaceous western Tethyan archipelago. The hylaeochampsid and the atoposaurid-like crocodilians (Ősi et al., 2007), the rhabdodontid ornithopod, the basal nodosaurid ankylosaur (Ősi, 2005), and the basal tetanuran theropod (Ősi et al., 2010) are

all closely related to much older taxa. The primitive osteological features detected in these forms from Iharkút unambiguously indicate their relict nature and strongly suggest that the Iharkút area functioned as a refugium during the Santonian, as other continental areas in the western Tethyan archipelago likely did during the Late Cretaceous. Indeed, primitive and endemic forms have been recognized in other Late Cretaceous European faunas (see Pereda-Suberbiola, 2009, for a review). The presumed insular habitat of the Maastrichtian Hațeg Basin fauna was noted early on by Nopcsa (1923) and was later supported by other workers (Weishampel et al., 1991, 1993; Csiki and Grigorescu, 2007). The presence of the basal nodosaurid *Struthiosaurus* (Pereda-Suberbiola and Galton, 2001), rhabdodontids (Sachs and Hornung, 2006), and basal tetanurans (Ősi et al., 2010) in the Early Campanian Muthmannsdorf fauna (eastern Austria) clearly support the existence of primitive forms in that region as well. In addition, the occurrence of primitive hadrosauroids, such as *Telmatosaurus* (Weishampel et al., 1993), *Tethyshadros* in Italy (Dalla Vecchia, 2001, 2009), and unnamed hadrosauroids in Spain (Casanovas et al., 1999) further emphasize the importance of endemism in the western Tethyan archipelago.

3. The fauna is a mixture of Euramerican and Gondwanan components. Besides the above-discussed relict forms that represent the survivors of a much older (?Late Jurassic–Early Cretaceous) Euramerican ecosystem, the Iharkút fauna also contains faunal elements that have Gondwanan origins. Among the latter, bothremydid turtles unambiguously indicate the immigration of Gondwanan components during the Late Cretaceous. On the basis of the available fossil record, the lepisosteid fish, the ranoid-like frogs, the bothremydid turtles, and borioteiioids from Iharkút represent the geologically earliest European occurrence of their groups. In abelisaurids, however, it is known that the group was already present in Europe by the late Early Cretaceous (Albian, *Genusaurus*; see Carrano and Sampson, 2008); thus, it can be supposed that at least this group had colonized Europe and perhaps the western Tethyan archipelago already in the Early Cretaceous. Nevertheless, as a result of the lack of fossil record over this time interval, the continuous existence of the group from the Albian to the Santonian is uncertain (Ősi et al., 2010).

4. Although most of the vertebrate families (e.g., lepisosteids, albanerpetontids, scincomorphs, bothremydids, ziphodont and eusuchian crocodiles, nodosaurids, rhabdodontids, abelisaurids and paravians, enantiornithines, azhdarchids) recognized in Iharkút are also known from other Campano-Maastrichtian localities in Europe, some taxa uniquely occur in the Hungarian locality. Among these are the pycnodontiform fishes and the mosasaurs. Both groups are known mainly from marine sediments (but see the description of the fish remains above), and they have been generally considered to be marine animals. The discovery of these groups in the Csehbánya Formation composed of fluvial and alluvial sediments deposited in a freshwater environment (Kocsis et al., 2009) is unexpected.

Pycnodontiforms and mosasaurs were not only able to colonize these territories and live there permanently or periodically, but they may have even reproduced in this habitat. The small-bodied hylaeochampsid is also unique within the European crocodilian record because, on the basis of numerous cranial adaptations, this animal was capable of efficient oral food processing by transverse jaw movements and had a specialized feeding preference. Similar to the presence of the other relicts discussed above, the presence of these unique endemic forms is presumably related to the long isolation of the area.

Finally, it is important to note that although extensive fieldwork and screen washing has been performed over the last 10 years, remains of mammals are still unknown from the locality. This might be explained by the small amount of screen-washed material. Nevertheless, it can also be supposed that mammals did not colonize the Iharkút area, at least during the Santonian. Whatever the case with the remains of mammals, we hope that future excavations including screen washing will answer this interesting question.

Acknowledgments

We thank Pascal Godefroit and Olivier Lambert for organizing the meeting Tribute to Charles Darwin and Bernissart *Iguanodon*s in Brussels. We are grateful to the reviewers Jim Gardner and Zoltán Csiki for their constructive comments and suggestions. We thank our colleagues who have helped us with their knowledge and provided literature over the years. We thank the Bakony Bauxite Mining Company and the Geovolán Company for logistical help and the 2000–2009 field crews for their assistance in the field. Fieldwork was supported by the Hungarian Natural History Museum, the National Geographic Society, the Jurassic Foundation, the National Scientific Research Fund (OTKAT-38045, PD 73021, NF 84193), the Hungarian Oil and Gas Company (MOL), the Eötvös University Department of Paleontology, and the Hantken Foundation, as well as several other institutions and companies. This is a contribution of the "Lendület Dinosaur Research Group" of the Hungarian Academy of Sciences.

References

Accarie, H., B. Beaudoin, J. Dejax, G. Fries, J.-G. Michard, and P. Taquet. 1995. Découverte d'un Dinosaure théropode nouveau (*Genusaurus sisteronis* n. g., n. sp.) dans l'Albien marin de Sisteron (Alpes de Haute-Provence, France) et extension au Crétacé inférieur de la lignée cératosaurienne. Comptes Rendus de l'Académie des Sciences (ser. IIA) Earth and Planetary Science 320: 327–334.

Allain, R. 2002. Discovery of megalosaur (Dinosauria, Theropoda) in the Middle Bathonian of Normandy (France) and its implications for the phylogeny of basal Tetanurae. Journal of Vertebrate Paleontology 22: 548–563.

Andrews, C. W. 1913. On the skull and part of the skeleton of a crocodile from the Middle Purbeck of Swanage, with the description of a new species (*Pholidosaurus laevis*), and a note on the skull of *Hylaeochampsa*. Annals and Magazine of Natural History 8: 485–494.

Astibia, H., E. Buffetaut, A. D. Buscalioni, H. Cappetta, C. Corral, R. Estes, F. Garcia-Garmilla, J. J. Jaeger, E. Jiménez-Fuentes, J. Le Loeuff, J. M. Mazin, X. Orue-Etexebarria, J. Pereda-

Suberbiola, J. E. Powell, J.-C. Rage, J. Rodriguez-Lazaro, J. L. Sanz, and H. Tong. 1990. The fossil vertebrates from Laño (Basque Country, Spain); new evidence on the composition and affinities of the Late Cretaceous continental faunas of Europe. Terra Nova 2: 460–466.

Astibia, H., X. Murelaga, X. Pereda-Suberbiola, J. J. Elorza, and J. J. Gómez-Alday. 1999. Taphonomy and palaeoecology of the Upper Cretaceous continental vertebrate-bearing beds of the Laño Quarry (Iberian Peninsula). Estudios del Museo de Ciencias Naturales de Alava 14: 43–104.

Báez, A. M., and C. Werner. 1996. Presencia de anuros ranoideos en el Cretácico de Sudan. Ameghiniana 33: 460.

Bardet, N., X. Pereda-Suberbiola, and N.-E. Jalil. 2003. A new mosasauroid (Squamata) from the Late Cretaceous (Turonian) of Morocco. Comptes Rendus Palevol 2: 607–616.

Baur, G. 1891. Notes on some little known American fossil tortoises. Proceedings of the Academy of Natural Sciences of Philadelphia 43: 411–430.

Behrensmeyer, A. K. 1982. Time resolution in fluvial vertebrate assembleges. Paleobiology 8: 211–227.

Bell, G. L., and M. J. Polcyn. 2005. *Dallasaurus turneri*, a new primitive mosasauroid from the Middle Turonian of Texas and comments on the phylogeny of the Mosasauridae (Squamata). Netherlands Journal of Geosciences 84: 177–194.

Benson, R., P. M. Barrett, H. P. Powell, and D. B. Norman. 2008. The taxonomic status of *Megalosaurus bucklandii* (Dinosauria: Theropoda) from the Middle Jurassic Oxfordshire, UK. Palaeontology 51: 419–424.

Bonaparte, J. F., and F. E. Novas. 1985. *Abelisaurus comahuensis*, n. g., n. sp., Carnosauria of the Late Cretaceous of Patagonia. Ameghiniana 21: 259–265.

Bossuyt, F., R. M. Brown, D. M. Hillis, D. C. Cannatella, and M. C. Milinkovitch. 2006. Phylogeny and biogeography of a cosmopolitan frog radiation: Late Cretaceous diversification resulted in continent-scale endemism in the family Ranidae. Systematic Biology 55: 579–594.

Brett-Surman, M. K., and G. Paul. 1985. A new family of bird-like dinosaurs linking Laurasia and Gondwanaland. Journal of Vertebrate Paleontology 5: 133–138.

Brochu, C. A. 1999. Phylogenetics, taxonomy, and historical biogeography of Alligatoroidea. Journal of Vertebrate Paleontology 19 (Memoir 6): 9–100.

———. 2004. Alligatorine phylogeny and the status of *Allognathosuchus* Mook,

1921. Journal of Vertebrate Paleontology 24: 857–873.

Buffetaut, E. 1979. Revision der Crocodylia (Reptilia) aus den Gosau-Schichten (Ober-Kreide) von Österreich. Beiträge zur Paläontologie von Österreich 6: 89–105.

———. 1989. Archosaurian reptiles with Gondwanan affinities in the Upper Cretaceous of Europe. Terra Nova 1: 69–74.

———. 1999. Pterosauria from the Upper Cretaceous of Laño (Iberian Peninsula): a preliminary comparative study. Estudios del Museo de Ciencias Naturales de Alava 14: 289–294.

Buffetaut, E., J. Le Loeuff, H. Tong, S. Duffaud, L. Cavin, G. Garcia, and D. Ward. 1999. Un nouveau gisement de vertébrés du Crétacé supérieur à Cruzy (Hérault, Sud de la France). Comptes Rendus de l'Académie des Sciences, Série IIA, Earth and Planetary Science 328: 203–208.

Bunzel, E. 1871. Die Reptilfauna der Gosau-Formation in der Neuen Welt bei Wiener-Neustadt. Abhandlungen der Geologische Reichsanstalt 5: 1–18.

Buscalioni, A. D., F. Ortega, and D. Vasse. 1997. New crocodiles (Eusuchia: Alligatoroidea) from the Upper Cretaceous of southern Europe. Comptes Rendus de l'Académie des Sciences, Sér. IIA, Earth and Planetary Science 325: 525–530.

———. 1999. The Upper Cretaceous crocodilian assemblage from Laño (Northcentral Spain): implications in the knowledge of the finicretaceous European faunas. Estudios del Museo de Ciencias Naturales de Alava 14: 212–233.

Buscalioni, A. D., F. Ortega, D. B. Weishampel, and C. M. Jianu. 2001. A revision of the Crocodyliform *Allodaposuchus precedens* from the Upper Cretaceous of the Haţeg Basin, Romania. Its relevance in the phylogeny of Eusuchia. Journal of Vertebrate Paleontology 21: 74–86.

Caldwell, M. W., and A. Palci. 2007. A new basal mosasauroid from the Cenomanian (U. Cretaceous) of Slovenia with a review of mosasauroid phylogeny and evolution. Journal of Vertebrate Paleontology 27: 863–880.

Camp, C. L. 1923. Classification of the lizards. Bulletin American Museum of Natural History 48: 289–481.

Carrano, M. T., and S. D. Sampson. 2008. The phylogeny of Ceratosauria (Dinosauria: Theropoda). Journal of Systematic Palaeontology 6: 183–236.

Carroll, R. L., and M. DeBraga. 1992. Aigialosaurs: mid-Cretaceous varanoid

lizards. Journal of Vertebrate Paleontology 12: 66–86.

Casanovas, M. L., X. Pereda-Suberbiola, J. V. Santafe, and D. B. Weishampel. 1999. A primitive euhadrosaurian dinosaur from the uppermost Cretaceous of the Ager syncline (Southern Pyrenees, Catalonia). Geologie en Mijnbouw 78: 345–356.

Cavin, L. 1999. Osteichthyes from the Upper Cretaceous of Laño (Iberian Peninsula). Estudios del Museo de Ciencias Naturales de Alava 14: 105–110.

Cavin, L., M. Martin, and X. Valentin. 1996. Découverte d'*Atractosteus africanus* (Actinopterygii, Lepisosteidae) dans le Campanien inférieur de ventabren (Brouches-du-Rhône, France). Implications paléobiogeographiques. Revue de Paléobiologie 15: 1–7.

Cavin, L., P. L. Forey, E. Buffetaut, and H. Tong. 2005. Latest European coelacanth shows Gondwanan affinities. Biology Letters 1: 176–177.

Cavin, L., U. Deesri, and V. Suteethorn. 2009. The Jurassic and Cretaceous bony fish record (Actinopterygii, Dipnoi) from Thailand; pp. 125–139 in E. Buffetaut, G. Cuny, J. Le Loeuff, and V. Suteethorn (eds.), Late Palaeozoic and Mesozoic ecosystems in SE Asia. Special Publications 315. Geological Society, London.

Chiappe, L. M., and C. A. Walker. 2002. Skeletal morphology and systematics of the cretaceous euenantiornithes (Ornithothoraces: Enantiornithes); pp. 240–267 in L. M. Chiappe and L. Witmer (eds.), Mesozoic birds: above the head of dinosaurs. University of California Press, Berkeley.

Cifelli, R. L., R. L. Nydam, J. D. Gardner, A. Weil, J. G. Eaton, J. I. Kirkland, and S. K. Madsen. 1999. Medial Cretaceous vertebrates from the Cedar Mountain Formation, Emery County, Utah: the Mussentuchit local fauna; pp. 219–242 in D. D. Gillette (ed.), Vertebrate paleontology in Utah. Utah Geological Survey Miscellaneous Publication 99-1. Utah Geological Survey, Salt Lake City.

Clark, J. M. 1994. Patterns of evolution in Mesozoic Crocodyliformes; pp. 84–97 in N. C. Fraser and H.-D. Sues (eds.), In the shadow of the dinosaurs. Cambridge University Press, New York.

Clark, J. M., and M. A. Norell. 1992. The Early Cretaceous crocodylomorph *Hylaeochampsa vectiana* from the Wealden of Isle of Wight. American Museum Novitates 3032: 1–19.

Company, J. 2004. Vertebrados continentales del Cretácico Superior

(Campaniense–Maastrichtiense) de Valencia. PhD dissertation, Universitad de València, Spain, 410 pp.

Company, J., X. Pereda-Suberbiola, J. I. Ruiz-Omenaca, and A. D. Buscalioni. 2005. A new species of *Doratodon* (Crocodyliformes: Ziphosuchia) from the Late Cretaceous of Spain. Journal of Vertebrate Paleontology 25: 343–353.

Cope, E. D. 1864. On the limits and relations of the Raniformes. Proceedings of the Academy of Natural Sciences of Philadelphia 16: 181–183.

———. 1868. On the origin of genera. Proceedings of the Academy of Natural Sciences of Philadelphia 20: 242–300.

Csiki, Z., and D. Grigorescu. 1998. Small theropods from the Late Cretaceous of the Hațeg Basin (western Romania)—an unexpected diversity at the top of the food chain. Oryctos 1: 87–104.

———. 2007. The dinosaur island—new interpretation of the Hațeg Basin vertebrate fauna after 110 years. Sargetia 20: 5–26.

Dalla Vecchia, F. M. 2001. Terrestrial ecosystems on the Mesozoic peri-Adriatic carbonate platforms: the vertebrate evidence. Abstract volume of the VIIth International Symposium on Mesozoic Terrestrial Ecosystems, Asociación Paleontológica Argentina, Publicación Especial 7: 77–83.

———. 2009. *Tethyshadros insularis*, a new hadrosauroid dinosaur (Ornithischia) from the Upper Cretaceous of Italy. Journal of Vertebrate Paleontology 29: 1100–1116.

Delfino, M., and B. Sala. 2007. Late pliocene Albanerpetontidae (Lissamphibia) from Italy. Journal of Vertebrate Paleontology 2: 716–719.

Delfino, M., V. Codrea, A. Folie, P. Dica, P. Godefroit, and T. Smith. 2008. A complete skull of *Allodaposuchus precedens* Nopcsa, 1928 (Eusuchia) and a reassessment of the morphology of the taxon based on the Romanian remains. Journal of Vertebrate Paleontology 28: 111–122.

Denton, R. K., and R. C. O'Neill. 1995. *Prototeius stageri*, gen. et sp. nov., a new teiid lizard from the Upper Cretaceous Marshalltown Formation of New Jersey, with a preliminary phylogenetic revision of the Teiidae. Journal of Vertebrate Paleontology 15: 235–253.

Dyke, G., and A. Ősi. 2010. Late Cretaceous birds from Hungary: implications for avian biogeography at the close of the Mesozoic. Geological Journal 45: 434–444.

Eaton, J. G., R. L. Cifelli, J. H. Hutchison, J. I. Kirkland, and J. M. Parrish. 1999.

Cretaceous vertebrate faunas from the Kaiparowits Plateau, south-central Utah; pp. 345–353 in D. D. Gillette (ed.), Vertebrate paleontology in Utah. Utah Geological Survey Miscellaneous Publication 99-1. Utah Geological Survey, Salt Lake City.

Estes, R. 1983. Sauria terrestria, Amphisbaenia. Handbook of Paleoherpetology. Gustav Fischer Verlag, Stuttgart, Part 10A, 249 pp.

Estes, R., and B. Sanchíz. 1982. Early Cretaceous lower vertebrates from Galve (Teruel), Spain. Journal of Vertebrate Paleontology 2: 21–39.

Estes, R., K. de Queiroz, and J. Gauthier. 1988. Phylogenetic relationships within Squamata; pp. 119–281 in R. Estes and G. Pregill (eds.), Phylogenetic relationships of the lizard families—essays commemorating Charles L. Camp. Stanford University Press, Stanford, Calif.

Feller, A. E., and B. S. Hedges. 1998. Molecular evidence for the early history of living amphibians. Molecular Phylogenetics and Evolution 9: 509–516.

Fischer von Waldheim, G. 1813. Zoognosia tabulis synopticis illustrata, in usum praelectionorum; Academiae Imperialis Medico-Chirurgicae Mosquensis edita. 3rd ed. Nicolai Sergeidis Vsevolozsky, Moscow.

Folie, A., and V. Codrea. 2005. New lissamphibians and squamates from the Maastrichtian of Hațeg Basin, Romania. Acta Palaeontologica Polonica 50: 57–71.

Fox, R. C., and B. G. Naylor. 1982. A reconsideration of the relationships of the fossil amphibian *Albanerpeton*. Canadian Journal of Earth Sciences 19: 118–128.

Friedman, M., J. A. Tarduno, and D. B. Brinkman. 2003. Fossil fishes from the high Canadian Arctic: further palaeobiological evidence for extreme climatic warmth during the Late Cretaceous (Turonian–Coniacian). Cretaceous Research 24: 615–632.

Fürbringer, M. 1900. Beitrag zur Systematik und Genealogie der Reptilien. Jenaischen Zeitschrift für Naturwissenschaften 34: 596–682.

Gaffney, E. S., and P. A. Meylan. 1988. A phylogeny of turtles; pp. 157–219 in M. J. Benton (ed.), The phylogeny and classification of the tetrapods. Vol. 1, Amhibians, reptiles, birds. Systematics Association Special Volume 35A.

———. 1992. The Transylvanian turtle, *Kallokibotion*, a primitive cryptodire of Cretaceous Age. American Museum Novitates 3040: 1–37.

Gaffney, E. S., H. Tong and P. A. Meylan. 2006. Evolution of the side-necked turtles: the families Bothremydidae, Euraxemydidae, and Araripemydidae. Bulletin of the American Museum of Natural History 300: 1–698.

Garcia, G., and X. Pereda-Suberbiola. 2003. A new species of *Struthiosaurus* (Dinosauria: Ankylosauria) from the Upper Cretaceous of Villeveyrac (southern France). Journal of Vertebrate Paleontology 23: 156–165.

Garcia, G., M. Pincemaille, M. Vianey-Liaud, B. Arandat, E. Lorenz, G. Cheylan, H. Cappetta, J. Michauxa, and J. Sudrea. 1999. Découverte du premier squelette presque complet de *Rhabdodon priscus* (Dinosauria, Ornithopoda) du Maastrichtien inferieur de Provence. Comptes Rendus de l'Académie des Sciences, Sér. IIA, Earth and Planetary Science 328: 415–421.

Garcia, A. P., F. Ortega, and X. Murelaga. 2009. A probable Pancryptodira turtle from the Upper Cretaceous of Lo Hueco (Cuenca, España) (in Spanish). Paleolusitana 1: 365–371.

Gardner, J. D. 2000. Albanerpetontid amphibians from the Upper Cretaceous (Campanian and Maastrichtian) of North America. Geodiversitas 22: 349–388.

Gardner, J. D., and M. Böhme. 2008. Review of the Albanerpetontidae (Lissamphibia), with comments on the paleoecological preferences of European Tertiary albanerpetontids; pp. 178–218 in J. T. Sankey and S. Baszio (eds.), Vertebrate microfossil assemblages: their role in paleoecology and paleobiogeography. Indiana University Press, Bloomington.

Gauthier, J. A. 1986. Saurischian monophyly and the origin of birds; pp. 1–55 in K. Padian (ed.), The origin of birds and the evolution of flight. Memoirs of the California Academy of Sciences 8.

Gayet, M., F. J. Meunier, and C. Werner. 2002. Diversification in Polypteriformes and special comparison with the Lepisosteiformes. Palaeontology 45: 361–376.

Grigorescu, D., M. Venczel, Z. Csiki, and R. Limberea. 1999. New latest Cretaceous microvertebrate fossil assemblages from the Hațeg Basin (Romania). Geologie en Mijnbouw 78: 301–314.

Haeckel, E. 1866. Generelle Morphologie der Organismen. Georg Reimer, Berlin, 574 pp.

Harzhauser, M., and P. M. Tempfer. 2004. Late Pannonian wetland ecology of the Vienna Basin based on molluscs and lower vertebrate assemblages (Late

Miocene, MN 9, Austria). Courier Forschungsinstitut Senckenberg 246: 55–68.

Hay, O. P. 1930. Second bibliography and catalogue of the fossil Vertebrata of North America. Publication 390. Carnegie Institute of Washington.

Hoffstetter, R., and J.-C. Gasc. 1969. Vertebrae and ribs of modern reptiles; pp. 201–310 in C. Gans (ed.), Biology of the Reptilia 1 (Morphology A). Academic Press, New York.

Holtz, T. R., Jr., and H. Osmólska. 2004. Saurischia; pp. 21–24 in D. B. Weishampel, P. Dodson, and H. Osmólska (eds.), The Dinosauria. 2nd ed. University of California Press, Berkeley.

Hussakof, L. 1916. A new pycnodont fish, *Coelodus syriacus*, from the Cretaceous of Syria. Bulletin of the American Museum of Natural History 35: 135–137.

Huxley, T. H. 1875. On *Stagonolepis Robertsoni*, and on the evolution of the Crocodilia. Quarterly Journal of the Geological Society London 3: 423–438.

———. 1880. On the application of the laws of evolution to the arrangement of the Vertebrata, and more particularly of the Mammalia. Proceedings of the Zoological Society London 1880: 649–662.

Jocha-Edelényi, E. 1988. History of evolution of the Upper Cretaceous Basin in the Bakony Mts at the time of the terrestrial Csehbánya Formation. Acta Geologica Hungarica 31: 19–31.

Kaup, J. 1834. Versuch einer Eintheilung der Säugethiere in 6 Stämme und der Amphibien in 6 Ordnungen. Isis von Oken 1834: 311–324.

Klein, E. F. 1885. Beiträge zur Bildung des Schädels der Knochenfische, 2. Jahreshefte Vereins Vaterlandischer Naturkunde in Würtenberg 42: 205–300.

Knauer, J., and Á. Siegl-Farkas. 1992. Palynostatigraphic position of the Senonian beds overlying the Upper Cretaceous bauxite formations of the Bakony Mountains. Annual Report of the Hungarica Geological Institute of 1990: 463–471.

Kocsis, L., A. Ősi, T. Vennemann, C. N. Trueman, and M. R. Palmer. 2009. Geochemical study of vertebrate fossils from the Upper Cretaceous (Santonian) Csehbánya Formation (Hungary): evidence for a freshwater habitat of mosasaurs and pycnodont fish. Palaeogeography, Palaeoclimatology, Palaeoecology 280: 532–542.

Kosma, R. 2004. The dentitions of recent and fossil scincomorphan lizards (Lacertilia, Squamata)—Systematics, Functional Morphology, Paleoecology. Ph.D. dissertation, Leibniz Universität Hannover, Germany, 187 pp.

Kriwet, J. 2002. *Anomoeodus pauciseriale* n. sp. (Neopterygii, Pycnodontiformes) from the White Chalk Formation (Upper Cretaceous) of Sussex, South England. Paläontologische Zeitschrift 76: 117–123.

———. 2004. Dental morphology of the pycnodontid fish †*Stemmatodus rhombus* (Agassiz 1844) (Neopterygii, †Pycnodontiformes) from the Early Cretaceous, with comments on its systematic position. Transactions of the Royal Society of Edinburgh, Earth Sciences 94: 145–155.

———. 2005. A comprehensive study of the skull and dentition of pycnodont fishes. Zitteliana A 45: 135–188.

Lapparent de Broin, F. de, and X. Murelaga. 1996. Une nouvelle faune de chéloniens dans le Crétacé Supérieur européen. Comptes Rendus de l'Académie des Sciences, Série IIA, Earth and Planetary Science 323: 729–735.

———. 1999. Turtles from Upper Cretaceous of Laño (Iberian Peninsula). Estudios del Museo de Ciencias Naturales de Alava 14: 135–212.

Lapparent de Broin, F. de, X. Murelaga Bereikua, and V. Codrea. 2004. Presence of Dortokidae (Chelonii, Pleurodira) in the earliest tertiary of the Jibou Formation, Romania: paleobiogeographical implications. Acta Paleontologica Romaniae 4: 203–215.

Laurenti, J. N. 1768. Specimen Medicurn, Exhibens Synopsen Reptilium Ernendatus curn Experimentis Circa Venea et Antidota Reptilium Austriacorum. J. T. Trattern, Vienna, 214 pp.

Lee, J.-N. 1996. A new nodosaurid ankylosaur (Dinosauria: Ornithischia) from the Paw Paw Formation (Late Albian) of Texas. Journal of Vertebrate Paleontology 16: 232–245.

Lehman, J. P. 1966. Actinopterygii; pp. 1–242, in J. Piveteau (ed.), Traité de Paléontologie. Vol. 4. Masson et Cie, Paris.

Le Loeuff, J., and E. Buffetaut. 1991. *Tarascosaurus salluvicus* nov. gen., nov. sp., dinosaure théropode du Crétacé supérieur du Sud de la France. Geobios 24: 585–594.

Linneaus, C. 1758. Systema naturae per regna tria naturae, secundum classes, ordines, genera, species, cum characteribus, differentiis, synonymis, locis. Editio decima, reformata. Holmiae. (Laurentii Salvii): (1–4): 1–824.

Maisey, J. G. 1991. Santana fossils. An illustrated atlas. T.F.H. Publications, Neptune City, N.J., 459 pp.

Makádi, L. 2005. A new aquatic varanoid lizard from the Upper Cretaceous of Hungary. Kaupia 14: 127.

———. 2006. *Bicuspidon* aff. *hatzegiensis* (Squamata: Scincomorpha: Teiidae) from the Upper Cretaceous Csehbánya Formation (Hungary, Bakony Mts). Acta Geologica Hungarica 49: 373–385.

Makádi L., G. Botfalvai, and A. Ősi. 2006. Late Cretaceous continental vertebrate fauna from the Bakony Mts. I: fishes, amphibians, turtles, lizards. Földtani Közlöny 136: 487–502. [In Hungarian]

Marsh, O. C. 1881. Principal characters of American Jurassic dinosaurs. Part V. The American Journal of Science and Arts (ser. 3) 21: 417–423.

———. 1890. Additional characters of the Ceratopsidae with notice of new Cretaceous dinosaurs. American Journal of Science (ser. 3) 39: 418–426.

Martin, J. E. 2010. *Allodaposuchus* Nopcsa, 1928 (Crocodylia, Eusuchia) from the Late Cretaceous of southern France and its relationships to Alligatoroidea. Journal of Vertebrate Paleontology 30: 756–767.

Martin, J. E., and E. Buffetaut. 2008. *Crocodilus affuvelensis* Matheron, 1869 from the Late Cretaceous of southern France: a reassessment. Zoological Journal of the Linnean Society 152: 567–580.

Martin, J. E., and M. Delfino. 2010. Recent advances on the comprehension of the biogeography of Cretaceous European eusuchians. Palaeogeography, Palaeoclimatology, Palaeoecology 293: 406–418.

Martin, J. E., Z. Csiki, D. Grigorescu, and E. Buffetaut. 2006. Late Cretaceous crocodylian diversity of the Hațeg Basin, Romania. Abstract volume of the 4th Annual Meeting of the European Association of Vertebrate Paleontologists. Hantkeniana 5: 31–37.

Martin, J. E., M. Rabi, and Z. Csiki, D. 2010. Survival of *Theriosuchus* (Mesoeucrocodylia: Atoposauridae) in a Late Cretaceous archipelago: a new species from the Maastrichtian of Romania. Naturwissenschaften 97: 845–854

Matsubara, K. 1955. Fish morphology and hierarchy. 3 vols. Ishizaki Shoten, Tokyo, 1605 pp.

Molnar, R. E., and E. Frey. 1987. The paravertebral elements of the Australian ankylosaur *Minmi* (Reptilia: Ornithischia, Cretaceous). Neues Jahrbuch für Geologie und Paläontologie Abhandlungen 175: 19–37.

Müller, J. 1845. Über den Bau und die Grenzen der Ganoiden, und über das natürliche System der Fische. Archiv für Naturgeschischte 1: 91–141.

Nessov, L. A. 1984. Upper Cretaceous pterosaurs and birds from Central Asia.

Paleontologicheskyy Zhurnal 1: 47–57. [In Russian]

Nopcsa, F. 1923. On the geological importance of the primitive reptilian fauna of the uppermost Cretaceous of Hungary; with a description of a new tortoise (*Kallokibotium*). Quarterly Journal of the Geological Society of London 79: 100–116.

———. 1928. Paleontological notes on Reptilia. 7. Classification of the Crocodilia. Geologica Hungarica, Series Paleontologica 1: 75–84.

———. 1929. Dinosaurierreste aus Siebenbürgen V. Geologica Hungarica, Series Palaeontologica 4: 1–76.

Norell, M. A., and P. Makovicky. 1999. Important features of the dromaeosaurid skeleton II: information from a newly collected specimens. American Museum Novitates 3282: 1–5.

Norell, M. A., J. M. Clark, and H. J. Hutchinson. 1994. The Late Cretaceous alligatoroid *Brachychampsa montana* (Crocodylia): new material and putitave relationships. American Museum Novitates 3116: 1–26.

Novas, F. E., and S. Bandyopadhyay. 2001. Abelisaurid pedal unguals from the Late Cretaceous of India; pp. 145–149 in Abstract volume of the VIIth International Symposium on Mesozoic Terrestrial Ecosystems. Publicación Especial 7. Asociación Paleontológica Argentina.

Nursall, J. R. 1996. The phylogeny of pycnodont fishes; pp. 125–152 in G. Arratia and G. Viohl (eds.), Mesozoic fishes: systematics and paleoecology. Verlag Dr. Friedrich Pfeil, Munich.

Nydam, R. L. 1999. Polyglyphanodontinae (Squamata: Teiidae) from the medial and Late Cretaceous: new taxa from Utah, USA, and Baja California del Norte, Mexico; pp. 303–317 in D. D. Gillette (ed.), Vertebrate paleontology in Utah. Utah Geological Survey Miscellaneous Publication 99-1. Utah Geological Survey, Salt Lake City.

———. 2002. Lizards of the Mussentuchit local fauna (Albian–Cenomanian boundary) and comments on the evolution of the Cretaceous lizard fauna of North America. Journal of Vertebrate Paleontology 22: 645–660.

Nydam, R. L., and R. L. Cifelli. 2002. A new teiid lizard from the Cedar Mountain Formation (Albian–Cenomanian boundary) of Utah. Journal of Vertebrate Paleontology 22: 276–285.

Nydam, R. L., J. G. Eaton, and J. Sankey. 2007. New taxa of transversely-toothed lizards (Squamata: Scincomorpha) and new information on the evolutionary history of "teiids." Journal of Paleontology 81: 538–549.

Oppel, M. 1811. Die Ordnungen, Familien und Gattungen der Reptilien, als Prodrom einer Naturgeschichte derselben. Joseph Lindauer, Munich, pp. 1–87.

Osborn, H. F. 1923. Two Lower Cretaceous dinosaurs from Mongolia. American Museum Novitates 95: 110.

Owen, R. 1874. Monograph on the fossil Reptilia of the Wealden and Purbeck Formations. Suppl. No. 6 (*Hylaeochampsa*). Paleontographical Society Monographs 27: 1–7.

Ősi, A. 2004a. The first dinosaur remains from the Upper Cretaceous of Hungary (Csehbánya Formation, Bakony Mts.). Geobios 37: 749–753.

———. 2004b. Dinosaurs from the Late Cretaceous of Hungary—similarities and differences with other European Late Cretaceous faunas. Revue de Paléobiologie 9: 51–54.

———. 2005. *Hungarosaurus tormai*, a new ankylosaur (Dinosauria) from the Upper Cretaceous of Hungary. Journal of Vertebrate Paleontology 25: 370–383.

———. 2008a. Cranial osteology of *Iharkutosuchus makadii*, a Late Cretaceous basal eusuchian crocodyliform from Hungary. Neues Jahrbuch für Geologie und Paleontologie Abhandlungen 248: 279–299.

———. 2008b. Enantiornithine bird remains from the Late Cretaceous of Hungary. Oryctos 7: 55–60.

Ősi, A., and S. Apesteguía. 2008. Non-avian theropod dinosaur remains from the Upper Cretaceous (Santonian) Csehbánya Formation (Iharkút, Bakony Mountains, western Hungary); pp. 78–79 in Abstract volume of the 6th Meeting of the European Association of Vertebrate Paleontologists.

Ősi, A., and L. Makádi. 2009. New remains of *Hungarosaurus tormai* (Ankylosauria, Dinosauria) from the Upper Cretaceous of Hungary: skeletal reconstruction and body mass estimation. Paläontologische Zeitschrift 83: 227–245.

Ősi, A., and A. Mindszenty. 2009. Iharkút, Dinosaur-bearing alluvial complex of the Csehbánya Formation; pp. 51–63 in E. Babinszky (ed.), Cretaceous sediments of the Transdanubian Range. Field guide of the geological excursion organized by the Sedimentological Subcommission of the Hungarian Academy of Sciences and the Hungarian Geological Society, Budapest, Hungary.

Ősi, A., and D. B. Weishampel. 2009. Jaw mechanism and dental function in the Late Cretaceous basal eusuchian *Iharkutosuchus*. Journal of Morphology 270: 903–920.

Ősi, A., and E. Buffetaut. In press. Additional non-avian theropod and bird remains from the early Late Cretaceous (Santonian) of Hungary and a review of the European abelisauroid record. Annales de Paleontologie.

Ősi, A., C. M. Jianu, and D. B.Weishampel. 2003. Dinosaurs from the Upper Cretaceous of Hungary. Advances in Vertebrate Paleontology, Hen to Pantha, Bucharest, pp. 117–120.

Ősi, A., D. B. Weishampel, and C. M. Jianu. 2005. First evidence of Azhdarchid pterosaurs from the Late Cretaceous of Hungary. Acta Palaeontologica Polonica 50: 777–787.

Ősi A., J. M. Clark, and D. B. Weishampel. 2007. First report on a new basal eusuchian crocodyliform with multicusped teeth from the Upper Cretaceous (Santonian) of Hungary. Neues Jahrbuch für Geologie und Paläontologie Abhandlungen 243: 169–177.

Ősi, A., S. Apesteguía, and M. Kowalewski. 2010. Non-avian theropod dinosaurs from the early Late Cretaceous of Central Europe. Cretaceous Research 31: 304–320

Ősi, A., E. Buffetaut, and E. Prondvai. 2011. New pterosaurian remains from the Late Cretaceous (Santonian) of Hungary (Iharkút, Csehbánya Formation). Cretaceous Research 32: 456–463.

Padian, K. 1986. A taxonomic note on two pterodactyloid families. Journal of Vertebrate Paleontology 6: 289.

Prasad, G. V. P., and F. Lapparent de Broin. 2002. Late Cretaceous crocodile remains from Naskal (India): comparisons and biogeographic affinities. Annales de Paléontologie 88: 19–71.

Pereda-Suberbiola, X. 2009. Biogeographical affinities of Late Cretaceous continental tetrapods of Europe: a review. Bulletin de la Société géologique de France 180: 57–71.

Pereda-Suberbiola, X., and P. Galton. 2001. Reappraisal of the nodosaurid ankylosaur *Struthiosaurus austriacus* Bunzel from the Upper Cretaceous Gosau Beds of Austria; pp. 173–210 in K. Carpenter (ed.), The armored dinosaurs. Indiana University Press, Bloomington.

Pereda-Suberbiola, X., H. Astibia, X. Murelga, J. J. Elorza, and J. J. Gómez-Alday. 2000. Taphonomy of the Late Cretaceous dinosaur-bearing beds of the Laño Quarry (Iberian Peninsula).

Palaeogeography, Palaeoclimatology, Palaeoecology 157: 247–275.

Poyato-Ariza, F. J., M. R. Talbot, M. A. Fregenal-Martínez, N. Meléndez, and S. Wenz. 1998. First isotopic and multidisciplinary evidence for nonmarine coelacanths and pycnodontiform fishes: palaeoenvironmental implications. Palaeogeography, Palaeoclimatology, Palaeoecology 144: 65–84.

Rabi, M. 2006. Do alligatoroids really derive from North America? Abstract volume of the 4th Annual Meeting of the European Association of Vertebrate Paleontologists. Hantkeniana 5: 102.

———. 2009. An update of the Late Cretaceous chelonian and crocodilian fauna of Central Europe. Journal of Vertebrate Paleontology 29 (supplement to 3): 168A

Rabi, M., and G. Botfalvai. 2006. A new bothremydid (Chelonia: Pleurodira) fossil assemblage from the Late Cretaceous (Santonian) of Hungary—additional studies in historical paleobiogeography of Late Cretaceous bothremydids. Abstract volume of the 4th Annual Meeting of the European Association of Vertebrate Paleontologists. Hantkeniana 5: 61–65.

Rabi, M., H. Tong, and G. Botfalvai. In press. A new species of the side-necked turtle Foxemys (Pelomedusoides: Bothremydidae) from the Late Cretaceous of Hungary and the historical biogeography of the Bothremydini. Geological Magazine.

Rafinesque, C. S. 1820. Ichthyolgia ohiensis, or natural history of the fishes inhabiting the river Ohio and its tributary streams, preceded by a physical descirption of the Ohio and its branches. Western Review and Miscellaneous Magazine 1: 361–377

Rage, J.-C. 2003. Oldest Bufonidae (Amphibia, Anura) from the Old World: a bufonid from the Paleocene of France. Journal of Vertebrate Paleontology 23: 462–463.

Rage, J.-C., and Z. Roček. 2003. Evolution of anuran assemblages in the Tertiary and Quaternary of Europe, in the context of palaeoclimate and palaeogeography. Amphibia-Reptilia 24: 133–167.

Roelants, K., and F. Bossuyt. 2005. Archeobatrachian paraphyly and Pangaean diversification of crown-group frogs. Systematic Biology 54: 111–126.

Sachs, S., and J. J. Hornung. 2006. Juvenile ornithopod (Dinosauria: Rhabdodontidae) remains from the Upper Cretaceous (Lower Campanian, Gosau Group) of Muthmannsdorf (Lower Austria). Geobios 39: 415–425.

Sánchez-Hernández, B., M. J. Benton, and D. Naish. 2007. Dinosaurs and other fossil vertebrates from the Late Jurassic and Early Cretaceous of the Galve area, NE Spain. Palaeogeography, Palaeoclimatology, Palaeoecology 249: 180–215.

Savage, J. 1973. The geographic distribution of frogs: patterns and predictions; pp. 351–445 in J. L. Vial (ed.), Evolutionary biology of the anurans: contemporary research on major problems. University of Missouri Press, Columbia.

Seeley, H. G. 1881. The reptile fauna of the Gosau Formation preserved in the Geological Museum of the University of Vienna. Quarterly Journal of the Geological Society London 37: 620–702.

———. 1888. The classification of the Dinosauria. Report of the British Association of Advancement of Science 1887: 698–699.

Sereno, P. C. 1997. The origin and evolution of dinosaurs. Annual Review of Earth and Planetary Sciences 25: 435–489.

Smith, J. B., D. R. Vann, and P. Dodson. 2005. Dental morphology and variation in theropod dinosaurs: implications for the taxonomic identification of isolated teeth. Anatomical Record A 285: 699–736.

Szalai, E. 2005. Paleomagnetic studies in Iharkút. Manuscript, Eötvös Loránd University, Department of Environmental Geology, Budapest, Hungary. [In Hungarian]

Szentesi, Z., and M. Venczel. 2010. An advanced anuran from the Late Cretaceous (Santonian) of Hungary. Neues Jahrbuch für Geologie und Paläontologie 256: 291–302.

Tong, H., and E. S. Gaffney. 2000. Description of the skull of Polysternon provinciale (Matheron, 1869), a side-necked turtle (Pelomedusoides: Bothremydidae) from the Late Cretaceous of Villeveyrac, France. Oryctos 3: 9–18.

Tong, H., E. S. Gaffney, and E. Buffetaut. 1998. Foxemys, a new side-necked turtle (Bothremydidae: Pelomedusoides) from the Late Cretaceous of France. American Museum Novitates 3251: 1–19.

Tuba, Gy., P. Kiss, M. Pósfai, and A. Mindszenty. 2006. Diagenesis history studies on the bone material of the Upper Cretaceous dinosaur locality in the Bakony Mts. Földtani Közlöny 136: 1–24. [In Hungarian]

Turner, A. H., D. Pol, J. A. Clarke, G. M. Erickson, and M. A. Norell. 2007. A basal dromaeosaurid and size evolution preceding avian flight. Science 317: 1378–1381.

Van der Meiden, A., M. Vences, S. Hoegg, and A. Meyer. 2005. A previously unrecognized radiation of ranid frogs in Southern

Africa revealed by nuclear and mitochondrial DNA sequences. Molecular and Phylogenetic Evolution 37: 674–685.

Venczel, M., and J. D. Gardner. 2005. The geologically youngest albanerpetontid amphibian, from the lower Pliocene of Hungary. Palaeontology 48: 1273–1300.

Vremir, M., and V. Codrea. 2009. Late Cretaceous turtle diversity in Transylvanian and Haţeg basins (Romania); pp. 1–5 in the Abstract volume of 7th National Sympoysum of Palaeontology of Romania.

Vullo, R., D. Néraudeau, and T. Lenglet. 2007. Dinosaur teeth from the Cenomanian of Charentes, western France: evidence for a mixed Laurasian–Gondwanan assemblage. Journal of Vertebrate Paleontology 27: 931–943.

Vullo, R., D. Néraudeau, R. Allain, and H. Cappetta. 2005. Un nouveau gisement à microrestes de vertébrés continentaux et littoraux dans le Cénomanien inférieur de Fouras (Charente-Maritime, Sud-Ouest de la France). Comptes Rendus Paleovol 4: 95–107.

Walker, A. D. 1970. A revision of the Jurassic reptile Hallopus victor (Marsh), with remarks on the classification of crocodiles. Philosophical Transactions of the Royal Society of London B 257: 323–372.

Walker, C. A. 1981. A new subclass of birds from the Cretaceous of South America. Nature 292: 51–53.

Weishampel, D. B., D. Grigorescu, and D. B. Norman. 1991. The dinosaurs of Transylvania. National Geographic Research and Exploration 7: 196–215.

———. 1993. Telmatosaurus transsylvanicus from the Late Cretaceous of Romania: the most basal hadrosaurid dinosaur. Palaeontology 36: 361–385.

Weishampel, D. B., C. M. Jianu, Z. Csiki, and D. B. Norman. 2003. Osteology and phylogeny of Zalmoxes (n. g.), an unusual euornithopod dinosaur from the latest Cretaceous of Romania. Journal of Systematic Palaeontology 1: 65–123.

Whetstone, K. N., and P. J. Whybrow. 1983. A "cursorial" crocodilian from the Triassic of Lesotho (Basutoland), southern Africa. Occasional Papers of the Museum of Natural History, University of Kansas 106: 1–37.

Wiley, E. O. 1976. The phylogeny and biogeography of fossil and recent gars (Actinopterygii: Lepisosteidae). Miscellaneous Publication 64. University of Kansas Museum of Natural History.

Winkler, D. A., P. A. Murry, and L. L. Jacobs. 1990. Early Cretaceous (Comanchean) vertebrates of central Texas. Journal of Vertebrate Paleontology 10: 95–116.

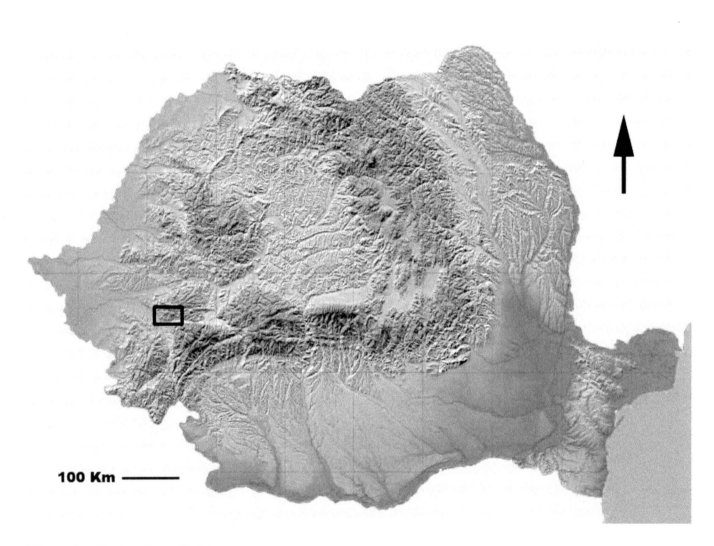

31.1. Location of the Rusca Montană Basin in Romania.

First Discovery of Maastrichtian (Latest Cretaceous) Terrestrial Vertebrates in Rusca Montană Basin (Romania)

Vlad A. Codrea*, Pascal Godefroit, and Thierry Smith

First mentioned by F. Nopcsa, Hațeg Island was a paleogeographical concept sketched by this famous paleontologist in order to explain the presence of small-sized dinosaurs in uppermost Cretaceous localities from Transylvania (western Romania), and particularly from the Hațeg Basin. However, this insularity is still debated, even after more than a century of research. In order to reconstruct the precise paleogeography of this area by Maastrichtian time, it is important to study all the coeval uppermost Cretaceous continental deposits from Transylvania. The westernmost region where these formations are exposed is the Rusca Montană Basin (western Romanian Carpathians). The geological history of this sedimentary basin appears similar to that of the Hațeg Basin. We report the first Maastrichtian vertebrates from the Rusca Montană Basin. These fossils include dinosaurs (ornithopod, sauropod, theropods), turtles (?*Kallokibotion*), indeterminate crocodiles, and multituberculate mammals (Kogaionidae). This fauna closely resembles that from the Hațeg Basin and is the first evidence of their presence to the west of Hațeg.

More than a century ago, the paleontologist Baron Francise von Nopcsa (1897) reported the first latest Cretaceous small-sized dinosaurs from Transylvania (western Romania) in the Hațeg Basin in the southern Carpathians. Soon, he also reported similar taxa in other areas from Transylvania, including Alba and Sălaj counties (Nopcsa, 1905), where Maastrichtian continental formations are widely exposed (Codrea and Dica, 2005; Codrea and Godefroit, 2008; Codrea et al., 2010).

Nopcsa (1914) was impressed by the remarkably small size of the dinosaurs in Transylvania and regarded it as an example of dwarfism on an island later called by Weishampel "Hațeg Island" (Weishampel et al., 1991). According with this idea, the island had a surface area of about 7,500 km² and was located 200 to 300 km from the rest of Europe. This topic remains controversial; some authors have supported Nopcsa's original hypothesis (Lapparent, 1947; Weishampel et al., 1991, 1993, 2003; Jianu and Weishampel, 1999; Dalla Vecchia, 2006, 2009), whereas others have challenged it (Jianu and Boekschoten, 1999; LeLoeuff, 2005; Pereda Suberbiola and Galton, 2009). In our opinion, the geological data at hand suggest that Europe was an archipelago during the Maastrichtian (see, e.g., Smith et al., 1994) rather than a wide land, so we are inclined to agree with Nopcsa's hypothesis. However, episodic connections between Hațeg Island and wider

Introduction

European emerged areas cannot be excluded (Folie and Codrea, 2005; Benton et al., 2010; Weishampel et al., 2010).

In order to reconstruct the paleogeography of the Transylvania area during the Maastrichtian, it is important to take all exploitable exposures into consideration (Codrea et al., 2010). The presence of the rhabdodontid *Zalmoxes* at Someş-Odorhei, a site located in the basin of Transylvania, confirms the northeastern extension for Haţeg Island in Transylvania (Nopcsa, 1905; Codrea and Godefroit, 2008). The Rusca Montană Basin is the westernmost region where continental Maastrichtian deposits can be observed in Romania. This sedimentary basin is located in the western Romanian Carpathians, on the southwestern side of the Poiana Ruscă Mountains (Fig. 31.1).

Although Romanian geologists have long observed that the geological history of Rusca Montană Basin appears similar to that of Haţeg Basin, vertebrate remains have not been found in the Rusca Montană Basin. Here, we briefly describe the first Maastrichtian terrestrial vertebrates ever found in the Rusca Montană Basin.

Institutional abbreviation. UBB, University Babeş-Bolyai, Cluj-Napoca, Romania.

Geological Setting

Rusca Montană Basin is one of the Late Cretaceous ("Senonian") synorogenic sedimentary basins located south of Mureş River, which were formed as a consequence of the Late Cretaceous tectogenesis that erected the Getic and Supragetic nappes in the Median Dacides (Săndulescu, 1984). Rusca Montană Basin and Haţeg Basin probably share the same geological evolution, from piggyback to collapse type (Willingshofer et al., 2001). The basement of Rusca Montană Basin is formed by metamorphic rocks covered by Jurassic (Lias, Malm), Lower Cretceous (?Albian; terrestrial environments with bauxite) and Upper Cretaceous (Cenomanian, Turonian) marine deposits (Dincă et al., 1972; Dincă, 1977; Bucur et al., 1985), all related to the Getic Nappe realm. A disconformity is present in the Middle–Late Turonian or within the Late Turonian (Strutinski, 1986; Strutinski and Hann, 1986). Therefore, "Senonian" evolution started with transgressing Coniacian deposits (calcarenites, marls) over older rocks (Mamulea, 1955). Santonian and Campanian deposits are represented by deep-water turbidites (marlstones and claystones, distal flysh deposits; Dincă, 1977) (Fig. 31.2).

A sedimentary turnover occurred when marine environments (Pop et al., 1972) were replaced by continental ones. As in Haţeg Basin (Melinte-Dobrinescu, 2010), continental sedimentation probably began during the Late Campanian and lasted through the Maastrichtian. The continental succession begins with siliciclastic rocks bearing coal beds (?Early Maastrichtian) and follows with siliciclastic beds interleaving Banatitic volcanoclastic strata (Kräutner et al., 1986; Grigorescu, 1992). In fact, volcanic activity in Rusca Montană Basin started even earlier in the Turonian–Coniacian, when basic tuffs associated with radiolarian rocks accumulated as a result of marine eruptions related to an east–west fracture bounding the northern basin margin (Strutinski and Bucur, 1985; Strutinski and Hann, 1986).

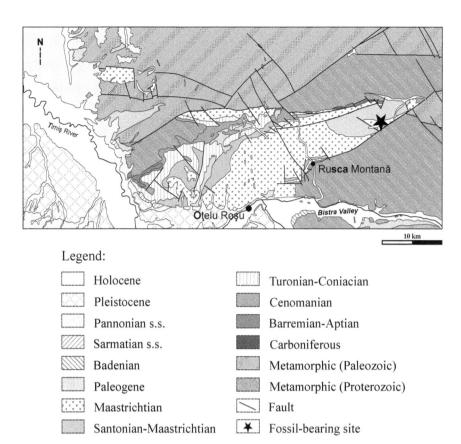

Legend:

☐	Holocene	⊞	Turonian-Coniacian
☐	Pleistocene	▨	Cenomanian
☐	Pannonian s.s.	▨	Barremian-Aptian
▨	Sarmatian s.s.	■	Carboniferous
▧	Badenian	▨	Metamorphic (Paleozoic)
☐	Paleogene	▨	Metamorphic (Proterozoic)
⋯	Maastrichtian	◺	Fault
▨	Santonian-Maastrichtian	★	Fossil-bearing site

The Upper Cretaceous deposits form an asymmetric syncline: the strata on the northern sector are more abruptly inclined than those on the southern side. The faults are not very numerous, and two main fault directions—southwest–northeast and southeast–northwest—can be observed.

The Terrestrial Maastrichtian in the Rusca Montană Basin

The Maastrichtian continental deposits are particularly well documented in Rusca Montană Basin, mainly because of coal mining works, including rich drilling data (Papp, 1915; Duşa, 1987). However, despite of the numerous boreholes and mining excavations, which stopped immediately under Coal Bed 1, the basin basement was never reached. Therefore, the base of the Maastrichtian continental sequence remains poorly known and is based on data restricted to only a few scarce outcrops.

Coal exploitation started at the beginning of the twentieth century and stopped around 1930. The geological survey continued until 1970, but the results were disappointing for profitable mining exploitation; although the quality of the coal is good, the coal beds are rather thin (Duşa, in Petrescu et al., 1987).

Grigorescu (1990, 22–23) briefly described the molass deposits from Rusca Montană Basin and correlated them with the "lower and median members of the Densuş-Ciula Formation" in the Haţeg Basin. Later, Therrien et al. (2002) mentioned the so-called Rusca Formation, a name devoid of formal lithostratigraphy. On the basis of drilling data, Duşa (1974, 1987) reported two Maastrichtian complexes, a lower clastic, coaly complex and an upper clastic, volcanic complex. The former is 320 m thick, with conglomerates, microconglomerates, sandstones, clay shale with few acid lava flows, tuffs, and volcanic agglomerate interbeds, bearing toward its base up to 20 bituminous coal beds and coaly intercalations, 1.4 m the thickest; the coal beds are of different lateral extent, some of them completely vanishing toward the basin central areas. The latter is 370 m thick, with coarser clastic rocks such as conglomerates, microconglomerates, and sandstones alternating with volcanic agglomerates and tuffs, but rarer lava flows; thermal metamorphism of coals at the contact with volcanic dikes was reported by Duşa and Bărilă (1973).

It is clear that an additional survey is necessary for a refined stratigraphy in Rusca Montană Basin. If the Maastrichtian flora is particularly poor in Haţeg Basin, the latest Cretaceous flora in Rusca Montană Basin is the richest in Romania. Several fossil macroflora were collected from the coal deposits, and 34 species have been described so far (Schafarzik, 1907; Tuzson, 1913; Cantuniari, 1937; Givulescu, 1966, 1968; Balteş, 1966; Petrescu and Duşa, 1970, 1980, 1984; Pop and Petrescu, 1983). However, Kvaček and Herman (2004), in a revision of the *Pandanites*, rejected some of the new species proposed by Petrescu and Duşa (1980). The plant assemblage includes ferns, monocotyledonous (palm trees), and dicotyledonous angiosperms.

The Late Cretaceous palaeoenvironment in Rusca Montană Basin was tentatively reconstructed by several authors. Tuzson (1913) illustrated a floodplain-dense forest, rich in *Pandanus* trees (Fig. 31.3A). Duşa (1974) also suggested a mangrovelike paleoenvironment, as figured in a picture from the collection of the Babeş-Bolyai University, Cluj-Napoca (Fig. 31.3B).

The continental deposits from Rusca Montană Basin are usually regarded as Maastrichtian (="Danian" in older contributions) in age. This age is based on similarities with the Maastrichtian formations in Haţeg Basin of the macroflora and especially of the pollen and spore assemblage, with *Pseudopapilopollis praesubhercynicus* (Antonescu et al., 1983). According to Antonescu et al. (1983), this latter marker indicates a Late Maastrichtian age for the latest Cretaceous continental formations in Haţeg Basin. As mentioned by Dalla Vecchia (2006, 2009), the dinosaur-bearing deposits from Transylvania were usually regarded as Late Maastrichtian during the 1970s and 1980s. However, more recent data instead suggest an Early Maastrichtian age for the dinosaur localities in Haţeg Basin (López-Martínez et al., 2001; Panaiotu and Panaiotu, 2002, 2010; Bojar et al., 2005; Van Itterbeeck et al., 2005; Therrien, 2005). Therefore, it is not clear whether the continental formations in Rusca Montană Basin are Early rather than Late Maastrichtian in age, or both.

Maastrichtian Land Vertebrates from the Rusca Montană Basin

We found the first Maastrichtian vertebrates from Rusca Montană Basin in the eastern part of the basin, near Lunca Cernii de Sus. The fossils formed a vertebrate microfossil assemblage (sensu Eberth and Currie, 2005) concentrated in red beds of fluvial origin both in the overbank silty clay and in the sandy channel fills. About a ton of sediments were screen washed. The red clay is rich in plant remains, including small pieces of yellowish amberlike resin. All the vertebrate remains are dark colored. Aquatic taxa (crocodiles, turtles) are typically most abundant, but dinosaur bone fragments and teeth are also well represented. Multituberculate mammals are also present. The bones and teeth of large reptiles are rare.

A

B

Sauropoda

Sauropods are represented by a single distal caudal vertebra (UBB NgS1; Fig. 31.4A) with a maximum preserved length of 105 mm. The centrum is eroded and broken, marked by intensive postmortem reworking. It is dorsoventrally compressed, with both articulation facets poorly preserved. Although the general aspect of the centrum evokes some Titanosauriformes, this centrum is too poorly preserved to warrant precise identification, and we regard it as an indeterminate sauropod.

Ornithopoda

Two isolated dental crowns (UBB NgO2 and NgO3; Fig. 31.4D,E) resemble the dentary teeth of the rhabdodontid *Zalmoxes*, abundantly represented in the Sânpetru Formation in the Haţeg Basin (Weishampel et al., 2003; Godefroit et al., 2009). The enamel is much thicker on the lingual side than on the labial side of the crown. The lingual side of the crown is asymmetrically divided by a strong primary ridge. High, slightly divergent vertical subsidiary ridges cover either side of the median ridge. Tiny crenulations are present along both edges of the crown. There is no real cingulum, but a thin enameled lip marks the base of the distal part of the crown. Contrary to typical *Zalmoxes* teeth, the basal lip bears tiny mammillations (Fig. 31.4, D1). Subsidiary ridges extend on the apical part of the labial side in UBB NgO3 (Fig. 31.4, E2), a character not described in *Zalmoxes* dentary teeth

31.3. Paleoenvironmental reconstructions of the Rusca Montană Basin (Romania) during the Maastrichtian.

A, After Tuzson (1913), modified. B, Oil on canvas by V. Svinţiu in the collection of Babeş-Bolyai University, Cluj-Napoca (Romania).

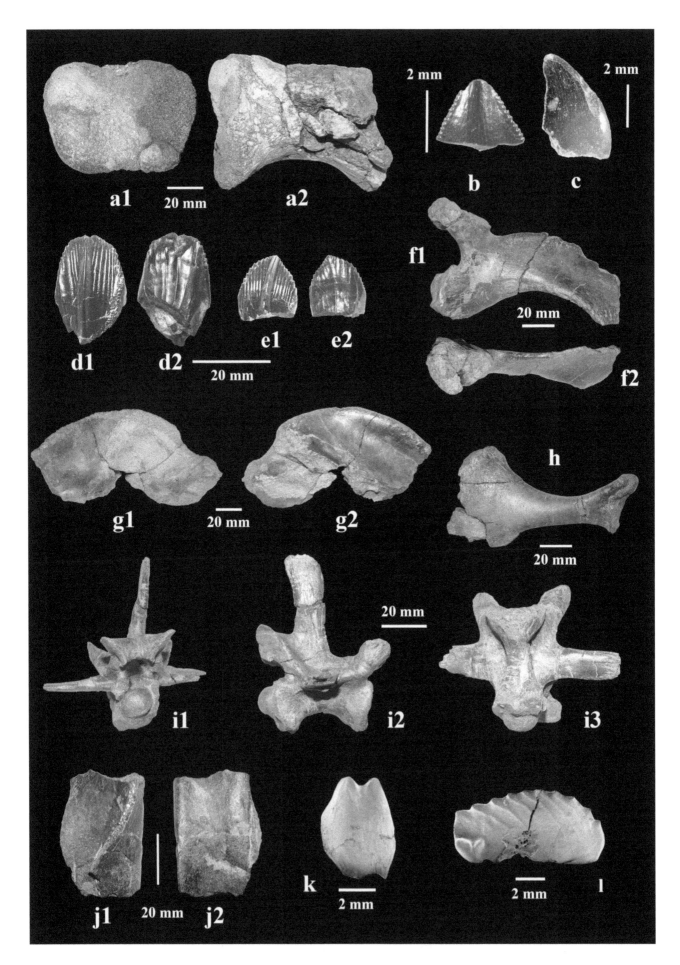

so far. The whole labial side of UBB NgO2 and the basal part the labial side of UBB Ngo3 are heavily worn. Therefore, because of small differences in the morphology of the dental crowns, it cannot be ascertained that these teeth really belong to *Zalmoxes*. One caudal vertebral centrum (UBB NgO1) may also tentatively be referred to an ornithopod dinosaur. Hadrosaurids are not represented in the Rusca Montană Basin fossil record; however, the collected sample is quite small, so this absence is not significant.

Theropoda

Theropods are represented by several isolated teeth. Three of them may be assigned to the velociraptorine morphotype (Currie et al., 1990; Sankey et al., 2002) because of the great disparity in size and distribution of the denticles along the mesial and distal carinae. In UBB NgTh2 (Fig. 31.4), the upper third of the distal carina bears 10 denticles per millimeter, whereas denticles are nearly completely missing on the mesial carina, with the exception of few denticles along its basal portion. The tooth crown is strongly compressed laterally, pointed, and sharply recurved.

UBB NgTh1 (Fig. 31.4B) is the apical part of a theropod tooth. It may be tentatively attributed to a troodontid-like theropod because it displays the following characters (Currie et al., 1990): the crown is less recurved than in teeth ascribed to velociraptorines, and both the mesial and distal denticles are well developed (six denticles per millimeter) and hooklike.

Both velociraptorine and troodontid dental morphotypes have already been described from the Sânpetru Formation of the Hațeg Basin (Csiki and Grigorescu, 1998; Codrea et al., 2002).

Crocodylia

Pelvic elements and vertebrae of crocodilians are rather abundant in the bonebed. UBB NGC1 (Fig. 31.4G) is a well-preserved iliac blade characterized by a reduced anterior process. UBB NGC1 (Fig. 31.4F) is a partial ischium; its posterior part is incomplete, but the iliac and pubic processes are preserved on the anterior portion of the bone. The pubis (UBB NgC3; Fig. 31.4H) is gracile and the articulation facet for the ischium is preserved. The anterior portion of pubis expands and becomes thin. Two crocodilian caudal vertebrae, probably from the middle part of the tail, have been discovered in the bonebed. UBB NgC4 (Fig. 31.4I) is procoelous, with an elongated centrum. The neural arch is much higher than the centrum, with well-developed prezygapophyses and postzygapophyses, and a robust neural spine. UBB NgC5 is less complete.

These fossils are too scarce to warrant a precise identification, and pending further information, we regard them as belonging to indeterminate crocodilians. Their size corresponds to similar fossils discovered in Hațeg Basin and attributed to the eusuchian *Allodaposuchus*, with a body longer than 3 m (Delfino et al., 2008)

The crocodilian bones are the most abundant fossils collected in this bonebed. However, they are less abundant than in the Maastrichtian locality of Oarda de Jos, in the southwestern part of the Transylvania Basin to the north (Codrea et al., 2010). There, crocodile teeth and bones can be

31.4. Late Cretaceous vertebrates from the Rusca Montană Basin. A, Caudal vertebra of Sauropoda indet. (UBB NgS1) in anterior (A1) and lateral (A2) views. B, Troodontid-like theropod tooth (UBB NgTh 1) in lingual or labial view. C, Velociraptorine-like theropod tooth (UBB NgTh 2) in lingual or labial view. D, E, Dentary teeth of rhabdodontid indet. (UBB NgO2, NgO3) in lingual (D1, E1) and labial (D2, E2) views. F, Left ischium of Crocodylia indet. (UBB NgC2) in lateral (F1) and ventral (F2) views. G, Left ilium of Crocodylia indet. (UBB NgC1) in medial (G1) and lateral (G2). H, Left pubis of Crocodylia indet. (UBB NgC3) in lateral view. I, Caudal vertebra (UBB NgC4) of Crocodylia indet. in anterior (I1), lateral (I2), and ventral (I3) views. J, Peripheral fragment of the carapace of a Chelonia indet. in ventral (J1) and dorsal (J2) views. K, P1 (UBB Ng-1-02) of Kogaionidae indet. in posterior view. L, p4 (UBB Ng-2-01) of Kogaionidae indet. in lateral view.

found at nearly all levels, with some of the collected bones even displaying bite marks.

Chelonia

Cryptodiran turtles are represented in the Maastrichtian assemblage from Rusca Montană Basin by scarce carapace fragments, including one peripheral fragment (UBB NgCh1; Fig. 31.4J), one costal plate fragment (UBB NgCh2), and three indeterminate other fragments. Bone thickness and ornamentation suggest the presence of the genus *Kallokibotion*, widespread in Maastrichtian localities from Transylvania (Codrea and Vremir, 1997; Vremir, 2004; Lapparent et al., 2009; Vremir and Codrea, 2009), but this identification needs to be confirmed by more complete material.

Mammalia

Multituberculate mammals are represented by isolated teeth. The simple tooth pattern and the reduced number of cusps in premolar and molar series are indicative for Kogaionidae. P2 (UBB NgCh1; Fig. 31.4K) has only two simple cusps, one on the labial side and the other on the lingual side. The damaged right p4 (NgCh1; Fig. 31.4L) has a bladelike crown with 10 cusps. Parallel labial ridges exist only under cusps 3–9, whereas sharp lingual ridges are present only under cusps 3–8. Under cusp 9, only an interrupted trace could mark the presence of a faint lingual ridge.

Two multituberculate taxa have been described from the Hațeg Basin, *Barbatodon transylvanicus* Rădulescu and Samson, 1986, and *Kogaionon ungureanui* Rădulescu and Samson, 1996; the second species is the type species for the family Kogaionidae. Csiki et al. (2005) recently discussed *B. transylvanicus* and other Kogaionidae from Hațeg Basin.

The Rusca Montană Basin multituberculate sample is too limited for precise identification. However, the discovery of kogaionid teeth in Rusca Montană Basin confirms that these multituberculates were widespread and probably diversified in Transylvania during the Maastrichtian. Indeed, Kogaionidae are rather abundant in Hațeg Basin, but also in the Basin of Transylvania (Codrea et al., 2010). Late Cretaceous kogaionids from Transylvania are apparently closely related to the later *Hainina*—for example, those reported from the Thanetian (Late Paleocene) of Jibou in the northwestern part of the Transylvania Basin (Gheerbrant et al., 1999). As in the other Maastrichtian localities from Transylvania, eutherians are apparently missing from the Rusca Montană Basin record.

Conclusion

The westernmost occurrences of continental Maastricthian formations in Romania can be observed in the Rusca Montană Basin. Westward, observation of these formations is particularly problematic because late Mesozoic rocks are hidden under molasse deposits in the Lugoj and Faget Cenozoic sedimentary basins.

Besides a rich Maastrichtian plant assemblage, vertebrate fossils are now reported from the Rusca Montană Basin. Preliminary observations suggest that the Maastrichtian vertebrate assemblage from Rusca Montană

Basin closely resembles assemblages described in other sedimentary basins (Haţeg Basin, Transylvania Basin) from western Romania (Codrea et al., 2001, 2010; Codrea and Godefroit, 2008). The Maastrichtian fauna appears globally similar in these different basins, reflecting a homogeneous paleoenvironment. The Rusca Montană Basin Maastrichtian assemblage demonstrates the extent of Haţeg Island toward the west. In spite of over a century's worth of research, this island still cannot be completely defined only on the basis of paleontological and sedimentological data as a result of the lack of information outside the Haţeg Basin. As already pointed out by Rage (2002) and Codrea et al. (2010), every piece of new data on the Upper Cretaceous continental deposits located outside Haţeg greatly aids our understanding of the paleobiology and paleogeography of the latest Cretaceous in this part of the continent. In this chapter, we show that the Rusca Montană Basin has great potential for future research.

Acknowledgments

We thank all our colleagues involved in the field missions in Rusca Montană Basin: Paul Dica, Emanoil Săsăran, Cristina Fărcaş, Cătălin Jipa (Cluj-Napoca), and Jimmy van Itterbeeck (Leuven). The activities of V.A.C. were partially supported by grant 1930/2009 from the CNCSIS (National University Research Council). M. J. Benton and D. B. Weishampel reviewed an earlier version of this chapter and made many useful comments.

References

Antonescu, E., D. Lupu, and M. Lupu. 1983. Correlation palynologique du Crétacé terminal du sud-est des Monts Metaliferi et des dépressions de Haţeg et de Rusca Montană. Anuarul Institutului de Geologie şi Geofizică 59: 71–77.

Balteş, N. 1966. Remarques sur la microflore de certains dépôts charbonneux daniens du bassin de Rusca Montană (Roumanie). Pollen et spores 8: 214–221.

Benton, M. J., Z. Csiki, D. Grigorescu, R. Redelstorff, P. M. Sander, K. Stein, and D. B. Weishampel. 2010. Dinosaurs and the island rule: the dwarfed dinosaurs from Haţeg Island. Palaeogeography, Palaeoclimatology, Palaeoecology 293: 438–454.

Bojar, A.-V., D. Grigorescu, F. Ottner, and Z. Csiki. 2005. Palaeoenvironmental interpretation of dinosaur- and mammal-bearing continental Maastrichtian deposits, Haţeg basin, Romania. Geological Quarterly 49: 205–222.

Bucur, I. I., C. Strutinski, and I. Cucuruzan. 1985. Formaţiunile mezozoice din sud-vestul Bazinului Rusca Montană. Dări de Seamă ale şedinţelor, Institutul de Geologie şi Geofizică 69: 57–76.

Cantuniari, M. S. 1937. Études géologiques dans les Monts Poiana Ruscă. I. Bassin de Rusca. Région de Rusca Montană. Comptes Rendus des Séances, Institut Géologique de Roumanie 21: 156–168.

Codrea, V., and P. Dica. 2005. Upper Cretaceous-lowermost Miocene lithostratigraphic units exposed in Alba Iulia-Sebeş-Vinţu de Jos area (SW Transylvanian basin). Studia Universitatis Babeş-Bolyai, Geologia 50: 19–26.

Codrea, V., and P. Godefroit 2008. New Late Cretaceous dinosaur findings from northwestern Transylvania (Romania). Comptes Rendus Palevol 7: 289–295.

Codrea, V., and M. Vremir. 1997. Kallokibotion bajazidi Nopcsa (Testudines, Kallokibotidae) in the red strata of Râpa Roşie (Alba County). Sargetia 17: 233–238.

Codrea, V., A. Hosu, S. Filipescu, M. Vremir, P. Dica, E. Săsăran, and I. Tanţău. 2001. Aspecte ale sedimentaţiei cretacic superioare din aria Alba-Iulia—Sebeş (jud. Alba). Studii şi cercetări, Geologie-Geografie 6: 63–68.

Codrea, V., T. Smith, P. Dica, A. Folie, G. Garcia, P. Godefroit, and J. Van Itterbeeck. 2002. Dinosaur egg nests, mammals and other vertebrates from a new Maastrichtian site of the Haţeg Basin (Romania). Comptes Rendus Palevol 1: 173–180.

Codrea, V., M. Vremir, C. Jipa, P. Godefroit, Z. Csiki, T. Smith, and C. Fărcaş.

2010. More than just Nopcsa's Transylvanian dinosaurs: a look outside the Haţeg Basin. Palaeogeography, Palaeoclimatology, Palaeoecology 293: 391–405.

Csiki, Z., and D. Grigorescu. 1998. Small theropods from the Late Cretaceous of the Haţeg Basin (Western Romania)—an unexpected diversity at the top of the food chain. Oryctos 1: 87–104.

Csiki, Z., D. Grigorescu and M. Rücklin. 2005. A new multituberculate specimen from the Maastrichtian of Pui, Romania and reassessement of *Barbatodon*. Acta Palaeontologica Romaniae 5: 73–86.

Currie, P. J., J. Keith Rigby Jr., and R. E. Sloan. 1990. Theropod teeth from the Judith River Formation of southern Alberta, Canada; pp. 107–125 in K. Carpenther and P. J. Currie (eds.), Dinosaur systematics: approaches and perspectives. Cambridge University Press, Cambridge.

Dalla Vecchia, F. M. 2006. *Telmatosaurus* and the other hadrosaurids of the Cretaceous European Archipelago. An overview. Natura Nascosta 32: 1–55.

———. 2009. European hadrosaurids; pp. 45–74 in Actas de las IV Jornadas Internacionales sobre Paleontologia de Dinosaurios y su Entorno. Colectivo Arqueológico-Paleontológico de Salas, Salas de los Infantes.

Delfino, M., V. Codrea, A. Folie, P. Dica, P. Godefroit, and T. Smith 2008. A complete skull of *Allodaposuchus precedens* Nopcsa, 1928 (Eusuchia) and a reassessement of the morphology of the taxon based on the Romanian remains. Journal of Vertebrate Paleontology 28: 111–122.

Dincă, A. 1977. Geologia Bazinului Rusca Montană. Partea de vest. Anuarul Institutului de Geologie şi Geofizică 52: 99–173.

Dincă, A., M. Tocorjescu, and A. Stilla. 1972. Despre vîrsta depozitelor continentale cu dinozaurieni din bazinele Haţeg şi Rusca Montană. Dări de Seamă 58: 83–94.

Duşa, A. 1974. Aspecte ale formării cărbunilor din Bazinul Rusca Montană. Studia Universitatis Babeş-Bolyai, Geologia-Geographia 19: 36–43.

Duşa, A. 1987. Zăcământul de la Rusca Montană; pp. 74–81 in I. Petrescu, G. Mărgărit, E. Nicorici, M. Nicorici, C. Biţoianu, A. Duşa, N. ţicleanu, I. Pătruţoiu, C. Todros, A. Munteanu, M. Ionescu, and A. Buda (eds.), Geologia zăcămintelor de cărbuni, vol. 2. Editura Tehnică, Bucharest.

Duşa, A., and M. Bărilă. 1973. Aspecte petrografice şi paleobotanice ale cărbunilor din bazinul Rusca Montană.

Studia Universitatis Babeş-Bolyai, Geologia-Mineralogia 18: 31–38.

Eberth, D. A., and P. J. Currie. 2005. Vertebrate taphonomy and taphonomic modes; pp. 453–477 in P. J. Currie and E. B. Koppelhus (eds.), Dinosaur Provincial Park, a spectacular ancient ecosystem revealed. Indiana University Press, Bloomington.

Folie, A., and V. Codrea. 2005. New lissamphibians and squamates from the Maastrichtian of Haţeg Basin, Romania. Acta Palaeontologica Polonica 50: 57–71.

Gheerbrant, E., V. Codrea, A. Hosu, S. Sen, C. Guernet, F. de Lapparent de Broin F., and J. Riveline. 1999. Découverte de vertébrés dans les Calcaires de Rona (Thanétien ou Sparnacien), Transylvanie, Roumanie: les plus anciens mammifères cénozoiques d'Europe Orientale. Eclogae geologicae Helvetiae 92: 517–535.

Givulescu, R. 1966. Sur quelques plantes fossiles du Danien de Roumanie. Comptes Rendus de l'Académie des Sciences de Paris 262: 1933–1936.

———. 1968. Nouvelles plantes fossiles du Danien de Roumanie. Comptes Rendus de l'Académie des Sciences de Paris 267: 880–882.

Godefroit, P., V. Codrea, and D. B. Weishampel. 2009. Osteology of *Zalmoxes shqiperorum* (Dinosauria, Ornithopoda), based on new specimens from the Upper Cretaceous of Nălaţ-Vad (Romania). Geodiversitas 31: 525–553.

Grigorescu, D. 1990. Nonmarine formations connected with the Laramian tectogenesis (Post-Early Maastrichtian formations in the Haţeg and Poiana Ruscă basins); pp. 18–23 in D. Grigorescu, E. Avram, G. Pop, M. Lupu, N. Anastasiu, and S. Radan (eds.), International geological correlation program Project 245: Nonmarine Cretaceous Correlation; Project 262: Tethyan Cretaceous Correlation, Guide to excursions. Institute of Geology and Geophysics, Bucharest.

———. 1992. Nonmarine Cretaceous formations of Romania; pp. 142–164 in N. Mateer and P.-J. Chen (eds.), Aspects of nonmarine cretaceous geology. Special volume, ICGP Project 245. China Ocean Press, Beijing.

Jianu, C.-M., and G. J. Boekschoten. 1999. The Haţeg—island or outpost? Deinsea 7: 195–198.

Jianu, C.-M., and D. B. Weishampel. 1999. The smallest of the largest: a new look at possible dwarfing in sauropod dinosaurs. Geologie en Mijnbouw 78: 335–343.

Kräutner, H. G., T. Berza, and R. Dimitrescu. 1986. K-Ar dating of the Banatitic magmatites from the southern Poiana

Ruscă Mountains (Rusca Montană Sedimentary Basin). Dări de Seamă ale şedinţelor, Institutul de Geologie şi Geofizică 70–71: 373–388.

Kvaček, J., and A. B. Herman. 2004. Monocotyledons from the Early Campanian (Cretaceous) of Grünbach, Lower Austria. Review of Palaeobotany and Palynology 128: 323–353.

Lapparent, A. F. de. 1947. Les dinosaures du Crétacé supérieur du Midi de la France. Bulletin de la Société Géologique de France 56: 1–54.

Lapparent, F. de, V. Codrea, T. Smith, and P. Godefroit. 2009. New turtle remains (Kallokibotionidae, Dortokidae) from the Upper Cretaceous of Transylvania (Romania); pp. 68–69 in 7th Romanian Symposium of Paleontology, Abstract book. Cluj-Napoca.

LeLoeuff, J. 2005. Romanian Late Cretaceous dinosaurs: big dwarfs or small giants? Historical Biology 17: 15–17.

López-Martínez, J. I. Canudo, L. Ardèvol, X. Pereda Superbiola, X. Orue-Extebarria, G. Cuenca-Bescós, J. I. Ruiz-Omeñaca, X. Murelaga, and M. Feis. 2001. New dinosaur sites correlated with Upper Maastrichtian pelagic deposits in the Spanish Pyrenees: implications for the dinosaur extinction pattern in Europe. Cretaceous Research 22: 41–61.

Mamulea, A. 1955. Cercetări geologice în regiunea Rusca Montană—Lunca Cernii. Dări de Seamă ale şedinţelor Comitetului Geologic 39: 172–178.

Melinte-Dobrinescu, M. C. 2010. Lithology and biostratigraphy of Upper Cretaceous marine deposits from the Haţeg region (Romania): palaeoenvironmental implications. Palaeogeography, Palaeoclimatology, Palaeoecology 293: 283–294.

Nopcsa, F. von. 1897. Vorläufiger Bericht über das Auftreten von oberer Kreide im Hatzeger Tal in Siebenbürgen. Verhandlungen der Kaiserlich-Königlichen Akademie der Wissenschaften, Wien 1897: 273–274.

———. 1905. A Gyulafehérvár, Déva, Ruszkabánya és a Romániai határ közé eső vidék geológiája. A magyar királyi Földtani Intézet Évkönyve 14: 82–254.

———. 1914. Über das Vorkommen der Dinosaurier in Siebenbürgen. Verhandlungen der zoologischen und botanischen Gessellschaft 54: 12–14.

Panaiotu, C., and C. Panaiotu. 2002. Paleomagnetic studies; p. 61 in 7th European Workshop on Vertebrate Paleontology, Abstracts volume and excursions field guide. Sibiu.

————. 2010. Palaeomagnetism of the Upper Cretaceous Sânpetru Formation (Haţeg Basin, South Carpathians). Palaeogeography, Palaeoclimatology, Palaeoecology 293: 343–352.

Papp, K. V. 1915. A Magyar birodalom vasérc-és kőszénkészlete. A Fraklintársulat nyomdája, Budapest, 964 pp.

Pereda Suberbiola, X., and P. M. Galton. 2009. Dwarf dinosaurs in the latest Cretaceous of Europe?; pp. 263–272 in Actas de las IV Jornadas Internacionales sobre Paleontología de Dinosaurios y su Entorno Colectivo. Arqueológico-Paleontologíco de Salas, Salas de los Infantes, Burgos, Spain.

Petrescu, I., and A. Duşa. 1970. Asupra unui punct paleofloristic din Cretacicul superior al bazinului Rusca Montană. Buletinul Societăţii Geologice, R.S. România 12: 165–172.

————. 1980. Flora din Cretacicul superior de la Rusca Montană—o raritate în patrimoniul paleobotanic naţional. Ocrotirea naturii şi mediului înconjurător 24: 147–155.

————. 1984. Paleoflora din Senonianul Bazinului Rusca Montană. Dări de Seamă ale şedinţelor Institutului de Geologie şi Geofizică 64: 107–124.

Pop, G., Th. Neagu, and L. Szász. 1972. Senonianul din regiunea Haţegului (Carpaţii Meridionali). Dări de Seamă ale şedinţelor Institutului de Geologie 58 (4): 95–118.

Pop, G., and I. Petrescu. 1983. Consideraţii paleoclimatice asupra vegetaţiei din Cretacicul superior de la Rusca Montană. Studia Universitatis Babeş-Bolyai, Geologia-Mineralogia 28: 49–54.

Rage, J.-C. 2002. The continental Late Cretaceous of Europe: toward a better understanding. Comptes Rendus Palevol 1: 257–258.

Rădulescu, C., and P.-M. Samson. 1986. Précisions sur les affinités des Multituberculés (Mammalia) du Crétacé supérieur de Roumanie. Comptes Rendus de l'Académie des Sciences Paris II 303: 1825–1830.

Rădulescu, C. and P.-M. Samson. 1996. The first multituberculate skull from the Late Cretaceous (Maastrichtian) of Europe (Haţeg Basin, Romania). Anuarul Institutului Geologic al României, Supplement 1, 69: 177–178.

Sankey, J. T., D. B. Brinkman, M. Guenther, and P. J. Currie. 2002. Small theropod and bird teeth from the Late Cretaceous (late Campanian) Judith River Group, Alberta. Journal of Paleontology 76: 751–763.

Săndulescu, M. 1984. Geotectonica României. Editura Tehnică, Bucharest, 336 pp.

Schafarzik, Fr. 1907. Über die geologische Verhältnisse der SW Poiana Ruska Gebirges im Komitate Krassó-Szörény. Mitteilungen aus dem Jahrbuch der königlich ungarischen geologischen Anstalt 1905: 84–95.

Smith, A. G., D. G. Smith, and B. M. Funnell. 1994. Atlas of Mesozoic and Cenozoic coastlines. Cambridge University Press, Cambridge, 99 pp.

Strutinski, C. 1986. Upper Cretaceous formations South of Ruşchiţa. Paleotectonic significance. Dări de Seamă ale şedinţelor Institutului de Geologie şi Geofizică 70–71: 247–254.

Strutinski, C., and I. I. Bucur. 1985. Basic tuffites in the Turonian and Coniacian of the Rusca Montană Basin (South Carpathians) and their paleogeographic significance. Studia Universitatis Babeş-Bolyai, Geologia-Geographia 31: 9–13.

Strutinski, C., and P. Hann. 1986. Reconsidération de la structure géologique de Rusca Montană et ses implications sur la tectonique du massif de Poiana Ruscă. Dări de Seamă ale şedinţelor Institutului de Geologie şi Geofizică 70–71: 255–268.

Therrien, F. 2005. Paleoenvironments of the Late Cretaceous (Maastrichtian) dinosaurs of Romania: insights from fluvial deposits and paleosols of the Transylvanian and Haţeg basins. Palaeogeography, Palaeoclimatology, Palaeoecology 218: 15–56.

Therrien, F., C.-M. Jianu, S. Bogdan, D. B. Weishampel, and J. W. King. 2002. Paleoenvironmental reconstruction of the latest Cretaceous dinosaur-bearing formations of Romania: preliminary results. Sargetia, Scienties Naturalis 19: 33–59.

Tuzson, J. 1913. Adatok Magyarország fosszilis flórájához (Additamenta ad floram fossilem Hungariae III), A magyar kiraly Földtani Intézet Évkönyve 21: 209–233.

Van Itterbeeck, J., S. V. Markevich, and V. Codrea. 2005. Palynostratigraphy of the Maastrichtian dinosaur- and mammal sites of the Râul Mare and Bărbat valleys (Haţeg Basin, Romania), Geologica Carpathica 56: 137–147.

Vremir, M. 2004. Fossil turtle found in Romania—overview. A Magyar Földtani Intézet Évi Jelentèse 2002: 143–152.

Vremir, M., and V. Codrea. 2009. Late Cretaceous turtle diversity in Transylvanian and Haţeg basins (Romania); pp. 122–124 in The 7th Romanian Symposium of Paleontology, Abstract book. Cluj-Napoca.

Weishampel, D. B., D. Grigorescu, and D. B. Norman. 1991. The dinosaurs of Transylvania. National Geographic Research and Exploration 7: 196–215.

Weishampel, D. B., D. B. Norman, and D. Grigorescu. 1993. *Telmatosaurus transsylvanicus* from the Late Cretaceous of Romania: the most basal hadrosaurid dinosaur. Palaeontology 36: 361–385.

Weishampel, D. B., C.-M. Jianu, Z. Csiki, and D. B. Norman. 2003. Osteology and phylogeny of *Zalmoxes* (n. g.), an unusual euornithopod dinosaur from the latest Cretaceous of Romania. Journal of Systematic Palaeontology 1: 65–123.

Weishampel, D. B., Z. Csiki, M. J. Benton, D. Grigorescu, and V. Codrea. 2010. Palaeobiogeographic relationships of the Haţeg biota—between isolation and innovation. Palaeogeography, Palaeoclimatology, Palaeoecology 293: 419–437.

Willingshofer, E., P. Andriessen, S. Cloetingh, and F. Neubauer. 2001. Detrital fission track thermochronology of Upper Cretaceous syn-orogenic sediments in the South Carpathians (Romania): inferences on the tectonic evolution of a collisional hinterland. Basin Research 13: 379–395.

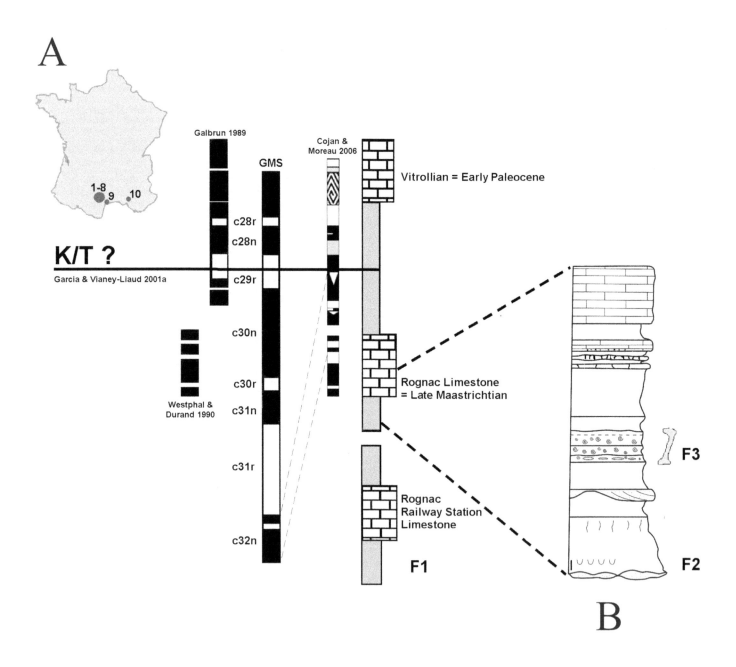

A

Galbrun 1989

GMS

Cojan &
Moreau 2006

1-8
9
10

Vitrollian = Early Paleocene

K/T ?

Garcia & Vianey-Liaud 2001a

c28r
c28n

c29r

c30n

c30r

c31n

c31r

c32n

Westphal &
Durand 1990

Rognac Limestone
= Late Maastrichtian

Rognac
Railway Station
Limestone

F1

F3

F2

B

First Late Maastrichtian (Latest Cretaceous) Vertebrate Assemblage from Provence (Vitrolles-la-Plaine, Southern France)

32

Xavier Valentin, Pascal Godefroit, Rodolphe Tabuce, Monique Vianey-Liaud, Wu Wenhao, and Géraldine Garcia*

A Late Maastrichtian locality from Provence (southwestern France) has yielded a diversified vertebrate fauna, including a "zhelestid" mammal, *Valentinella vitrollense*, in association with lacertilians, cryptodiran chelonians, alligatoroid crocodilians, and a varied dinosaur fauna including Neoceratosauria, Coelurosauria, Titanosauria, basal Iguanodontian, and Hadrosauridae. It is the first noticeable report of the occurrence of hadrosaurids in Provence. The morphology of the dentary and dentary teeth suggests that the hadrosaurid material belongs to a noneuhadrosaurian hadrosaurid, close to *Telmatosaurus transsylvanicus* and *Tethyshadros insularis*, also known from the Late Cretaceous of Europe. It is clearly different from hadrosaurid material previously described in Europe. However, the material discovered so far is still inadequate to erect a new taxon. This new site offers new perspectives on the diversity and evolution of the European vertebrate ecosystems during the Maastrichtian. Indeed, the association of hadrosaurids with titanosaurids and *Rhabdodon* is unusual in the latest Cretaceous of southern France. It questions the hypothesis of the replacement in Western Europe of an Early Maastrichtian fauna dominated by titanosaurid sauropods and *Rhabdodon* by a Late Maastrichtian assemblage dominated by hadrosaurids.

Introduction

Late Cretaceous continental deposits are widely exposed in southern France. Nevertheless, in spite of intensive field investigations during the last 20 years, the main localities that have yielded a diverse Late Maastrichtian continental fauna were restricted so far to the western part of southern France. All of them (Fig. 32.1) are located in the foothills of the Pyrenees, from the eastern Corbières in the east to both sides of the Garonne Valley in the west, via the Plantaurel in Ariège (see, e.g., Paris and Taquet, 1973; Le Lœuff et al., 1994; Laurent et al., 1999, 2002).

Vitrolles-La-Plaine is the first Late Maastrichtian site from Provence that contains diverse articulated and scattered reptile bones associated with mammal remains. This chapter is a preliminary description of the fauna collected during different fieldwork sessions in 1998–1999 and two main excavation campaigns in spring 2007 and 2008.

Institutional abbreviations. ISEM, Institut des Sciences de l'Évolution, Université Montpellier II, Montpellier, France; UP, Université de Poitiers, Poitiers, France.

32.1. A, Map showing the location of the main French sites with hadrosaurid remains dated from the Late Maastrichtian. 1, Ausseing. 2, Auzas. 3, Cassagnau. 4, Le Jadet. 5, Lestaillats. 6, Mérignon. 7, Peyrecave. 8, Tricouté. 9, Le Bexen, 10, the locality of Vitrolles-La-Plaine. B, Schematic stratigraphic column of Vitrolles section with paleomagnetic data and global magnetostratigraphic scale (GMS). Log of Vitrolles-la-Plaine with fossiliferous levels: F1, Vitrolles-Couperigne locality, where one skeleton of *Rhabdodon priscus* was collected (Garcia et al., 1999); F2, dinosaur eggs organized in clutches; F3, location of the studied assemblage. No precise correlation have been proposed by Galbrun (1989) and Westphal and Durand (1990) for the Rognac Limestone in this part of Basin. The results (dotted lines on B) suggested by Cojan and Moreau (2006) are inexact because they did not take into account the diachronism and the large thickness variation of this formation across the Basin, as already demonstrated by Garcia and Vianey-Liaud (2001a).

Geological Setting

The Vitrolles-La-Plaine site is located in the eastern part of the Aix Basin, close to Lake Berre. It was discovered during a geological survey by one of us (X.V.) in 1997, in Upper Cretaceous deposits near the town of Vitrolles (Garcia, 1998; Garcia and Vianey-Liaud, 2001a). It corresponds to a succession of clays and mottled marls with some interbedded sandstone lenses, topped by the thick lacustrine Limestone of Rognac. By using the stratigraphic distribution of the egg species, Garcia and Vianey-Liaud (2001a) have demonstrated that the Limestone of Rognac, present in the western and eastern areas of the Aix Basin, is clearly diachronous throughout the basin and corresponds to a Late Maastrichtian age in its western part.

The vertebrate remains are scattered through a 1.5-m-thick lignite marl layer containing freshwater and terrestrial gastropods (*Pyrgulifera armata, Pupilla* sp., *Lychnus bourguignati, L. matheroni*, and *Cyclophorus heliciformis*) and a single bivalve (*Unio cuvieri*).

The co-occurrence of typical "Rognacian" gastropods (*Lychnus matheroni* and *Pyrgulifera armata*), charophytes (*Peckichara sertulata*, M. Feist, pers. comm.) and dinosaur eggs belonging to the Maastrichtian oospecies *Megaloolithus mamillare* (Fig. 32.3), as well as the position of the site just below the Rognac Limestone (Fig. 32.1), clearly indicate that the Vitrolles-La-Plaine locality is Late Maastrichtian in age (Garcia and Vianey-Liaud, 2001a).

The Vitrolles-La-Plaine fossil locality is clearly an allochtonous assemblage of elements belonging to numerous animals of different sizes. The rare associated skeletal elements indicate that the vertebrate carcasses were disarticulated before reworking (Fig. 32.2). Dense elements, such as limb bones, appear overrepresented at Vitrolles-La-Plaine, although lighter elements (vertebrae, skull bones), are proportionally rarer. This suggests hydraulic sorting during transportation (Voorhies, 1969). Moreover, most of the long bones are broken off at both ends, and the fractured edge is always quite rounded (abrasion levels 2–3; Fiorillo, 1988). This feature indicates that the bones were significantly reworked after being broken (Fiorillo, 1988). Behrensmeyer (1988) observed that fresh limb bones from large mammals often showed no evidence of breakage during vigorous hydraulic reworking. According to Ryan et al. (2001) and Eberth and Getty (2005), large numbers of broken limb bones indicate a destructive history before or during final reworking. Thus, it is more likely that many limb elements discovered at Vitrolles-La-Plaine experienced an earlier taphonomic episode, such as the breakdown of trabecular bone and collagen (Eberth and Getty, 2005), that weakened the specimens and increased their susceptibility to hydraulically induced breakage.

Biodiversity of Vitrolles-la-Plaine Locality

Eggs

Eggshells were collected in three levels of the section (Fig. 32.1, F1 [see also Garcia et al., 1999]; F2 and F3) and include prismatic (Fig. 33.3 D), ratite (Fig. 33.3E), and geckonoid (Fig. 33.3F) morphotypes. Complete spherical eggs organized in small clutches containing five to eight eggs (a total of 30 eggs were found in the same nesting layer; Fig. 33.3A) were unearthed in level F2, 4 m under the vertebrate level F3. This egg spatial

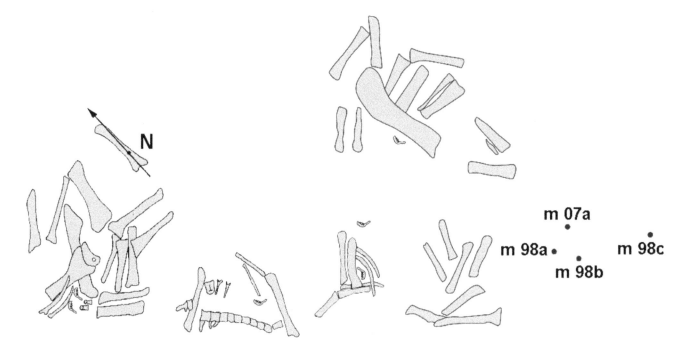

m 07a

m 98a • • m 98c

m 98b

_10 cm

arrangement with randomly structures has already been described for this ootaxa (Garcia, 1998; Vianey-Liaud and Garcia, 2000; Cousin, 2002). One pathological egg with two superimposed eggshell layers (Fig 33.3, B1–B4) was discovered with the normal eggs, suggesting a reproduction system closer to modern avians that laid their eggs, one ovum per oviduct, over a long period of time (Varicchio et al., 1997). All belong to the megaloolithid oospecies *Megaloolithus mamillare* (Vianey-Liaud et al., 1994, Garcia and Vianey-Liaud, 2001a, 2001b), present in several sites of southern Europe (Vianey-Liaud and Lopez-Martinez, 1997; Garcia, 1998) and typically representing an ootaxa of the Maastrichtian continental deposits (Garcia and Vianey-Liaud, 2001a). Classically, the megaloolithid eggs are assigned to titanosaurian sauropods (Grellet-Tinner et al., 2006; Sander et al., 2008); the most convincing evidence for this assignment is the discovery in Argentina of embryonic remains inside eggs (Chiappe et al., 1998). However, this attribution was recently questioned by Grigorescu et al. (2010).

Fishes

Rare fish remains are exclusively represented by isolated lepisosteid scales; all these scales are particularly thick, rhomboidal in shape, and covered by ganoine. Lepisoteid scales compose the majority of fish remains discovered in Late Cretaceous continental deposits from southern France (Laurent et al., 1999).

Squamates

Only one fragmentary dentary containing some incomplete straight teeth (Fig. 32.4A) indicates the presence of an indeterminate lacertilian at Vitrolles-La-Plaine (J.-C. Rage, pers. comm.). Squamates are also rare in other Maastrichtian assemblages from Europe (Tremp Basin from Spain:

32.2. Map indicating the repartition of vertebrate specimens collected during three field missions since 1998. The mammal remains (labeled "m") are concentrated in the same area of the excavation.

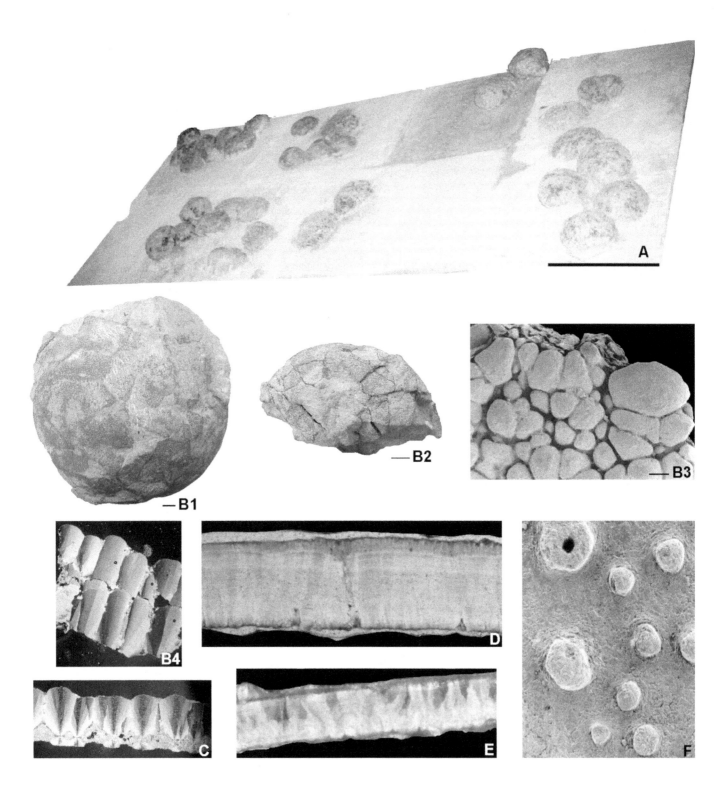

Lopez-Martinez et al., 2001; localities from of the Garonne Valley in France: Laurent et al., 2002; Hateg and Transylvanian basins from Romania: Folie and Codrea, 2005; Codrea et al., 2010), with the occurrence of Iguanidae, Scincomorpha, and Anguimorpha (Pereda-Suberbiola, 2009).

Chelonia

Turtle remains are not abundant, and they are mostly represented by isolated carapace plates with a typical reticulate ornamentation. They are tentatively referable to the cryptodiran turtle *Solemys*, originally described from the Campanian of Laño in Spain (Lapparent de Broin and Murelaga, 1996) and also recorded in several localities of the southern France (Lapparent de Broin and Murelaga, 1999; Pereda-Suberbiola, 2009).

Crocodylia

Isolated crocodilian teeth are rather abundant at Vitrolles-La-Plaine. They have lanceolate and robust crowns, with a circular base, two prominent carinae, and a blunt apex. Variations in their morphology clearly depend on their position in the jaws and on the size of the animal (Fig. 33.4B,C). They closely resemble the teeth of the alligatoroids *Musturzabalsuchus buffetauti*, from the Campanian of Laño in Spain (Buscalioni et al., 1997), and those of *Massaliasuchus affuveliensis*, from the Santonian–Campanian of Bouches-du-Rhône (Martin and Buffetaut, 2008). Nevertheless, the enamel of the teeth from Vitrolles-La-Plaine appears smooth, although it is profusely ridged with the margins crenulated in *Musturzabalsucuhus* (Buscalioni et al., 1999); the ornamentation is more discrete in *Massaliasuchus*, consisting of small wrinkles on the surface of the enamel (Martin and Buffetaut, 2008). Of course, precise identification of crocodilian taxa on the basis of isolated teeth remains conjectural. However, the presence of alligatoroids at Vitrolles-La-Plaine is likely because these crocodilians were common components of Late Cretaceous deposits in North America and Europe (Martin and Buffetaut, 2008).

Neoceratosauria

Several limb bones belonging to medium-sized theropods have been unearthed from Vitrolles-la-Plaine locatilty. Because of the poor preservation of this material, most diagnostic characters are unfortunately not preserved.

A right femur from Vitrolles-la-Plaine (Fig. 32.5A,B) resembles the holotype of *Tarascosaurus salluviscus* Le Lœuff and Buffetaut, 1991. Originally described as belonging to the family Abelisauridae (Le Lœuff and Buffetaut, 1991), *Tarascosaurus* was subsequently regarded as a nomen dubium (Rauhut, 2003) or a potential Abelisauroidea incertae sedis (Tykosky and Rowe, 2004). Like in *Tarascosaurus*, the cranial side of the proximal portion of UP-VLP-98C-001 is narrow, virtually reduced to a prominent ridge starting for the lesser trochanter and extending along the proximal third of the bone (Fig. 32.5A). The lesser trochanter is positioned low on the femur, probably well below the level of the femoral head. Like in *Tarascosaurus*, it is mediolaterally wide but not prominently developed, and in cranial view, it occupies a median position above the proximocranial ridge. The presence of a nutritive foramen, characteristic for *Tarascosaurus* (Le Lœuff and Buffetaut, 1991), cannot be ascertained in UP-VLP-98C-001. The femoral head is not preserved in UP-VLP-98C-001. In cranial view, the femoral shaft is convex externally, but it may also be a consequence of postmortem deformation. The fourth trochanter forms a large, bladelike structure on the caudomedial shaft of the femur (Fig. 32.5B); its apex is located at the level of the proximal third of the bone. The distal end of the femur is incompletely preserved. The medial epicondyle is exceptionally developed (Fig.

32.3. A–C, *Megaloolithus mamillare*. A, Eggs exposed in the level F2 of Vitrolles-La-Plaine. Scale = 1 m. Pathological egg in dorsal (B1) and lateral (B2) views. Scale = 1 cm. B3, Outer ornamentation with abnormal nodes. Scale = 1 mm. B4, Radial view showing the two pathological layers (original magnification, ×40). C, Radial view of normal eggshell *M. mamillare* (original magnification, ×40). D, Prismatic morphotype in radial view with pore canal (original magnification, ×40). E, Ratite morphotype in thin section (original magnification, ×100). F, Outer surface with a dispersituberculate ornamentation with pore opening in node (scanning electron micrograph, original magnification ×35).

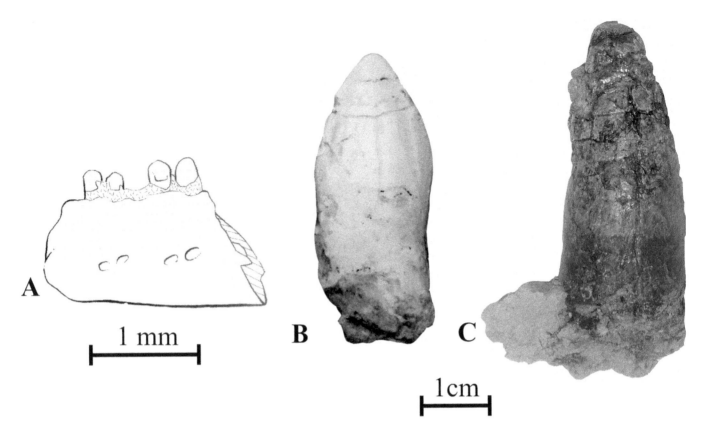

32.4. A, Lacertilian left dentary in labial view. B, UP-VLP-07-00, alligatoroid tooth in lingual view. C, UP-VLP-07-002, alligatoroid tooth in labial view.

32.5A,B), as it is usually observed in abelisauroids (Sampson et al., 2001; Tykosky and Rowe, 2004). In UP-VLP-98C-001, it is oriented quite cranially above the tibial condyle, although it is usually oriented medially in typical abelisauroids (see, e.g., Carrano et al., 2002, fig. 14a).

UP-VLP-98A-002 is a left tibiotarsus (Fig. 32.5C–E) that closely resembles another specimen previously described from the Late Campanian of La Boucharde (Allain and Pereda-Suberbiola, 2003). Although it is incompletely preserved and deformed, its cnemial crest is clearly craniocaudally longer than its articular condyles (Fig. 32.5C,D), which is regarded as a synapomorphy for Neoceratosauria (Tykosky and Rowe, 2004). The tibial shaft is transversely compressed and convex laterally. The distal end of the tibiotarsus is better preserved in UP-VLP-98A-002. It is triangular in distal view and slightly enlarged transversely. The cranial side of the tibia is reduced to a rounded ridge on the distal quarter of the bone. As it is usual in Neoceratosauria (Tykosky and Rowe, 2004), there is a well-developed crista fibularis on the lateral side of the distal tibia (Fig. 32.5E). The astragalus is fused to the distal end of the tibia. It forms a high caudal ascending process along the distal quarter of the tibia (Fig. 32.5E).

Even if they are poorly preserved, these limb bones can be confidently referred to as neoceratosaurian theropods. Tortosa et al. (2010) also recently reported the presence of abelisaurid neoceratosaurians in latest Cretaceous deposits from Provence.

Coelurosauria

Isolated theropod teeth have also been discovered at Vitrolles-La-Plaine. The *Richardoestesia* morphotype, known from latest Cretaceous localities

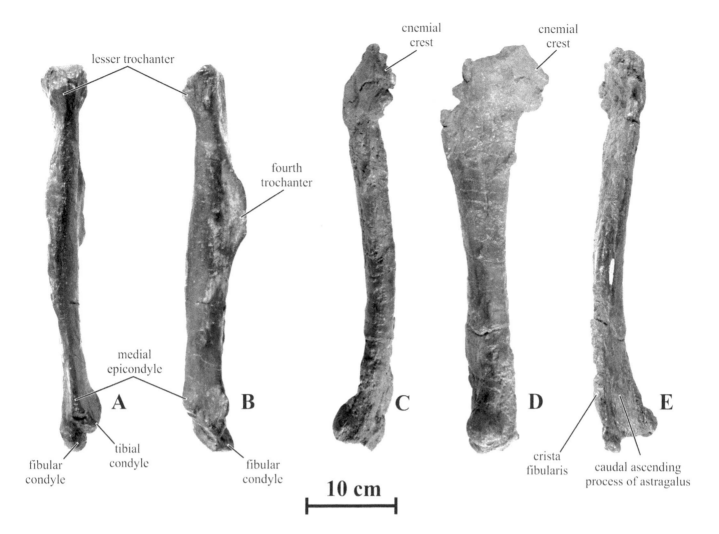

lesser trochanter

fourth trochanter

cnemial crest

cnemial crest

medial epicondyle

A   **B**   **C**   **D**   **E**

fibular condyle   tibial condyle   fibular condyle   crista fibularis   caudal ascending process of astragalus

10 cm

in western North America (e.g., Currie et al., 1990; Sankey, 2001) and Romania (Codrea et al., 2002), is represented by several laterally compressed teeth, characterized by their slight curvature and by the presence of minute serrations along the distal carina only (Fig. 32.6A). Similar isolated teeth were also collected in the lower part in the section (Fig. 32.1, Level 1). Maastrichtian representatives of the *Richardoestesia* morphotype, as discovered at Vitrolles-La-Plaine, are usually less recurved distally than the Campanian ones (Currie et al., 1990; Codrea et al., 2002).

32.5. A, B, UP-VLP-98C-001, right femur of a Neoceratosauria indet. in cranial (A) and medial (B) views. C–E, UP-VLP-98A-002, left tibiotarsus of a Neoceratosauria indet. in cranial (C), medial (D), and caudal (E) views.

Titanosauria

Some compressed limb bones (humerus [Fig. 32.6B], femur, and tibia) were found during the third field campaigns. Their proximal and distal ends are not well preserved, but their size and robustness indicate that they belonged to fairly large animals. Unfortunately, any diagnostic characters allowing it to be referred to one of the three valid titanosaur species currently recognized on the Iberian–Armorican island—*Ampelosaurus atacis* (Le Lœuff, 1995, 2005), *Lirainosaurus astibiae* (Sanz et al., 1999), and *Atsinganosaurus velauciensis* (Garcia et al., 2010)—have not been preserved on this material. The material collected in 2007 and 2008 needs to be prepared in laboratory before being studied in detail for accurate identification.

32.6. A, UP-VLP-08-001, Coelurosauria tooth (*Richarddoestesia* morphotype) in lateral view. B, UP-VLP-98B-001, left humerus of a Titanosauria indet. in cranial view. C, UP-VLP-08-002, basal Iguanodontia ("*Rhabdodon*") dentary tooth in labial view.

Basal Iguanodontia

Massive isolated teeth (Fig. 32.6C) closely resemble those of the basal Iguanodontia *Rhabdodon*, the most common dinosaur of Late Cretaceous vertebrate assemblages from southern France (Allain and Pereda-Suberbiola, 2003). The enamel is distributed on both sides of the crowns. The lingual side of the dentary teeth is characterized by a strong median primary ridge and by more than eight slightly divergent on either sides of the primary ridge. There is no real cingulum, but a thin enameled lip marks the base of the distal part of the crown. The labial side is heavily worn. The crown of the maxillary teeth is devoid of a primary ridge, but numerous subequal, slightly vertically divergent ridges cover the buccal surface of the crown. The lingual side is heavily worn.

Systematic revisions of the *Rhabdodon* material from southern France are currently in progress in order to clarify the intrageneric variability (how many species?) and the individual differences linked to sexual dimorphism (Chanthasit and Buffetaut, 2007; Goussard, 2009).

Hadrosauridae

Five dentaries were collected in the Vitrolles-La-Plaine locality (Fig. 32.7). All are crushed and incomplete. Because of their small size, they probably belong to juvenile specimens. The dentary ramus appears long and only slightly recurved ventrally. The dental battery is formed by narrow parallel-sided grooves (Fig. 32.7A,D); this character is synapomorphic for advanced Hadrosauroidea, including *Batyrosaurus, Probactrosaurus, Eolambia, Protohadros*, and Hadrosauridae (see Appendix 19.2 in Chapter 19 in this book, character 53). Even in juveniles, the dental battery is formed by more than 25 tooth families and is proportionally higher than in *Batyrosaurus, Probactrosaurus, Eolambia*, and *Protohadros*, suggesting that the hadrosauroid from Vitrolles-La-Plaine was more derived than these taxa and was probably a Hadrosauridae (the most recent common ancestor of *Bactrosaurus* and *Parasaurolphus*, plus all the descendants of this common ancestor; see Chapter 20 in this book). Moreover, the dental battery extends far caudally to the level of the apex of the coronoid process (Fig. 32.7A,D), a character found in Euhadosauria (sensu Weishampel et al., 2003: Hadrosaurinae + Lambeosaurinae) and *Telmatosaurus transylvanicus*. The coronoid process of the dentary is narrow and much higher than the dentary ramus (Fig. 32.7C,D). Although this character is unusual among hadrosaurids, it may be partially explained by ontogeny: juveniles are usually characterized by a proportionally higher coronoid process than older specimens (P.G., pers. obs.). The coronoid process is laterally offset and separated from dentition by a wide shelf (Fig. 32.7B), as is usual in hadrosauroids except *Bolong* and *Jinzhousaurus* (Appendix 19.2, character 48). It is distinctly curved rostrally (Fig.32.7C,D), like in Euhadrosauria and *Tethyshadros insularis* (Dalla Vecchia, 2009). Like in basal hadrosauroids, the apex of the coronoid process is only slightly expanded rostrally, and the surangular apparently formed much of the caudal margin of the coronoid process. In Euhadrosauria, on the contrary, the dentary forms nearly all of the rostrocaudally greatly expanded apex, and the surangular is reduced to a thin sliver along caudal margin and does not reach the dorsal end of the coronoid process (Appendix 19.2, character 49).

The dentary teeth are lanceolate and narrow (Fig. 33.7E). Their lingual side bears a strong primary ridge that divides the crown into two subequal halves. This character is shared by Euhadrosauria and *Telmatosaurus transsylvanicus*. A secondary ridge is developed on the mesial half of the crown and reaches the apex of the tooth, like in basal hadrosauroids, but also in *Tethysahdros insularis* (Dalla Vecchia, 2009). In Euhadrosauria and *Telmatosaurus transsylvanicus*, on the other hand, the secondary ridge is usually absent or faintly developed (Weishampel et al., 1993). The margins of the crown are denticulate; denticulations are usually less developed in Euhadrosauria.

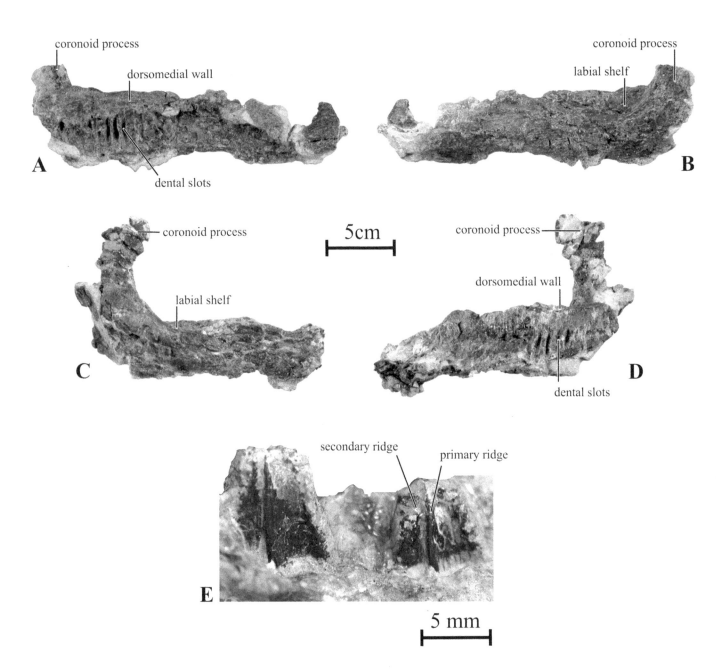

32.7. Hadrosaurid dentaries. A, B, UP-VLP-07 P15-001, left dentary in lingual (A) and labial (B) views. C, D, UP-VLP-07 P03-001, right dentary in labial (C) and lingual (D) views. E, UP-VLP-08E-001, detail of dentary teeth.

Therefore, the mosaic of characters observed on the dentaries and dentary teeth from Vitrolles-La-Plaine suggests that these fossils belong to a noneuhadrosaurian Hadrosauridae, close to *Telmatosaurus transsylvanicus* and *Tethyshadros insularis*, also from uppermost Cretaceous deposits of Europe. Moreover, these dentaries display a potential autapomorphy that we did not observe in any other hadrosauroid: along at least the caudal third of the dentary, the dental slots do not reach the dorsal border of the dentary ramus but are bordered dorsally by a medial wall that raises medially from the shelf at the base of the coronoid process (Fig. 32.7A,C). The biomechanical significance of this character remains enigmatic.

Mammalia

In 1998–1999, and more recently in 2007, the excavations at Vitrolles-La-Plaine yielded some fragmentary remains of a eutherian mammal,

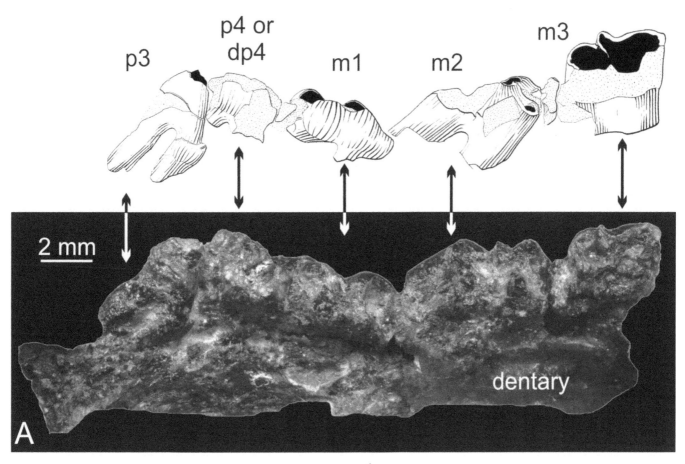

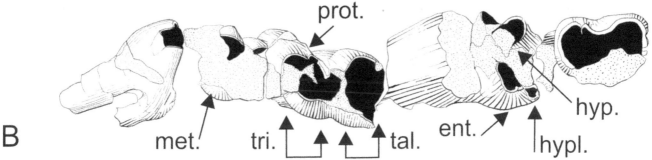

B

met.   tri.   tal.   ent.   hyp.

prot.

hypl.

Valentinella vitrollense (Tabuce et al., 2004). The specimen discovered in 2007, still undescribed, corresponds to a lingual part of an upper molar retaining a prominent protocone and a distinct low cingular hypocone.

The Upper Cretaceous fossil record of eutherians is scanty in Europe: only two genera, *Labes* and *Lainodon*, were known in France and Spain before the discovery of *Valentinella* (Pol et al., 1992; Sigé et al., 1997; Gheerbrant and Astibia, 1999). On the basis of several dental traits, such as the hypoconulid–entoconid twinning and the anteroposteriorly short trigonid on m1–3, *Labes, Lainodon*, and *Valentinella* are ascribed to the "zhelestids."

The "zhelestids" are a paraphyletic unit well documented from the Cenomanian through Turonian of Uzbekistan and Kazakhstan. Other species from the Coniacian through Paleocene are also known in North America and Japan (Archibald, 2007). Most "zhelestids" are known only from fragmentary dental remains, but several taxa from Uzbekistan are now documented by associated cranial and dental remains in addition to

32.8 *Valentinella vitrollense*, ISEM/VLP-2 (holotype) in lingual (A) and occlusal (B) views. *Abbreviations:* met, metaconid; pro, protoconid; tri, trigonid; tal, talonid; ent, entoconid; hyp, hypoconid; hypl, hypoconulid.

ear regions (Ekdale et al., 2004) and referred postcranial elements (Chester et al., 2007, 2008, 2010). The "zhelestids" were initially regarded as being at the origin of both "condylarths" and Cenozoic ungulates within the Ungulatomorpha clade (Archibald et al., 2001). A recent broader-scale phylogenetic analysis argues, however, that the morphological similarities observed in "zhelestids" and ungulate placentals are only due to convergences. "Zhelestids" are now positioned near the base of the Eutheria clade in the mammalian tree (Wible et al., 2007).

Conclusions

Vitrolles–La-Plaine is currently among the most diversified Late Maastrichtian sites in southern France. The vertebrate fauna, composed of at least 11 taxa, includes a "zhelestid" mammal, *Valentinella vitrollense*, in association with lacertilians, cryptodyran chelonians, alligatoroid crocodilians, and a diversified dinosaur fauna including Neoceratosauria, Coelurosauria, Titanosauria, basal Iguanodontia, and Hadrosauridae.

For the first time, hadrosaurid dinosaurs are formally recognized in Provence. The morphology of the dentary and dentary teeth suggests that this hadrosaurid material belongs to a basal form, close to *Telmatosaurus transsylvanicus* and *Tethyshadros insularis*, also from the Late Cretaceous of Europe. It apparently differs from the hadrosaurid material hitherto described, but the material at hand is too fragmentary and too poorly preserved to justify the erection of a new taxon.

The association of hadrosaurids with titanosaurids and *Rhabdodon* is unusual in the latest Cretaceous of southern France. Previous works based on sites in southern France have concluded that an important faunal replacement related to environmental changes occurred in southern Europe during the Late Cretaceous: a Late Campanian–Early Maastrichtian fauna dominated by titanosaurid sauropods and *Rhabdodon* was replaced by a Late Maastrichtian assemblage dominated by hadrosaurids (e.g., Le Lœuff et al., 1994). This hypothesis has already been challenged by recent discoveries in northern Spain (Lopez-Martinez et al., 2001): the hadrosaurid *Pararhabdodon isonensis* (including *Koutalisaurus kohlerorum*) from the Tremp Basin is considered as Early Maastrichtian to early Late Maastrichtian in age (Pereda-Suberbiola et al., 2009). The presence of hadrosaurids, titanosaurs, and *Rhabdodon* in the allochtonous assemblage from Vitrolles-La-Plaine suggests that these animals lived together in southern France during the latest Cretaceous. However, it is highly probable that they lived in different environments. Their co-occurrence in the allochtonous assemblage from Vitrolles-La-Plaine can easily be explained by taphonomical processes (hydraulic transport) that concentrated, in the same level, bones of animals living in various environments. However, it cannot be excluded that older Late Campanian–Early Maastrichtian sediments, containing dinosaur fossils (titanosaurs, basal Iguanodontia), were reworked and that the Vitrolles-la-Plaine assemblage is consequently a mixing of faunas of different ages.

It may be concluded Vitrolles-La-Plaine potentially offers new perspectives on the diversity and evolution of vertebrate ecosystems in southern Europe during the Maastrichtian. However, further research is needed in this site in order to clarify its taphonomy.

We gratefully acknowledge the Vitrolles Municipality, particularly the mayor, G. Obino, for permission to conduct our research in this part of the Aix Basin. Logistical and field assistance were provided by L. Blein, R. Dumont, S. Jiquel, R. Licand, B. Marandat, F. Mebrouk, D. Roggero, S. Rafa, J. Sudre, and E. Turini.

Acknowledgments

References

Allain, R., and X. Pereda-Suberbiola. 2003. Dinosaurs of France. Comptes Rendus Palevol 2: 27–44.

Archibald, D. J. 2007. Zhelestids: stem eutherians or basal laurasiatherians, but no evidence for placental orders in the cretaceous. Journal of Vertebrate Paleontology 27: 41A.

Archibald, J. D., A. O. Averianov, and E. G. Ekdale. 2001. Late Cretaceous relatives of rabbits, rodents, and other extant eutherian mammals. Nature 414: 62–65.

Behrensmeyer, A. K. 1988. Vertebrate preservation in fluvial channels. Palaeogeography, Palaeoclimatology, Palaeoecology 63: 183–199.

Buscalioni, A. D., F. Ortega, and D. Vasse. 1997. New crocodiles (Eusuchia: Alligatoroidea) from the Upper Cretaceous of southern Europe. Comptes rendus de l'Académie des Sciences de Paris (ser. 2) 325: 525–530.

———. 1999. The upper Cretaceous crocodilian assemblage from Laño (northcentral Spain): implications in the knowledge of the finicretaceous European faunas. Estudios Museo de Ciencia Naturales de Alava 14: 213–233.

Carrano, M. T., S. D. Sampson, and C. A. Foster. 2002. The osteology of *Masiakasaurus knopfleri*, a small abelisauroid (Dinosauria: Theropoda) from the Late Cretaceous of Madagascar. Journal of Vertebrate Paleontology 22: 510–534.

Chanthasit, P., and Buffetaut E. 2007. The distribution and new localities of the Late Cretaceous ornithopod dinosaur *Rhabdodon* in southern France. 12th European Workshop of Vertebrate Palaeontology, 12

Chester, S., E. Sargis, F. Szalay, J. D. Archibald, and A. O. Averianov. 2007. A functional analysis of mammalian humeri from the Late Cretaceous of Uzbekistan. Journal of Vertebrate Paleontology 27 (supplement to 3): 58A.

Chester, S., E. Sargis, F. Szalay, J. D. Archibald, and A. O. Averianov. 2008. Therian femora from the Late Cretaceous of Uzbekistan. Journal of Vertebrate Paleontology 28 (supplement to 3): 63A.

Chester, S. G. B., Sargis, E. J., Szalay, F. S., J. D. Archibald, and A.O. Averianov. 2010. Mammalian distal humeri from the Late Cretaceous of Uzbekistan. Acta Palaeontologica Polonica 55: 199–211.

Chiappe, L. M., L. M. Coria, L. Dingus, F. Jackson, A.Chinsamy, and M. Fox. 1998. Sauropod embryos from the Late Cretaceous of Patagonia. Nature 396: 258–261.

Codrea, V., T. Smith, P. Dica, A. Folie, G. Garcia, P. Godefroit, and J. Van Itterbeeck. 2002. Dinosaur egg nests, mammals and other vertebrates from a new Maastrichtian site of the Haţeg Basin (Romania). Comptes rendus Palevol 1: 173–180.

Codrea, V., O. Barbu, and C. Jipa-Murzea. 2010. Upper Cretaceous (Maastrichtian) land vertebrate diversity in Alba District (Romania). Bulletin of the Geological Society of Greece 43: 594–601.

Cojan, I., and M. G. Moreau. 2006. Correlation of terrestrial climatic fluctuations with global signals during the Upper Cretaceous–Danian in a compressive setting (Provence, France). Journal of Sedimentary Research 76: 589–604.

Cousin, R. 2002. Organisation des pontes des Megaloolithidae Zhao, 1979. Bulletin trimestriel de la Société géologique de Normandie et des Amis du Muséum du Havre, Éditions du Muséum d'Histoire Naturelle du Havre 89: 1–176.

Currie, P. J., J. K. Rigby Jr., and R. E. Sloan. 1990. Theropod teeth from the Judith River Formation of southern Alberta, Canada; pp. 107–125 in: K. Carpenter and P. J. Currie (eds.), Dinosaur systematics: approaches and perspectives. Cambridge University Press, Cambridge.

Dalla Vecchia, F. M. 2009. *Tethyshadros insularis*, a new hadrosauroid dinosaur (Ornithischia) from the Upper Cretaceous of Italy. Journal of Vertebrate Paleontology 29 (4): 1100–1116.

Eberth, D. A., and M. A. Getty. 2005. Ceratopsian bonebeds: occurrence, origins, and significance; pp. 501–506 in P. J. Currie and E. B. Koppelhus (eds.), Dinosaur Provincial Park, a spectacular ancient ecosystem revealed. Indiana University Press, Bloomington.

Ekdale, E. G., D. J. Archibald, and A. O. Averianov. 2004. Petrosal bones of placental mammals from the Late Cretaceous of Uzbekistan. Acta Palaeontologica Polonica 49: 161–176.

Fiorillo, A. R. 1988. Taphonomy of the Hazard Homestead Quarry (Ogallala Group), Hitchcock County, Nebraska. University of Wyoming Contributions to Geology 26: 57–98.

Folie, A., and V. Codrea. 2005. New lissamphibians and squamates from the Maastrichtian of Hateg Basin, Romania. Acta Palaeontologica Polonica 50: 57–71.

Galbrun, B., 1989. Résultats magnéto-stratigraphiques à la limite Rognacien-Vitrollien: précisions sur la limite Crétacé-Tertiaire dans le Bassin d'Aix-en-Provence. Cahiers de la Réserve géologique de Haute-Provence, Digne 1: 34–37.

Garcia, G. 1998. Les coquilles d'œufs de dinosaures du Crétacé supérieur du sud de la France: Diversité, paléobiologie, biochronologie et paléoenvironnements. Ph.D. thesis, University of Montpellier II, Montpellier, 313 pp.

Garcia, G., and M. Vianey-Liaud. 2001a. Dinosaur eggshells as new biochronological markers in Late Cretaceous continental deposits. Palaeogeography, Palaeoclimatology, Palaeoecology 169: 153–164.

———. 2001b. Nouvelles données sur les coquilles d'œufs de dinosaures de Megaloolithidae du sud de la France: systématique et variabilité intraspécifique. Comptes Rendus de l'Académie des Sciences de Paris 332: 185–191.

Garcia, G., M. Pincemaille, B. Marandat, M. Vianey-Liaud, E. Lorenz, H. Cappetta, G. Cheylan, J. Michaux, and J. Sudre. 1999. Découverte du premier squelette presque complet de Rhabdodon priscus (Dinosauria, Ornithopoda) dans le Maastrichtien inférieur de Provence. Comptes Rendus de l'Académie des Sciences de Paris 328: 415–421.

Garcia, G., S. Amico, F. Fournier, E. Thouand, and X. Valentin. 2010. A new titanosaur genus (Dinosauria, Sauropoda) from the Late Cretaceous of southern France and its paleobiogeographic implications. Bulletin de la Société Géologique de France 181: 269–277.

Gheerbrant, E., and H. Astibia. 1999. The Upper Cretaceous mammals from Lano (Spanish Basque Country). Estudios del Museo de Ciencias naturales de Alava 14: 295–323.

Grellet-Tinner, G., L. Chiappe, M. Norell, and D. Bottjer. 2006. Dinosaur eggs and nesting behaviors: a paleobiological investigation. Palaeogeography, Palaeoclimatology, Palaeoecology 232: 294–321.

Goussard, F. 2009. Zalmoxes, Rhabdodon and Rhabdodontidae: how many genera and how many species?; p. 51 in P. Godefroit and O. Lambert (eds.), Tribute to Charles Darwin and Bernissart iguanodons: new perspectives on vertebrate evolution and early cretaceous ecosystems. Royal Belgian Institute of Natural Sciences, Brussels.

Grigorescu, D., G. Garcia, Z. Csiki, V. Codrea, and A.-V. Bojar. 2010. Uppermost Cretaceous megaloolithid eggs from the Hațeg Basin, Romania, associated with hadrosaur hatchlings: search for explanation. Palaeogeography, Palaeoclimatology, Palaeoecology 293: 360–374.

Lapparent de Broin, F. de, and X. Murelaga. 1996. Une nouvelle faune de chéloniens dans le Crétacé supérieur européen. Comptes Rendus de l'Académie des Sciences de Paris 323: 729–735.

———. 1999. Turtles from the Upper Cretaceous of Laño (Iberian Peninsula). Estudios del Museo de Ciencias Naturales de Alava 14: 135–211.

Laurent, Y., L. Cavin, and M. Bilotte. 1999. Découverte d'un gisement à vertébrés dans le Maastrichtien supérieur des Petites-Pyrénées. Comptes Rendus de l'Académie des Sciences de Paris 328: 781–787.

Laurent, Y., M. Bilotte, and J. Le Loeuff. 2002. Late Maastrichtian continental vertebrates from southwestern France: correlation with marine faunas. Palaeogeography, Palaeoclimatology, Palaeoecology 187: 121–135.

Le Lœuff, J. 1995. Ampelosaurus atacis (nov. gen., nov. sp.), un nouveau Titanosauridae (Dinosauria, Sauropoda) du Crétacé supérieur de la Haute Vallée de l'Aude (France). Comptes Rendus de l'Académie des Sciences de Paris 321: 693–699.

———. 2005. Osteology of Ampelosaurus atacis (Titanosauridae) from southern France; pp. 115–137 in V. Tidwell and K. Carpenter (eds.), Thunder-lizards: the sauropodomorph dinosaurs. Indiana University Press, Bloomington.

Le Lœuff, J., and E. Buffetaut. 1991. Tarascosaurus salluvicus nov. gen., nov. sp., un nouveau dinosaure théropode (Abélisauridé) du Crétacé supérieur (Campanien inférieur) du synclinal du Beausset (Bouches-du-Rhône, France). Geobios 25: 585–594.

Le Lœuff, J., E. Buffetaut, and M. Martin. 1994. The last stages of dinosaur faunal history in Europe: a succession of Maastrichtian dinosaur assemblages from the Corbières (southern France). Geological Magazine 131: 625–630.

Lopez-Martinez, N., J. I. Canudo, L. Ardevol, X. Pereda-Suberbiola, X. Orue-Extbarria, G. Cuenca-Bescos, J. I. Ruiz-Omenaca, X. Murelaga, and M. Feist. 2001. New dinosaur sites correlated with Upper Maastrichtian pelagic deposits in the Spanish Pyrenees: implications for the dinosaur extinction in Europe. Cretaceous Research 22: 41–61.

Martin, J. E., and E. Buffetaut. 2008. Crocodilus affuvelensis Matheron, 1869 from the Late Cretaceous of southern France: a reassessment. Zoological Journal of the Linnean Society 152: 567–580.

Paris, J. P., and P. Taquet. 1973. Découverte d'un fragment de dentaire d'hadrosaurien (reptile dinosaurien) dans le Crétacé supérieur des Petites Pyrénées (Haute Garonne). Bulletin du Muséum d'Histoire Naturelle de Paris 130: 17–27.

Pereda-Suberbiola, X. 2009. Biogeographical affinities of Late Cretaceous continental tetrapods of Europe: a review. Bulletin de la Société Géologique de France 180: 57–71.

Pereda-Suberbiola, X., J. I. Canudo, J. Company, P. Cruzado-Caballero, and J. I. Ruiz-Omeñaca. 2009. Hadrosauroid dinosaurs from the Latest Cretaceous of the Iberian Peninsula. Journal of Vertebrate Paleontology 29: 946–951.

Pol, C., A. D. Buscalioni, J. Carballeira, V. Francés, N. Lopez-Martinez, B. Marandat, J. J. Moratalla, J. L. Sanz, B. Sigé, and J. Villatte. 1992. Reptiles and mammals from the Late Cretaceous new locality Quintanilla del Coco (Burgos province, Spain). Neues Jahrbuch für Geologie und Paläontologie 184: 279–314.

Rauhut, O. W. M. 2003. The interrelationships and evolution of basal theropod dinosaurs. Special Papers in Palaeontology 69: 1–213.

Ryan, M. J., A. P. Russell, D. A. Eberth, and P. J. Currie. 2001. The taphonomy of a Centrosaurus (Ornithischia: Ceratopsidae) bone bed from the Dinosaur Park Formation (Upper Campanian), Alberta, Canada, with comments on cranial ontogeny. Palaios 16: 482–506.

Sampson, S. D., M. T. Carrano, and C. A. Forster. 2001. A bizarre predatory dinosaur from the Late Cretaceous of Madagascar. Nature 409: 504–506.

Sander, P. M., C. Peitz, F. Jackson, and L. M. Chiappe. 2008. Upper Cretaceous titanosaur nesting sites and their implication for sauropod dinosaur reproductive biology. Palaeontographica A 284: 69–107.

Sanz, J. L., J. E. Powell, J. Le Lœuff, R. Martínez, and J. Pereda Suberbiola. 1999. Sauropod remains from the Upper Cretaceous of Laño (northcentral Spain). Titanosaur phylogenetic relationships. Estudios del Museo de Ciencias Naturales de Alava 14 (Num. Esp. 1): 235–255.

Sankey, J. T. 2001. Late Campanian southern dinosaurs, Aguja Formation, Big Bend, Texas. Journal of Paleontology 75: 208–215.

Sigé, B., A. D. Buscalioni, S. Duffaud, M. Gayet, B. Orth, J.-C. Rage, and J. L. Sanz. 1997. Etat des données sur le gisement crétacé supérieur continental de Champ-Garimond (Gard, Sud de la France). Münchner Geowissenschaftliche Abhandlungen 34: 111–130.

Tabuce, R., M. Vianey-Liaud, and G. Garcia. 2004. A eutherian mammal in the latest Cretaceous of Vitrolles, southern France. Acta Paleontologica Polonica 49: 347–356.

Tortosa, T., E. Buffetaut, Y. Dutour, and G. Cheylan. 2010. Abelisaurid remains from Provence (Southeastern France): phylogenetic and paleobiogeographic implications; p. 82 in 8th meeting of the European Association of Vertebrate Palaeontologists, Aix-en-Provence, June 7–12, 2010.

Tykosky, R. S., and T. Rowe. 2004. Ceratosauria; pp. 47–70 in D. B. Weishampel, P. Dodson, and H. Osmólska (eds.), The Dinosauria. 2nd ed. University of California Press, Berkeley.

Varicchio, D. J., F. Jackson, J. J. Borkowski, and J. H. Horner. 1997. Nest and clutches of the dinosaur Troodon formosus and the evolution of avian reproductive traits. Nature 385: 247–250.

Vianey-Liaud, M., and G. Garcia. 2000. The interest of french Late Cretaceous dinosaur eggs and eggshells; pp. 165–176 (extended abstracts) in A. M. Bravo and T. Reyes (eds), 1st International symposium on dinosaurs eggs and babies. Isona i Conca Della, Catalonia, Spain.

Vianey-Liaud, M., and N. Lopez-Martinez. 1997. Late Cretaceous dinosaur eggshells from the Tremp Basin, southern Pyrenees, Lleida, Spain. Journal of Paleontology 71: 1157–1171.

Vianey-Liaud, M., Mallan, P. O. Buscail, and C. Montgelard. 1994. Review of French dinosaur eggshells: morphology, structure, mineral, and organic composition; pp. 151–183 in K. Carpenter, K. F. Hirsch and J. R. Horner (eds.), Dinosaur eggs and babies. Cambridge University Press, New-York.

Voorhies, M. R. 1969. Taphonomy and population dynamics of an Early Pliocene vertebrate fauna, Knox County, Nebraska. Contributions to Geology, University of Wyoming Special Paper 1.

Weishampel, D. B., D. B. Norman, and D. Grigorescu. 1993. Telmatosaurus transsylvanicus from the Late Cretaceous of Romania: the most basal hadrosaurid dinosaur. Palaeontology 36: 361–385.

Weishampel, D. B., C.-M. Jianu, Z. Csiki, and D. B. Norman. 2003. Osteology and phylogeny of Zalmoxes (n. g.), an unusual euornithopod dinosaur from the latest Cretaceous of Romania. Journal of Systematic Palaeontology 1: 65–123.

Westphal, M., and J. P. Durand. 1990. Magnétostratigraphie des séries continentales fluvio-lacustres du Crétacé supérieur dans le synclinal de l'Arc (Région d'Aix-en-Provence, France). Bulletin de la Société Géologique de France 8: 609–620.

Wible, J. R., G. W. Rougier, M. J. Novacek, and R. J. Asher. 2007. Cretaceous eutherians and Laurasian origin for placental mammals near the K/T boundary. Nature 447: 1003–1006.

33.1. Dorsal view of endocranial casts in three Mesozoic mammals. A, Triconodont *Triconodon.* B, Multituberculate *Chulsanbaatar.* C, Eutherian *Barunlestes.* A and B represent the cryptomesencephalic type (vermis and the paraflocculi very large, no apparent cerebellar hemispheres, no dorsal midbrain exposure). C represents the eumesencephalic type (following this interpretation, all extant mammals are derived from type C). From Kielan-Jaworowska (1997), modified from Kielan-Jaworowska (1986). Not to scale.

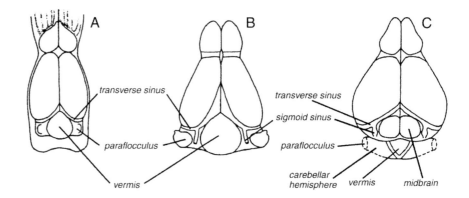

Reassessment of the Posterior Brain Region in Multituberculate Mammals

33

Emmanuel Gilissen* and Thierry Smith

Two distinct types of morphology have been suggested to characterize the brains of Mesozoic mammals. In some primitive mammals such as *Triconodon* and multituberculates, the vermis and the paraflocculi are large, with no apparent cerebellar hemispheres and no dorsal midbrain exposure (cryptomesencephalic type). In the other type of brain morphology (eumesencephalic type), the cerebral hemispheres are actually well developed, the presence of cerebellar hemispheres is apparent, and there is a large dorsal midbrain exposure. The terms *cryptomesencephalic* and *eumesencephalic brains* may have to be abandoned in light of a new interpretation of multituberculate and eutriconodontan endocasts. In this new interpretation, the region described as the vermis is rather an impression of the superior cistern covering both the midbrain and the cerebellar vermis itself. A cistern is an expansion of the subarachnoid space containing cerebrospinal fluid. This interpretation suggests that the superior cistern must have been large enough in eutriconodonts and multituberculates to press on the internal table of the cranial cavity during development in order to create the triangular bulge evident on the endocasts. The term *vermis* to refer to the triangular bulge visible on multituberculate and eutriconodontan endocasts therefore ought to be replaced by *superior cistern*. In support of this new interpretation, it was suggested that a similar "enlarged vermis" exists in extant species such as the marsupial koala, in which this enlargement is due to the expansion of the superior cistern, which makes a bulging impression on the skull. Further, the midbrain (tectum mesencephali) is dramatically exposed in the koala brain but is covered on the endocast. In the current study, we assessed this new interpretation of the endocranial cast of Mesozoic mammals with new observations of the endocranial cast morphology in extant monotremes (*Zaglossus, Tachyglossus, Ornithorhynchus*), marsupials (*Macropus, Thylacinus, Protemnodon, Sarcophilus, Perameles, Phascolomys, Dasyurus, Phascolarctos, Didelphis, Lutreolina, Caluromys, Chironectes, Marmosa, Metachirus*), Tenrecidae, primates, megabats, and fossil mammalian taxa. For the same genus or species examined, we compared structures on endocasts with structures on the exteriors of brains via dissections or illustrated descriptions of brains. In all the specimens we analyzed, the cerebellar vermis is clearly visible on endocranial casts. In koalas (genus *Phascolarctos*), the triangular space covering the midbrain between the posterior portions of the cerebral hemispheres is evident but cannot be confused with the cerebellar vermis, which is clearly apparent between the cerebellar hemispheres. The bulging aspect of the triangular space between the cerebral hemispheres is mainly due to the presence of the venous transverse sinuses, located anterior to the vermis.

An interpretation alternative to vermis or superior cistern could also be given. In the megachiropteran *Dobsonia praedatrix*, this bulging structure is the pineal organ. It is prominent on *D. praedatrix* endocranial casts and covers the anterior part of the cerebellar vermis. Overall, our observations indicate that it is not the superior cistern but most probably the cerebellar vermis that makes a clearly visible impression on the posterior part of the endocranial casts of all extant and fossil mammals examined, including multituberculates and eutriconodonts.

Introduction

Multituberculates are among the most important representatives of the Mesozoic mammalian fauna. When considering the phylogenetic cladogram including both extinct and extant mammalian groups, they currently occupy a place between the monotremes and the marsupials (Rougier and Novacek, 1998). The first multituberculates known with certainty appeared during the Middle Jurassic Period (Bathonian, 165 Ma). They constituted major elements of Mesozoic mammalian faunas in the Northern Hemisphere and were especially widespread and diverse during the Late Cretaceous (Kielan-Jaworowska et al., 2004).

They survived into the Cenozoic together with three extant mammalian taxa (monotremes, marsupials, and eutherians) and remain among the most common fossil mammals in well-sampled Paleocene localities of the Northern Hemisphere. Multituberculates reached their peak of diversity during the Paleocene and then became extinct during the Late Eocene (Priabonian, 35 Ma) (Luo, 2007).

It therefore appears that multituberculates persisted for at least 130 Ma, which is almost as long as the nonavian dinosaurs, and were the longest-lived mammalian order. Brain function was likely a key element for the success of this group because brains can be considered as buffers against environmental variability and changes. In the chaotic natural world, animals face the challenge of finding the resources necessary for their survival, and they also face the hazards that threaten their survival. Brains are informed by the senses about the presence of resources and hazards. They evaluate and store this input to generate appropriate responses (Allman, 1999). This may be a reason why brain size and anatomy differs between different animals. It is therefore of paramount interest to gather information about brain morphology in such a successful mammalian group.

Nervous systems are generally poorly represented in mammal phylogenetic analyses (Kielan-Jaworowska, 1997; Macrini et al., 2006). However, this topic was the subject of a breakthrough with the recognition by Kielan-Jaworowska (1986, 1997) of two distinct types of brain morphology that characterize Mesozoic mammals. According to Kielan-Jaworowska (1986, 1997), in some primitive mammals such as *Triconodon* and multituberculates, the vermis and the paraflocculi are large, with no apparent cerebellar hemispheres and no dorsal midbrain exposure (cryptomesencephalic type). In the other type of brain morphology (eumesencephalic type), the cerebral hemispheres are actually well developed, the presence of cerebellar hemispheres is apparent, and there is a large dorsal midbrain exposure (Fig. 33.1).

This terminology used for describing endocasts of Mesozoic mammals is challenged, however, by more recent phylogenetic analyses. Macrini et al. (2007b) observed that although the cryptomesencephalic endocast character is plesiomorphic in multituberculates and tricondonts, the presence of broad dorsal midbrain exposure on endocasts described by the eumesencephalic condition occurs independently in several mammalian lineages and some nonmammalian cynodonts (Macrini et al., 2007b). However, midbrain exposure is not an archaic character of placental mammals. It also varies widely between closely related taxa, even within archaic mammals such as the living Didelphimorphia (Starck, 1963), and midbrain itself can widely vary in size, independently from the whole brain, as exemplified by Kaas and Collins (2001) for the superior colliculus of the ground squirrel (genus *Spermophilus*) and the laboratory rat (genus *Rattus*).

In a new interpretation of Mesozoic mammal brains, Kielan-Jaworowska and Lancaster (2004) and later Kielan-Jaworowska et al. (2004) suggested that the terms *cryptomesencephalic* and *eumesencephalic brains* should be abandoned in light of a new interpretation of multituberculate and eutriconodontan endocasts. In this new interpretation, the region described as the vermis is rather an impression of the superior cistern covering both the midbrain and the cerebellar vermis itself. A cistern is an expansion of the subarachnoid space containing cerebrospinal fluid. This interpretation suggests that the superior cistern must have been large enough in eutriconodonts and multituberculates to press onto the internal table of the cranial cavity during development in order to create the triangular bulge evident on the endocasts. Kielan-Jaworowska and Lancaster (2004) and Kielan-Jaworowska et al. (2004) no longer use the term *vermis* for the triangular bulge visible behind the cortex on multituberculate and eutriconodontan endocasts; they replace it with the term *superior cistern*. In support of this new interpretation, previous unpublished observations suggest that a similar "enlarged vermis" exists in extant species such as the marsupial koala, in which this enlargement is due to the expansion of the superior cistern, making a bulging impression on the skull. Kielan-Jaworowska and Lancaster (2004) further indicate that the midbrain (tectum mesencephali) is dramatically exposed in the koala brain but is covered on the endocast.

In this study, we assess this new interpretation of the endocranial cast of Mesozoic mammals with new observations of the endocranial cast morphology in fossil and extant species. The cerebellar vermis is here defined as the middle part of the cerebellum, stretching over the entire anteroposterior length of the cerebellum (cf. Nieuwenhuys et al., 1998, fig. 22.31). Endocasts were compared with brains of the same genus or species.

Institutional abbreviations. RBINS, Royal Belgian Institute of Natural Sciences, Brussels, Belgium; RMCA, Royal Museum for Central Africa, Tervuren, Belgium.

Material

We observed the morphology of the posterior portion of the brain in 33 silicone endocranial casts made from the cranial material of a sample of representative extant species (Tables 33.1 and 33.2). Except when mentioned ($n = 2$), all species are represented by one silicone endocast. Most of the cranial material is housed at the Royal Belgian Institute of Natural Sciences

Table 33.1. List of Extant Noneutherian Mammal Species in This Study[a]

Taxonomic group	Species
Monotremata (n = 3)	*Ornithorhynchus anatinus*[j]
	Tachyglossus aculeatus[j,k]
	Zaglossus bruijni (cf. *Tachyglossus*)[j]
Marsupialia (Australidelphia) (n = 11)	*Dasyurus maculatus*[c,e,j]
	Macropus rufus[b,c,e,h,j,k]
	Perameles nasuta[e,h,i,j]
	Phascolarctos cinereus (n = 2)[d]
	Protemnodon bicolour (cf. *Macropus*)[b,c,e,h,j,k]
	Sarcophilus harrisii[c,d,e,j]
	Thylacinus cynocephalus[h]
	Vombatus ursinus (n = 2)[d,e,h,j]
	Wallabia bicolor (cf. *Macropus*)[b,c,e,h,j,k]
Marsupialia (Ameridelphia, Didelphimorphia) (n = 10)	*Caluromys lanatus*[e,j]
	Chironectes minimus (cf. *Didelphis*)[e,h,j]
	Didelphis marsupialis (n = 2)[e,f,g,h,j]
	Lutreolina crassicaudata (cf. *Didelphis*)[e,h,j]
	Marmosa sp.[e,j]
	Marmosa elegans[e,j]
	Metachirus nudicaudatus (cf. *Didelphis*)[e,h,j]
	Monodelphis domestica[f]
	Philander opossum[e,h,j]

[a] Except when mentioned in parentheses, each species is represented by one specimen (i.e., one endocranial cast). For each genus, references are provided for comparison with actual brains or descriptions of brains. In cases where there was no available brain, reference is given for the closest sister genus following Wilson and Reeder (2005).
[b] Gervais (1869).
[c] Haight and Murray (1981).
[d] Haight and Nelson (1987).
[e] Johnson (1977).
[f] Macrini et al. (2007c).
[g] Nieuwenhuys et al. (1998).
[h] Brain specimen (formalin fixed) housed at the Museum National d'Histoire Naturelle, Paris, France.
[i] The brain of another Bandicoot, *Isoodon obesulus*, is illustrated in Johnson (1977) and Haight and Murray (1981).
[j] Comparative Mammalian Brain Collections (http://brainmuseum.org/index.html).
[k] Brauer and Schober (1970).

(RBINS), Brussels, Belgium. The afrosoricid and the macroscelid specimens (Table 33.2) are housed in the Royal Museum for Central Africa (RMCA), Tervuren, Belgium. The *Galago demidovii* silicone endocast was made at the Anthropology Institute, University of Zürich-Irchel, Switzerland. The endocast of *Monodelphis domestica* was taken from Macrini et al. (2007a). New reconstructions of fossil monotreme, marsupial, and eutherian mammal endocranial casts were taken from the literature (Gingerich and Gunnell, 2005; Macrini et al., 2006, 2007a, 2007b; Meng et al., 2003; Novacek, 1982; Silcox et al., 2010) (Table 33.3).

For the same genus or species examined, we compared structures on endocasts with structures on the exteriors of brains via dissections or illustrated descriptions of brains. For these comparisons, details of the references are given in Tables 33.1 and 33.2. The anatomical nomenclature is taken from Rauber and Kopsch (Leonhardt et al., 1987).

Methods

Endocranial casts of extant mammals were created by pouring silicone (DC 3481; Mida, Brussels, Belgium) mixed with 5% catalyst (81R; Mida) into the skull through the foramen magnum after stopping up the optic canal and

Taxonomic group	Species
Afrosoricida (n = 3)	Tenrec ecaudatus (n = 2)[b,d,f]
	Potamogale velox[b,d]
Soricomorpha (n = 2)	Solenodon cubanus[b,d]
	Galemys pyrenaicus[d]
Erinaceomorpha (n = 1)	Echinosorex sp.[d]
Macroscelidea (n = 2)	Rhynchocyon cirnei stuhlmanni[b,e]
	Elephantulus brachyrhynchus[b,e,f,g]
Chiroptera (n = 1)	Dobsonia praedatrix[c]
Primates (n = 1)	Galago demidovii[e,g]

Table 33.2. List of Extant Eutherian Mammal Species in This Study[a]

[a] Except when mentioned in parentheses, each species is represented by one specimen (i.e., one endocranial cast). For each genus, references are given for comparison with actual brains or descriptions of brains.
[b] Bauchot and Stephan (1967).
[c] Bhatnagar et al. (1990).
[d] Stephan et al. (1991).
[e] Brain dissection (EG) of ethanol preserved specimens housed at the Royal Museum for Central Africa, Tervuren, Belgium.
[f] Comparative Mammalian Brain Collections (http://brainmuseum.org/index.html).
[g] Brauer and Schober (1970).

Taxonomic group	Species
Monotremata	Obdurodon dicksoni[b]
Stem marsupials	Pucadelphys andinus[c]
	Herpetotherium fugax[d]
	Wynyardia bassiana[e]
Theriiform	Vincelestes neuquenianus[f]
Early placental mammals	Leptictis dakotensis[g]
	Rhombomylus[h]
	Ignacius graybullianus[i]
	Microsyops annectens[i]
	Plesiadapis cookei[j]

Table 33.3. List of Fossil Mammalian Species in This Study[a]

[a] Reconstructions are taken from the literature.
[b] Macrini et al. (2006).
[c] Macrini et al. (2007a).
[d] Sanchez-Villagra et al. (2007).
[e] Haight and Murray (1981).
[f] Macrini et al. (2007b).
[g] Novacek (1982).
[h] Meng et al. (2003).
[i] Silcox et al. (2010).
[j] Gingerich and Gunnell (2005).

other cranial foramina with plastic material and after covering the tabula interna of the skull with Vaseline spray (Mida) to facilitate the removal of the endocast. The skull was then positioned to allow the silicone to run out of the skull, thus leaving a layer that covered the whole endocranial surface. This operation was repeated two or three times until the resulting stack of silicone layers generated an endocranial cast solid enough to be extracted by hand from the skull through the foramen magnum without damaging the skull. Once extracted, the hollow silicone endocranial cast was hardened with two-component synthetic resin (F31a and b, Axson Technologies, Saint-Ouen l'Aumône, France).

On all the endocranial casts observed in this study as well as on all the illustrations of endocasts of fossil specimens taken from the literature,

Results and Discussion

33.2. Dorsal view of the endocranial cast of the koala, *Phascolarctos cinereus*. The triangular space between the cerebral hemispheres (H1) is cisterna venae cerebri magnae (cvcm). It is slightly bulging because of the presence of the transverse sinus (ts) (see Fig. 33.1 for its location). All the major cerebellar subdivisions are distinct and visible: vermis (V), cerebellar hemispheres (H2), and parafloculli (p). The vermis is clearly distinct from the triangular space between the cerebral hemispheres. Specimen RBINS IG 4957 Reg 468 (adult female). Scale bar = 1 cm.

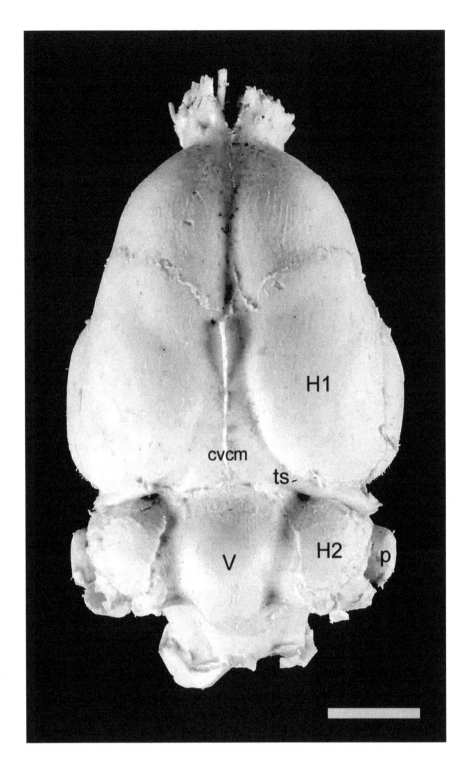

the cerebellar vermis and the cerebellar hemispheres are always clearly visible and distinct. The identification of these structures is confirmed by the observation of corresponding brains. This observation is of particular interest when considering the Didelphimorphia (Table 33.1) because these marsupials are considered to reflect the ancestral mammalian state more than most other present-day mammals (Karlen and Krubitzer, 2007). In koalas (*Phascolarctos cinereus*), the triangular space covering the midbrain between the posterior portions of the cerebral hemispheres is wide and cannot be confused with the cerebellar vermis. The latter is clearly apparent between the cerebellar hemispheres (Figs. 33.2–33.3). The cerebellar

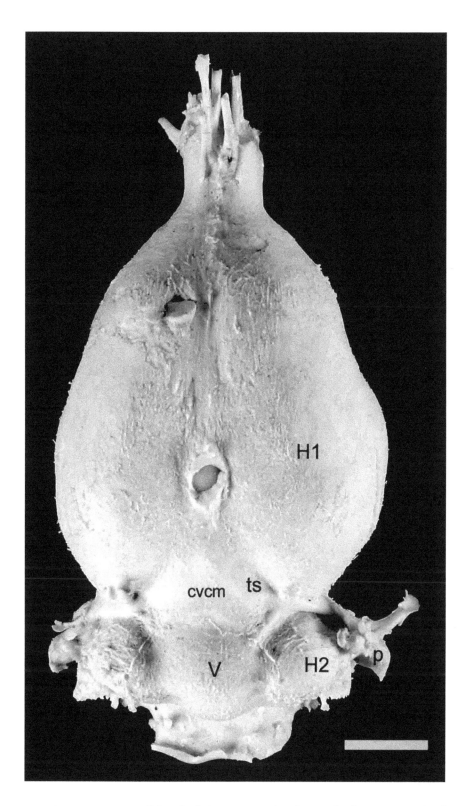

33.3. Dorsal view of the endocranial cast of the koala, *Phascolarctos cinereus*. See Figure 33.2 for a description. The triangular space between the cerebral hemispheres is cisterna venae cerebri magnae (cvcm). It is bulging because of the presence of the transverse sinus (ts) and is clearly distinct from the cerebellar vermis (V). Specimen RBINS IG 4943 Reg 46 (adult). Scale bar = 1 cm.

vermis presents a striking bulging aspect in the sagittal or parasagittal planes, as clearly appears in De Miguel and Henneberg (1998, fig. 1) (Fig. 33.4) and Taylor et al. (2006). The triangular space between the cerebral hemispheres is called the *cisterna venae cerebri magnae* (Figs. 33.2–33.3). It is the subarachnoid space that appears in black above the midbrain (M) on Figure 33.4. It could also show a slightly bulging aspect, but this is mainly due to the presence of the venous transverse sinuses, located just behind the cerebral hemispheres (Figs. 33.2–33.3; see also Figs. 33.1 and 33.7). The

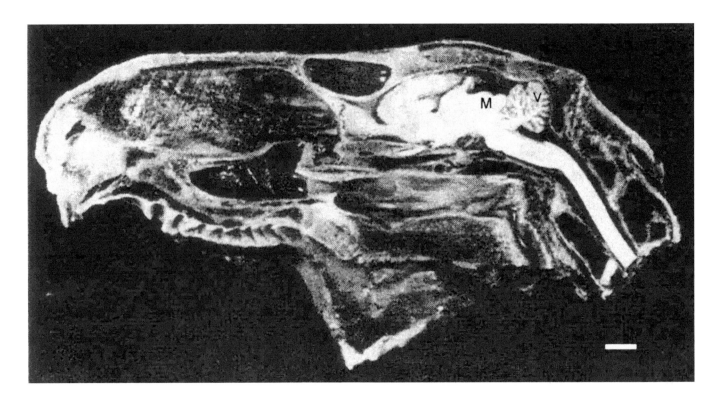

33.4. Midsagittal section of the head of the koala, *Phascolarctos cinereus,* with the brain in situ. Note the subarachnoid space (cisterna venae cerebri magnae, black space) above the midbrain (M) and the bulge due to the expansion of the cerebellar vermis (V). The superior cistern itself (cisterna cerebelli superior) is the subarachnoid space covering the whole cerebellum. Modified from De Miguel and Henneberg (1998). Scale bar = 1 cm.

superior cistern itself (cisterna cerebelli superior) is the subarachnoid space covering the whole cerebellum and could not anatomically be a bulge.

The koala endocast therefore shows a common mammalian pattern, but as summarized by Ashwell (2008), the koala is notorious for having an oddly small and lissencephalic brain. Peculiar features include enlarged cerebral ventricles and subarachnoid spaces (Haight and Nelson, 1987; De Miguel and Henneberg, 1998; Tyndale-Biscoe, 2005; Taylor et al. 2006). The koalas form with the common wombats (*Vombatus ursinus*) a distinct group of Australian diprotodonts (Haight and Nelson, 1987). Interestingly, the common wombat is reported as the most highly encephalized marsupial and the koala as the least in the study of Haight and Nelson (1987). The koala endocast is lissencephalic (Figs. 33.2–33.3), and the wombat endocast shows a highly convoluted brain (Fig. 33.5). The cerebellar structure appears to be similar in both species. The characteristics of the koala brain could be linked to the peculiar terpene- and phenol-rich eucalyptus leaf diet of this animal and the need to handle a high load of dietary phytotoxins (Flannery, 1994; Tyndale-Biscoe, 2005). The hydrocephalic appearance of the koala brain with its enlarged ventricular space could therefore be a specific adaptation that allows energy conservation in response to the poor nutritional quality of its diet.

Our observations of endocranial surface anatomy on the current available sample are in accordance with previous studies. Cerebellar hemispheres indicate the lateral expansion of the cerebellum and are visible in extant and fossil monotremes (Macrini et al., 2006) and in crown therians (Kielan-Jaworowska, 1984, 1986; Novacek, 1982; Macrini et al., 2007a), as well as in the theriiform *Vincelestes* (Macrini et al., 2007b). They therefore must have been present in the most recent common ancestor of the clade Mammalia, and the observation that cerebellar hemispheres are notably absent from endocasts of multituberculates as well as of eutriconodonts

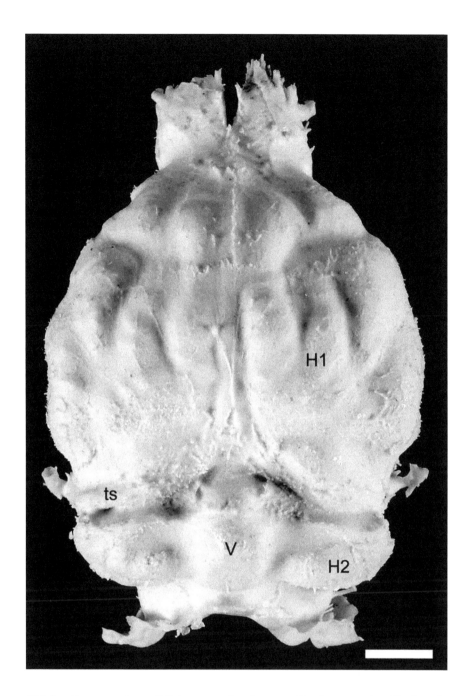

33.5. Dorsal view of the endocranial cast of the common wombat, *Vombatus ursinus*. The triangular space between the cerebral hemispheres (H1) is reduced compared to the koala. The transverse sinus (ts) and the cerebellar components, vermis (V), and hemispheres (H2) are clearly distinct. The parafloculli are not visible on this endocast. Specimen RBINS IG 3081 Reg 42B (adult). Scale bar = 1 cm.

(Kielan-Jaworowska, 1986) is a peculiar feature. Dechaseaux (1958) already recognized the brains of Mesozoic mammals such as the multituberculate *Ptilodus* or the eutriconodontan *Triconodon* as clearly different from the brains of archaic placental mammals and described the cerebellum of these Mesozoic mammal brains as constituted by the archeocerebellum only.

From the above data, it appears that cerebellar hemispheres either independently evolved in monotremes and therian mammals or were independently lost in multituberculates and eutriconodonts (Fig. 33.6). Lack of cerebellar hemispheres is a condition that occurs independently in multituberculates and nonmammalian cynodonts (see Macrini et al., 2007b, for a literature review), indicating that this condition is possibly a reversal in the former group. However, although the mammalian cerebellar cortex is histologically uniform with a cytoarchitectural organization similar through all lobules of the vermis and hemispheres (Sillitoe et al., 2003), the white

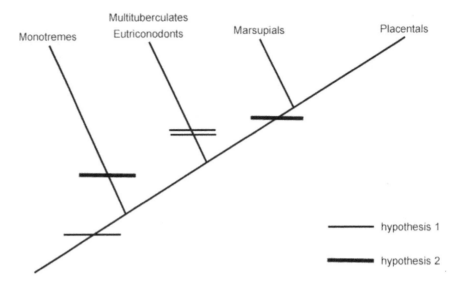

33.6. Simplified phylogenetic tree illustrating the two competing hypotheses that could be suggested from the present study. In hypothesis 1 (thin lines), cerebellar hemispheres were independently lost (double line) in multituberculates and eutriconodonts. In hypothesis 2 (thick lines), cerebellar hemispheres independently evolved in monotremes and therian mammals (marsupials and placentals).

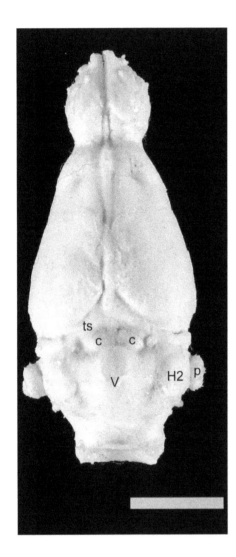

33.7. Dorsal view of the endocranial cast of the didelphid marsupial *Lutreolina crassicaudata*. The superior or the inferior colliculus, the twin structures of the exposed part of the midbrain (C), is clearly visible between the transverse sinus (ts) and the cerebellum. *Abbreviations:* V, vermis; H2, cerebellar hemisphere; p, parafloculus. The vermis, the cerebellar hemispheres, and the parafloculli are clearly visible. Specimen RBINS IG 10520 Reg 326e (adult female). Scale bar = 1 cm.

matter of the cerebellar hemispheres is characterized by the presence of nuclei that are connected to other brain structures (Stephan et al., 1991). The loss of cerebellar hemispheres would therefore involve an important reorganization of brain connectivity. Moreover, if a possible independent evolution of cerebellar hemispheres is considered, independent origins of complex structures in different mammal groups is not uncommon. Striking examples are the homoplasies in mammalian middle ear evolution and the possible independent origin of middle ear bones in monotremes and therians (Rich et al., 2005). Conversely, there are examples of unexpectedly simple brain structure patterns within mammalian taxa for which sister taxa show strikingly different patterns for the same structures. The simple lamination pattern in the lateral geniculate nucleus of gibbons and siamangs is such an example. Gibbons and siamangs are apes with a *Tarsius*-like lateral geniculate nucleus lamination pattern, whereas other hominoid primates, including great apes as well as Old and New World monkeys, have more complex patterns (Tigges and Tigges, 1987; Kaas and Huerta, 1988). Another example is the cerebellum-like structure of the dorsal cochlear nucleus in marsupials and eutherian mammals, but not in monotremes when the latter have a cerebellum structure comparable to the ones of marsupials and placentals (Bell, 2002).

An alternative interpretation for the bulging structure as a vermis or as a superior cistern could not involve the midbrain. When exposed on the endocast, the superior or the inferior colliculus (midbrain) is always represented by twin structures that cannot be mistaken for other structures, as it appears to be clearly visible on the examples in Figure 33.7 (*Lutreolina crassicaudata*, a didelphid marsupial) and Figures 33.8 and 33.9 (*Elephantulus brachyrhynchus* and *Rhynchocyon cirnei stuhlmanni*, macroscelids or elephant shrews).

However, another interpretation of the bulging structure could be suggested after observation of chiropteran endocasts. In microchiropterans such as *Peropteryx macrotis*, *Rhinolofus trifoliatus*, *Hypsugo imbricatus*, or *Mops mops*, pineal organs are large and superficially located (Bhatnagar et al., 1986). It is also the case for megachiropterans such as *Dobsonia moluccensis*, *D. inermis*, and especially *D. praedatrix*, which have a large pineal

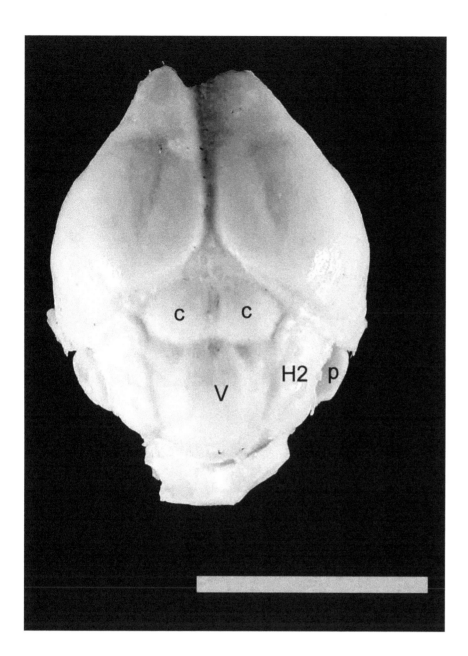

33.8. Dorsal view of the endocranial cast of the macroscelid *Elephantulus brachyrhynchus*. The superior or the inferior colliculus (midbrain) (C) is clearly visible. The vermis (V), the cerebellar hemispheres (H2), and the parafloculli (p) are also clearly distinct. Specimen RMCA not registered (adult). Scale bar = 1 cm.

organ, both in absolute and relative size (Bhatnagar et al., 1990). In the megachiropteran *Dobsonia praedatrix*, the pineal organ is a bulging structure covering the midbrain and the anterior part of the vermis (Fig. 33.10). The usual location of the pineal organ among other megachiropteran bats (family Pteropodidae) is deep into the brain, either under or between the cerebral hemispheres. In *Dobsonia praedatrix*, it is prominent and extends onto the brain surface. Both on the brain (see Bhatnagar et al., 1990, fig. 2) and on the endocranial cast, the pineal organ can be seen to cover the anterior part of the cerebellar vermis (Fig. 33.10). Although this interpretation cannot be discarded, it is certainly not the most parsimonious one to interpret brain structure in large groups of Mezosoic mammals such as the eutriconodonts and the mutituberculates.

Conclusion

On the basis of our current observations, it appears that the cerebellar vermis always makes a bulging impression on the posterior part of endocranial

33.9. Dorsal view of the endocranial cast of the macroscelid *Rhynchocyon cirnei stuhlmanni*. See Figure 33.7 for a description. Specimen RMCA 90-042-M-198 (adult). Scale bar = 1 cm.

33.10. Dorsal view of the brain (left inset) and endocranial cast (right inset) of the megabat *Dobsonia praedatrix*. The view of the brain is modified from Bhatnagar et al. (1990). On the endocranial cast, the pineal organ is bulging between the transverse sinus and the cerebellar vermis. It covers the anterior part of the vermis (cf. Bhatnagar et al. 1990). *Abbreviations:* p, pineal organ, ts, transverse sinus; V, cerebellar vermis. The endocranial cast was made from specimen RBINS IG 5070 Reg 189 (adult male). Scale bar = 1 cm.

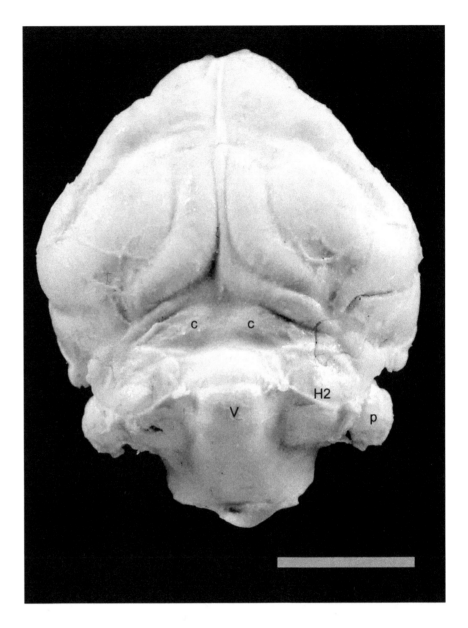

casts and cannot be mistaken for the superior cistern. If accepted, this view means that the cerebellum of multituberculates has no hemispheres and is mainly represented by the vermis and the paraflocculi on endocasts (Fig. 33.1). These structures would not have been the only parts of the multituberculate cerebellum; they are the only major features of the cerebellum that are visible on endocasts. This cerebellar structure is different from the one of the three extant mammalian taxa (monotremes, marsupials, and eutherians) with which multituberculates coexisted at least until the Late Eocene, 35 Ma. This feature, already recognized by Dechaseaux (1958), certainly deserves further investigation as more fossil material becomes available.

Acknowledgments

We extend our thanks to George Lenglet (Royal Belgian Institute of Natural Sciences, Brussels) for the loan of most of the specimen used in this study; Eric Dewamme (Royal Belgian Institute of Natural Sciences, Brussels) for making the silicone endocranial casts of all the taxa used in this study

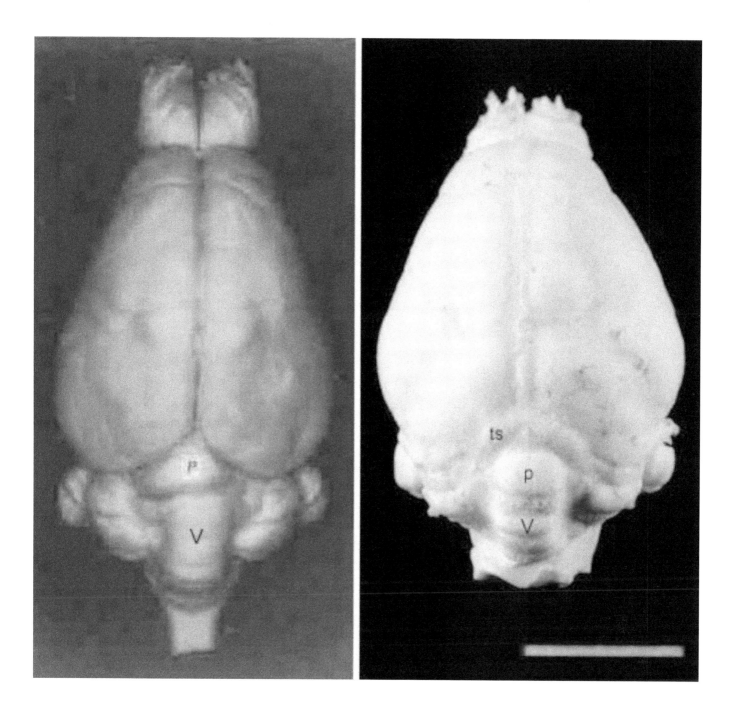

except *Galago demidovii*; Jean-Marc Vandyck (Royal Museum for Central Africa, Tervuren) for taking the photographs of the endocasts; Marc Herbin (Museum National d'Histoire Naturelle, Paris) for access to brain specimens; Matthew Tocheri (Smithsonian Institution, Washington, D.C.) for insightful discussion; and Ted Macrini (St. Mary's University, San Antonio, Tex.) for helpful comments on a first draft of this chapter.

Allman, J. M. 1999. Evolving brains. Scientific American Library/W. H. Freeman, New York, 224 pp.

Ashwell, K. W. S. 2008. Encephalization of Australian and New Guinean marsupials. Brain, Behavior and Evolution 71: 181–199.

Bauchot, R., and H. Stephan. 1967. Encéphales et moulages endocrâniens de quelques insectivores et primates actuels. Coll. Intern. CNRS 163: 575–586

Bell, C. C. 2002. Evolution of cerebellum-like structures. Brain, Behavior and Evolution 59: 312–326.

Bhatnagar, K .P., H. D. Frahm, and H. Stephan. 1986. The pineal organ of bats: a comparative morphological and volumetric investigation. Journal of Anatomy 147: 143–161.

———. 1990. The megachiropteran pineal organ: a comparative morphological and volumetric investigation with special emphasis on the remarkably large pineal of *Dobsonia praedatrix*. Journal of Anatomy 168: 143–166.

Brauer, K., and W. Schober. 1970. Katalog der Säugetiergehirne. Veb Gustav Fischer Verlag, Jena, 150 pp.

Dechaseaux, C. 1958. Encéphales de Condylarthres; pp. 28–30 in J. Piveteau (ed.), Traité de Paléontologie, Tome 6, Vol. 2. Masson, Paris.

De Miguel, C., and M. Henneberg. 1998. Encephalization of the koala, *Phascolarctos cinereus*. Australian Mammalogy 20: 315–320.

Flannery, T. F. 1994. The future eaters: an ecological history of the Australasian lands and people. Reed Books, Sydney, 367 pp.

Gervais, P. 1869. Mémoire sur les formes cérébrales propres aux marsupiaux. Nouvelles Archives du Muséum d'Histoire Naturelle de Paris 5: 229–251.

Gingerich, P. D., and G. F. Gunnell. 2005. Brain of *Plesiadapis cookei* (Mammalia, Proprimates): surface morphology and encephalization compared to those of Primates and Dermoptera. University of Michigan Contributions from the Museum of Paleontology 31: 185–195.

Haight, J. R., and P. F. Murray. 1981. The cranial endocast of the Early Miocene marsupial, *Wynyardia bassiana*: an assessment of taxonomic relationships based upon comparisons with recent forms. Brain Behavior and Evolution 19: 17–36

Haight, J. R., and J. E. Nelson. 1987. A brain that doesn't fit its skull: a comparative study of the brain and endocranium of the koala, *Phascolarctos cinereus* (Marsupialia: Phascolarctidae); pp. 331–352 in M. Archer (ed.), Possums and opossums. Royal Zoological Society of New South Wales, Sydney.

Johnson, J. J. 1977. Central nervous system of marsupials; pp. 157–278 in D. Hunsaker II (ed.), The biology of marsupials. Academic Press, New York.

Kaas, J. H., and C. E. Collins. 2001. Variability in the sizes of brain parts. Behavioral and Brain Sciences 24: 288–290.

Kaas, J. H., and M. F. Huerta. 1988. The subcortical visual system of primates; pp. 327–391 in H. D. Steklis and J. Erwin (eds.), Comparative primate biology. Vol. 4, Neurosciences. Alan R. Liss, New York.

Karlen, S. J., and L. Krubitzer. 2007. The functional and anatomical organization of marsupial neocortex: evidence for parallel evolution across mammals. Progress in Neurobiology 82: 122–141

Kielan-Jaworowska, Z. 1984. Evolution of the therian mammals in the Late Cretaceous of Asia. VI. Endocranial casts of eutherian mammals. Palaeontologica Polonica 46: 157–171.

———. 1986. Brain evolution in Mesozoic mammals. Contributions to Geology, University of Wyoming, Special Paper 3: 21–34.

———. 1997. Characters of multituberculates neglected in phylogenetic analyses of early mammals. Lethaia 29: 249–266.

Kielan-Jaworowska, Z., and T. E. Lancaster. 2004. A new reconstruction of multituberculate endocranial casts and encephalization quotient of *Kryptobaatar*. Acta Palaeontologica Polonica 49: 177–188.

Kielan-Jaworowska, Z., R. L. Cifelli, and Z.-X. Luo. 2004. Mammals from the Age of Dinosaurs: origins, evolution, and structure. Columbia University Press, New York, 630 pp.

Leonhardt, H., G. Töndury, and K. Zilles. 1987. Rauber/Kopsch, Anatomie des Menschen. Band III, Nervensystem Sinnesorgane. Georg Thieme Verlag, Stuttgart, 678 pp.

Luo, Z.-X. 2007. Transformation and diversification in early mammal evolution. Nature 450: 1011–1019.

Macrini, T. E., T. Rowe, and M. Archer. 2006. Description of a cranial endocast from a fossil platypus, *Obdurodon dicksoni* (Monotremata, Ornithorhynchidae), and the relevance of endocranial characters to monotreme monophyly. Journal of Morphology 267: 1000–1015.

Macrini, T. E., C. de Muizon R. L. Cifelli, and T. Rowe. 2007a. Digital cranial endocast of *Pucadelphys andinus*, a Paleocene metatherian. Journal of Vertebrate Paleontology 27: 99–107.

Macrini, T. E., G. W. Rougier, and T. Rowe. 2007b. Description of a cranial endocast from the fossil mammal *Vincelestes neuquenianus* (Theriiformes) and its relevance to the evolution of endocranial characters in Therians. Anatomical Record 290: 875–892.

Macrini, T. E., T. Rowe, and J. L. VandeBerg 2007c. Cranial endocasts from a growth series of *Monodelphis domestica* (Didelphidae, Marsupialia): a study of individual and ontogenetic variation. Journal of Morphology 268: 844–865.

Meng, J., Y. Hu, and C. Li. 2003. The osteology of *Rhombomylus* (Mammalia,

Glires): implications for phylogeny and evolution of Glires. Bulletin of the American Museum of Natural History 275: 1–247.

Nieuwenhuys, R., H. J. Ten Donkelaar, and C. Nicholson. 1998. The central nervous system of vertebrates. Vol. 3. Springer-Verlag, New York: 1525–2219.

Novacek, M. J. 1982. The brain of *Leptictis dakotensis*, an Oligocene leptictid (Eutheria: Mammalia) from North America. Journal of Paleontology 56: 1177–1186.

Rich, T. H., J. A. Hopson, A. M. Musser, T. F. Flannery, and P. Vickers-Rich. 2005. Independent origins of middle ear bones in monotremes and therians. Science 307: 910–914.

Rougier, G. W., and M. J. Novacek. 1998. Early mammals: teeth, jaws and finally . . . a skeleton! Current Biology 8: R284–R287.

Sanchez-Villagra, M., S. Ladevèze, I. Horovitz, C. Argot, J. J. Hooker, T. E. Macrini, T. Martin, S. Moore-Fay, C. de Muizon, T. Schmelzle, and R. J. Asher. 2007. Exceptionally preserved North American Paleogene metatherians: adaptations and discovery of a major gap in the opossum fossil record. Biology Letters 3: 318–322

Silcox, M. T., A. E. Benham, and J. I. Bloch. 2010. Endocasts of *Microsyops* (Microsyopidae, Primates) and the evolution of the brain in primitive primates. Journal of Human Evolution 58: 505–521.

Sillitoe, R. V., H. Künzle, and R. Hawkes. 2003. Zebrin II compartmentation of the cerebellum in a basal insectivore, the Madagascan hedgehog tenrec *Echinops telfairi*. Journal of Anatomy 203: 283–296.

Starck, D. 1963. "Freiliegendes Tectum mesencephali" ein Kennzeichen des primitiven Säugetiergehirns? Zoologischer Anzeiger 171: 350–359.

Stephan, H., G. Baron, and H. D. Frahm. 1991. Insectivora. Comparative brain research in mammals. Vol. 1. Springer-Verlag, New York, 573 pp.

Taylor, J., F. J. Rühli, G. Brown, C. De Miguel, and M. Henneberg. 2006. MR imaging of brain morphology, vascularisation and encephalization in the koala. Australian Mammalogy 28: 243–247.

Tigges, J., and M. Tigges. 1987. Termination of retinofugal fibrs and lamination pattern in the lateral geniculate nucleus of the gibbon. Folia Primatologica 48: 186–194.

Tyndale-Biscoe, H. 2005. Life of Marsupials. CSIRO Publishing, Sydney, 442 pp.

Wilson, D. E., and D. M. Reeder (eds.). 2005. Mammal species of the world: a taxonomic and geographic reference. 3rd ed. Johns Hopkins University Press, Baltimore, Md., 2142 pp.

Index

306, 400; size and posture; skull, 295, 297, 299; splenial, 295, 304; squamosal, 292, 299, 300; surangular, 292, 299, 304; tail, 308; teeth, 302, 303, 304; tibia, 292, 315, 315; ulna, 292, 309, 311

bolus, 148

bonebed: at Baudour, 149; iguanodon at Bernissart, 41, 155, 156, 158, 159–160, 159, 164, 165, 166, 167, 169; in Iharkút, 535–536, 541, 577; in northeastern Iberia, 410, 413, 417

Bon-Tsagan, 452, 458, 459, 463

Borioteiioidea, 546, 548, 549

Bornhardt, Wilhelm, 23

Bornholm, Isle of, 435, 436, 437, 439, 442, 445, 448

Bornholmian, 439, 441, 446, 447

Bornholmian-Scanian, 441, 447

Bostobe Formation, 335

Bostobinskaya Svita, 335, 336, 355

Bothremydidae, 543, 544, 545

Bothremydina, 544

Botrychites reheensis, 462

boudinage, 65, 158

Boulenger, Georges Albert, xiii, 14, 15, 178, 199

Brachylophosaurus, 309, 318, 329, 353

Brachylophosaurus canadensis, 298

Brachyoxylon, 103

Brachyphyllum, 97, 102, 103, 105, 107, 108, 109, 110

Brachyphyllum longispicum, 462

Bracquegnies Formation, 69, 71, 72, 73, 75

brain, 168, 219, 350; *Batyrosaurus rozhdestvenskyi*, 350–352; cavity endocasts, 215–216, 221; encephalization quotient, 219–221; *Hexing qingyi*, 469, 471; *Hypsilophodon foxii*, 239; *Iguanodon bernissartensis*, 216–218; lambeosaurines, 222; *Leaellynasaura amicagraphica*, 497; mammal, 599–610; *Mantellisaurus atherfieldensis*, 213, 218–219

braincase, 216, 264, 323; *Batyrosaurus rozhdestvenskyi*, 337, 338, 339, 340, 349, 350; *Bolong yixianensis*, 301; *Hypsilophodon foxii*, 238, 239, 240, 245

breccia, 51, 52, 55, 88, 169, 535, 536, 541

brine, 169

British Museum (Natural History), 4, 15, 179, 230

Brooke (Isle of Wight), 213

Brook Chine (Isle of Wight), 196, 210, 269, 274

Buenache de la Sierra, 379, 381, 416, 417, 523

Bukachacha, 454, 455

Bukukun River, 454

Bulun Formation, 456

Bureau de Recherches Géologiques et Minières, 149

Bureya Basin, 457

Burgos: Ankylosauria indet. from, 388; baryonychine teeth from, 389; basal ornithopods from, 385, 387; *Dacentrurus* sp. from, 388; *Demandasaurus darwini* from, 379, 400; dinosaur localities from, 379, 380, 381, 382, 383; dromaeosaurid from, 391; *Iguanodon*-like iguanodontoids from, 384; sauropods from, 396, 397; theropods from, 389; titanosauriforms from, 395; thyreophorans from, 387

Buteo, 561

Buxaceae, 80

Byrka River, 453

Cabezón de la Sierra, 381

Café Dubruille (Bernissart), 5

Calamosaurus, 400

Calamosaurus foxi, 398

calcaneum: *Bolong yixianensis*, 315, 316; *Hexing qingyi*, 468, 480; *Hypsilophodon foxii*, 255

Caledonian highs, 66

Calizas de la Huergina Formation, 482

Callovosaurus leedsi, 320

Caluromys, 599

Caluromys lanatus, 602

Camarasaurus, 372, 395

Camarillas Formation: geological framework, 382, 408, 414, 415; multituberculates from, 415, 420, 427

Cambridge Greensand, 365, 366, 374, 375, 376

Cameros, 413

Cameros Basin, 378, 380, 382

Camptosaurus, 182, 183, 228, 253, 255, 260, 262, 263, 264, 266, 271, 273, 275, 310, 311, 318, 320, 328, 353

Camptosaurus aphanoectes, 273

Camptosaurus dispar, 182, 183, 268, 272, 273, 294

Camptosaurus hoggii, 181, 227, 266

Camptosaurus prestwichii, 320

Camptosaurus valdensis, 225, 257–259, 387, 400

Cantabria, 379, 381, 383, 523

Cantalera, 419

Cantalera abadi, 419, 426, 427 428

Capart, André, 11

carbon isotopes, 18, 79, 80, 81, 83

Carcharhinus leucas, 520, 525

Carcharodontosauria, 399

Carcharodontosaurus, 221, 375

Carpolithus multiseminalis, 462

Carpolithus pachythelis, 462

carpus, 191, 203, 251, 311, 327

Carrascosa de la Sierra, 381

Cassagnau, 583

Castellar Formation, 408

Castellón, 379, 382; Allosauroidea indet. from, 390; carcharodontosaurid from, 389; dinosaur localities from, 380, 381, 383; dromaeosaurids from, 391; cf.

Hypsilophodon from, 385; *Iguanodon bernissartensis* from, 384; spinosauroids from, 389, 390; titanosauriforms from, 396; thyreophorans from, 387, 388

Castellote, 379, 381, 388, 390, 415

Castrillo de la Reina Formation, 382

Catillon Formation, 69, 71, 72, 73, 75

caudal vertebrae, 157, 186, 192, 197, 245, 249, 308, 309, 363, 372–374, 373, 374

Caulier, Jules, 25

Cearachelys, 544

Cedar Montain Formation, 320, 321, 354

Cedarosaurus, 151

Cedrorestes, 320

Cedrorestes crichtoni, 354

celestine, 113, 118, 119, 121, 129, 130

Cellosize (hydroxyethyl cellulose), 70, 79

Central Iberian Range, 378, 382, 409, 410, 412, 413, 417, 433

Ceratodus africanus, 285

Ceratodus tiguidensis, 285

Ceratosaurus, 221

cerebellar hemispheres, 598, 599, 600, 604, 607, 608; *Elephantulus brachyrhynchus*, 609; evolution in mammals, 608; *Lutreolina crassicaudata*, 608; *Phascolarctos cinereus*, 604, 606; *Vombatus ursinus*, 607

Cerebropollenites, 83

Cerebrum relative volume (CRV), 221, 222

Cerrada Roya-Mina, 408, 415, 427

cervical ribs, 245, 307, 475 476

cervical vertebrae, 227, 292, 304, 307, 326, 391, 393, 475–477, 477, 549, 562

Chamopsiinae, 548

cheeks, 232, 237, 243–244

Chegdomyn Formation, 457

Cheirolepidiaceae, 97, 108

Chelonia, 577, 578, 586

Chemchukin, 457

chemostratigraphy, 18, 40, 79, 80, 81, 101

Chernovskoe coal mine, 454

Cherves-de-Cognac, 439

Chikoi River, 454

Chironectes, 599

Chironectes minimus, 602

Chita-Ingoda Basin, 457

Chonkogor Formation, 457

Chulsanbaatar, 598

Cimolodonta, 409, 410

Cinctorres, 379, 381

Cismon, 81, 82, 83

cisterna cerebelli superior, 606

cisterna venae cerebri magnae, 604, 605, 606

Citipati, 221

Cladophlebidium, 456

Claosaurus agilis, 355

Clift, William, 3, 176

clinochlore, 119, 125

Cloverly Formation, 179, 375

Coccolepis woodwardi, 492

"*Coelodus*" *syriacus*, 539

This book was designed by Jamison Cockerham at Indiana University Press, set in type by Jamie McKee at MacKey Composition, and printed by Sheridan Books, Inc.

The fonts are Electra, designed by William A. Dwiggins in 1935, Frutiger, designed by Adrian Frutiger in 1975, and Futura, designed by Paul Renner in 1927. All were issued by Adobe Systems.

Milton Keynes UK
Ingram Content Group UK Ltd.
UKHW050329161223
434452UK00003B/79

9 780253 357212